COMPUTING METHODS IN APPLIED SCIENCES AND ENGINEERING

Fourth International Symposium on Computing Methods in Applied Sciences and Engineering

Organized by

Institut de Recherche d'Informatique et d'Automatique (IRIA)

Sponsored by

AFCET, GAMNI, IFIP TC-7

Scientific Secretaries

J. F. Bourgat

A. Marrocco

Secretary

Th. Bricheteau

ROCKET PROPULSION ESTABLISHMENT
LIBRARY 1/5

Please return this publication, or request a renewal, by the date stamped below.

Name	Date
~~A. WITHEY~~ AW.	~~-3 JUL 1981~~ B
~~1. J. Schnuler~~	~~-8 JUL 1981~~ S

(4/64) L23964 442077 Wt29280 D7061 10/64 10M T&Co G871. R.P.E. Form 243

COMPUTING METHODS IN APPLIED SCIENCES AND ENGINEERING

Proceedings of the Fourth International Symposium on Computing Methods in Applied Sciences and Engineering

Versailles, France, December 10-14, 1979

Edited by

R. GLOWINSKI

J. L. LIONS

Institut de Recherche d'Informatique et d'Automatique
Paris, France

1980

NORTH-HOLLAND PUBLISHING COMPANY
AMSTERDAM · NEW YORK · OXFORD

ISBN: 0 444 86008 8

Published by:

North-Holland Publishing Company
Amsterdam · New York · Oxford

Sole distributors for the U.S.A. and Canada:

Elsevier North-Holland, Inc.
52 Vanderbilt Avenue
New York, NY 10017

Library of Congress Cataloging in Publication Data

International Symposium on Computing Methods in Applied Sciences and Engineering, 4th, Versailles, 1979.
Computing methods in applied sciences and engineering.

English or French
1. Engineering--Data processing--Congresses.
2. Science--Data processing--Congresses.
I. Glowinski, R. II. Lions, Jacques Louis.
III. Institut de recherche d'informatique et d'automatique. IV. Association française de cybernétique économique et technique. V. Groupe pour l'avancement des méthodes numériques dans les sciences de l'ingénieur. VI. International Federation for Information Processing. Technical Committee on Optimization (TC7)
TA345.I57 1979 620'.0028'54 80-16250
ISBN 0-444-86008-8

Printed in The Netherlands

PREFACE

This book contains the texts of the lectures which were presented during the Fourth International Symposium on Computing Methods in Applied Sciences and Engineering, December 10-14, 1979, organized by IRIA-LABORIA under the sponsorship of Association Française pour la Cybernétique Economique et Technique (AFCET), Groupe pour l'Avancement des Méthodes Numériques de l'Ingénieur (GAMNI), and International Federation for Information Processing (IFIP WG7-2).

More than 400 scientists and engineers from many countries attended this meeting.

The organizers which to express their gratitude to Mr. A. Danzin, Director of IRIA, and address their thanks to each session chair-person who directed very interesting discussions and also to all speakers.

Sincere gratitude is also expressed to the IRIA Public Relation Office whose help contributed greatly to the success of this Symposium and also to the Scientific Secretaries, Drs. Bourgat and Marrocco.

R. GLOWINSKI
and
J.L. LIONS

IRIA-LABORIA
Institut de Recherche d'Informatique et d'Automatique
IRIA Research Laboratory

TABLE OF CONTENTS

SESSION I

NUMERICAL METHODS IN NON LINEAR MECHANICS

COMPUTING METHODS IN APPLIED SCIENCES AND ENGINEERING
R. Glowinski, J.L. Lions (editors)
North-Holland Publishing Company

ON THE ANALYSIS OF FORMING PROBLEMS IN NATURAL FORMULATION

J.H. Argyris, J.St. Doltsinis, H. Wüstenberg

Institut für Statik und Dynamik der Luft- und Raumfahrtkonstruktionen
University of Stuttgart

Abstract

A concise presentation of the natural formulation for large strain inelastic problems is given. The adopted stress-strain relations account also for the thermomechanical coupling due to dissipation of mechanical work. Depending on the properties activated by the deformation process, the plastic or the viscoplastic description of inelastic flow may be used. Algorithms are discussed for the numerical treatment in the quasistatic and the dynamic cases.

1. Introduction

The natural presentation of the strain and stress state yields a transparent description of large strain inelastic phenomena. Following [1] which confirms the equivalence of the natural and the standard formulations in the large strain domain, the former may simply be considered as the combination of uniaxial strain applications in six natural material directions. Furthermore, the time rates of natural strains and stresses assume particular simple forms.

The general theory of Green and Naghdi [2] is an appropriate foundation for the description of a material behaviour containing elastic and inelastic properties. The specification for the description of metal materials follows the proposals made by Lee (c.f. e.g. [3]). Within this framework the elastic constants remain unaffected by any inelastic deformations if referred to the local stress-free configuration. The major part of the mechanical work dissipated during inelastic flow is converted into heat affecting the stress-strain relations for the elastic and the inelastic material components. Two limiting cases may be considered without any need of examining the heat flow explicitly. While the isothermal condition applies to slow deformations, the adiabatic condition is appropriate for rapid motions occurring in dynamic deformations. Depending on the properties activated by the process either the plastic or the viscoplastic description of inelastic flow may be used in the description of the inelastic material component.

Various procedures are feasible for the numerical solution of the deformation problem of the discretized system [4]. In the quasistatic case an algorithm derives from a combination of the

Newton method with respect to the total deformation and the normal iteration with respect to inelastic flow. The procedure may also be applied in the dynamic case as a matrix method of solution. Complex schemes for the approximate integration of the equations of motion can more easily be incorporated into vector iteration algorithms [5]. Applications of the theory shown in Section 5 concern various metal forming problems and impact phenomena.

2. Natural strains and stresses; time rates

Following [1] the state of strain is represented by the natural strain vector

$$\gamma_N = \{ \tfrac{1}{2} ({}^c_r\lambda^2 - {}^o_r\lambda^2)^\alpha \} \qquad \alpha = 1, \ldots, 6 \tag{1}$$

which is associated with the deformation from the initial unstressed configuration ${}^oC$ to the current state ${}^cC$ (fig. 1). For each natural material direction α at the point considered the extension ratio

$$\lambda = \frac{l}{{}^r l} \tag{1a}$$

is taken with respect to the length ${}^r l$ of the line element at the reference configuration ${}^rC$, as yet not specified. The elastic part of the deformation may, as usual, be recovered by local unloading up to the current stress-free state ${}^fC$ which differs from ${}^oC$ as a result of the inelastic part of the deformation. This imaginary, purely elastic, stress removal allows the additive decomposition of the natural strain vector

$$\gamma_N = \varepsilon_N + \eta_N \tag{2}$$

with an elastic component

$$\varepsilon_N = \{ \tfrac{1}{2} ({}^c_r\lambda^2 - {}^f_r\lambda^2) \} \tag{2a}$$

and an inelastic component

$$\eta_N = \{ \tfrac{1}{2} ({}^f_r\lambda^2 - {}^o_r\lambda^2) \} \tag{2b}$$

Taking the reference configuration at the initial state eq. (1) furnishes the natural Green strains γ_{GN} and eqs. (2a, 2b) their elastic part ε_{GN} and inelastic part η_{GN} respectively. If the reference configuration is taken at the current state, eq. (1) defines the natural Almansi strains γ_{AN} and eqs. (2a, 2b) their elastic and inelastic parts ε_{AN}, η_{AN} respectively. Finally, taking the local stress-free configuration as reference, the natural strain vector γ_{FN} may also be considered as the sum of a Green strain ε_{FN} and an Almansi strain η_{FN} defined for the elastic and the inelastic parts of the deformation respectively, each part being considered independently. The different strain measures are interrelated through

$$\gamma_{FN} = {}^{c}_{f}\lambda^{2}\,\gamma_{AN} = {}^{o}_{f}\lambda^{2}\,\gamma_{GN} \tag{3}$$

$$\lambda = \lceil \lambda^{\alpha} \rfloor \tag{3a}$$

which holds also for the elastic and the inelastic strain components. When considering strain rates the fact that the strains refer to a system of material coordinates is of importance. If the material coordinates are defined at a configuration varying with time we have to distinguish between the standard derivative with respect to time and that based on material coordinates. To this purpose we use the symbol $(\ )^{\bullet}$ in the latter case. Then the time derivative of eq. (1) may be expressed as

$$\dot{\gamma}_{N} = {}^{c}_{r}\lambda^{2}\,(\gamma_{AN})^{\bullet} \tag{4}$$

where the Almansi strain rate

$$(\gamma_{AN})^{\bullet} = \{({}_{c}\lambda^{\alpha})^{\bullet}\} \quad , \qquad ({}_{c}\lambda^{\alpha})^{\bullet} = \left(\frac{{}^{c}\dot{\ell}}{{}^{c}\ell}\right)^{\alpha} \tag{4a}$$

simply describes the instantaneous rate of deformation in natural terms. Considering on the other hand the standard derivative of the Almansi strain with respect to time we may write

$$\dot{\gamma}_{AN} = (\gamma_{AN})^{\bullet} - 2({}_{c}\lambda)^{\bullet}\gamma_{AN} \; , \quad ({}_{c}\lambda)^{\bullet} = \lceil ({}_{c}\lambda^{\alpha})^{\bullet} \rfloor \tag{5}$$

the difference between the standard and the material time derivative involving the ignoration of the convective term in the latter case. The time rate of the inelastic part of strain eq. (2b) follows as

$$\dot{\eta}_{N} = {}^{f}_{r}\lambda^{2}\,(\eta_{FN})^{\bullet} \tag{6}$$

in which

$$(\eta_{FN})^{\bullet} = \{({}_{f}\lambda^{\alpha})^{\bullet}\} \quad , \qquad ({}_{f}\lambda^{\alpha})^{\bullet} = \left(\frac{{}^{f}\dot{\ell}}{{}^{f}\ell}\right)^{\alpha} \tag{6a}$$

represents the rate of deformation for the local inelastic motion. The elastic strain rate deriving from eq. (2a) assumes then the form

$$\dot{\varepsilon}_{N} = {}^{c}_{r}\lambda^{2}\,(\gamma_{AN})^{\bullet} - {}^{f}_{r}\lambda^{2}\,(\eta_{FN})^{\bullet} \tag{7}$$

and is given as the difference between the overall and the inelastic strain rate. For completeness we note that for the rates there holds also

$$\dot{\gamma}_{N} = \dot{\varepsilon}_{N} + \dot{\eta}_{N} \tag{8}$$

and that the time rates of the different strain measures are interrelated through

$$(\gamma_{FN})^{\bullet} = {}^{c}_{f}\lambda^{2} (\gamma_{AN})^{\bullet} = {}^{o}_{f}\lambda^{2} \dot{\gamma}_{GN} \tag{9}$$

which also holds for the elastic and inelastic parts.

The natural stress vector

$$\boldsymbol{\sigma}_{N} = \{ \sigma^{\alpha} \} \qquad \alpha = 1, \ldots, 6 \tag{10}$$

is defined as corresponding to the rate of natural strain if their product expresses the virtual work over the reference volume ${}^{r}V$

$$\dot{W} = {}^{r}V \boldsymbol{\sigma}_{N}^{t} \dot{\boldsymbol{\gamma}}_{N} \tag{11}$$

Hence the true Cauchy stress $\boldsymbol{\sigma}_{CN}$ defined at the current configuration corresponds to the Almansi strain rate (rate of deformation) $(\boldsymbol{\gamma}_{AN})^{\bullet}$ eq. (4a). The natural Piola-Kirchhoff stress $\boldsymbol{\sigma}_{PN}$ accords with the second (symmetric) Piola-Kirchhoff definition of stress with reference to the initial unstressed configuration and corresponds to the Green strain rate $\dot{\boldsymbol{\gamma}}_{GN}$. The stress measure $\boldsymbol{\sigma}_{FN}$ corresponding to the strain rate $(\boldsymbol{\gamma}_{FN})^{\bullet}$ of the partitioned deformation process is referred to the current local stress-free configuration. The interrelation of the different stress measures obtained from the invariance of the rate of elementary work eq. (11) and application of eq. (9) follows as

$$\boldsymbol{\sigma}_{FN} = \frac{{}^{c}V}{{}^{f}V} {}^{f}_{c}\lambda^{2} \boldsymbol{\sigma}_{CN} = \frac{{}^{o}V}{{}^{f}V} {}^{f}_{o}\lambda^{2} \boldsymbol{\sigma}_{PN} \tag{12}$$

Discussing now stress rates we note that the time derivatives of the natural stresses in eq. (12) constitute themselves objective stress rates as they represent the change of stress in the natural material directions. Considering the derivative of the natural Piola-Kirchhoff stress in terms of the natural Cauchy stress from eq. (12), we obtain

$$\dot{\boldsymbol{\sigma}}_{PN} = \frac{{}^{c}V}{{}^{o}V} {}^{o}_{c}\lambda^{2} \left[\dot{\boldsymbol{\sigma}}_{CN} + \frac{{}^{c}\dot{V}}{{}^{c}V} \boldsymbol{\sigma}_{CN} - 2 ({}_{c}\lambda)^{\bullet} \boldsymbol{\sigma}_{CN} \right] \tag{13}$$

where

$$\frac{{}^{c}\dot{V}}{{}^{c}V} = \mathbf{e}^{t} \, {}^{c}\mathbf{A}^{-1} (\boldsymbol{\gamma}_{AN})^{\bullet} \tag{14}$$

$$\mathbf{e} = \{ 1 \; 1 \; 1 \; 1 \; 1 \; 1 \} \tag{14a}$$

$${}^{c}\mathbf{A} = {}^{c}[\, c^{2}_{\alpha,\beta} \,] \quad , \qquad c_{\alpha,\beta} = \cos(\alpha,\beta) \tag{14b}$$

is nothing but the volumetric part of the Almansi strain rate (rate of deformation).

The expression within the brackets in eq. (13) is Truesdell's rate of natural Cauchy stress

$$(\sigma_{CN})^{\bullet}_{T} = \dot{\sigma}_{CN} + \frac{{}^{c}\dot{V}}{{}^{c}V}\sigma_{CN} - 2({}_{c}\lambda)^{\bullet}\sigma_{CN} \tag{15}$$

Using the relation between Truesdell's and Oldroyd's rate of natural Cauchy stress (c.f. [1]) one obtains the expression

$$(\sigma_{CN})^{\bullet}_{O} = (\sigma_{CN})^{\bullet}_{T} - \frac{{}^{c}\dot{V}}{{}^{c}V}\sigma_{CN} = \dot{\sigma}_{CN} - 2({}_{c}\lambda)^{\bullet}\sigma_{CN} \tag{16}$$

The time rate of σ_{PN} in terms of σ_{FN} from eq. (12) assumes the form

$$\dot{\sigma}_{PN} = \frac{{}^{f}V}{{}^{o}V}\,{}^{o}_{f}\lambda^{2}\left[\dot{\sigma}_{FN} + \frac{{}^{f}\dot{V}}{{}^{f}V}\sigma_{FN} - 2({}_{f}\lambda)^{\bullet}\sigma_{FN}\right] \tag{17}$$

where the volumetric change of the local stress-free configuration

$$\frac{{}^{f}\dot{V}}{{}^{f}V} = e^{t}\,{}^{f}A^{-1}(\eta_{FN})^{\bullet} \quad , \qquad {}^{f}A = {}^{f}[\,c^{2}_{\alpha,\beta}\,] \tag{18}$$

is that caused by inelastic flow. Comparison of eq. (17) with eqs. (13) and (15) justifies the following notation for the expression within the brackets of eq. (17)

$$(\sigma_{FN})^{\bullet}_{T} = \dot{\sigma}_{FN} + \frac{{}^{f}\dot{V}}{{}^{f}V}\sigma_{FN} - 2({}_{f}\lambda)^{\bullet}\sigma_{FN} \tag{19}$$

which defines Truesdell's rate of σ_{FN}. In the case of isochoric inelastic deformation eq. (19) reduces to an expression which compares with Oldroyd's rate of Cauchy stress

$$(\sigma_{FN})^{\bullet}_{O} = \dot{\sigma}_{FN} - 2({}_{f}\lambda)^{\bullet}\sigma_{FN} \tag{20}$$

The stress rate in eq. (20) can be derived from σ_{FN} with respect to inelastic deformation in the same manner as Oldroyd's rate of Cauchy stress can with respect to the total deformation. Thus the adopted notation is justified. We further obtain the relation

$$(\sigma_{FN})^{\bullet}_{O} = (\sigma_{FN})^{\bullet}_{T} - \frac{{}^{f}\dot{V}}{{}^{f}V}\sigma_{FN} \tag{21}$$

We finally note that an equation interrelating the rates of the different stress measures may be written in the form

$$(\sigma_{FN})^{\bullet}_{T} = \frac{{}^{c}V}{{}^{f}V}\, {}^{f}_{c}\lambda^{2} (\sigma_{CN})^{\bullet}_{T} = \frac{{}^{o}V}{{}^{f}V}\, {}^{f}_{o}\lambda^{2}\, \dot{\sigma}_{PN} \tag{22}$$

which is analogous to eq. (12).

Concerning Jaumman's co-rotational rate of the natural Cauchy stress we remark that by its very nature it cannot be expressed in purely natural terms and refer to [1].

3. Stress-strain relations

Following Green and Naghdi [2] we base the description of material behaviour on thermodynamic considerations. Expressing the local energy balance and the inequality of local entropy production in terms of Helmholtz's free energy function f for which we assume the dependence

$$f = f(\varepsilon_{GN}, \eta_{GN}, T) \tag{23}$$

we obtain

$$\sigma_{PN} = {}^{o}\rho\, f^{t}_{,\varepsilon GN} \tag{24a}$$

$$s = - f_{,T} \tag{24b}$$

$$\frac{1}{{}^{o}\rho}\, \sigma^{t}_{PN}\, \dot{\eta}_{GN} \geqq f_{,\eta GN}\, \dot{\eta}_{GN} \tag{24c}$$

Eq. (24a) relates the natural stress to the natural strains and the temperature T with reference to the initial state while eq. (24c) compares the rate of inelastic work per unit mass with the rate of the specific free energy arising from inelastic deformation. The specific entropy s is not a convenient measure in mechanical applications; however eq. (24b) comes into its own when considering thermomechanical coupling.

If the deformation proceeds slowly isothermal conditions can be assumed and the temperature need not be considered further. In rapid motions occurring in dynamic situations the assumption of adiabatic deformation is appropriate. In this case the entropy production is only due to the dissipation of mechanical work and we may write in accordance with [3]

$$T\dot{s} = \xi \frac{1}{{}^{o}\rho}\, \sigma^{t}_{PN}\, \dot{\eta}_{GN} \tag{25}$$

where $\xi = 1 - \delta$ is a number less than but very near to unity indicating the fraction of inelastic work dissipated and converted into heat. From eq. (25) the local temperature change can be derived as

$$\dot{T} = -\frac{T}{c}\left[s_{,\varepsilon GN}\,\dot{\varepsilon}_{GN} + \left(s_{,\eta GN} - \frac{\xi}{T\,{}^{o}\rho}\,\sigma^{t}_{PN} \right)\dot{\eta}_{GN} \right] \qquad (26)$$

where

$$c = T\, s_{,T} \qquad (26a)$$

is the specific heat of the material.

As pointed out by Lee (c.f. e.g. [3]) the behaviour of common metals can be specified in a simple manner with reference to the stress-free configuration considered as varying with inelastic deformation. To this end we assume the free energy function in the form

$$f = f(\varepsilon_{FN}, \eta_{FN}, T) \qquad (28)$$

and considering eqs. (12) and (3) we deduce from eq. (24a)

$$\sigma_{FN} = {}^{f}\rho\, f^{t}_{,\varepsilon FN} \qquad (28)$$

If we next require the elastic behaviour to be independent of inelastic strains, f must for

$${}^{f}\rho = {}^{o}\rho = const \qquad (29a)$$

satisfy the non-coupling conditions

$$\sigma_{FN,\eta FN} = {}^{f}\rho\, f^{t}_{,\varepsilon FN,\eta FN} = 0 \;, \qquad s_{,\eta FN} = -f_{,T,\eta FN} = 0 \qquad (29b)$$

leading to the form

$$f = f^{e}(\varepsilon_{FN}, T) + f^{i}(\eta_{FN}) \qquad (29)$$

for the free energy function.

We assume in what follows isotropic materials and use a free energy expansion restricted to the second order in elastic strains and third order in temperature

$${}^{f}\rho\, f^{e} = a_1 J_1^2 + a_2 J_2 + b_1 J_1 \vartheta + c_1 \vartheta^2 + c_2 \vartheta^3 \qquad (30)$$

In accordance with [3] eq. (30) goes back to Murnaghan [6].

In eq. (30) the first and second invariants of elastic strain can be expressed in natural terms

$$J_1 = (\varepsilon_I + \varepsilon_{II} + \varepsilon_{III})_F = \mathbf{e}^t\, {}^f\!\mathbf{A}^{-1}\, \boldsymbol{\varepsilon}_{FN} \tag{31}$$

$$J_2 = (\varepsilon_I \varepsilon_{II} + \varepsilon_{II}\varepsilon_{III} + \varepsilon_{III}\varepsilon_I)_F = \frac{1}{2}\, \boldsymbol{\varepsilon}^t_{FN} ({}^f\!\mathbf{A}^{-1} \mathbf{E}\, {}^f\!\mathbf{A}^{-1} - {}^f\!\mathbf{A}^{-1})\, \boldsymbol{\varepsilon}_{FN} \tag{32}$$

$\varepsilon_{I,II,III}$ denoting principal strains and

$$\mathbf{E} = \mathbf{e}\,\mathbf{e}^t \tag{33a}$$

while

$${}^f\!\mathbf{A} = {}^f[\, c^2_{\alpha,\beta}] \qquad c_{\alpha\beta} = \cos(\alpha,\beta) \tag{33b}$$

fixes the natural directions relative to each other at the current stress-free state. The temperature dependence on eq. (30) is expressed by the quantity

$$\vartheta = T - {}^\circ T \tag{34}$$

representing the deviation of the temperature from the reference temperature ${}^\circ T$ at the initial unstressed state.

Using eq. (30) in eq. (28) and identifying the material constants for the limiting case of small strains, the natural stress-strain relations follow as

$$\boldsymbol{\sigma}_{FN} = {}^f\mathbf{k}_N (\boldsymbol{\varepsilon}_{FN} - \alpha\vartheta\mathbf{e}) \tag{35}$$

in which the natural elastic stiffness

$${}^f\mathbf{k}_N = 2G\left({}^f\!\mathbf{A}^{-1} + \frac{\nu}{1-2\nu}\, {}^f\!\mathbf{A}^{-1}\mathbf{E}\, {}^f\!\mathbf{A}^{-1}\right) \tag{35a}$$

is to be taken at the current stress-free state.

If the process is assumed isothermal the temperature term in eq. (35) is ignored. Considering the temperature change in an adiabatic deformation we express eq. (25) as

$$T\dot{s} = \xi \frac{1}{{}^f\rho}\, \boldsymbol{\sigma}^t_{FN} (\boldsymbol{\gamma}_{FN})^{\cdot} \tag{36a}$$

and derive $\dot{s}$ using immediately the dependence

$$s = -f^e_{,T} = s(J_1, T) \tag{36b}$$

resulting from the particular form of the free energy in eq. (30). We thus arrive at

$$\dot{T} = -\frac{1}{f\rho c}\left[3K\alpha T\,\dot{j}_1 - \xi\,\boldsymbol{\sigma}^t_{FN}(\boldsymbol{\eta}_{FN})^{\cdot}\right] \tag{37}$$

the form of f in eq. (30) permitting a quadratic variation of the specific heat with temperature

$$c = \frac{2}{f\rho}\left[c_1 + 3c_2(T - {}^{\circ}T)\right]T \tag{37a}$$

The evaluation of the stress-strain relations eq. (35) requires the knowledge of the inelastic deformation on which ${}^f\boldsymbol{K}_N$, $\boldsymbol{\varepsilon}_{FN}$ and ϑ depend. Temperature dependent plastic deformation may be governed by the von Mises yield condition appropriate for metals

$$\Phi = \bar{\sigma} - \sigma_Y(\bar{\eta}, T) \leqq 0 \tag{38}$$

In eq. (38) the von Mises equivalent stress is expressed in terms of $\boldsymbol{\sigma}_{FN}$. The uniaxial yield stress σ_Y is assumed as a function of the temperature and the equivalent plastic strain $\bar{\eta}$ which is given as the time integral of its rate $\dot{\bar{\eta}}$ expressed in terms of $(\boldsymbol{\eta}_{FN})^{\cdot}$. The inelastic strain rate may conveniently be written as

$$(\boldsymbol{\eta}_{FN})^{\cdot} = \dot{\bar{\eta}}\,\boldsymbol{S}_N \tag{39}$$

where the natural direction of plastic flow $\boldsymbol{S}_N$ derives from eq. (38) by the normality condition

$$\boldsymbol{S}_N = \frac{3}{2}\frac{1}{\bar{\sigma}}\left({}^f\boldsymbol{A} - \frac{1}{3}\boldsymbol{E}\right)\boldsymbol{\sigma}_{FN} \tag{40}$$

If we next describe the instantaneous uniaxial inelastic response of the material by

$$\dot{\sigma}_Y = \sigma_{Y,\bar{\eta}}\,\dot{\bar{\eta}} + \sigma_{Y,T}\,\dot{T} = \zeta\,\dot{\bar{\eta}} + \psi\,\dot{T} \tag{41}$$

and substitute for the temperature rate eq. (37) due to adiabatic deformation we deduce from the yield condition the equivalent plastic strain rate

$$\dot{\bar{\eta}} = \frac{1}{\zeta + \frac{\psi}{f\rho c}\xi\bar{\sigma}}\left(\dot{\bar{\sigma}} + \frac{\psi}{f\rho c}3K\alpha T\,\dot{j}_1\right) \geqq 0 \tag{42}$$

We reproduce the isothermal case by simply ignoring in eq. (42) the terms containing ψ which normally assumes negative values. While

$$\dot{\bar{\sigma}} = \boldsymbol{S}^t_N\,\dot{\boldsymbol{\sigma}}_{FN} \tag{42a}$$

is expressed in eq. (42) in terms of the rate of stress, an alternative expression for $\dot{\bar{\eta}}$ may be derived in the form

$$\dot{\bar{\eta}} = \frac{1}{\zeta + 3G^* + \frac{\psi}{f\rho c}\xi\bar{\sigma}} \left(\dot{\bar{\sigma}}^* + \frac{\psi}{f\rho c} 3K\alpha T \dot{J}_1 \right) \tag{43}$$

Eq. (43) contains the rate of strain as (c.f. [1])

$$\dot{\bar{\sigma}}^* = \mathbf{s}_N^t \, \dot{\boldsymbol{\sigma}}_{FN}^* \tag{43a}$$

where

$$\dot{\boldsymbol{\sigma}}_{FN}^* = {}^f\boldsymbol{\kappa}_N (\boldsymbol{\gamma}_{FN} - \alpha\vartheta \mathbf{e})^{\cdot} + {}^f\dot{\boldsymbol{\kappa}}_N (\boldsymbol{\varepsilon}_{FN} - \alpha\vartheta \mathbf{e}) \tag{43b}$$

Also

$$3G^* = \mathbf{s}_N^t \, {}^f\boldsymbol{\kappa}_N \, {}^c_f\boldsymbol{\lambda}^2 \, \mathbf{s}_N \tag{43c}$$

As in eq. (42) the isothermal case implies the ignoration of the term in eq. (43) associated with ψ.

If the material possesses viscous properties which are supposed to be activated by the particular deformation process the combined viscoplastic description of inelastic flow may be appropriate. We adopt for this purpose the viscoplastic material model as discussed in [7], which combines a viscous and a plastic element in parallel and seems to introduce in effect a yield condition dependent on the rate of inelastic deformation. We retain the presentation of eq. (39) for the inelastic strain rate also for the present viscoplastic case. We follow in this reference [7] but use natural quantities referred to the current stress-free configuration. It turns out that the natural direction of viscoplastic flow $\mathbf{s}_N$ coincides with eq. (40) for inviscid plasticity. The equivalent inelastic strain rate is expressed in the viscoplastic case as

$$\dot{\bar{\eta}} = \mu F(\varphi) \geqq 0 \tag{44}$$

Here μ accounts for the material fluidity and φ is defined in terms of the static yield condition eq. (38) as

$$\varphi = \frac{1}{\sigma_Y(\bar{\eta}, T)} \phi = \frac{\bar{\sigma}}{\sigma_Y(\bar{\eta}, T)} - 1 \tag{45}$$

The discontinuous function $F(\varphi)$ ensures that no viscoplastic flow occurs below $\varphi = 0$ and must therefore vanish in this case. Taking the special case

$$F(\varphi) = \varphi^n \tag{46}$$

we obtain from eq. (44) the relation

$$\bar{\sigma} - [1 + (\frac{\dot{\bar{\eta}}}{\mu})^{1/n}]\,\sigma_Y(\bar{\eta},T) = 0 \tag{47}$$

Eq. (47) may be interpreted as expressing the influence of viscous phenomena in the yield condition and indicates the viscoplastic behaviour ranging between elastic and inviscid plastic response (fig. 2).

4. Large strain inelastic analysis

We assume the body under consideration to be discretized into an assembly of finite elements which are conveniently described in a natural manner [1]. For example, in the three-dimensional case the natural directions of the strain and stress presentation are taken to coincide with those of the edges of a tetrahedron element (fig. 3). Simple extension of the local continuum to a homogeneous finite element domain leads to finite elements (urelements [8]) which may be used as the subelements for the derivation of higher order elements.

In the quasistatic case the resultants of the internal stresses $\boldsymbol{S}$ at the nodal points of the discretized body are equivalent to the applied loads $\boldsymbol{R}$. Thus for isothermal conditions

$$\boldsymbol{R} - \boldsymbol{S}(\boldsymbol{X}, {}^f\boldsymbol{\ell}) = 0 \tag{48}$$

According to the stress-strain relations eq. (35), $\boldsymbol{S}$ is a function of the nodal point coordinates $\boldsymbol{X}$ and the stress-free element lengths ${}^f\boldsymbol{\ell}$. Following eq. (6a) the stress-free lengths ${}^f\boldsymbol{\ell}$ depend on inelastic deformation and may be related to the actual element lengths $\boldsymbol{\ell}$ compatible with $\boldsymbol{X}$ by an expression of the form

$$\boldsymbol{f}(\boldsymbol{\ell}, {}^f\boldsymbol{\ell}) = 0 \tag{49}$$

Eq. (49) requires the approximate stepwise integration of the equations governing plastic flow.

Eqs. (48) and (49) controlling the quasistatic motion are solved iteratively. Starting with $\boldsymbol{X}_i, {}^f\boldsymbol{\ell}_i$ as estimates we obtain a new estimate for $\boldsymbol{X}$ by developing eq. (48) into a Taylor series for constant ${}^f\boldsymbol{\ell}$, as

$$\boldsymbol{X}_{i+1} = \boldsymbol{X}_i + \boldsymbol{S}_{,X}^{-1}(\boldsymbol{X}_i, {}^f\boldsymbol{\ell}_i)[\boldsymbol{R} - \boldsymbol{S}(\boldsymbol{X}_i, {}^f\boldsymbol{\ell}_i)] \tag{50a}$$

and solve eq. (49) for the corresponding $(i+1)$ th estimate of ${}^f\boldsymbol{\ell}$ satisfying

$$\boldsymbol{f}(\boldsymbol{\ell}_{i+1}, {}^f\boldsymbol{\ell}_{i+1}) = 0 \tag{50b}$$

If we next set

$$\delta \mathbf{R} = \mathbf{R} - \mathbf{S}(\mathbf{X}_i, {}^f\boldsymbol{\ell}_i) \qquad (51a)$$

$$\delta \mathbf{r} = \mathbf{X}_{i+1} - \mathbf{X}_i \qquad (51b)$$

we may write eq. (50a) in the familiar form

$$\delta \mathbf{r} = \mathbf{K}^{-1} \delta \mathbf{R} \qquad (52)$$

in which the stiffness matrix

$$\mathbf{K} = \mathbf{S}_{,X} = \mathbf{K}_E + \mathbf{K}_G \qquad (52a)$$

is composed of the elastic stiffness $\mathbf{K}_E$ and the geometric stiffness $\mathbf{K}_G$ the explicit form of which is to be found in [1] for the present large strain inelastic case.

The resulting procedure represents a Newton iterative method with respect to $\mathbf{X}$ combined with a normal iteration with respect to the stress-free lengths ${}^f\boldsymbol{\ell}$ of the elements. The calculation of the latter requires the approximate integration of the inelastic flow equations and thus dictates an incremental treatment of the deformation process.

In the dynamic case the external loads inducing displacements $\mathbf{r}$ velocity $\dot{\mathbf{r}}$ and acceleration $\ddot{\mathbf{r}}$, are equivalent to the nodal resultants of the internal stresses $\mathbf{S}$ and to the inertia forces operating at the nodes of the discretized body. The equations of motion thus read

$$\mathbf{R}(t) = \mathbf{S}(\mathbf{r}, \dot{\mathbf{r}}) + \mathbf{M}\ddot{\mathbf{r}} \qquad (53)$$

In eq. (53) $\mathbf{R}$ has been assumed to be solely dependent on the time t while $\mathbf{S}$ requires a knowledge of $\mathbf{r}, \dot{\mathbf{r}}$ and in particular the dependence arising in the large strain range. The mass matrix $\mathbf{M}$ kinematically consistent or lumped, is unaffected by any geometry changes and need therefore be calculated only once as long as conservation of mass can be presumed.

The numerical integration of eq. (53) calls for a step by step procedure, whereby the unknown displacements and velocities required for the calculation of the stresses at the end of each time increment are expressed in terms of the accelerations at the beginning and the end of the time interval. The acceleration for its part is controlled by eq. (53). Instead of the acceleration and higher order derivatives of the motion it is more convenient to work with the inertia forces

$$\mathbf{D} = \mathbf{M}\ddot{\mathbf{r}} = \mathbf{R} - \mathbf{S} \qquad (54)$$

and their time derivatives. The velocity of the inertia forces e.g. must satisfy the equation

$$\dot{\mathbf{D}} = \dot{\mathbf{R}} - \dot{\mathbf{S}} \tag{55}$$

With regard to the right-hand side of eq. (55) it seems not appropriate to include higher order derivatives of the inertia forces in our considerations because their accurate determination is hardly feasible.

Bearing this aspect in mind, approximations for the velocities and displacements at the end of a typical time increment

$$\tau = {}^{b}t - {}^{a}t \tag{56a}$$

may be presented in the general scheme

$${}^{b}\dot{\mathbf{r}} = {}^{a}\dot{\mathbf{r}} + \tau \mathbf{M}^{-1}(c_1\,{}^{a}\mathbf{D} + c_2\,{}^{a}\dot{\mathbf{D}} + c_3\,{}^{b}\mathbf{D} + c_4\,{}^{b}\dot{\mathbf{D}}) \tag{56b}$$

$${}^{b}\mathbf{r} = {}^{a}\mathbf{r} + \tau\,{}^{a}\dot{\mathbf{r}} + \tau^2 \mathbf{M}^{-1}(c_5\,{}^{a}\mathbf{D} + c_6\,{}^{a}\dot{\mathbf{D}} + c_7\,{}^{b}\mathbf{D} + c_8\,{}^{b}\dot{\mathbf{D}}) \tag{56c}$$

Assuming a certain variation of the inertia force $\mathbf{D}$ within the time step the coefficients c_i of $\mathbf{D}$, $\dot{\mathbf{D}}$ in eqs. (56b, c) may be chosen so that ${}^{b}\dot{\mathbf{r}}$, ${}^{b}\mathbf{r}$ follow as consistent approximations. Taylor expansions start always from values at the beginning of the time interval and lead therefore to explicit integration algorithms. Hermitean approximations on the other hand make use of the set of values at the beginning and the end of the time interval leading to implicit schemes. Table 1 summarizes the coefficients c_i of $\mathbf{D}$, $\dot{\mathbf{D}}$ at ${}^{a}t$, ${}^{b}t$ in eqs. (56b, c) appertaining to consistent approximations based either on Taylor expansions or on Hermitean approximations [5]. Alternatively, these coefficients may be determined from different considerations involving, for example, requirements on stability and/or convergence of the solution scheme. In particular, they may reproduce the unconditional stable algorithms derived and discussed in [9].

	$\mathbf{D}$	c_1	c_2	c_3	c_4	c_5	c_6	c_7	c_8
Taylor expansion	const.	1	0	0	0	1/2	0	0	0
	linear	1	$\tau/2$	0	0	1/2	$\tau/6$	0	0
Hermitean approx.	linear	1/2	0	1/2	0	1/3	0	1/6	0
	cubic	1/2	$\tau/12$	1/2	$-\tau/12$	21/60	$\tau/20$	3/20	$-\tau/30$

Table 1 Coefficients c_i of $\mathbf{D}$, $\dot{\mathbf{D}}$ in eqs. (56b, c) for consistent approximations of ${}^{b}\dot{\mathbf{r}}$, ${}^{b}\mathbf{r}$

If the incremental solution is to be achieved by vector iteration we start with ${}^{b}\boldsymbol{D}_i\,,\,{}^{b}\dot{\boldsymbol{D}}_i$ as estimates and obtain

$$ {}^{b}\dot{\boldsymbol{r}}_{i+1} = {}^{a}\dot{\boldsymbol{r}} + \tau\,\boldsymbol{M}^{-1}(c_1\,{}^{a}\boldsymbol{D} + c_2\,{}^{a}\dot{\boldsymbol{D}} + c_3\,{}^{b}\boldsymbol{D}_i + c_4\,{}^{b}\dot{\boldsymbol{D}}_i) \tag{57a} $$

$$ {}^{b}\boldsymbol{r}_{i+1} = {}^{a}\boldsymbol{r} + \tau\,{}^{a}\dot{\boldsymbol{r}} + \tau^2\boldsymbol{M}^{-1}(c_5\,{}^{a}\boldsymbol{D} + c_6\,{}^{a}\dot{\boldsymbol{D}} + c_7\,{}^{b}\boldsymbol{D}_i + c_8\,{}^{b}\dot{\boldsymbol{D}}_i) \tag{57b} $$

The vectors $\boldsymbol{S}\,,\,\dot{\boldsymbol{S}}$ equivalent to the stresses and their time rates at time ${}^{b}t$ can now be computed from the stress-strain relations using the derived kinematic quantities [1] . Next the inertia forces and their time rate at time ${}^{b}t$ are obtained from eqs. (54, 55) as

$$ {}^{b}\boldsymbol{D}_{i+1} = \boldsymbol{R}({}^{b}t) - \boldsymbol{S}({}^{b}\boldsymbol{r}_{i+1}\,,{}^{b}\dot{\boldsymbol{r}}_{i+1}) \tag{58a} $$

$$ {}^{b}\dot{\boldsymbol{D}}_{i+1} = \dot{\boldsymbol{R}}({}^{b}t) - \dot{\boldsymbol{S}}({}^{b}\boldsymbol{r}_{i+1}\,,{}^{b}\dot{\boldsymbol{r}}_{i+1}) \tag{58b} $$

and are used as new estimates for the subsequent iteration cycle. We arrive at a matrix scheme for the solution of the dynamic problem in each step by rewriting eq. (53) at time ${}^{b}t$ in the form

$$ \boldsymbol{R}(t) - [\boldsymbol{S}(\boldsymbol{X},{}^{f}\boldsymbol{\ell},\boldsymbol{T}) + \boldsymbol{D}(\boldsymbol{X})] = \boldsymbol{0} \tag{59} $$

where the index b has been omitted.

For the internal stresses the dependence indicated in eq. (48) is extended to include the temperatures $\boldsymbol{T}$ arising from the thermomechanical coupling in the adiabatic case. The latter are taken to satisfy in conjunction with the stress-free lengths ${}^{f}\boldsymbol{\ell}$ a relation of the form

$$ \tilde{\boldsymbol{f}}(\boldsymbol{\ell},{}^{f}\boldsymbol{\ell},\boldsymbol{T}) = \boldsymbol{0} \tag{60} $$

which will differ from eq. (49) used in the quasistatic case. In particular the approximate integration of the inelastic flow equations may also be based on the rate of deformation derivable from the velocity. This latter interpretation is covered by the dependence of $\boldsymbol{S}$ indicated in eq. (53). The dependence of the inertia force $\boldsymbol{D}$ from the nodal point positions $\boldsymbol{X}$ follows from eq. (56c) omitting time derivatives in the approximation, as

$$ \boldsymbol{D} = \frac{1}{c_7\tau^2}\,\boldsymbol{M}(\boldsymbol{r} - {}^{a}\boldsymbol{r} - \tau\,{}^{a}\dot{\boldsymbol{r}}) - \frac{c_5}{c_7}\,{}^{a}\boldsymbol{D} = \boldsymbol{D}(\boldsymbol{X}) \tag{61} $$

because of

$$ \boldsymbol{X} = {}^{0}\boldsymbol{X} + \boldsymbol{r} \tag{61a} $$

in which ${}^{o}\mathbf{X}$ denotes the nodal point positions at the initial unstressed state. The iterative solution of eqs. (59), (60) proceeds now analogously to that of eqs. (48), (49) governing the quasistatic case. The Newton step for constant ${}^{f}\boldsymbol{\ell}, T$ furnishes here

$$\mathbf{X}_{i+1} = \mathbf{X}_i + \tilde{\mathbf{K}}^{-1}\,\delta\tilde{\mathbf{R}} \tag{62}$$

where

$$\tilde{\mathbf{K}} = \mathbf{S}_{,X} + \mathbf{D}_{,X} = \mathbf{K} + C\,\mathbf{M} \tag{63}$$

$$C = \frac{1}{c_7\,\tau^2} \tag{63a}$$

is to be taken at $\mathbf{X}_i, {}^{f}\boldsymbol{\ell}_i, T_i$ and

$$\delta\tilde{\mathbf{R}} = \mathbf{R}(t) - [\mathbf{S}(\mathbf{X}_i, {}^{f}\boldsymbol{\ell}_i, T_i) + \mathbf{D}(\mathbf{X}_i)] \tag{63b}$$

As a next step eq. (60) is solved for the corresponding values of ${}^{f}\boldsymbol{\ell}, T$ satisfying

$$\tilde{f}(\boldsymbol{\ell}_{i+1}, {}^{f}\boldsymbol{\ell}_{i+1}, T_{i+1}) = 0 \tag{64}$$

and with $\mathbf{X}_{i+1}, {}^{f}\boldsymbol{\ell}_{i+1}, T_{i+1}$ as new estimates to be entered into eq. (62) the procedure is repeated until $\delta\tilde{\mathbf{R}}$ approaches zero.

5. Applications

The examples of this section have been computed with the programming system LARSTRAN developed at the Institut für Statik und Dynamik (ISD) in Stuttgart.

a) Steel projectile striking a rigid plate

The impact of a steel projectile on a rigid plate has been treated numerically in [8]. The present study concerns the behaviour of different integration schemes used in dynamic analysis as well as of the solution method within each increment of time. Furthermore, the influence of thermomechanical coupling on the results is examined. The problem is described in fig. 4, where the simple mesh of TRIAX3 elements is shown. The adopted material of the projectile is a nickel-chrome steel for which the rate dependence of the stress-strain relations is negligible. The material characteristics for uniaxial compression indicated in fig. 4 for the reference temperature ${}^{o}T$ = 293 K justify the assumption of an elastic-ideally plastic behaviour with respect to the current stress-free configuration. The yield stress is assumed to vary with $\vartheta = T - {}^{o}T$ in a parabolic manner described by $\sigma_\ell(\vartheta^2)$. In order to demonstrate also a stronger dependence, a linear variation $\sigma_\ell(\vartheta)$ is alternatively applied.

The deformation behaviour of the projectile possessing a striking velocity of V_0 = 300 m/s is analysed using the different algorithms described in section 4. The impact process is pursued numerically up to a response time of t = 20 μs. In the study of algorithmic behaviour isothermal conditions are assumed, i.e. a thermomechanical coupling is not taken into account. Results are shown in fig. 5 and represent the lateral and longitudinal movement of two characteristic points. A reference solution is obtained first by operating the implicit schemes with a time step of $\tau = 0.01\,\mu s$ and a strong criterion of termination for the iteration within each time interval. Both implicit schemes - linear and cubic Hermitean - yield the same results, requiring approximately the same computational effort in the case of the vector iteration. The linear scheme is also pursued using the matrix method of solution described in section 4. This technique converges more rapidly than the vector iterative solution, but the computational effort is somewhat higher due to the associated matrix operations. In all cases a high rate of convergence is achieved in the early stage of the iteration process. The same techniques are subsequently operated with a limit of three iterations and yield practically identical results for a word length of four to five digits. The higher convergence rate of the cubic Hermitean approximation is also confirmed. With reference to the explicit schemes we find that the results of the Taylor expansion based on the inertia force and its time derivative at the beginning of each step are approximately the same as the results of the reference solution. The explicit scheme ignoring the time derivative of the inertia force furnishes substantially different results. Operating the explicit algorithms with $\tau = 0.1\,\mu s$, i.e. ten times the former step length, the higher explicit scheme remains stable even in this case but with a reduced accuracy. The lower explicit scheme behaves in an unstable manner already in the early stages of the motion. An analogous behaviour takes place for the Hermitean schemes if used explicitly. Already after three iterations both approximations yield almost the same results, which in the present case ($\tau = 0.1\,\mu s$) do not reproduce however the reference solution because of the low number of iterations. Finally, the linear implicit scheme with $c_1 = c_2 = 1/2$, $c_5 = c_7 = 1/4$, $c_2 = c_4 = c_6 = c_8 = 0$ in eqs. (56b), (56c) (which is known to be unconditionally stable for linearly elastic problems [10]) is studied in connection with the matrix iterative method of incremental solution. It is found that in the present case of large strain inelastic deformations the algorithm behaves more stably than the consistent linear approximation. The influence of the numerical integration on the plastic strains is demonstrated in fig. 6.

Consider next the behaviour under adiabatic conditions when thermomechanical coupling becomes significant. Here the calculation is performed with the linear implicit scheme using a time step of τ = 0.01 μs . Fig. 7 indicates how the temperature dependence of the yield condition influences the deformation of the projectile during impact. Likewise fig. 8 shows the effect of thermomechanical coupling on the permanent strain distribution. A considerable increase of the temperature is obtained near the face of impact (fig. 9). This however is not sufficient to activate a strong variation of the yield limit within the parabolic dependence $\sigma_\ell(\vartheta')$. The temperature increase along the axis of the projectile is shown in fig. 10 at the beginning and end of the impact. In fig. 11 the final shape of the projectile is compared with the stage of deformation at t = 1 μs after the initiation of the impact.

b) Impact of an aluminium bar

While in the former example the overall behaviour of the body is considered, the present study focuses attention on more detailed local phenomena accompanying the impact of an aluminium bar. In addition, the influence of an assumed rate dependence of the yield stress is examined.

The initial geometry of the bar is reproduced in fig. 12 together with the finite element discretization. The TRIAX3 mesh is considerably finer in the present case, in accordance with the required detailed information. The aluminium bar is assumed to be subjected to an impact on the lower face, imposing there an initial velocity of V_o = 180 m/s while its upper face is fixed in the axial direction. In this case thermomechanical coupling may be assumed not to have a significant influence and is hence neglected.

The impact process is calculated up to a response time somewhat in excess of t = 15 μs using a time step of τ = 0.02 μs with the consistent linear Hermitean scheme applied explicitly. For this time step the implicit vector iteration scheme converges. Thus, in accordance with the discussion of algorithmic behaviour in the foregoing example, the results are expected to be accurate enough. Inviscid material is first considered for which fig. 13 illustrates the distribution of the equivalent plastic strain within the bar at t = 7.5 μs and at t = 15 μs after impact. The figure clearly demonstrates the high permanent deformation at the face under impact, where the strain concentration near the edge is seen to be higher than that near the axis. The opposite is true at a distance of approximately 1/8 times the diameter from the face under impact; here the permanent strain is concentrated near the axis.

Subsequently, the viscoplastic description for the inelastic part of the deformation is applied in place of the inviscid plastic flow theory. To this end the yield stress of the material is assumed to exhibit an increase of 50 % for a certain permanent strain rate. Taking this strain rate as 10^{-4}, 10^{-3}, 10^{-2} $(\mu s)^{-1}$, it follows that the fluidity μ attains twice these values. Fig. 14 compares the evolution of the permanent strain during impact near the edge at the impacted end of the bar both for the elasto-plastic and the elasto-viscoplastic materials. The figure demonstrates also that the inelastic wave is reaching the fixed end of the bar at approximately $t = 15\,\mu s$; note that the plastic region indicated in fig. 13 is associated with higher values of permanent strain. Due to the high inelastic deformation rates the dynamic yield stress introduced by the viscoplastic approach increases considerably (fig. 15), yielding a corresponding increase of the von Mises equivalent stress. With increasing fluidity μ the dynamic and static yield stresses become closer as the former decreases and the latter increases; the opposite is true for decreasing fluidity. In the plastic case the equivalent stress is of course restricted by the static yield condition and during the impact process remains close to the initial yield limit of the material as shown in fig. 15.

c) Dynamic heading of a steel bolt

As a final investigation the forming process concerning the dynamic heading of a steel bolt is performed. The problem is described in fig. 16, where the elasto-plastic properties of the steel material exhibiting a temperature-dependent yield stress are also given. The bottom of the cylinder is constrained while the head is formed during the impact by a ram - this causes the part of the cylinder which is initially unconstrained to be reduced by 60 %. As the deformation develops, the free cylinder surface comes into contact with the ram as well as with the surrounding material, both assumed rigid. This fact requires flexible boundary conditions. A frictional force per unit area acts on the impacted top surface and is assumed to vary linearly with the radius. The intensity of the frictional forces shown in fig. 17 is such that the average frictional stress equals 0.24 times the initial yield stress of the material in shear and is kept constant over the entire process. Fig. 17 illustrates the finite element mesh of TRIAX3 elements.

The desired 60 % reduction of the cylinder corresponding to 18 mm downward movement of the top surface is imposed by a ram velocity of $V = 9$ m/s that remains constant over the forming process. In order to study high inertia effects, the computations are also carried out for a ram velocity of $V = 90$ m/s. In both cases the linear implicit scheme is applied in its unconditionally stable form (cf. the first example in this section). In this connection the matrix method of solution presented in section 4 as a modification of the static algorithm for large strain inelastic analysis in the presence of inertia effects is used. The computation implies 600 equal increments of prescribed displacements of the top with three iteration cycles in each step. The computational effort is approximately twice that for the purely static case. Fig. 18 shows the development of the equivalent plastic strain at four characteristic points of the cylinder during the dynamic forming process. The higher forming velocity clearly shifts the permanent deformation to the regions near the impacted surface mainly in the early stages of the process; the difference diminishes as time proceeds. Fig. 19 compares the deformed meshes resulting at an overall reduction of 10 % and 60 %, respectively, for the two forming velocities. While essential differences occur at the beginning of the forming process, the final configurations ultimately appear almost identical. For this reason the distribution of $\bar{\sigma}$ and $\vartheta = T - {}^{\circ}T$, ${}^{\circ}T = 293$ K at the final stage of 60 % reduction is reproduced in fig. 20 only for the process velocity of $V = 9$ m/s, the results for $V = 90$ m/s being not substantially different.

References

[1] J.H. Argyris, J.St. Doltsinis, On the large strain inelastic analysis in natural formulation, Part I, Quasistatic problems, Comp. Meths. Appl. Mech. Eng. 20, Nr. 2, Part II, Dynamic problems, Comp. Meths. Appl. Mech. Eng. 21, Nr. 1.

[2] A.E. Green, P.M. Naghdi, A general theory of an elastic-plastic continuum, Arch. Rat. Mech. An. 18 (1965) 251-281

[3] E.H. Lee, T. Wierzbicki, Analysis of the propagation of plane elastic-plastic waves at finite strain, J. Appl. Mechs. 34 (1967) 931-935

[4] J.H. Argyris, J.St. Doltsinis, W.C. Knudson, L.E. Vaz, K.J. Willam, Numerical solution of transient nonlinear problems, Comp. Meths. Appl. Mech. Eng. 17 (1979) 341-409

[5] J.H. Argyris, P.C. Dunne and Th. Angelopoulos, Nonlinear oscillations using the finite element technique, Comp. Meths. Appl. Mech. Eng. 2 (1973) 203-250

[6] F.D. Murnaghan, Finite deformations of an elastic solid, Am. J. Mathcs. 59 (1937) 235-260

[7] P. Perzyna, The constitutive equations for rate sensitive plastic materials, Q. Appl. Math. 20 (1962) 321-332

[8] J.H. Argyris, H. Balmer, J.St. Doltsinis, P.C. Dunne, M. Haase, M. Kleiber, G.A. Malejannakis, H.-P. Mlejnek, M. Müller, D.W. Scharpf, Finite element method - the natural approach, Comp. Meths. Appl. Mech. Eng. 17/18 (1979) 1-106

[9] J.H. Argyris, P.C. Dunne and Th. Angelopoulos, Dynamic response by large step integration, Earthq. Eng. Str. Dyn. 2 (1973) 185-203

[10] N.M. Newmark, A method of computation for structural dynamics, Proc. ASCE 85 (1959) 67-94

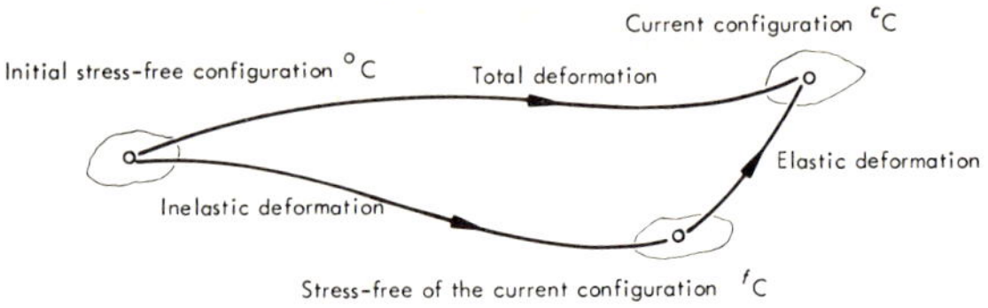

Fig. 1 Elastic and inelastic parts of deformation

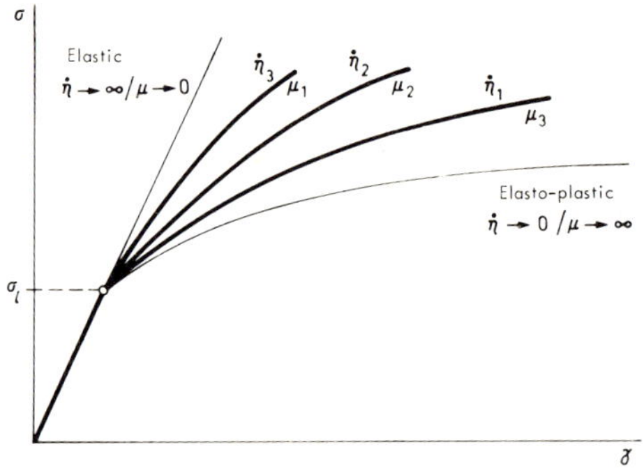

Fig. 2 Range of elasto-viscoplastic behaviour

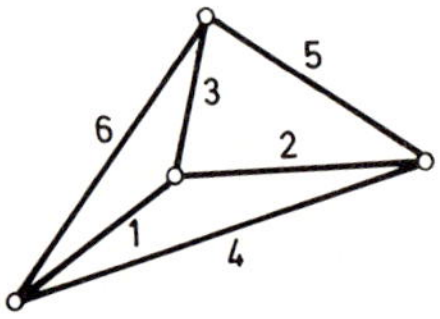

Fig. 3 Definition of natural directions 1 to 6 on the tetrahedron element

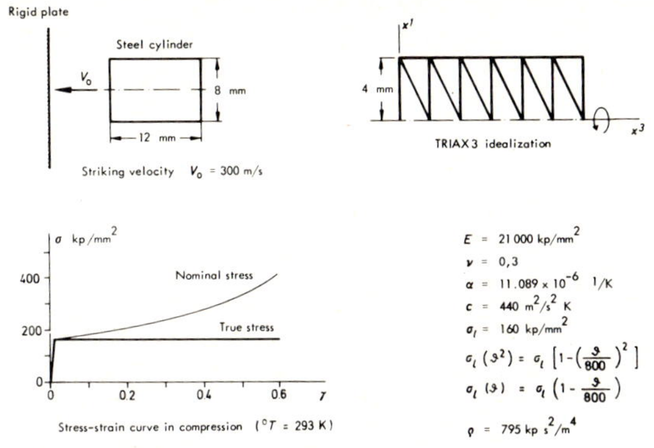

Fig. 4 Steel projectile striking a rigid plate

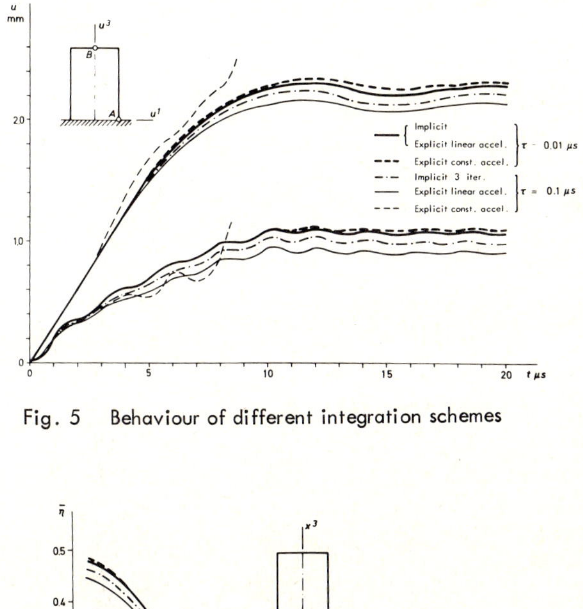

Fig. 5 Behaviour of different integration schemes

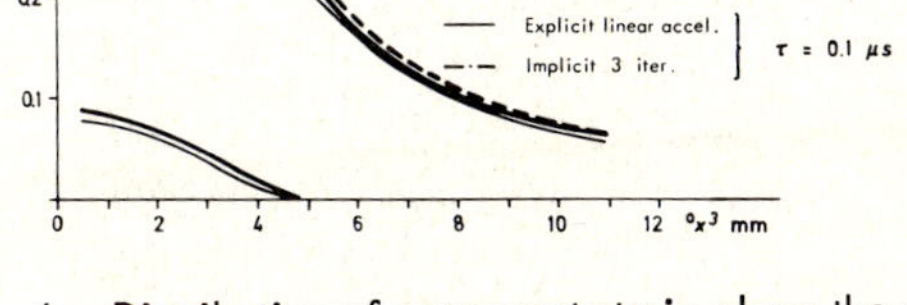

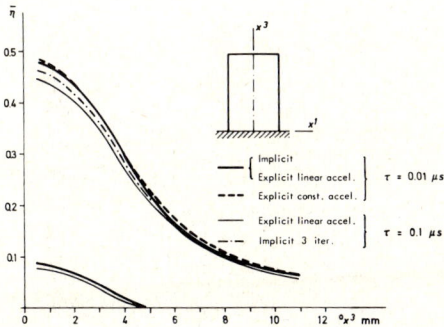

Fig. 6 Distribution of permanent strain along the axis

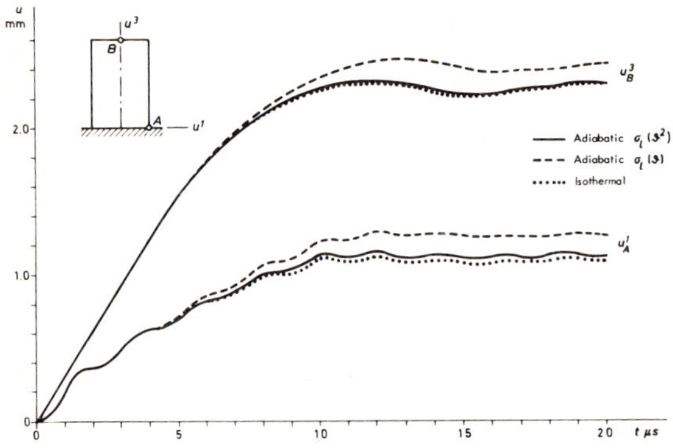

Fig. 7
Influence of thermomechanical coupling on the motion

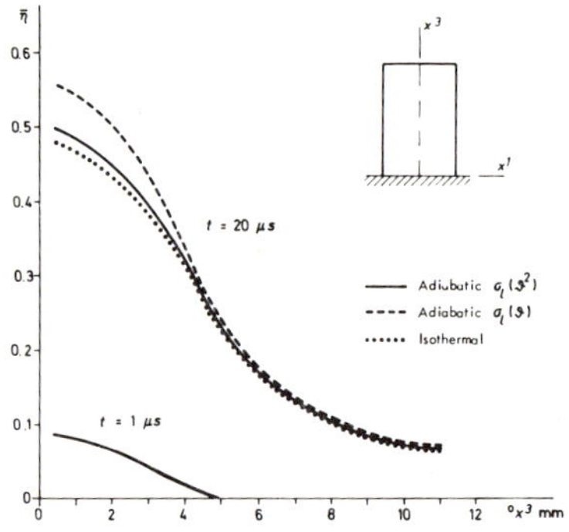

Fig. 8
Influence on the permanent strain

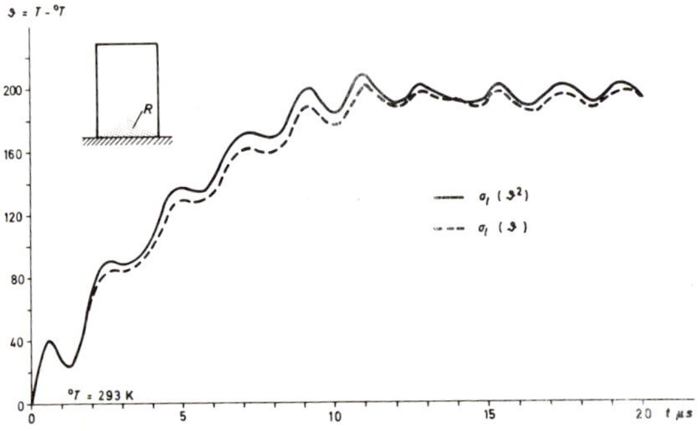

Fig. 9
Variation of temperature near the face of impact (region R)

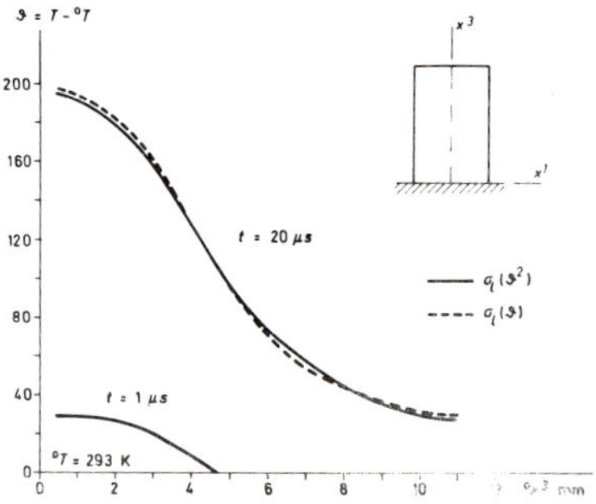

Fig. 10
Distribution of temperature

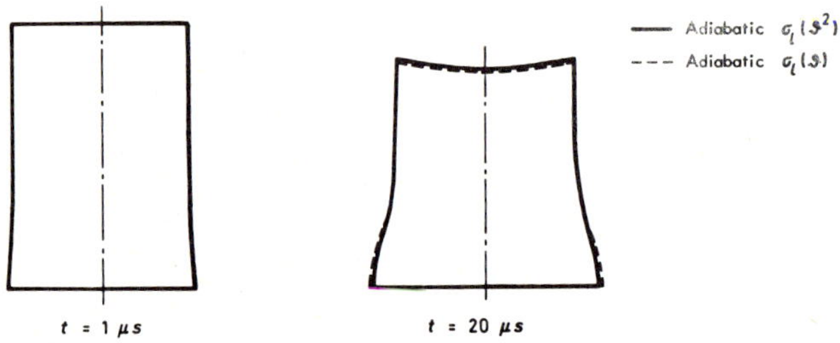

Fig. 11 Initial and final stages of deformation

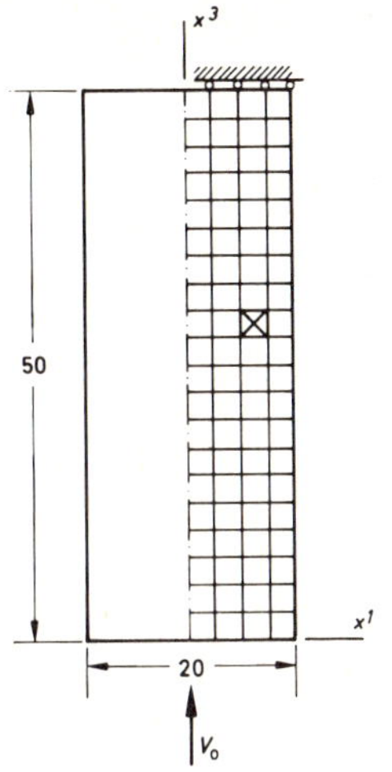

$E = 7000\ \mathrm{kp/mm^2}$

$\nu = 0.3$

$\sigma_l = 20\ \mathrm{kp/mm^2}$

$\sigma_y = \sigma_l + 115.5\ \eta$

Impact velocity

$V_0 = 180\ \mathrm{m/s}$

Fig. 12 Impact of an aluminium bar

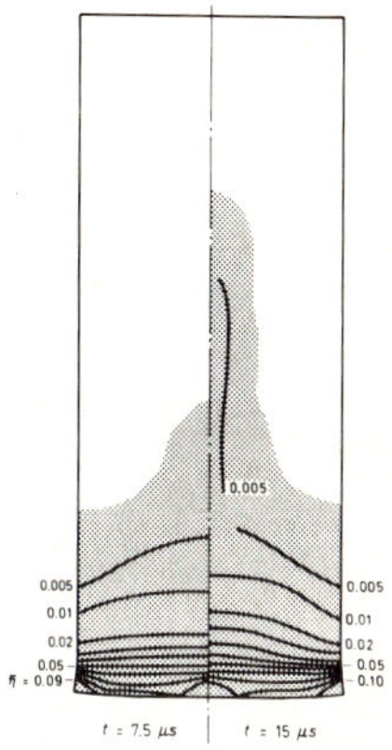

Fig. 13 Distribution of permanent strain

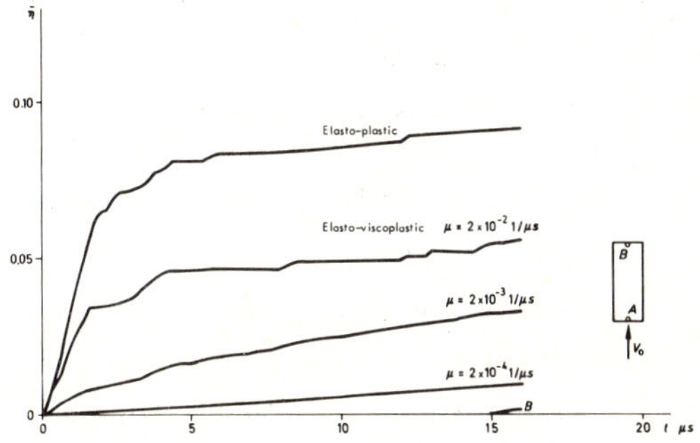

Fig. 14 Development of permanent strain during impact (point A)

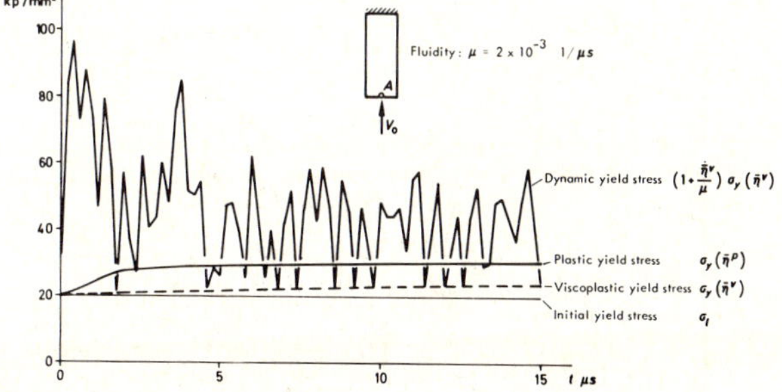

Fig. 15 Increase of initial yield stress during impact (point A)

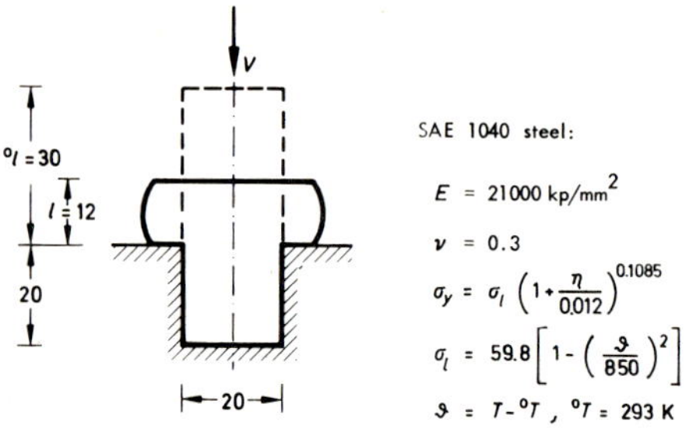

Fig. 16 Dynamic heading of a steel bolt

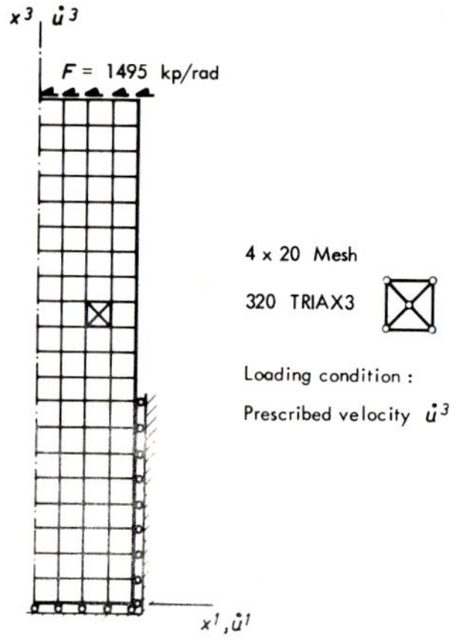

Fig. 17 Finite element discretization

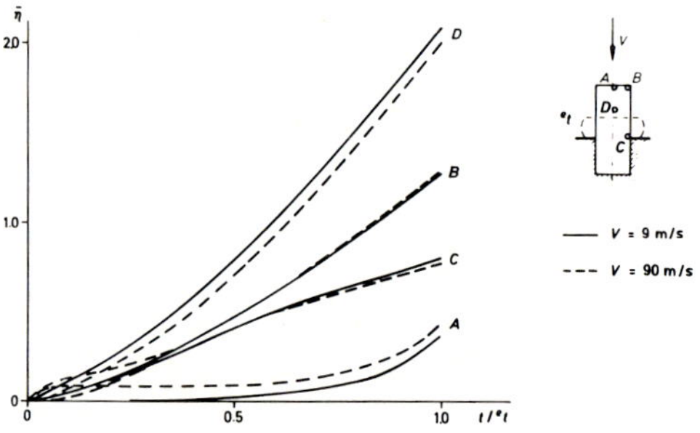

Fig. 18 Development of permanent strain

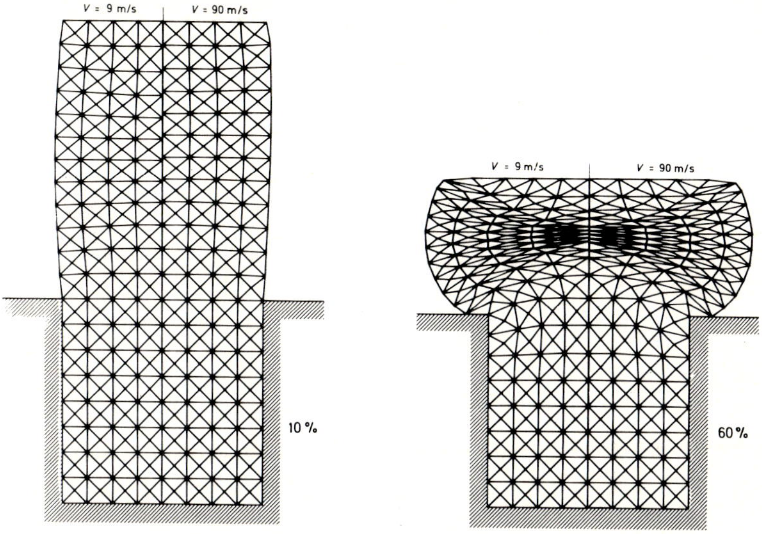

Fig. 19 Deformed configurations: comparison of two forming velocities

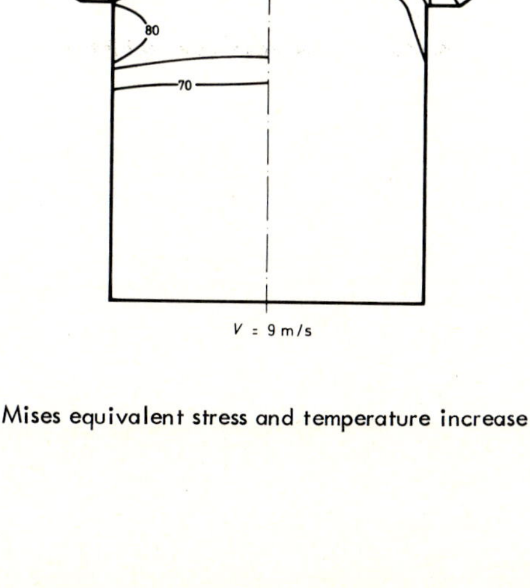

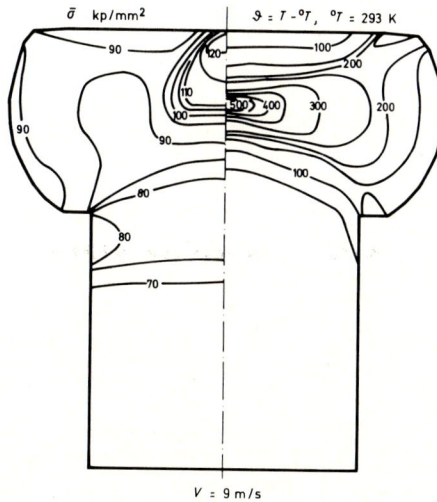

Fig. 20 Von Mises equivalent stress and temperature increase at 60 % reduction

COMPUTING METHODS IN APPLIED SCIENCES AND ENGINEERING
R. Glowinski, J.L. Lions (editors)
North-Holland Publishing Company

THE FICTITIOUS REGIONS METHOD IN PROBLEMS OF MATHEMATICAL PHYSICS

A.N.Konovalov
Institute of pure and applied mechanics
Novosibirsk, USSR

The paper deals with the results obtained by the author,his collaborators by application and foundation of the fictitious regions method in problems of mathematical physics.

Main difficulties of the numerical solution of problems of mathematical physics are connected with a construction of a mesh near the curvelinear boundary γ of a region D . The fictitious regions method is expected to allow to a certain degree to avoid this difficulty.The main idea of this method is the following. By means of a fictitious region D_1 we shall supplement D to some other region D_0 with the boundary Γ . Usually D_1 is chosen as complement of D to the rectangle in view of the fact that this choice has an irrefutable practical advantage for numerical solution of problems by the finite difference method. Of course, in theoretical consideration such a choice of D is not obligatory in the least.

Someway, dependence of the type of boundary conditions on γ the coefficients of the initial problem are continued to D_1 and some agreement conditions on γ are assumed. As a rule these conditions have a real physical sense, for example, equality of saturation and expenditure on the boundary γ in the problem of porous flow; continuity of displacement vector on the boundary γ , equality of forces with respect to their value and their contrusts with respect to their direction on the boundary γ in elasticity problems.

On the external boundary we can always consider the Direchlet boundary problem. Now finding a solution of an initial problem in D is reduced to finding a solution of an auxiliary problem with discontinuous coefficients but in a simpler region.

At once note, that as it followes from agreement conditions, already the first derivatives of a solution of the auxiliary problem will be discontinuous on γ. So we have to consider only the problem of finding the generalized solution of an auxiliary problem.From different (equivalent) definitions of generalized solution more constructive for practical purposes is that associated with execution of some integral conservation law for any "reserved contour" entirely contained in a solution region. Approximation of this integral conservation law leads to conservative difference scheme.

Construction of a conservative difference scheme in those cases when the line of discontinuity coincides with a coordinate line of mesh provides execution of agreement conditions on lines of discontinuity, in particular, those which have been mentioned above. It permits to do a numerical calculation of auxiliary problem by uniform manner without special distinguishing of discontinuities. So foundation of fictitious regions method in the first approach can be reduced to foundation of proximity in D of solutions of initial and auxiliary problems.

It should be mentioned at once that such a consideration is especially convenient for the creation of the programmes for solution of class problems (packet of programmes). The fictitious regions method in given lower wordings allows to consider each problem from a class of initial problems only in standard regions (restangle, restangular parallelepiped), leaving in quality of individual properties of the problem the collection of its coefficients and right parts. As it is imagined such (geometrical) a description has undoubted advantage to an logical description of an arbitrary region for concrete programming.

Although the results knownto author relating to fictitious regions method are connected with the names of a lot of authors (rather detailed bibliography one can find in the papers which are quoted lower), but exactly from the works of V.K.Sauliev and V.I.Lebedev fictitious regions method is considered as a means of a solving of practical problems.

Thus according to fictitious regions method instead of solving an initial problem "u" in a region D we find a solution of an auxiliary problem "u_ε" in D_0. Then the difference problem finding $u_{\varepsilon h}$ in D_{0h} is formulated.

But, so far as

$$\|u_{\varepsilon h} - u\|_D \leq \|u_{\varepsilon h} - u_h\|_D + \underline{\|u_h - u\|_D},$$

or

$$\|u_{\varepsilon h} - u\|_D \leq \|u_{\varepsilon h} - u\|_D + \underline{\|u_{\varepsilon h} - u_\varepsilon\|_D},$$

and for considered lower problems we have either estimation methods or estimations of the emphasized members themselves, then the main attention is paid to obtaining estimations of a type

$$\|u_\varepsilon - u\|_D = O(\varepsilon^\alpha), \quad \|u_{\varepsilon h} - u_h\| = O(\varepsilon^\beta)$$
$$0 < \varepsilon \ll 1, \quad \alpha > 0, \quad \beta > 0 \tag{1}$$

in such or other norms. Usually, it is norms of Sobolev spaces W_2^p. In stationary problems $u_{\varepsilon h}$ itself as a rule is found by means of iterative methods. Therefore we specially research a rate of convergence of iterative methods for finding $u_{\varepsilon h}$.

Problem 1 is torsion of homogeneous isotropic cylindrical beam. In a plane one-connected domain D with a boundary γ a solution of Neumann problem is found:

$$\Delta\varphi = 0, \qquad \frac{\partial\varphi}{\partial n}\Big|_\gamma = y\cos(nx) - x\cos(ny) \tag{2}$$

where φ is torsion function. Problem (2) allows the following wording:

$$\frac{1}{\mu}\Delta R + 2 = 0, \qquad R_{|\gamma} = 0.$$

Here

$$R = \mu\varphi, \quad \varphi = \psi - \tfrac{1}{2}(x^2 + y^2), \tag{3}$$

$$\frac{\partial\varphi}{\partial x} = \frac{\partial\psi}{\partial y}, \qquad \frac{\partial\varphi}{\partial y} = -\frac{\partial\psi}{\partial x}$$

and $\mu > 0$ is Lame's constant. The auxiliary problem for (3) is determined in the following way:

$$\begin{gathered}\frac{1}{\mu}\Delta R_\varepsilon + 2 = 0, \quad M \in D, \qquad \frac{1}{\mu\varepsilon^2}\Delta R_\varepsilon + 2 = 0, \quad M \in D_1, \\ R_{\varepsilon|\Gamma} = 0, \quad R^+_{\varepsilon|\gamma} = R^-_{\varepsilon|\gamma}, \quad \frac{\partial R^+_\varepsilon}{\partial n}\Big|_\gamma = \frac{1}{\varepsilon^2}\frac{\partial R^-_\varepsilon}{\partial n}\Big|_\gamma .\end{gathered} \tag{4}$$

Signs plus or minus in agreement conditions mean that the corresponding value is a limit value while a point M approaches to γ within or outside the domain D.

Theorem 1. Let $R \in W_2^2(D) \cap \overset{\circ}{W}{}_2^1(D)$ be a solution of the problem(3), $R_\varepsilon \in \overset{\circ}{W}{}_2^1(D_0)$ be a solution of the problem (4). Then

$$\|R - R_\varepsilon\|_{W_2^1(D)} \leq K_1\varepsilon. \tag{5}$$

Here and further K_m means independent of ε constant.

The other construction of an auxiliary problem is possible, exactly

$$\begin{gathered}\frac{1}{\mu}\Delta R_\varepsilon + 2 = 0, \quad M \in D, \quad \frac{1}{\mu}\Delta R_\varepsilon - \frac{R_\varepsilon}{\varepsilon^2} = 0, \quad M \in D_1, \\ R_{\varepsilon|\Gamma} = 0.\end{gathered} \tag{6}$$

Theorem 2. Let $R \in W_2^2(D) \cap \overset{\circ}{W}{}_2^1(D)$ be a solution of the problem (3), $R_\varepsilon \in W_2^1(D_0)$ be a solution of the problem (6). Then

$$\|R - R_\varepsilon\|_{W_2^1(D)} \leq K_2\varepsilon. \tag{7}$$

Problem II is static elasticity problem.

$$A\vec{u} = \vec{f}, \qquad A\vec{u} = -\mu\Delta\vec{u} - (\lambda + \mu)\,\text{grad div}\,\vec{u}$$

$$a)\ \vec{u}_{/\gamma}=0 \quad ; \quad b)\ T\vec{u}\cdot\vec{n}_{/\gamma}=0 \tag{8}$$

$\lambda>0$, $\mu>0$ are Lame's constants, $T\vec{u}$ is stress tensor, corresponding to a displacement vector $\vec{u}$, $\vec{n}$ is a vector of an external normal to γ. The problem (8a) is an equilibrium problem of an elastic body with fixed border. The problem (8b) is not always solvable. For its solvability it is necessary to assume

$$\int_D \vec{f}\,d\mathcal{M}=0, \qquad \int_D \vec{z}\times\vec{f}\,d\mathcal{M}=0. \tag{9}$$

The only solution of the problem (8b), (9) we shall share by means of conditions

$$\int_D \vec{u}\,d\mathcal{M}=\vec{C}_1, \qquad \int_D \operatorname{rot}\vec{u}\,d\mathcal{M}=\vec{C}_2, \tag{10}$$

where C_1, C_2 - some vector constants, fixing a progressive transference of the body and its rotation as an indivisible whole. The auxiliary problem for (8b), (9), (10) problems is formulated in the following way:

$$\begin{aligned} &A\vec{u}_\varepsilon=\vec{f}, \quad \mathcal{M}\in D, \qquad A_\varepsilon\vec{u}_\varepsilon=0, \quad \mathcal{M}\in D_1, \\ &\vec{u}_{\varepsilon/\Gamma}=0, \quad \vec{u}^+_{\varepsilon/\gamma}=\vec{u}^-_{\varepsilon/\gamma}, \quad T\vec{u}^+_\varepsilon\cdot\vec{n}_{/\gamma}=T_\varepsilon\vec{u}^-_\varepsilon\cdot\vec{n}_{/\gamma}. \end{aligned} \tag{11}$$

Theorem 3. Let $\vec{u}\in W_2^2(D)\cap \overset{o}{W}{}_2^1(D)$ be a solution of the problem (8a), $\vec{u}_\varepsilon\in \overset{o}{W}{}_2^1(D_0)$ be a solution of the problem (11), in which A_ε, T_ε are calculated with constans $\lambda_\varepsilon=\frac{\lambda}{\varepsilon^2}$, $\mu_\varepsilon=\frac{\mu}{\varepsilon^2}$. Then

$$\|\vec{u}-\vec{u}_\varepsilon\|_{W_2^1(D)} \le K_3\,\varepsilon. \tag{12}$$

Theorem 4. Let $\vec{u}\in W_2^1(D)$ be a solution of the problem (8b), $\vec{u}_\varepsilon\in \overset{o}{W}{}_2^1(D_0)$ be a solution of a problem (11), in which A_ε, T_ε are calculated with constants $\lambda_\varepsilon=\lambda\varepsilon^2$, $\mu_\varepsilon=\mu\varepsilon^2$. Then there exist such constants $\vec{C}_1$, $\vec{C}_2$, that

$$\|\vec{u}-\vec{u}_\varepsilon\|_{W_2^1(D)} \le K_4\,\varepsilon. \tag{13}$$

For the problem (8a) analogous to (6) construction of auxiliary problem is possible

$$A\vec{u}_\varepsilon=\vec{f}, \quad \mathcal{M}\in D, \quad A\vec{u}_\varepsilon+\frac{\vec{u}_\varepsilon}{\varepsilon^2}=0, \quad \mathcal{M}\in D_1, \quad \vec{u}_{\varepsilon/\Gamma}=0. \tag{14}$$

Theorem 5. Let $\vec{u}\in W_2^1(D)\cap \overset{o}{W}{}_2^1(D)$ be a solution of a problem (8a), $\vec{u}_\varepsilon\in \overset{o}{W}{}_2^1(D_0)$ be a solution of the problem (14).
Then

$$\|\vec{u}-\vec{u}_\varepsilon\|_{W_2^1(D)} \le K_5\,\varepsilon. \tag{15}$$

The structure of the auxiliary problems of the type (6), (14)for the first time was suggested by V.I.Lebedev.

The obtained results are generalized for the case of the mixed boundary conditions in the problem 2 (A.N.Bugrov, A.N.Konovalov, V.I.Kramarenko)

$$\vec{u}_{/\gamma_1}=0, \qquad T\vec{u}\cdot\vec{n}_{/\gamma_2}=0\,, \qquad \gamma=\gamma_1+\gamma_2\,. \tag{16}$$

Problem III is the model problem of porous flow of two-phase incompressible fluid with capillary forces. In cylinder $Q=\{D\times [0<t\leqslant T]\}$ with the side surface $S=\{\gamma\times[0\leqslant t\leqslant T]\}$ a solution of Cauchy mixed problem is found.

$$\begin{aligned} \frac{\partial u_1}{\partial t}-\frac{\partial u_2}{\partial t} &= \lambda_1\Delta u_1+f_1 \\ -\frac{\partial u_1}{\partial t}+\frac{\partial u_2}{\partial t} &= \lambda_2\Delta u_2+f_2 \end{aligned} \qquad \lambda_i>0 \tag{17}$$

$$u_1(M,0)-u_2(M,0)=\psi(M)\,, \quad M\in D\,,$$

a) $u_{i/S}=0\,,$ b) $\dfrac{\partial u_i}{\partial n}\Big|_S=0\,.$

In this case the most simple auxiliary problem is the following:

$$\begin{aligned} \frac{\partial u_{1\varepsilon}}{\partial t}-\frac{\partial u_{2\varepsilon}}{\partial t} &=\lambda_1\Delta u_{1\varepsilon}+f_1 \\ -\frac{\partial u_{1\varepsilon}}{\partial t}+\frac{\partial u_{2\varepsilon}}{\partial t} &=\lambda_2\Delta u_{2\varepsilon}+f_2 \end{aligned} \quad M\in D \qquad \begin{aligned} \frac{\partial u_{1\varepsilon}}{\partial t}-\frac{\partial u_{2\varepsilon}}{\partial t} &=\lambda_{1\varepsilon}\Delta u_{1\varepsilon} \\ -\frac{\partial u_{1\varepsilon}}{\partial t}+\frac{\partial u_{2\varepsilon}}{\partial t} &=\lambda_{2\varepsilon}\Delta u_{2\varepsilon} \end{aligned} \quad M\in D_1 \tag{18}$$

$$u_{i\varepsilon/S_0}=0\,, \qquad u_{1\varepsilon}(M,0)-u_{2\varepsilon}(M,0)=\begin{cases}\psi(M)\,, & M\in D\\ 0\,, & M\in D_1\end{cases}$$

$$u^+_{i\varepsilon/S}=u^-_{i\varepsilon/S}\,, \qquad \lambda_i\frac{\partial u^+_{i\varepsilon}}{\partial n}\Big|_S=\lambda_{i\varepsilon}\frac{\partial u^-_{i\varepsilon}}{\partial n}\Big|_S\,.$$

Where $S_0=\{\Gamma\times[0<t\leqslant T]\}$, $\lambda_{i\varepsilon}=\lambda_i\,\varepsilon^{2\alpha}$ and $\alpha=-1$ for the problem (17a) and $\alpha=1$ for the problem (17b).

Theorem 6. Let $u_i\in W_2^{2,0}(Q)\cap \overset{\circ}{W}{}_2^{1,0}(Q)$ be a solution of the problem (17a), $u_{i\varepsilon}\in \overset{\circ}{W}{}_2^{1,0}(Q_0)$ be a solution of the problem (18). Then

$$\|u_{i\varepsilon}-u_i\|_{W_2^{1,0}(Q)}\leqslant K_6\,\varepsilon. \tag{19}$$

For the problem (17a) the structure of the auxiliary problem of type (6), (14) is possible.

Theorem 7. Let $u_i\in W_2^{1,0}(Q)$ be the solution of the problem (17b), $u_{i\varepsilon}\in \overset{\circ}{W}{}_2^{1,0}(Q_0)$ be a solution of the problem (18). Then

$$\|u_{i\varepsilon}-u_i\|_{W_2^{1,0}(Q)}\leqslant K_7\,\varepsilon. \tag{20}$$

Constants K_6, K_7 are uniformly restricted for all $t \in [0<t\leq T]$.

Obtained results are also true for the problem (17) with boundary conditions (J.L.Korobicina, K.Utegenov)

$$u_{1/s}=0, \qquad \frac{\partial u_2}{\partial n}\Big|_s=0. \tag{21}$$

Problem IY is a problem about cross bending of a thin plate.

$$\lambda\Delta\Delta u=f, \quad \mathcal{M}\in D\subset R^2, \quad \lambda>0.$$

$$\text{a)} \quad u_{/\gamma}=0, \quad \frac{\partial u}{\partial n}\Big|_\gamma=0, \tag{22}$$

$$\text{b)} \quad \frac{\partial^2 u}{\partial n^2}+\nu\left(\frac{\partial^2 u}{\partial \tau^2}-\frac{1}{\rho}\frac{\partial u}{\partial n}\right)_{/\gamma}=0, \quad \frac{\partial}{\partial n}\Delta u+(1-\nu)\frac{\partial}{\partial \tau}\frac{\partial^2 u}{\partial n\partial \tau}\Big|_\gamma=0.$$

Here $\lambda>0$ - rigidity, $\frac{\partial}{\partial \tau}$ means differentiation along a tangent, ν - Poisson coefficient, ρ - a radius of curvature γ.

For the problem (22b) it must be executed the solvability conditions

$$\int_D f\,d\mathcal{M}=0, \quad \int_D x_i f\,d\mathcal{M}=0 \quad ; \quad i=1,2. \tag{23}$$

In this case the only solution is standed out by conditions

$$\int_D u\,d\mathcal{M}=C_0, \quad \int_D x_i u\,d\mathcal{M}=C_i, \quad i=1,2, \tag{24}$$

where C_0, C_1, C_2 are constants.

We shall formulate an auxiliary problem in the following way

$$\frac{\partial^2}{\partial x_1^2}\left[\lambda_\varepsilon\left(\frac{\partial^2 u_\varepsilon}{\partial x_1^2}+\nu\frac{\partial^2 u_\varepsilon}{\partial x_2^2}\right)\right]+2\frac{\partial^2}{\partial x_1\partial x_2}\left[\lambda_\varepsilon(1-\nu)\frac{\partial^2 u_\varepsilon}{\partial x_1\partial x_2}\right]+$$

$$+\frac{\partial^2}{\partial x_2^2}\left[\lambda_\varepsilon\left(\nu\frac{\partial^2 u_\varepsilon}{\partial x_1^2}+\frac{\partial^2 u_\varepsilon}{\partial x_2^2}\right)\right]+\omega_\varepsilon u_\varepsilon=f_\varepsilon, \quad \mathcal{M}\in D_0, \tag{25}$$

$$u_{\varepsilon/\Gamma}=0, \qquad \frac{\partial u_\varepsilon}{\partial n}\Big|_\Gamma=0.$$

In (25)

$$\lambda_\varepsilon, \omega_\varepsilon, f_\varepsilon = \begin{cases} \lambda, & 0, & f, & \mathcal{M}\in D \\ \varepsilon^\alpha\lambda, & \frac{\omega}{\varepsilon^\beta}, & 0, & \mathcal{M}\in D_1 \end{cases} \tag{26}$$

and $\omega=0$, or $\omega=1$. On γ agreement conditions are laied down

$$u_\varepsilon^+=u_\varepsilon^-, \quad \frac{\partial u_\varepsilon^+}{\partial n}=\frac{\partial u_\varepsilon^-}{\partial n}, \quad \ell_i u_\varepsilon^+=\varepsilon^\alpha\ell_i u_\varepsilon^- \qquad i=1,2 \tag{27}$$

where

$$\ell_1 v\equiv\lambda\left[\frac{\partial^2 v}{\partial n^2}+\nu\left(\frac{\partial^2 v}{\partial \tau^2}-\frac{1}{\rho}\frac{\partial v}{\partial n}\right)\right],$$

$$\ell_2 v = \lambda \left[\frac{\partial}{\partial n}\Delta v + (1-\nu)\frac{\partial}{\partial \tau}\,\frac{\partial^2 v}{\partial n \partial \tau})\right].$$

For the problem (22a) following continuations are possible: $\omega = 0$, $\alpha < 0$; $\omega = 1, \beta > 0, \alpha < 0$; $\alpha = 0$, $\omega = 1, \beta > 0$. In the last case agreement conditions become unnecessary. For the problem (22b) the continuation $\alpha > 0, \omega = 0$ is possible.

Omiting formulations of theorems, we notice, that all above-mentioned continuations were studied by A.N.Bugrov and for each of them the first estimation of type (1) was obtained.

He also obtained the first estimation of type (1) for a system of equations of Navey-Stoks both in a formulation $\vec{v}$, p

$$\frac{\partial \vec{v}}{\partial t} + v_\kappa \frac{\partial \vec{v}}{\partial x_\kappa} = \nu \Delta \vec{v} - \operatorname{grad} p + \vec{f} \qquad \kappa = 1,2$$

$$\operatorname{div} \vec{v} = 0 \tag{28}$$

$$\vec{v}_{/t=0} = \vec{a}(\mathcal{U}), \qquad \vec{v}_{/\gamma \times [0,T]} = 0, \quad \operatorname{div} \vec{a} = 0, \quad \vec{a}_{/\gamma} = 0,$$

and in formulation ψ, ψ is a function of current.

$$\frac{\partial}{\partial t}(-\Delta \psi) + \nu \Delta\Delta \psi + \frac{\partial}{\partial x_2}\left(\frac{\partial \psi}{\partial x_1}\Delta \psi\right) - \frac{\partial}{\partial x_1}\left(\frac{\partial \psi}{\partial x_2}\Delta \psi\right) = g, \tag{29}$$

$$\psi_{/t=0} = \psi_0(\mathcal{U}), \qquad \psi_{/\gamma \times [0,T]} = \frac{\partial \psi}{\partial n}{}_{/\gamma \times [0,T]} = 0.$$

Now some words about the second estimation (1).

While obtained theoretical results are concerned with the regions D_h of types "corner" and "triangle". Such a restriction is connected with used methods of proof.

Let $D = D_0 - D_1$ and $D_0 = (0 \le x_\kappa \le 1)$, $D_1 = (0.5 \le x_\kappa \le 1)$. In D, D_1, D_0 we shall introduce a rectangular mesh concordant with γ, Γ (possibly irregular). The difference solution u_h of the main problem

$$Lu \equiv -\sum_{\kappa,m=1}^{2} \frac{\partial}{\partial x_\kappa}\left(a_{\kappa m}\frac{\partial u}{\partial x_m}\right) = f, \quad u_{/\gamma} = 0 \tag{30}$$

where $a_{\kappa m}$ satisfy conditions of strong ellipticity, we shall define by means of summarised identities (O.A.Ladyzhenskaya, A.D.Lyashko). For problem (30) we consider a continuation of type (4). The difference solution $u_{\varepsilon h}$ of the auxiliary problem we also define by means of summarised identity. Then

$$\| u_{\varepsilon h} - u_h \|_{W_2^1(D_h)} = O(\varepsilon). \tag{31}$$

If we shall use a continuation of type (6), then we can obtain

$$\| u_{\varepsilon h} - u_h \|_{W_2^1(D_h)} = O(\sqrt{\varepsilon}). \tag{32}$$

These estimations are obtained if we suppose, that

$$\| \nabla_h u_h \|_{\gamma_h^*} \le const , \tag{33}$$

where γ_h^* is a general part of boundaries D_h and D_{1h}, and $\nabla_h u_h$ is mesh analog of a conormal derivative. Without this supposition we have to change the right part in (31), (32) by $O(\varepsilon/\sqrt{h})$, $O(\sqrt{\varepsilon/h})$ accordingly.

For the problem

$$- Lu + cu = f, \qquad \sum_{k,m=1}^{2} a_{km} \frac{\partial u}{\partial x_m} \cos(n, x_k)_{/\gamma} = 0 \tag{34}$$

estimation (31) is obtained without supposition (33).

For problem II estimations (31), (32) are obtained by A.N.Bugrov. Now about iterative processes for defining $u_{\varepsilon h}$. Constructing iterative methods whose asymptotic rates of convergence are independent of ε is based on a general theory of convergence by A.A.Samarsky. It is known (A.A.Samarsky), that if for finding the solution of an operational equation

$$Au = f , \qquad A^* = A > 0 . \tag{35}$$

we use an iterative method (κ is a number of iteration)

$$B \frac{y^{\kappa+1} - y^{\kappa}}{\tau} + A y^{\kappa} = f , \tag{36}$$

$$B^* = B > 0 , \quad \gamma_1 B < A < \gamma_2 B , \quad \gamma_1 > 0$$

then for $\tau = \frac{2}{\gamma_1 + \gamma_2}$ we have

$$\| u - y^{\kappa+1} \|_C \le \frac{1 - \xi}{1 + \xi} \| u - y^{\kappa} \|_C , \quad \xi = \frac{\gamma_1}{\gamma_2} . \tag{37}$$

In (37) $C = A$, or $C = B$.

For application of fictitious regions method, we have

$$\tilde{A} u_{\varepsilon h} = f_\varepsilon , \qquad \tilde{A} = \begin{cases} A , & u \in D_h \\ A_\varepsilon , & u \in D_{1h} . \end{cases} \tag{38}$$

The main problem is therefore to choose such an operator $\tilde{B}$, that $\tilde{\xi} = \tilde{\gamma}_1 / \tilde{\gamma}_2$ will be independent of ε. It turns out, that such a choice is possible for considering continuations. We shall limit ourselves by a simple example (A.N.Bugrov).

Let

$$\tilde{A} = A + \frac{1}{\varepsilon^2} \mathcal{P} , \tag{39}$$

where $\mathcal{P}$ is a narrowing operator on D_{1h}. As

$$\gamma_1 E \le A \le \gamma_2 E , \quad Eg = g$$

and

$$\frac{\gamma_1}{\gamma_2}\frac{1}{\varepsilon^2}\mathcal{P} \le \frac{1}{\varepsilon^2}\mathcal{P} \le \frac{\gamma_2}{\gamma_1}\frac{1}{\varepsilon^2}\mathcal{P},$$

then

$$\gamma_1\left(E+\frac{1}{\gamma_1\varepsilon^2}\mathcal{P}\right) \le \tilde{A} \le \gamma_2\left(E+\frac{1}{\varepsilon^2\gamma_1}\mathcal{P}\right).$$

It is therefore sufficient to take

$$\tilde{B} = E + \frac{1}{\varepsilon^2\gamma_1}\mathcal{P}$$

for $\tilde{B}$ in order to assert, that rate of convergence of iterative processes

$$\tilde{B}\,\frac{u_{\varepsilon h}^{k+1} - u_{\varepsilon h}^{k}}{\tau} + \tilde{A}u_{\varepsilon h}^{k} = f_\varepsilon \qquad (40)$$

$$\tau = \frac{2}{\gamma_1+\gamma_2}$$

$$\frac{u_h^{k+1} - u_h^{k}}{\tau} + Au_h^k = f \qquad (41)$$

is the same and is therefore independent of ε.

So far as an implicit iterative scheme is reduced to an explicit one by the known way, then constructing similar iterative schemes in the implicit case will be a problem of techniques.

In particular, we note, that iterative schemes for the fictitious regions method with asymptotic rate of convergence independent of ε, have been constructed for schemes of the type of variable directions and for other, than in (39) methods of continuation (A.N.Konovalov , A.N.Bugrov). These schemes have been turned out close to those considered by O.B.Widlund.

Application of the fictitious regions method is not restricted only by problems, whose solution is found in complicated, not standard regions.

Worth while to applicate this method also in such cases, when a region, in which a solution is found, is standard, but boundary condition for it is approximate, moreover it is often difficult to estimate a degree of approximation.

As a simple example we shall consider a problem of cooling of an infinite ingot, which occupies a region D with boundary γ in a plan. Let us have to define temperature of the ingot as a function of coordinates and the time t . The problem is reduced to solving the mixed Cauchy problem for a heat conduction equation. Usually in similar problems an assumption taken is the following: there is convection heat exchange with environment according to the Newton law on ,and temperature of environment is known. Then we have

$$\left[\frac{\partial u}{\partial n}+\frac{\alpha}{\lambda}(u-u_o)\right]_{/\gamma\times[0,T]}=0\;. \qquad (42)$$

In (42) λ is a coefficient of heat conduction of ingot material, α is a coefficient of heat exchange, whose real value depends on heat physical properties of environment and on temperature on the surface of the interfoice of two media. The latter circumstance does not allow one to define degree of conformity of the problem cooling of the ingot with boundary condition (42) to real physical problem. Usually a problem of such a type is worth while to be replaced by the problem

$$c\rho\frac{\partial u}{\partial t}=\lambda\Delta u\,,\quad \mathcal{M}\in D\,,\qquad c_1\rho_1\frac{\partial u}{\partial t}=\lambda_1\Delta u\,,\quad \mathcal{M}\in D_1$$

$$u_{/S_o}=u_o\,,\quad u_{/S^+}=u_{/S^-}\,,\qquad \lambda\frac{\partial u}{\partial n}_{/S^+}=\lambda_1\frac{\partial u}{\partial n}_{/S^-} \qquad (43)$$

$$u(\mathcal{M},0)=\begin{cases}\varphi(\mathcal{M})\;, & \mathcal{M}\in D\\ u_o\;, & \mathcal{M}\in D_1\;.\end{cases}$$

In (43) C is heat capacity, ρ is density, region D_1 is filled with environment. So far as heat physical prorperties of the domain D and D_1 are essentially different, we have a standard problem of the fictitious regions method. Problem (43) will describe real physical process the better.,the further from ingot a boundary γ is, as in this case boundary condition $u_{/S_o}=u_o$ is exact.

Other examples can be connected with impenetrable boundary conditions in problems of porous flow of two-phase incompressible fluid with capillary forces (A.N.Konovalov, J.L.Korobicina), or with condition of free border in the problem of cross bending of a thin plate (A.N.Konovalov, B.I.Moskalenko).

Both theoretical researches made and numerous numerical calculations, frequently for real problems, allow one with sufficient certainty to recommend the fictitious regions method for the solution of problems of mathematical physics.

LITERATURE

1. Bugrov A.N., Konovalov A.N., Scherbak B.A. Fictitious regions method in plane static elasticity problems. Sb."Chislennye metody mekhaniki sploshnoy sredy", Novosibirsk, 1974, v.5,N 1,p.20-29.
2. Bugrov A.N., Konovalov A.N., Kramarenko V.I. Fictitious regions method for elliptic problems with boundary conditions of different

types. Sb."Aeromekhanika", M.,"Nauka", 1976, p.275-282.

3. Bugrov A.N. Foundation of fictitious regions method for finite difference methods. Materialy XIY Vsesojuznoy studencheskoy konferentsii, Matematika, Novosibirsk, 1976, p.19-25.

4. Bugrov A.N. Fictitious regions method for cross bending of a thin plate problem., Sb."Chislennye metody mekhaniki sploshnoy sredy", Novosibirsk, 1977, v.8, N 4, p.45-58.

5. Bugrov A.N. Fictitious regions method for elliptic partial derivative equations. Trudy Y Vsesojuznoy konferentsii po chislennym metodam resheniya zadach teorii uprugosti i plastichnosti, Novosibirsk, 1978, p.28-35.

6. Il'in V.P., Korotkevich V.A. On solving Poisson's equation in non--rectangular regions. Sb."Chislennye metody mekhaniki sploshnoy sredy", Novosibirsk, 1976, v.7, N 7, p.30-44.

7. Konovalov A.N. Fictitious regions method for problems of porous flow of two-phase incompressible fluid with capillary forces. Sb. "Chislennye metody mekhaniki sploshnoy sredy", 1972,v.3,N 5, p.52--67.

8. Konovalov A.N. Fictitious regions method for torsion problems.Sb. "Chislennye metody mekhaniki sploshnoy sredy", Novosibirsk, 1973, v.4, N 2, p.109-115.

9. Konovalov A.N. On one version of fictitious regions method. Sb. "Nekotorye problemy vychislitelnoy i prikladnoy matematiki",Nauka, Novosibirsk, 1975.

10. Konovalov A.N., Korobicina J.L. Modelling of boundary conditions in porous flow problems, by fictitious regions method. Sb."Chislennoe reshenie zadach filtratsii mnogofaznoy nesjimaemoy jidkosti", Novosibirsk, 1977, p.115-120.

11. Korobitsina J.L. Fictitious regions method for problems of porous flow of twophase in compressible fluid. Sb."Chislennye metody mekhaniki sploshnoy sredy", Novosibirsk, 1976, v.7,N 3,p.103-111.

12. Kopchenov B.D. On one variational problem with small parameter.
   1. Dif. uravneniya 1966, v.2, N 6, p.709-815.
   2. Dif. uravneniya 1966, v.2, N 7, p.967-987.

13. Konovalov A.N., Korobitsina J.L., Kuznetsov S.B. Fictitious regions method for elasticity problems in stress functions.Sb."Modulnyj analiz", Novosibirsk, 1978.

14. Konovalov A.N., Moskalenko V.I. Modelling free border conditions for bending plate problem. Sb."Modulnyj analiz", Novosibirsk,1978.

15. Kopchenov B.D. Fictitious regions method for the second and third boundary problem. Trudy MIAN, USSR, 1975,v.131, p.119-127.

16. Kuznetsov Yu.A., Matsokin A.M. Solving Helmholtz's equation by fictitious regions method. Sb."Vychislitelnye metody lineinoy algebry". VC SO AN SSSR, Novosibirsk, 1972, p.127-144.
17. Lebedev V.I. Difference analogies of orthogonal decomposition of basic differential operators and some boundary-value problems of mathematical physics. JVM i MF, 1964, v.4, N 3, p.449-465.
18. Lions J.L. Sur d'approximation des solution de certains problemes aux limites. Rendiconti del seminario mathematico della Universita di Padova, 32, 1962, p.1-54.
19. Marchuk G.I. Methods of computing mathematics. M., Nauka,1977.
20. Matsokin A.M. On development of fictitious regions method.Sb."Vychiclitelnye metody lineinoy algebry. Novosibirsk, VC SO AN SSSR, 1973, p.48-56.
21. Mignot A.L. Methods d'approximation des solutions des certans problemes aux limites lineares. These, Paris, 1967.
22. Osmolovsky V.G., Rivkind V.Y. Fictitious regions method for creeping metals problem. Sb."Chislennye metody mekhaniki sploshnoy sredy", Novosibirsk, v.8, N 2, 1977, p.89-93.
23. Rivkind R.Y. About estimations of rates of convergence of solutions of difference equations to solutions of elliptic equations with discontinuous coefficients and about one numerical method for Dirichlet problems. DAN SSSR, v.149, N 6, p.1264-1267.
24. Rukhovec L.A. Observation to fictitious regions method.Dif.equations, v.3, N 4, 1967.
25. Samarsky A.A. Theory of difference schemes. M., Nauka, 1977.
26. Saul'ev V.K. On one method of automation for solving boundary-value problems by fast computers. DAN SSSR, 1962, v.144,N 3,p.497-500.
27. Saul'ev V.K. On solving some boundary-value problems by fictitious regions method on fast computers. Sib.Mat.jurnal, 1963, v.IY,N 4, p.912-925.
28. Widlund O.B. On the effects of scaling of the Peasceman-Rachford Method. Mathematics of computation. v.25, N 113, January, 1971, p.33-41.

COMPUTING METHODS IN APPLIED SCIENCES AND ENGINEERING
R. Glowinski, J.L. Lions (editors)
North-Holland Publishing Company

METHOD TO INCREASE THE ACCURACY OF LOW-DEGREE ELEMENT SOLUTION IN NONLINEAR PROBLEMS

LIN QUN

Institute of Mathematics
Academia Sinica
Peking
China

SUMMARY:

The iterated finite element solution and the extrapolation from the iterated collocation solution are used to increase the accuracy of low-degree elements.

The accuracy of finite element method using piecewise linear elements with mesh size h is of $O(h^2)$ for the displacement but only $O(h)$ for the stress ([8] [16]).

The first problem is how to increase the accuracy of stress, especially for some nonlinear problems.

§1 For ease of understanding, we begin with the eigen-problem

$$(1) \quad \begin{cases} \Delta u_i = \lambda_i u_i & \text{in } \Omega \\ u_i = 0 & \text{on } \partial\Omega \end{cases}$$

Let (λ_i^h, u_i^h) denote the piecewise linear finite element solution to (1), then it follows from [16] that

$$||u_i^h - u_i||_o = O(h^2), \qquad ||u_i^h - u_i||_1 = O(h), \; |\lambda_i^h - \lambda_i| = O(h^2).$$

We show that if (λ_i^h, u_i^h) are used to calculate the iterated finite element solution $\hat{u}_i$ from Poisson equation

$$(1') \quad \begin{cases} \Delta \hat{u}_i = \lambda_i^h u_i^h & \text{in } \Omega \\ \hat{u}_i = 0 & \text{on } \partial\Omega, \end{cases}$$

then, following the suggestion of STRANG - FIX ([16] p.50), the order of accuracy is doubled to

$$(2) \qquad ||\hat{u}_i - u_i||_1 = 0(h^2), \quad \left| \frac{||\nabla \hat{u}_i||_2^2}{||\hat{u}_i||_2^2} - \lambda_i \right| = 0(h^4).$$

The numerical experiments coincide with the theorical result (2). Furthermore, it suggests that the accuracy in L_2-norm, $||\hat{u}_i - u_i||_o$, can be increased too. For the detail we refer to [14] and the table in the end of this paper.

We remark that a complete theory about the approximate solution of eigen-problem has been given by CHATELIN in [5].

§2 Let us turn to nonlinear problem

$$(3) \qquad \begin{cases} -\nabla.(a(u)\nabla u) = f(x,u,\nabla u) & \text{in } \Omega \\ u = 0 & \text{on } \partial\Omega \end{cases}$$

which DOUGLAS and DUPONT [7] have studied deeply, where $a(u)$ and $f(x,u,v)$ satisfy some local boundedness and local Lispichitz conditions.

The high accuracy method for (3) is as that for the eigen-problem (1). The piecewise linear finite element solution u^h of (3) is used to calculate the iterated solution $\hat{u}$ from linear equation

$$\begin{cases} -\nabla.(a(u^h)\nabla\hat{u}) = f(x,u^h,\nabla u^h) & \text{in } \Omega \\ \hat{u} = 0 & \text{on } \partial\Omega, \end{cases}$$

then the accuracy of stress is doubled to

$$(4) \qquad ||\hat{u} - u||_1 = 0(h^2).$$

For the proof of (4), we refer to [11].

§3 Specifically we consider in [12] the equation

$$(5) \qquad \begin{cases} -\Delta u + \sum_1^3 u_i \frac{\partial u}{\partial x_i} = f(x) & \text{in } \Omega \\ u = 0 \quad \text{on } \partial\Omega, \ u = (u_1, u_2, u_3). \end{cases}$$

An existence theorem to (5) is proved, (if $f \in C^{(\alpha)}(\bar{\Omega})$ then there exists a solution $u \in C^{(2+\alpha)}(\bar{\Omega})$). Furthermore, it can be shown that there eixists a unique finite element solution u^h (using piecewise linear elements) in a neighborhood of the isolated solution u of (5) such that

$$||u^h - u||_o = O(h^2), \quad ||u^h - u||_1 = O(h).$$

If u^h is used to calculate the iterated solution $\hat{u}$ from Poisson equation

$$\begin{cases} -\Delta\hat{u} = f(x) - \sum_1^3 u_j^h \dfrac{\partial u^h}{\partial x_j} & \text{in } \Omega \\ \hat{u} = o & \text{on } \partial\Omega, \end{cases}$$

then the accuracy of stress is doubled to $||\hat{u} - u||_1 = O(h^2)$.

§4 For the nonlinear problem

$$(5') \qquad \begin{cases} -\nabla.(a(|\nabla u|)\,\nabla u) = f(x,u,\nabla u) & \text{in } \Omega \\ u = o & \text{on } \partial\Omega \end{cases}$$

which GLOWINSKI et al (see [9]) have studied deeply, it is possible to increase the accuracy of finite element solution u^h by calculating the iterated solution $\hat{u}$ from linear equation

$$(5'') \qquad \begin{cases} -\nabla.(a(|\nabla u^h|)\nabla\hat{u}) = f(x,u^h,\nabla u^h) & \text{in } \Omega \\ \hat{u} = o & \text{on } \partial\Omega. \end{cases}$$

Some numerical experiments have been given in [14]. See also the table in the end of this paper.

§5 Consider the Galerkin method for solving the nonlinear integral equation

$$(6) \qquad u(x) = \int_0^1 K[x,y,u(y)]dy,$$

where K satisfies some regularity conditions. It can be shown that there exists a unique Galerkin solution u^h (using piecewise linear elements) in a neighborhood of the isolated solution u of (6) such that

$$||u^h - u||_C = O(h^2).$$

If u^h is used to calculate the iterated Galerkin solution

$$\hat{u}(x) = \int_0^1 K[x,y,u^h(y)]dy,$$

then the order of accuracy is doubled to

$$(7) \qquad ||\hat{u} - u||_C = O(h^4).$$

For the proof of (7), we refer to [11].

§6 Consider the collocation method for solving integral equation

$$(8) \qquad u(x) = \int_0^1 K(x,y)\, f[y,u(y)]dy$$

with non-smooth Kernel K which allows a discontinuity along the diagonal $x = y$ which the Green function posseses. Let u^h denotes the piecewise linear collocation solution to (8). It is easy to show that

$$||u^h - u||_C = 0(h^2),$$

and the iterated collocation solution

$$\hat{u}^h(x) = \int_0^1 K(x,y)\, f[y,u^h(y)]dy$$

does not give higher order approximation to (8) than u^h. However, the method of extrapolation from iterated collocation solution will give a higher order approximation to (8) :

$$(9) \qquad ||\frac{1}{3}\left(4\hat{u}^{\frac{h}{2}} - \hat{u}^h\right) - u||_C = 0(h^4).$$

For the proof of (9), we refer to [13] *)

We remark that a complete theory about the extrapolation method for integral equation with smooth Kernel has been given by BAKER in [2]. We also mention that the methods of iterated collocation solution have been studied deeply by ATKINSON [1] CHANDLER [4] CHATELIN [6] SLOAN [15] and all.

§7 Consider the collocation method for solving two point boundary value problem

$$(10) \qquad \begin{cases} -u'' = f(x,u) \\ u(o) = u(1) = 0 \end{cases}$$

Let u^h denotes the cubic spline collocation solution to (10). It is easy to show that

$$||u^h - u||_{C(2)} = 0(h^2).$$

---

*) We have shown by an example $u(x) = \int_0^1 |x-y|u(y)dy + 2 - x^2 - (1 - x)^2$ that the extrapolation from Nyström solution can not generate a global superconvergent approximation.

If u^h is used to calculated the iterated collocation aolurion $\hat{u}^h$ from

$$\begin{cases} -(\hat{u}^h)'' = f(x,u^h) \\ \hat{u}^h(o) = \hat{u}^h(1) = 0, \end{cases}$$

then the extrapolation from $\hat{u}^h$ and $\hat{u}^{\frac{h}{2}}$ will give a higher order approximation to (10) than u^h or $\hat{u}^h$:

$$\left|\left|\frac{1}{3}\left(4\hat{u}^{\frac{h}{2}} - \hat{u}^h\right) - u\right|\right|_{C(2)} = O(h^4).$$

For some important references about the extrapolation procedure of two point boundary value problem, we refer to BIRKHOFF-GULATTI [3] KELLER [10] STRANG-FIX [16] and al.

Final Remark. The most important task in nonlinear finite element methods is how to solve the discrete nonlinear equations. Fortunatelly, there is a lecture on this topic by STRANG in [17].

NUMERICAL EXPERIMENTS

1. Using iterated finite element solution $\hat{u}_1$ of (1') to improve the finite element solution u_1^h of (1), where $\Omega = (0,1) \times (0,1)$, $h = \frac{1}{4}$ and $\hat{u}_1$ is calculated by the quadric finite element solution $\hat{u}^h$, and

$$\hat{\lambda}^h = \frac{||\nabla \hat{u}^h||^2}{||\hat{u}^h||^2} .$$

$u_1^h(\frac{1}{2}, \frac{1}{2})$	$\hat{u}^h(\frac{1}{2}, \frac{1}{2})$	$u_1(\frac{1}{2}, \frac{1}{2})$
2.2019	2.0193	2
u_{1x}^h	$\hat{u}_x^h$	u_{1x}
1.9900	2.7294	2.7206
u_{1y}^h	$\hat{u}_y^h$	u_{1y}
0.6633	2.7294	2.7207
λ_i^h	$\hat{\lambda}^h$	λ_1
22.865	19.817	19.739

2. Using iterated finite element solution $\hat{u}$ of (5") to improve the finite element solution u^h of magnetic field distribution equation (5'), where $a(\lambda)$ is shown in the graph

$$\frac{||\nabla u^h|| - ||\nabla u||}{||\nabla u||} \approx 24\ \%, \quad \frac{||\nabla \hat{u}|| - ||\nabla u||}{||\nabla u||} \approx 2\ \%$$

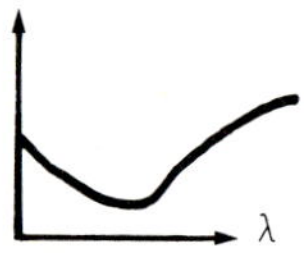

REFERENCES

[1] ATKINSON K.E. : A survey of numerical methods for..., SIAM 1976.

[2] BAKER C.T.H. : The numerical treatment of integral equations, Oxford University. Press 1977.

[3] BIRKHOFF G., GULATI S. : Optimal few-point discretization, SIAM J. Numer. Anal. 11 (1974).

[4] CHANDLER G.A. : Superconvergence of numerical solutions to second kind integral equations, A thesis submitted to A.N.U. (1979).

[5] CHATELIN F. : The spectral approximation of linear operator...., to appear in SIAM review.

[6] CHATELIN F.: Linear spectral approximation in Banach spaces with applications to integral and differential operators, Birkhäuser, Basel (to appear).

[7] DOUGLAS J.,Jr., DUPONT T. : A Galerkin method for a nonlinear Dirichlet problem, Math. Comput. 29 (1975).

[8] FENG KANG : Appl. and Comp. Math. 3 (1965) in Chinese.

[9] GLOWINSKI R. : Numerical analysis of some nonlinear elliptic boundary value problems, Press de l'Univ. de Montreal (to appear).

[10] KELLER H.B. : Numerical solution of two point boundary value problem, Phila. SIAM 1976.

[11] LIN QUN : Acta Math. Sinica 2 (1979)

[12] LIN QUN , JIANG LISHANG : Tech. Rep. Peking (1979).

[13] LIN QUN , LIU JIAQUAN : Tech. Rep. Peking (1979).

[14] LIN QUN XIE GANQAN : Tech. Rep. Peking (1979).

[15] SLOAN I.H. : A review of numerical methods for Fredholm equations, Seminar on Appl. and Numer. solution of integral equations, Canberra, Nov. 29 (1978).

[16] STRANG G., FIX G. : An analysis of the finite element method, Prentice Hall 1973.

[17] STRANG G.: Quasi-Newton methods in nonlinear mechanics 1979.

SESSION II

NUMERICAL METHODS IN BIFURCATION AND APPLICATIONS

COMPUTING METHODS IN APPLIED SCIENCES AND ENGINEERING
R. Glowinski, J.L. Lions (editors)
North-Holland Publishing Company

CALCULATION OF FLOWS BETWEEN ROTATING DISKS

H.B. KELLER and R.K.-H. SZETO

Applied Mathematics, Cal Tech 101-50, Pasadena, Calif. 91125, U.S.A.

INTRODUCTION and FORMULATION.

Similarity solutions of the Navier-Stokes equations for the steady viscous flow of an incompressible fluid between two rotating disks have been introduced by von Karman [4] and Batchelor [1]. Subsequently numerous studies of such flows have been made both numerically [2,3,8,10,11,12,13,14] and analytically [9,15,18]. We present here part of very extensive calculations on this problem [16,19]. Our results show bifurcation, perturbed bifurcation, many limit point curves, cusp behavior, isola formation and at least 19 nonunique solutions for some parameter ranges. Finite difference methods are used along with pseudoarclength continuation methods to circumvent limit point difficulties and to easily find bifurcated branches [6] and some information on stability [7,16].

In terms of dimensionless variables (lengths scaled by disk separation, d, and velocities scaled by $d\Omega_o$ where Ω_o is the angular velocity of the "lower" disk) we seek velocity components (u,v,w) in cylindrical coordinates (r,θ,z) in the form :

$$(1) \qquad w = f(z)\ ,\ u = -r\,f'(z)/_2\ ,\ v = r\,g(z)$$

These von Karman similarity variables reduce the Navier-Stokes equations to the system of ordinary differential equations :

$$(2) \qquad \left.\begin{cases} f'''' = R[ff''' + 4gg'] \\ g'' = R[fg' - gf'] \end{cases}\right\} 0 < z < 1 .$$

The boundary conditions are

$$(3) \qquad \begin{cases} f(o) = f'(o) = 0\ ,\ g(o) = 1\ ; \\ f(1) = f'(1) = 0\ ,\ g(1) = \gamma\ ; \end{cases}$$

since the fluid moves with the disks. Here we have introduced a Reynolds number, R, and disk speed ratio, γ, by :

(4) $$R \equiv \frac{\Omega_o d^2}{\nu} \quad , \quad \gamma \equiv \frac{\Omega_1}{\Omega_o} \quad ,$$

where Ω_1, is the angular velocity of the upper disk (at z=1).

Every solution with $|\gamma|\geq 1$ has a corresponding solution with $|\gamma|\leq 1$. This is equivalent to interchanging the disks. The precise correspondence is given as

Lemma 5 : Let a solution of (2,3) with $\gamma \neq 0$ be given by

$$S \equiv [f(z), g(z), R, \gamma, \text{ on } 0 \leq z \leq 1].$$

Then a solution is also given by

$$\hat{S} \equiv [\hat{f}(\zeta), \hat{g}(\zeta), \hat{R}, \hat{\gamma}, \text{ on } 0 \leq \zeta \leq 1] \, ,$$

where

(5) $$\zeta \equiv 1-z, \; \hat{R} \equiv |\gamma| R, \; \hat{\gamma} \equiv \frac{1}{\gamma} \, , \; \hat{f}(\zeta) \equiv \frac{-f(z)}{|\gamma|} \, , \; \hat{g}(\zeta) \equiv \frac{g(z)}{\gamma} \, .$$

Proof : The proof easily follows by substitution of (5) into (2,3).

Two important special cases of Lemma 5 occur for $\gamma = \pm 1$. The "second" solution, $\hat{S}$, is distinct from S for γ=1 unless f(z) is antisymmetric and g(z) is symmetric about z = 1/2. For γ=-1 the corresponding pair of solutions are distinct unless both f(z) and g(z) are antisymmetric about z=1/2. These observations aide in finding the multiplicity of solutions that we report.

Some trivial solutions useful for starting our calculations are given by :

(6) $$\begin{cases} \text{a) } f(z) \equiv 0, \; g(z) = 1+(\gamma-1)z \; , \; \gamma = \text{arbitrary}, \; R=0 \; ; \\ \text{b) } f(z) \equiv 0, \; g(z) \equiv 1, \; \gamma = 1 \; , \; R = \text{arbitrary}. \end{cases}$$

It can be shown that these solutions are isolated ; that in (6b) is just a rigid body rotation.

For the numerical work we reformulate (2,3) in the first order form :

(7) $$\text{a) } \underset{\sim}{u}' - \underset{\sim}{F}(\underset{\sim}{u};R) = 0 \quad ; \quad \text{b) } \begin{cases} B_{\underset{\sim}{o}}\underset{\sim}{u}(o) - \underset{\sim}{e}_3 = 0 \; , \\ B_{\underset{\sim}{o}}\underset{\sim}{u}(1) - \gamma \underset{\sim}{e}_3 = 0. \end{cases}$$

Here we have introduced :

$$
(8)\quad \begin{cases}
\text{a)}\ \underset{\sim}{u}(z) \equiv (f,f',f'',f''',g,g')^T \equiv (u_1,u_2,\ldots,u_6)^T\ ; \\
\text{b)}\ \underset{\sim}{F}(\underset{\sim}{u},R) \equiv (u_2,u_3,u_4,R[u_1u_4+4u_5u_6]\ ,u_6,R[u_1,u_6-u_2u_5])^T\ ; \\
\text{c)}\ B_o \equiv \begin{pmatrix} 1&0&0&0&0&0 \\ 0&1&0&0&0&0 \\ 0&0&0&0&1&0 \end{pmatrix},\quad \underset{\sim}{e}_3 = \begin{pmatrix} 0\\0\\1 \end{pmatrix}.
\end{cases}
$$

We note that our problem (7) depends on two real parameters, (R,γ), and by virtue of Lemma 5 we need only study this problem in the strip

$$(9)\qquad \mathcal{S} \equiv \{(R,\gamma)\ :\ R \geq 0\ ,|\gamma| \leq 1\}\ .$$

NUMERICAL METHODS AND CONTINUATION.

We use centered finite differences and Newton's method to compute approximate solutions of (7). Specifically on a nonuniform net $\{z_j\}_1^J$, with $z_{j+1}=z_j+h_j$, $z_1=0$, $z_J=1$ we use the box or centered Euler scheme :

$$
(10)\quad \begin{cases}
\text{a)}\ \underset{\sim}{g}_o(u_h) \equiv B_o\underset{\sim}{u}_1-\underset{\sim}{e}_3 = 0\ ; \\
\text{b)}\ \underset{\sim}{g}_j(u_h,R) \equiv \dfrac{\underset{\sim}{u}_{j+1}-\underset{\sim}{u}_j}{h_j} - \underset{\sim}{F}\left(\dfrac{\underset{\sim}{u}_{j+1}+\underset{\sim}{u}_j}{2},R\right) = 0,\ 1 \leq j \leq J-1\ ; \\
\text{c)}\ g_J(u_h,\gamma) \equiv B_o\underset{\sim}{u}_J-\gamma\underset{\sim}{e}_3 = 0
\end{cases}
$$

Here we use the notation

$$u_h \equiv (\underset{\sim}{u}_1,\underset{\sim}{u}_2,\ldots,\underset{\sim}{u}_J)^T$$

and the system (10) of 6J nonlinear algebraic equations can be written as

$$(11)\qquad G_h(u_h,\lambda) = 0.$$

Of course $\underset{\sim}{u}_j$ is to approximate $\underset{\sim}{u}(z_j)$ and λ in (11) represents either R or γ. The above scheme is justified for any smooth isolated solution $\underset{\sim}{u} = \underset{\sim}{u}(z;R,\gamma)$ of (7) by means of Theorems 2.27 and 2.31 of [5]. These results imply that : for $h = \max_j h_j$ sufficiently small (10) has a unique solution, u_h, in some sphere about $U_h \equiv (\underset{\sim}{u}(z_1), \underset{\sim}{u}(z_2),\ldots,\underset{\sim}{u}(z_J))^T$; Newton's method converges quadratically to this solution for any initial guess in the sphere ; the error in the solution is $\mathcal{O}(h^2)$.

The approximation of certain nonisolated solutions of (7) can also be justified when (10) is solved by appropriate continuation techniques. These include limit points, see [6], which seem to be the generic type of nonisolated solution. That is all the nonisolated solutions, save a set of measure zero, are of limit point type. Our calculations clearly support this conjecture.

The continuation procedures play a crucial role in assuring that we have an appropriate initial guess for Newton's method (or any other iterative method). In terms of (11) Newton's method defines the iterates $\{u_h^\nu\}$ by means of :

$$(12)\qquad \left\{\begin{array}{ll} \text{a)} & G^\nu_{h,u_h}\,\delta u^\nu_h = -G^\nu_h \\ \text{b)} & u^{\nu+1}_h = u^\nu_h + \delta u^\nu_h \end{array}\right\} \qquad \nu = 0,1,2,\ldots$$

where

$$G^\nu_h \equiv G_h(u^\nu_h,\lambda)\ ,\quad G^\nu_{h,u_h} \equiv D_{u_h} G_h(u^\nu_h,\lambda).$$

From (10) and (11) we see that G_{h,u_h} is the matrix of order 6J given by :

$$(13)\qquad \text{a)}\quad G_{h,u_h} \equiv \begin{bmatrix} B_o & & & \\ L_2 & R_2 & & \\ & \ddots & \ddots & \\ & & L_J & R_J \\ & & & B_o \end{bmatrix} \equiv \mathbb{A}_h$$

where

$$(13)\quad \text{b)}\quad \left\{\begin{array}{l} L_j \equiv -\left[\dfrac{1}{h_{j-1}}\, I + \dfrac{1}{2}\, A\left(\dfrac{\underset{\sim}{u}_j+\underset{\sim}{u}_{j-1}}{2}, R\right)\right] ; \\ R_j \equiv \left[\dfrac{1}{h_{j-1}}\quad I - \dfrac{1}{2}\, A\left(\dfrac{\underset{\sim}{u}_j+\underset{\sim}{u}_{j-1}}{2}, R\right)\right] ; \\ A(\underset{\sim}{u},R) \equiv \dfrac{\partial \underset{\sim}{F}}{\partial \underset{\sim}{u}}(\underset{\sim}{u},R)\ ;\ I \equiv 6\times 6 \text{ identity} . \end{array}\right.$$

We use block-elimination or band-elimination with a mixed partial-pivoting to solve the linear systems in (12a). For nonsingular $\mathbb{A}_h$ in the form (13) these techniques are justified in [5] and do not enlarge the nonzero block structure in a factorization (neglecting pivots for rotational simplicity) :

$$(14)\qquad \mathbb{A}_h = \mathbb{L}_h\,\mathbb{U}_h\ ,\ \text{diag}\ \mathbb{L}_h = \mathbb{I}\ .$$

Furthermore these techniques can be used if $\mathbb{A}_h$ has a one dimensional null space until the last 6×6 block in $\mathbb{A}_h$ is processed. Then we easily employ full pivoting and still obtain the triangular or block triangular form (14). Of course limit point nonisolated solutions lead to singular $\mathbb{A}_h$ but with a one dimensional null space (so do most bifurcations).

The initial guess, $u^o_h(\lambda+\delta\lambda)$, to employ in (12) is obtained by continuation in λ (i.e. either R or γ) from a solution obtained for λ. We use

$$(15)\qquad \text{a)}\quad u^o_h(\lambda+\delta\lambda) = u_h(\lambda)+\delta\lambda D_\lambda u_h(\lambda),$$

where $D_\lambda u_h(\lambda)$ is the solution of the linear variational problem :

(15) b) $$G_{h,u_h} D_\lambda u_h(\lambda) = - D_\lambda G_h(u_h,\lambda) \ .$$

If, in the factorization (14) the matrix $\mathbb{A}_h$ may be approaching singularity, which is easily detected by considering the ratio of the last two diagonal elements in U_h, we switch over to pseudoarclength continuation [6].

This new procedure determines branches or arcs of solutions, $[u_h(s),\lambda(s)]$, parametrized in terms of an arclength - like parameter, s, rather than $[u_h(\lambda),\lambda]$ with $\lambda = R$ or $\lambda=\gamma$ as the parameter. In brief rather than solve (10) or (11) for $u_h(\lambda)$ with λ fixed and known we solve (11), with u_h and λ unknown, together with the additional constraint :

(16) $$\langle \dot{u}_h(s_o), [u_h(s)-u_h(s_o)] \rangle + \dot{\lambda}(s_o)[\lambda(s)-\lambda(s_o)] = s-s_o \ .$$

We can solve the system (11,16) by Newton's method with quadratic convergence even when $\mathbb{A}_h$ is singular at a limit point. The reason is that the coefficient matrix of the enlarged system,

$$\mathcal{A}_h \equiv \begin{pmatrix} \mathbb{A}_h & G_{h,\lambda} \\ \dot{u}_h^T & \dot{\lambda} \end{pmatrix} ,$$

is nonsingular, see [6].

At bifurcation points some additional computations are required to switch from the tangent vector $[\dot{u}_h(s_o),\dot{\lambda}(s_o)]^I$ on the branch just computed to the new tangent vector, say $[\dot{u}_h,\dot{\lambda}]^{II}$, on the bifurcating branch [6,7]. Then we simply use this new tangent in (16) in place of the old tangent and proceed as before to continue on the new branch.

SOME COMPUTATIONAL RESULTS

We seek to determine the "sheet" of solutions say $S(R,\gamma)$ that lies above the strip $\mathbb{S}$ of (9). That is for each $(R,\gamma) \in \mathbb{S}$ we seek a solution $S(R,\gamma)$ of (7). It turns out that this sheet is multivalued over most of $\mathbb{S}$ and we do not have a complete picture of its structure. It has many simple folds several cusps, hyperbolic points, isola centers and other singular structures which we have not yet completely established. In the brief space alotted to us here we can only sketch some of these results and indicate how they were obtained.

Starting from the exact trivial solution (6a) with twenty-one different values :

$$\gamma=\gamma_k = 0.1k-1, \ k=0,1,\dots,20 \ ;$$

we use continuation in $\lambda=R$ to generate 21 different arcs or paths of solutions, call them $S(R,\gamma_k)$, over $0 \leq R \leq 1000$. All of these solution arcs were able to be continued this far and even to larger R values. However some of these paths contained limit points, denoting simple folds in $S(R,\gamma)$, and one in particular, $S(R,\gamma_o)$, contained a bifurcation point. At several fixed R values, say R_j, continuation was done in $\lambda=\gamma$ to generate arcs $S(R_j,\gamma)$ eminating from those previsouly computed. These new γ-continuation arcs, orthogonal to the original R-continuation arcs, also contained limit points denoting other folds in the $S(R,\gamma)$ surface.

Furthermore these arcs frequently traversed parameter values (R_i,γ_k) at which the original arcs contained <u>different</u> solutions. This also indicates that several "sheets" of solutions lie above some regions of $\mathbb{S}$ and they can be detected without knowing just what folds, cusps or other singularities in $S(R,\gamma)$ give rise to them. When solutions at $|\gamma|= 1$ are obtained we apply Lemma 5 to get other solutions at the same γ value ; then new branches are traced from them.

One way to summarise many of these results is shown in Figure 1 where a number of simple folds in the solution surface $S(R,\gamma)$ are projected onto the strip $\mathbb{S}$. The integers in each region are the number of distinct solutions that have been computed for (R,γ) in the corresponding region. Note that integers in adjacent regions differ by two. This is because the boundaries of the regions represent folds across which the number of solutions changes by two. We have not completely traced out all the folds and thus some simply end in the figure.

Somewhat surprising is the suggested region of uniqueness : $|\gamma|\leq 1$, $R < R_U = 55$, which seems to be uniform in γ. The fold which occurs along $R = R_U$ is not on the sheet which directly eminates form the trivial solution at R=0. Thus it may not be easy to establish this result analytically (i.e. by contractions).

Five special critical points are marked on some of the folds shown in Figure 1. The two saddle point bifurcations (marked X, S and T) and the isola center (marked -ф-) are easily visualized in $\mathbb{R}^3$ as simple folds which are either convex (X) or concave (-ф-) with respect to the pair of solutions meeting at the fold. The two cusp bifurcations (marked ●,A and B) are more difficult to visualize and other figures can be used to show their behavior. In particular we show "slices" of the surface $S(R,\gamma)$ in planes orthogonal to the strip $\mathbb{S}$ and parallel to the γ-direction (Figures 2,3,4) or parallel to the R-direction (Figures 5,6).

In Figures 2,3,4 we plot $g(z=1/2)$ vs γ for the three R values : 1000, 500 and 400, respectively. Not all of the computed solutions are shown in these figures in part to keep them simple and in part due to scale difficulties. Near $\gamma=1$ in Figures 2 and 3 at least 7 or 9 solution branches are indicated along with the fold labelled $\overline{JK}$ in Figure 1. Three folds in $|\gamma| < 1/2$ are clear in these figures, they are labelled $\overline{AL}$, $\overline{A'L'}$ and $\overline{SR}$ in Figure 1.

As R is reduced from 500 to 400 (Figures 3 and 4) the "slicing" plane in the γ-direction passes the hyperbolic bifurcation point, T in Figure 1, and two new folds appear, both on $\overline{BTJ}$. The branches meeting on the folds $\overline{AL}$ and $\overline{A'L'}$ have been eliminated in Figure 4, to better show the perturbed hyperbolic bifurcation.

In Figure 5 we show the bifurcation that occurs from the anti-symmetric solution that exists for the counter-rotating disks, $\gamma=-1$. On the anti-symmetric solution $g(z=1/2)=0$ for all R and thus this figure suggests bifurcation from a trivial solution (but of course no analytic representation of this solution is known). The bifurcation occurs at $R=R_A= 119.4$ and again our figure is misleading in suggesting that the bifurcation is symmetric. For the full solution it is not ; but for $g(z=1/2)$ it is. Although this bifurcation phenomenon for counter-rotating flows has not been known previously, solutions on the "antisymmetric branch" have been determined [10,12,14,15] and a solution on one of the bifurcating branches has been computed at R=1000 [12].

When the disk speed ratio is altered the bifurcation is perturbed and we show the results at $\gamma=-0.6$ in Figure 6. The bifurcation at A in Figures 1 and 5 now goes over into the fold $\overline{AL}$. The axis in Figure 6 no longer represents a solution but an entire branch of solutions still extends over $0 \leq R \leq 500$, now they are not anti-symmetric however. Clearly the segment from $\overline{OA}$ and the lower bifurcating branch have joined to give the continuous branch over $0 \leq R \leq 500$. There is also an exchange of stability at this bifurcation. We have determined some features of the standard linearized stability analysis in all of our calculations. These details and many other results of our calculations are contained in [16,19].

All computed solutions known to us are contained in our results. We have been able to determine what part of the sheet $S(R,\gamma)$ they lie on and how they can be continued to other known solutions [16].

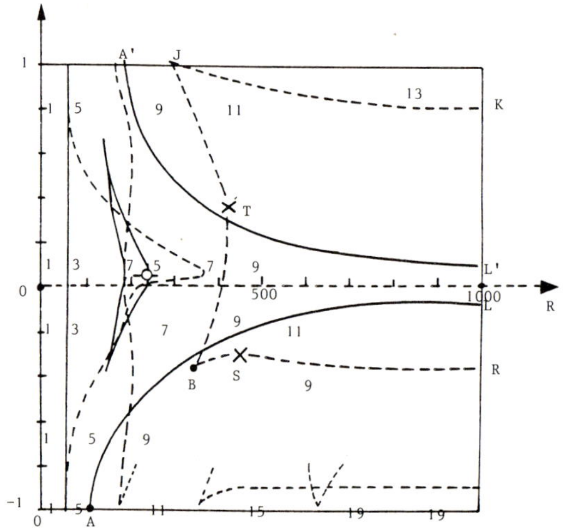

FIGURE 1

Folds and special critical points of the solution surface $S(R,\gamma)$ in the region: $|\gamma|\leq 1$, $0 \leq R \leq 500$. The integers, 1,3,5,7,9,11 etc... tell how many solutions have been found in the indicated areas.

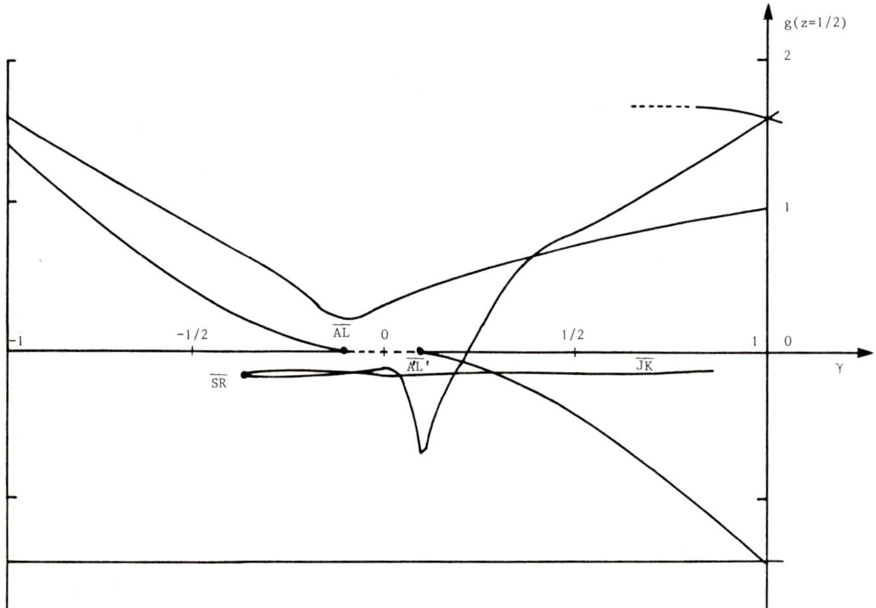

FIGURE 2

Section of the solution surface for R=1000, $|\gamma|\leq 1$. The angular velocity midway between the plates, $g(z=1/2)$, is plotted vs γ.

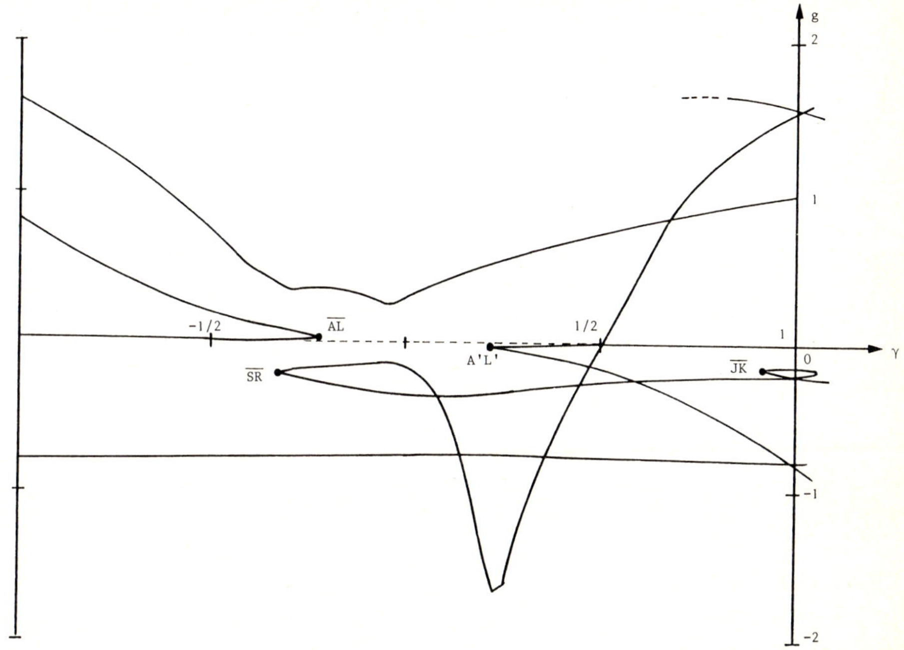

FIGURE 3

Section of the solution surface for R=500, $|\gamma| \leq 1$. The angular velocity midway between the plates, $g(z=1/2)$, is plotted vs γ.

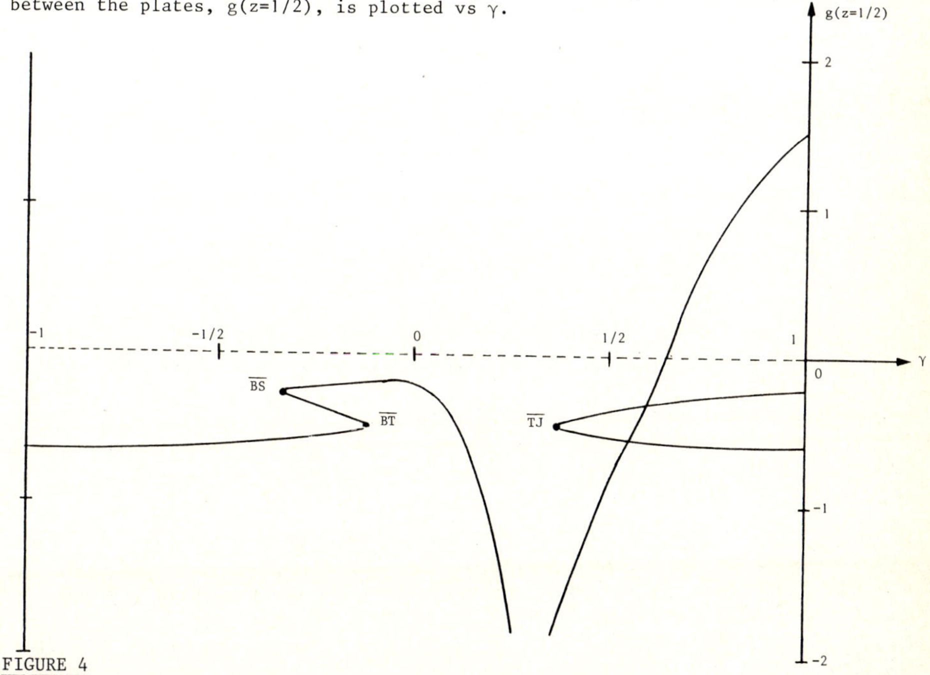

FIGURE 4

Section of the solution surface for R=400, $|\gamma| \leq 1$. The angular velocity midway between the plates, $g(z=1/2)$, is plotted vs γ.

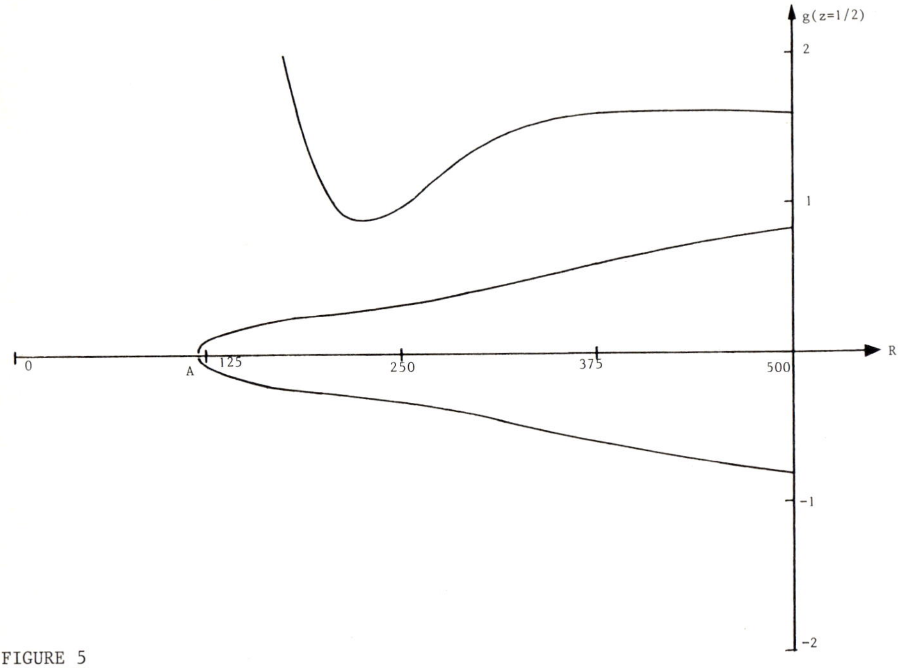

FIGURE 5

Section of the solution surface for $\gamma=-1$, $0 \le R \le 500$. Plot is of $g(z=1/2)$ vs R.

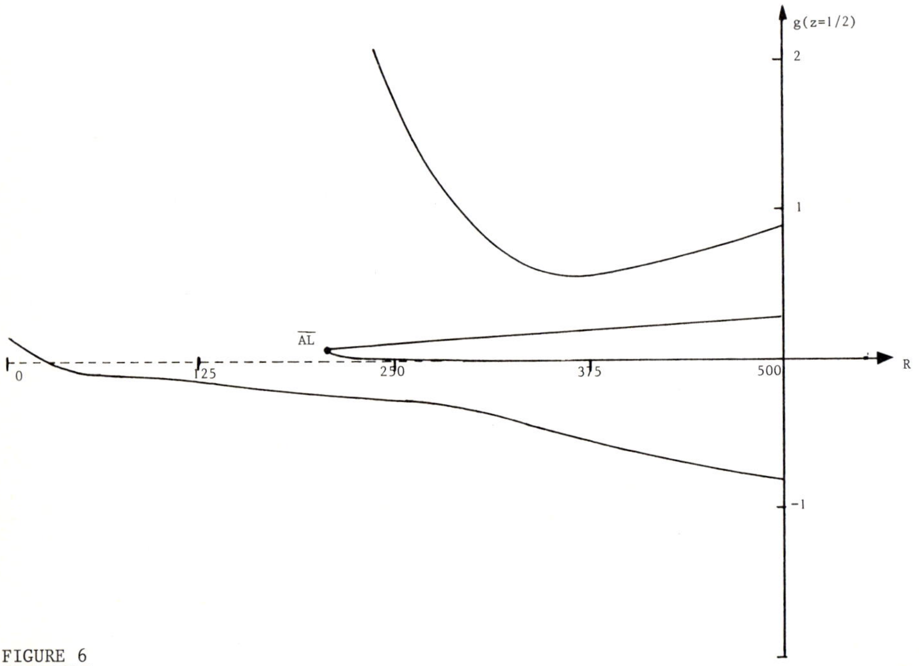

FIGURE 6

Section of the solution surface for $\gamma=-0.6$, $0 \le R \le 500$. Plot is of $g(z=1/2)$ vs R.

ACKNOWLEDGEMENTS

This work was supported by D.O.E. under contract EY-76-S-03-0767 Project Agreement N° 12 and by the U.S. A.R.O. under contract N° DAAG 29-78-C-0011. I also thank the Guggenheim Foundation and the Institut de Recherche d'Informatique et d'Automatique for support during a nine month stay at IRIA where this paper was written.

REFERENCES

[1] Batchelor, G.K., Quart. J. Mech. 4 (1951) 29-41.

[2] Greenspan, D., J. Int. Math. Appl. 9 (1977) 370-377.

[3] Holodniok, M., Kubicek, M. and Hlavacek, V., J.F.M. 81 (1977) 689-699.

[4] Karman, T. von, Z.A.M.M. 1 (1921) 232-252.

[5] Keller, H.B., SIAM Reg. Conf. Ser. # 24 (1976) 61 pp.

[6] Keller, H.B., Applications of Bifurcation Theory (ed. P. Rabinowitz), Academic Press, New York (1977) 359-384.

[7] Keller H.B., Lecture Notes in Math. # 704, Springer-Verlag, Berlin (1979) 241-251.

[8] Lance, G.N. and Rogers, M.H., Proc. Roy. Soc. A 266 (1962) 109-121.

[9] McLeod, J.B. and Parter S.V., A.R.M.A. 54 (1974) 301-327.

[10] Mellor, G.L., Chapple, P.J. and Stokes, V.K., J.F.M. 31 (1968) 95-112.

[11] Nguyen, N.D., Ribault, J.P. and Florent, P. ; J.F.M. 68 (1975) 369-

[12] Pearson, C.E.; J.F.M. 21 (1965) 623-633.

[13] Roberts, S.M. and Shipman, J.S.; J.F.M. 73 (1976) 53-

[14] Rogers, M.H. and Lance G.N.; J.F.M. 7 (1960) 617-631.

[15] Stewartson, K., Proc. Conf. Phil. Soc. 49 (1953) 333-341.

[16] Szeto, R.K.-H., Ph.D. Thesis, Cal Tech. Pasadena, CA. (1977).

[17] Tam, K.K., SIAM J. Appl. Math. 17 (1969) 1305-1310.

[18] Matkowski, B.J. and Siegmann, W.L. ; SIAM J. Appl. Math. 30 (1976) 720-727.

[19] Keller, H.B. and Szeto, R.K.-H., in preparation.

COMPUTING METHODS IN APPLIED SCIENCES AND ENGINEERING
R. Glowinski, J.L. Lions (editors)
North-Holland Publishing Company

NUMERICAL PATTERN FORMATION AND GROUP THEORY

Hiroshi Fujii
Institute of Computer Sciences
Kyoto Sangyo University, Kyoto 603, Japan

1. INTRODUCTION

The phenomenon of instability in nonlinear mechanics has a close and intrinsic relation with the existence of singularities such as snap-through or bifurcation bucklings. In understanding the mechanism of instability, the knowledge of global *bifurcation diagram* may have the primary significance. Here, we use "global" in the sense of parametric dependency of solutions.

In mathematical ecology, two interacting populations of prey-predator systems display sharp spatial heterogeneity - even in a homogeneous environment and with a diffusion process. This phenomenon of pattern formation is known as *patchiness* in ecology. The onset of such patterns are generally related to bifurcation phenomena ([21]). As we have found (in Sec.3), several *stable* steady state patterns (=solutions) may co-exist in these systems. Again it seems highly desirable to have information about the global bifurcation diagram to gain insight into the structure of such phenomena.

Similar phenomena of instability and pattern formation are widely observed in a variety of fields - for instance, the Bénard and Taylor problems in fluid mechanics, morphogenesis in biology, reaction-diffusion problems in bio-chemistry, and so on.

It is these nonlinear phenomena that motivate our work to construct *numerically* global bifurcation diagrams of nonlinear systems. The purpose of this paper is thus to provide a basic theory to the numerical analysis of global bifurcation problems.

We present here two model problems, one is from the classical nonlinear elasticity and the other from mathematical ecology.

The von Kármán-Donnell Shell Model :

Let $\Omega \subset \mathbb{R}^2$ be a polygon, with the boundary $\partial\Omega$.

$$(P)_1 \qquad \begin{aligned} \Delta^2\psi &= -\frac{1}{2}[w,w] - [w,w_0] \\ \Delta^2 w &= [w,\psi] + [w_0,\psi] + [w,\psi_0] + \lambda\cdot p, \end{aligned} \qquad \text{in } \Omega, \qquad (0.1)_1$$

with the boundary condition

$$w = \frac{\partial w}{\partial n} = 0 = \psi = \frac{\partial \psi}{\partial n} \qquad \text{on } \partial\Omega, \qquad (0.1)_2$$

where Δ^2 is the bi-harmonic operator, and the bracket [,] is

$$[u,v] = u_{xx}v_{yy} + u_{yy}v_{xx} - 2u_{xy}v_{xy}. \qquad (0.1)_3$$

Here, w_0 and ψ_0 denote, respectively, the known deflection of the shell and the Airy function of the applied edge force; p is the external normal load on the shell, with the loading parameter $\lambda \in \mathbb{R}^1$.

A Diffusive Prey Predator Model :

Let $\Omega \subset \mathbb{R}^N$ ($N \leq 3$) be a bounded, sufficiently smooth domain with the boundary $\partial\Omega$. Let $\underline{u} = (u_1, u_2)$ denote the population densities of a prey species and its predator.

$$(P)_2 \qquad \left| \; D\cdot\Delta\underline{u} + \underline{f}(\underline{u}) = 0 \qquad \text{in } \Omega, \right.$$

subject to the zero-flux boundary condition :

$$\left| \; \frac{\partial}{\partial n}\underline{u} = 0 \qquad \text{on } \partial\Omega. \right.$$

Here, D is a diagonal matrix $\begin{bmatrix} d_1 & \\ & d_2 \end{bmatrix}$ $(d_1, d_2 \geq d_0 > 0)$, and Δ is the Laplace operator. $\underline{f}(\underline{u})$ is a smooth function $\mathbb{R}^2 \to \mathbb{R}^2$. Following Mimura, Nishiura and Yamaguti [15], it is generally assumed that there is a constant stationary state $\underline{\bar{u}} = (\bar{u}_1, \bar{u}_2)$, with $\underline{f}(\underline{\bar{u}}) = 0$, which is stable to small homogeneous perturbations. (I.e., if $B = (b_{ij}) = f_{\underline{u}}(\underline{\bar{u}})$, $\det B > 0$ and $\operatorname{tr} B < 0$ is assumed.)

Among the pattern formation and singularity, there is a key which links these two notions, that is, the symmetry group of the system. If a physical system is independent of the observer, we may expect that this system is covariant under a group operation G - either finite or compact.

We note that there are mainly two aspects in the application of group theory to bifurcation problems. The first one is based on the principle that the symmetry group of the system is inherited to its bifurcation equations. See, e.g., Sattinger [19], [20] and Ruelle [18]. The second, which is more significant in the study of numerical analysis, concerns the "structural stability" of the criticalities in the presence of a symmetry group. By "structural stability" we mean the uniform existence of a criticality, say, of a simple, symmetry breaking bifurcations, with respect to small changes of the equation.

This structural stability is the key to the numerical realization of the whole bifurcation structure, and to error estimate of numerical solutions. At the last IRIA symposium, we have suggested such a situation, but only in the case of the reflection symmetry $C_s \cong \{e,s\}$ $(s^2=e)$ [23]. The theory was then fully extended to a finite or a compact group G in Fujii and Yamaguti [7]. In this paper, we focus principally on "discrete" structural stability problem. We discuss, in the light of the symmetry group, how a simple symmetry breaking bifurcation can be realized numerically. The argument here is, in essence, parallel to [7], but includes an extension to the case of non self-adjoint linearized operators.

We give also a numerical result on the global bifurcation diagram in a diffusive prey predator model proposed by Mimura and Murray. Various interesting phenomena, such as snap-through, and primary or secondary bifurcation criticalities, disappearing of a bifurcating path, successive recovery of stability of bifurcating paths, and so on, are observed there.

Our basic tool from the group representation theory is the *standard decomposition* of a Hilbert space with respect to a symmetry group G.

The outline of this paper is as follows. We first discuss bifurcations in G-covariant systems in Sec.1. Sec.2 is devoted to numerical aspects of group symmetry in bifurcation problems. We obtain a discrete analogy of the structural stability theorem. Also, we give some remarks from group-theoretical viewpoints to numerical algorithms which solves the discrete nonlinear systems. In Sec.3, we give results of our numerical study of global bifurcation diagram.

1. BIFURCATIONS IN G-COVARIANT SYSTEMS. STRUCTURAL STABILITY OF SYMMETRY BREAKING CRITICALITIES

Suppose we are given a Fredholm mapping $F: \Lambda \times V \to V$, where $\Lambda \subset \mathbb{R}^1$, and V is a complex Hilbert space. We consider the problem

$$(P) \qquad F(\lambda,u) = 0 \tag{1. 1}$$

under a physically natural situation that (P) is covariant with respect to a group operation G.

Instead of studying only one (P), we consider a family ($\varepsilon \in E \subset \mathbb{R}^1$) of perturbed mappings:

$$F: E \times \Lambda \times V \to V. \tag{1. 2}$$

F is Fredholm, and is sufficiently smooth. The problem is to seek the triplet $(\varepsilon;\lambda, u) \in E \times \Lambda \times V$ which satisfies

$$(P)_\varepsilon \qquad F(\varepsilon;\lambda,u) = 0. \tag{1. 3}$$

Here, we suppose that $F_\varepsilon(0;\lambda,u)$ does not identically vanish(*), and that

$$F(0;\lambda,u) = F(\lambda,u), \qquad (\lambda,u) \in \Lambda \times V. \tag{1. 4}$$

Throughout in this paper, F is a real mapping $\overline{F(\varepsilon;\lambda,u)} = F(\varepsilon;\lambda,\bar{u})$.

Let $T: G \to GL(V)$ be a unitary representation of G on V, where G is a finite or a compact group. $(P)_\varepsilon$ is assumed to be covariant under G, that is,

$$T_g F(\varepsilon;\lambda,u) = F(\varepsilon;\lambda,T_g u), \qquad (\varepsilon;\lambda,u) \in E \times \Lambda \times V, \text{ and } \quad g \in G. \tag{1. 5}$$

Our discussions on the G-covariant systems are based on the following two principles. The first is that a G-covariant system is *enclosed* in the G-symmetric space V_+ (as defined below). The second one, due to Sattinger [20], is that the G-covariance of the mapping F is inherited to bifurcation equations.

The following is the basic tool for our future arguments.

Lemma 1.1 (The standard decomposition of V with respect to G)

(*) We denote by $F_\varepsilon(\varepsilon;\lambda,u)$ the Frechet derivative of F with respect to ε at $(\varepsilon;\lambda,u)$, and similarly for $F_u(\varepsilon;\lambda,u)$, $F_\lambda(\varepsilon;\lambda,u)$, $F_{u\lambda}(\varepsilon;\lambda,u)$, and so on.

The Hilbert space V is decomposed into a direct sum

$$V = \sum_{k=1}^{q} \oplus V_k \tag{1. 6}$$

where $q \leq +\infty$ is the number of conjugacy classes of G. (q is finite if G is finite.)

See, for details [14] or [22]. A characterization of V_k is that to each V_k there is associated a subgroup $G_k \subseteq G$, such that every element of V_k is invariant under G_k. One may thus call G_k the symmetry group of V_k. There exists always a subspace V_1 of V, which has G itself as the symmetry group.(*)

The projections P_k: $V \to V_k$ ($k=1,2,..,q$) are (uniquely) given by

$$P_k = \frac{n_k}{n(G)} \sum_{g\in G} \overline{\chi_k(g)}\, T_g, \quad (**) \tag{1. 7}$$

which are self-adjoint, commute with T_g ($\forall g \in G$) and satisfy the relations

$$P_k P_m = P_k\, \delta_{k,m} \quad \text{and} \quad \sum_k P_k = I. \tag{1. 8}$$

Here, $n(G)$ denotes the order of G; χ_k($k=1,2,..,q$) the simple characters of irreducible representations τ_k ($k=1,2,..,q$), and n_k ($k=1,..,q$) the dimension of τ_k($k=1,..,q$), respectively.

Whenever it is convenient, we use the symbols P_+ and P_-, where

$$P_+ = P_1 \quad \text{and} \quad P_- = P_2 + P_3 + \cdots + P_q, \tag{1. 9}$$

and write $V = V_+ \oplus V_-$; V_+ is the G-symmetric space, while V_- is the G-asymmetric space.

A fundamental property of the G-covariant system (P) is that it is *enclosed* in V_+ in the following sense.

Lemma 1.2. Suppose F is G-covariant. Then, it holds that

(i) $P_k F(\varepsilon;\lambda,u_+) = 0$, $k=2,3,..,q$, for $\forall\, (\varepsilon,\lambda,u_+) \in E \times \Lambda \times V_+$, (1.10)

(ii) $P_k F_u(\varepsilon;\lambda,u_+) P_m = 0$, $k \neq m$, $k,m=1,2,..,q$, for $\forall (\varepsilon;\lambda,u_+) \in E \times \Lambda \times V_+$. (1.11)

Proof. Use the fact $T_g u_+ = u_+$ ($\forall\, g \in G$), and $F_u(\varepsilon;\lambda,T_g u)T_g = T_g F_u(\varepsilon;\lambda,u)$ ($\forall u \in V$, $\forall g \in G$.).

Now, let us consider the problem (P). If $F_u(\lambda_o,u_o)$ is invertible, then by the implicit function theorem there exists (locally) a smooth path $u=u(\lambda)$ of solutions through the point (λ_o,u_o). Moreover, if u_o is G-invariant, i.e., $u_o \in V_+$, then so is $u(\lambda)$. This is a direct consequence of the preceding lemma and the (local) uniqueness of the solution path. This means that *a G-symmetric path continues to be G-symmetric until it "arrives" at a critical point* $C_+ = (\lambda_c,u_c) \in \Lambda \times V$. Note that u_c is itself G-symmetric due to the completeness of the space V_+.

(*) The projection onto the G-symmetric space V_+ is given by

$$P_1 = \frac{1}{n(G)} \sum_{g \in G} T_g, \quad \text{and hence } T_g P_1 = P_1 \ (\forall g \in G).$$

(**) When G is a compact continuous group, the sum is to be replaced by the invariant integral on G. See, Serre [22] or Miller [14].

Suppose $L_c = F_u(\lambda_c, u_c)$ is a Fredholm operator of index zero. Assume that $N \equiv \ker L_c$ is of dimension 1, and $R \equiv$ range L_c is of codimension 1. Let $N = \{\phi_c\}$, and $R = \{\phi_c^*\}^{\perp}$, where ϕ_c and ϕ_c^* are the critical eigenvectors of L_c and its adjoint L_c^*, respectively.

A simple, symmetry preserving criticality is the case that u_c, ϕ_c and $\phi_c^* \in V_+$. This generically corresponds to a *snap-through* (=limit) point, characterized by the conditions:

$$\langle F_\lambda(\lambda_c, u_c), \phi_c^* \rangle \neq 0 \quad \text{and} \quad \langle F_{uu}(\lambda_c, u_c)(\phi_c, \phi_c), \phi_c^* \rangle \neq 0. \tag{1.12}$$

For more discussions, see [7].

A simple, symmetry breaking criticality is the case that $u_c \in V_+$, while ϕ_c and $\phi_c^* \in V_k \subset V_-$. ($\exists\, k \in \langle 2,3,..,q \rangle$.)

Here, we note two properties of a simple, symmetry breaking criticality.

(i) By definition, it holds generally that $(T|V_k) \cong \tau_k$. If $N \subset V_k$, we know that τ_k is necessarily a one-dimensional representation. This is because that dim (N) should be a multiple of $\dim(\tau_k)$, since N is invariant under G. (Note that $T_g L_c = L_c T_g$, $\forall g \in G$.)

(ii) Since $u_c \in V_+$, the linearized operator $L_c = F_u(\lambda_c, u_c)$ is block diagonal (Lemma 1.2). Hence, $N \subset V_k \subset V_-$ implies that $L_{c,+} = P_+ L_c P_+$ is *invertible as an operator on* V_+.

By virtue of (ii) above and Lemma 1.2, it is clear that *at least in a neighborhood* Λ_o *of* $\lambda = \lambda_c$, *there is a G-symmetric path* $(\lambda, u_+(\lambda)) \in \Lambda_0 \times V_+$ *which crosses* C_+. (*Proof*. Restrict the problem to V_+, and use the implicit function theorem.)

We come now at the classical situation of « *bifurcations from a "known" path* $(\lambda, u_+(\lambda))$, $\lambda \in \Lambda_0$ ». We let consider the problem

$$H(\lambda, v) = F(\lambda, u_+(\lambda)+v) - F(\lambda, u_+(\lambda)) \quad \text{for } \lambda \in \Lambda_0. \tag{1.13}$$

If $L(\lambda) = H_u(\lambda, 0) = F_u(\lambda, u_+(\lambda))$ has a simple isolated eigenvalue $\zeta_c(\lambda)$, which crosses the origin at $\lambda = \lambda_c$, i.e., $\zeta_c(\lambda_c) = 0$ and $\dot{\zeta}_c(\lambda_c) \neq 0$, there is a bifurcating path of solutions $(\lambda_b(\alpha), u_b(\alpha)) \in \Lambda_0 \times V$, for $\alpha \in A_0 \subset \mathbb{R}^1$. See, e.g., [4].

Since $\dot{\zeta}_c(\lambda_c) = \langle H_{u\lambda}(\lambda_c, 0)\phi_c, \phi_c^* \rangle$, the second requirement is equivalent to

$$B_c \equiv \langle F_{u\lambda}(\lambda_c, u_+(\lambda_c))\phi_c, \phi_c^* \rangle + \langle F_{uu}(\lambda_c, u_+(\lambda_c))(\dot{u}_+(\lambda_c), \phi_c), \phi_c^* \rangle \neq 0 \tag{1.14}$$

where $\dot{u}_+(\lambda_c) = -L_c^\dagger \omega_c F_\lambda(\lambda_c, u_+(\lambda_c))$; ω_c is the projection onto $N^\perp$, and $L_c^\dagger$ is the bounded inverse of $(L_c|N^\perp)$.

We call Eq. (1.14) the *non-degeneracy* condition. We conclude that a simple, non-degenerate symmetry breaking critical point is always a bifurcation point, which we abbreviate to a *SNSBB*. With regards to the symmetry property of the bifurcating path $(\lambda_b(\alpha), u_b(\alpha))$, $\alpha \in A_0$, we have the following

Lemma 1.3. $u_b(\alpha)$, $\alpha \in A_o$ is G_k-invariant, that is $u_b(\alpha) \in V_+^{(k)}$, $\alpha \in A_o$, where $V_+^{(k)}$ is the G_k-symmetric space:

$$V_+^{(k)} = \left(\frac{1}{n(G_k)} \sum_{g \in G_k} T_g \right) V. \tag{1.15}$$

Proof. $u_b(\alpha)$ is obtained in the form $u_b(\alpha) = u_+(\lambda_b(\alpha)) + \alpha(\phi_c + \Psi(\alpha))$, $\lambda_b(\alpha) = \lambda_c + \nu_b(\alpha)$, where $<\Psi(\alpha), \phi_c> = 0$. Both $u_+(\lambda_b(\alpha))$ and ϕ_c are G_k-invariant, it suffices to show the G_k-invariance of $\Psi(\alpha)$. Since $\Psi(\alpha)$ is G-covariant (see, [7] or [19]), $(T_g | N) = 1$, $\forall g \in G_k$ proves the lemma.

Thus, a *secondary* bifurcation problem from the bifurcating path $(\lambda, u_b(\lambda))$ can be considered as a G_k-covariant problem.

As a corollary that the bifurcation equation is also covariant under G, we have

Lemma 1.4. A SNSBB *cannot* be a fold (= trans-critical, = asymmetric point of) bifurcation. In other words, a fold bifurcation is, if exists, symmetry preserving.

See, [7].

Finally, we claim that a SNSBB is stable with respect to small changes of the equation, provided they keep G-covariant. (*)

Proposition 1.5. Suppose $F(0;\lambda,u)$ possesses a SNSBB $(C_+;\phi_c,\phi_c^*) = ((\lambda_c,u_{+,c});\phi_c,\phi_c^*) \in \Lambda \times V_+ \times V_k \times V_k$ $(V_k \subset V_-)$. Then, there is a ε-family of SNSBBs $(C_+(\varepsilon);\phi_c(\varepsilon), \phi_c^*(\varepsilon)) \in \Lambda \times V_+ \times V_k \times V_k$, exists for each $|\varepsilon| < \varepsilon_o$.

Proof. The standard decomposition (1.6) 6) being taken in mind, we have as the G-symmetric component:

$P_+F(\varepsilon;\lambda,u_+) = 0$, where $u_+ \in V_+$.

When $\varepsilon=0$, there is a G-symmetric path $(\lambda,u_+(\lambda)) \in \Lambda_o \times V_+$ such that $P_+F(0;\lambda, u_+(\lambda))=0$. Since $P_+F_u(0;\lambda,u_+(\lambda)), \lambda \in \Lambda_o$ is invertible on V_+, there is a unique $u_+=u_+(\varepsilon;\lambda)$ for each (fixed) ε, $\forall |\varepsilon| < \varepsilon_o$, such that

$\|u_+(\varepsilon;\lambda) - u_+(\lambda)\|_V < C_o|\varepsilon|$, $\forall |\varepsilon| < \varepsilon_o$.

The pair $(\lambda,u_+(\varepsilon;\lambda))$ satisfies $(P)_\varepsilon$.

We study next a (ε,λ)-family of eigenproblems in $V_k \subset V_-$: $(\phi_c(\varepsilon;\lambda) \in V_k)$,

$L_k(\varepsilon;\lambda)\phi_c(\varepsilon;\lambda) = \zeta_c(\varepsilon;\lambda)\phi_c(\varepsilon;\lambda)$,

where $L_k(\varepsilon;\lambda) = P_kF_u(\varepsilon;\lambda,u_+(\varepsilon;\lambda))$.

By hypothesis, $\zeta_c(0;\lambda_c) = 0$ at

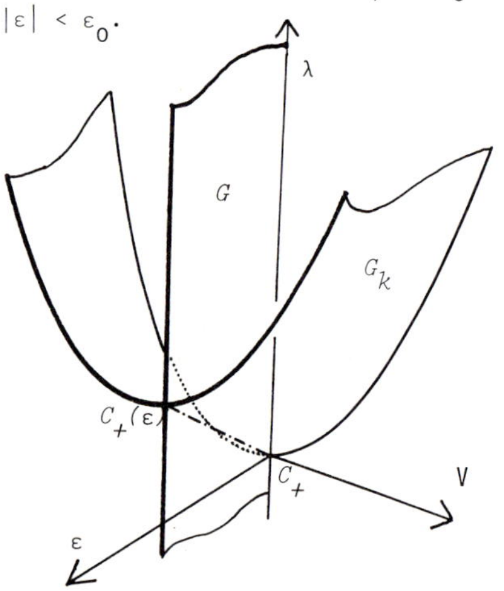

"Two sheets of solutions intersecting transversally. One sheet is G-invariant, while the other (bifurcating) is G_k-invariant."

Fig.1

(*) If both $N(L_c) \subset V_k$ and the corresponding τ_k are m (≥ 1) dimensional, C_+ is a «group theoretical» m-multiple criticality. A «group theoretical» multiple critical point is also stable under G-covariant perturbations. See, [5] and the discussions in Sec.3.

$\lambda = \lambda_c$. We want to seek a ε-family of $\lambda=\lambda_c(\varepsilon)$ such that $\zeta_c(\varepsilon;\lambda) = 0$. Thanks to the non-degeneracy condition (1.14), i.e., $(\partial/\partial\lambda)\zeta_c(0;\lambda_c) \neq 0$, we have the unique existence of $\lambda = \lambda_c(\varepsilon)$ such that $\zeta_c(\varepsilon;\lambda_c(\varepsilon)) = 0$, and that $|\lambda_c(\varepsilon) - \lambda_c(0)| \le C_0|\varepsilon|$ for $\forall$ $|\varepsilon| \le \varepsilon_0$. It is clear by construction that $\phi_c(\varepsilon;\lambda_c(\varepsilon)) \in V_k \subset V_-$.

Together with Lemma 1.4, we have near a SNSBB of (P) two sheets of solutions which are intersecting transversally. One sheet is G-invariant, and the other is G_k -invariant. See, Fig.1.

2. NUMERICAL ASPECTS OF GROUP SYMMETRY IN BIFURCATION PROBLEMS

The aim of this section is, firstly to establish a discrete analogy of the structural stability theorem (Prop.1.5) for a class of numerical schemes, and secondly, to give a group theoretical remark to algorithms for solving numerical schemes.

2.1 Numerical Realization of Symmetry Breaking Bifurcations in Approximate Space V^h

We study numerical approximation of solutions of an operator equation of the form:

$$\text{(P)} \qquad u = G(\lambda,u), \qquad (\lambda,u) \in \Lambda \times V. \tag{2. 1}$$

G is a smooth and compact mapping $\Lambda \times V \to V$. Throughout in this section, G is assumed to be covariant under G - a finite or a continuous compact group. As before, G is real $\overline{G(\lambda,u)} = G(\lambda,\bar{u})$.

Let $V^h \subset V$ be a sequence ($h\to 0$) of finite dimensional subspaces of V, and P^h the orthogonal projection onto V^h:

$$\langle P^h u, v^h\rangle = \langle u, v^h\rangle, \ \forall v^h \in V^h. \tag{2. 2}$$

$$\text{Let} \qquad Q^h \equiv I - P^h. \tag{2. 3}$$

We investigate a class of approximate scheme (P^h) given in the form

$$(P^h) \qquad u^h = P^h G(\lambda,u^h), \quad (\lambda,u^h) \in \Lambda \times V^h. \tag{2. 4}$$

The scheme (P^h) may be regarded as an *internal approximation* with respect to (P), including certainly the conforming finite element scheme. (*)

As for the group theoretical property, we assume that (P^h) is covariant under G *or* its subgroup $G' \subseteq G$. (**) This is equivalent to say that the projection P^h and T' - the unitary representation of G' on V - commute each other.

In lieu of studying a general mapping G, we discuss here two particular examples within the setting (P) and (P^h).

(*) See, also Brezzi and Fujii [2], where a mixed class of scheme is considered for the von Kármán equation.

(**) In some situation, this latter case may be inevitable. Suppose (P) is a D_∞-covariant problem defined on a circle. A finite difference or element approximation (P^h), even with a uniform mesh, pollutes the D_∞-symmetry to D_K-symmetry.

The von Kármán-Donnell Shell Model:

The formulation here is in the form of systems of equations, similar to Brezzi and Fujii[2]. This seems simpler than Fujii and Yamaguti[7] at the cost of losing the self-adjointness of linearized operators.

Let
$\underline{V} = (H_0^2(\Omega))^2$, with the inner product
$\langle \underline{u},\underline{v}\rangle = \sum_{i=1}^{2} (\Delta u_i, \Delta v_i)_{L_2(\Omega)}$, for $u = (u_1,u_2)$, $\underline{v} = (v_1,v_2) \in \underline{V}$.
$\underline{U} = (L_2(\Omega))^2$ and $\underline{W} = (H_0^2(\Omega) \cap H^{3+\sigma}(\Omega))^2$, $(0 < \sigma \le \sigma_0(\Omega))$.(*)

Definition 2.1: Let

$$\underline{B}(\underline{u},\underline{v}) = (B(u_2,v_2), -B(u_1,v_2)-B(u_2,v_1)), \quad \text{for } \underline{u}, \underline{v} \in \underline{V}, \tag{2. 5}$$

where $B(u,v) = \Delta^{-2}[u,v]$, for $u,v \in H_0^2(\Omega)$.

Lemma 2.2: $\underline{B}$ is a symmetric, bilinear mapping $\underline{V} \times \underline{V} \to \underline{V}$, and is continuous $\underline{B} : \underline{V} \times \underline{W} \to \underline{W}$ and $\underline{U} \times \underline{W} \to \underline{V}$.

Proof. Direct from the corresponding properties of $B(u,v)$. See, [7].

Note that $B(u,\cdot)$ is a linear self-adjoint operator $V \to V$, $(V = H_0^2(\Omega))$, due to the property $\langle B(u,v),w\rangle = \langle B(u,w),v\rangle$. However, $\underline{B}(\underline{u},\cdot)$ is no more self-adjoint as an operator $\underline{V} \to \underline{V}$.

The equivalence of the von Kármán-Donnell model (0.1) to the following equation is an exercise.

$$(P)_1 \qquad \underline{u} = -[\tfrac{1}{2}\underline{B}(2\underline{u}_0+\underline{u},\underline{u}) + \lambda\underline{A}\, p\,], \tag{2. 6}$$

where $\underline{u} = (\psi,w)$, $\underline{u}_0 = (\psi_0,w_0)$ and $\underline{A} = \begin{bmatrix} 0 \\ \Delta^{-2} \end{bmatrix}$.

If a conforming class of subspaces $\underline{V}^h$ $(h \to 0)$ is given, our scheme may be:

$$(P^h)_1' \qquad \langle \underline{u}^h,\underline{v}^h\rangle = -\tfrac{1}{2}\int_\Omega [\underline{u}^h + \underline{u}_0^h, \underline{u}^h]\,\underline{v}^h - \lambda \int_\Omega p v_2^h\,, \quad \text{for all } \underline{v}^h \in \underline{V}^h \tag{2. 7}$$

where $[\underline{u},\underline{v}] = ([u_2,v_2], -[u_1,v_2]-[u_2,v_1])$.

Thus, $(P^h)_1'$ is rewritten in the form :

$$(P^h)_1 \qquad \underline{u}^h = -P^h[\tfrac{1}{2}\underline{B}(2\underline{u}_0+\underline{u},\underline{u}) + \lambda\underline{A}p\,], \quad \underline{u}^h \in \underline{V}^h. \tag{2. 8}$$

Some remarks about $(P)_1$. It is known that if $\underline{u}_0 \in \underline{W}$, then $\underline{u} \in \underline{W}$. See, e.g., [7]. Next, since the form $[u,v]$ is covariant under D_∞ - the group of reflections and rotations in the plane, the von Kármán-Donnell shell is covariant under $G \subseteq D_\infty$, where G is the symmetry group of the domain Ω, defined as $G = \{ g \mid g\cdot\Omega = \Omega, g \in D_\infty\}$.

That $(P^h)_1$ is covariant under $G' \subsetneq G$ implies that the element pattern of $\underline{V}^h$ is G'-invariant.

A Diffusive Prey-Predator Model: (Eq.(0.2))

Let $\underline{V} = (H^1(\Omega))^2$, $\underline{U} = (L_2(\Omega))^2$ and $\underline{W} = (H^2(\Omega))^2$.

(*) There is a $\sigma_0 = \sigma_0(\Omega)$ such that for any real $\sigma \in [0,\sigma_0[$, Δ^{-2} is continuous from $H^{-1+\sigma}(\Omega)$ into $H^{3+\sigma}(\Omega)$. See, Grisvard[8].

The inner product of $\underline{V}$ is defined as, for $\underline{u}=(u_1,u_2)$, $\underline{v}=(v_1,v_2) \in \underline{V}$,

$$<\underline{u},\underline{v}> = \sum_{i=1}^{2} [(\nabla u_i,\nabla v_i)_{L_2(\Omega)} + (u_i,v_i)_{L_2(\Omega)}]$$

By the Riesz theorem, there is $\underline{L} \in \mathcal{B}(\underline{U},\underline{V})$ (*) such that $<\underline{L}\underline{u},\underline{v}> = (\underline{u},\underline{v})_{\underline{U}}$. In fact, it holds that $\underline{L} \in \mathcal{B}(\underline{U},\underline{W})$.

We write the diffusive prey-predator equation (0.2) in the form:

$$\Delta\underline{u} + \Xi(\lambda)\cdot\underline{f}(\underline{u}) = 0 \quad \text{in } \Omega, \qquad \frac{\partial}{\partial n}\underline{u} = 0 \quad \text{on } \partial\Omega, \tag{2. 9}$$

and assume that $\underline{f}(\underline{u})$ is given by

$$\underline{f}(\underline{u}) = A\underline{u} + \frac{1}{2}\underline{b}(\underline{u},\underline{u}) + \frac{1}{3!}\underline{c}(\underline{u},\underline{u},\underline{u}), \tag{2.10}$$

where $\underline{b}$ and $\underline{c}$ are symmetric, bi- and tri-linear mappings on $\mathbb{R}^2$. $\Xi(\lambda)$ is a diagonal matrix

$$\Xi(\lambda) = \begin{bmatrix} d_1^{-1}(\lambda) & \\ & d_2^{-1}(\lambda) \end{bmatrix} \tag{2.11}$$

with a bifurcation parameter $\lambda \in \Lambda \subset \mathbb{R}^1$. We have then the operator equation on $\underline{V}$:

$$(P)_2 \qquad \underline{u} = \underline{L}\,\underline{u} + \Xi(\lambda)\cdot\underline{L}\underline{f}(\underline{u}). \tag{2.12}$$

Define the symmetric, bi and tri-linear mappings $\underline{B}: \underline{V} \times \underline{V} \to \underline{V}$ and $\underline{C}: \underline{V} \times \underline{V} \times \underline{V} \to \underline{V}$, respectively, by

$$\underline{B}\,(\underline{u},\underline{v}) = \underline{L}\,\underline{b}(\underline{u},\underline{v}), \qquad (\underline{u},\underline{v} \in \underline{V}), \tag{2.13}$$

and

$$\underline{C}\,(\underline{u},\underline{v},\underline{w}) = \underline{L}\,\underline{c}(\underline{u},\underline{v},\underline{w}) \qquad (\underline{u},\underline{v},\underline{w} \in \underline{V}). \tag{2.14}$$

It is easily seen that $\underline{B}$ and $\underline{C}$ are continuous mappings:

$$\underline{B} : \underline{U} \times \underline{V} \to \underline{V}, \text{ and } \underline{V} \times \underline{V} \to \underline{W}, \tag{2.15}$$

$$\underline{C} : \underline{U} \times \underline{V} \times \underline{V} \to \underline{V}, \text{ and } \underline{V} \times \underline{V} \times \underline{V} \to \underline{W}. \tag{2.16}$$

If $\underline{V}^h$ = {piecewise linear, continuous finite element functions} $\subset \underline{V}$, we have a conforming finite element scheme $(P^h)_2$, in the form $\underline{u}^h = P^h G(\lambda,\underline{u}^h)$, where G is the right-hand side of Eq.(2.12).

On the group theoretical property of $(P)_2$ and of $(P^h)_2$, see Sec.3.

As results from linear finite element approximations, we refer to

Lemma 2.3: For $(P^h)_1$ and $(P^h)_2$, we have that

$$\|Q^h\underline{u}\|_{\underline{V}} \le C\,h^{1+\sigma}\|\underline{u}\|_{\underline{W}} \tag{2.17}$$

and

$$\|Q^h\underline{u}\|_{\underline{U}} \le C'h^{2(1+\sigma)}\|\underline{u}\|_{\underline{W}} \tag{2.18}$$

where $\sigma \in [0,\sigma_0[$ for $(P^h)_1$, and $\sigma = 0$ for $(P^h)_2$.

See, e.g., Ciarlet[3]. The second inequality is due to Nitsche's trick.

Now, we shall show that *a SNSBB - symmetry breaking bifurcation C_+ with respect to G' is uniformly realized in V^h for all $h < h_o$ (small).* Proofs are parallel to

---

(*) $\mathcal{B}(X,Y)$ denotes the set of bounded linear mappings $X \to Y$, and $\mathcal{B}_0(X,Y)$ the set of bounded, compact operators $X \to Y$.

to [7], except Lemma 2. 5. We sketch the proofs, which are devided in three steps.

In the following, $G=G(\lambda,u)$ is the right-hand side of Eq.(2.6) or Eq.(2.12). We suppose the following situation *(A)*. *(P) has a SNSBB with respect to G' - namely, (P) has a* $C_+ \equiv (\lambda_c, \underline{u}_+(\lambda_c)) \in \Lambda \times (\underline{V}_+ \cap \underline{W})$, *with* $N(I-G_u(\lambda_c,\underline{u}_+(\lambda_c)) = \{\underline{\phi}_c\} \subset \underline{V}_k \subset \underline{V}_-$.

First Step: Show the realization of G'-symmetric path in $\underline{V}^h$.

By assumption, P^h and the standard decomposition projectors (with respect to *G'*) P_k' (k=1,2,..,q') commute. Thus, $P_k'P^hG = P^hP_k'G$ clearly holds.

Lemma 2.4. Under the hypothesis *(A)*, there exists a *G'*-symmetric path $(\lambda,\underline{u}_+(\lambda)) \in \Lambda_0 \times \underline{V}_+^h$ of (P^h), at least in a neighborhood Λ_0 of $\lambda = \lambda_c$, where $\Lambda_0 \subset \Lambda$ is independent of $h < h_0$. Moreover,

$$\|\underline{u}_+^h(\lambda) - P^h u_+(\lambda)\|_{\underline{V}} \le C(|\lambda|,\|\underline{u}_+(\lambda)\|_{\underline{W}})\, h^{2(1+\sigma)}, \tag{2.19}$$

$$\|\dot{\underline{u}}_+^h(\lambda) - P^h\dot{u}_+(\lambda)\|_{\underline{V}} \le C'(|\lambda|,\|\underline{u}_+(\lambda)\|_{\underline{W}})\, h^{2(1+\sigma)}. \tag{2.20}$$

Proof. Restrict (P) and (P^h) to $\underline{V}_+ = P_+'\underline{V}$ and $\underline{V}_+^h = P_+'\underline{V}^h = P^h\underline{V}_+$, respectively. If $[I - P_+'G_u(\lambda,\underline{u}_+(\lambda))]$ is invertible on $\underline{V}_+$, then $[I - P_+'P^hG_u(\lambda,P^h\underline{u}_+(\lambda))]$ is *equi-invertible* on $\underline{V}_+$ for $h < h_0$. Next, note the estimate

$$\|G(\lambda,u) - G(\lambda,P^hu)\|_{\underline{V}} \le C(|\lambda|, \|\underline{u}\|_{\underline{W}})[\|Q^h\underline{u}\|_{\underline{V}}^2 + \|Q^h\underline{u}\|_{\underline{U}}] \tag{2.21}$$

and the approximation property Lemma 2.3; use the *equi*-implicit function theorem (see, [7]) to the problem $P_+'(P^h)$ with $\underline{v}^h = \underline{u}_+^h(\lambda) - P^h\underline{u}_+(\lambda)$ for each fixed $\lambda \in \Lambda_0$.

Second Step: Show the unique existence of a critical point on the G'-symmetric path $(\lambda,\underline{u}_+^h(\lambda))$, $\lambda \in \Lambda_0$.

Define a family $(\lambda \in \Lambda_0)$ of linear, compact operators:

$$K\quad(\lambda) = P_k'G_u(\lambda,\ \underline{u}_+(\lambda)), \tag{2.22}$$

and a sequence $(h \to 0)$ of families $(\lambda \in \Lambda_0)$ of linear bounded operators:

$$K^{(h)}(\lambda) = P_k'G_u(\lambda,\ \underline{u}_+^h(\lambda)). \tag{2.23}$$

$K(\lambda)$ and $K^{(h)}(\lambda)$ are clearly smooth with respect to $\lambda \in \Lambda_0$; $K(\lambda) \in \mathcal{B}_0(V,V) \cap \mathcal{B}(U,V) \cap \mathcal{B}(V,W)$ (since $\underline{u}_+(\lambda) \in W$), and $K^{(h)}(\lambda) \in \mathcal{B}(V,V)$, for $\lambda \in \Lambda_0$. It is not hard to show the following estimates:

$$\|\Sigma^{(h)}(\lambda)\|_{\underline{V}\to\underline{V}} \le C\,(|\lambda|,\|\underline{u}_+(\lambda)\|_{\underline{W}})\, h^{1+\sigma}, \tag{2.24}$$

$$\|\dot{\Sigma}^{(h)}(\lambda)\|_{\underline{V}\to\underline{V}} \le C'(|\lambda|,\|\underline{u}_+(\lambda)\|_{\underline{W}})\, h^{1+\sigma}, \tag{2.25}$$

$$\|\Sigma^{(h)}(\lambda)\|_{\underline{W}\to\underline{V}} \le C''(|\lambda|,\|\underline{u}_+(\lambda)\|_{\underline{W}})\, h^{2(1+\sigma)}, \tag{2.26}$$

and

$$\|\Sigma^{(h)*}(\lambda)\|_{\underline{W}\to\underline{V}} \le C'''(|\lambda|,\|\underline{u}_+(\lambda)\|_{\underline{W}})\, h^{2(1+\sigma)}, \tag{2.27}$$

where

$$\Sigma^{(h)}(\lambda) = K(\lambda) - K^{(h)}(\lambda). \tag{2.28}$$

$\Sigma^{(h)*}(\lambda)$ is the adjoint operator of $\Sigma^{(h)}(\lambda)$, defined as $\langle\Sigma^{(h)}(\lambda)u,v\rangle = \langle u,\Sigma^{(h)*}(\lambda)v\rangle$. Note that the last estimate does not necessarily follow from Eq.(2.26). For the von Kármán-Donnell equation, we come back to the definition of B(u,v), Eq.(2.5), and use the symmetry $\langle B(u,v),w\rangle = \langle B(u,w),v\rangle$, for $u,v,w \in V$.

Lemma 2.5. Consider two eigenproblems ($\lambda \in \Lambda_0$):

$$(E) \qquad K(\lambda)\,\underline{\phi}(\lambda) = \zeta(\lambda)\cdot\underline{\phi}(\lambda), \qquad (\underline{\phi}(\lambda) \in \underline{V}_k \cap \underline{W}), \tag{2.29}$$

$$(E^h) \qquad P^h K^{(h)}(\lambda)\underline{\phi}^h(\lambda) = \zeta^h(\lambda)\cdot\underline{\phi}^h(\lambda) \qquad (\underline{\phi}^h(\lambda) \in \underline{V}_k^h), \tag{2.30}$$

and their adjoint problems, (E^*) and (E^{h*}). If an eigenpair $(\zeta(\lambda), \underline{\phi}(\lambda)) \in \mathbb{C}^1 \times (\underline{V}_k \cap \underline{W})$ is simple, isolated in $\lambda \in \Lambda_0$, then there is a pair $(\zeta^h(\lambda), \underline{\phi}^h(\lambda)) \in \mathbb{C}^1 \times \underline{V}_k^h$, $(h < h_0)$ of (E^h), which is simple and isolated. We have that

$$|\zeta^h(\lambda) - \zeta(\lambda)| \le C\,(|\lambda|, \|\underline{\phi}(\lambda)\|_{\underline{W}}, \|\underline{\phi}^*(\lambda)\|_{\underline{W}})\, h^{2(1+\sigma)}, \tag{2.31}$$

$$|\dot{\zeta}^h(\lambda) - \dot{\zeta}(\lambda)| \le C'(|\lambda|, \|\underline{\phi}(\lambda)\|_{\underline{W}}, \|\underline{\phi}^*(\lambda)\|_{\underline{W}})\, h^{1+\sigma}, \tag{2.32}$$

and

$$\|\underline{\phi}^h(\lambda) - P^h\underline{\phi}(\lambda)\|_{\underline{V}} \le C''(|\lambda|, \|\underline{\phi}(\lambda)\|_{\underline{W}}, \|\underline{\phi}^*(\lambda)\|_{\underline{W}})\, h^{2(1+\sigma)}. \tag{2.33}$$

(Here, $\underline{\phi}^*(\lambda)$ is the eigenfunction of (E*) corresponding to $\overline{\zeta(\lambda)}$.)

For a proof, see [5].

Proposition 2.6. Assume (A), i.e., $N(I - G_u(\lambda_c, \underline{u}_+(\lambda_c))) = \{\underline{\phi}_c\} \subset \underline{V}_k \subset \underline{V}_{-}$. Then, there exists a unique $\lambda_c^h \in \Lambda_0$, such that

$$N(I - P^h G_u(\lambda_c^h, \underline{u}_+^h(\lambda_c^h))) = \{\underline{\phi}_c^h\} \subset \underline{V}_k^h \subset \underline{V}_{-}^h ,$$

and that

$$|\lambda_c - \lambda_c^h| \le C_0\, h^{2(1+\sigma)}, \tag{2.34}$$

$$\|P^h\underline{\phi}_c - \underline{\phi}_c^h\|_{\underline{V}} \le C_0'\, h^{2(1+\sigma)}, \tag{2.35}$$

$$\|P^h\underline{\phi}_c^* - \underline{\phi}_c^{h*}\|_{\underline{V}} \le C_0''\, h^{2(1+\sigma)}. \tag{2.36}$$

Proof. We apply the previous lemma. Taking note of the reality $\overline{K(\lambda)} = K(\lambda)$, $\lambda \in \Lambda_0$, and since $\zeta_c(\lambda_c) = 1$ is real, simple, $K(\lambda)$ has a real, simple isolated eigenvalue $\zeta_c(\lambda)$ (by taking a smaller $\Lambda_0' \subset \Lambda_0$, if necessary). Note also that $\zeta_c^h(\lambda)$ is *real* and simple. Since $\dot{\zeta}_c(\lambda_c) \ne 0$, one can assume that $\zeta_c(\lambda)$ is strictly monotone in $\lambda \subset \Lambda_0$ (or, if necessary, in $\lambda \in \Lambda_0'' \subset \Lambda_0'$.) By Eq.(2.32), $\zeta_c^h(\lambda)$ is strictly monotone in λ Λ_0, for $h < h_0$ (sufficiently small). Thus, two curves $\zeta = \zeta_c^h(\lambda)$ and $\zeta = 1$ crosses transversally, only once in $\lambda \in \Lambda_0$, at $\lambda = \lambda_c^h$. The estimate (2.34) is left as an exercise.

<u>*Third Step*</u>: *Uniform existence and convergence of bifurcating paths.*

<u>*Proposition 2.7.*</u> For each $h \in \,]0,h_0]$ (fixed), there bifurcates from $C_+^h \equiv (\lambda_c^h, \underline{u}_+^h(\lambda_c^h)) \in \mathbb{R}^1 \times \underline{V}_+^h$ a path $(\lambda_b^h(\alpha), \underline{u}_b^h(\alpha)) \in \mathbb{R}^1 \times \underline{V}^h$, for $\alpha \in A_0' = \{\alpha;\ |\alpha| < \delta'\}$, ($\exists\,\delta' > 0$, independent of h).

A proof may be obtained if we consider the discrete version of the mapping $H(\lambda, \underline{v})$ defined by Eq. (1.13), and use the equi-implicit function theorem. We adopt here another strategy which makes use of the G'-covariance of the discrete bifurcation equation.

We write (P^h) as

$$F^h(\lambda,\underline{u}^h) \equiv \underline{u}^h - P^h G(\lambda,\underline{u}^h) = 0. \tag{2.37}$$

Let $L_c^h \equiv F_u^h(\lambda_c^h, \underline{u}_+^h(\lambda_c^h))$. Then, by Prop. 2.5, $N(L_c^h) = \{\underline{\phi}_c^h\}$, $N(L_c^{h*})^\perp = \{\underline{\phi}_c^{h*}\}^\perp = R(L_c^h)$. We denote by ω_c^h the projection of $\underline{V}$ onto $R(L_c^h)$, and $\pi_c^h = I - \omega_c^h$. Also, $\tilde{\pi}_c^h$ is the projection of $\underline{V}$ onto $N(L_c^h)$ and $\tilde{\omega}_c^h = I - \tilde{\pi}_c^h$.

We seek a α-family of solutions of Eq.(2.37) in the form:

$$\lambda_b^h = \lambda_c^h + \nu_b^h \quad \text{and} \quad \underline{u}_b^h = \underline{u}_+^h(\lambda_c^h) + \alpha\underline{\phi}_c^h + \underline{\Psi}_b^h, \tag{2.38}$$

where $\tilde{\omega}_c^h \underline{\Psi}_b^h = \underline{\Psi}_b^h$. Through the Lyapounov-Schmidt decomposition, $\omega_c^h F^h(\lambda_b^h, \underline{u}_b^h) = 0$ is written as:

$$\begin{aligned} L_c^h \underline{\Psi}_b^h &+ \nu_b^h \omega_c^h F_{\lambda,c}^h + \frac{1}{2}(\nu_b^h)^2 \omega_c^h F_{\lambda\lambda,c}^h + \nu_b^h \omega_c^h F_{u\lambda,c}^h(\alpha\underline{\phi}_c^h + \underline{\Psi}_b^h) + \\ &+ \frac{1}{2}\omega_c^h F_{uu,c}^h(\alpha\underline{\phi}_c^h + \underline{\Psi}_b^h, \alpha\underline{\phi}_c^h + \underline{\Psi}_b^h) + \omega_c^h R_c^h(\nu_b^h, \alpha\underline{\phi}_c^h + \underline{\Psi}_b^h) = 0, \end{aligned} \tag{2.39}$$

where R_c^h is the remainder term with $\|R_c^h(\nu, \underline{v})\|_{\underline{V}} = O(|\nu|^3, \|\underline{v}\|_{\underline{V}}^3)$. *)

Thanks to the equi-invertibility of L_c^h from $N(L_c^h)^\perp$ into $R(L_c^h)$, $\underline{\Psi}_b^h$ can be solved uniquely as a function of α and ν_b^h:

$$\begin{aligned} \Psi_b^h &= \Psi_b^h(\alpha, \nu_b^h) \\ &= - L_c^{h\dagger}\omega_c^h[\nu_b^h F_{\lambda,c}^h + \alpha\nu_b^h F_{u\lambda,c}^h \underline{\phi}_c^h + \frac{1}{2}\alpha^2 F_{uu,c}^h(\underline{\phi}_c^h, \underline{\phi}_c^h) + (\text{h.o.t.})]. \end{aligned} \tag{2.40}$$

We note that Ψ_b^h is G'-covariant, and of equi-C^p class (whenever F^h is of equi-C^p, and in fact, $p = +\infty$ in our case).

Substituting $\Psi_b^h = \Psi_b^h(\alpha, \nu_b^h)$ to $\pi_c^h F^h(\lambda_b^h, \underline{u}_b^h) = 0$, we have the discrete bifurcation equation $\Gamma^h(\alpha, \nu_b^h) = 0$. Again, Γ^h is covariant under G'. As a consequence, both the zero-th and second order terms in α vanish. After a short calculation, it turns out that

$$\begin{aligned} \Gamma^h(\alpha, \nu_b^h) &= \alpha \cdot \tilde{\Gamma}(\alpha, \nu_b^h) \\ &= \alpha \cdot [B_c^h \cdot \nu_b^h + \frac{1}{3!} D_c^h \cdot \alpha^2 + \cdots]. \end{aligned} \tag{2.41$_1$}$$

Here, $$B_c^h \equiv \langle F_{u\lambda,c}^h \underline{\phi}_c^h, \underline{\phi}_c^{h*}\rangle + \langle F_{uu,c}^h(\underline{u}_+^h(\lambda_c^h), \underline{\phi}_c^h), \underline{\phi}_c^{h*}\rangle, \tag{2.41$_2$}$$

and $$D_c^h \equiv \langle F_{uuu,c}^h(\underline{\phi}_c^h, \underline{\phi}_c^h, \underline{\phi}_c^h), \underline{\phi}_c^{h*}\rangle - 3\langle F_{uu,c}^h(\underline{\phi}_c^h, L_c^{h\dagger}\omega_c^h F_{uu}^h(\underline{\phi}_c^h, \underline{\phi}_c^h)), \underline{\phi}_c^{h*}\rangle. \tag{2.41$_3$}$$

It is noted that $\alpha \equiv 0$ corresponds the G'-symmetric path.

Γ^h and consequently, $\tilde{\Gamma}^h$ are of equi-C^p, and $|B_c^h|$ is bounded below since $|B_c^h - B_c| \le Ch^{1+\sigma}$. Thus, the equi-implicit function theorem as applied to $\tilde{\Gamma}^h(\alpha, \nu_b^h) = 0$ guarantees the existence of a constant $\delta' > 0$, independent of h, and a family of solutions $\nu_b^h = \nu_b^h(\alpha) = -(D_c^h/3!B_c^h)\cdot\alpha^2 + \cdots$, for $\alpha \in A_0' = \{\alpha;\ |\alpha| < \delta'\}$. Hence, follows Prop. 2.7.

To show the convergence of $(\lambda_b^h(\alpha), \underline{u}_b^h(\alpha))$ to $(\lambda_b(\alpha), \underline{u}_b(\alpha))$, the bifurcating path of (P), where

$$\lambda_b(\alpha) = \lambda_c + \nu_b(\alpha) \quad \text{and} \quad \underline{u}_b(\alpha) = \underline{u}_+(\lambda_c) + \alpha\phi_c + \underline{\Psi}_b(\alpha),\ \alpha \in A_0, \tag{2.42}$$

we let, for $\alpha \in A_0'' \equiv A_0 \cap A_0'$,

*) Here, $F_{u,c}^h$, $F_{\lambda u,c}^h, \cdots$ denote $F_u^h(\lambda_c^h, \underline{u}_+^h(\lambda_c^h))$, $F_{\lambda u}^h(\lambda_c^h, \underline{u}_+^h(\lambda_c^h))$, $\cdots$, respectively.

$$\lambda_b^h(\alpha) = \lambda_b(\alpha) + \varepsilon_b^h(\alpha) \quad \text{and} \quad \underline{\Psi}_b^h(\alpha) = \tilde{\omega}_c^h P^h \underline{\Psi}_b(\alpha) + \underline{\eta}_b^h(\alpha), \tag{2.43}$$

with the condition $\varepsilon_b^h(0) = 0$ and $\underline{\eta}_b^h(0) = 0$.

Lemma 2.8.

$$|\varepsilon_b^h(\alpha)| + \|\eta_b^h(\alpha)\|_{\underline{V}} \le C|\alpha|^2\, h^{2(1+\sigma)}. \tag{2.44}$$

We only sketch the proof. After the substitution of Eq. (2.43) into (P^h), and after the Lyapounov-Schmidt decomposition, we have an equation in terms of $\underline{\eta}_b^h$, similar to Eq. (2.39). $\underline{\eta}_b^h$ can be solved as a function of α and ε_b^h as:

$$\underline{\eta}_b^h\,(\alpha,\varepsilon_b^h) = -L_c^h\, \omega_c^h[\, \tilde{F}_c^h(\alpha) + \varepsilon_b^h F_{\lambda,c}^h + \cdots], \tag{2.45}$$

with the estimate

$$\|\underline{\eta}_b^h\|_{\underline{V}} = O(|\varepsilon_b^h|, \|\tilde{F}_c^h(\alpha)\|_{\underline{V}}). \tag{2.46}$$

Here,

$$\tilde{F}_c^h(\alpha) \equiv F^h(\lambda_c^h + \nu_b(\alpha),\ \underline{u}_+^h(\lambda_c^h) + \alpha\underline{\phi}_c^h + \tilde{\omega}_c^h P^h \Psi_b(\alpha)). \tag{2.47}$$

The key in the above is the following estimate:

$$\|\tilde{F}_c^h(\alpha)\|_{\underline{V}} \le C|\alpha|^2\, h^{2(1+\sigma)}, \tag{2.48}$$

which follows from the approximation property Eqs.(2.17)-(2.18) and

$$\|\tilde{\pi}_c^h P^h \underline{\Psi}_b(\alpha)\|_{\underline{V}} \le C h^{2(1+\sigma)} \|\underline{\Psi}_b(\alpha)\|_{\underline{W}} \le C'|\alpha|^2\, h^{2(1+\sigma)}, \tag{2.49}$$

in view of the relations $F_c(\alpha) \equiv F(\lambda_c + \nu_b(\alpha),\ \underline{u}_+(\lambda_c) + \alpha\underline{\phi}_c + \underline{\Psi}_b(\alpha)) = 0$.

Then, the kernel component of the Lyapounov-Schmidt decomposition becomes

$$\begin{aligned} \Xi^h(\alpha,\varepsilon_b^h) &= \alpha\cdot\tilde{\Xi}^h(\alpha,\varepsilon_b^h) \\ &= \alpha\cdot[B_c^h\cdot\varepsilon_b^h + \alpha^{-1}\langle\tilde{F}_c^h(\alpha),\underline{\phi}_c^{h*}\rangle - \langle F_{uu,c}^h(L_c^{h\dagger}\omega_c^h\tilde{F}_c^h(\alpha),\underline{\phi}_c^h),\underline{\phi}_c^{h*}\rangle + (\text{h.o.t.})] \end{aligned} \tag{2.50}$$

Again, we have made an essential use of the G'-covariance of Ξ^h and $\tilde{F}_c^h(\alpha)$. We should note also that from the G'-covariance of $\tilde{F}_c^h(\alpha)$, $|\langle\tilde{F}_c^h(\alpha),\underline{\phi}_c^h\rangle| \le C|\alpha|^3\, h^{2(1+\sigma)}$ holds. Then, Eq. (2.44) is again a consequence of the equi-implicit function theorem applied to $\Xi^h(\alpha,\varepsilon_b^h) = 0$, and of Eqs.(2.46)-(2.48).

Thus, noting Eq. (2.49),

$$\|P^h\underline{\Psi}_b(\alpha) - \underline{\Psi}_b^h(\alpha)\|_{\underline{V}} \le \|\underline{\eta}_b^h(\alpha)\|_{\underline{V}} + \|\tilde{\pi}_c^h P^h \underline{\Psi}_b(\alpha)\|_{\underline{V}} \le C|\alpha|^2\, h^{2(1+\sigma)}, \tag{2.51}$$

we are at our final conclusion.

Proposition 2.9. Under the situation (*A*), the G_k'-symmetric bifurcating paths of (P^h), $(\lambda_b^h(\alpha), \underline{u}_b^h(\alpha)) \in \mathbb{R}^1 \times \underline{V}^h$, converge to $(\lambda_b(\alpha),\underline{u}_b(\alpha)) \in \mathbb{R}^1 \times \underline{W}$, uniformly with respect to $\alpha \in A_0''$, in the following sense:

$$|\lambda_b^h(\alpha) - \lambda_b(\alpha)| + \|\underline{u}_b^h(\alpha) - P^h\underline{u}_b(\alpha)\|_{\underline{V}}$$

$$\le C_0\, h^{2(1+\sigma)}, \quad \text{for all } \alpha \in A_0''.$$

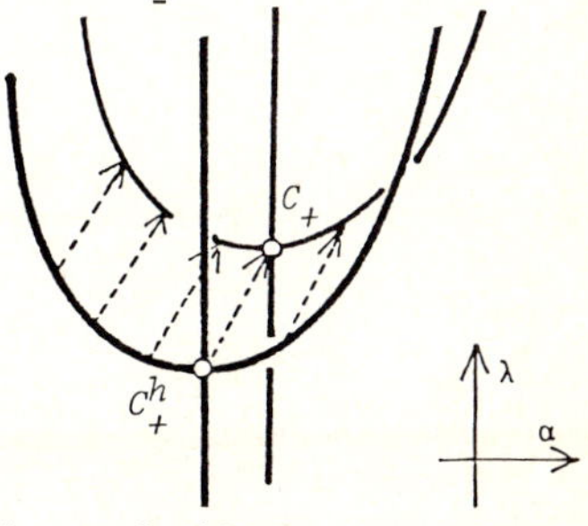

Fig. 2 "Uniform (in α) convergence of bifurcating paths"

2.2 Some Remarks on Algoritms - A Group Theoretical Decomposition

A few remarks about the question how to actually construct *paths of solutions* of a given discrete equation (P^h):

$$(P^h) \qquad u^h = P^h G(\lambda, u^h) \equiv G^h(\lambda, u^h), \qquad u^h \in V^h.$$

We first refer to the *static perturbation method* of Hangai and Kawamata[10], and the *pseudo-arclength continuation* of H.B. Keller[12] (presented at the last IRIA symposium).

The key is the *parametrization* - the choice of local incremental parameters in the continuation of paths. At an ordinary point, λ is the first candidate for the local parameter, as the implicit function theorem shows. At, or in a neighborhood of, a snap-through (=limit) point, the parameter $\alpha = \langle u_c^h, \phi_c^h \rangle$ is eligible, and which is the procedure of Hangai and Kawamata. An alternative is to adopt the inflation procedure and use the *arclength* s as the local parameter. See, [12]. In both methods, snap-through criticalities are eliminated.

Our next question is, in a neighborhood of a SNSBB - *a simple, non-degenerate symmetry breaking bifurcation* point C_+, how we can reach and locate it.

The key lies in the structure of SNSBBs. Assuming the standard decomposition in mind (with the help of the *character table* of the group G), we know that

(i) a G-symmetric path $(\lambda, u_+^h(\lambda))$ continues to be G-symmetric, until it arrives at a critical point C_+; (Note: At a snap-through point, it does not lose its symmetry.)

(ii) there is a G-symmetric path which *traverses* the SNSBB point C_+. This path is *ordinary* in V_+^h-space.

The above note (ii) implies that if we *mask* the V_--part of the equation, the SNSBB point is eliminated. Thus, the G-symmetric path can be obtained using the continuation method restricted to the space V_+ without difficulty - even in a neighborhood of C_+.

In order to locate a SNSBB point on this G-symmetric path $(\lambda, u_+^h(\lambda))$, it is enough to "check" the submatrices $L_k^h(\lambda) = P_k L^h(\lambda) P_k$, where $L^h(\lambda) = I - G_u^h(\lambda, u_+^h(\lambda))$, $k = 2,3,..,q$. If $L_k^h(\lambda_c)\cdot\phi_c = 0$, then $\phi_c^h \in V_k^h$ and the bifurcating path is to be traced in the G_k-symmetric space $V_+^{h(k)}$.

The construction of the projectors could be done by the use of *the symmetry adopted basis* of V^h ([9]). In many cases, however, it seems more convenient to check simply the matrix $L^h(\lambda)$. In this case, however, it is important to detect the symmetry group $G_k \subsetneq G$ of ϕ_c^h, if we want to compute further possible *secondary* or *tertiary* SNSBB points.

It is intuitively clear that an equivalent equation of the V_+-part of the scheme (P^h) may be obtained *usually* by restricting the domain Ω^h to $P_+\Omega^h$, and imposing the symmetric boundary condition on internal boundaries.

We may summarize the above procedure as *the group theoretical decomposition*. Clearly, this algorithm is more economical in many situations. In the next section, we show an actual example solved by this algorithm, where we have obtained secondary and tertiary bifurcations stably.

3. NUMERICAL STUDY OF GLOBAL BIFURCATION DIAGRAM - AN EXAMPLE OF REACTION-DIFFUSION EQUATIONS

To illustrate the group theoretical aspects of bifurcation theory and its applications to numerical analysis in an actual problem, we show a result of numerical study of global bifurcation diagram for the diffusive prey-predator model proposed by Mimura and Murray.

For the general setting of the problem, see Mimura, Nishiura and Yamaguti[15]. Global existence of bifurcating paths for the case $d_2 \gg 1$ is discussed in Nishiura [17]. As for the singular perturbation limit $d_1 \to 0$, we refer the work of Mimura, Tabata and Hosono[16]. The picture of complete bifurcation diagram which summarize those analytical and numerical studies, together with group theoretical discussions, will be reported in Fujii, Mimura and Nishiura[6].

We study the following system of equations:

$$(P)_{mm} \quad \left| \begin{array}{l} d_1 u_{xx} + \{f(u) - v\}u = 0 \\ d_2 v_{xx} - \{g(v) - u\}v = 0 \end{array} \right. \qquad x \in I = (-\frac{\pi}{2}, \frac{\pi}{2}) \qquad (3.1)_1$$

with the zero-flux boundary condition at $x = \pm\,\pi/2$, where $f(u)$ and $g(v)$ are

$$\left| \; f(u) = \frac{1}{9}(35 + 16u - u^2) \quad \text{and} \quad g(v) = 1 + \frac{2}{5}v. \right. \qquad (3.1)_2$$

Remark 3.1. Due to the zero-flux boundary condition (the zero Neumann condition), a remark from the group theoretical viewpoint may be necessary. In appearance, the reflection at the origin $C_s \cong \{e,s\}$ $(s^2 = e)$, is the only symmetry group of $(P)_{mm}$. However, a deeper insight may be obtained , if we identify the set of solutions S of $(P)_{mm}$ with the reflection symmetric cross-section of $\tilde{S}$, the set of solutions of the virtual problem $(\tilde{P})_{mm}$; $(\tilde{P})_{mm}$ is defined on a circle $\tilde{I} = [-\pi,\pi]$, and satisfy the periodic boundary condition. $(\tilde{P})_{mm}$ *is* D_∞*-covariant*. It is clear that $S \subset \tilde{S}$, and in fact, $S = \tilde{S}|\tilde{V}^+$, where $\tilde{V}^+$ is the reflection invariant subspace of $\tilde{V}$, i.e., $\tilde{V}^+ = \frac{1}{2}(I+T_s)\tilde{V}$. Thus, our strategy is to look for all bifurcating solutions $\tilde{S}$ of $(\tilde{P})_{mm}$, and then take the cross-section S of $\tilde{S}$. (It is noted that a simple criticality of $(P)_{mm}$ generally corresponds to a *group theoretical double criticality* of $(\tilde{P})_{mm}$. See, the footnote of Prop.1.5. Thus, we need to perform bifurcation analysis of double criticalities. For details, see [5] and [6].) With the understanding of this, *we can still label each bifurcating path of* $(P)_{mm}$ *by its symmetry group* $D_k \subset D_\infty$.

Following [15], we show in Fig. 4 the bifurcation curves Γ_k's in the two dimensional parameter space (d_1,d_2). Γ_k's are the set of critical parameters of $(P)_{mm}$

from which primary bifurcations, with respect to the constant steady state $(\bar{u},\bar{v})$ = (5,10), of D_k-symmetry occur.

We consider the problem for a fixed $d_2=d_2^0 > 0$, and regard $d_1=\lambda$ as the bifurcation parameter. There exist an infinite number of primary bifurcation points $\mathcal{C}_k^p = \{\lambda_k^{c,p},(\bar{u},\bar{v})\}$, $k=k_0,k_0+1,\cdots$.(k_0 : depends on d_2^0.) From each of $\mathcal{C}_k^p$, there bifurcates a D_k-invariant primary path. All of $\mathcal{C}_k^p$'s are simple, and of cusp type [15].

We are interested in the global behavior of these bifurcating paths.

Fig.3 shows numerical results about the D_1- and D_2-paths.

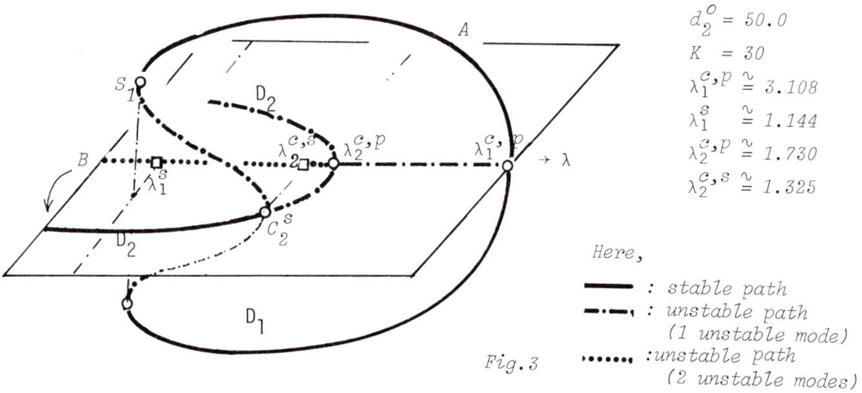

Fig.3

As $\lambda \to 0$, the D_1-path turns back at S_1 - *a snap-through (=limit) point*, and then is absorbed to the D_2-path at $\mathcal{C}_2^s$. The D_2-path, originally unstable, *recovers its stability* at $\mathcal{C}_2^s$ - a symmetry breaking *secondary* bifurcation point, and continues to exist for $\lambda \to 0$.(*) We show the numerically obtained patterns of (u,v) at the points specified on the diagram. (Fig. 5)

Now, we give some comments on the scheme $(P^h)_{mm}$.

(i) Let K be the number of mesh (element) division of I. Then, the scheme $(\tilde{P}^h)_{mm}$ is D_{2K}-covariant. (This looks like the situation that we approximate a circle by a 2K-sided regular polygon.) The number of possible primary bifurcations is now finite: $\mathcal{C}_{k,h}^p = (\lambda_{k,h}^c, (\bar{u},\bar{v}))$, $k=k_0+1,..,K$.

Since it is only the subgroups $D_k \subset D_{2K}$ that $(\tilde{P}^h)_{mm}$ is covariant, the bifurcating path from $\mathcal{C}_{j,h}$ is not in general D_j-invariant, *except for those* D_k*-paths such that* $D_k \subset D_{2K}$. (Namely, except for the case that k is a divisor of 2K. This is obviouly satisfied if we adopt the group theoretical decomposition.)

(ii) Suppose then we have a D_k-invariant bifurcating path. The secondary bifurcation on this path is regarded as a problem of D_k-covariant system. (Now, equi-

(*) Although the numerical solution seems to exist always to the limit $\lambda \to 0$, the analysis of Sec.2 does not guarantee the accuracy of solutions for the case that $\lambda \lesssim h$ where the solution path enters into the region of singular perturbation theory.

valent to a problem on a k-polygon.)

(iii)The orders of convergence for the critical values of the primary and secondary bifurcation points $\lambda^{c,p}_{2,h}$, $\lambda^{c,s}_{2,h}$ ... were found to be $O(h^2)$, for the piecewise linear finite element basis, and also with lumpings in the semi-linear terms. This result supports the analysis of Sec.2. Detailed results will appear in [5] and [6].

Fig.5 exhibits the *secondary* bifurcations on the D_3-path. C^p_3 ($C^p_{3,h}$) is of "fold" (=transcritical) type. From the viewpoint of the reflection symmetry C_s, D_3-path has only the trivial group E={e} as its symmetry. Numerical realization of such simple, symmetry preserving bifurcations is not included in the theory of Sec. 2. However, this can be justified by considering $(\tilde{P})_{mm}$, and regarding $C^p_{3,h}$ as a *group theoretical double criticality* on a D_3-invariant path. See the footnote of Prop. 1.5. See, also [5].

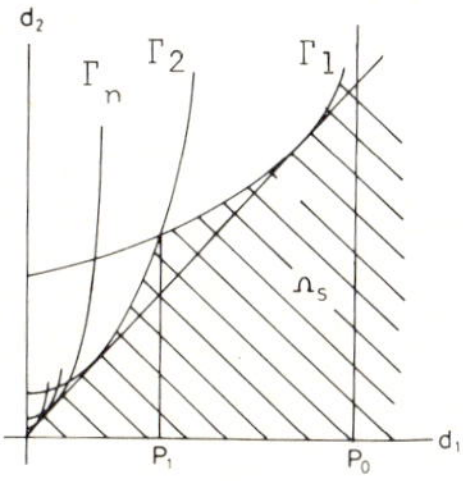

Fig.4 (left)
"Bifurcation curve in (d_1,d_2)-space".

Fig.5 (below)

(a) at A ($\lambda \cong 2.800$)
(b) at S_1 ($\lambda \cong 1.144$)
(c) at C_2 ($\lambda \cong 1.325$)
(d) at B ($\lambda \cong 0.100$)

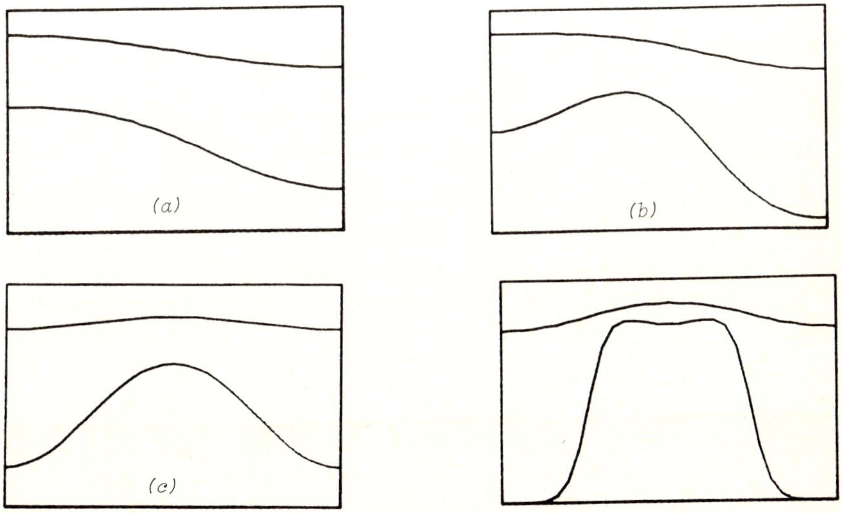

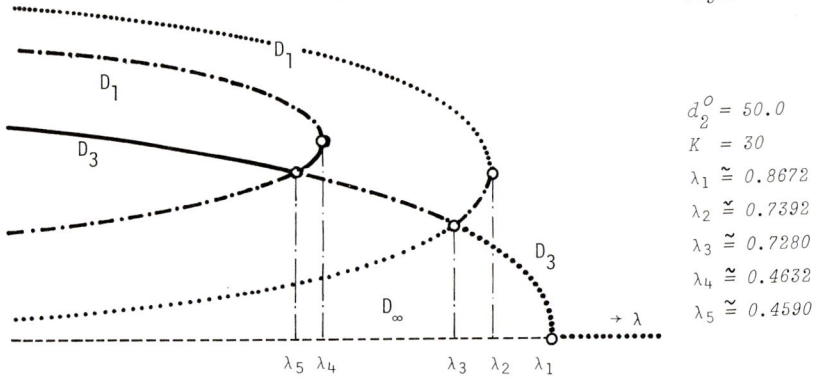

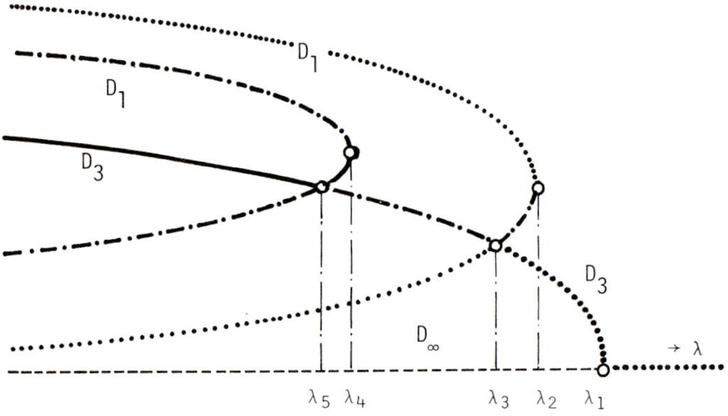

Fig.6

$d_2^o = 50.0$
$K = 30$
$\lambda_1 \cong 0.8672$
$\lambda_2 \cong 0.7392$
$\lambda_3 \cong 0.7280$
$\lambda_4 \cong 0.4632$
$\lambda_5 \cong 0.4590$

A tertiary bifurcation was also observed from the D_2-secondary bifurcated path, which had bifurcated from the D_4-path. See, [5] and [6].

We have thus observed here a number of interesting behaviors of bifurcating solutions; appearance of a snap-through point, of secondary and tertiary bifurcation points, disappearance of a bifurcated path, global existence of a bifurcated path, and so on.

An interesting phenomenon is the *successive recovery of stability* of originally unstable paths. This phenomenon was generally observed for other D_k-paths. It is also reported by Herschkowitz-Kaufman and Erneux [11] for a model system of biochemical reaction-diffusion equations. As a consequence, *several* non-constant and *stable* stationary solutions co-exist. See, Fig.3. For λ, $\lambda_1^s > \lambda > \lambda_2^{c,s}$, both the D_1- and D_2-paths are stable.

This paper concludes with a remark that the co-existence of several stable solutions may pose an interesting question on the pattern selection of initial value problems in a Hilbert space: $\partial \underline{u}/\partial t = F(\lambda, \underline{u})$, $\underline{u}(t) \in \underline{V}$.

ACKNOWLEDGEMENTS

The author would like to express his appreciation to Professor M. Yamaguti of Kyoto University for continuous encouragement and discussions. He expresses his thanks to his colleagues M. Mimura, Y. Nishiura and M. Tabata for their discussions.

REFERENCES

[1] Atkinson, K.E.,The numerical solution of a bifurcation problem, SIAM J. Numer. Anal. 14, (1977), n°4.

[2] Brezzi, F. and Fujii, H., Mixed finite element approximations of the von Kármán equations, IV th LIBLICE Conference on Basic Problems of Numerical Analysis, Chechoslovakia, 1978.

[3] Ciarlet, P.G., The finite element method for elliptic problems, North Holland, Amsterdam, 1978.

[4] Crandall, M.G. and Rabiniwitz, P.H., Bifurcation from simple eigenvalues, J. Funct. Anal. 8 (1971).

[5] Fujii, H., Numerical analysis of global bifurcations in reaction-diffusion equations. (in preparation)

[6] Fujii, H., Mimura, M. and Nishiura, Y., A picture of global bifurcation diagram of ecological interacting and diffusing systems. (manuscript)

[7] Fujii, H. and Yamaguti, M., Structure of singularities and its numerical realization in nonlinear elasticity, to appear in J. Math. Kyoto Univ.

[8] Grisvard, P., Singularité des solutions du problèmes de Stokes dans un polygone, Publications de l'Univ. de Nice, 1978.

[9] Hamermesh, M., Group theory and its applications to physical problems. Readings: 1962, Addison-Wesley.

[10] Hangai, Y. and Kawamata, S., Analysis of geometrically nonlinear and stability problems by static perturbation method, Report of Inst. Industrial Sci., 22 (1973), The Univ. of Tokyo.

[11] Herschkowitz-Kaufman, M. and Erneux, T., The bifurcation diagram of model chemical reactions, Annals of the New York Academy of Sciences 316 (1979).

[12] Keller, H.B., Constructive methods for bifurcation and nonlinear eigenvalue problems, 3 ème Colloque International sur les Méthodes de Calcul Scientifique et Technique, 1977.

[13] Keener, J.P., Secondary bifurcation in nonlinear diffusion reaction equations, Studies in Appl. Math., 55 (1976).

[14] Miller Jr., W., Symmetry groups and their applications, Academic Press, 1972.

[15] Mimura, M., Nishiura, Y. and Yamaguti, M., Some diffusive prey and predator systems and their bifurcation problems, Annals of the New York Academy of Sciences, 316 (1979).

[16] Mimura, M, Tabata, M. and Hosono, Y., Multiple solutions of two-point boundary value problems of Neumann type with a small parameter, to appear in SIAM J. Math. Anal.

[17] Nishiura, Y., Global branching theorem for spatial patterns of reaction-diffusion system, Proc. Japan Academy, 55, Ser.A, No.6 (1979).

[18] Ruelle, D., Bifurcations in the presence of a symmetry group, Arch. Rat. Mech. Anal. 51 (1975).

[19] Sattinger, D.H., Group representation theory and branch points of nonlinear functional equations, SIAM J. Math. Anal., 8 (1977).

[20] Sattinger, D.H., Group theoretic methods in bifurcation theory, Lec. Notes in Math., Univ. of Chicago, 1978.

[21] Segal, L. and Jackson, J., Dissipative structure: an explanation and an ecological example., J. Theor. Biol. 37 (1972).

[22] Serre, J.-P., Représentations linéaires des groupes finis, Hermann s.a., Paris 1971.

[23] Yamaguti, M. and Fujii, H., On numerical deformation of singularities in nonlinear elasticity, 3 ème Colloque International sur les Méthodes de Calcul Scientifique et Technique, 1977.

COMPUTING METHODS IN APPLIED SCIENCES AND ENGINEERING
R. Glowinski, J.L. Lions (editors)
North-Holland Publishing Company

A NEW APPROACH TO BIFURCATION THEORY WITH NUMERICAL APPLICATIONS TO CONVECTION INSTABILITIES IN PLASMAS

H. Haken and H. Klenk
Institut für theoretische Physik
Universität Stuttgart

Abstract

We treat nonlinear stochastic differential equations and study the branching of new solutions out of a homogeneous time-independent state when external control parameters are changed. From a linear stability analysis we derive critical points in parameter space. The newly developing solutions are expanded into stable (damped) and unstable (undamped) modes of the linear problem. We derive equations for the fully nonlinear problem expressed in slowly varying amplitudes. By an elimination of the time-dependent stable modes we find a closed set of equations for the amplitudes of the unstable modes ("order parameters"). We apply this method explicitly to the convection instability in plasmas, i.e. a plasma heated from below in the presence of a magnetic field. Beyond a critical temperature gradient the formerly homogeneous plasma shows the formation of convection cells. In this case both the linear problem as well as the elimination procedure is treated by means of a computer.

§ 1 Introduction

Many phenomena in physics and mechanical and electrical engineering can be described by equations of the form

$$\dot{\underset{\sim}{u}} = \underset{\sim}{N}(\underset{\sim}{u}, \sigma) \tag{1.1}$$

In (1.1) the vector $\underset{\sim}{U}$ comprises a set of variables which describe the system, for instance densities, velocity fields, currents etc. $\underset{\sim}{N}$ is a nonlinear function of the variables $\underset{\sim}{U}$. It depends also on external control parameters σ , for instance the temperature, the heatflux through a system etc. There are numerous examples in physics, e.g. plasmas, lasers, fluids, but also electric networks and mechanics where the following situation appears. For a certain range of parameter values σ a certain solution $\underset{\sim}{U} = \underset{\sim}{U}^0$ is stable. When σ is changed beyond a critical value that solution becomes unstable and qualitatively new solutions arise. In many cases several new stable solutions become possible. Such a phenomenon is often called bifurcation. In spatially extended systems bifurcation is often connected with the occurrence of spatial structures. We have developed a rather general method to cope with this phenomenon. In it the socalled slaving principle plays a crucial role. It allows one to reduce the originally high dimensional problem to one containing only very few variables.

§ 2 Our bifurcation method

We assume the equations for $\underset{\sim}{U}$ of the following type:

$$\frac{\partial}{\partial t} U_\mu = G_\mu(\nabla, \underset{\sim}{U}) + D_\mu \nabla^2 U_\mu + F_\mu(t) \;; \mu = 1, 2, \ldots n \tag{2.1}$$

In (2.1) G_μ are nonlinear functions of $\underset{\sim}{U}$ and perhaps of a gradient. In most applications like hydrodynamics, G is a linear or bilinear function of $\underset{\sim}{U}$, though in certain cases a cubic coupling term may equally well occur. The next term in (2.1) describes diffusion (D real) or wave type propagation (D imaginary). In this latter case the second order time derivative of the wave equation has been replaced by the first derivative by use of the "slowly varying amplitude approximation". The $F_\mu(t)$'s are fluctuating forces which are caused by external reservoirs and internal dissipation and which are connected with the damping terms occurring in (2.1).

We shall not be concerned with the derivation of equations (2.1). Rather our goal will it be to derive from (2.1) equations for the "unstable" modes which aquire a macroscopic size and determine the dynamics of the system in the vicinity of the instability point. These modes form a mode skeleton which grows out from fluctuations above the instability and thus describes the "embryonic" state of the evolving spatio-temporal structure. We now start with a treatment of (2.1). We assume that the functions G_μ in (2.1) depend on external parameters (e.g. energy pumped into the system). First we consider such values of σ so that $\underset{\sim}{U} = \underset{\sim}{U}_0$ (= independent of space and time) is a stable solution of (2.1). We decompose $\underset{\sim}{U}$

$$\underset{\sim}{U} = \underset{\sim}{U}_0 + \underset{\sim}{q} \tag{2.2}$$

with

$$\underset{\sim}{q} = \begin{pmatrix} q_1(\underset{\sim}{x}, t) \\ \vdots \\ q_n(\underset{\sim}{x}, t) \end{pmatrix} \tag{2.3}$$

Splitting the right hand side of (2.1) into a linear part, $K\underset{\sim}{q}$, and a nonlinear part $\underset{\sim}{g}$ we obtain (2.1) in the form

$$\left(\frac{\partial}{\partial t} - K(\nabla^2)\right)\underset{\sim}{q} = \underset{\sim}{g}(\underset{\sim}{q}) + \underset{\sim}{F}(t) \tag{2.4}$$

In it the matrix

$$K = (\hat{K}_{\mu\nu}) \tag{2.5}$$

has the form

$$\hat{K}_{\mu\nu} = K_{\mu\nu} + \delta_{\mu\nu} D_\mu \nabla_\mu^2 \qquad K_{\mu\nu} = \left.\frac{\partial G_\mu}{\partial u_\nu}\right|_{u_{\nu,0}} \tag{2.6}$$

Our whole procedure applies, to a matrix K which depends in a general way on ∇ . g is assumed in the form

$$g_i(\underset{\sim}{q}) = \sum_{\mu\nu} q_\mu \, g^{(2)}_{i\mu\nu}(\nabla) \, q_\nu \tag{2.7}$$

$g^{(2)}$ may or may not depend on ∇. We have included only quadratic terms in (2.7), but higher order terms could be treated as well by our procedure. We first deal with K and introduce operators

$$\underset{\sim}{O}^{(j)} = \underset{\sim}{O}^{(j)}(\nabla) \tag{2.8}$$

which still depend on ∇ and which are defined as eigenvectors satisfying the equation

$$\underset{\sim}{K}(\nabla)\underset{\sim}{O}^{(j)} = \lambda_j(\nabla)\,\underset{\sim}{O}^{(j)} \tag{2.9}$$

When ∇ is replaced by $i\cdot\underset{\sim}{k}$ (2.9) becomes a linear algebraic equation so that (2.9) can easily be solved without resorting to any operator techniques. We furthermore introduce eigenfunctions of the wave equation

$$\nabla^2 \chi_{\underset{\sim}{k}}(\underset{\sim}{x}) = -k^2 \chi_{\underset{\sim}{k}}(\underset{\sim}{x}) \tag{2.10}$$

provided K depends on ∇^2. $\chi_{\underset{\sim}{k}}$ can be chosen lateron adequately. We shall use a notation which suggests that the $\chi_{\underset{\sim}{k}}$'s are plane-wave solutions, but it may be advantageous to use other representations as well, e.g. Bessel functions or spherical wave functions, depending on the problem. If K depends on ∇ in odd powers (and in even powers), $\chi_{\underset{\sim}{k}}$'s are taken as complex plane wave solutions of (2.10). We represent $\underset{\sim}{q}(x)$ as superposition

$$\underset{\sim}{q}(\underset{\sim}{x},t) = \sum_{\underset{\sim}{k},j} \underset{\sim}{O}^{(j)} \xi_{\underset{\sim}{k},j} \, \chi_{\underset{\sim}{k}}(\underset{\sim}{x}) \tag{2.11}$$

Since our procedure has been described elsewhere in detail [1] we present here a special case of it. In particular we neglect so called finite band excitations of unstable modes and we neglect fluctuations, i.e. F = 0. Our first goal is to derive a general set of equations for the mode amplitudes ξ. To do so we insert (2.11) into (2.4), multiply from the left hand side by $\chi^*_{k'}(x)\,\bar{O}^{(j')}$ and integrate over space. After some analysis the basic set of equations has the following structure

$$\dot{\xi}_{k,j} - \lambda_j(k)\,\xi_{k,j} = H_{k,j}(\{\xi(x)\}) \tag{2.12}$$

where H_{kj} is a polynomial of ξ's up to second order. So far no approximations have been made, but to cast (2.12) into a practicable form we have to <u>eliminate</u> the unwanted or uninteresting modes which are the <u>damped ("stable") modes.</u> Accordingly we put

$$j = u \quad \text{unstable if} \quad \mathrm{Re}\,\lambda_\mu(k) \gtrsim 0 \tag{2.13}$$

and

$$j = s \quad \text{stable} \quad \text{if } \mathrm{Re}\,\lambda_s(k) < 0 \tag{2.14}$$

An important point should be observed at this stage: Though ξ bears two indices, k and u, these indices are not independent of each other. Indeed the instability usually occurs only in a small region of $k = k_c$. Thus we must carefully distinguish between the k values at which (2.19) (see below) is evaluated. When we treat boundaries with finite dimensions k runs over a set of discrete values with $|k| = k_c$. The basic idea of our further procedure is this. Because the undamped modes may grow unlimited provided the nonlinear terms are neglected, we expect that the amplitudes of the undamped modes are considerably bigger than those of the damped modes. Since, on the other hand, close to the "phase transition" point the relaxation time of the undamped modes tends to infinity, i.e. the real part of λ tends to zero, the damped modes must adiabatically follow the undamped modes. Though the amplitudes of the damped modes are small, they must not be neglected completely. This neglect would lead to a catastrophe if the cubic terms are lacking. As one convinces oneself very quickly, quadratic terms in the order parameter equations cannot lead to a globally stable situation. Thus cubic terms are necessary for a stabilization. Such cubic terms are introduced even in the absence of those in the original equations by the elimination of the damped modes. To exhibit the main features of our elimination procedure more clearly, we put for the moment

$$\xi_{\underset{\sim}{k},j} \longrightarrow (\underset{\sim}{k},j) \tag{2.15}$$

and drop all coefficients in (2.11). We assume $|\xi_s| << |\xi_u|$ and, in a selfconsistent way, $\xi_s \propto \xi_u^2$. Keeping in (2.11) only terms up to third order in ξ_u , we obtain

$$\left(\frac{d}{dt} - \lambda_u\right)(\underset{\sim}{k},u) = \sum_{\underset{\sim}{k}'\underset{\sim}{k}''u's} (\underset{\sim}{k}',u')\cdot(\underset{\sim}{k}'',s) + \sum_{\underset{\sim}{k}'\underset{\sim}{k}''u'u''} (\underset{\sim}{k}',u')\cdot(\underset{\sim}{k}'',u'')$$

$$+ \sum_{\substack{\underset{\sim}{k}'\underset{\sim}{k}''\underset{\sim}{k}'''\\ u'u''u'''}} (\underset{\sim}{k}',u')(\underset{\sim}{k}'',u'')(\underset{\sim}{k}''',u''') \tag{2.16}$$

Consider now the corresponding equation for j = s. In it we keep only terms necessary to obtain an equation for the unstable modes up to third order

$$\left(\frac{d}{dt} - \lambda_s\right)(\underset{\sim}{k},s) = \sum_{\underset{\sim}{k}'\underset{\sim}{k}''u'u''} (\underset{\sim}{k}',u')(\underset{\sim}{k}'',u'') + \dots \tag{2.17}$$

If we adopt an iteration scheme using the inequality $|\xi_s| \ll |\xi_u|$ one readily convinces oneself that ξ_s is at least proportional to ξ_u^2 so that the only relevant terms in (2.17) are those exhibited explicitly. We now use our second hypothesis, namely, that the stable modes are damped much more quickly than the unstable ones which is well fulfilled for the soft mode instability. In the case of a hard mode, we must be careful to remove the oscillatory part of (k', u'), (k",u") in (2.17). This is achieved by keeping the time derivative in (2.17). We therefore write the solution of (2.17) in the form

$$(\underset{\sim}{k},s) = \left(\frac{d}{dt} - \lambda_s\right)^{-1} \sum_{\underset{\sim}{k}'\underset{\sim}{k}''u'u''} (\underset{\sim}{k}',u')(\underset{\sim}{k}'',u'') \tag{2.18}$$

The presciption to evaluate d/dt is this: In the soft mode case it can be neglected, whereas in the case $(\underset{\sim}{k},u) \propto \exp(i\omega_{\underset{\sim}{k}} t)$, d/dt is to be replaced by $i(\omega_{\underset{\sim}{k}'} + \omega_{\underset{\sim}{k}''})$. The equation of type (2.17) can now be readily solved. If time and space derivatives are neglected, the solution of (2.17) is a purely algebraic problem so that one could keep also higher order terms in (2.17) without difficulties, at least in principle. Inserting the results into (2.16) we obtain the fundamental set of equations for the order parameters

$$\left(\frac{d}{dt} - \lambda_u(\underset{\sim}{k})\right)\xi_{\underset{\sim}{k},u} = \sum_{\underset{\sim}{k}'\underset{\sim}{k}''u'u''} a_{\underset{\sim}{k}\underset{\sim}{k}'\underset{\sim}{k}''u'u''u'''}\,\xi_{\underset{\sim}{k}',u'}\,\xi_{\underset{\sim}{k}'',u''}$$

$$+ \sum_{\substack{\underset{\sim}{k}'\underset{\sim}{k}''\underset{\sim}{k}''' \\ u'u''u'''}} C_{\underset{\sim}{k}\underset{\sim}{k}'\underset{\sim}{k}''\underset{\sim}{k}'''uu'u''u'''}\,\xi_{\underset{\sim}{k}',u'}\,\xi_{\underset{\sim}{k}'',u''}\,\xi_{\underset{\sim}{k}''',u'''} \qquad (2.19)$$

where in the soft mode case the C's are certain constants, which can be calculated explicitly. The procedure to express the stable modes by the order parameters (undamped modes) can be pushed to arbitrary order (compare [1]). The equations (2.19) are still formidable but due to symmetries, for instance imposed by boundary conditions often only few combinations of $\xi_{\underset{\sim}{k},u}$ need to be kept.

§ 3 Application to the convection instability of plasmas

We have applied the procedure described in § 2 to the following problem: A horizontal layer of a homogeneous one-component plasma, exposed to a homogeneous magnetic field in vertical direction is heated from below. More specifically, the temperatures at the lower and upper boundaries are kept fixed. For a small temperature gradient, the plasma stays spatially homogeneous. When the temperature gradient is increased, at a certain critical value, the plasma acquires spatial or spatio-temporal structures. We have calculated these emerging structures. Our approach starts from the following set of equations for the velocity field $\underset{\sim}{v}$, the magnetic induction $\underset{\sim}{B}$, density ϱ and temperature T in the conventional dimensionless units.

$$\operatorname{div} \underset{\sim}{v} = 0 \quad ; \quad \operatorname{div} \underset{\sim}{B} = 0$$

$$\varrho \frac{\partial \underset{\sim}{v}}{\partial t} + \varrho(\underset{\sim}{v}\nabla)\underset{\sim}{v} - \frac{1}{R_e}\nabla^2 \underset{\sim}{v} = -\frac{1}{\gamma}\nabla P - \frac{A_0^2}{2}\nabla \underset{\sim}{B}^2 + A_0^2(\underset{\sim}{B}\nabla)\underset{\sim}{B} - \varrho g_0 \underset{\sim}{e}_z$$

$$\frac{\partial \underset{\sim}{B}}{\partial t} - \frac{1}{R_m}\nabla^2 \underset{\sim}{B} = \operatorname{rot}(\underset{\sim}{v}\times\underset{\sim}{B})$$

$$\frac{\partial T}{\partial t} + (\underset{\sim}{v}\nabla)T - \frac{\gamma}{P_e}\nabla^2 T = \frac{A_0^2}{R_m}\gamma(\gamma-1)(\operatorname{rot}\underset{\sim}{B})^2 - A_0^2\gamma(\gamma-1)(\underset{\sim}{v}\times\underset{\sim}{B})\operatorname{rot}\underset{\sim}{B}$$

$$+ \frac{\gamma(\gamma-1)}{2R_e}\sum_{i,j=1}^{3}\left(\frac{\partial v_i}{\partial x_j} + \frac{\partial v_j}{\partial x_i}\right)^2$$

The constants occurring these equations have the following meaning:

R_e :	Reynolds number
γ :	adiab. exponent
A_0 :	Alfvén number
g_0 :	gravity acceleration constant (dimensionless)
$R_m = v_s d/\eta$:	magnetic Reynolds number
$P_e = v_s d/\varkappa$:	Peclit number

The equations were solved under the following boundary conditions. We treat a plasma layer with periodic boundary conditions along the x and y - directions(horizontal directions). The vertical component of the velocity field is assumed to vanish at z = 0 and z = d.

Figs. 1 to 6 show explicit cases which have been obtained by our procedure described above.

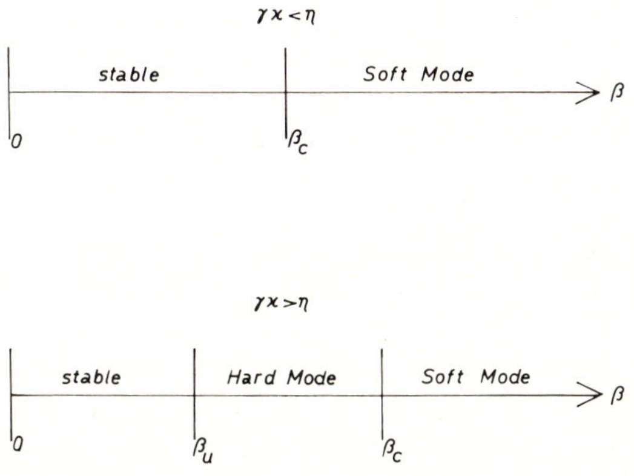

Fig. 1
Upper part: for $\gamma\varkappa < \eta$ i.e. $\gamma/P_e < 1/R_m$
With increasing temperature gradient β (given in the dimensionless units) at $\beta = \beta_c$ the system becomes unstable and a soft mode appears.
Lower part: for $\gamma\varkappa > \eta$ i.e. $\gamma/P_e > 1/R_m$
Two instability points appear, the first one giving rise to a hard mode instability (Hopf bifurcation)and at still higher β the hard mode disappears and is replaced by a soft mode.

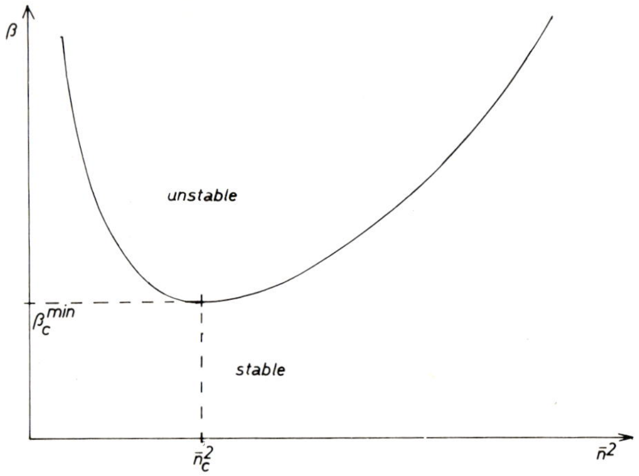

Fig. 2
The dependence of the critical temperature gradient β_c on the wave number in horizontal direction, $\bar{n}$.

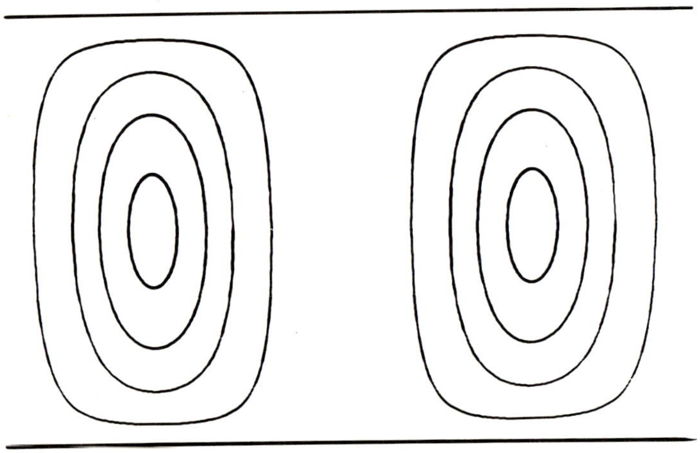

Fig. 3
Velocity field in the x, z-plane. The horizontal bars indicate the boundaries. Single mode case.

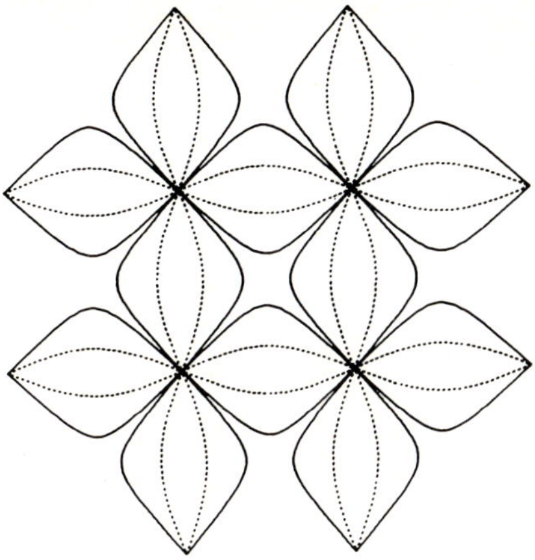

Fig. 4
Velocity field seen from above. Coexistence of two modes in perpendicular direction.

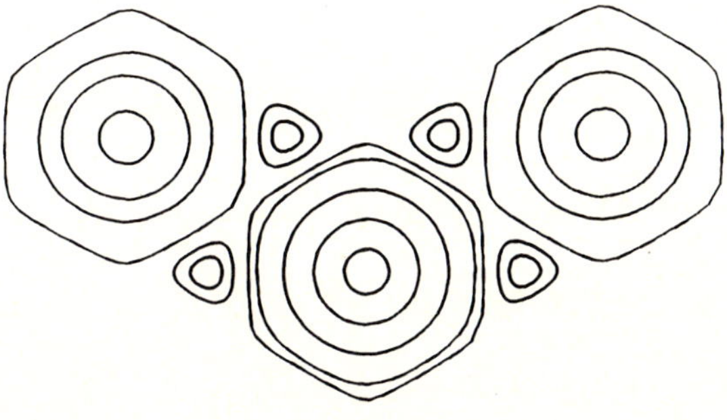

Fig. 5
The velocity field of the hexagonal structure seen from above (coexistence of three modes).

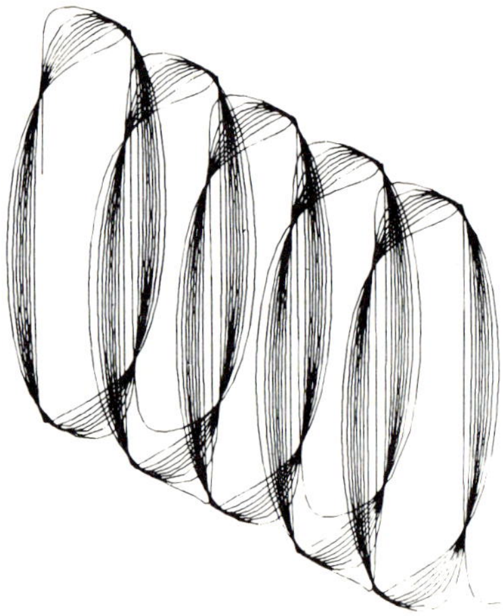

Fig. 6
Three-dimensional trajectory of a volume element in x, y, z-direction in perspective representation in the case of a hard mode.

References

[1] H. Haken: Synergetics. An Introduction, 2nd enlarged edition Springer Verlag, Heidelberg, Berlin, New York 1978

[2] F.F. Cap: Handbook on Plasma Instabilities, Vol. 1, Academic Press,New York 1976

COMPUTING METHODS IN APPLIED SCIENCES AND ENGINEERING
R. Glowinski, J.L. Lions (editors)
North-Holland Publishing Company

ETUDE NUMERIQUE DES BIFURCATIONS ET DE LA STABILITE DES SOLUTIONS DES EQUATIONS DE GRAD-SHAFRANOV.

Michel SERMANGE
IRIA-LABORIA
Domaine de Voluceau
78150 LE CHESNAY (France)

La configuration d'équilibre d'un plasma confiné dans une machine du type Tokamak est modélisée par une équation elliptique non linéaire et non continûment différentiable. On étudie ici des méthodes numériques nouvelles pour déterminer les branches de solutions de tels problèmes ainsi que leurs bifurcations (cf. [12]). D'autre part, la stabilité de ces équilibres (par rapport à des perturbations axisymétriques) est déterminée par un problème aux valeurs propres non local (faisant intervenir des intégrales curvilignes le long des lignes magnétiques). Le problème approché est obtenu par une double discrétisation : des éléments finis et une famille de lignes magnétiques. Les résultats numériques présentés mettent en évidence des bifurcations et des changements de stabilité. Signalons qu'une étude théorique de l'existence de bifurcation a fait l'objet de [13] .

1. LES EQUATIONS

1.1. Equations d'équilibre.

Les équations standard d'équilibre en M H D sont

$$(1.1)\qquad \begin{cases} \mathrm{rot}\; B \times B = \mathrm{grad}\; p\ , \\ \mathrm{div}\; B = 0\ , \end{cases}$$

où B est le champ magnétique et p la pression.

Dans les cas d'équilibres cylindriques indépendants de z ou d'équilibres toriques axisymétriques, il est possible de réduire le système (1.1) à une seule équation portant sur le flux poloïdal (Ψ) ; c'est l'équation de GRAD-SHAFRANOV, qui s'écrit [1]

$$(1.2)\qquad -\Delta\Psi = p'(\Psi)$$

[1] Par commodité, on écrit seulement les expressions associées au cas cylindrique.

où $\Delta\Psi = \sum_{i=1}^{2} \frac{\partial^2\Psi}{\partial x_i^2}$ et $p(\Psi)$ représente la somme de la pression et de $\frac{1}{2} \; B_z(\Psi)^2$.

Le plasma étant maintenant considéré à frontière libre dans un domaine de section Ω, il est possible de formuler le problème de la façon suivante (cf. R. TEMAM [16]).

Soient Ω un domaine de $\mathbb{R}^2$, p une fonction de $\mathbb{R}$ dans $\mathbb{R}$ donnée, I une constante réelle positive donnée, on cherche $\Psi \in H^1(\Omega)$ (espace de Sobolev d'ordre 1 sur $L^2(\Omega)$) tel que

$$(1.3) \qquad \begin{cases} -\Delta\Psi = p'(\Psi) & \textit{p.p dans } \Omega \\ \Psi = \textit{cte (inconnue) p.p. sur } \Gamma(=\partial\Omega), \\ -\displaystyle\int_\Gamma \frac{\partial\Psi}{\partial n}\, d\sigma = I. \end{cases}$$

La fonction p est supposée vérifier les conditions suivantes

(1.4) $\quad p : \mathbb{R} \longrightarrow \mathbb{R}$ est une fonction C^1 telle que

$$p(z) = o \quad \forall\, z \leq o \quad \text{et} \quad p'(o) = o,$$

(1.5) $\quad p$ est convexe

(1.6) $\quad \exists\ \alpha > 1$ tel que $\lim\limits_{z \to \infty} \dfrac{p'(z)}{z^\alpha} = o,$

(1.7) $\quad \lim\limits_{z \to \infty} \ mes(\Omega)\, p'(z) > I.$

Lorsque la solution Ψ est connue, la région Ω_p occupée par le plasma s'en déduit par (cf. Figure 1)

$$\Omega_p = \left\{ x \in \Omega \mid \Psi(x) > o \right\}.$$

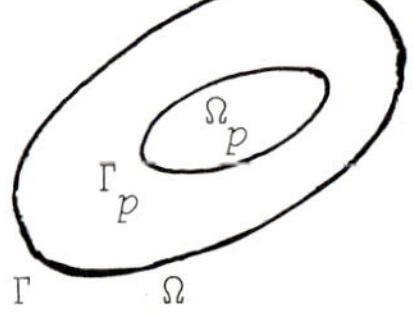

Figure 1.

L'existence de solution Ψ pour le problème (1.3) a d'abord été démontrée par R. TEMAM [16], puis J.P. PUEL [10], dans le cas modèle où $p(z) = \frac{1}{2}\lambda(z_+)^2$, $\lambda > o$ quelconque ; ce résultat a été étendu par H. BERESTYCKI et H. BREZIS [2] au cas général où p vérifie (1.4)-(1.7).

L'unicité de solution a été démontrée (cf. [2]) dans le cas où p' est strictement monotone et k -lipschitzienne avec k suffisamment petit.

1.2. Solutions variationnelles et solutions linéairement stables.

On introduit la variété K,

$$K = \left\{ \Psi \in H^1(\Omega) \mid \Psi = cte \text{ (quelconque) } p.p. \text{ sur } \Gamma, \int_\Omega p'(\Psi)\, dx = I \right\},$$

et on définit sur cette variété une fonctionnelle e par

$$e(\Psi) = \frac{1}{2}\int_\Omega |\nabla\Psi|^2\, dx - \int_\Omega p(\Psi)\, dx + I\,\Psi(\Gamma)$$

où $\Psi(\Gamma)$ désigne la valeur (constante) de Ψ sur Γ.

Les points critiques de e sur K sont les solutions de (1.3) (cf. [2] [16]). On appelera solution variationnelle de (1.3) les solutions réalisant le minimum de e sur K.

On trouvera dans l'ouvrage récent de G. BATEMAN [1] les références de plusieurs méthodes pour déterminer la stabilité linéaire d'équilibres quelconques de la M H D. L'originalité du travail présenté ici (cf. [14]) est la relative simplicité des expressions, qui a été obtenue en se plaçant dans le cadre particulier de la dimension 2 (y compris pour les perturbations).

En effet, dans le cas d'un équilibre cylindrique indépendant de z (rep. axisymétrique) on peut étudier un type particulier de stabilité : la stabilité par rapport à des déplacements indépendants de z (resp. axisymétrique). Ce cadre a été étudié par E. REBHAN et A. SALAT [11]. Si on ajoute encore deux conditions, à savoir que la composante verticale (resp. toroïdale) du champ magnétique à l'équilibre est partout nulle et que les déplacements admissibles ont des composantes verticales (resp. toroïdales) nulles, on est ramené au problème en dimension 2 suivant .

Avec l'espace des déplacements admissibles

$$E = \left\{ u \in (L^2(\Omega))^2 \mid u.\nabla\Psi \in H^1_o(\Omega) , \quad div\, u|_{\Omega_p} \in L^2(\Omega_p) \right\},$$

et la fonctionnelle (déduite de J. BERNSTEIN, E.A. FRIEDMAN, N. KRUSKAL, R.M. KULSRUD [3])

$$J(u) = \frac{1}{2} \int_\Omega |\nabla u^1|^2 \, dx - \frac{1}{2} \int_\Omega p''(\Psi)|u^1|^2 + \frac{\gamma}{2} \int_\Omega p(\Psi)|div\ u|^2 \, dx$$

où $u^1 = u.\nabla\Psi$

et $\gamma = 5/3$ est la constante de chaleur spécifique,

l'équilibre (p,Ψ) est dit linéairement stable (pour la dim.2) si et seulement si

$$\operatorname*{Inf}_{\substack{u \in E \\ \int_\Omega u^2 \, dx = 1}} J(u) \geq 0 \ ;$$

l'équilibre est dit linéairement instable sinon, c'est-à-dire

$$\operatorname*{Inf}_{\substack{u \in E \\ \int_\Omega u^2 \, dx = 1}} J(u) < 0.$$

2. DETERMINATION DES BRANCHES DE SOLUTIONS ET DES BIFURCATIONS.

2.1. Deux algorithmes généraux.

Le premier algorithme que nous présentons concerne la minimisation de la différence de deux fonctions convexes.

Soient V, H deux espaces de Banach ; on suppose que V s'injecte de façon compacte dans H et qu'il est dense dans H. On s'intéresse à la minimisation de fonctionnelles de la forme

$$e(v) = \frac{1}{2} a(v,v) - f(v)$$

où a est une forme bilinéaire continue coercive sur V, f est une fonctionnelle convexe et C^1 sur H, e est minorée sur V et $\lim_{||v||_V \to \infty} e(v) = +\infty$.

Les solutions du problème de minimisation

(2.1) $u \in V$ tel que $e(u) \leq e(v) \quad \forall\, v \in V$

sont aussi des solutions de l'équation d'Euler

(2.2) $u \in V$ tel que $a(u,v) = f'(u).v \quad \forall\, v \in V.$

On définit un algorithme par la relation de récurrence

(2.3) $u^{n+1} \in V$ tel que $a(u^{n+1},v) = f'(u^n).v \quad \forall\, v \in V.$

On démontre, généralisant ainsi un résultat de H. BERESTYCKY et H. BREZIS (cf. th.2.3),

Théorème 2.1 . Sous les hypothèses précédentes, pour tout $u_o \in V$, il existe une sous-suite $u^{n'}$ de la suite définie par (2.3) qui converge fortement dans H vers une solution du problème (2.2). De plus, toute sous-suite convergente de (u^n) converge vers une solution de (2.2).

Principe de la démonstration. Grâce à la convexité de f, on obtient directement que $e(u^{n+1}) \leq e(u^n) \quad \forall\, n \geq 1$. La suite $e(u^n)$ est décroissante et minorée. Elle est donc convergente. On en déduit facilement le théorème.

On remarque que si le problème (2.2) admet au plus une solution, alors toute la suite définie par (2.3) converge vers cette solution.

Le deuxième algorithme concerne la détermination des points fixes d'opérateurs non contractants et non continûment différentiables.

Soient V un espace de Banach réel et T une application continue de V dans lui-même dont on cherche les points fixes.

On se donne un réel positif α et on construit par récurrence une suite (u^n) d'éléments de V. On part d'un élément $u^o \in V$ quelconque et, l'élément u^n étant connu, on détermine u^{n+1} par trois pas intermédiaires :

$$(2.4)_\alpha \qquad \begin{cases} u^{n+\frac{1}{3}} = T(u^n), \\ u^{n+\frac{2}{3}} = T(u^{n+\frac{1}{3}}), \\ u^{n+1} = (1-\alpha)\, u^n + 2\alpha\, u^{n+\frac{1}{3}} - \alpha\, u^{n+\frac{2}{3}} \,. \end{cases}$$

La continuité de T entraîne le résultat de consistance suivant. S'il existe $u^* \in V$ tel que, pour $n \to \infty$, $u^n \to u^*$ et $u^{n+\frac{1}{3}} \longrightarrow u^*$ dans V, alors u^* est un point fixe de T . Par contre, la seule hypothèse : $u^n \to u^*$ dans V ne permet pas en général de conclure que u^* est un point fixe de T.

On démontre le résultat de convergence suivant :

Théorème 2.1. Supposons que

i) $u^* \in V$ est un point fixe de T ;

ii) T est Fréchet-différentiable en u^*, de dérivée notée A ;

iii) le complexifié de A est compact, de valeurs propres réelles et différentes de 1 ;

Alors, pour tout α , $0 < \alpha < \alpha^*$, α^* assez petit, il existe un voisinage V_α de u^* tel que, si $u^o \in V_\alpha$ et si (u^n) est défini par $(2.4)_\alpha$,

$$u^{n+\frac{i}{3}} \longrightarrow u^* \quad \text{dans } V, \qquad i = 0,\ 1,\ 2.$$

Principe de la démonstration. On remarque que la suite u^n est obtenue par itération de l'opérateur

$$U_\alpha = (1-\alpha)\ I + 2\ \alpha\ T - \alpha\ T^2.$$

On vérifie que, pour α plus petit que α^* donné par

$$(2.5) \qquad \alpha^* = 2/_{b^2} \quad \text{où} \quad b = \sup_{y \in Sp(A)} |1-y|,$$

le rayon spectral de l'opérateur U_α est strictement plus petit que 1. On en déduit la convergence locale de la suite (u^n) par application d'un théorème de J.W. KITCHEN [8].

L'algorithme $(2.4)_\alpha$ convient principalement au calcul d'une branche de solutions passant par une solution connue. Plus précisément, soit à calculer une branche de solutions de

$$T(\lambda,u) = u \qquad \underline{\lambda} \leq \lambda < \bar{\lambda}$$

passant par une solution (λ^*,u^*) donnée.

On choisit une subdivision croissante (resp. décroissante) $(\lambda_i)_{i=1}^N$ de $[\lambda^*,\bar{\lambda}]$ (resp. $[\underline{\lambda}\ ,\lambda^*]$ et on calcule les solutions (λ_i,u_i) de proche en proche : supposant (λ_i,u_i) calculé, on évalue α^*_{i+1} grâce à α^*_i (donné par (2.5)) et on calcule la suite $(u^n_{i+1})_n$ par $(2.4)_\alpha$ $(\alpha < \alpha^*_{i+1})$ avec l'initialisation $u^o_{i+1} = u_i$.

2.2. Application aux équations de GRAD-SHAFRANOV avec frontière libre.

On cherche à déterminer les solutions de

$$(2.6)\qquad \begin{cases} -\Delta u = p'(u) \text{ dans } \Omega, \\ u = cte(inconnue) \text{ sur } \Gamma \\ -\displaystyle\int_\Gamma \frac{\partial u}{\partial n}\, d\sigma = I , \end{cases}$$

avec $p = \lambda\, p_o$, où p_o est donnée et λ est un paramètre que l'on fera varier. On ne s'intéressera qu'aux solutions u de (2.6) vérifiant aussi

$$(2.7)\qquad mes\left\{x \in \Omega \mid u(x) = o\right\} = o \ ;$$

on pourra alors introduire la première valeur propre ν_1 du problème linéarisé

$$(2.8)\qquad \begin{cases} -\Delta\varphi = \nu\, p''(u)\varphi \quad p.p. \text{ dans } \Omega , \\ \varphi = cte(inconnue)\ p.p. \text{ sur } \Gamma, \\ \displaystyle\int_\Omega p''(u)\varphi \ dx = o , \end{cases}$$

où $\varphi \in H^1(\Omega)$ et $\varphi \not\equiv o$.

On déduit de l'étude précédente deux algorithmes pour déterminer des branches de solutions de (2.6).

Premier algorithme

$$(2.9)\qquad \begin{cases} -\Delta u^{n+1} = \lambda\, p_o'(u^n) \text{ dans } \Omega, \\ u^{n+1} = cte(inconnue) \text{ sur } \Gamma, \\ \displaystyle\int_\Omega \lambda\, p_o'(u^{n+1})\, dx = I. \end{cases}$$

Deuxième algorithme

$$(2.10)\qquad \begin{cases} -\Delta u^{n+1/3} = \lambda\, p_o'(u^n) \text{ dans } \Omega, \\ u^{n+1/3} = cte\ (inconnue) \text{ sur } \Gamma, \\ \displaystyle\int_\Omega \lambda\, p_o'(u^{n+1/3})\, dx = I , \end{cases}$$

$$(2.11)\quad \begin{cases} -\Delta u^{n+2/3} = \lambda\, p_o'(u^{n+1/3}) \text{ dans } \Omega, \\ u^{n+2/3} = cte(inconnue) \text{ sur } \Gamma, \\ \int_\Omega \lambda\, p_o'(u^{n+2/3})\, dx = I, \end{cases}$$

$$(2.12)\quad u^{n+1} = (1-\alpha)\, u^n + 2\alpha\, u^{n+1/3} - \alpha\, u^{n+2/3}.$$

Le problème (2.9) (et donc aussi les problèmes (2.10) et (2.11)) admettent une unique solution dans $H^2(\Omega)$. En effet, posant

$$u^{n+1} = \tilde{u}^{n+1} + C^{n+1}, \quad \tilde{u}^{n+1} \in H_0^1(\Omega), \; C^{n+1} \in \mathbb{R},$$

la fonction $\tilde{u}^{n+1}$ est l'unique solution d'un problème de Dirichlet usuel et la constante C^{n+1} est l'unique solution du problème non linéaire monotone

$$(2.13)\quad \int_\Omega \lambda\, p_o'(\tilde{u}^{n+1} + C^{n+1})\, dx = I.$$

On a le résultat de convergence suivant, dû à H.BERESTYCKI et H. BREZIS [2] et du même type que le th. 2.1 :

<u>Théorème 2.3.</u> Supposons que p vérifie (1.4)-(1.7) ; quelque soit $u^o \in H^1(\Omega)$, il existe une sous-suite de (u^n) définie par (2.9) qui converge fortement dans $H^1(\Omega)$ vers une solution du problème (2.6). De plus, toute sous-suite convergente de (u^n) converge vers une solution de (2.6).

D'autre part, par application du théorème 2.2, on peut démontrer

<u>Théorème 2.4</u>. Supposons que p vérifie (1.4)-(1.7), que p' est dérivable en tout point $z > o$ et que $p'(z) > o \;\; \forall\, z > o$.

Soit u^* une solution de (2.6) vérifiant en outre (2.7) et supposons que 1 n'est pas valeur propre de (2.8). Alors

i) pour tout α, $o < \alpha < \alpha^*$, α^* assez petit, il existe un voisinage V_α de u dans $H^2(\Omega)$ tel que, si $u^o \in V_\alpha$ et (u^n) est défini par (2.12), pour $n \to \infty$,

$$u^{n+i/3} \longrightarrow u^* \text{ dans } H^2(\Omega), \quad i = 0, 1, 2$$

ii) si de plus u^* est une solution variationnelle [1], il existe un voisinage V de u^* dans $H^2(\Omega)$ tel que, si $u^o \in V$ et (u^n) est défini par (2.9) alors, pour $n \to \infty$,

$$u^n \longrightarrow u \text{ dans } H^2(\Omega).$$

2.3 Quelques remarques sur la mise en oeuvre des algorithmes.

Les algorithmes précédents sont appliqués après discrétisation du problème (2.6) par la méthode des éléments finis d'ordre 1. On a vérifié que le problème discrétisé possède toujours une solution.

Pour chaque étape d'un des algorithmes, le calcul du second membre requiert l'évaluation de

$$(2.14) \qquad \int_T p'(u^n)\, w_i\, dx = \int_{T \cap u^n > o} p'(u^n)\, w_i\, dx$$

où T est un triangle et w_i une fonction de base . Puisque la fonction p' n'est pas régulière en zéro, l'approximation de l'intégrale (2.14) est effectuée sur l'expression de droite, c'est-à-dire qu'on calcule les surfaces exactes des ensembles $\left\{ x \in T \mid u^n(x) > o \right\}$ (cf. Figure 2).

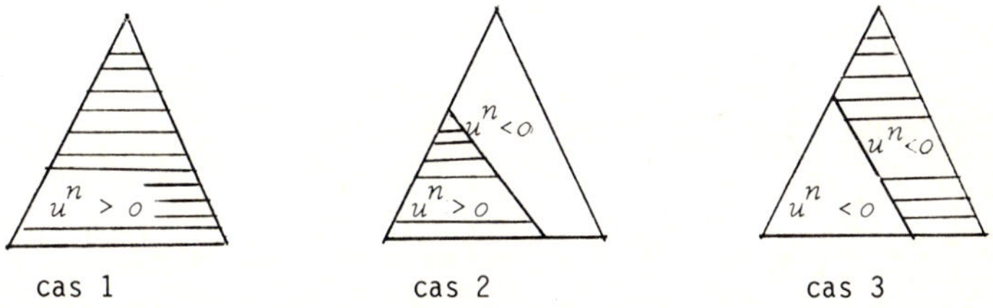

Figure 2.

Les systèmes linéaires (2.9), (2.10) où (2.11) sont résolus par la méthode itérative de relaxation avec paramètre optimal.

L'équation (2.13) est résolue par la méthode de Newton.

Le calcul de ν_1 est effectué par la méthode des puissances itérées.
Quant au choix des algorithmes, on applique de préférence le premier, plus performant, le second n'étant utilisé que lorsqu'on prédit une valeur propre $\nu_1 < 1$,

[1] ou , plus généralement, si $\nu_1 > 1$.

auquel cas le premier algorithme peut ne pas converger.

Enfin, lorsque le domaine Ω possède un axe de symétrie, le problème (2.6) possède au moins une solution symétrique ; elle peut être déterminée par application de l'algorithme (2.9) sur le demi-ouvert avec condition de Neumann sur l'axe de symétrie.

Remarque 2.1 . On trouvera d'autres méthodes numériques pour déterminer les solutions d'un problème très voisin de (2.6) dans [6] et [9].

2.4. Une remarque sur la détermination numérique des points de bifurcation.

Lors du calcul des branches de solutions, il est possible de reconnaître un point de bifurcation grâce à la condition suffisante suivante.

Soient $\mathfrak{J}$ un intervalle de $\mathbb{R}_+$ et u une application continue de $\mathfrak{J}$ dans $H^2(\Omega)$ telle que, pour tout $\lambda \in \mathfrak{J}$, $(\lambda, u(\lambda))$ soit solution de (2.6) et (2.7). On peut définir $\nu_1(\lambda)$ par (2.8).

Supposons qu'il existe $\lambda^* \in \mathfrak{J}$ et $\varepsilon > 0$ tels que

i) $\nu_1(\lambda^*) = 1$

ii) $\nu_1(\lambda) > 1$ pour $\lambda^* - \varepsilon < \lambda < \lambda^*$ (resp. $\nu_1(\lambda) < 1$)

iii) $\nu_1(\lambda) < 1$ pour $\lambda^* < \lambda < \lambda^* + \varepsilon$ (resp. $\nu_1(\lambda) > 1$) ;

alors $(\lambda^*, u(\lambda^*))$ est un point de bifurcation de (2.6) dans $\mathfrak{J} \times H^2(\Omega)$ par rapport à la branche $\left\{(\lambda, u(\lambda)) , \lambda \in \mathfrak{J}\right\}$.

Principe de la démonstration . Supposant qu'il n'y a pas de bifurcation, on montre que le degré topologique de LERAY-SCHAUDER de fonctionnelles homotopes bien choisies n'est pas constant.

Rappelons enfin qu'il a été démontré par R. TEMAM, puis T. GALLOUET [5] , que la condition i) ci-dessus est une condition nécessaire de bifurcation.

3. DETERMINATION NUMERIQUE DE LA STABILITE LINEAIRE.

3.1. Un problème de valeur propre.

Nous allons d'abord montrer que le calcul de la stabilité linéaire (cf. § 1), c'est-à-dire la recherche du signe de l'infimum de $J(\vec{u})$,

$$J(\vec{u}) = J_o(\vec{u}) + \gamma J_1(\vec{u})$$

où

$$J_o(u) = \frac{1}{2} \int_\Omega |\nabla u^1|^2 dx - \frac{1}{2} \int_{\Omega_p} p''(\Psi)|u^1|^2 dx , \quad u^1 = \vec{u}.\vec{\nabla}\Psi,$$

$$J_1(\vec{u}) = \frac{1}{2} \int_{\Omega_p} p(\Psi)|div \vec{u}|^2 dx ,$$

sur l'espace

$$E = \left\{ \vec{u} \in (L^2(\Omega))^2 | \quad \vec{u}.\vec{\nabla}\Psi \in H_0^1(\Omega), \quad div \vec{u}|_{\Omega_p} \in L^2(\Omega_p) \right\} ,$$

peut être ramené au problème de la détermination de la plus grande valeur propre d'un opérateur compact, ce qui est un problème classique.

Grâce au théorème de Federer, on a

$$J_1(\vec{u}) = \int_o^\infty p(t) \left(\int_{\Psi=t} |div \vec{u}|^2 \frac{d\sigma}{|\nabla\Psi|} \right) dt,$$

où $d\sigma$ désigne l'élément de longueur sur la courbe $\Psi = t$.

Soient $(\vec{\tau}, \vec{n})$ les vecteurs tangents et perpendiculaires à la courbe $\Psi = t$ passant par un point x quelconque ; on note

$$\vec{u}(x)=u_{/\!/}(x)\vec{\tau}(x) + u_\perp(x) \vec{n}(x).$$

Dans ce système de coordonnées, la divergence est donnée par l'expression

$$div \vec{u} = |\nabla\Psi| \frac{\partial}{\partial\tau} \left(\frac{u_{/\!/}}{|\nabla\Psi|} \right) + \frac{\partial}{\partial n} u_\perp,$$

d'où

$$J_1(\vec{u}) = \int_o^\infty p(t) \left[\int_{\Psi=t} \left(|\nabla\Psi| \frac{\partial}{\partial\tau} \left(\frac{u_{/\!/}}{|\nabla\Psi|} \right) + \frac{\partial}{\partial n} u_\perp \right)^2 \frac{d\sigma}{|\nabla\Psi|} \right] dt$$

Minimisant cette expression par rapport à $u_{/\!/}$, on obtient une fonctionnelle J_2 qui ne dépend que de $u^1 = u_\perp |\nabla\Psi|$.

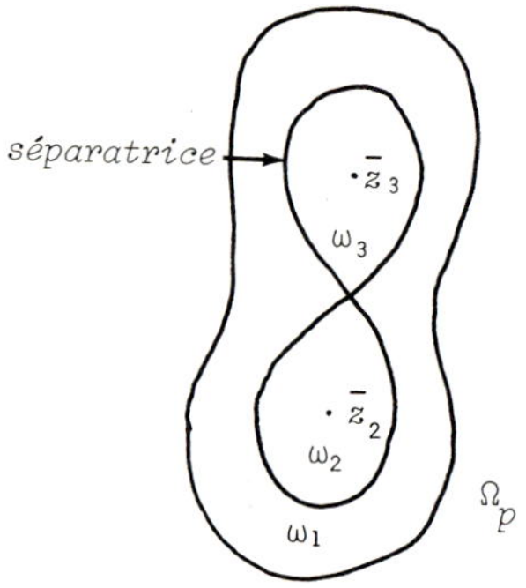

On appelle séparatrice une ligne $\Psi(x) = t$ possédant au moins un point double (cf. Figure 3). Soient $\omega_i, i=1,\dots,N,$ les composantes connexes de Ω_p privé des séparatrices. On note $\underline{z_i}$ (resp. $\overline{z_i}$) le minimum (resp. le maximum) de Ψ sur $\overline{\omega_i}$. On note $\Gamma_i(t)$ la ligne $\Psi(x) = t$ pour $x \in \omega_i$; c'est une ligne fermée sans point double.

Notant, pour $i = 1,\dots, N$ et $\underline{z_i} \le t \le \overline{z_i}$

$$q_i(t) = p(t) \int_{\Gamma_i(t)} |\nabla\Psi| d\sigma . \left(\int_{\Gamma_i(t)} d\sigma\right)^{-2} , \tag{3.1}$$

$$\beta_i v(t) = \int_{\Gamma_i(t)} \left[\frac{\nabla\Psi \cdot \nabla v}{|\nabla\Psi|^3} - \frac{\nabla\Psi\cdot\nabla(|\nabla\Psi|^2)}{2|\nabla\Psi|^5} v\right] d\sigma \tag{3.2}$$

on a

$$b(v_1 , v_2) = \sum_{i=1}^{N} \int_{\underline{z_i}}^{\overline{z_i}} q_i(t)\ \beta_i\ v_1(t)\ \beta_i\ v_2(t)\ dt, \tag{3.3}$$

et finalement

$$J_2(v) = b(v, v).$$

On s'est donc ramené (<u>formellement</u> dans le travail actuel) au problème de minimisation de $J_o(v) + \gamma J_2(v)$ sur un espace F de <u>fonctions scalaires</u>, où

$$F = \left\{ v \in H_0^1(\Omega) , \sum_{i=1}^{N} \int_{\underline{z}_i}^{\overline{z}_i} q_i(t)\ (\beta_i\ v(t))^2\ dt < \infty \right\} .$$

On démontre que cet espace F, muni du produit scalaire

$$\int_\Omega \nabla v_1 . \nabla v_2\ dx + b(v_1, v_2) \tag{3.4}$$

est un <u>espace de Hilbert</u>.

Puisque l'opérateur A, qui à $u \in F$ associe l'unique solution $v \in F$ de

$$\int_\Omega \nabla v.\nabla w \, dx + \gamma \, b(v, w) = \int_{\Omega_p} p''(\Psi) \, u \, w \, dx \qquad \forall w \in F,$$

est un opérateur compact, on en déduit qu'il existe une plus petite valeur propre μ_1 de : $u \in F$ tel que

$$\text{(3.5)} \quad \int_\Omega \nabla u.\nabla v \, dx + \gamma \, b(u,v) = \mu \int_{\Omega_p} p''(\Psi) \, u \, v \, dx \qquad \forall v \in F.$$

La stabilité linéaire est liée à μ_1 par :

l'équilibre est linéairement stable $\Longleftrightarrow$ $\mu_1 \geq 1$
l'équilibre est linéairement instable $\Longleftrightarrow$ $\mu_1 < 1$.

Remarque 3.1. La forme bilinéaire b est non locale : pour certaines fonctions v_1 , v_2 à supports disjoints mais coupés par une même ligne $\Psi(x) = t$, on a $b(v_1, v_2) \neq o$.

Remarque 3.2. Heuristiquement, les fonctions de F sont les fonctions de $H^1_0(\Omega)$ qui sont "voisines de o" autour des points x où $\nabla\Psi(x) = o$.

3.2. Approximation du problème aux valeurs propres.

Il s'agit en fait d'une double approximation : on calcule d'abord un équilibre d'une façon approché (Ψ_h) puis on cherche à calculer une stabilité approchée.

On choisit une seule triangulation $\mathcal{T}_h$ de l'ouvert Ω (supposé polygonal) pour les deux approximations. L'équilibre Ψ_h est calculé suivant les méthodes du § 2, donc avec des éléments finis d'ordre 1.

On choisit d'autre part des subdivisions $\{z_{i,1}, \ldots, z_{i,n_i}\}$ des segments $[\underline{z_i}, \overline{z_i}]$ définis précédemment (et dépendant de Ψ_h).

On introduit l'espace V_h des éléments finis d'ordre 1 sur $\mathcal{T}_h$, nuls sur le bord de Ω. C'est une approximation externe de F (cf. remarque 3.2).

On définit la forme bilinéaire b_h (par analogie avec (3.3)) par

$$b_h(v_1 , v_2) = \sum_{i=1}^{N} \sum_{j=1}^{n_i-1} \frac{1}{2}\left(z_{i,j+1} - z_{i,j}\right) \cdot \Big[q_i \, \beta_{ih} v_1 \, \beta_{ih} v_2(z_{i,j+1}) + q_i \, \beta_{ih} v_1 \, \beta_{ih} v_2(z_{i,j}) \Big] .$$

l'expression $\beta_i v$ contient le terme

$$\nabla \left(|\nabla \Psi_h|^2 \right)$$

qui est une distribution localisée sur les côtés des triangles. L'approximation $\beta_{ih}\, v$ est obtenue par une méthode employée par M.O. BRISTEAU dans [4] et qui consiste à faire une analogie avec la formule de GREEN.

On définit donc une plus petite valeur propre μ_{1h} de : $u_h \in V_h$ tel que

$$\int_\Omega \nabla u_h . \nabla v_h \, dx + \gamma \, b_h(u_h, v_h) = \mu_h \int_{\Omega_p} p''(\Psi_h) \, u_h \, v_h \, dx \quad \forall \, v_h \in V_h. \tag{3.6}$$

Le problème précédent s'écrit encore

$$Bx = \mu_h \, Cx \tag{3.7}$$

où la matrice B a pour coefficients

$$b_{ij} = \int_\Omega \nabla w_i . \nabla w_j \, dx + \gamma \, b_h(w_i, w_j) \, .$$

On effectue une factorisation de CHOLEWSKY de la matrice B et on calcule μ_{1h} par la méthode des puissances itérées.

Signalons que la matrice B peut être assez remplie (50% dans l'exemple du § 4).

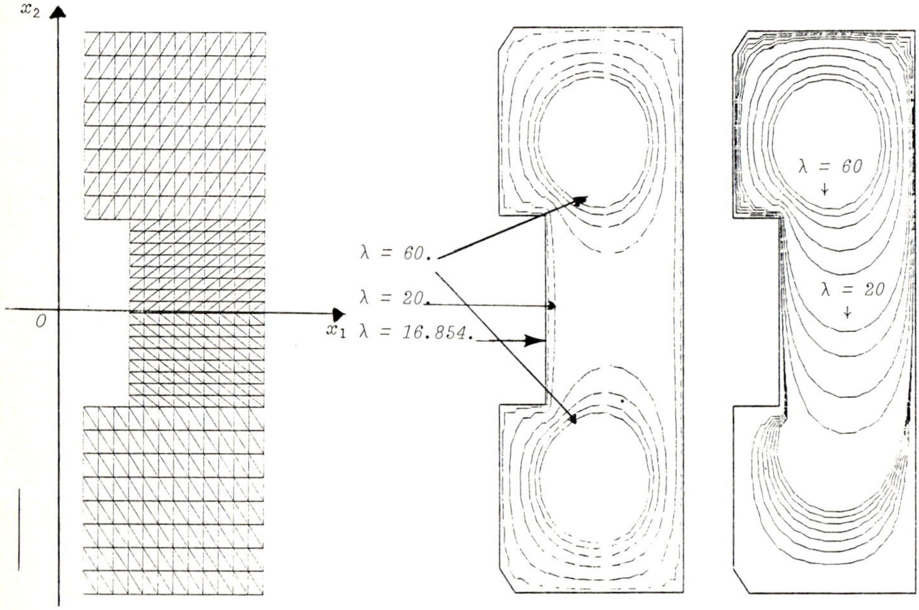

Figure.4 Triangulation

Figure. 5 Configurations d'équilibres symétriques

Figure. 6 Configurations d'équilibres dissymétrique

4. RESULTATS NUMERIQUES.

Les résultats numériques présentés ici concernent un cylindre droit de section Ω ; ce domaine Ω possède l'axe de symétrie $x_2 = 0$ (cf. fig. 4) ; la fonction p est $p(z) = \frac{1}{2} \lambda(z_+)^2$.

La triangulation du domaine Ω comprend 384 sommets et 672 triangles.

Pour calculer la branche des solutions symétriques, on a appliqué le premier algorithme sur le demi-ouvert, en suivant le sens des λ croissant. On a ensuite cherché à calculer une solution dissymétrique pour λ grand $(\lambda = 60)$. On a employé le premier algorithme avec une initialisation dissymétrique et on a effectivement obtenu une solution dissymétrique. On a alors calculé une branche de solutions (dissymétriques) en employant le premier algorithme et en faisant décroître λ. Cette branche de solutions s'est raccordée à la première branche pour $\lambda = 16{,}854$. La solution symétrique correspondante est donc un point de bifurcation.

On avait d'ailleurs la preuve de cette bifurcation dès le calcul de la branche des solutions symétriques, car la courbe de $1/\nu_1$ en fonction de λ traverse l'axe $1/\nu_1 = 1$ pour la valeur $\lambda = 16{,}854$ (cf. § 2.4).

La bifurcation est illustrée par la fig. 7 où, pour chaque solution Ψ calculée, on représente la valeur de

$$\int_{x_2 > 0} \Psi \, dx - \int_{x_2 < 0} \Psi \, dx$$

qui renseigne sur la symétrie des solutions.

D'autre part, on a déterminé la stabilité de ces solutions en calculant la valeur de $1/\mu_1$ (cf. fig. 8).

Voici quelques données pratiques pour un ordinateur IBM 370/168. Pour le calcul d'une solution, la place mémoire est 180 K-octets, le temps d'exécution varie entre 10 et 30 s. suivant l'initialisation. Pour le calcul de μ_1, la place mémoire est 550 K-octets, le temps d'éxécution est de 30 s.

Remarque 4.1 : On trouvera d'autres résultats numériques de bifurcation pour des problèmes très voisins de (2.6) dans [7] [16].

En conclusion, on peut dire que, d'une part, les algorithmes précédents sont très performants et, d'autre part, qu'ils ont permis de tracer un diagramme de bifurcation et de changement de stabilité, ce qui est un résultat nouveau.

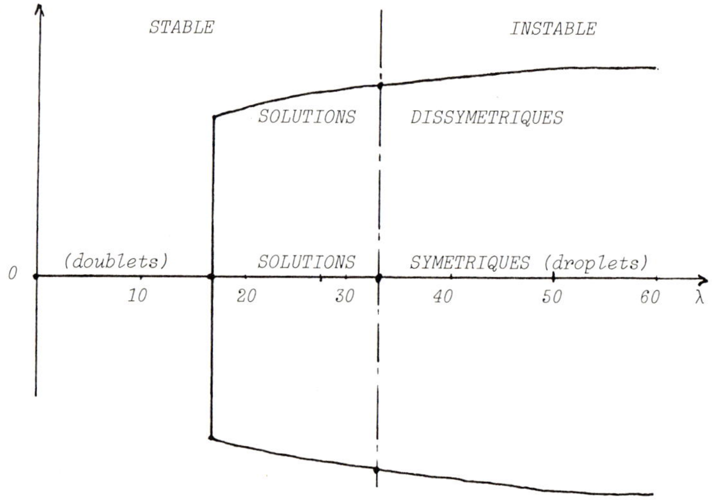

Figure 7

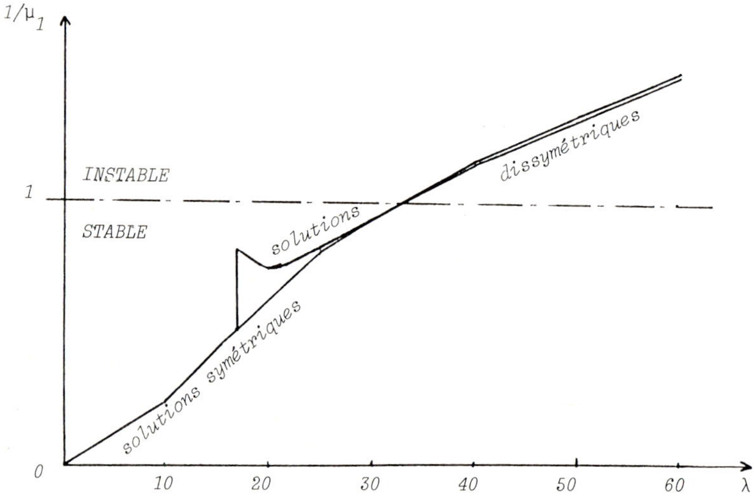

Figure 8

i) les solutions symétriques sont des doublets (c.-à-d. le plasma est connexe) jusqu'à $\lambda = 27{,}3$; ce sont des droplets (le plasma est non connexe) au-delà ;

ii) au-delà de $\lambda = 16{,}854$, les solutions variationnelles sont les solutions dissymétriques ;

iii) le changement de stabilité a lieu pour $\lambda = 33$.

REMERCIEMENTS

L'auteur voudrait remercier ici Monsieur R. TEMAM pour avoir dirigé ce travail, Monsieur R. GLOWINSKI pour en avoir permis l'exécution, Messieurs H. GRAD, B. GROSSMANN et H. WEITZNER pour l'avoir invité au sein de leur équipe au Courant Institute.

REFERENCES

[1] G. BATEMAN, MHD instabilities, the MIT Press, Massachusetts, 1978.

[2] H. BERESTYCKI et H. BREZIS, On a free boundary problem arising in plasma physics, Rapport 78017, Université de Paris, 1978.

[3] I.B. BERNSTEIN, E.A. FRIEMAN, M.D. KRUSKAL and R.M. KULSRUD, An energy principle for hydromagnetic stability problems, Proc. Royal Soc., A 244, 1958, 17-40.

[4] M.O. BRISTEAU, Application of optimal control theory to transonic flow computations by finite element methods, Lecture Notes in Physics, Computing Methods in Applied Sciences, 91, Third International Symposium, Versailles, 77, Springer-Verlag, 1979.

[5] T. GALLOUET, Contribution à l'étude d'une équation apparaissant en physique des plasmas, thèse de 3ème cycle, Université de Paris VI, 1978.

[6] C. GUILLOPE, Sur un problème à frontière libre intervenant en physique des plasmas, thèse de 3ème cycle, Univ. de Paris XI, 1977.

[7] F.J. HELTON, T.S. WANG, MHD equilibrium in non-circular Tokamaks with field-shaping coil systems, Nuclear Fusion 18 11, 1978, 1523-1533.

[8] J.W. KITCHEN, Concerning the convergence of iterates to fixed points, Stud. Math., 27, 1966, 247-249.

[9] J. LAMINIE, Determination numérique de la configuration d'équilibre du plasma d'un Tokamak, Thèse de 3ème cycle, Univ. Paris XI, 1977.

[10] J.P. PUEL, Un problème de valeurs propres non linéaires et de frontière libre, Rapport 76016, Université de Paris, 1976.

[11] E. REBHAN, A. SALAT, Axisymmetric MHD stability of sharp-boundary Tokamaks, Nuclear Fusion, 17 2, 1977, 251-260.

[12] M. SERMANGE, Une méthode numérique en bifurcation, Application à un problème à frontière libre de la physique des plasmas, Appl. Math. and Optim. 5, 1979, 127-151.

[13] M. SERMANGE, Bifurcation of free boundary plasma equilibria, Rapport Laboria, 385 (1979).

[14] M. SERMANGE, Détermination numérique de la stabilité des solutions des équations de Grad-Shafranov, à paraître.

[15] J. SIJBRAND, Computations of the bifurcation solutions of a class of problems with a free boundary, Bifurcation analysis for a class of problems with a free boundary, lecture notes 90, 91, University of Utrecht, (1978).

[16] R. TEMAM, Remarks on a free boundary value problem arising in plasma physics, Comm. in P.D.E., 2(6), 1977, 563-585.

SESSION III

FLUID MECHANICS

COMPUTING METHODS IN APPLIED SCIENCES AND ENGINEERING
R. Glowinski, J.L. Lions (editors)
North-Holland Publishing Company

ON THE NONLINEAR ACCELERATION OF ITERATIVE SCHEMES

N.N. Yanenko, Yu.I. Shokin, Yu.N. Zaharov
Institute on Pure and Applied Mechanics
Academic of Sciences of the USSR Siberian
Branch Novosibirsk, 630090 USSR

At the present time for the solution of stationary problems the relaxation method is widely used. As a rule in this case one stand the way of physics analogies, i.e. they find the physics process, which gives us the stationary solution as $t \to \infty$. For example, for the solution of Poisson equation the heat flow equation is used, and for the solution of the steady-state Navier-Stokes equations - nonstationary system. But if we interested only the final solution it does not necessary for the solution of non-stationary problem to have any physical sence. The convergence with the best estimation is of great importance.

In this paper the question of constructing of nonstationary systems without necessary physics interpretation is investigated. This nonstationary systems possesses the higher convergence speed than some of well known nonstationary systems.

§ 1. The using of artificial viscosity for the solution of Navier-Stokes equations.

1. Let's consider the interior problem in the domain Ω

$$\bar{u}\cdot\nabla\bar{u} + \nabla p_u = \mu\Delta\bar{u} + g,$$
$$\operatorname{div}\bar{u} = 0 \quad , \quad \bar{u}|_\Gamma = \varphi , \qquad (1)$$

where $\bar{u} = (u_1, u_2, u_3)$ is the velosity vector, $p_u = p(x_1, x_2, x_3)$ - the pressure, $g = (g_1, g_2, g_3)$ - the mass forces vector, $\varphi = \varphi(x_1, x_2, x_3)$ - some given function, $u_i = u_i(\bar{x}) = u_i(x_1, x_2, x_3)$, μ - the positive viscosity coeffissient, Γ - the boundary of Ω .

It is supposed that problem (1) has the only solution from $\mathcal{L}_2$ and slow steady-state flows ($\alpha = \mu/\ell^2 - m > 0$, ℓ - diameter of Ω , m is the maximum of the module of the tensor of velosity strain of the problem (1)) the solution on nonstationary problem

$$\frac{\partial \bar{v}}{\partial t} + \bar{v}\cdot\nabla\bar{v} + \nabla p_v = M\Delta\bar{v} + \bar{g},$$
$$\operatorname{div}\bar{v} = 0 \quad, \bar{v}|_{\Gamma} = \varphi \quad, \bar{v}(\bar{x},0) = \bar{v}_0(\bar{x}) \tag{2}$$

converges to the solution of (1) with the rate equal to exp $(-\alpha t)$, that is for the error vector $\bar{\omega} = \bar{u} - \bar{v}$ the estimate [1]

$$\|\bar{\omega}(\bar{x},t)\|_2 \le e^{-\alpha t}\|\bar{\omega}(\bar{x},0)\|_2 \tag{3}$$

is justified. Here $\bar{v} = \bar{v}(\bar{x},t) = (v_1, v_2, v_3)$, $p_v = p_v(\bar{x},t)$ - the pressure, $\bar{v}_0(\bar{x})$ - some given function, $\|\cdot\|_2$ - the norm of $\mathcal{L}_2$.

Let's consider the next boundary problem in the same domain Ω :

$$\frac{\partial \bar{w}}{\partial t} + \bar{w}\cdot\nabla\bar{w} + \nabla p_w = M\Delta\bar{w} + \sum_{i=1}^{3} M_i \frac{\partial^2 \bar{w}}{\partial x_i^2} + g,$$
$$\operatorname{div}\bar{w} = 0 \quad, \bar{w}|_{\Gamma} = \varphi \ , \bar{w}(\bar{x},0) = \bar{w}_0(\bar{x}) = \bar{v}_0(\bar{x}), \tag{4}$$

where $\bar{w} = \bar{w}(\bar{x},t) = (w_1, w_2, w_3)$, $p_w = p_w(\bar{x},t)$ - pressure, M_i $(i = 1,2,3)$ - some nonnegative matrices.

Let's suppose that the problem (4) has the only solution from $\mathcal{L}_2(\Omega)$ and consider the question of choosing of matrices M_i so that the solution of (4) as $t \to \infty$ converge faster than the solution of (2). Thus the estimate (3) will be some better.

Theorem 1. If $\alpha > 0$ and $\|M_i(t)\| \le K_1 \exp(-a_1 t)$, $a_1 \ge \alpha$ $(i = 1,2,3)$, than $\|\bar{\omega}_1\|_2 = \|\bar{w} - \bar{u}\|_2 \to 0$ as $t \to \infty$ and if

$$(M_i(t)y, y) \ge K_2 e^{-a_2 t}(y,y) \qquad \forall y \in \mathcal{L}_2(\Omega),$$

where K_1, K_2, a_1, a_2 are nonnegative constants, than for all times t as

$$\|\bar{w} - \bar{u}\|_2 > \frac{K_1}{K_2} e^{(a_2 - a_1)t}\|u\|_2 \tag{5}$$

the estimate

$$\|\bar{\omega}_1(\bar{x},t)\|_2 < e^{-\alpha_1 t}\|\bar{\omega}_1(\bar{x},0)\|_2 \qquad (\alpha_1 > \alpha)$$

is justified and hence

$$\|\bar{\omega}_1(\bar{x},t)\|_2 \le e^{-[\alpha t + (\alpha_1 - \alpha)t_1]}\|\bar{\omega}_1(\bar{x},0)\|_2 .$$

Here t_1 is the time moment up to which the inequality (5) is valid, that is $\bar{w}$ converges to $\bar{u}$ faster than $\bar{v}$.

Notes, that if (5) is not valid, that is we are in the neighbourhood of the required solution, we must to set matrices M_i equal to zero identically.

Let's consider the structure of matrices M_i in the two-dimensional case. The system (4) is not the system of Cauchy-Kovalevskaya type and because for the numerical search of the pressure it is possible to use the equation with the small parameter [2] . Approximate the equation of coutinuity by the equation

$$\varepsilon \frac{\partial p}{\partial t} + \operatorname{div} \bar{w} = \varepsilon \mu \Delta p \quad , \ \varepsilon > 0,$$

we obtain the system

$$\frac{\partial \bar{w}}{\partial t} + A_1 \frac{\partial \bar{w}}{\partial x_1} + A_2 \frac{\partial \bar{w}}{\partial x_2} = A_3 \cdot \Delta \bar{w} + \sum_{i=1}^{2} M_i \frac{\partial^2 \bar{w}}{\partial x_i^2} , \tag{6}$$

where

$$A_j = \begin{pmatrix} u_j & 0 & \delta_{j1}/\varepsilon \\ 0 & u_j & \delta_{j2}/\varepsilon \\ \delta_{j1}/\varepsilon & \delta_{j2}/\varepsilon & 0 \end{pmatrix} \ (j=1,2) , \ A_3 = \mu E , \quad \bar{w} = \begin{pmatrix} u_1 \\ u_2 \\ \varepsilon p \end{pmatrix} ,$$

E -identical matrix, $\varepsilon > 0$ - the small parameter.

Split the system (6):

on odd steps

$$\frac{1}{2} \frac{\partial \bar{w}}{\partial t} + A_1 \frac{\partial \bar{w}}{\partial x_1} = (A_3 + M_1) \frac{\partial^2 \bar{w}}{\partial x_1^2} , t_n \le t \le t_n + \frac{\tau_n}{2} , \tag{7}$$

on even steps

$$\frac{1}{2} \frac{\partial \bar{w}}{\partial t} + A_2 \frac{\partial \bar{w}}{\partial x_2} = (A_3 + M_2) \frac{\partial^2 \bar{w}}{\partial x_2^2} , t_n + \frac{\tau_n}{2} \le t \le t_n + \tau_n$$

and demand that matrices M_i lead to diagonal forms simultaneously with matrices A_i $(i=1,2)$ we find the structure of this matrices:

$$M_1 = \begin{pmatrix} f_{11} \cos^2 \gamma_1 + f_{13} \sin^2 \gamma_1 & 0 & \frac{1}{2}(f_{11} - f_{13}) \sin 2\gamma_1 \\ 0 & f_{12} & 0 \\ \frac{1}{2}(f_{11} - f_{13}) \sin 2\gamma_1 & 0 & f_{11} \sin^2 \gamma_1 + f_{13} \cos^2 \gamma_1 \end{pmatrix},$$

$$M_2 = \begin{pmatrix} f_{21} & 0 & 0 \\ 0 & f_{22} \cos^2 \gamma_2 + f_{23} \sin^2 \gamma_2 & \frac{1}{2}(f_{22} - f_{23}) \sin 2\gamma_2 \\ 0 & \frac{1}{2}(f_{22} - f_{23}) \sin 2\gamma_2 & f_{22} \sin^2 \gamma_2 + f_{23} \cos^2 \gamma_2 \end{pmatrix},$$

where $\gamma_i = -0{,}5 \arcsin (2\sqrt{u_i^2 + 4})$, f_{ij} - arbitrary non-negative functions $(i=1,2;\ j=1,2,3)$.

Notes, that if $f_{i1} = f_{i3}$ $(i=1,2)$ than matrices

M_i are independent of velocity and have diagonal form.

2. Let's the splitting system (7) is approximated by the splitting scheme

$$\frac{u_1^{n+\frac12}-u_1^n}{\tau}+\frac{\Delta_1+\Delta_{-1}}{2h}[(u_1^n)^2+p^n]=\mu(1+f_1^n)\frac{\Delta_1\Delta_{-1}}{h^2}u_1^{n+\frac12},$$

$$\frac{u_2^{n+\frac12}-u^n}{\tau}+\frac{\Delta_1+\Delta_{-1}}{2h}(u_1^n u_2^n)=\mu(1+f_2^n)\frac{\Delta_1\Delta_{-1}}{h^2}u_2^{n+\frac12},$$

$$\frac{\varepsilon}{2}\frac{p^{n+\frac12}-p^n}{\tau}+\frac{\Delta_1+\Delta_{-1}}{2h}u_1^{n+\frac12}=f_3\frac{\Delta_1\Delta_{-1}}{h^2}p^{n+\frac12};$$

$$\frac{u_1^{n+1}-u_1^{n+\frac12}}{\tau}+\frac{\Delta_2+\Delta_{-2}}{2h}(u_1^{n+\frac12}u_2^{n+\frac12})=\mu(1+f_1^n)\frac{\Delta_2\Delta_{-2}}{h^2}u_1^{n+1},$$

$$\frac{u_2^{n+1}-u_2^{n+\frac12}}{\tau}+\frac{\Delta_2+\Delta_{-2}}{2h}[(u_2^{n+\frac12})^2+p^{n+\frac12}]=\mu(1+f_2^n)\frac{\Delta_2\Delta_{-2}}{h^2}u_2^{n+1},$$

$$\frac{\varepsilon}{2}\frac{p^{n+1}-p^{n+\frac12}}{\tau}+\frac{\Delta_2+\Delta_{-2}}{2h}u_2^{n+1}=f_3\frac{\Delta_2\Delta_{-2}}{h^2}p^{n+1},$$

where f_j^n $(j=1,2,3)$ - some nonnegative functions which are tend to zero as $n\to\infty$, $\Delta_i=T_i-E$, $\Delta_{-i}=E-T_{-i}$, $T_{\pm1}u(x_1,x_2)=$ $=u(x_1\pm h,x_2)$, $T_{\pm2}u(x_1,x_2)=u(x_1,x_2\pm h)$.

Let's, first of all, consider the difference scheme, which is a linear analog of the one of the equation of the system (8):

$$\frac{u^{n+\frac12}-u^n}{\tau}+a_1\frac{\Delta_1+\Delta_{-1}}{2h}u^n=(\mu+f^n)\frac{\Delta_1\Delta_{-1}}{h^2}u^{n+\frac12},$$

$$u^n|_{\Gamma_{1h}}=u^{n+\frac12}|_{\Gamma_{1h}}=\varphi(x_1,x_2)|_{\Gamma_{1h}},\quad u^0=\varphi_0(x_1,x_2);$$

$$\frac{u^{n+1}-u^{n+\frac12}}{\tau}+a_2\frac{\Delta_2+\Delta_{-2}}{2h}u^{n+\frac12}=(\mu+f^n)\frac{\Delta_2\Delta_{-2}}{h^2}u^{n+1},\tag{9}$$

$$u^{n+\frac12}|_{\Gamma_{2h}}=u^{n+1}|_{\Gamma_{2h}}=\varphi(x_1,x_2)|_{\Gamma_{2h}},\quad a_1,a_2=\mathrm{const},$$

f^n - is nonnegative function, depending on the number of iteration. This scheme we shall use for the solution the following steady-state boundary problem

$$a_1 \frac{\Delta_1+\Delta_{-1}}{2h} u_h + a_2 \frac{\Delta_2+\Delta_{-2}}{2h} u_h = \mu\left(\frac{\Delta_1\Delta_{-1}}{h^2} + \frac{\Delta_2\Delta_{-2}}{h^2}\right) u_h,$$

$$u_h|_{\Gamma_h} = \varphi(x_1, x_2) \tag{10}$$

in the square grid domain $\Sigma_h = \{0 < ih, jh < 1; i,j = 1, \ldots, N-1\}$ with the boundary $\Gamma_h = \Gamma_{1h} \cup \Gamma_{2h}, \Gamma_{1h} = \{i=0,N; 0 \le j \le N\}, \Gamma_{2h} = \{0 \le i \le N; j=0,N\}$.

As the iterative scheme (9) is not complete approximation scheme [2] , the parameter of iteration τ must be necessary choosen as $O(h^2)$, hence, the rate of convergency of this scheme, when $f^n \equiv 0$ is not high. Let us consider the case when $f^n \not\equiv 0$.

Suppose that R is $(N-1)^2$ - dimensional vector space with elements $u_h, u^n, u^{n+\frac{1}{2}}, u^{n+1}$ and these vectors are defined on the grid Σ_h .

We can perfome the scheme (9) into entire steps as

$$u^{n+1} = G_n^2 G_n^1 u^n,$$

where

$$G_n^1 = (E + \tau\mu_n\Omega_1)^{-1}(E - \tau A_1), G_n^2 = (E + \tau\mu_n\Omega_2)^{-1}(E - \tau A_2),$$

$$A_i = a_i \frac{\Delta_i + \Delta_{-i}}{2h}, \quad \Omega_i = -\frac{\Delta_i\Delta_{-i}}{h^2}, \quad \mu_n = \mu + f^n.$$

The solution of the scheme (9) and the error vector $\omega^n = u^n - u_h$ satisfy to schemes

$$u^{n+1} = G_n u^n, \quad \omega^{n+1} = G_n(\omega^n - \tau f^n \Omega u_h),$$

$$G_n = (E + \tau\mu_n\Omega)^{-1}(E - \tau A), \; A = A_1 + A_2, \; \Omega = \Omega_1 + \Omega_2 \tag{11}$$

with the accuracy up to order $O(\tau^2)$. The error vector $\omega_1^n = u_1^n - u_h$ where u_1^n is the solution of the scheme (9) when $f^n \equiv 0$, satisfy to scheme

$$\omega_1^{n+1} = G\omega_1^n, \quad G = (E + \tau\mu\Omega)^{-1}(E - \tau A).$$

Further using the equality $G_n = G(E + \tau f^n \Omega G)^{-1}$ the next proposition may be easy proved.

Lemma 1. If $f^n > 0$ then

$$\|G_n\| \le K_n \|G\|,$$

where

$$K_n = \left(1 + \frac{2\tau f^n \delta + \tau^2 (f^n)^2 \delta^2}{\|G^{-1}\|^2}\right)^{-\frac{1}{2}} < 1 ,$$

δ is the minimun eigenvalue of the operator Ω .

Let us introduce the operator χ_n which transform the error vector ω^n into the error vector $-\Omega u_h$. The operator χ_n may be represented as $(N-1)^2 \times (N-1)^2$ matrix in R space. For instance

$$\chi_n = \frac{1}{\omega_i^n} \begin{pmatrix} 0 & 0 & \cdots & 0 & \overset{i}{} & 0 & \cdots & 0 \\ \vdots & \vdots & & \vdots & & & & \\ \vdots & \vdots & & \vdots & -\Omega u_h & & & \\ 0 & 0 & \cdots & 0 & & 0 & \cdots & 0 \end{pmatrix} ,$$

ω_i^n is a maximum elements of error vector ω^n . Rewrite (11) in the next form

$$\omega^{n+1} = \bar{G}_n \omega^n ,$$

where $\quad \bar{G}_n = G_n (E - \tau f^n \chi_n)$.

Theorem 2. Let $\|\chi_n\| < \delta K / \|G^{-1}\|$, $0 < K = \text{const} \le 1$. If f^n satisfy to inequality

$$f^n > r_1 + \sqrt{r_1^2 + r_2^2} ,$$

where

$$r_1 = \frac{\|\chi_n\| \|G^{-1}\|^2 - \delta K^2}{\tau(\delta^2 K^2 - \|\chi_n\|^2 \|G^{-1}\|^2)} , \quad r_2 = \frac{(1 - K^2) \|G^{-1}\|^2}{\tau(\delta^2 K^2 - \|\chi_n\|^2 \|G^{-1}\|^2)} ,$$

then the inequality is valid and the convergent rate of the scheme (9) with $f^n \neq 0$ is greater than the same for the scheme (9) with $f^n = 0$.

In practical calculations it accurs that it is more convinient to choose f^n in the next form

$$f^n = \eta_n \|u^n - u^{n-1}\| , \tag{12}$$

where η_n is some positive constant.

Theorem 3. If the sequence $\{\eta_n\}$ is limited from above for all $n \geq n_0$ by some constant $B = B(\mu, \tau, h) > 0$ then solution of (9) convergents as $n \to \infty$ to solution (10).

In conclusion of this section we shall give some results of

our calculations, which were carried out by two iterative schemes with nonlinear "viscosity" and obtained from complete approximation schemes.

For solving the problem (10) with the boudary function $\varphi(x_1,x_2)=\sin x_1 \sin x_2$ next schemes are used:

$$\frac{u^{n+\frac{1}{2}}-u^n}{\tau}+A_{1n}u^{n+\frac{1}{2}}+A_{2n}u^n=0,$$

$$\frac{u^{n+1}-u^{n+\frac{1}{2}}}{\tau}+A_{1n}u^{n+\frac{1}{2}}+A_{2n}u^{n+1}=0, \qquad (13)$$

$$u^n|_{\Gamma_h}=u^{n+\frac{1}{2}}|_{\Gamma_h}=\varphi(x_1,x_2);$$

$$\frac{u^{n+\frac{1}{2}}-u^n}{\tau}+A_{1n}u^{n+\frac{1}{2}}+A_{2n}u^n=0,$$

$$\frac{u^{n+1}-u^{n+\frac{1}{2}}}{\tau}+A_{2n}(u^{n+1}-u^n)=0, \qquad (14)$$

$$u^n|_{\Gamma_h}=u^{n+\frac{1}{2}}|_{\Gamma_h}=\varphi(x_1,x_2),$$

where

$$A_{in}=\frac{\Delta_i+\Delta_{-i}}{2h}-\mu_n\frac{\Delta_i\Delta_{-i}}{h^2}\quad(i=1,2),\quad \mu_n=\mu(1+f^n),$$

$$f^n=\gamma_1\max_{i,j}|u_{ij}^n-u_{ij}^{n-1}|/n^{\gamma_2}+\gamma_3\max_{i,j}|u_{ij}^{n-1}-u_{ij}^{n-2}|/n^{\gamma_4}.$$

These schemes may be obtained from the usual alternating direction scheme (AD) and stabilizing corrections scheme (SC) by the addition of the „artificial viscosity " μf^n to the "natural viscosity" μ .

Numerical results are performed in tables 1,3 and for comparison one can see the table 2. Which follows from tables 1,2 the schemes with nonlinear viscisity convergent 1,4-1,5 faster than AD and SC schemes. From the table 3 it is obvious that if all iterative parameters in schemes (13), (14)are choosen to be optimal for $N=10$ than the same parameters will be optimal for $N=20,30$ and if N increases the number of iteration does not increase significantly. This fact will be discussed further.

Let's note that in test computations splitting schemes with artificial viscosity (12) have the convergent rate two order heigher than schemes without this artificial viscisity.

3. Let's consider the dependence of rate of convergence of the

scheme	N	τ	γ_1	γ_2	γ_3	γ_4	n $\varepsilon=10^{-5}$	n $\varepsilon=h^5$
(13)	10	0.023	8	1	1	1	10	-
	20	0.01	20	1	10	1	15	25
	30	0.007	20	1	0	-	25	46
(14)	10	0.023	8	1	0	-	21	-
	20	0.01	20	2	1	2	21	42
	30	0.007	20	1	1	1	29	71

Table 1

N	τ	A.D.		S.C.	
		$n(\varepsilon=10^{-5})$	$n(\varepsilon=h^5)$	$n(\varepsilon=10^{-5})$	$n(\varepsilon=h^5)$
10	0.023	15	-	28	-
20	0.01	28	39	45	65
30	0.007	42	66	60	116

Table 2

scheme	τ	N	γ_1	γ_2	γ_3	γ_4	n $\varepsilon=10^{-3}$	n $\varepsilon=10^{-4}$	n $\varepsilon=10^{-5}$
(13)	0.023	10	8	1	1	1	4	7	10
			100	2	10	2	6	8	12
		20	8	1	1	1	5	10	19
			100	2	10	2	8	12	20
		30	8	1	1	1	6	14	30
			100	2	10	2	9	15	31
(14)	0.023	10	8	1	0	-	7	12	21
			100	2	10	2	12	18	25
		20	8	1	0	-	9	23	46
			100	2	10	-	12	19	39
		30	8	1	0	-	9	31	72
			100	2	10	2	12	27	70

Table 3

scheme (8) on the artificial viscosity in the two physical problems: 1) Puaseil flow in a rectangular plane channel with unit whidth; 2) the flow in square box with two holes. For the first problem functions f_i^n have the next form

$$f_i^n = \beta_i \exp(-\gamma_i n)\,\varepsilon_n \qquad (i=1,2,3),$$

where

$$\beta_j = 2Re \quad (j=1,2), \quad \beta_3 = 20,$$

$$\gamma_i = \begin{cases} 0.05, & \tau \le 0.2\cdot 10^{-2}, \\ 0.1, & \tau > 0.2\cdot 10^{-2}, \end{cases}$$

$$\varepsilon_n = \max_{i,j} |u_{ij}^n - U|,$$

U is the analytical exact solution at the exit of the channel.

In the table 4 results of these computations are shown.

Re	τ	grid points	the number of iterations	$\|u^n - U\|$
25	0.01	15x90	455	0.02
100	0.001	15x40	1500	0.15
100	0.001	15x40	900	0.15
100	0.004	15x40	350	0.1
200	0.004	15x40	350	0.1
500	0.002	15x40	500	0.13
1000	0.002	15x60	900	0.13

Table 4

The same problem was solved in paper [4] with some splitting scheme. Those results show that the 0.025 accuracy was reached after 2000 iterations for $Re = 25$ (to compare with the first line of the table 4). Moreover, if $Re > 100$, the scheme without artificial viscosity does not converges during the real computing time. The second problem cannot be solved by splitting scheme without additional viscisity. The complete approximate scheme converges after 3000 iterations for $Re = 1000$. In our computations the steady-state flow qualitatively coiwiding with the same [5] was obtained after 1200 iterations for $Re = 1000$ (fig.1). The method based on stabilization by Reinolds number is used for the numerical solving of problem (2) by some difference

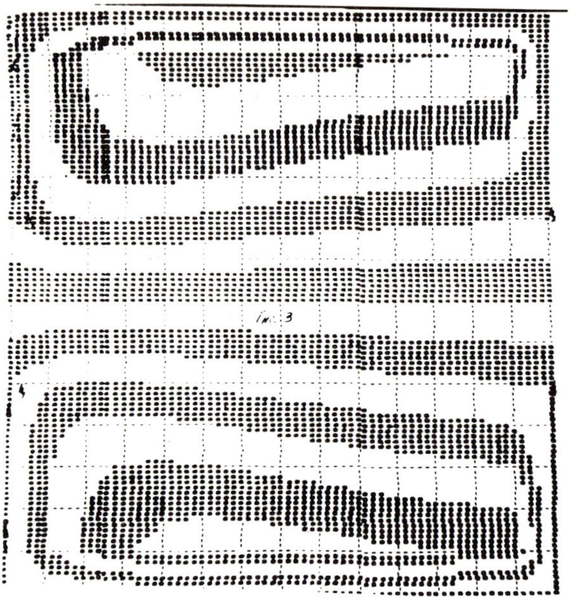

FIGURE 1

scheme for $Re \sim 1000$. That is, for begining the system (2) is solved up to stabilisation for small Re , than this system is solved up to stabilisation for some greater Re and so on. This method allow us to decrease the number of iterations for stabilisation [6] , [7] , but it must be used carefully because the uniqueness may be broken [8] . It follows from the theorem 1 that we must increase the Reynolds number faster than some exponent.

§ 2 The noncomplete approximation schemes using for solving linear equations.

1. As it follows from upper the nonlinear viscosity using allow us to increase the convergent rate of iteration schemes for nonlinear equations. The artificial viscosity introducing breakes the complete approximation condition for these schemes. Thus it is worth-while to treat this question in detail.

The division of iteration schemes on complete and noncomplete approximation schemes was made initially in [2] and practically all wellknown schemes, based on the approximation of the nonsteady equation

$$\frac{du}{dt} + Au = g$$

with the positive operator A were comlete approximation schemes. Noncomplete approximation schemes were known apriori as slowly convergent schemes and does not attract mathematician attention. Our test computation carried out with some noncomplete approximation schemes show their high effectivity even for nonlinear equations [9], [10]. Moreover it may be demonstrated that among these schemes it is easy to choose anyone, which converge faster than stationary complete approximation schemes.

2. Let's H is real Hilbert space with the scalar product (u, v) and the norm $\|u\| = (u, u)^{1/2}$. Let's consider also the space H_D which was given rise by selfconjugate positive operator D defined in the space H $(D = D^* > 0)$. In this space H_D the scalar product and the norm are defined as follows:

$$(u, v)_D = (Du, v) \quad , \quad \|u\|^2_D = (u, u)_D \qquad , \quad u, v \in H .$$

Let's assume that $A \le B$ if $(Ay, y) \le (By, y) \quad \forall y \in H$.

Let's consider the operator equation

$$Au = g , \tag{1}$$

where A is a linear positive operator defined in H, g and u are defined and unknown vectors from H. Let's consider two iteration schemes

$$B \frac{u^{n+1} - u^n}{\tau} + Au^n = g \qquad , \; n = 0, 1, 2, \dots , \tag{2}$$

$$B_{1n} \frac{u_1^{n+1} - u_1^n}{\tau} + Au_1^n + R_n u_1^n = g \quad , \quad n = 0, 1, 2, \dots . \tag{3}$$

for approximate solving the problem (1). Where B is defined positive operator acted in H, B_{1n} is the positive operator defined in H and R_n is some linear operator acted in H. The scheme (3) is a complete approximation scheme if $R_n \not\equiv 0$ for only n. It is obvious that for the investigation of the rate of convergence of scheme (2), (3) we can take initial values equal to zero. It is easy to obtain explicit schemes

$$v^{n+1} = S^{n+1} v^0, \tag{4}$$

$$S = E - \tau C \quad , C = D^{-1/2}(DB^{-1}A)D^{-1/2},$$

$$v_1^{n+1} = L_n v_1^0, \tag{5}$$

$$L_k = G_{1k}L_k + G_{2k} \quad , G_{1k} = S_k - G_{2k},$$

$$S_k = E - \tau_k D^{-\frac{1}{2}}(DB_{1k}^{-1}A)D^{-\frac{1}{2}} \quad , L_0 = S_0(\tau_0),$$

$$G_{2k} = \tau_k D^{-\frac{1}{2}}(DB_{1k}^{-1}A)D^{-\frac{1}{2}} \quad , k = 1,2,\dots,n$$

from (2),(3) with the next substitutions: $v^n = D^{\frac{1}{2}} w^n = D^{\frac{1}{2}}(u^n - u)$, $v_1^n = D^{\frac{1}{2}} w_1^n = D^{\frac{1}{2}}(u_1 - u)$. From the equality $\|v^n\| = \|w^n\|_D$ follows that estimating the convergence rate of schemes (4),(5) in H we can estimate the convergence of schemes (2),(3) in H_D

Let's suppose further that operators G_{2k}, G_{1k}, L_{k-1} are selfconjugate and commutative than it is easy to show that if G_{2k} satisfy to the inequality

$$-[\alpha_k \|S\|^{k+1} E + S_k L_{k-1}] \le G_{2k}(E - L_{k-1}) \le \le [\alpha_k \|S\|^{k+1} - S_k L_{k-1}], \tag{6}$$

where α_k is some positive constant then

$$\|L_k\| \le \alpha_k \|S\|^{k+1} \quad , k = 1,2,\dots . \tag{7}$$

Let constans $\ell_{k-1}^{(1)}$, $\ell_{k-1}^{(2)}$ from inequality

$$\ell_{k-1}^{(1)} E \le S_k L_{k-1} \le \ell_{k-1}^{(2)} E \quad , \ell_{k-1}^{(1)} \le \ell_{k-1}^{(2)} \tag{8}$$

are known. Than the estimate (7) is valid if G_{2k} satisfies to inequality

$$-[\alpha_k \|S\|^{k+1} + \ell_{k-1}^{(1)}]E \le G_{2k}(E - L_{k-1}) \le [\alpha_k \|S\|^{k+1} - \ell_{k-1}^{(2)}]E . \tag{9}$$

From (9) we get that

$$\alpha_k \geqslant \frac{\ell_{k-1}^{(2)} - \ell_{k-1}^{(1)}}{2\|S\|^{k+1}} \quad , k = 1,2,\dots . \tag{10}$$

and if $\ell_{k-1}^{(2)} \neq |\ell_{k-1}^{(1)}|$, than $\alpha_k < \alpha_{k-1}$ $(k=1,2,\dots)$ and the right side of (10) is always less than 1.

Let's suppose further that $B_{1k} = B$, $\tau_k = \tau$ $(k=0,1,2,\dots)$ and operator $DB^{-1}A$ is selfconjugate, than we get

$$0 \le S_1 L_0 = S L_0 = S^2 \le \|S\|^2 E.$$

Hence $\ell_0^{(1)} = 0$, $\ell_0^{(2)} = \|S\|^2$. In this case $\alpha_0 = 1$, $0.5 \le \alpha_1 < 1$ the scheme (3) will converge faster than (2) for any operator B and optimal iterative parameter τ independetly of the norm of step operator S of the scheme (4). Within this the difference between convergent rates of schemes (2),(3) may be significant. Thus the next theorem is proved.

Theorem 1. If constants $\ell_{k-1}^{(1)}$, $\ell_{k-1}^{(2)}$ from (8) are known and the operator G_{2k} satisfy to (9), than (7) is valid for any operators B, A such as $DB^{-1}A$ is selfconjugate and for α_k (10) is valid and $\alpha_n \le \alpha_{n-1} \le \ldots \le \alpha_1 < 1$.

Suppose $G_{2k} = \tau_k F_k$ where F_k is positive operator, τ_k is some constant and

$$0 \le f_k E \le F_k \le \|F_k\| E \quad , f_k = \mathrm{const}.$$

It is easy to show that if $S > 0$ than $|\ell_{k-1}^{(1)}| < \ell_{k-1}^{(2)}$ and from the fact that τ_k satisfy to unequality

$$-\frac{1}{\|F_k\|(1+\|L_{k-1}\|)}[\alpha_k \|S\|^{k+1} + \ell_{k-1}^{(1)}] \le \tau_k \le$$

$$\le \frac{1}{f_k(1-\|L_{k-1}\|)}[\alpha_k \|S\|^{k-1} - \ell_{k-1}^{(2)}],$$

where

$$1 > \alpha_k \geqslant \frac{\bar{\varkappa}_{k-1}\ell_{k-1}^{(2)} - \ell_{k-1}^{(1)}}{(1+\bar{\varkappa}_{k-1})\|S\|^{k+1}}$$

$$\bar{\varkappa}_{k-1} = (\|F_k\|/f_k)\varkappa_{k-1} \quad , \varkappa_{k-1} = (1+\|L_{k-1}\|)/(1-\|L_{k-1}\|)$$

it follows (7) and $\alpha_k < \alpha_{k-1} < \ldots < \alpha_1 < 1$. If $F_k = E$ than we can set

$$G_{2k} = \tau_k E = -0.5\,(\ell_{k-1}^{(2)} + \ell_{k-1}^{(1)}) \qquad (11)$$

and in this case

$$\alpha_k = \frac{\varkappa_{k-1}\ell_{k-1}^{(2)} - \ell_{k-1}^{(1)}}{(1+\varkappa_{k-1})\|S\|^{k+1}} < 1 \qquad (k = 1, 2, \ldots).$$

According to $\ell_0^{(1)} = 0$, $\ell_0^{(2)} = \|S\|^2$ than $\alpha_1 = 0.5(1+\|S\|) < 1$ for $G_{2k} = -0.5\,\|S\|^2$.

It is easy to constract the iterative scheme with the positive step operator. Really, if in the scheme (3)

$$B_{1n} = B(E+S)^{-1} \quad , \quad S = E - \tau B^{-1}A \tag{12}$$

than

$$S = S_1^2 > 0 \quad , \quad S_1 = E - \tau D^{-\frac{1}{2}}(DB^{-1}A)D^{-\frac{1}{2}} .$$

Hence, always $\rho_0^{(1)} = 0$, $\rho_0^{(2)} = \| S_1 \|^4$ and for $G_{21} = -0.5 \|S_1\|^4$, $\alpha_1 = 0.5\ (1 + \| S_1 \|^4) < 1$ that is the scheme (3) will converge in H faster than (2).

Thus we can conclude that if $DB^{-1}A$ is selfconjugate then for any operator B the scheme (3) will converge faster than scheme (2) for the constant optimal parameter τ . It is suffisient for this to set in (3) B_{1n} from (12) and to define G_{2k} by equality (11), for example.

Notes, if $G_{2k} = \tau_n E$, then $R_n = \tau_n B$.

3. In this section we shall give results of our computations which were carried out for the investigation of the convergent rate of noncomplete approximation schemes. These computations were made for the model finit-difference Dirichle problem

$$\Lambda u_h(\bar{x}_h) = (\Lambda_1 + \Lambda_2) u_h(\bar{x}_h) = 0 \ , \ \bar{x}_h = P_h \ , \tag{13}$$

$$u_h|_{\Gamma_h} = \varphi_h(\bar{x}_h) , \bar{x}_h \in \Gamma_h \ , \ \Lambda_i = \frac{\Delta_i \Delta_{-i}}{h^2} \quad (i = 1,2) .$$

As a region of computation we used the unit square P . In this region the uniform square grid with the step h was constructed. Initial values for all iterative schemes were choosen equal to zero and iterations were continued up to $\max_{i,j} |u_{ij}^n - u_{ij}^{n-1}| \le \varepsilon$, $\varepsilon = 10^{-5}$ or $1/N^5$.

Next iterative schemes were used to solve the problem (13)

$$\frac{u^{n+\frac{1}{2}} - u^n}{\tau} = (1+f_n)\Lambda_1 u^{n+\frac{1}{2}} + \Lambda_2 u^n ,$$
$$\frac{u^{n+1} - u^{n+\frac{1}{2}}}{\tau} = (1+f_n)\Lambda_2 u^{n+1} + \Lambda_1 u^{n+\frac{1}{2}} ; \tag{14}$$

$$\frac{u^{n+\frac{1}{2}} - u^n}{\tau} = (1+f_n)\Lambda_1 u^{n+\frac{1}{2}} + \Lambda_2 u^n ,$$
$$\frac{u^{n+1} - u^{n+\frac{1}{2}}}{\tau} = (1+f_n)\Lambda_2 u^{n+1} - \Lambda_2 u^n ; \tag{15}$$

$$\frac{u^{n+\frac{1}{2}}-u^n}{\tau} = (1+f_n)\Lambda_1 u^{n+\frac{1}{2}} + \Lambda_2 u^n + \omega_0 \Lambda u^n,$$
$$\frac{u^{n+1}-u^{n+\frac{1}{2}}}{\tau} = (1+f_n)\Lambda_2 u^{n+1} - \Lambda_2 u^n, \qquad (16)$$

where

$$f_n = \gamma_1 \| u^n - u^{n-1} \|_c^{\gamma_2} / n^{\gamma_3} + \gamma_4 \| u^{n-1} - u^{n-2} \|_c^{\gamma_5} / n^{\gamma_6}, \qquad (17)$$

$\| u^n \|_c = \max_{i,j} | u_{ij}^n |$, γ_k $(k = 1, \dots, 6)$ are nonnegative constants and ω_0 is some positive constant.

It is obvious that there is a constant for such as if $\| \Lambda_1 \Lambda_2 u_h \| \le \varepsilon$ then $\| u^1 - u_h \| \le 2\varepsilon$.

Some results of solving the problem (13) with the boundary function $\varphi(\bar{x}_h) = \sin x_{1h} \sin x_{2h}$ are perfomed in table 5. In this case in (17) $\gamma_2 = 1$, $\gamma_3 = \gamma_4 = 0$ and in the scheme (16) $\omega_0 = \rho_D^2 / (2 - \rho_D^2)$ where ρ_D is a constant which is same lower estimate of Douglas-Rachford (D.-R.) scheme's step operator [2].

The results of computations with the boundary function $\varphi(\bar{x}_h) = \sin x_{1h} \sin x_{2h}$ when τ , γ_i $(i = 1, \dots, 6)$ are constant for each scheme and all $N = 10, 20, 30$ and $\gamma_2 = \gamma_5 = 1$ are performed in the table 6. In the table 7 one can see for comparison, the results of solving of the some problem by different complete approximation schemes.

The next table show the results of computation by schemes (15),(16) of the problem (13) with another boundary conditions. The tables 8,9 represent the results of computation by scheme (16) where $\omega_0 = \rho_D / 2$ (in the table 10 $\rho_D = 0.535 \ln 1/N$ is the estimate of the norm of the step operator of the D.-R. scheme [11] and in the table 11 $\rho_D = 1 - 1/N$ is the assimptotic estimate of the step operator of D.-R. scheme). In the table 10 the results of computation for $\varphi(\bar{x}_h) = \sin x_{1h} \cos x_{2h}$ and in the table 11 - for $\varphi(\bar{x}_h) = \sin x_{1h}\, e^{x_{2h}}$ are shown. Here f_n has the form (17) where $\gamma_2 = \gamma_5 = 1$. In tables 10,11 the results of computations by scheme (15) for boundary function $\varphi(\bar{x}_h) = \sin x_{1h} \sin x_{2h}$ and $\varphi(\bar{x}_h) = 3 - \cos x_{1h}\, e^{x_{2h}}$ are shown, respectively. In the table 13 the results of computations by scheme (15) for $\varphi(\bar{x}_h) = \sin x_{1h} \cos x_{2h}$ and $N = 40$ with different iterative parameters are shown. In the table 13 the results of computations by scheme (15) for the same boundary problem and $N = 40$ are shown, where f_n is some function dependent on the number of iteration n . Note, that $f_0 = 0$ during all computations.

Numerical experiments show:

1) All three noncomplete approximation schemes converge faster than schemes from which they were constracted, and faster than 2-cycle implicit Richardson scheme (se tables 5,7,11),

2) The choosing of f_n and the value of parameter have small influence on the rate of convergrnce (see tables 12,13),

3) It is interesting that if one choose the parameters τ, γ_i $(i=1,\ldots,6)$ in schemes (14)-(16) to be optimal for $N=10$,than for $\varepsilon \sim 10^{-4} - 10^{-5}$ these parameters will be optimal for the other N (see tables 6,8,11), and the number of iterations slightly increase or don't increase at all unlike the stationary complete approximation schemes. The same results take place and for nonselconjugate case (see table 3 from § 1).

sche-mes	N	ρ_D	γ_1	n $\varepsilon=10^{-5}$	N	ρ_D	γ_1	n $\varepsilon=10^{-5}$	n $\varepsilon=h^5$	N	ρ_D	γ_1	n $\varepsilon=10^{-5}$	n $\varepsilon=h^5$
14	10	-	8	9	20	-	30	20	29	30	-	20	25	47
15	10	-	40	15	20	-	150	19	29	30	-	265	22	35
16	10	0,7	30	10	20	0,86	60	16	22	30	0,91	15	20	36

Table 5.

scheme	τ	N	γ_1	γ_3	γ_4	γ_6	n $\varepsilon=10^{-5}$
(14)	0,023	10	8	0	0	-	8
			100	2	10	2	9
		20	8	0	0	-	14
			100	2	10	2	15
		30	8	0	0	-	16
			100	2	10	2	17
(15)	0,023	10	40	0	0	-	15
			100	2	10	2	14
		20	40	0	0	-	17
			100	2	10	2	15
		30	40	0	0	-	19
			100	2	10	2	18
(16)	0,023	10	30	0	0	-	10
			100	2	10	2	8
		20	30	0	0	-	14
			100	2	10	2	12
		30	30	0	0	-	16
			100	2	10	2	14

Table 6.

scheme	Zeidel		Upper relaxation		D.-R.		A.D.		Implicit Richardson	
	n $\varepsilon=10^{-5}$	n $\varepsilon=h^5$	n $\varepsilon=10^{-5}$	n $\varepsilon=h^5$	n $\varepsilon=10^{-5}$	n $\varepsilon=h^5$	n $\varepsilon=10^{-5}$	n $\varepsilon=h^5$	n $\varepsilon=10^{-5}$	n $\varepsilon=h^5$
10	83	-	36	-	26	-	14	-	12	-
20	275	415	72	98	45	65	26	36	22	28
30	542	1042	100	161	66	109	34	62	30	44

Table 7.

N	τ	γ_1	γ_3	γ_4	γ_6	n $\varepsilon=10^{-5}$
10	0.23	0	-	0	-	17
10	0.023	100	2	10	2	12
40	0.006	0	-	0	-	58
40	0.012	100	2	10	2	14
40	0.023	100	2	10	2	17

Table 8

scheme	N	τ	γ_1	γ_3	γ_4	γ_6	n $\varepsilon=10^{-5}$
	10	0.023	0	-	0	-	30
(16)	10	0.023	100	2	10	2	18
	40	0.006	0	-	0	-	91
(16)	40	0.023	100	2	10	2	23

Table 9

N	τ	γ_1	γ_2	γ_3	γ_4	γ_5	γ_6	n $\varepsilon=10^{-4}$
10	0.005	10^4	2	1	10	0	1	12
20	0.025	100	2	1	10	2	1	10
20	0.005	100	2	1	10	0	1	10

Table 10

N	τ	γ_1	γ_2	γ_3	γ_4	γ_5	γ_6	n $\varepsilon=10^{-4}$
20	0.025	100	2	1	10	1	1	20
40	0.025	100	2	1	10	1	1	20
70	0.025	100	2	1	10	1	1	20

Table 11

τ	γ_1	γ_2	γ_3	γ_4	γ_5	γ_6	n $\varepsilon=10^{-4}$
0.001	100	2	1	0	-	-	19
0.005	100	2	1	1	0	1	14
0.005	10^8	0	8	0	-	-	18
0.005	10	0	1	100	0	1	15
0.01	100	1	1	10	0	1	11

Table 12

f_n	$\gamma_1 \operatorname{arctg}(\gamma_2 \|u^n-u^{n-1}\|_c / n^{\gamma_3})$	$1-\exp\left(\frac{\gamma_1 \|u^n-u^{n-1}\|_c^{\gamma_2}}{n^{\gamma_3}} + \frac{\gamma_4 \|u^{n-1}-u^{n-2}\|_c^{\gamma_5}}{n^{\gamma_6}}\right)$	$\gamma_1 \exp(-n^{\gamma_2})$	$\exp(\gamma_1 - n^{\gamma_2})$
τ	0.005	0.005	0.005	0.005
γ_1	1	10	10^6	5
γ_2	10^6	2	4	1
γ_3	1	1	-	-
γ_4	0	1	-	-
γ_5	-	0	-	-
γ_6	-	1	-	-
$n(\varepsilon=10^{-4})$	8	14	16	15

Table 13

References

I. Ладыженская О.А. Математические вопросы динамики вязкой несжимаемой жидкости. 2-е издание, М., "Наука", I970.

2. Яненко Н.Н. Метод дробных шагов решения многомерных задач математической физики. Новосибирск, "Наука", Сибирское отделение АН СССР, I967.

3. Владимирова Н.Н., Кузнецов Б.Г., Яненко Н.Н. Численный расчёт симметричного обтекания пластинки плоским потоком вязкой несжимаемой жидкости. В сб. "Некоторые вопросы прикладной и вычислительной математики", I966.

4. Тейлор Т.А., Ндэфо Э. Расчёт течения вязкой жидкости в канале при помощи метода расщепления. В сб. "Численные методы в механике жидкости", М., "Мир", I973, 2I8-229.

5. Громов В.П., Кузнецов Б.Г. Об одном методе расчёта задач вязкой несжимаемой жидкости. В сб. "Труды 2-го Всесоюзного семинара по численным методам по механике вязкой жидкости"., Новосибирск, "Наука", I969.

6. Кузнецов Б.Г., Смагулов Ш. Об аппроксимации уравнений Навье-Стокса. В сб. "Численные методы механики сплошной среды", I975, т.6, №2, 70-79.

7. Расщупкин В.И. О применении метода дробных шагов к численному решению уравнений Навье-Стокса. В сб. "Численные методы механики сплошной среды", I976, т.7, №6, I27-I35.

8. Джакупов К.Б., Кузнецов Б.Г. Об одном методе расчёта задач вязкой несжимаемой жидкости. В сб. "Труды 2-го Всесоюзного семинара по численным методам вязкой жидкости". , Новосибирск, "Наука", I969, 96-I06.

9. Захаров Ю.Н., Шокин Ю.И., Яненко Н.Н. Об одном методе ускорения сходимости итерационных схем. В сб. "Численные методы механики сплошной среды", I974, т.5, №5, 57-62.

I0. Захаров Ю.Н. Ускорение сходимости итерационных схем. В сб. "Численные методы механики сплошной среды", I976, т.7, №7, I2-22

II. Вазов В., Форсайт Дж. Разностные методы решения дифференциальных уравнений в частных производных. М., "ИЛ", I963.

COMPUTING METHODS IN APPLIED SCIENCES AND ENGINEERING
R. Glowinski, J.L. Lions (editors)
North-Holland Publishing Company

TRANSPORT APPROACH FOR COMPRESSIBLE FLOW

S. KANIEL
Department of Mathematics
The Hebrew University of Jerusalem
Jerusalem, Israel

J. FALCOVITZ
A.D.A.
P.O.B. 2250
Haifa, Israel

In this paper a novel approach for the numerical solution of the equations governing the motion of compressible flow, in the Eulerian formulation is described.

The equations are:

$$(1.1)\qquad \frac{\partial \rho}{\partial t} + \sum_i \frac{\partial}{\partial x_i} (\rho u_i) = 0$$

$$(1.2)\qquad \frac{\partial u_i}{\partial t} + \sum_j u_j \frac{\partial u_i}{\partial x_j} + \frac{1}{\rho} \frac{\partial p}{\partial x_i} = 0$$

$$(1.3)\qquad \frac{\partial S}{\partial t} + \sum_i u_i \frac{\partial S}{\partial x_i} = 0$$

$$(1.4)\qquad p = F(\rho, S)$$

The scalar ρ denotes the density, the vector u denotes the velocity, the scalar p the pressure and the scalar S the entropy. The equations are formulated for points (x,t) where x is an n-dimensional vector, $n = 1,2,3,.$

Equation (1.2) may be replaced by an equivalent equation for the momentum m

$$(1.2a)\qquad \frac{\partial m_i}{\partial t} + \sum_j \frac{\partial}{\partial x_j} \left(\frac{m_i m_j}{\rho}\right) + \frac{\partial p}{\partial x_i} = 0$$

Equation (1.3) may be replaced by equivalent equations for the internal energy or the pressure. The equation of state (1.4) is general.

The basic reasoning behind the numerical method is similar to that in the kinetic theory of gases. A distribution function $f(x,\xi,t)$, where ξ is a vector in velocity space is considered. It is the distribution of "molecules" located at x and having velocity ξ.

If such distribution function is defined then

$$(1.5)\qquad \rho(x,t) = \int f(x,\xi,t)\,d\xi$$

(1.6) $m_k(x,t) = \rho(x,t) \cdot u_k(x,t) = \int \xi_k \; f(x,\xi,t)d\xi$

There is a wide choice of distribution functions that, for given $\rho(x,t)$ and $m(x,t)$ satisfy equations (1.5) - (1.6).
Since the "phase space" (x,ξ) and the distribution function $f(x,\xi,t)$ are introduced expressly for the solution of the system (1.1) - (1.4) the determination will be such that a simple evolution in time of $f(x,\xi,t)$ will result in approximation to the system to be computed.
The evolution in time of $f(x,\xi,t)$ be that of free streaming i.e., each molecule keeps its own velocity. Thus

(1.7) $f(x,\xi,t) = f(x - \xi t,\xi,0)$.

The differential equation that f satisfies is

(1.8) $$\frac{\partial f}{\partial t} + \sum_i \xi_i \frac{\partial f}{\partial x_i} = 0$$

The equation (1.7) is meaningfull even for discontinuous initial values $f(x,\xi,0)$. Indeed, (1.7) is a "weak solution" of (1.8).

Integrate now equation (1.8)

(1.9) $$\frac{\partial}{\partial t} \int f d\xi + \sum_i \frac{\partial}{\partial x_i} \int \xi_i \; f d\xi = 0$$

By (1.5) , (1.6) it follows that

$$\frac{\partial \rho}{\partial t} + \sum_i \frac{\partial}{\partial x_i} m_i = 0$$

which is equation (1.1).

Multiply equation (1.8) by ξ_k and integrate $d\xi$.

(1.10) $$\frac{\partial}{\partial t} \int \xi_k \; f d\xi + \sum_i \frac{\partial}{\partial x_i} \int \xi_i \xi_k \; f d\xi = 0$$

By (1.6)

(1.11) $$\frac{\partial m_k}{\partial t} + \sum_i \frac{\partial}{\partial x_i} \int \xi_i \xi_k \; f d\xi = 0$$

The second term in (1.11) is not equal in general, to the second term of (1.2).

Nevertheless, it will be seen that, by a judicious choice of $f(x,\xi,0)$ equation (1.2) will, indeed, be satisfied. In so doing it will turn out that the variable ξ will no longer be the real velocity. For example, in a three dimensional space, the second moment $\frac{1}{2}\int |\xi|^2 f(x,\xi,t)d\xi$ (where $|\xi|^2 = \sum_i |\xi_i|^2$) will approximate the sum of kinetic energy and 3/2 times the pressure[1] (not the total energy which would have been the case, had f been the distribution of molecules having "true velocity" ξ).

2. $f(x,\xi,0)$ is constructed by a function $g_S(|\xi|)$. $g_S(|\xi|)$ itself is not connected to any particular flow. It is a general thermodynamic function. As such it has to be dependent on other thermodynamic variables (like ρ,p,S etc.). This dependence is best described by the equations, in n dimensional space

(2.1) $$\rho(\alpha) = \int_{|\xi| \le \alpha} g_S(|\xi|)d\xi$$

(2.2) $$p(\alpha) = \frac{1}{n}\int_{|\xi| \le \alpha} |\xi|^2 g_S(|\xi|)d\xi$$

(2.3) $$p(\alpha) = F(\rho(\alpha),S).$$

i.e., if, for some α , the integral in (2.1) is a density and the integral in (2.2) is a pressure then, for a fixed value of the entropy, the density and pressure should be connected by the equation of state (2.3).

For three dimensional space, using polar coordinates

(2.4) $$\rho(\alpha) = 4\pi \int_{r\le\alpha} r^2 g_S(r)dr$$

(2.5) $$p(\alpha) = \frac{4\pi}{3} \int_{r\le\alpha} r^4 g_S(r)dr$$

Differentiate (2.4) and (2.5)

(2.6) $$\frac{\partial\rho}{\partial\alpha} = 4\pi\alpha^2 g_S(\alpha)$$

(2.7) $$\frac{\partial p}{\partial\alpha} = \frac{4\pi}{3}\alpha^4 g_S(\alpha)$$

Divide (2.7) by (2.6) to get

(2.8) $$\frac{\partial p}{\partial\alpha} = \frac{\alpha^2}{3}\frac{\partial\rho}{\partial\alpha} .$$

On the other hand, from the equation of state, $\frac{\partial p}{\partial \alpha} = \frac{\partial p}{\partial \rho} \cdot \frac{\partial \rho}{\partial \alpha}$ where $\frac{\partial p}{\partial \rho} = c^2(\rho, S)$, a known function. Hence

$$\alpha^2 = 3\, c^2(\rho, S) \tag{2.9}$$

Hence α, the limit of integration, being a multiple of the speed of sound, is also a thermodynamic variable. By (2.6) and (2.9)

$$g_S(\alpha) = \frac{1}{4\pi} \cdot \frac{1}{\alpha^2} \cdot \frac{\partial \rho}{\partial \alpha} = \tag{2.10}$$

$$= \frac{1}{4\pi} \cdot \frac{1}{\alpha^2} \left\{ \frac{\partial \alpha}{\partial \rho} \right\}^{-1} = \frac{1}{4\pi} \cdot \frac{1}{\alpha^2} \cdot \frac{\partial}{\partial \rho} \left\{ \sqrt{3}\, \frac{\partial p}{\partial \rho}^{1/2} \right\}^{-1} =$$

$$= \frac{1}{2\sqrt{3}\,\pi} \cdot \frac{1}{\alpha^2} \cdot \frac{\partial^2 p}{\partial \rho^2}^{-1} \cdot \frac{\partial p}{\partial \rho}^{1/2} .$$

Since, for "reasonable" equations of state, $\frac{\partial^2 p}{\partial \rho^2} > 0$, it follows that $g_S(\alpha) > 0$. An explicit derivation for the equation of state of an ideal gas $p = A(S)\rho^\gamma$ follows.

In this case, for three dimensional space

$$\frac{\partial p}{\partial \rho} = A\gamma p^{\gamma-1} \qquad \alpha = \sqrt{3A\gamma} \cdot \rho^{\frac{\gamma-1}{2}}$$

$$\frac{\partial \rho}{\partial \alpha} = (3A\gamma)^{-\frac{1}{\gamma-1}} \cdot \frac{2}{\gamma-1} \cdot \alpha^{\frac{3-\gamma}{\gamma-1}} \tag{2.11}$$

$$g_S(\alpha) = \frac{1}{4\pi} (A(S) \cdot 3 \cdot \gamma)^{-\frac{1}{\gamma-1}} \cdot \frac{2}{\gamma-1} \cdot \alpha^{\frac{5-3\gamma}{\gamma-1}} \tag{2.12}$$

In particular, for $\gamma = \frac{5}{3}$, $g_S(\alpha)$ is a constant depending on S.

The analysis for one and two dimensional distributions is similar. For two dimensional distribution the result is

$$g_S(\alpha) = \frac{1}{2\pi} (A(S) \cdot 2 \cdot \gamma)^{-\frac{1}{\gamma-1}} \cdot \frac{2}{\gamma-1} \cdot \alpha^{\frac{4-2\gamma}{\gamma-1}} \qquad n = 2 \tag{2.13}$$

For one dimensional distribution

$$g_S(\alpha) = \frac{1}{\gamma-1} \cdot (A(S) \cdot \gamma)^{-\frac{1}{\gamma-1}} \cdot \alpha^{\frac{3-\gamma}{\gamma-1}} \qquad n = 1. \tag{2.14}$$

For $n=2$, $\gamma = 2$ or $n = 1$, $\gamma = 3$ it turns out that $g_S(\alpha) = K(S)$

3. Define $f(x,\xi,0)$ by

$$f(x,\xi,0) = g_{S(x)}(|\xi - u(x)|) \qquad |\xi - u(x)| \le \sqrt{n}\, c(x) \tag{3.1}$$

$$(3.2) \qquad f(x,\xi,0) = 0 \qquad\qquad |\xi - u(x)| > \sqrt{n}\ c(x)$$

Proposition: Define $f(x,\xi,t)$ by (1.7), define $\rho(x,t)$, $u(x,t)$ and $m(x,t)$ by (1.5) and (1.6) respectively. Define

$$(3.3) \qquad \hat{p}(x,t) = \frac{1}{n} \int \sum_i (\xi_i - u_i(x,t))^2\, f(x,\xi,t)\,d\xi.$$

Then, for $t = 0$, $\hat{p}(x,0) = p(x,0)$ and equation (1.2) is satisfied.

Proof: Evaluate, for $t = 0$, the integrals in (1.11)

$$\int \xi_i \xi_k\, f(x,\xi,0)\,d\xi = \int_{|\xi - u(x)| \le \sqrt{3}\, c(x)} \xi_i \xi_k\, g_{S(x)}(|\xi - u(x)|)\,d\xi$$

$$= \int_{|\eta| \le \sqrt{n}\, c(x)} (\eta_i + u_i(x))\ (\eta_k + u_k(x))\ g_{S(x)}(|\eta|)\,d\eta$$

where $\eta = \xi - u$.

The last expression is composed of four integrals

$$\int_{|\eta| \le \sqrt{n} c(x)} u_i \cdot u_k\, g_{S(x)}(|\eta|)\,d\eta = u_i \cdot u_k \int_{|\eta| \le \sqrt{n}\, c(x)} g_{S(x)}(|\eta|)\,d\eta = \rho\, u_i \cdot u_k$$

$$u_i \int_{|\eta| \le \sqrt{n} c} \eta_k\, g_S(|\eta|)\,d\eta = u_k \int_{|\eta| \le \sqrt{n} c} \eta_i\, g_S(|\eta|)\,d\eta = 0$$

because of symmetry.

If $i \neq k$ then also $\int \eta_i \eta_k\, g_S(|\eta|)\,d\eta = 0.$

For $i = k$

$$\int_{|\eta| \le \sqrt{n} c} \eta_k^2\, g_S(|\eta|)\,d\eta = \frac{1}{n} \int_{|\eta| \le \sqrt{n} c} |\eta|^2 g_S(|\eta|)\,d\eta = p$$

by the definition of $g_S(|\eta|)$

substitution in (1.11) results in (1.2). The result $\hat{p}(x,0) = p(x,0)$ was also proved, during the computation above.

4. One dimensional isentropic computation.

The scheme developed so far is already suitable for isentropic computations (as assumed, say, in transonic flow).

For computation in conservation law form the space is divided into cells. Let the i'th cell be bounded by x_i and x_{i+1}. Denote for $t = 0$, the total mass and total momentum in the i'th cell by $\rho^{(i)}(0)$ and $m^{(i)}(0)$ respectively. These quantities are used in order to specify approximate density $\rho(x,0)$ and approximate momentum $m(x,0)$ for each point x. (The specification is not unique). These, in turn, determine approximate velocity $u(x,0)$ and speed of sound $c(x,0)$. The actual algorithm is based on piecewise linear $u(x,0)$ and $c(x,0)$, with possible discontinuities at the cells boundaries.

By the fundamental construction

$$(4.1)\quad \rho^{(i)}(0) = \int_{x_i}^{x_{i+1}} \rho(x,0)\,dx = \int_{x_i}^{x_{i+1}} \int f(x,\xi,0)\,d\xi dx =$$

$$= \int_{x_i}^{x_{i+1}} \int_{u(x,0)-c(x,0)}^{u(x,0)+c(x,0)} g_S(|\xi - u(x,0)|)\,d\xi dx$$

$$(4.2)\quad m^{(i)}(0) = \int_{x_i}^{x_{i+1}} m(x,0)\,dx = \int_{x_i}^{x_{i+1}} \int \xi f(x,\xi,0)\,d\xi dx =$$

$$= \int_{x_i}^{x_{i+1}} \int_{u(x)-c(x)}^{u(x)+c(x)} \xi\, g_S(|\xi - u(x)|)\,d\xi dx$$

Since $f(x,\xi,t) = f(x - \xi t,\xi,0)$ it follows that

$$(4.3)\quad \rho^{(i)}(\Delta t) = \int_{x_i}^{x_{i+1}} \int f(x,\xi,\Delta t)\,d\xi dx =$$

$$= \iint\limits_{x_i \le x+\xi\Delta t \le x_{i+1}} f(x,\xi,0)\,d\xi dx = \iint\limits_{\substack{x_i \le x+\xi\Delta t \le x_{i+1} \\ u(x,0)-c(x,0) \le \xi \le u(x,0)+c(x,0)}} g_S(|\xi - u(x,0)|)\,d\xi dx$$

$$= \iint\limits_{\substack{\frac{x_i}{\Delta t} - u(x,0) \le \eta \le \frac{x_{i+1}}{\Delta t} - u(x,0) \\ -c(x,0) \le \eta \le c(x,0)}} g_S(|\eta|)\,d\eta dx$$

$$m^{(i)}(\Delta t) = \int_{x_i}^{x_{i+1}} \int \xi f(x,\xi,\Delta t)\,d\xi dx = \iint\limits_{x_i \le x+\xi t \le x_{i+1}} \xi f(x,\xi,0)\,d\ dx$$

(4.4)

$$= \iint_{\substack{x_i \le x+\xi\Delta t \le x_{i+1} \\ u(x,0)-c(x,0) \le \xi \le u(x,0)+c(x,0)}} \xi g_S(|\xi - u(x)|)\,d\xi\,dx =$$

$$= u(x,0) \iint g_S(|\eta|)\,d\eta \;+\; \iint_{\substack{\frac{x_i}{\Delta t} - u(x,0) \le \eta \le \frac{x_{i+1}}{\Delta t} - u(x,0) \\ -c(x,0) \le \eta \le c(x,0)}} \eta g_S(|\eta|)\,d\eta$$

The main part of the algorithm consists of the computation of the integrals in the right hand side of (4.3) and (4.4). Since $g_S(|\xi|)$ is a known thermodynamic function the integrals are dependent only on t, $u(x,0)$ and $c(x,0)$. These, in turn, are dependent on $\rho^{(i)}(0)$ and $m^{(i)}(0)$. The outcome is $\rho^{(i)}(\Delta t)$ and $m^{(i)}(\Delta t)$, which serve as initial values for the subsequent cycle. The process is described, schematically, in the following figure.

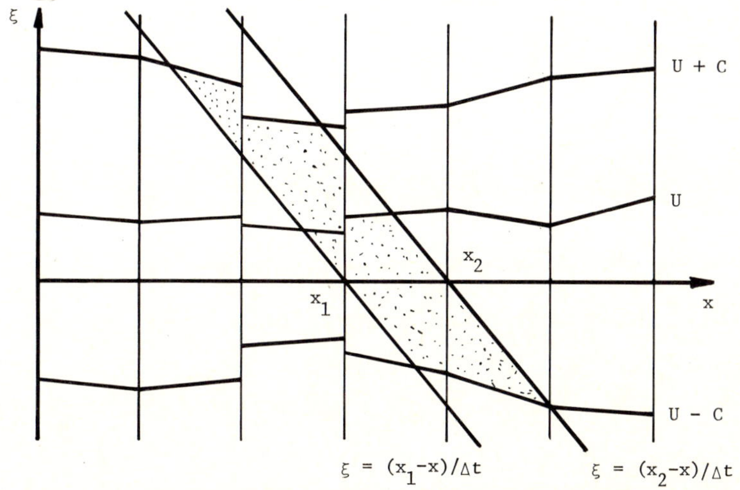

Figure 4.1 Phase Space Integration (One-Dimensional)

The integration can be performed analytically as follows:

Denote

$$G_S(|\eta|) = \int_0^{|\eta|} g_S(|\xi|)\,d\xi$$

Then the right hand side of (4.3) is equal to

$$(4.6)\qquad \int_{x_{max}}^{x_{min}} \{G_S(\eta_1(x)) - G_S(\eta_2(x))\}dx$$

where

$$(4.7)\qquad x_{min} + \{u(x_{min}) + c(x_{min})\}\Delta t = x_i$$

$$(4.8)\qquad x_{max} + \{u(x_{max}) - c(x_{max})\}\Delta t = x_{i+1}$$

$$(4.9)\qquad \eta_1(x) = \min\{u(x) + c(x), \frac{x_{i+1}-x}{\Delta t} - u(x)\}$$

$$(4.10)\qquad \eta_2(x) = \max\{u(x) - c(x), \frac{x_i-x}{\Delta t} - u(x)\} .$$

Equations (4.7) and (4.8) may have more than one solution. Thus the integral (4.6) may consist of several intervals. Each function on the right hand side of (4.9) and (4.10) is piecewise linear. Therefore $\eta_1(x)$ and $\eta_2(x)$ are also piecewise linear. So the integral (4.6) breaks into a finite sum of integrals of the form $\int_{x_k}^{x_{k+1}} G_S(\alpha_k x + \beta_k)dx$.

Each of the last integrals is analytically computable in terms of the indefinite integral of $G_S(|\eta|)$.

If the Courant condition $\frac{(|u| + c)\Delta t}{\Delta x} \leq 1$ is satisfied then, by (4.7) and (4.8) $x_{min} \geq x_{i-1}$ while $x_{max} \leq x_{i+2}$. This means that the contribution from each cell is only from the neighbouring ones. In this case the algorithm consists of the analytic evaluation of at most two integrals for cell.

The violation of Courant condition does not result in instability (there is no proof yet, but, heuristically, there is no way for the density to become negative). In this case the number of integral evaluations increases, but the actual algorithm used is still fast, though logically intricate.

If the flow is supersonic, say $u \geq c$ then $x_{max} \leq x_{i+1}$ i.e. the contribution is only from the left.

It is easy to see that the computational scheme is Galilei invariant i.e., the addition of a constant velocity to the flow does not affect, essentially, the computations. This can clearly be seen by looking at figure 4.1.The addition of a constant velocity will result in a rigid motion of the whole shape. The authors are not aware of any other Eulerian scheme that has this property.

For the particular case of the equation of state $p = A(S)\rho^3$ the function $g_S(|\xi|)$ is constant (cf (2.14)). Assume, without loss of generality $g \equiv 1$. It follows that, in figure 4.1 the mass in each cell is equal to its area while the

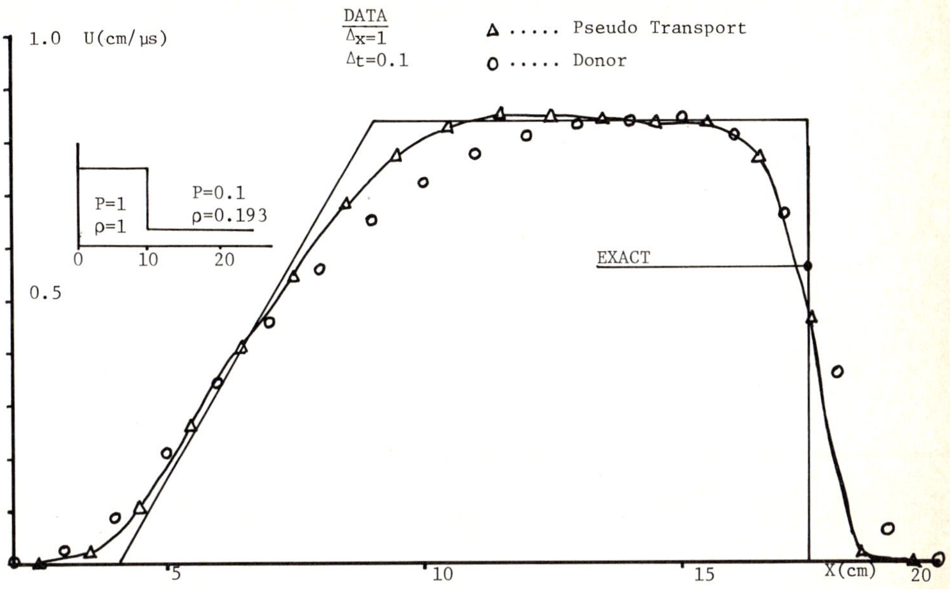

Figure 4.2 Shock Tube Velocity Distribution (γ = 1.4)

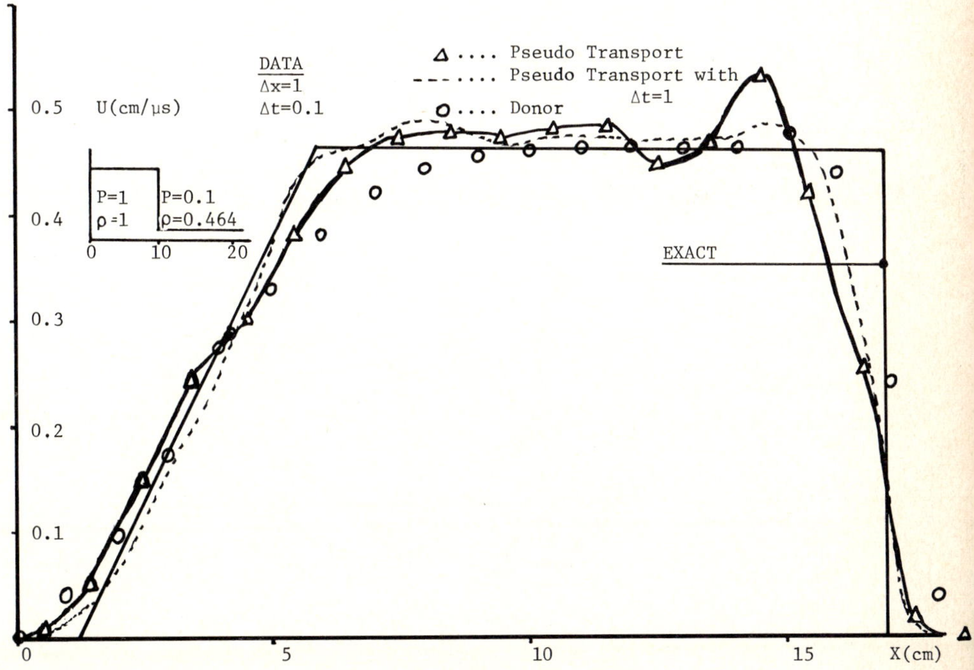

Figure 4.3 Shock Tube Velocity Distribution (γ = 3)

momentum is equal to the moment about the x-axis. The algorithm, in this case, consists of the computation of areas and moments of shapes as drawn between the two skew lines.

The computation of flow in a shock tube is exhibited in figures 4.2 and 4.3. It is for an ideal gas with $\gamma = 1.4$ (air) and $\gamma = 3$. Observe that for $\gamma = 3$, there is an excellent fit for $\Delta t = 1$. For this time step the Courant condition is violated. Indeed

$$\frac{(|u| + c)\Delta t}{\Delta x} \sim 2.15 \; .$$

For $\gamma = 1.4$ a large time step did not furnish a good approximation. The reason for the good fit when $\gamma = 3$ will be given in section 6.

5. Two dimensional isentropic Computations

Let the (i,j) cell be bounded by $x_1 = x_1^{(i)}, x_1 = x_1^{(i+1)}, x_2 = x_2^{(j)}, x_2 = x_2^{(j+1)}$. Denote it by $C^{(i,j)}$. Let $\rho^{(i,j)}$ and $m^{(i,j)}$ denote the total mass and momentum in $C^{(i,j)}$. As in the one dimensional case, these quantities determine approximate velocity and speed of sound, taken to be piecewise linear with possible discontinuities at the cells boundaries. Thus by (3.1) and (3.2) $\rho^{(i,j)}(0)$ and $m^{(i,j)}(0)$ determine the distribution function $f(x,\xi,0)$ which, by (1.7) determines $f(x,\xi,\Delta t)$.

The algorithm consists of the computation of $\rho^{(i,j)}(\Delta t)$ and $m^{(i,j)}(\Delta t)$.

It is convenient, in this case, to consider it with "donor viewpoint" i.e., to ask what is the total mass and momentum that for $t = 0$ is in $C^{(k,\ell)}$ and for $t = \Delta t$ is in $C^{(i,j)}$. This, in turn, is determined by the transport of "molecules" from $C^{(k,\ell)}$ to $C^{(i,j)}$.

The velocity ξ of a molecule located at x can be decomposed into $\xi = u(x) + (\xi - u(x))$. Consider first the movement of $C^{(i,j)}$ by $\Delta t \cdot u(x,0)$. Since $u(x,0)$ is assumed to be linear in each cell, a rectangular cell $C^{(k,\ell)}$ will move, by the linear transformation thus defined, to a parallelogram $U^{(k,\ell)}$. Then the molecules will be dispersed by $\Delta t \cdot \eta$ where $\eta = \xi - u(x)$ and $|\eta| \le \sqrt{2}\, c(x)$. Since $c(x)$ is assumed to be linear in $C^{(k,\ell)}$ it will be also a linear function of the new coordinates, defining the parallelogram $U^{(k,\ell)}$.

In order to determine the total mass and momentum in each cell it is enough to determine these in half spaces (say $x_1 \ge x_1^{(i)}$) and quarter-spaces (say $x_1 \ge x_1^{(i)}, x_2 \le x_2^{(j)}$).

The contribution of $C^{(k,\ell)}$ to the mass located right of $x_1 = x_1^{(i)}$ will be

$$\iint\limits_{\substack{x_1 \ge x_1^{(i)} \\ x - \xi t \in C^{(k,\ell)}}} f(x,\xi,\Delta t)\, d\xi dx = \iint\limits_{\substack{x \ge x_1^{(i)} \\ x - \xi t \in C^{(k,\ell)}}} f(x - \xi\Delta t, \xi, 0)\, d\xi dx = \tag{5.1}$$

$$(5.1) \quad = \iint\limits_{\substack{x_1+\xi_1\Delta t \geq x_1^{(i)} \\ |\xi - u(x)| \leq \sqrt{2}\, c(x) \\ x \in C^{(k,\ell)}}} g_S(x\,|\,\xi - u(x)|\,)\,d\xi\,dx \quad = \iint\limits_{\substack{y_1+\eta_1\Delta t \geq x_1^{(i)} \\ |\eta| \leq \sqrt{2}\, c(y) \\ y \in U^{(k,\ell)}}} g_S(y,|\eta|)\,d\eta\,dy$$

The integration in η can be carried out analytically (or can be precomputed numerically) It amounts to the integration of $g_s(|\eta|)$ over a spherical cap. This, in turn, is a function of the radius of the sphere (i.e., $\sqrt{2}\ c(y)$) and the quantity $d_1 = \dfrac{x_1^{(i)} - x_1}{\Delta t}$.

A similar computation holds for a quarter-space. One has to add a constraint (say $y_2 + \eta_2\Delta t \leq x_2^{(j)}$) to get a known function depending on three variables. The components of the momentum are evaluated by a similar procedure.

This way the ξ-integration is eliminated from the numerical computation. In fact, no function depending on ξ is explicitly constructed. The x integration has to be performed numerically. Methods for fast evaluation of the integrals involved are being developed. These methods are similar, in spirit, to some integration techniques used in F.E.M.

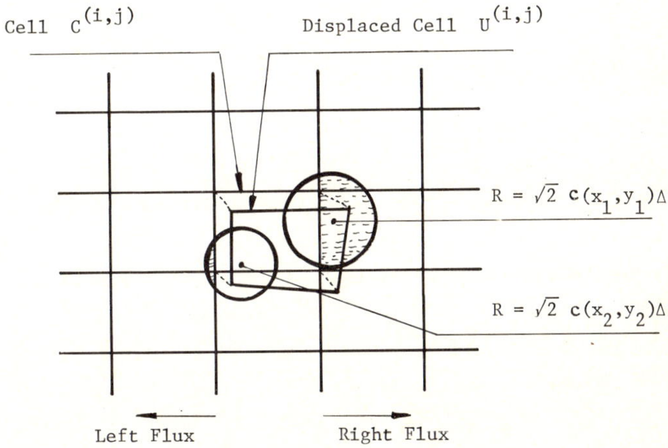

Figure 5.1 Integration Domain for Side Flux

As an illustration to the techniques and procedures used we exhibit a detailed analytical computation for an ideal gas $p = A(S)\rho^2$, where the mass and momentum are assumed to be piecewise constant. The problem is flow into a wedge.

There are some further simplifications in this sample 2-D problem. The flow field chosen is that of a plane shock generated by a shock tube, that impinges on a 45° wedge, as depicted in Figure 5.2. The boundary conditions are specular reflections of the pseudo molecules from the walls. Since the cells are squares, this can be achieved for the 45° wedge by introducing mirror cells along the boundaries. Thus, the ensuing code is of a lesser complexity than a general code for the computation of compressible flow by the pseudo-transport scheme.

At this point we introduce a change of notation. The space coordinates will be (x,y) and the pseudo molecular velocity will be (ξ,η).

For clarity, we reiterate the basic assumptions on which the simplified scheme is based:

(a) The fluid is an ideal gas having $\gamma = 2$.

(b) The time step is limited by

$$[|u| + \sqrt{2}\, c]\, \Delta t < \Delta x$$
$$[|v| + \sqrt{2}\, c]\, \Delta t < \Delta y$$

(c) The flow field ρ, u, c is considered piecewise constant in each cell.

(d) The cells are squares. $(\Delta x = \Delta y)$. The wedge coincides with cell diagonals (45° wedge).

The time-step limitation (b) justifies consideration of nearest-neighbor exchange of mass and momentum. We consider the cell $C^{(i,j)}$ and its eight neighbors, as depicted in Figure 5.3. Let us denote the flux from $C^{(i,j)}$ to its neighbor q, by

$$(5.2) \qquad F^q_{m,n} \qquad q = 1,2,\dots,8, \quad (m,n) = \begin{cases} (0,0) & \text{mass} \\ (1,0) & \text{x-momentum} \\ (0,1) & \text{y-momentum} \end{cases}$$

(unlike the conventional terminology, we denote by "flux", the time-integral of the mass or momentum flux over the time interval Δt).

We now consider in some detail the evaluation of the flux terms (5.2). The grand-strategy is to exploit the following obvious relation

$$(5.3) \qquad S^k_{m,n} = F^{2k-1}_{m,n} + F^{2k}_{m,n} + F^{2k+1}_{m,n} \qquad k = 1,2,3,4$$

where $S^k_{m,n}$ is the flux through the entire (infinite) side k of $C^{(i,j)}$. ($k = 1$ is the side $y = 0$, and the order is anticlockwise). The side flux $S^k_{m,n}$ is relatively easy to evaluate; the main effort is the evaluation of the corner (quarter-space) flux $F^q_{m,n}$ with $q = 1,3,5,7$. Since the domain of integration for the corner flux is rather complex, we proceed in the following manner: first we evaluate two basic flux functions: a side flux W_m and a corner flux $H_{m,n}$. Then we use these basic functions in some combinations which will be briefly outlined later, to obtain $S^k_{m,n}$, $F^{2k-1}_{m,n}$, $k = 1,2,3,4$.

The basic side flux function $W_m(a)$ is defined as the total flux of the ξ^m moment that emmanates from the region $0 \leq x \leq a$ $(a \geq 0)$ crossing the line $x = 0$. The integration domain for (ξ,η) is a circular cap, and for (x,y) it is the strip $0 \leq x \leq a$. The results are expressed by elementary functions.

$$W_0(a) = \frac{\rho R}{\pi} [\sigma \text{ arc } \cos(\sigma) - \sqrt{1-\sigma^2} + \frac{1}{3}(1-\sigma^2)^{3/2} + \frac{2}{3}]$$

(5.5) $$W_1(a) = -\frac{\sqrt{2}\rho RC}{\pi}[\frac{1}{12}\sigma(5-2\sigma^2)\sqrt{1-\sigma^2} + \frac{1}{4}\text{ arc } \sin(\sigma)]$$

$$R = \sqrt{2}\ c\Delta t \qquad \sigma = \frac{a}{R}$$

When $\sigma > 1$, $W_m(\sigma) = W_m(1)$

The evaluation of the basic corner flux function involves two steps. First we perform the (ξ,η) integration, with the integration domain Ω_1 shown in Figure 5.4. The results are

(5.6)
$$G_{0,0}(x,y) = \frac{\rho}{2\pi}[\text{arc } \cos(\sigma) - \text{arc } \sin(\theta) + 2\sigma\theta - \theta\sqrt{1-\theta^2} - \sigma\sqrt{1-\sigma^2}]$$
$$G_{1,0}(x,y) = -\frac{\rho c\sqrt{2}}{6\pi}(\theta - \sqrt{1-\sigma^2})^2(\theta + 2\sqrt{1-\sigma^2})$$
$$G_{0,1}(x,y) = -\frac{\rho c\sqrt{2}}{6\pi}(\sigma - \sqrt{1-\theta^2})(\sigma + 2\sqrt{1-\theta^2})$$

$$\sigma = \frac{x}{R},\ \theta = \frac{y}{R},\quad R = \sqrt{2}\ c\Delta t,\quad \sigma^2 + \theta^2 \leq 1$$

When $\sigma^2 + \theta^2 > 1$, $G_{m,n}(x,y) = 0$

Where $G_{m,n}(x,y)$ is the moment $\xi^m\eta^n$ of molecules emanating from point (x,y) in quarter I $(x > 0,\ y > 0)$, into quarter III $(x < 0,\ y < 0)$.

For the second step we consider the displaced cell $U^{(i,j)}$ as shown in Figure 5.5. The integration of $G_{m,n}(x,y)$ on the domain Ω_2 of Figure 5.5 results in the total flux $H_{m,n}(\alpha,\beta)$ of molecules that originated from the cell $C^{(i,j)}$ whose bottom left corner $(x,y) = (0,0)$ moved to $(\alpha,\beta) = (u\Delta t, v\Delta t)$, with $\alpha \gtrless 0$ and $\beta \gtrless 0$, and stream into the opposite quarter $(x < 0, y < 0)$. The results of this integration are:

$$H_{0,0}(\alpha,\beta) = \frac{\rho R^2}{24\pi}\{12\bar{\alpha}\bar{\beta}[\text{arc } \cos(\bar{\beta}) - \text{arc } \sin(\bar{\alpha})] -$$
$$- (\bar{\alpha}^4 + \bar{\beta}^4) + 6(\bar{\alpha}^2 + \bar{\beta}^2 + \bar{\alpha}^2\bar{\beta}^2) + 3 -$$
$$- 12(\bar{\alpha}\sqrt{1-\bar{\beta}^2} + \bar{\beta}\sqrt{1-\bar{\alpha}^2}) +$$
$$+ 4\bar{\alpha}(1-\bar{\beta}^2)^{3/2} + 4\bar{\beta}(1-\bar{\alpha}^2)^{3/2}\}$$

(5.7) $$H_{1,0}(\alpha,\beta) = \frac{\sqrt{2}\rho R^2 c}{120\pi}\{\sqrt{1-\bar{\beta}^2}(2\bar{\beta}^4 - 9\bar{\beta}^2 - 8) - 5\sqrt{1-\bar{\alpha}^2}\,\bar{\alpha}\bar{\beta}(5 - 2\bar{\alpha}^2) -$$
$$- \bar{\alpha}(10\bar{\alpha}^2\bar{\beta}^2 - 3\bar{\alpha}^4 + 5\bar{\beta}^4 - 30\bar{\beta}^2 + 10\bar{\alpha}^2 - 15)$$
$$+ 15\bar{\beta}[\text{arc } \cos(\bar{\beta}) - \text{arc } \sin(\bar{\alpha})]\}$$

(5.7)
$$H_{0,1}(\alpha,\beta) = H_{1,0}(\beta,\alpha)$$
$$\bar{\alpha} = \mathrm{Min}[\mathrm{Max}\,(\tfrac{\alpha}{R},0),1]$$
$$\bar{\beta} = \mathrm{Min}[\mathrm{Max}\,(\tfrac{\beta}{R},0),1]$$

The evaluation of $F^q_{m,n}$ from the basic flux functions W_m and $H_{m,n}$ is done as follows. We consider the general case shown in Figure 5.6, where the displaced cell $U^{(i,j)}$ has moved through time step Δt so that the point $(x,y) = (0,0)$ is at $(\alpha < 0, \beta < 0)$. Thus the cell $U^{(i,j)}$ is divided by the axes (x,y) into four parts. We have to evaluate $F^1_{m,n}$ and $S^1_{m,n}$. These will suffice, since by shifting the origin $(x,y) = (0,0)$ to the other vertices of $C^{(i,j)}$ (while rotating the axes by $\frac{\pi}{2}$ at each vertex change), and repeating the computation of $F^1_{m,n}$, $S^1_{m,n}$, we get all side and corner fluxes. Then, by (5.3) we get the flux to the four side-neighbors of $C^{(i,j)}$. The manner by which $F^1_{m,n}$ is obtained from the basic flux functions W_m, $H_{m,n}$, will now be briefly outlined. The flux contributed to quarter III $(x < 0, y < 0)$ is the sum of four contributions each corresponding to one of the four parts of $U^{(i,j)}$ (see Figure 5.6). Thus, each one of these sub-fluxes is a result of integration on the relatively simple domain Q_n $(n = 1,2,3,4)$, and can readily be expressed by suitable combinations of the basic flux functions $W_{m,n}$, $H_{m,n}$. (While so doing, we exploit both the identity (5.3) and the symmetry relative to point $(x,y) = (0,0)$). Consequently, the resulting code involves just the computation of W_m and $H_{m,n}$ in the appropriate combinations.

6. The first order approximations for ρ and u were derived in the general (variable entropy) case. In this case one still needs to compute a third thermodynamic variable.

It is possible to integrate equation (1.3) separately. Some methods like the F.C.T. of Boris and Book are suitable for such an equation.
An alternative approach, consistent with the ideas developed in this paper, is to consider the pressure as the third thermodynamic variable.

It can be shown that for $\hat{p}$ as defined in (3.3) and $t = 0$

$$\frac{\partial \hat{p}}{\partial t} - \frac{\partial p}{\partial t} = -\frac{1}{n}\left((n+2)p - n\rho\frac{\partial p}{\partial \rho}\right)\sum_{j=1}^{n}\frac{\partial u_j}{\partial x_j}\ . \tag{6.1}$$

So, if $\hat{p}$ is taken to be an approximation to the pressure, a correction term has to be added.
Furthermore, if one checks the order of approximation it turns out that, for

$t = 0$, $\dfrac{\partial^2\hat{\rho}}{\partial t^2} = \dfrac{\partial^2\rho}{\partial t^2}$ while

(6.2) $$\frac{\partial^2 \hat{m}_k}{\partial t^2} - \frac{\partial^2 m_k}{\partial t^2} = \sum_i \frac{\partial}{\partial x_i} \left\{ \left(\frac{\partial u_i}{\partial x_k} + \frac{\partial u_k}{\partial x_i} \right) p \right\} + \frac{\partial}{\partial x_k} \left\{ \left(p - \frac{\partial p}{\partial \rho} \cdot \rho \right) \sum_i \frac{\partial u_i}{\partial x_i} \right\}$$

Here again, the hats denote computed variables.

It turns out that a second order approximation is possible. Define

(6.3) $$f^*(x,\xi,t) = f(x - \xi t,\xi,0) + th(x - \frac{\xi}{2} t,\xi,0)$$

Then, for an appropriate $h(x, ,0)$

(6.4) $$\frac{\partial^3 \hat{\rho}}{\partial t^3} = \frac{\partial^3 \rho}{\partial t^3} , \quad \frac{\partial^2 \hat{m}}{\partial t^2} = \frac{\partial^2 m}{\partial t^2} , \quad \frac{\partial \hat{p}}{\partial t} = \frac{\partial p}{\partial t}$$

$h(x, ,0)$ is constructed in a way similar to the construction of $f(x,\xi,0)$.

Observe that for one dimensional flow and $\gamma = 3$ it follows that second order accuracy for the momentum equation is achieved already in the basic approximation. This explains the good results for $\gamma = 3$ and large time step.

The details of the results reported in this section are somewhat lengthy. They will be carried out elsewhere.

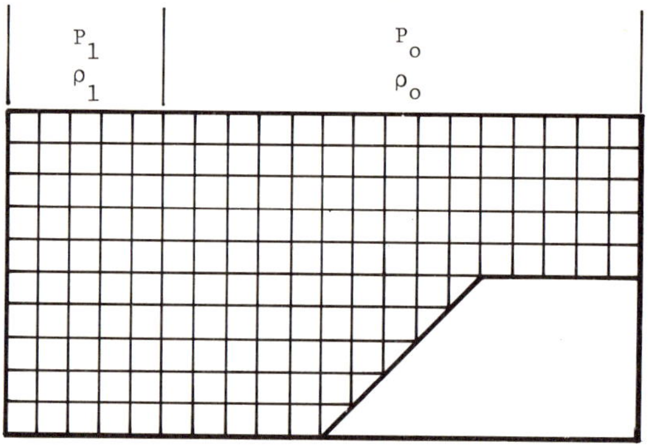

Figure 5.2 45° Wedge Shock Tube

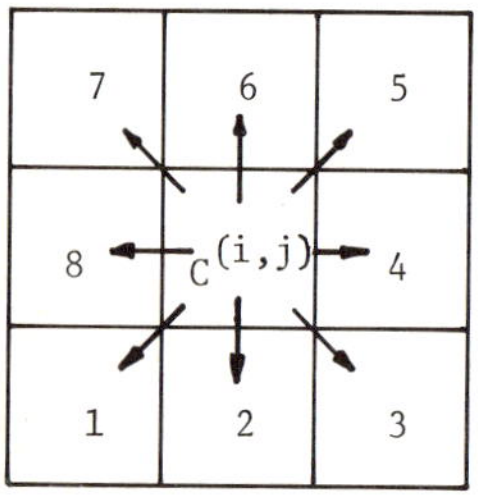

Figure 5.3
Neighbor cell Flux $F^q_{m,n}$
$(q = 1,2,\ldots,8)$

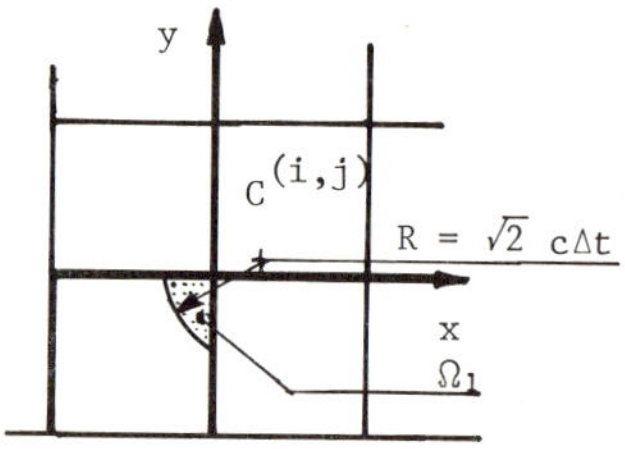

Figure 5.4
(ξ, η) Integration Domain for Basic Corner Flux $G_{m,n}(x, y)$

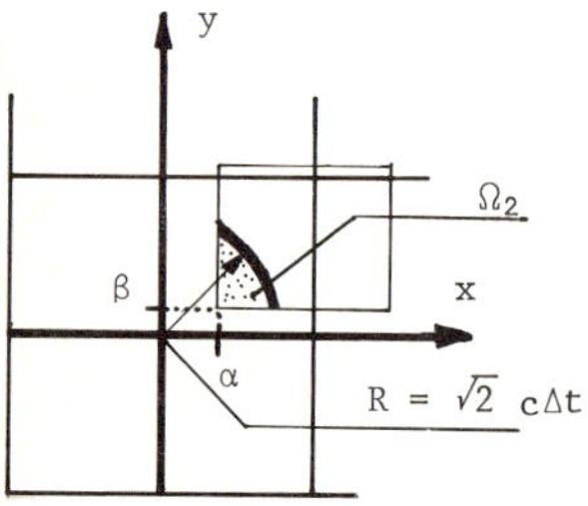

Figure 5.5
(x, y) Integration Domain for Basic Corner Flux $H_{m,n}(\alpha, \beta)$

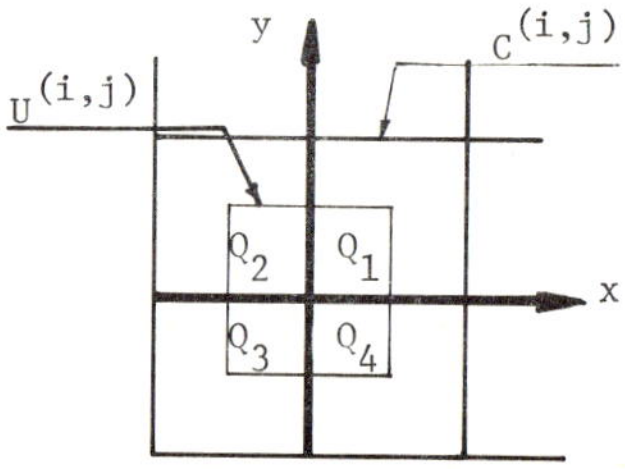

Figure 5.6
The Sub-Domains of Displaced Cell $U^{(i,j)}$

7. Numerical Results

At the end of Section 5 we described how the neighbor flux terms $F^q_{m,n}$ are evaluated. We have to add that the momentum flux is now obtained by combining a "molecular" contribution with a mass-flow contribution, as follows:

$$(7.1)\qquad \begin{aligned} (\text{Flux})^q_{1,0} &= F^q_{1,0} + u\, F^q_{0,0} \\ (\text{Flux})^q_{0,1} &= F^q_{0,1} + v\, F^q_{0,0} \end{aligned} \qquad q = 1,2,\ldots,8$$

The test problem for numerical computation is a plane shock wave of Mach number $M_s = 1.75$, that propagates in an ideal gas with $\gamma=2$, $\rho_o = 0.316$, $P_o = 0.1$. The shock wave is reflected from a 45° compressive corner. For comparison with known methods, we used a finite-difference code employing a conventional donor scheme, with shock smoothing by the Von-Neuman artificial viscosity. The results are shown in figures 7.2 and 7.3. The familiar pattern of the reflected bow-shock is clearly visible in the results. However the spacial spread of the shock front obtained by the transport code is considerably larger than that of the donor code. A known remedy to that is to carry out the flux integration (section 5) with interpolated field variables, rather than assume them to be piecewise constant in each cell as we presently do. We plan to do that in the future. However, as a demonstration of this effect, we consider the same shock wave computed in a 1-D code (section 4). The pressure distribution is shown in figure 7.1. Compare the two types of computation: (a) with interpolation, (b) piecewise constant. Clearly, the spread of the shock front in case (b) is much larger than that of (a), while the latter is comparable to the spread obtained by a donor code with artifical viscosity (2-3 cells).

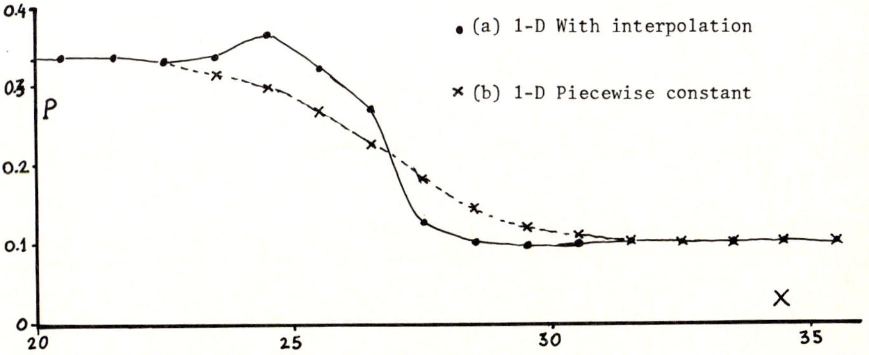

Figure 7.1 - Pressure distribution in a plane shock. ($\gamma=2$, $M_s=1.75$, $\rho_o=0.316$, $P_o=0.1$)

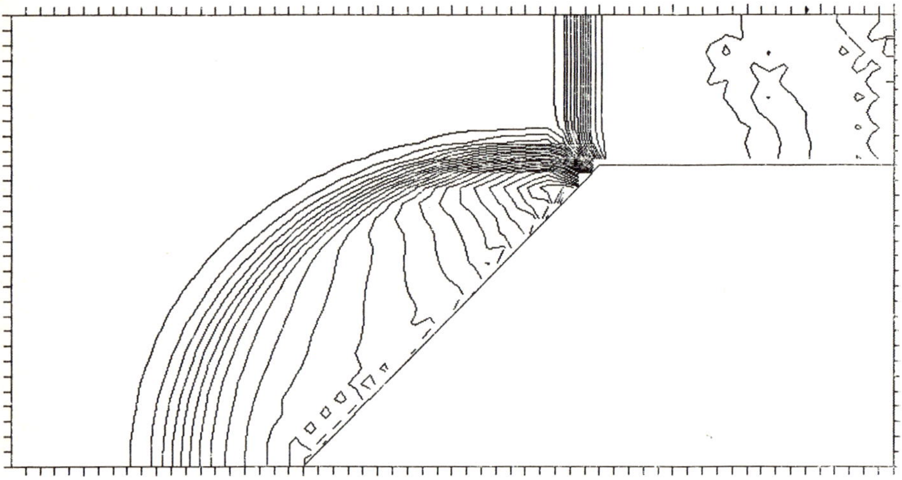

(a) Donor Scheme

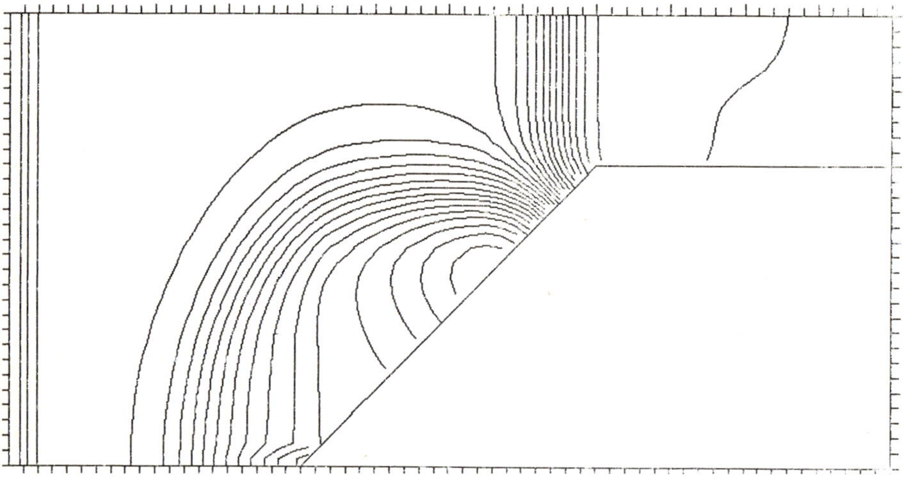

(b) Transport Scheme

Figure 7.2: Shock Reflection by a Wedge (45°) Time = 27.5
Isobars with $\Delta p = 0.02$ ($P_o = 0.1$, $\rho_o = 0.316$, $\gamma = 2$, $M_s = 1.75$)

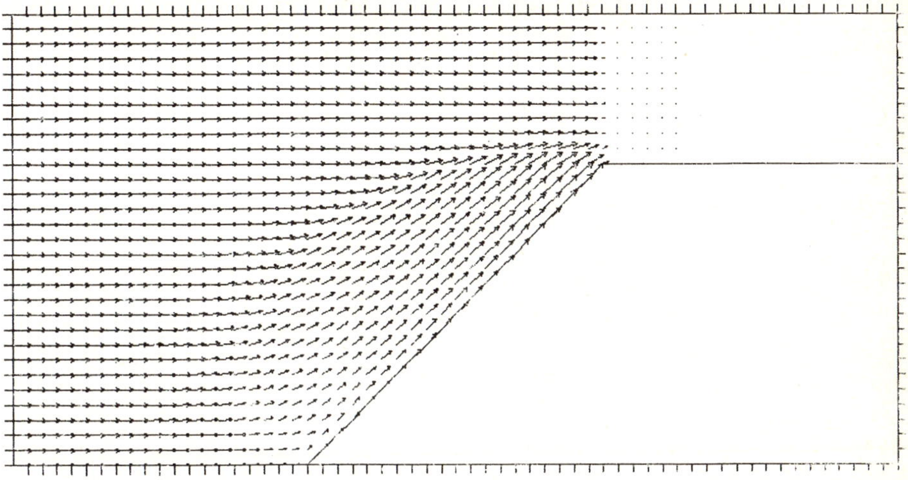

(a) Donor Scheme

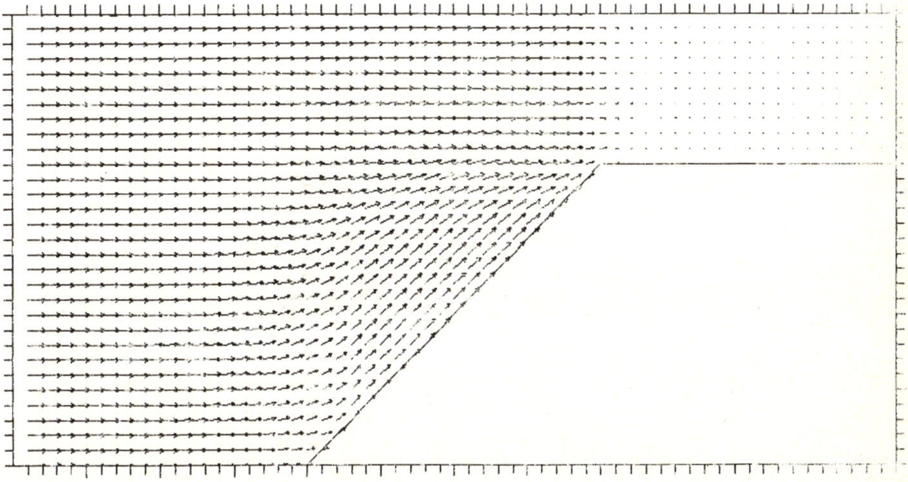

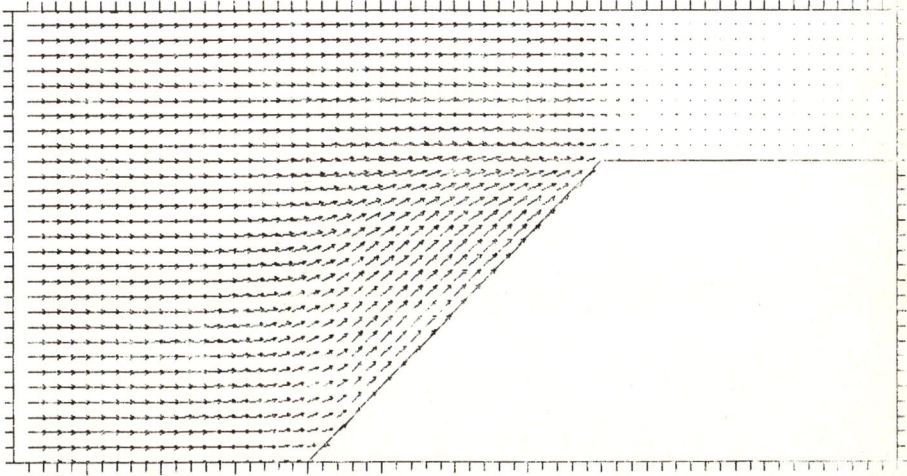

(b) Transport Scheme

Figure 7.3: Shock Reflection by a Wedge (45°) Time = 27.5
Velocity field. Scale = 0.5 (cm/ms.)/cm.
(P_o = 0.1, ρ_o = 0.316, γ = 2, M_s = 1.75)

COMPUTING METHODS IN APPLIED SCIENCES AND ENGINEERING
R. Glowinski, J.L. Lions (editors)
North-Holland Publishing Company

One-Sided Difference Schemes and Transonic Flow

Stanley Osher

Mathematics Department
University of California
Los Angeles, California
90024

In recent years one sided difference approximations have been frequently used in the numerical solution of transonic flow problems, both for the small disturbance approximation and the full potential equation e.g. [1], [6], [7], [8]. Björn Engquist and I have developed a simple second order accurate one sided method for the small disturbance equation which is nonlinearly stable, converges only to physically correct solutions, and appears to give excellent shock resolution. This extends our earlier work in which we simplified the first order Cole-Murman algorithm to rule out physically incorrect shocks, and instabilities, merely by changing the switch.

The main problems in computational fluid dynamics for inviscid compressible flow involve devising efficient algorithms which have some of the following properties:

(1) High order accuracy in smooth regions of the flow.

(2) Shocks which must satisfy the unique jump conditions.

(3) Physically correct, i.e. nonexpansive, shocks satisfying the entropy condition.

(4) Nonlinear stability.

(5) Good shock resolution-minimal overshoot and smearing.

We have recently devised algorithms based on upwind or one-sided differencing and splitting which have the above five properties for the small disturbance equations of transonic flow. These methods are very simple computationally. We have successfully performed many numerical experiments for the small disturbance equation and hope to extend these algorithms to apply to the full potential equation.

Consider first the low frequency time dependent small disturbance equation of transonic flow:

$$2\Phi_{tx} = \left(K\Phi_x - \tfrac{1}{2}(\gamma+1)\Phi_x^2\right)_x + \Phi_{yy} + \Phi_{zz}$$

for $t \geq 0$ and x,y,z in some region.

Begin by numerically splitting the equation

$$2\Phi_{tx} = \left(K\Phi_x - \tfrac{1}{2}(\gamma+1)\Phi_x^2\right)_x \qquad \text{Step 1}$$

$$2\Phi_{tx} = \Phi_{yy} \qquad \text{Step 2}$$

$$2\Phi_{tx} = \Phi_{zz} \qquad \text{Step 3}$$

Numerically let $D_-\Phi_j = \dfrac{\Phi_j - \Phi_{j-1}}{\Delta x} = \dfrac{1}{\Delta x}(\Delta_-\Phi_j) = u_j$, which approximates $\Phi_x(x_j) = u(x_j)$, and proceed as follows:

Step 1: $\quad u_t + f(u)_x = 0 \ , \ f(u) = -\tfrac{1}{2}(Ku - \tfrac{1}{2}(\gamma+1)u^2)$

The Cole-Murman conservative first order differencing, [8], approximates

$$f(u)_x \to \frac{1}{\Delta x}(\Delta_+ f(u_j) - \Delta_- \theta_j \Delta_+ f(u_j))$$

with the switch

$$\theta_j \equiv \begin{cases} 1 & \text{if } u_{j+1} + u_j \geq 2\bar{u} \\ 0 & \text{if } u_{j+1} + u_j < 2\bar{u} \end{cases}$$

for $\bar{h} = K/\gamma+1$, the sonic point.

This admits nonphysical entropy violating shocks as shown by Jameson [8]. In [3] a new switch is introduced.

$$f(u)_x \to \frac{1}{\Delta x}[\Delta_+ f_-(u_j) + \Delta_- f_+(u_j)]$$

$$\begin{cases} f_+(u_j) = f(\max(u_j,\bar{u})) \\ f_-(u_j) = f(\min(u_j,\bar{u})) \end{cases}$$

which agrees with Cole-Murman away from sonic points. The full difference scheme:

$$u_j^{n+1} = u_j^n - \frac{\Delta t}{\Delta x}[\Delta_+ f_-(u_j^n) + \Delta_- f_+(u_j^n)]$$

with CFL number 1, is shown theoretically in [3], [4], [5] to have properties (2)-(5) above.

Step 2: $\quad 2\Phi_{tx} - \Phi_{yy}$.

Let

$$\Delta x\, u_{jk\ell}^n = \Delta_-^x \Phi_{jk\ell}^n \ , \ \Delta y\, v_{jk\ell}^n = \Delta_-^y \Phi_{jk\ell}^n$$

and use the differencing:

$$u^{n+1}_{jk\ell} = u^n_{jk\ell} + \tfrac{1}{4}\frac{\Delta t}{\Delta y}\,\Delta^y_+\left(v^{n+1}_{jk\ell} + v^n_{jk\ell}\right) \ .$$

where $\Phi^n_{jk\ell} \approx \Phi(k_j, y_k, z_\ell, t^n)$.

Similarly for Step 3.

Both of these linear implicit steps are unconditionally stable and easily invertible.

For second order scheme, in Step 1, approximate

$$\partial_x f(u) \rightarrow (\Delta_+ f_-(u_j) - \tfrac{1}{2}\Delta_+(f'_-(z_j)\Delta_+ u_j)$$

$$+ \Delta_- f_+(u_j) + \tfrac{1}{2}\Delta_-(f'_+(w_j)\Delta_- u_j))/\Delta x$$

$$z_j = \begin{cases} \max(u_j, u_{j+1}) & \text{if } u_{j-1} < \bar{u} \\ \bar{u} & u_{j-1} \geq \bar{u} \end{cases}$$

$$w_j = \begin{cases} \min(u_j, u_{j+1}) & \text{if } u_{j+1} > \bar{u} \\ \bar{u} & u_{j+1} \leq \bar{u} \ . \end{cases}$$

Properties 1 - 5 inclusive are now valid for the full scheme in Step 1, where a Lax-Wendroff upwind type of time discretization is used in order to preserve second order accuracy.

Steps 2 and 3 are easily modified to be second order accurate and a Strang type of splitting [9] preserves full second order accuracy in many dimensions.

The design principle behind the construction of these schemes is intimately connected to the proof of the associated stability and entropy theorems. We thus expect to be able to extend these results to the full unsteady potential equation.

It also appears that the method which approximates time dependent problems serves as a relaxation method for the steady problem, i.e. if the boundary conditions are time independent, the numerical time iteration converges to a solution of the steady problem.

REFERENCES

[1] Ballhaus, W. F., Jameson A. and Albert J., (1977). "Implicit Approximate-Factorization Schemes for the Efficient Solution of Steady Transonic Flow Problems", NASA Technical Memorandum NASA TM X-73, 202.

[2] Engquist, B. and Majda A., (1979). "Radiation Boundary Conditions for Acoustic and Elastic Wave Calculations", Comm. Pure Appl. Math., V. 32, 313-357.

[3] Engquist B. and Osher S. "Stable and Entropy Satisfying Approximations for Flow Calculations", Math. Comp., (to appear).

[4] Engquist, B. and Osher, S. "One Sided Difference Schemes and Transonic Flow", Submitted to Proc. Nat. Acad. Sciences.

[5] Engquist, B. and Osher, S. "One Sided Difference Approximations for Nonlinear Conservation Laws, (in preparation).

[6] Goorjian, P., (1979). "Computations of Unsteady Transonic Flow Governed by the Conservative Full Potential Equation Using an Alternating Direction Implicit Algorithm", Informatics Inc. Preprint. CR - 152274.

[7] Holst, T. and Albert, J., (1978). "An Implicit Algorithm for the Conservative, Transonic Full Potential Equation Using an Arbitrary Mesh", AIAA paper 78 - 1113.

[8] Jameson, A., (1976). "A Numerical Solution of Nonlinear Partial Differential Equations of Mixed Type", Numerical Solutions of Partial Differential Equations, III., (Academic Press), 275-320.

[9] Strang, G., (1968). "On the Construction and Comparison of Difference Schemes", SINUM, V. 5, 506-517.

COMPUTING METHODS IN APPLIED SCIENCES AND ENGINEERING
R. Glowinski, J.L. Lions (editors)
North-Holland Publishing Company

AN IMPLICIT ALGORITHM FOR THE TRANSONIC FULL-POTENTIAL EQUATION IN CONSERVATIVE FORM

Terry Holst
Ames Research Center, NASA
Moffett Field, California 94035, U.S.A.

I. INTRODUCTION

Implicit approximate factorization (AF) algorithms for solving the low-frequency (unsteady) transonic small-disturbance equation were investigated in reference 1. Among the techniques studied were versions of the classical Peaceman-Rachford alternating direction implicit (ADI) algorithm (referred to as AF1 in ref. 1) and another type of AF algorithm referred to as AF2. Both two- and three-dimensional schemes were discussed. The AF2 algorithm has been subsequently applied to the solution of steady flows for several different formulations, including the two-dimensional transonic small-disturbance potential equation (ref. 2), the two-dimensional transonic full-potential equation (refs. 3-5), and the three-dimensional transonic full-potential equation (ref. 6). For the steady formations, all of which utilized a relaxation form of the AF2 algorithm, significant improvement in convergence speed has been experienced relative to the standard transonic solution procedure, successive-line overrelaxation (SLOR).

Several guidelines for the construction of the implicit AF schemes can be conveniently introduced by considering the general two-level interaction scheme

$$NC^n + \omega L\phi^n = 0 \tag{1}$$

where C^n is the correction $(\phi^{n+1} - \phi^n)$, L is the residual operator, and ω is a relaxation parameter. The iteration scheme given by equation (1) can be considered as an iteration in pseudotime, where the n superscript indicates the time-step level of the solution; that is, $(\ )^{n+1} - (\ )^n \sim \Delta t(\ )_t$. The nth level residual $(L\phi^n)$ is an indication of how well the finite-difference equation is satisfied by the nth level velocity-potential solution (ϕ^n). The residual operator is a particularly good indicator of high-frequency error content in the solution but is a poor measure of low-frequency error content (refs. 3, 7). More discussion of this point will be presented in the section on computed results.

The N operator of equation (1) determines the type of iterative procedure and, therefore, determines the rate at which the solution procedure converges. Classical successive overrelaxation schemes (SOR) or SLOR schemes effectively use only a portion of the L operator in forming the N operator. As a consequence, the iteration scheme is relatively simple, but the convergence rate may be very slow. In the present AF approach, the philosophy is to choose for N a form that closely approximates L. This in theory will produce a scheme with good convergence characteristics. The procedure for obtaining N consists of two steps: (1) linearize L and (2) factor the linearized result. There are usually two factors for two-dimensional algorithms and three factors for three-dimensional algorithms. The resulting scheme retains the simplicity of having to perform only narrow-banded matrix inversions (for the present case bidiagonal and tridiagonal). The effects of both the error terms resulting from the AF and the linearization are removed from the solution simultaneously by means of the iteration scheme. Because each grid point is influenced by every other grid point during each iteration, much faster convergence can be obtained.

Stability in the present full-potential formulation for supersonic regions of flow has been achieved by an upwind bias of the density coefficient. This procedure is effectively the same as the addition of an artificial viscosity term similar to that introduced in reference 8. Use of the upwind biased density coefficient greatly simplifies the solution procedure and effectively allows the simple two- and three-banded matrix form of the AF scheme to be retained over the entire flow field, even in regions of supersonic flow. Other studies (refs. 9-11) have used similar steady-state differencing procedures in a wide variety of problems to further substantiate this differencing procedure as being both reliable and flexible.

In the development of the new three-dimensional algorithm, special attention has been given to the treatment of "exact" wing surface boundary conditions (i.e., transformation of the wing surface to a constant coordinate surface of the computational domain). The elliptic-solver grid generation technique of reference 12 has been adapted to the present algorithm because of its flexibility. This technique has very few limitations and, for instance, can be used to directly model wind-tunnel walls.

II. GOVERNING EQUATIONS

The three-dimensional full-potential equation written in strong conservation-law form is given by

$$(\rho\phi_x)_x + (\rho\phi_y)_y + (\rho\phi_z)_z = 0 \tag{2a}$$

$$\rho = \left[1 - \frac{\gamma - 1}{\gamma + 1}(\phi_x{}^2 + \phi_y{}^2 + \phi_z{}^2)\right]^{1/\gamma-1} \tag{2b}$$

The density (ρ) and velocity components (ϕ_x, ϕ_y, and ϕ_z) are nondimensionalized by the stagnation density (ρ_s) and the critical sound speed (a_*), respectively; x, y, and z are Cartesian coordinates in the streamwise, spanwise, and vertical directions, respectively; and γ is the ratio of specific heats. The two-dimensional conservation-law form of the full-potential equation is simply obtained by dropping all y-derivative terms from equation (2).

Equation (2) expresses mass conservation for flows that are steady, isentropic, and irrotational. The corresponding shock-jump conditions are valid approximations to the Rankine-Hugoniot relations for many transonic flow applications. A comparison of isentropic and Rankine-Hugoniot shock polars is given in reference 13.

Equation (2) is transformed from the physical domain (Cartesian coordinates) to the computational domain by using a general independent variable transformation. This transformation $[(\xi,\eta,\zeta) \leftrightarrow (x,y,z)$, see fig. 1], maintains the strong conservation-law

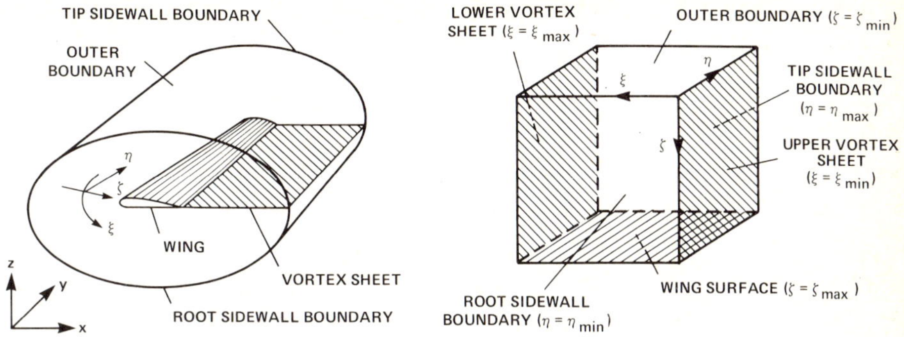

(a) Physical domain. (b) Computational domain.

Figure 1. Schematic of general $(x,y,z) \leftrightarrow (\xi,\eta,\zeta)$ transformation.

form of equation (2) as discussed in references 14-17. The full-potential equation written in the computational domain (ξ-η-ζ coordinate system) is given by

$$\left(\frac{\rho U}{J}\right)_\xi + \left(\frac{\rho V}{J}\right)_\eta + \left(\frac{\rho W}{J}\right)_\zeta = 0 \tag{3}$$

$$\rho = \left[1 - \frac{\gamma - 1}{\gamma + 1}\,(U\phi_\xi + V\phi_\eta + W\phi_\xi)\right]^{1/\gamma-1} \tag{4}$$

where

$$\left.\begin{aligned}
&U = A_1\phi_\xi + A_4\phi_\eta + A_5\phi_\zeta\ , \quad V = A_4\phi_\xi + A_2\phi_\eta + A_6\phi_\zeta\ , \quad W = A_5\phi_\xi + A_6\phi_\eta + A_3\phi_\zeta\\
&A_1 = \xi_x{}^2 + \xi_y{}^2 + \xi_z{}^2\ , \quad A_2 = \eta_x{}^2 + \eta_y{}^2 + \eta_z{}^2\ , \quad A_3 = \zeta_x{}^2 + \zeta_y{}^2 + \zeta_z{}^2\\
&A_4 = \xi_x\eta_x + \xi_y\eta_y + \xi_z\eta_z\ , \quad A_5 = \xi_x\zeta_x + \xi_y\zeta_y + \xi_z\zeta_z\ , \quad A_6 = \eta_x\zeta_x + \eta_y\zeta_y + \eta_z\zeta_z\\
&\text{and}\\
&J = \xi_x\eta_y\zeta_z + \xi_y\eta_z\zeta_x + \xi_z\eta_x\zeta_y - \xi_z\eta_y\zeta_x - \xi_y\eta_x\zeta_z - \xi_x\eta_z\zeta_y
\end{aligned}\right\} \tag{5}$$

U, V, and W are contravariant velocity components along the ξ, η, and ζ directions, respectively; $A_1 - A_6$ are metric quantities; and J is the Jacobian of the transformation. The two-dimensional form of the full-potential equation written in the computational domain (ξ-ζ coordinates) is obtained by dropping all y and η terms in equations (3)-(5); that is, all y and η derivatives as well as all derivatives of y and η are set equal to zero. An exception to this is that y_η and η_y must be set equal to 1.

The transformed full-potential equation (eqs. (3) and (4)) is only slightly more complicated than the original Cartesian form (eq. (2)), provided the quantities A_1-A_6 and J are considered as known constants. Several significant advantages are offered by this very general form. The main advantage is that boundaries associated with the physical domain are transformed to boundaries of the computational domain. This aspect is illustrated in figure 1 where the physical and computational domains for a typical transformation are shown. The computational coordinates, ξ, η, and ζ, are in the wraparound, spanwise, and radial-like directions, respectively. The inner wing boundary transforms to $\zeta = \zeta_{max}$ and the outer physical boundary transforms to $\zeta = \zeta_{min}$. Note that no restrictions have been placed on the shape of the outer boundary. Arbitrarily shaped outer boundaries, including wind-tunnel walls, may be used. The symmetry-plane boundary transforms to $\eta = \eta_{min}$, and the wing-tip boundary transforms to $\eta = \eta_{max}$. The last two sides of the computational domain are formed from the upper and lower cuts along the vortex sheet. Additional details concerning the general $(x,y,z) \rightarrow (\xi,\eta,\zeta)$ transformation used in the present study are given in reference 7.

III. GRID GENERATION

The grid-generation scheme used in the present three-dimensional formulation is a simple extension of the two-dimensional scheme presented in reference 4, and is described in reference 7. Basically, this scheme uses numerically generated solutions of Laplace's equation (or in some cases, Poisson's equation) to establish regular and smooth finite-difference meshes around arbitrary bodies. In the present case, the finite-difference mesh for each spanwise plane (η = constant plane) is generated using the standard two-dimensional algorithm. This requires solution of the following two Laplace equations in each spanwise plane:

$$\xi_{xx} + \xi_{zz} = 0 \ , \qquad \zeta_{xx} + \zeta_{zz} = 0 \tag{6}$$

These equations are transformed to (and solved in) the computational domain, that is, ξ and ζ are the independent variables, and x and z are the dependent variables. A fast approximate factorization relaxation algorithm is used to solve the resulting transformed equations (ref. 4). This establishes values for x and z in each spanwise plane. Coordinate values in the spanwise direction (y values) are established by a simple stretching formula. Given the values of x, y, and z at each grid point, the values of the metric quantities (eq. (5)) are easily computed using standard finite-difference formulas. Details of this procedure are given in reference 7.

IV. FULL-POTENTIAL EQUATION ALGORITHM

Spatial Differencing

A finite-difference approximation to equation (3), suitable for both subsonic and supersonic flow regions is given by

$$\overleftarrow{\delta}_\xi\left(\frac{\tilde{\rho}U}{J}\right)_{i+1/2,j,k} + \overleftarrow{\delta}_\eta\left(\frac{\bar{\rho}V}{J}\right)_{i,j+1/2,k} + \overleftarrow{\delta}_\zeta\left(\frac{\hat{\rho}W}{J}\right)_{i,j,k+1/2} = 0 \tag{7}$$

The operators

$$\overleftarrow{\delta}_\xi(\ ) \ , \quad \overleftarrow{\delta}_\eta(\ ) \ , \quad \text{and} \quad \overleftarrow{\delta}_\zeta(\ ) \tag{8}$$

are the first-order-accurate backward-difference operators in the ξ, η, and ζ directions, respectively, and are defined by

$$\overleftarrow{\delta}_\xi(\ )_{i,j,k} = (\ )_{i,j,k} - (\ )_{i-1,j,k}\ , \qquad \overleftarrow{\delta}_\eta(\ )_{i,j,k} = (\ )_{i,j,k} - (\ )_{i,j-1,k}$$
$$\overleftarrow{\delta}_\zeta(\ )_{i,j,k} = (\ )_{i,j,k} - (\ )_{i,j,k-1} \tag{9}$$

The standard $\Delta\xi$, $\Delta\eta$, and $\Delta\zeta$ quantities are all equal to 1 and therefore have been omitted. The density coefficients, $\tilde{\rho}$, $\bar{\rho}$, and $\hat{\rho}$ are upwind evaluations of the density and are defined by

$$\tilde{\rho}_{i+1/2,j,k} = [(1 - \nu)\rho]_{i+1/2,j,k} + \nu_{i+1/2,j,k}\rho_{i+r+1/2,j,k} \tag{10a}$$
$$\bar{\rho}_{i,j+1/2,k} = [(1 - \nu)\rho]_{i,j+1/2,k} + \nu_{i,j+1/2,k}\rho_{i,j+s+1/2,k} \tag{10b}$$
$$\hat{\rho}_{i,j,k+1/2} = [(1 - \nu)\rho]_{i,j,k+1/2} + \nu_{i,j,k+1/2}\rho_{i,j,k+t+1/2} \tag{10c}$$

The density is computed in a straightforward manner by using equation (4). The quantity ν is a switching function controlling the level of upwinding in the spatial difference scheme. Further discussion about the calculation of density and the definition of ν will be presented subsequently. The r, s, and t indices control the upwind direction and are defined by

$$\begin{aligned} r &= \mp 1 \quad \text{when} \quad U_{i+1/2,j,k} \gtrless 0 \\ s &= \mp 1 \quad \text{when} \quad V_{i,j+1/2,k} \gtrless 0 \\ t &= \mp 1 \quad \text{when} \quad W_{i,j,k+1/2} \gtrless 0 \end{aligned} \tag{11}$$

The differencing scheme, given by equations (7)-(11), maintains an upwind influence for supersonic regions anywhere in the general ξ-η-ζ finite-difference mesh for any orientation of the velocity vector. Thus, the effect of rotated differencing is closely approximated (ref. 5). This aspect greatly contributes to the stability and reliability of the present algorithm for many difficult strong shock-wave cases.

Two variations for computing the density have been studied using the two-dimensional version of the algorithm. (The two-dimensional algorithm is obtained by eliminating all η terms and dropping j subscripts in eqs. (7)-(11).) The first variation involves the calculation of the density at mesh points (i,k) (refs. 4,5). Simple averages are then used to obtain the density at the required half points i+1/2,k and i,k+1/2. The second variation involves the calculation of density at the mesh cell centers (i+1/2, k+1/2) (refs. 7,18). Simple averages are used again to obtain the density at the required half points. In each case, values of ϕ_ξ, ϕ_ζ, U, and W required in the density calculation are computed with standard second-order-accurate finite-difference formulas. Calculation of the density at the cell centers produces sharper shock waves and better resolution of the re-expansion singularity at the foot of the shock wave. This is because the computational module for the latter version extends over fewer grid points and thereby causes less smearing of the solution. The first variation for computing the density, by virtue of its increased dissipation, is more easily stabilized for strong shock-wave calculations. Density values in the three-dimensional scheme are computed and stored at i+1/2,j,k. Values of the density required at $i,j+1/2,k$ and $i,j,k+1/2$ are determined by simple four-element averages. Values of the density required in the ξ-derivative term of equation (7) (i.e., $\rho_{i+1/2,j,k}$) are computed with the smallest possible computational module in the ξ direction. Therefore, shock waves should be optimally captured in regions where the ξ coordinate is nearly streamwise.

Use of the present upwinded density coefficients in an otherwise centrally differenced scheme, is equivalent to adding an artificial-viscosity-type term to a standard centrally differenced scheme. The equivalent artificial viscosity term is given by

$$-\Delta\xi\left(\nu\rho_\xi \frac{|U|}{J}\right)_\xi - \Delta\eta\left(\nu\rho_\eta \frac{|V|}{J}\right)_\eta - \Delta\zeta\left(\nu\rho_\zeta \frac{|W|}{J}\right)_\zeta \tag{12}$$

To obtain the scheme of equation (7), the ρ_ξ, ρ_η, and ρ_ζ derivatives of equation (12) must be differenced with a backward (forward) difference when U, V, and W are positive (negative), respectively.

The artificial viscosity coefficient ν strongly affects the stability of the present scheme and is defined as follows:

$$\nu_{i+1/2,j,k} = \begin{cases} \max[(M^2_{i,j,k} - 1)C,0] & \text{for } U_{i+1/2,j,k} > 0 \\ \max[(M^2_{i+1,j,k} - 1)C,0] & \text{for } U_{i+1/2,j,k} < 0 \end{cases} \tag{13}$$

The parameter C is a user-specified constant and is usually set between 1.0 and 2.0. Use of larger values of C increases the amount of upwinding and, therefore, the effective amount of artificial viscosity added to the difference scheme. For subsonic regions ν is zero, making the spatial difference scheme entirely second-order-accurate and centrally differenced. In supersonic regions, as the Mach number increases, the density coefficients and, therefore, the spatial-differencing scheme are increasingly retarded in the upwind direction. An additional constraint placed on ν is $\nu \le 1$. This improves stability and, in some cases, improves the convergence rate. Expressions for ν at $i,j+1/2,k$ and $i,j,k+1/2$ are required in equations (10b) and (10c) and are defined similarly to $\nu_{i+1/2,j,k}$.

AF2 Iteration Scheme

The AF2 fully implicit approximate factorization scheme applied to the three-dimensional full-potential equation (refs. 1,6), can be expressed by choosing the N operator of equation (1) as follows:

$$\alpha N C^n_{i,j,k} = -\left[\left(\alpha - \frac{1}{A_k}\overleftarrow{\delta}_\eta A_j \overleftarrow{\delta}_\eta\right)\left(A_k - \frac{1}{\alpha}\overrightarrow{\delta}_\xi A_i \overleftarrow{\delta}_\xi\right) - \alpha E^{+1}_\zeta A_k\right](\alpha + \overleftarrow{\delta}_\zeta) C^n_{i,j,k} \tag{14}$$

where

$$A_i = \left(\frac{\tilde{\rho}A_1}{J}\right)^n_{i-1/2,j,k}, \quad A_j = \left(\frac{\bar{\rho}A_2}{J}\right)^n_{i,j-1/2,k}, \quad A_k = \left(\frac{\hat{\rho}A_3}{J}\right)^n_{i,j,k-1/2} \tag{15}$$

The $\tilde{\rho}$, $\bar{\rho}$, and $\hat{\rho}$ coefficients are defined by equations (10a) to (10c), α is a free parameter defined subsequently, and the operator E^{+1}_ζ is a shift operator given by

$$E_\zeta^{+1}(\ )_{i,j,k} = (\ )_{i,j,k+1} \tag{16}$$

Note that one form of the two-dimensional AF2 iteration scheme is obtained from equation (14) by simply setting the η difference equal to zero (ref. 4).

Multiplying out the three factors of equation (14) yields an approximation to the L operator defined by equation (7) plus a number of error terms. This approximate L operator does not have the mixed-derivative terms contained in the exact L operator and has been effectively linearized; that is, all the coefficients used in equation (14), A_i, A_j and A_k, are evaluated at the nth iteration level. In spite of these approximations, unconditional linear stability exists for this scheme; it is discussed in the appendix. (See also refs. 1 and 4 for additional discussion on the stability of the present algorithm.) Because of the implicit construction of this scheme, each point in the finite-difference mesh influences every other point during each iteration. As a result, evolution of the solution proceeds at a much faster rate, relative to explicit or semi-implicit algorithms.

Implementation of the AF2 scheme is achieved by writing it in a three-step form given by

Step 1

$$\left(\alpha - \frac{1}{A_k}\vec{\delta}_\eta A_j \overleftarrow{\delta}_\eta\right) g_{i,j}^n = \alpha\omega L\phi_{i,j,k}^n + \alpha A_{k+1} f_{i,j,k+1}^n \tag{17}$$

Step 2

$$\left(A_k - \frac{1}{\alpha}\vec{\delta}_\xi A_i \overleftarrow{\delta}_\xi\right) f_{i,j,k}^n = g_{i,j}^n \tag{18}$$

Step 3

$$(\alpha + \overleftarrow{\delta}_\zeta) C_{i,j,k}^n = f_{i,j,k}^n \tag{19}$$

Here, ω is a relaxation factor equal to 1.8 for all cases presented; $g_{i,j}^n$ is an intermediate result stored at each grid point in a given k plane, that is, g requires only a two-dimensional array of storage; and $f_{i,j,k}^n$ is an intermediate result stored at each point in the finite-difference mesh. In step 1, the g array is obtained by solving a tridiagonal matrix equation for each ξ = constant line in the kth plane. In step 2, the f array is obtained from g by solving a tridiagonal matrix equation for each η = constant line, again for just the kth plane. Next, step 1 is used to obtain the g array for the k + 1 plane, and then step 2 is used to obtain the f array for the k + 1 plane, etc. This process continues until all values of f in the three-dimensional mesh are established. Then, by using step 3, the correction array is obtained from the f array by solving a simple bidiagonal matrix equation for each ζ = constant line in the entire finite-difference mesh. With this sweeping procedure all coefficients, A_i, A_j, and A_k are required only once per iteration and, therefore, do not have to be stored in three-dimensional arrays. The nature of this AF2 factorization places a sweep-direction restriction on the step 1-2 combination and on step 3. The step 1-2 combination must be swept in the direction of the decreasing k subscript, that is, from the wing boundary toward the outer boundary (see fig. 1). The step 3 sweep must proceed in just the opposite direction, that is, from the outer boundary toward the wing. There are no sweep-direction limitations placed on any of the three sweeps due to flow direction.

Initiation of step 1 at the wing boundary ($k = NK$) requires knowledge of f at $NK + 1$ which is generally unobtainable. A simple solution is to set f at $NK + 1$ equal to zero. Because the present iteration scheme is written in the correction form, f must approach zero as the solution converges. This boundary condition is therefore consistent with the steady-state solution and seems to provide acceptable performance. A similar boundary condition is required for g at $\eta = \eta_{min}(j = 1)$ and $\eta = \eta_{max}(j = NJ)$ and is implemented by imposing $(g_\eta)_{i,1} = (g_\eta)_{i,NJ} = 0$.

Temporal Damping

For the AF2 factorization, the N operator must be written so that either the ξ-, η-, or ζ-difference approximation to the full-potential equation is split between two factors. This construction generates either a $\phi_{\xi t}$-, $\phi_{\eta t}$-, or $\phi_{\zeta t}$-type term, and if it is properly upwind-differenced (ref. 19) provides time-dependent dissipation to the convergence process. When a particular coordinate direction is split (e.g., the ξ direction), the resulting $\phi_{\xi t}$ difference direction is fixed by the construction of the AF2 algorithm; that is, the term is either backward or forward differenced over the entire mesh. Due to the wraparound ξ coordinate, a backward- (forward-) differenced $\phi_{\xi t}$ is upwind (downwind) differenced below the wing and downwind (upwind) differenced above. Therefore, a problem with $\phi_{\xi t}$ arises either above or below the wing. Following the two-dimensional algorithm development (ref. 4) the ζ-difference approximation is split between two factors. This allows control over the other more important coordinate directions (ξ and η) because the $\phi_{\xi t}$ and $\phi_{\eta t}$ terms are added to the iteration scheme explicitly and are not part of the factorization construction. The $\phi_{\eta t}$ and $\phi_{\xi t}$ terms are included by adding

$$\mp\beta_\eta \left|V_{i,j,k}\right| \overleftrightarrow{\delta}_\eta \quad \text{and} \quad \mp\alpha\beta_\xi \overleftrightarrow{\delta}_\xi \tag{20}$$

inside the brackets of the first and second sweeps, equations (17) and (18), respectively. The parameter β_ξ is fixed to 0.0 in subsonic regions and specified as needed in supersonic regions. The parameter β_η is a user-specified constant, fixed over the entire mesh. For all cases presented in the present study $\beta_\eta = 0$. The double arrow notation on the ξ- and η-difference operators indicates that the difference is always upwind. For the η direction, a backward difference is used when the η contravariant velocity component ($V_{i,j,k}$) is positive; a forward difference is used when V is negative. The sign is chosen in each case so that the addition of $\phi_{\eta t}$ and $\phi_{\xi t}$ increases the magnitude of the first- and second-sweep diagonal coefficients, respectively. More discussion about temporal damping can be found in references 4 and 6.

Selection of α for the AF2 Algorithm

The quantity α appearing in equation (14) is an as yet undefined free parameter. If α was chosen to be Δt^{-1}, then the AF2 scheme could be considered as an iteration in pseudotime. As shown in the appendix, unconditional linear stability exists for the AF2 algorithm providing α remains positive. Thus, one strategy for obtaining fast convergence is to advance time as fast as possible with large time steps (i.e., small α's). In practice, however, a nonlinear stability limitation on the time-step size does exist, and an optimal value of α must be determined by numerical experiment. As pointed out in reference 2, use of large time steps is only effective for treating the low-frequency errors, not the high-frequency errors. An improved approach is to use an α sequence containing several values of α. The small values are effective for reducing the low-frequency errors, and the large values are effective for reducing the high-frequency errors. End points for a suitable α sequence can be approximated analytically (ref. 2), $\alpha_H \simeq 1/\Delta$, $\alpha_L \simeq 1$, where Δ is a representative mesh cell spacing in the finite-difference mesh near the airfoil or wing. The α sequence endpoints should only be considered as order-of-magnitude estimates. Refinement of these estimates by numerical experiment is required before optimal efficiency of the AF2 iteration procedure is obtained. The α sequence used presently is given by

$$\alpha_k = \alpha_H(\alpha_L/\alpha_H)^{(k-1)/(M-1)} \tag{21}$$

where $k = 1, 2, \ldots, M$, and M is the number of elements in the sequence. This sequence has been found to be effective (refs. 2-6), although other sequences perhaps could provide equivalent or improved convergence performance.

Boundary Conditions

The wing surface boundary condition is that of flow tangency (i.e., no flow through the wing surface), and requires the ζ contravariant velocity component at the wing surface to be zero ($W = 0$). This boundary condition is implemented by applying

$$\left(\frac{\rho W}{J}\right)_{i,j,NK+1/2} = -\left(\frac{\rho W}{J}\right)_{i,j,NK-1/2} \tag{22}$$

where $k = NK$ is the wing surface. In expressions where ϕ_ζ is required at the wing surface, the $W = 0$ boundary condition is used again to obtain

$$\phi_\zeta\Big|_{wing} = -\frac{A_5}{A_3}\phi_\xi - \frac{A_6}{A_3}\phi_\eta \tag{23}$$

Thus, a value of ϕ_ζ at the wing surface can be obtained without using a one-sided difference on ϕ.

In the present study, a special wing geometry has been chosen to evaluate the new three-dimensional AF2 algorithm, namely, flow past an arbitrary wing mounted between parallel walls. The purpose of this model problem is to simulate the flow past a wing in a wind tunnel. The parallel side walls are treated with the same tangency boundary condition used for the wing surface ($V = 0$). Details about the implementation of this boundary condition are given in reference 6.

V. COMPUTED RESULTS

The implicit algorithm discussed in the previous section has been coded into a transonic airfoil analysis computer code (TAIR) and a transonic wing analysis computer code (TWING). Details of the TAIR and TWING computer codes are given in references 4 and 6, respectively. Several numerically computed examples using these codes are presented in this section. The two-dimensional results from TAIR were all computed in the default mode. This simply means that all parameters affecting the convergence rate including the relaxation factor (ω), the acceleration parameters (α_k), and the temporal damping coefficient (β_ξ) are either held fixed or are adjusted automatically by internal computer code logic. This feature greatly simplifies operation of TAIR and improves reliability, especially for inexperienced users. However, the convergence speed of TAIR in the default mode is below optimum by about 10% to 50%, depending on the particular case. For more details about the default mode feature see reference 6.

The first two-dimensional test case involves the Korn airfoil (airfoil 75-06-12, ref. 20) at a free-stream Mach number of 0.74 and zero degrees angle of attack. The pressure coefficient distribution for this slightly off-design case is compared in figure 2 with a result from the GRUMFOIL computer code (ref. 21). Both calculations are in excellent agreement. The GRUMFOIL computer code is similar to TAIR in that both codes solve the conservative full-potential equation, but different in that TAIR uses the AF2 iteration scheme and GRUMFOIL uses a hybrid direct-solver/SLOR iteration scheme (ref. 8). This hybrid iteration scheme is composed of one direct-solver iteration (which is very effective for reducing low-frequency errors but is unstable for supersonic regions) followed by several (10 is the default) SLOR iterations. The purpose of the SLOR iterations is to smooth high-frequency errors generated by the direct-solver step in regions of supersonic flow. The boundary-layer option which is available in GRUMFOIL has not been used in any of the results presented herein. All GRUMFOIL results are computed on a 148 × 32 mesh; the TAIR results are computed on a 149 × 32 mesh.

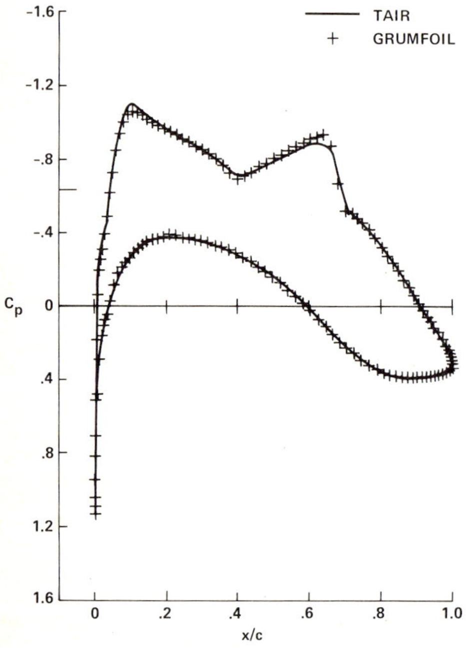

Figure 2. Pressure coefficient comparison (Korn airfoil, M_∞ = 0.74, α = 0°).

The rms error (E_{rms}) convergence-history curves for the Korn airfoil calculation are presented in figure 3. The E_{rms} at iteration n (E^n_{rms}) is defined by

$$E^n_{rms} = \left[\frac{\sum_{i=1}^{NI}\left(C^n_{P_i} - C_{P_i}\right)^2}{\sum_{i=1}^{NI} C^2_{P_i}}\right]^{1/2} \qquad (29)$$

where $C^n_{P_i}$ is the surface pressure coefficient at the ith grid point and the nth iteration, C_{P_i} is the surface pressure coefficient at the ith grid point taken from the converged solution, and NI is the total number of surface grid points. Using E_{rms} to compare convergence performance is a much more quantatively correct procedure than using the standard maximum residual quantity. This situation is the result of two factors (refs. 3,7). First, AF schemes (as well as other fully implicit schemes) treat all error components equally well (approximately), whereas SOR and SLOR schemes (as well as other explicit or semi-implicit schemes) perform efficiently on only high-frequency errors. Second, it can be shown that the residual is a weighted sum of errors, over the entire error frequency spectrum, weighted by the eigenvalue of the finite-difference scheme. The eigenvalue for high-frequency errors is $O(\Delta X^{-2})$ and for the low-frequency errors is O(1) (ref. 7). Hence, the maximum residual operator is heavily influenced by the high-frequency errors. Therefore, the maximum residual operator should not be used for comparing iteration schemes with basically different characteristics. Root-mean-square errors are much better suited for this purpose and will be used here. More discussion on this topic can be found in references 3 and 7.

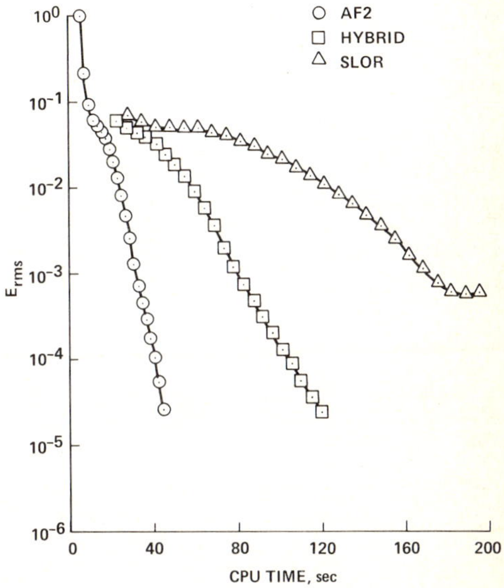

Figure 3. Two-dimensional convergence histories (Korn airfoil, $M_\infty = 0.74$, $\alpha = 0°$).

The three curves shown in figure 3 correspond to the following iteration schemes: (1) AF2, (2) hybrid, and (3) SLOR. The SLOR scheme is simply the hybrid iteration scheme without benefit of the direct-solver step. Convergence for the SLOR scheme has been approximately optimized by a trial-and-error adjustment of the relaxation parameter. Each convergence-history curve is constructed by plotting E_{rms} versus CPU time (Ames CDC 7600 computer). The hybrid case has been computed with default values for all relaxation parameters. Setup times, that is, the CPU time required for grid generation, initialization, and coarse- and medium-mesh calculations, are included in each convergence-history curve. The AF2 curve includes 6 sec for grid generation and initialization. The hybrid and SLOR curves both use coarse-medium-fine mesh sequences. Converged results from the coarse mesh are interpolated onto the medium mesh, and then from the medium mesh onto the fine mesh, thus providing a good initial guess for the fine-mesh calculation. The setup times for these cases are 23 sec for the hybrid case and 28 sec for the SLOR case. For this calculation the AF2 scheme is about 2.5 times faster than the hybrid scheme and about 5 times faster than SLOR.

The second two-dimensional test case involves the NACA 0012 airfoil at a free-stream Mach number of 0.82 and zero degrees angle of attack. The pressure coefficient distribution for this case is compared in figure 4 with a result from the GRUMFOIL computer code. The two results are in excellent agreement. The rms error convergence-history curves for this calculation are presented in figure 5. The three curves correspond to the AF2, hybrid, and SLOR iteration schemes. Setup times for each iteration scheme have been included as before. For this calculation AF2 is about 5 times faster than the hybrid scheme and about 8 times faster than SLOR.

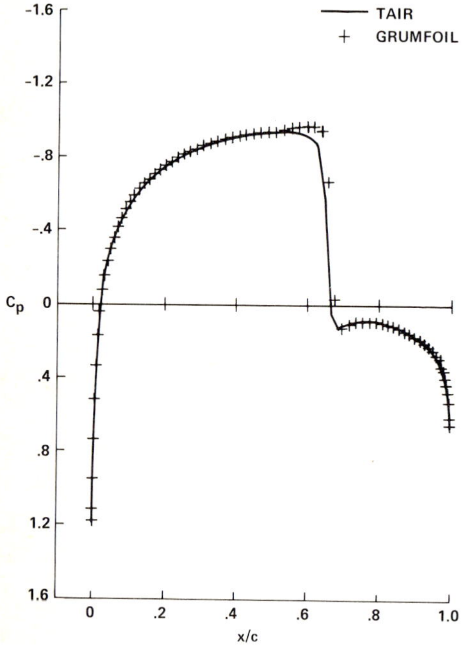

Figure 4. Pressure coefficient comparison (NACA 0012 airfoil, $M_\infty = 0.82$, $\alpha = 0°$).

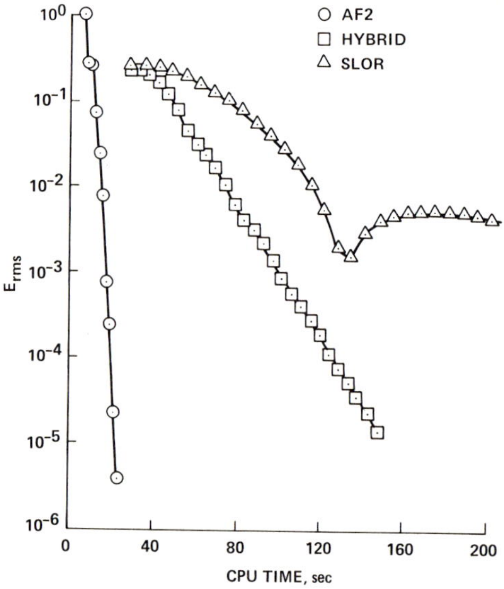

Figure 5. Two-dimensional convergence histories (NACA 0012 airfoil, $M_\infty = 0.82$, $\alpha = 0°$).

Solutions with M_∞ Approaching Unity

As the free-stream Mach number approaches unity, interesting airfoil shock-wave patterns develop. These solutions were the subject of discussion in references 5, 6, 9, and 22. An example calculation about an NACA 0012 airfoil at $M_\infty = 0.95$ and $\alpha = 4°$ is shown in figure 6. The Mach number contours clearly show the existence of a so-called "fishtail" shock-wave pattern downstream of the airfoil trailing edge. The angle of attack produces lift ($C_L = 0.43$), and therefore, desymmetrizes the fishtail shock-wave structure. The oblique shock emanating from the trailing edge upper surface has been strengthened while the oblique shock emanating from the trailing edge lower surface has been weakened and is almost nonexistent. The normal shock above the airfoil is much stronger than the normal shock below the airfoil. This difficult calculation demonstrates the convergence reliability of the present transonic solution procedure.

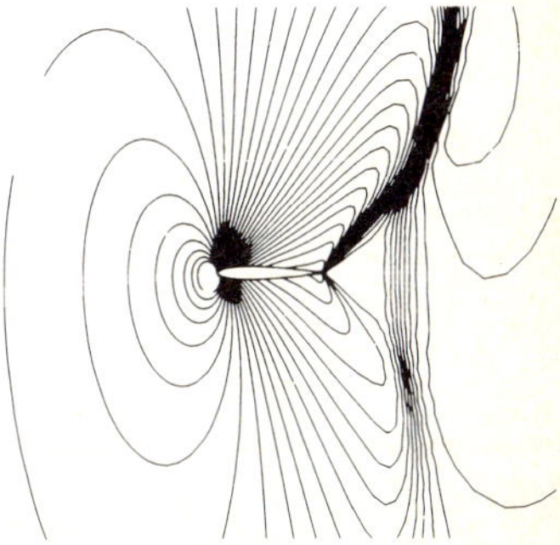

Figure 6. Mach number contours (NACA 0012 airfoil, $M_\infty = 0.95$, $\alpha = 4°$).

Convergence history curves for this case — including E_{rms}, maximum residual ($|R|_{max}$), number of supersonic points (NSP), and C_L convergence histories — are presented in figure 7. Convergence is achieved for this case in approximately 20 sec of CPU time (Ames CDC 7600 computer) (116 iterations) and is indicated by constant values of both NSP and C_L. The steady-state value of NSP is just over 2,000 which represents about 45% of all mesh points. At this point, $|R|_{max}$ has dropped only slightly, while E_{rms} has dropped by over 2.5 orders of magnitude. This discrepancy is partially due to the fact that the airfoil surface solution, which is effectively monitored by E_{rms}, converges before the downstream fishtail shock-wave pattern. Another cause is that during the initial phase of convergence in which the residual does not drop, the position of the shock sonic line is rapidly being adjusted. This excites high-frequency errors, and therefore keeps the

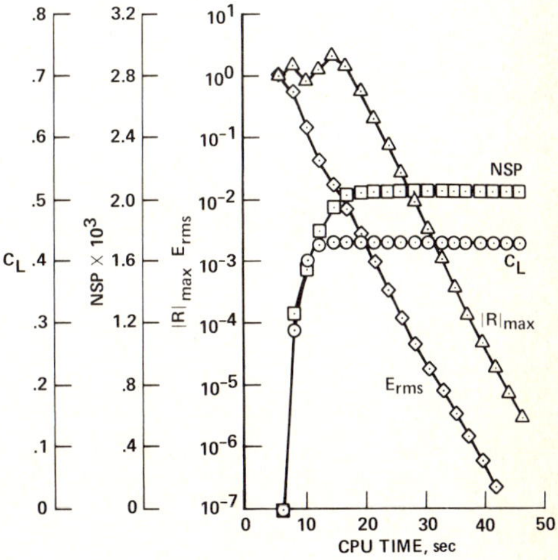

Figure 7. Two-dimensional convergence histories (NACA 0012 airfoil, $M_\infty = 0.95$, $\alpha = 4°$).

residual artificially high even though E_{rms} is being reduced. Establishment of convergence for such a small reduction in the maximum residual is a characteristic behavior of many strong shock calculations using the present algorithm. Use of a three-order-of-magnitude reduction in $|R|_{max}$ as a convergence criteria for the present case would require E_{rms} to be reduced by 5 orders of magnitude, which represents a factor of 2 more iterations than necessary.

Three-Dimensional Solutions

Results from the transonic wing analysis code (TWING) are presented in this section. All calculations have been computed with the density coefficients upwinded in only the ξ and η directions. For the cases presented herein, this was sufficient; however, for cases with stronger shocks at the trailing edge, density upwinding along the ζ direction would probably be required. All results have been computed with $\alpha_L = 0.4$, $\alpha_H = 4.0$, and β_ξ ranging from 0.1 to 0.5. The larger values of β_ξ were required for the larger aspect ratio cases.

To evaluate the three-dimensional code, several infinite-aspect-ratio results have been compared with two-dimensional results using the concept of simple sweep theory. For three-dimensional infinite-aspect-ratio calculations, the solution in the wing-normal plane is the same as a two-dimensional solution with free-stream Mach number, pressure coefficient, and airfoil thickness scaled appropriately, as follows:

$$M_{\infty,2D} = \cos\lambda\, M_{\infty,3D} \;, \qquad C_{P,2D} = C_{P,3D}/\cos^2\lambda \;, \qquad T(x)_{2D} = T(x)_{3D}/\cos\lambda$$

where $T(x)$ is the airfoil thickness distribution. Both subcritical and supercritical results using this concept have been compared and are in excellent agreement. For an actual comparison of two-dimensional and three-dimensional infinite aspect ratio solutions see reference 6.

Figures 8-10 show the results of a nonlifting wing calculation with the following characteristics: NACA 0015 wing, AR = 1.9, $M_\infty = 0.86$, and $\lambda = 30°$. Figures 8 and 9 show the wing planform Mach number contours at 20 and 80 iterations, respectively. In just 20 iterations, a reasonable approximation to the final solution is already established. As expected, the shock-wave approaches both sidewalls perpendicularly. Near center span the shock wave is swept, approximately parallel to the wing leading and trailing edges. The spanwise shock-strength gradient is quite large. This is indicated by both the Mach contours of figure 9 and the surface Cp distributions shown in figure 10. The maximum local Mach numbers in the root, center span, and tip planes are 1.26, 1.33, and 1.78, respectively. A large part of this spanwise shock-strength gradient is caused by the tip sidewall/wing interaction, which is essentially the opposite of a three-dimensional relief effect. The existence of the tip sidewall constrains the streamlines to remain in the tip plane. The wing sweep induces a spanwise component of velocity which in effect squeezes the streamlines toward the tip plane. This increases the Mach number and therefore the shock strength in the tip plane region.

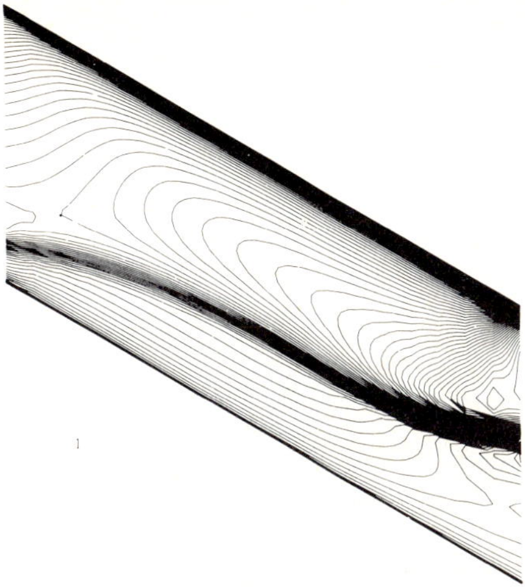

Figure 8. Wing planform Mach contours (20 iterations) (NACA 0015 wing, $M_\infty = 0.86$, $\alpha = 0°$, $\lambda = 30°$, AR = 1.9).

Figure 9. Wing planform Mach contours (fully converged) (NACA 0015 wing, $M_\infty = 0.86$, $\alpha = 0°$, $\lambda = 30°$, AR = 1.9).

The TWING calculation presented herein was computed with a 23,200 (=58 × 20 × 20) point mesh. Because of the wraparound mesh, all 58 points in the ξ direction lie on the wing surface. Each iteration of the three-dimensional algorithm requires about 2.1 sec of CPU time on the Ames CDC 7600 computer. For the case of figure 10, this equates to a total run time of less than 3 min to achieve plottable accuracy. Cases with weaker shock waves are faster. For subcritical cases the run time is faster still, requiring only about a minute of CPU time. These times include CPU time for metric recalculation every iteration. Another version of TWING using disk to store all metric quantities actually saves time by requiring only about 1.8 sec CPU time per iteration. However, the latter case is actually slightly more expensive, because of the extra I/O charges incurred by accessing the disk repeatedly during each iteration.

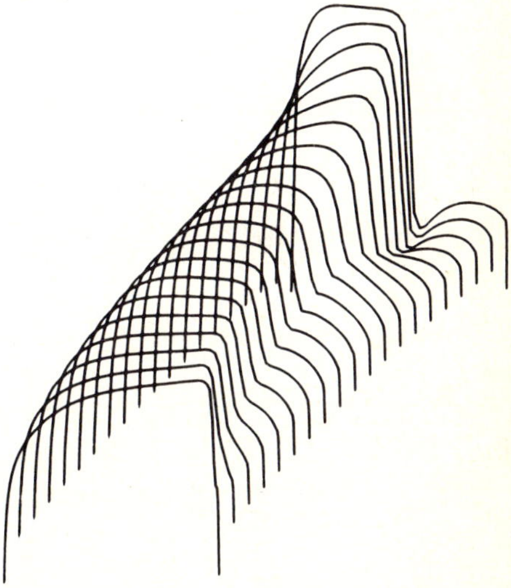

Figure 10. Three-dimensional pressure coefficient distribution (NACA 0015 wing, $M_\infty = 0.86$, $\lambda = 30°$, AR = 1.9, $\alpha = 0°$).

VI. CONCLUSIONS

A fast, implicit algorithm for solving the conservative full-potential equation in both two and three dimensions is presented. Stability in supersonic regions is maintained by using an upwind evaluation of the density coefficient along all coordinate directions. This provides an effective upwind difference of the streamwise terms for any orientation of the velocity vector (i.e., rotated differencing), and thereby greatly enhances the reliability of the present algorithm. The present scheme has been used to compute a number of difficult two-dimensional test cases, including cases with "fishtail" shock-wave patterns. The rapid convergence of these difficult cases demonstrates the reliability and efficiency of the present transonic flow solution procedure.

The present fully implicit AF2 algorithm has been compared with both the standard transonic-solution procedure, successive-line overrelaxation (SLOR), and a hybrid (direct-solver/SLOR) scheme. The surface C_p distributions produced by these schemes are in good agreement. Based on CPU time, the rms error in the surface pressures is reduced from 5 to 8 times faster by the AF2 algorithm relative to SLOR, and from 2.5 to 5 times faster by the AF2 algorithm relative to the hybrid scheme.

A three-dimensional solution for a swept wing mounted between parallel walls is also presented. For this calculation a strong shock wave extends across the entire wingspan indicating a high degree of reliability for the three-dimensional AF2 algorithm. Convergence histories indicate that the convergence rate improvement experienced in two dimensions carries over to the three-dimensional version of the fully implicit algorithm.

ACKNOWLEDGMENTS

The author expresses gratitude to William F. Ballhaus for many helpful suggestions during the course of this study. Many thanks are also extended to Juanita Frick and Alan Fenquist for writing the computer codes described herein and to Karen Gundy for developing the computer graphics shown in this report.

APPENDIX

AF2 STABILITY ANALYSIS

The linearized three-dimensional AF2 stability analysis is presented in this appendix. Consider the following model equation:

$$\phi_{xx} + \phi_{yy} + \phi_{zz} = 0 \tag{A1}$$

The AF2 algorithm applied to equation (A1) yields

$$\left[\left(1 - \frac{1}{\alpha}\vec{\delta}_x\overleftarrow{\delta}_x\right)(\alpha - \vec{\delta}_y\overleftarrow{\delta}_y) - \alpha E_z^{+1}\right](\alpha + \overleftarrow{\delta}_z)C^n_{i,j,k} = \alpha\omega L\phi^n_{i,j,k} \tag{A2}$$

where ω is the standard relaxation parameter, C is the correction $(\phi^{n+1} - \phi^n)$, α is a free convergence acceleration parameter, and $L\phi^n_{i,j,k}$ is the residual defined by

$$L\phi^n_{i,j,k} = (\vec{\delta}_x\overleftarrow{\delta}_x + \vec{\delta}_y\overleftarrow{\delta}_y + \vec{\delta}_z\overleftarrow{\delta}_z)\phi^n_{i,j,k} \tag{A3}$$

Substitution of a single Fourier component into equation (A2) yields, after simplification, the following growth factor

$$G = \frac{\phi^{n+1}}{\phi^n} = \frac{(\omega - 1)\alpha a_1 + a_2}{a_2 - \alpha a_1} \tag{A4}$$

where

$$a_1 = -2 + 2 \cos r - 2 + 2 \cos s - 2 + 2 \cos t$$

$$\begin{aligned} a_2 = & -\alpha^2(e^{it} - 1) + (-2 + 2 \cos r)(-2 + 2 \cos t) \\ & - (-2 + 2 \cos r)(1 - e^{-it}) - (-2 + 2 \cos s)(1 - e^{-it}) \\ & + \frac{1}{\alpha} (-2 + 2 \cos s)(-2 + 2 \cos r)(1 - e^{-it}) \end{aligned}$$

and r, s, and t are wave numbers. For stability, the magnitude of the growth factor must be less than unity. Since a_1 is real and a_2 is complex, an equivalent expression for stability is given by

$$\frac{(\omega - 1)\alpha a_1 + a_3}{a_3 - \alpha a_1} \leq 1 \tag{A5}$$

where $a_3 = \mathrm{Re}(a_2)$. Since a_3 is always positive and a_1 is always negative, unconditional stability is obtained providing $0 \leq \omega \leq 2$ and $\alpha \geq 0$. For more discussion on AF2 stability see references 1 and 4.

REFERENCES

1. Ballhaus, W. F. and Steger, J. L.: Implicit Approximate Factorization Schemes for the Low-Frequency Transonic Equation. NASA TM X-73,082, 1975.

2. Ballhaus, W. F.; Jameson, A.; and Albert, J.: Implicit Approximate Factorization Schemes for the Efficient Solution of Steady Transonic Flow Problems. AIAA Journal, vol. 16, no. 6, 1978, pp. 573-579.

3. Holst, T. L. and Ballhaus, W. F.: Fast Conservative Schemes for the Full Potential Equation Applied to Transonic Flows. NASA TM-78469, 1978. (Also AIAA Journal, vol. 17, no. 2, Feb. 1979, pp. 145-152.)

4. Holst, T. L.: An Implicit Algorithm for the Conservative, Transonic Full Potential Equation Using an Arbitrary Mesh. AIAA Paper 78-1113, July 1978.

5. Holst, T. L. and Albert, J.: An Implicit Algorithm for the Conservative, Transonic Full Potential Equation with Effective Rotated Differencing. NASA TM-78570, Apr. 1979.

6. Holst, T. L.: A Fast, Conservative Algorithm for Solving the Transonic Full-Potential Equation. AIAA Fourth Computational Fluid Dynamics Conference Proceedings, July 1979, pp. 109-121.

7. Ballhaus, W. F.: A Fast Implicit Solution Procedure for Transonic Flows. Lecture Notes in Physics, Computing Methods in Applied Sciences and Engineering, Springer-Verlag, 1977, pp. 90-102.

8. Jameson, A.: Transonic Potential Flow Calculations Using Conservative Form. AIAA Second Computational Fluid Dynamics Conference Proceedings, June 1975, pp. 148-155.

9. Hafez, M. M.; Murman, E. M.; and South, J. C.: Artificial Compressibility Methods for Numerical Solution of Transonic Full Potential Equation. AIAA Paper 78-1148, July 1978.

10. Eberle, A.: A Finite Volume Method for Calculating Transonic Potential Flow Around Wings from the Pressure Minimum Integral. Technical Translation, NASA TM-75,324, 1978.

11. Eberle, A.: Transonic Potential Flow Computations by Finite Elements: Airfoil and Wing Analysis, Airfoil Optimization. Lecture at the DGLR/GARTEUR 6 Symposium, Transonic Configurations, Bad Harzburg, Germany, June 1978.

12. Thompson, J. F.; Thames, F. C.; and Mastin, C. M.: Automatic Numerical Generation of Body-Fitted Curvilinear Coordinate System for Field Containing Any Number of Arbitrary Two-Dimensional Bodies. J. Comput. Phys., vol. 15, 1974, pp. 299-319.

13. Steger, J. L. and Baldwin, B. S.: Shock Waves and Drag in the Numerical Calculation of Isentropic Transonic Flow. NASA TN D-6997, 1972.

14. Lapidus, A.: A Detached Shock Calculation by Second-Order Finite Differences. J. Comput. Phys., vol. 2, 1967, pp. 154-177.

15. Viviand, H.: Conservative Forms of Gas Dynamic Equations. La Recherche Aerospatiale, no. 1, Jan.-Feb. 1974, pp. 65-68.

16. Vinokur, M.: Conservative Equations of Gas Dynamics in Curvilinear Coordinate Systems. J. Comput. Phys., vol. 14, Feb. 1974, pp. 105-125.

17. Steger, J. L.: Implicit Finite Difference Simulation of Flow About Arbitrary Geometries with Application to Airfoils. AIAA Paper 77-665, June 1977. (See also AIAA Journal, vol. 16, no. 7, July 1978, pp. 679-686.)

18. Jameson, A.: Transonic Flow Calculations. VKI Lecture Series 87, Mar. 15-19, 1976.

19. Jameson, A.: Iterative Solution of Transonic Flows over Airfoils and Wings, Including Flows at Mach 1. Commun. Pure and Appl. Math., vol. 27, 1974, pp. 283-309.

20. Bauer, F.; Garabedian, P.; Korn, D.; and Jameson, A.: Supercritical Wing Sections. Lecture Notes in Economics and Mathematical Systems, 108, Springer Verlag, 1975.

21. Melnik, R. E.: Wake Curvature and Trailing Edge Interaction Effects in Viscous Flow over Airfoils. Advanced Technology Airfoil Research, NASA CP-2045, Mar. 1978, pp. 255-270.

22. Newman, P. A. and South, J. C.: Conservative Versus Nonconservative Differencing: Transonic Streamline Shape Effects. NASA TM X-72,827, 1976.

COMPUTING METHODS IN APPLIED SCIENCES AND ENGINEERING
R. Glowinski, J.L. Lions (editors)
North-Holland Publishing Company

SUBSONIC AND TRANSONIC COMPUTATION OF CASCADE FLOWS

Herman DECONINCK
Research assistant IWONL

Charles HIRSCH
Professor

Vrije Universiteit Brussel
Dept. of Fluid Mechanics
Brussels , Belgium

ABSTRACT

The computation of cascade flows is discussed with main emphasis on a Finite Element formulation of the full potential equation in conservative form. Subsonic calculations are straightforward with a Galerkin method and comparisons are presented between linear triangular, bilinear and biquadratic quadrilateral elements. Transonic calculations with artificial compressibility, bilinear elements and relaxation as well as ADI iterative solutions are presented.

INTRODUCTION

The increasing demand for higher performance in the components of gas engines and turbomachines is supported by the continuous development of computational tools allowing a reliable prediction of detailed flow properties. The basic geometrical engine configuration is constituted by the two-dimensional plane cascade and its quasi three-dimensional extension whereby variations of radius and streamtube thickness are taken into account in a basic 2D formulation.

Compared to flows around isolated airfoils, cascade flows present some particularities mainly connected to the particular boundary conditions which have to be imposed along the in- and outlet boundaries of the computational domain, along the periodic parts of the domain and, in transonic regime, to the choking phenomenon leading to a solution dependent mass flow.

With regard to the numerical techniques, all methods developed in external aerodynamics can be applied to cascade flows and some were originally introduced for cascade calculations. A brief summary of applications of the Euler equations will be given in the first section together with a discussion of the boundary conditions. More attention will be given to the potential equation formulation and its application to subsonic and transonic flows in a Finite Element formulation.

REVIEW OF EULER FORMULATIONS

1. Equations and solution procedure

Inviscid flows are governed by the Euler equations expressing conservation of mass, momentum and energy :

$$\frac{\partial U}{\partial t} + \vec{\nabla}\vec{H} = 0 \tag{1}$$

where in two dimensional flows

$$U = \begin{Bmatrix} \rho \\ \rho u \\ \rho v \\ \rho t \end{Bmatrix} \qquad H_x = F = \begin{Bmatrix} \rho u \\ \rho u^2+p \\ \rho uv \\ \rho u(E+\frac{p}{\rho}) \end{Bmatrix} \qquad H_y = G = \begin{Bmatrix} \rho v \\ \rho uv \\ \rho v^2+p \\ \rho v(E+\frac{p}{\rho}) \end{Bmatrix} \tag{2}$$

E is the total internal energy per unit volume and is given by

$$E = \frac{p}{\rho(\gamma - 1)} + \frac{1}{2}(u^2 + v^2) \tag{3}$$

This system is written in conservation form in order to obtain correct weak solutions[1], i.e. solutions with discontinuities . A weak formulation with arbitrary weight functions W is given by

$$\int_S \frac{\partial U}{\partial t} W\, dS + \int_S \vec{\nabla}\vec{H}\, W\, dS = 0 \quad \text{where S is the flow domain} \tag{4}$$

Both equations (1) and (4) are used by several authors, the differential form (1) leading to Finite Difference (FD) methods[2,3] and the integral form (4) to finite volume[4,5,6] and Finite Element[7] (FE) methods

The stationary solution is obtained by time integration over a sufficient long time until the transient solution is damped out. This is the most common solution procedure in transonic calculations because the time dependent form of the Euler equations is hyperbolic in subsonic as well as in supersonic regions allowing the application of the same discretization scheme in the whole domain. However, the time step in the explicit time integration is limited by severe stability bounds, the Courant-Friedrichs-Lewi conditions depending on the time and space discretization scheme and on the mesh size. In general several hundreds of time steps are necessary to obtain the stationary solution.

Even when satisfying the CFL-conditions non linear oscillations are generated and amplified with progressing time. Therefore all methods include a damping mechanism going from a simple spatial smoothing procedure[4] or explicit addition of dissipative terms[3] to more elaborated methods such as Couston's "Damping Surface Technique"[6] and Denton's "Opposed Difference Technique"[5].

2. Boundary Conditions

The flow domain S is a blade to blade surface of a turbomachine, delimited by

ABCDHGFE, with x the meridional coordinate and y the tangential or pitchwise coordinate and b the streamtube thickness in radial direction due to variations in axial velocity ratio.

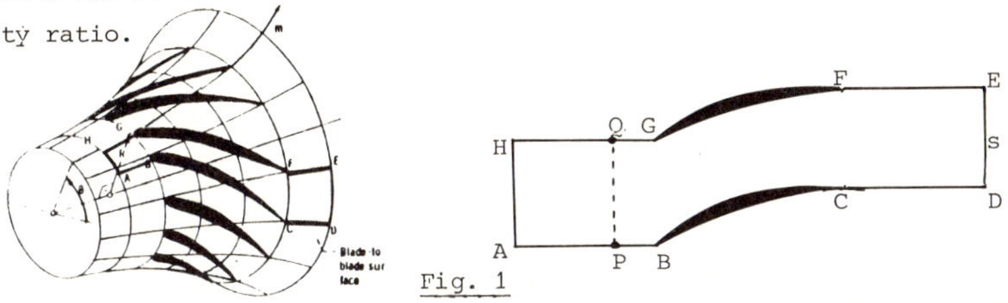

Fig. 1

System (1) is a well posed problem provided the boundary conditions are specified in a way compatible with the equations and the physics of the problem.

On the blade walls the normal velocity is required to be zero. For the periodic boundary AB, EF, CD and GH all flow quantities must be equal in corresponding periodic points such as P and Q. Analysis of the inlet and exit boundary conditions on AH and DE with help of the compatibility relations along the characteristic directions associated with the hyperbolic Euler system leads to the following result[8] : For subsonic axial flows, three variables can be specified and one variable must be computed as a function of time at the inlet boundary, while at the exit boundary three variables must be computed and one must be specified. If the axial velocity is supersonic all four variables must be specified at the inlet and none at the exit.

Based on physical considerations one arrives at the some conclusions : at the downstream boundary of the cascade the static pressure is usually specified in an experimental configuration (1 condition), while at the upstream boundary total temperature and pressure are fixed. Now the inlet flow direction can only be held fixed if the Mach number at inlet is subsonic. If the inlet Machnumber is supersonic the flow direction must be allowed to very f.e. by keeping the pitchwise velocity V_y fixed and allowing the mass flow to vary until the unique incidence condition is satisfied.

If the axial velocity is supersonic at inlet, downstream information cannot reach the upstream boundary and all flow conditions must be specified at inlet and none at outlet.

Numerical boundary conditions, under the form of mass flow in pure subsonic flows or rules for extrapolation at downstream boundaries have to be added in order to make the computation possible.

The circulation or equivalently the outlet angle is part of the solution[6], although some authors apply a Kutta condition at the trailing edge[2,3].

3. Conclusion

The main advantage of the methods based on the Euler equations compared to the potential formulation described below is their ability to take into account entropy variations induced by shocks and rotational effects downstream shocks. These effects are not represented in the potential flow model.

Further they make use of the physical boundary conditions while the discretization can be completely type independent due to the fully hyperbolic character although better convergence properties are likely if the distretization takes into account the type of the flow, supersonic or subsonic[9]. The main disadvantage is the high computer cost due to the slow convergence properties of explicit time integrating schemes, although, the implicit scheme of Beam & Warmig seems promising[10].

Some gain in computer cost can be obtained by replacing the energy equation by the assumption of constant stagnation enthalpy which is exact in stationary adiabatic flows.

The need for more efficient computer programs in routine applications made it necessary to consider the potential flow model by adding to the Euler model the assumption of irrotationality.

POTENTIAL FLOW FORMULATIONS

1. Basic Equations

For irrotational flows the velocity vector is the gradient of a scalar potential function ϕ and the Euler system (1) is then reduced to the potential equation retaining only the continuity equation

$$(\rho\, b\, \phi_x)_x + (\rho\, b\, \phi_y)_y = 0 \tag{5}$$

where b is the streamtube thickness in radial direction (Fig. 1).

Since no entropy can be created in irrotational flows the density is obtained from the isentropic relation

$$\rho = \rho_t [(1 - \frac{\gamma - 1}{2\,\gamma r T_t} (\phi_x^2 + \phi_y^2)]^{1/(\gamma-1)} \tag{6}$$

where the stagnation temperature T_t and stagnation pressure ρ_t are constant for non rotating cascade flow.

The quasilinear equation (5) is elliptic in subsonic points and hyperbolic in supersonic points of the flow domain S. A weak formulation allowing for solutions which contain discontinuous velocities ϕ_x and ϕ_y is given by

$$\int_S [\rho\, \phi_x W_x + \rho\, \phi_y W_y]\, dS - \int_s \rho\, W \frac{\partial \phi}{\partial n}\, ds = 0 \tag{7}$$

for any arbitrary weightfunction W continuous over S.s is the part of the boundary where a Neumann boundary conditions $\rho(\partial\phi/\partial n)$ is specified.

For inviscid transonic flows the only sources of entropy production are the shocks and since the entropy variation is neglectible for weak shocks, ($M \leqslant 1.5$), the potential formulation should be a valid model for moderate Mach number flows. A more fundamental problem caused by the absence of entropyproduction in shocks is the fact that the integral form (7) may contain non physical solutions with expansion shocks. Therefore a criterium is needed to select the physically relevant solutions of (7). The discontinuous solutions with physical meaning can be represented as the solution in the limit of vanishing viscosity $(\nu_1, \nu_2 > 0)$[11] of

(8) $$\frac{\partial}{\partial x}\left(\frac{\rho\, b\, \partial\phi}{\partial x}\right) + \frac{\partial}{\partial y}\left(\frac{\rho\, b\, \partial\phi}{\partial y}\right) = -\nu_1\, \phi_{xxx} - \nu_2\, \phi_{yyy}$$

The right hand side of this equation is called artificial viscosity term for its analogy to the physical viscosity expressed by similar terms in the Navier Stokes equation. Obviously, no artificial viscosity is needed in subsonic flow regions. An alternative potential flow formulation but only valid in two dimensions is the streamfunction formulation. Here the continuity equation is fullfilled by definition and the irrotationality condition is the equation to be solved

(9) $$\frac{\partial}{\partial x}\left(\frac{1}{\rho\, b}\frac{\partial\psi}{\partial x}\right) + \frac{\partial}{\partial y}\left(\frac{1}{\rho\, b}\frac{\partial\psi}{\partial y}\right) = 0$$

and the density expression is now implicit

(10) $$\rho = \rho_t \left[1 - \frac{\gamma - 1}{2\, \gamma r T_t}\frac{1}{\rho^2}(\psi_x^2 + \psi_y^2)\right]^{1/(\gamma-1)}$$

Both formulations with potential function and with streamfunction are commonly used in subsonic calculations while convergence problems with the implicit density relation (10) exclude the streamfunction formulation in transonics.

2. Boundary Conditions

Boundary conditions for elliptic quasi linear p.d.e. are well established from the mathematical point of view. They consist in Dirichlet $(\beta = 0)$, Neumann $(\alpha = 0)$ or mixed $(\alpha \neq 0,\ \beta \neq 0)$ conditions

(11) $$\alpha\, \phi + \beta\, \rho\, b\, \frac{\partial\phi}{\partial n} = q \qquad \text{resp. } \alpha\, \psi + \beta\, \frac{1}{\rho\, b}\, \frac{\partial\psi}{\partial n} = q$$

Restricting ourselves to the potential function formulation, physical considerations lead to the following conclusions.

At the inlet boundary and exit boundary Neumann boundary conditions

(12) $$\rho\frac{\partial\phi}{\partial n} = \rho_1(w_{x1}\ ,\ w_{y1})\ b_1\ w_{x1} \qquad \rho\frac{\partial\phi}{\partial n} = \rho_2(w_{x2}\ ,\ w_{y2})\ b_2\ w_{x2}$$

are obvious as well as the zero flux boundary condition on the blade walls

$$\rho\, \frac{\partial\phi}{\partial n} = 0$$

Subscripts 1 and 2 denote the inlet and exit boundary. For the periodic boundaries the periodicity results in a constant difference in potential between two corresponding periodic points (f.e. P - Q, Fig. 1).

(13) $$\phi_Q = \phi_P + s\, w_{y1} \text{ upstream} \qquad \phi_Q = \phi_P + s\, w_{y2} \text{ downstream}$$

where s is the pitch.

So from equation (12) and (13), it appears that the knowledge of w_{x1}, w_{x2}, w_{y1} and w_{y2} allow complete mathematical specification of the downstream and upstream boundary conditions, including the periodicity conditions. These four quantities are related by one obvious expression, the mass flow conservation :

$$\rho_1(w_{x1}, w_{y1})\, b_1\, w_{x1} = \rho_2(w_{x2}, w_{y2})\, b_2\, w_{x2}$$

So only three independent quantities can be specified for both upstream and downstream boundaries.

Hence specification of the inlet angle and Machnumber, together with the outlet angle or alternatively outlet Machnumber completely determines the potential flow trough a cascade, including the circulation. When the outlet angle or Machnumber is not known from external data a Kutta-Joukowsky condition must be imposed on the trailing edge fixing in this way the circulation

$$s\ (w_{y2} - w_{y1})$$

and then no independent flow quantity is specified at exit. It is the experience of the authors that simulation of a Kutta condition by requiring equal velocities for suction and pressure side near the trailing edge still offers an arbitrary estimation of the circulation in the real flow, especially for turbine cascades with rounded trailing edges.

For transonic flows and more particularly supersonic inlet or outlet flows the choking condition implies that the massflow is part of the solution, hence only one independent quantity.can be imposed f.e. the inlet Machnumber or inlet tangential velocity.

An important practical problem is the introduction of Neumann boundary conditions in the discretized equations. Methods starting from an integral formulation such as eq. (7), namely FE and FV methods are clearly more favorable than differential methods, since Neumann B.C. is treated automatically by the formulation.

3. Numerical solution procedures

For subsonic flows both the formulation with potential function and streamfunction have been widely used in streamline curvature methods[12,13], singularity methods[14] and field methods discretized with FD[15] and FE[16,17,18,19]. In the transonic range only the field methods with the potential function have met with some success : some of which are the Murman & Cole type calculation with type dependent finite differencing [20,21], the optimal control method with finite element discretization proposed by Glowinski and Pironneau, applied to cascade flow by Hirsch and Deconinck[22] and recently the Galerkin FE method with artificial compressibility[23]. We restrict ourselves to a discussion of field methods, the recent progress being mainly in this area. Considering the following general non linear p.d.e. to be solved in an iterative method and residual R

(14) $A(\phi) = f$ $\qquad R(\phi) = A(\phi) - f$

A linearized operator denoted A^n is obtained with help of the previous iteration number n and the linearized equation is given by

(15) $$A(\phi) \approx A^n \phi = f$$

A Newton iterative process obtained by Taylor expansion of $A(\phi^{n+1})$ leads to the following linear equation which must be solved in each iteration

(16) $$A_T^n \, \delta\phi = - R(\phi^n)$$ where

(17) $$A_T^n = \left.\frac{\partial A(\phi)}{\partial \phi}\right|_{\phi=\phi^n}$$ (18) $$\delta\phi = \phi^{n-1} - \phi^n$$

It is clear that operator A_T^n can be replaced by any other linear operator provided the iterative process converges, namely the residual tends to zero for n going to infinity

(19) $$C^n \, \delta\phi = - R\,(\phi^n)$$

for all operators C^n the choice A_T^n assures the best convergence speed although a simpler choice of C^n can result in a lower cost. A good approximation to A_T^n is obtained by using the linearized operator :

(20) $$A_T^n \approx \left.\frac{\partial\,[A^n \phi\,]}{\partial \phi}\right|_{\phi=\phi_n} = A^n$$

leading to the following scheme termed variable stiffness scheme in FE applications while the Newton scheme is called tangential stiffness scheme

(21) $$A^n \, \delta\phi = - R(\phi^n)$$

This scheme still requires the inversion of a matrix in each iteration. This can be avoided by using the same operator at each iteration or during a certain number of iterations (constant stiffness scheme)

(22) $$A^o \, \delta\phi = - R(\phi^n)$$

For a linear equation obviously $A_T^n = A^n = A^o = A$. In the case of the potential function formulation one obtains

(23) $$A(\phi) = - \partial_x [\rho(\phi)\, \partial_x \phi\,] - \partial_y [\rho(\phi)\, \partial_y \phi\,]$$

(24) $$A^n = - \partial_x \rho^n \partial_x - \partial_y \rho^n \partial_y$$

while the tangential stiffness matrix obtained by analytic differentiation of (22) takes the following form which is also hyperbolic in supersonic points

$$(M_x = \frac{\phi_x}{a}\,,\ M_y = \frac{\phi_y}{a}) :$$

(25)
$$A_T^n = - \partial_x \{\rho\,[\,1 - (M_x^n)^2\,]\,\partial_x\} + \partial_x\,(\rho\, M_x^n M_y^n \partial_y) + \partial_y\,(\rho\, M_x^n M_y^n \partial_x) - \partial_y \{\rho\,[\,1 - (M_y^n)^2]\,\partial_y\}$$

This is the form used by Prince[18] in his FE method applying Galerkin's technique to eq. (16) requiring very few iterations (4 to 10). In their subsonic applications the authors[16,17] preferred the simpler scheme of eq. (21) and even the constant stiffness form of eq. (22) where A^o simply was the Laplacian operator $\partial^2_{xx} + \partial^2_{yy}$ requiring still only 4 to 10 iterations depending on the Machnumber. However, the inversion of the operator C^n is only needed in the first iteration. This remarkable result is due to the fact that in subsonic cases the incompressible Laplacian operator is still a good approximation to the elliptic compressible A^n_T or A^n operator.

In FD applications a similar method exists, based on the use of fast Poisson Solvers, hence requiring a uniform orthogonal mesh.

In the methods described above, the operator C^n leads to a system implicit in the unknowns of all meshpoints explaining intuitively the very fast convergence properties. However, in transonic flows the operator A^n_T or A^n is no longer positive definite in supersonic points and standard direct inversion methods fail. Problems in this area were circumvented in FD methods due to the widespread use of relaxation methods, even in subsonic applications[15]. Relaxation methods are more easy to program with FD and require less computer storage but are also slower due to the lower degree of implicitness.

Jameson[11] proves that for the Laplace equation ($A = A^n = A^n_T = \partial^2_{xx} + \partial^2_{yy}$) on a uniform grid point overrelaxation is equivalent to the following form for C

(26) $$C = \alpha\, \partial^{\leftarrow}_x + \beta\, \partial^{\leftarrow}_y + \gamma$$

where $\partial^{\leftarrow}_x$ and $\partial^{\leftarrow}_y$ are backward differences allowing the scheme to be fully explicit. It is clear that only very slow convergence can be obtained with such a bad approximation of the Laplace operator. With line overrelaxation(SLOR) one obtains the following form for C

(27) $$C = \alpha\, \partial^{\leftarrow}_x + \beta\, \partial^2_{yy} + \gamma$$

which is implicit in the y-direction leading to a tridiagonal system for each $x = c^t$ line. In this case operator C is a better approximation of the Laplace operator and faster convergence is obtained. The introduction of a backward first derivative in the x-direction increases the numerical stability in supersonic regions[11] and this explains the common use of SLOR in external transonic flow applications. A better approximation of C with respect to A^n but still requiring only tridiagonal solution steps is obtained in the Alternating Direction Implicit (ADI) or Approximate Factorization method (AF) by posing

(28) $$C = C_1\, C_2 \ldots C_R$$

where each operator C_i is tridiagonal. In the case of the potential function a possible choice is the following (AF1 scheme)

(29) $$C^n = C_1^n C_2^n = (1 - \sigma \partial_x \rho^n \partial_x)(1 - \sigma \partial_y \rho^n \partial_y) \text{ giving } C^n \approx 1 + \sigma A^n$$

when σ is small enough. Equation (19) is now splitted up it two steps involving only tridiagonal operators

(30) $$(1 - \sigma \partial_x \rho^n \partial_x) g = - R(\phi^n) \qquad (1 - \sigma \partial_y \rho^n \partial_y) \delta\phi = g$$

Both the line overrelaxation and approximate factorization methods are used in section 5 for the solution of transonic flows with finite elements.

4. Solution of subsonic flows with direct Galerkin FE methods

The constant stiffness scheme eq. (22) discussed in the previous section proved to be the most efficient solution method for the fully elliptic equation. In weak form it is given by (for any W)

(31) $$- \int_S \rho^o \nabla W \nabla \delta \phi \, dS = \int_S \rho^n \nabla \phi^n \nabla W \, dS - \int_s (\rho \frac{\partial \phi}{\partial n}) W \, ds$$

A system of linear equations in the unknown values of the potential function at each meshpoint is obtained by choosing a weightfunction for each meshpoint, namely its shape function (Galerkin procedure) and by approximating the potential function as a linear combination of the meshpoint values with help of the shape functions N_J

(32) $$W = N_I \, , \; \phi(x,y) = \sum_J \phi_J N_J (x,y)$$

leading to the following system

(33) $$\sum_J K_{IJ}^o \, \delta\phi_J = R_I^n \qquad \text{where}$$

(34) $$K_{IJ}^o = - \rho^o \int_S \nabla N_I \nabla N_J \, dS$$

is the constant stiffness matrix depending only on the geometry and therefore its inversion is only needed in the first iteration. The residual R_I^n is given by

(35) $$R_I^n = \sum_K \phi_K^n \int \rho^n \nabla N_K \nabla N_I \, dS - \int_s (\rho \frac{\partial \phi}{\partial n})^o N_I \, ds$$

and contains the contribution of the boundary condition when I is situated on a Neumann boundary. For solid walls this contribution is zero. Except for linear elements where the integration can be carried out analytically all integrations are performed numerically in the orthogonal $\xi - \eta$ coordinate system obtained by isoparametric transformation of the physical $x - y$ coordinate system given by[24]

(36) $$x = \sum_I x_I N_I (\xi,\eta) \qquad y = \sum_I y_I N_I (\xi,\eta)$$

Hence the shape functions need only to be specified in the $\xi - \eta$ coordinates.

Computational results

Three different types of finite elements[24] were compared (Fig. 2), namely the 8-node biquadratic quadrilateral serendipity element, the 4-node bilinear quadrilateral element and the 3-node linear triangular element. The following table summarizes the relative CP times obtained for a calculation of the VKI-Gasturbine cascade[25] and the DCA-48° camber compressor cascade[26] for the same convergence criterium and number of iterations.

	Triangular	Bilinear	Biquadratic
VKI ($\pm$ 300 mesh points)	0.56	0.79	1.00
DCA ($\pm$ 300 mesh points)	0.58	0.77	1.00

With approximately the same number of nodes the number of elements in the bilinear case is about four times and in the triangular case about eight times the number of elements in the biquadratic case. The higher cost for the bilinear and biquadratic element is situated in the numerical integration. No numerical integration is needed for the triangular elements. The quality of the solution is very similar except for the leading and trailing edge where the curvature is important. The blade Machnumber distribution is given in Fig. 3 and 4, where the experiments[25,26] are denoted by 0,X.

The computational results are in good agreement with the experiments except for the peaky behaviour on the pressure side in the trailing edge region, which is a potential effect not present in physical flow due to boundary layer separation.

For triangular elements a clustering of the gridpoints towards the blade walls is necessary to obtain an accurate solution, while the other elements are very insensitive to the quality of the mesh (Fig. 5).

Conclusion

Accurate solutions are obtained with very little cost on coarse grids. No attention to smoothness or clustering of the grid is needed when the biquadratic element is used.

5. F.E. Computation of transonic cascade flows. Artificial compressibility form

Transonic calculations require the solution of the potential equation (8) with additional artificial viscosity terms which can be included implicitly in the discretization scheme by using central differences in subsonic and upwind differences in supersonic flow regions as was originally suggested by Murman and Cole. An attractive way of introducing artificial viscosity terms in an explicit way, which is the only possibility in FE methods is to include them in the density relation. This is realised by choosing the following form[27] for the artificial viscosity terms in the right hand side of equation (8)

$$(\nu \, \phi_x \, \rho^{\leftarrow}_x \, \Delta x)_x + (\nu \, \phi_y \, \rho^{\leftarrow}_y \, \Delta y)_y \tag{37}$$

where $\rho_x^{\leftarrow}$ and $\rho_y^{\leftarrow}$ are upwind differenced and ν is a switching function assuring zero viscosity in the subsonic flow below a cut-off Mach number M_c

$$(38) \qquad \nu = \max\left(o\ ,\ 1 - \frac{M_c^2}{M^2}\right)$$

equation (8) is then rearranged giving

$$(39) \qquad (\tilde{\rho}\ \phi_x)_x + (\bar{\rho}\ \phi_y)_y = 0 \qquad \text{with}$$

$$(40) \qquad \tilde{\rho} = \rho - \nu\ \rho_x^{\leftarrow}\ \Delta x \qquad\qquad \bar{\rho} = \rho - \nu\ \rho_y^{\leftarrow}\ \Delta y$$

Formal identity with the unmodified potential equation is obtained by the choice[28]

$$(41) \qquad \tilde{\rho} = \bar{\rho} = \rho - \nu\ \rho_s^{\leftarrow}\ \Delta s$$

where $\rho_s^{\leftarrow}$ is the upwind difference along a streamline.

The artificial compressibility formulation allows to include the artificial viscosity effects completely in the density updating which is performed after each iteration step according to equation (19). Hence the solution process itselves is not type dependent and equation (39) can be treated in a standard way with the iterative methods (SLOR or ADI) discussed in section 3, while the residual of eq. (39) is computed with the classical Galerkin procedure giving exactly the form of eq. (35) if (41) is used.

Line overrelaxation method

Considering bilinear elements the line overrelaxation leads to the following tridiagonal system which must be solved for each column i (in finite difference notation with row and column indices)

$$(42) \qquad \sum_{l=j-1}^{j+1} \delta\phi_{il}\ K_{ij}^{il} = -\ \omega\left[R_{ij}^{(n)} - \sum_{l=j-1}^{j+1} K_{ij}^{i-1,l}\ \delta\phi_{i-1,l}\right]$$

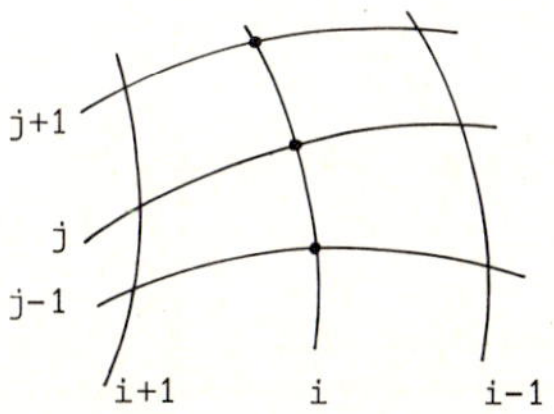

where ω is a relaxation factor and the residual R is given by eq. (35) after replacing ρ by $\tilde{\rho}$. The iteration matrix K is defined as

$$(43) \qquad K_{ij}^{kl} = \int_S \tilde{\rho}^n\ \nabla N_{ij}\ \nabla N_{kl}\ dS$$

which is zero for $(i - 1) \leqslant k \leqslant (i + 1)$ and $(j - 1) \leqslant l \leqslant (j + 1)$

The boundary conditions are included in the residual and no special treatment is needed. Further the method is valid for an arbitrary curvilinear mesh although in supersonic regions it is preferable that the grid is aligned with the flow direction in order to introduce correct stabilizing upstream differences mentioned in eq. (27) requiring also fixed sweep direction from upstream to downstream.

Compared to subsonic flows a higher number of meshpoints together with a higher degree of smoothness of the gridlines is required. Therefore a body fitted coordinate generating package was developed based on elliptic partial differential equations and exponential clustering to the blade walls in the line of the Sörenson and Steger method for single airfoil geometries[29]. This method allows a userspecified distance between the blade wall and the next coordinate line. A result of the geometry generated for the VKI turbine cascade is given in Fig. 6. All computations presented in the paper are run with a grid containing 1825 mesh points.

In order to speed up the convergence a grid refinement technique is applied going from 40, 133, 481 to 1825 mesh points. On each mesh 50 iterations are needed for a maximal residual drop of 3 orders of magnitude.

The excellent shock capturing properties of the method are demonstrated in a channel flow calculation with a circular bump. Mach distribution along the bump and isomach plot are shown in figure 7 and 8.

The first transonic cascade case is the DCA 9.5° compressor cascade[26] for an inlet Machnumber of 1.05. Experimental investigations show a shock from suction to pressure side moving downstream with increasing outlet Machnumber.

Numerical solution for an outlet Machnumber of .76 at the unique incidence of 58° taken from the experiments showed a purely subsonic flow for the pressure side (Fig.9).

However, for an incidence of 55° the shock was recovered showing also a weak detached shock in front of the leading edge (Fig. 10). The Mach distribution is in reasonable agreement with the experimental result (Fig. 11) taking into account the high viscous effects in the real flow.

The second transonic cascade is the VKI gas turbine cascade also used in the subsonic calculations, but this time different supersonic outlet Machnumbers were imposed. For an outlet Machnumber of .975 excellent agreement with the experiments is achieved without any modification of the trailing edge in order to reduce the high Machnumber peak on the pressure side, discussed in section 4 (Fig. 12).

For higher Machnumber the program fails due to these extremely high velocities and modification of the trailing edge is necessary in order to simulate the boundary layer separation and the wake. More work in this area is in progress.

Approximate Factorization Method

AF methods are known as very efficient tools for the solution of nonlinear partial differential equations discretized with FD, particularly in transonic flow calculations where direct methods (f.e. fast Poisson Solvers) are unapplicable, and it is expected that this is also true for FE applications.

The AF1 scheme was already developed in section 3 for a rectangular mesh leading to the splitted equation (30). In the case of a body fitted curvilinear mesh the operators $\partial_x \rho \partial_x$ and $\partial_y \rho \partial_y$ contain mixed $\xi - \eta$ derivatives when transformed to the computational plane and hence the resulting system will contain contributions of nodes on different coordinate lines.

In order to obtain systems implicit in unknowns of only one coordinate line it is necessary to perform the splitting along the $\xi - \eta$ directions : Starting from the potential equation transformed to the $\xi - \eta$ plane (with J and

$$\begin{vmatrix} A_1 & A_2 \\ A_2 & A_3 \end{vmatrix}$$

resp. the Jacobian and metric tensor).

$$A(\phi) = \partial_\xi \left[\frac{\tilde{\rho}}{|J|} (A_1 \partial_\xi + A_2 \partial_\eta)\phi \right] + \partial_\eta \left[\frac{\bar{\rho}}{|J|} (A_2 \partial_\xi + A_3 \partial_\eta)\phi \right] = 0 \tag{44}$$

the following AF1 scheme is obtained by neglecting the mixed ξ,η derivatives in the operator $A(\phi)$:

$$\left(1 - \sigma \partial_\xi \frac{\tilde{\rho}^n}{|J|} A_1 \partial_\xi\right) \left(1 - \sigma \partial_\eta \frac{\bar{\rho}^n}{|J|} A_3 \partial\eta\right)\delta\phi = - R(\phi^n) \tag{45}$$

which is splitted as before in :

$$\begin{aligned} \left(1 - \sigma \partial_\xi \frac{\tilde{\rho}^n}{|J|} A_1 \partial_\xi\right) g &= - R(\phi^n) \\ \left(1 - \sigma \partial_\eta \frac{\bar{\rho}^n}{|J|} A_3 \partial_\eta\right)\delta\phi &= g \end{aligned} \tag{46}$$

Applying FE Galerkin method to both equations and for factorizable shapefunctions

$$N_{ij}(\xi,\eta) = N_i(\xi)\, N_j(\eta)$$

one obtains for each line

$$(M_\xi + \sigma K_\xi) g = -R(\phi^n) \tag{47a}$$

and for each column :

$$(M_\eta - \sigma K_\eta)\delta\phi = g \tag{47b}$$

where

$$\begin{aligned} (M_\xi)_{ik} &= \int N_i(\xi)\, N_k(\xi) d\xi \\ (M_\eta)_{ik} &= \int N_i(\eta)\, N_k(\eta) d\eta \end{aligned} \tag{48}$$

$$\begin{aligned} (K_\xi)_{ik} &= \int \frac{\tilde{\rho}^n}{|J|} A_1\, \partial_\xi N_i\, \partial_\xi N_k\, d\xi \\ (K_\eta)_{ik} &= \int \frac{\bar{\rho}^n A_3}{|J|} \partial_\eta N_i\, \partial_\eta N_k\, d\eta \end{aligned} \tag{49}$$

while the residual in eq. (47) remains fully standard and identical to eq. (35) if $\bar{\rho} = \tilde{\rho}$. For bilinear elements the explicit form of the discretized eq. (47) is given in ref[30], showing that except for the mass matrices (48) the FE discretization of the L.H.S. of eq. (47) is idential to its FD discretization using central differences[27].

This AF1 scheme with FE discretization converges very fast in subsonic cases when compared to the SLOR solution[23]. However, it is less efficient than the direct methods discussed in a previous section. In transonic problems stability problems may occur in the shock region which are not present in the SLOR method due to the implicit backward $\overleftarrow{\partial}_x \delta\phi$ difference in the operator C. Therefore a variant of the AF1 scheme has been proposed, namely the AF2 scheme[31] replacing eq. (45) by

$$(48) \qquad \left(1 - \sigma\, \overrightarrow{\partial}_\xi \frac{\tilde{\rho}^n}{|J|} A1\right)\left(\overleftarrow{\partial}_\xi - \sigma\, \partial_\eta \frac{\bar{\rho}^n}{|J|} A_3\, \partial_\eta\right)\delta\phi = -\, R(\phi^n)$$

More details concerning the implementation of this scheme with FE discretization are given in ref[23].

A Machdistribution for the NACA0012 airfoil with a free stream Machnumber of .80 and zero incidence calculated with the AF1-FE scheme is given in figure 13 showing good agreement with the SLOR calculation of the same flow, except for a pre-shock overshoot caused by the numerical instability mentioned before.

CONCLUSION

Due to the artificial compressibility concept FE methods are now well established for the computation of transonic flows. Furthermore the FE methods have the intrincic advantage compared to FD methods of automatic Neumann boundary condition treatment and computation on a curvilinear body fitted grid without the need of explicit transformation of the equation to a computational orthogonal coordinate system.

Extension to higher order approximations is possible by the use of higher order elements.

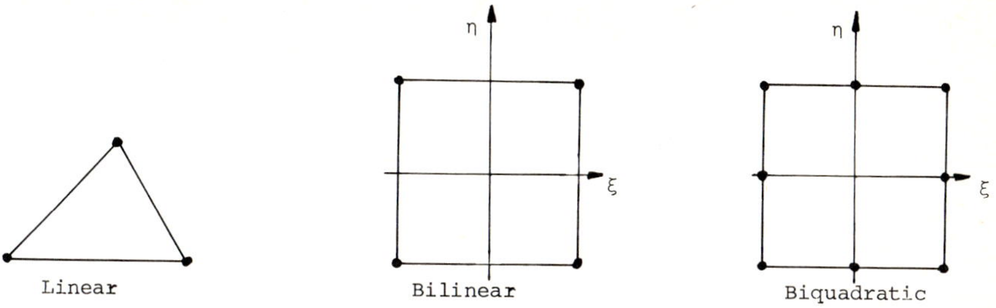

Fig. 2

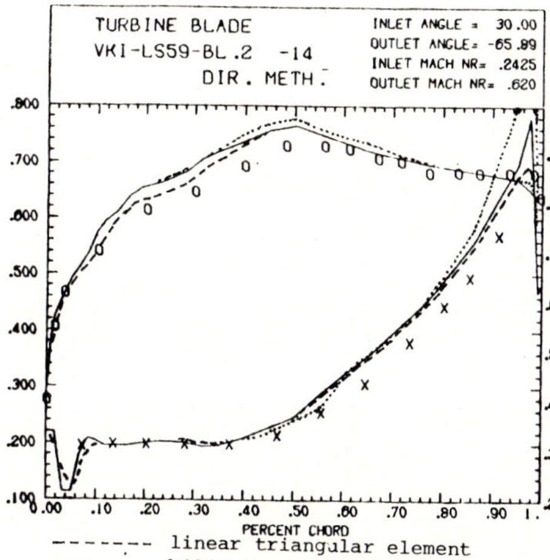

------- linear triangular element
——— bilinear rectangular element
.... biquadratic rectangular el.
o,x experimental result

Fig. 3 : Machdistribution

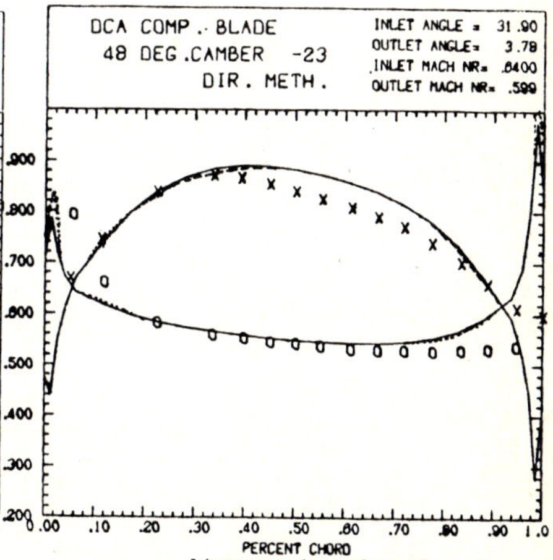

------- linear triangular element
——— bilinear rectangular element
.... biquadratic rectangular el.
o,x experimental result

Fig. 4 : Machdistribution

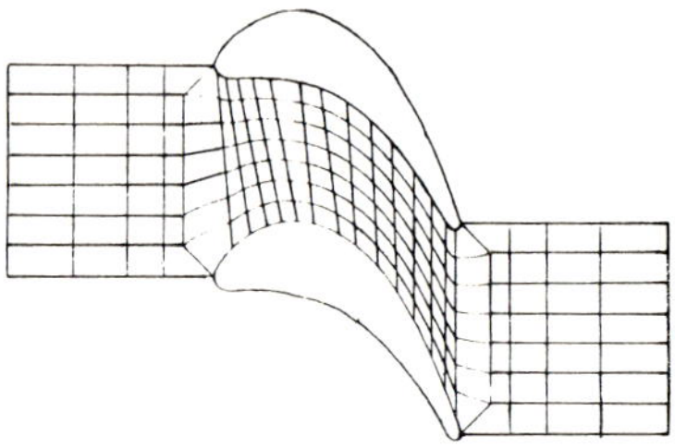

Fig. 5 : Mesh for subsonic calculation with biquadratic elements VK1-LS-50 turbine blade

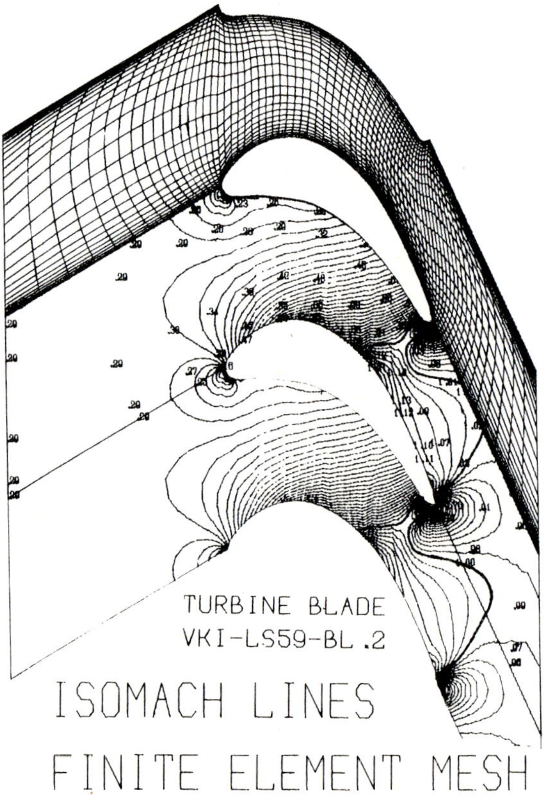

Fig. 6

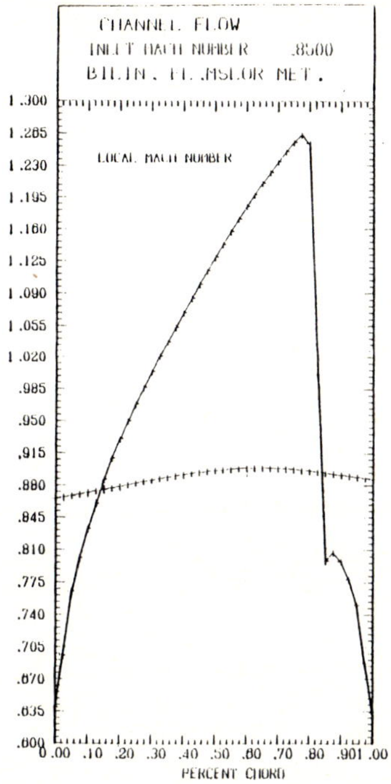
CHANNEL FLOW
INLET MACH NUMBER .8500
LOCAL MACH NUMBER
PERCENT CHORD

Fig. 7

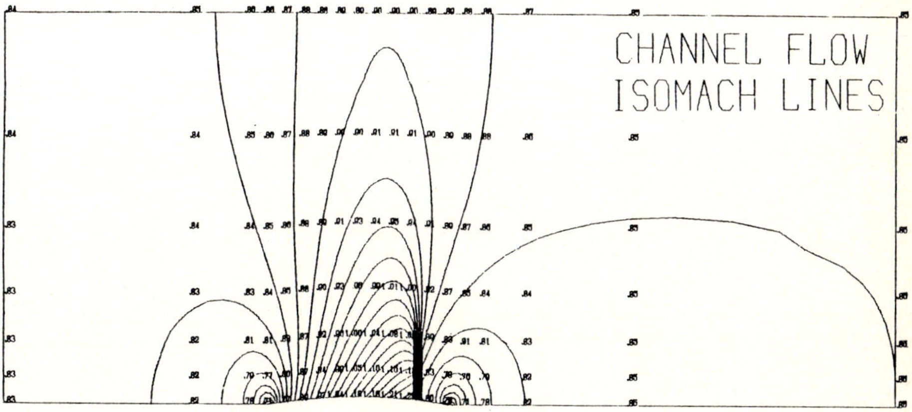
CHANNEL FLOW
ISOMACH LINES

Fig. 8

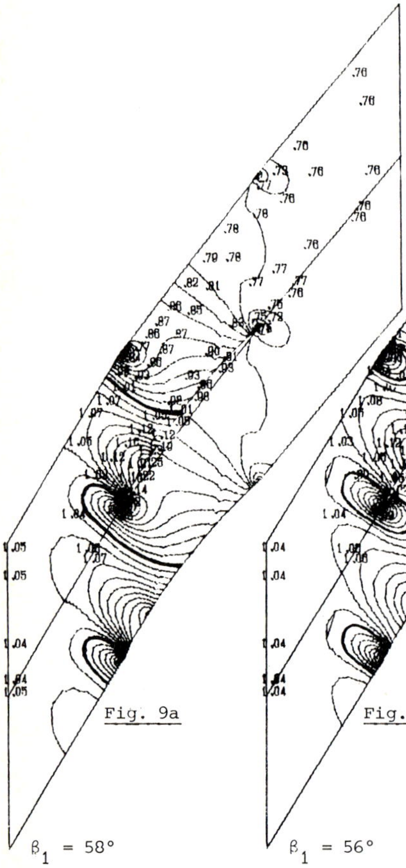

Fig. 9a

$\beta_1 = 58°$

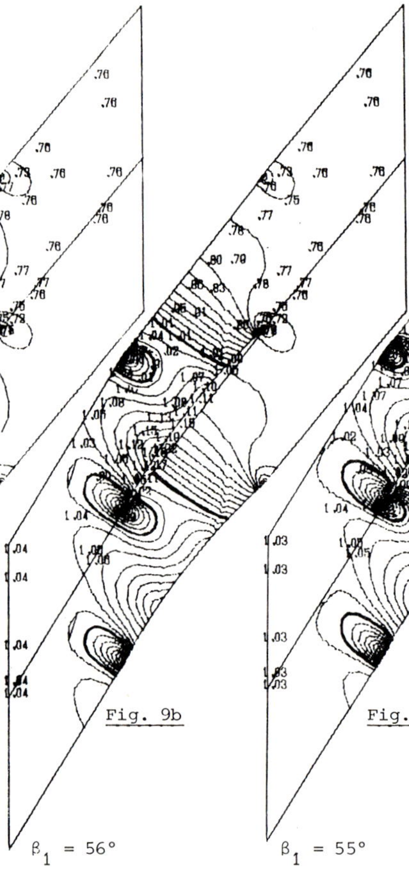

Fig. 9b

$\beta_1 = 56°$

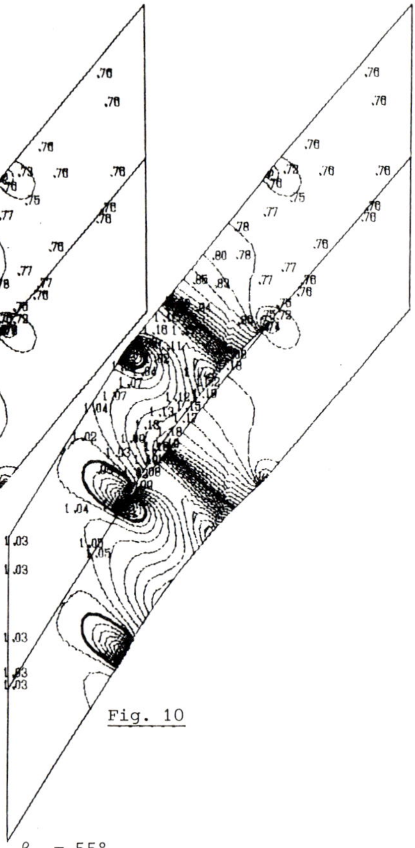

Fig. 10

$\beta_1 = 55°$

DCA 9.5° Camber Compressor Cascade

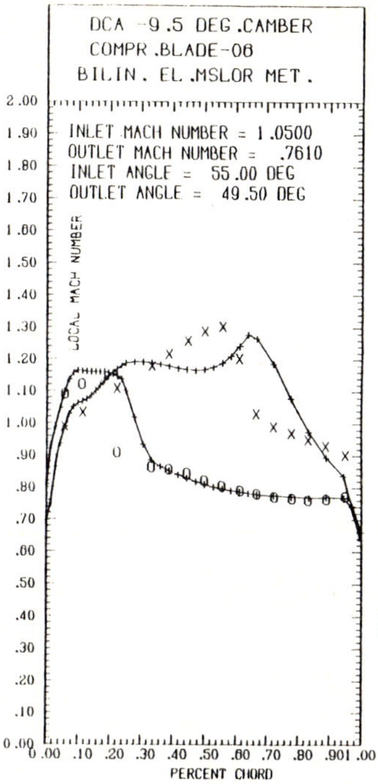
DCA -9.5 DEG.CAMBER
COMPR.BLADE-06
BILIN. EL.MSLOR MET.
INLET MACH NUMBER = 1.0500
OUTLET MACH NUMBER = .7610
INLET ANGLE = 55.00 DEG
OUTLET ANGLE = 49.50 DEG
LOCAL MACH NUMBER
PERCENT CHORD

Fig. 11

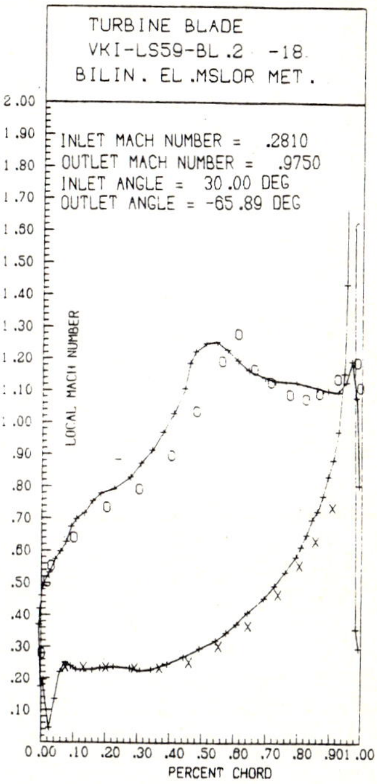
TURBINE BLADE
VKI-LS59-BL.2 -18
BILIN. EL.MSLOR MET.
INLET MACH NUMBER = .2810
OUTLET MACH NUMBER = .9750
INLET ANGLE = 30.00 DEG
OUTLET ANGLE = -65.89 DEG
LOCAL MACH NUMBER
PERCENT CHORD

Fig. 12

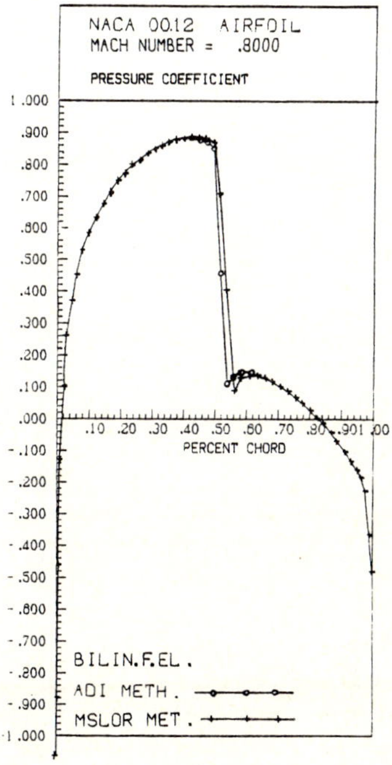
NACA 00.12 AIRFOIL
MACH NUMBER = .8000
PRESSURE COEFFICIENT
PERCENT CHORD
BILIN.F.EL.
ADI METH.
MSLOR MET.

Fig. 13

REFERENCES

1 LAX, P.D., "Weak Solutions of nonlinear hyperbolic equations and their numerical computation", Comm. Pure & Appl. Math. VII, 1954

2 GOPALAKRISHNAN S., BOZZOLA R., "A Numerical Technique for the Calculation of Transonic Flows in Turbomachinery Cascades", ASME Paper No. 71-GT-42

3 VEUILLOT J.P., "Calcul Numérique de l'écoulement transsonique d'un fluide parfait dans une grille d'aubes", La Recherche Aérospatiale n0 1975, pp. 327-338

4 Mac DONALD P.W., "The Computation of Transonic Flow trough two dimensional Gas Turbine Cascades", ASME paper 71-GT-89, 1971

5 DENTON J.D., SINGH U.K., "Time Marching Methods for Turbomachinery Flow Calculation", LS 1979-7, Von Karman Institute for Fluid Mechanics

6 COUSTON M., "Time marching-finite Area Method", VKI-LS 84, 1976, Von Karman Institute for Fluid Mechanics

7 HIRSCH Ch., WARZEE G., "An orthogonal Finite Element Method for Transonic Flow Calculations", Proc. 6th Int. Conf. on Numerical Methods in Fluid Dynamics, Springer Verlag 1979

8 GOPALAKRISHNAN S., BOZZOLA R., "Numerical Representation of Inlet and Exit Boundary Conditions in Transient Cascade Flow", ASME paper 73-GT-55

9 WARMING R.F., BEAM R.M., "Upwind Second-Order Difference schemes and Applications in Unsteady Aerodynamic Flows", AIAA Journal, Vol. 14, Sept. 1976, pp. 1241-1249

10 BEAM R.M., WARMING R.F., "An Implicit Finite-Difference Algorithm for Hyperbolic Systems in Conservation Law Form", Journal of Comp. Physics, Vol. 22, sept. 1976 pp. 87-110

11 JAMESON A., "Numerical Computation of Transonic Flows with Shocks Waves", Symposium Transsonicum II, Springer Verlag 1975

12 NOVAK R.A., "Streamline Curvature Computing Procedures", ASME J. Eng. Power 89, 4, 1967

13 WILKINSON D.H., "Calculation of blade to blade Flow in a Turbomachine by Streamline Curvature", ARCR & M.3704, 1972

14 VAN DEN BRAEMBUSSCHE R.A., "Calculation of Compressible subsonic Flows in Cascades with varying Blade Height", VKI-LS-59, 1973, Von Karman Institute for Fluid Mechanics

15 KATSANIS T., Mac NALLY W., "A Fortran Program for Calculating Velocities and Streamlines on a Blade to Blade Stream Surface of a Tandem Blade Turbomachine"; NASA TN D-5044, 1969

16 HIRSCH Ch., WARZEE G., "Finite Element Computation of Subsonic Cascade Flows", Proc. of 6th Canadian Congress an Applied Mechanics, Vancouver 1977

17 HIRSCH Ch., WARZEE G., "An Integrated Quasi-3D Finite Element Calculation Program for Turbomachinery Flows" Journal of Engineering for power, Vol. 101, pp. 141-148, 1979 also ASME paper No. 78-GT-56

18 PRINCE T.C., "Prediction of Steady Inviscid Compressible Flow on a Blade-to-Blade Surface by Finite Element Method", AIAA paper 78-244, 1978

19 PERIAUX J., "Three-Dimensional Analysis of Compressible Potential Flows with the Finite Element", Int. Journ. Num. Methods in Engineering 9, 1975

20 DODGE P., "Transonic Flows in Axial Turbomachinery-Transonic Relaxation", VKI-LS 84, 1976 Von Karman Institute for Fluid Mechanics

21 IVES D.C., LIUTERMOZA J.F., "Second Order Accurate Calculation of Transonic Flow over Turbomachinery Cascades", AIAA paper 78-1149, 1978

22 DECONINCK H., HIRSCH Ch., "A Finite Element Method solving the Full Potential Equation with Boundary Layer Interaction in Transonic Cascade Flow", AIAA paper

79-0132, 1979

23 DECONINCK H., HIRSCH Ch., "Finite Element Methods for Transonic Flow Calculations" Proc. III. GAMM Conf. on Numerical Methods in Fluid Mechanics, Köln 1979

24 ZIENKIEWICS O.C., "The Finite Element Method in Engineering Science", Mac Graw Hill, London 1971

25 SIEVERDING C., "Base Pressure Measurements in Transonic Turbine Cascades", Von Karman Institute for Fluid Dynamics, LS 84, 1976

26 BREUGELMANS F.A.E., in : Von Karman Institute for Fluid Dynamics LS 59, 1973

27 HOLST T.L., "An Implicit Algorithm for the Conservative Transonic Full Potential Equation using an Arbitrary Mesh, AIAA paper 78-1113, 1978 also AIAA Journal Vol. 17, No. 10, 1979

28 HAFEZ M., MURMAN E.M., SOUTH J.C., "Artificial Compressibility Methods for Numerical Solution of Transonic Full Potential Equation", AIAA paper 78-1148, 1978

29 SORENSON R.L., STEGER J.L., "Simplified Clustering of Nonorthogonal Grids Generated by Elliptic Partial Differential Equations", NASA TM 73252, Aug. 1977

30 DECONINCK H., HIRSCH Ch., "Transonic Flow Calculations with Finite Elements", GAMM workshop on numerical methods for the computation of inviscid transonic flow with shock waves", Stockholm, 1979

31 HOLST T.L., BALLHAUS W.F., "Conservative Implicit Schemes for the Full Potential Equation Applied to Transonic Flows", NASA TM 78469, 1978

COMPUTING METHODS IN APPLIED SCIENCES AND ENGINEERING
R. Glowinski, J.L. Lions (editors)
North-Holland Publishing Company

METHODE VARIATIONNELLE POUR LES PROBLEMES HYPERBOLIQUES ET MIXTES DU PREMIER ORDRE

par J.J. CHATTOT
OFFICE NATIONAL D'ETUDES ET DE RECHERCHES AEROSPATIALES
29, avenue de la Division Leclerc
92320 CHATILLON - FRANCE

RESUME

Les méthodes de moindres carrés ont été souvent utilisées pour résoudre des problèmes de mécanique des fluides. Nous montrons ici leur intérêt spécifique pour la résolution des problèmes hyperboliques et mixtes du premier ordre. D'une part la formulation moindres carrés permet de transformer le problème du premier ordre en un problème "équivalent" du second ordre, mieux adapté aux méthodes de discrétisation et à la résolution par relaxation. D'autre part la fonctionnelle ainsi construite favorise l'utilisation de techniques nouvelles de pénalisation et de contrôle optimal. Les discrétisations de type éléments finis sont alors directement applicables. Des exemples concrets d'utilisation de cette approche sont présentées pour des problèmes stationnaires et instationnaires.

INTRODUCTION -

Bien que historiquement, pour le calcul des écoulements transsoniques stationnaires, les méthodes instationnaires aient été les premières méthodes mises en oeuvre en vue de fournir des solutions complètes avec ondes de choc, (voir par exemple les travaux de Singleton [1], Laval [2] , etc...), les méthodes stationnaires se sont développées rapidement à la suite de la publication par Murman et Cole [3] d'une méthode de relaxation pour l'équation du potentiel des petites perturbations transsoniques. L'originalité de cette méthode réside dans l'utilisation d'un schéma mixte de discrétisation, centré en subsonique et décentré "amont" en supersonique, et d'une méthode de relaxation par colonne ayant de bonnes propriétés de convergence. Les développements ultérieurs ont abouti à des méthodes de résolution de l'équation complète du potentiel pour des profils et des ailes en écoulement transsonique [4 à 6] .

Il faut noter que les méthodes de relaxation couramment employées utilisent un modèle à une équation aux dérivées partielles du second ordre pour le potentiel des vitesses de petites perturbations ou pour le potentiel complet. Il semble en fait que le succès de ces méthodes soit lié à l'utilisation de la variable potentiel. Certes, si l'on compare avec le système des équations d'Euler, nécessitant la résolution de 4 équations pour les 4 inconnues ρ , u , v , p en deux dimensions d'espace, l'équation complète du potentiel pour la seule variable Φ correspond à un gain substantiel en volume de calcul pour une discrétisation comparable de l'espace physique. Mais cet aspect ne parvient pas à lui seul à expliquer la différence d'efficacité qui existe entre les méthodes de résolution de l'équation du potentiel et les méthodes de résolutions des équations d'Euler. Par contre il convient de remarquer qu'il n'existe pas de méthode de relaxation pour les équations d'Euler stationnaires. Les solutions "Euler" sont obtenues à partir des équations instationnaires pour de grandes valeurs de la variable temps. Malheureusement les schémas explicites ont une limitation sur le pas de temps Δt pour maintenir la stabilité (condition de Courant - Friedriechs - Lewy, CFL) et les propriétés de convergence vers la solution stationnaire ne sont pas aussi favorables que pour les schémas de relaxation. L'introduction de schémas implicites tels que ceux de Beam et Warming [7] n'a pas encore changé de façon notable cette conclusion.

En fait la discrétisation d'équations du premier ordre au moyen de schémas centrés conduit à des matrices mal conditionnées. L'utilisation d'un schéma implicite, s'il n'est pas accompagné d'un bon conditionnement de la matrice itérative, ne permettra pas en pratique de s'affranchir de la condition de stabilité du schéma explicite.

Nous allons montrer, dans une première partie, que l'utilisation d'une formulation variationnelle de type moindres carrés, pour les équations stationnaires, permet d'élever l'ordre du système et de formuler un "problème équivalent" aux dérivées partielles secondes conduisant, après discrétisation au moyen de schémas centrés, à des matrices bien conditionnées. Cette approche sera illustrée par des exemples d'application aux équations d'Euler monodimensionnelles (problème mixte) et à un problème de convection (problème hyperbolique).

Dans une deuxième partie une application récente de cette approche à une équation non linéaire de propagation d'une onde indique qu'il est possible d'étendre la formulation moindres carrés aux problèmes instationnaires.

1 - PROBLEMES STATIONNAIRES -

Pour les problèmes stationnaires elliptiques et mixtes du premier ordre, les méthodes de relaxation n'ont pas produit d'algorithmes d'efficacité comparable à ceux utilisés pour les équations du second ordre. Pour les problèmes purement hyperbolique en revanche on sait résoudre les équations d'Euler stationnaires à partir d'une donnée amont correspondant à un écoulement supersonique (réf. [8]). Si le vecteur vitesse est sensiblement aligné avec l'axe Ox , les équations sont intégrées pas à pas dans cette direction, la variable x étant alors analogue à une variable de temps. Les schémas numériques sont décentrés vers l'amont et ne permettent évidemment pas de traiter des régions, mêmes limitées, d'écoulement subsonique (équation elliptique), l'information ne pouvant avec de tels schémas se transmettre d'aval en amont. Avec ces schémas les matrices du système discrétisé sont bien conditionnées.

Lorsque le système est de type elliptique ou mixte, il serait naturel d'utiliser, au moins en subsonique, des schémas centrés afin de tenir compte du domaine de dépendance des équations. Malheureusement la discrétisation des dérivées partielles premières avec un schéma centré, par exemple $\left(\frac{\partial u}{\partial x}\right)_i = \frac{u(x_i+\Delta x) - u(x_i-\Delta x)}{2\Delta x}$ conduit à une matrice mal conditionnée ayant des zéros sur la diagonale principale. Les solutions obtenues ainsi sont fortement oscillatoires avec un découplage des lignes d'indice pair et impair.

Des méthodes de tir s'inspirant de la procédure utilisée en hyperbolique ont pu être utilisées. Il faut alors introduire des conditions aux limites surabondantes sur une frontière afin de procéder à l'intégration en direction de la frontière opposée, où les conditions aux limites ne seront pas satisfaites, mais où l'écart détecté permettra une remise en cause des conditions fictives. Mentionnons à ce propos les travaux de Chushkin [9] , Euvrard et Tournemine [10] , Chattot [11] , et le livre de Holt [12] dans lequel une revue d'ensemble de ces méthodes peut être trouvée.

Afin de tenter d'approfondir les difficultés inhérentes à la résolution des systèmes du premier ordre, il peut être instructif de comparer les deux formulations équivalentes du point de vue mathématique de l'écoulement transsonique autour d'un demi-profil dans un canal non bloqué en théorie des petites perturbations.

$$\begin{cases} \dfrac{\partial \sigma(u)}{\partial x} + \dfrac{\partial v}{\partial y} = 0 \;, \quad \sigma(u) = \left(1 - M_\infty^2 - \dfrac{\gamma+1}{2} M_\infty^2 u\right) u & (1.a) \\ -\dfrac{\partial v}{\partial x} + \dfrac{\partial u}{\partial y} = 0 & (1.b) \end{cases}$$

u et v sont les composantes de la vitesse de perturbation, M_∞ est le nombre de Mach de référence.

$$(2)\qquad \frac{\partial}{\partial x}\left[\sigma\left(\frac{\partial\varphi}{\partial x}\right)\right]+\frac{\partial}{\partial y}\left[\frac{\partial\varphi}{\partial y}\right]=0$$

φ est le potentiel de perturbation.

Les conditions aux limites pour un canal à section constante sont les suivantes:

$$(3)\quad\begin{cases} u_0=\frac{\partial\varphi}{\partial x}{}_0 = u_1=\frac{\partial\varphi}{\partial x}{}_1 & \text{donnés à l'amont et à l'aval}\\ v_0=\frac{\partial\varphi}{\partial y}{}_0=f'(x) & \text{où } f'(x) \text{ est la pente de la paroi équipée du demi-profil}\\ v_1=\frac{\partial\varphi}{\partial y}{}_1=0 & \text{sur la paroi supérieure}\end{cases}$$

La résolution du problème (2) (3) est maintenant bien établie [3] . Un schéma mixte à quatre opérateurs doit être utilisé pour la discrétisation du terme non-linéaire et un schéma centré est utilisé pour le second terme [13] . La matrice du système discret possède 5 diagonales et est bien conditionnée par construction du schéma mixte. La résolution est effectuée par une méthode itérative de relaxation par colonne, dont la matrice est tridiagonale à diagonale dominante. Les conditions aux limites sont prises en compte dans cette résolution. Les résultats obtenus par Collins et Krupp [14] mettent en évidence les excellentes propriétés de cette méthode.

La résolution numérique du problème (1) (3) n'est pas classique. Par analogie avec la formulation utilisant le potentiel, une méthode de relaxation par colonne conduit à calculer v à partir de (1.a) et u à partir de (1.b) La question se pose alors du choix du schéma permettant d'approcher les dérivées $\frac{\partial}{\partial y}$ le long de la colonne, et du traitement des conditions aux limites (on rappelle que v est imposé à chaque extrémité, et qu'il n'existe aucune condition sur u). Comme nous l'avons vu, un schéma centré conduit à une matrice mal conditionnée. Steger et Lomax [15] ont proposé un schéma décentré alternativement vers le haut et vers le bas. Nous avons montré dans [16] que ce schéma n'est pas conservatif et produit des oscillations (sur la variable v) associées à la résolution, le long d'une colonne, d'une équation différentielle ordinaire du premier ordre assujettie à une condition de Dirichlet à chaque extrémité. Pour ces raisons un "système équivalent" du second ordre à été introduit (réf. [16]) en dérivant les équations (1) par rapport à y (équations (4)) et en imposant deux conditions aux limites complémentaires (équations (1) discrétisées aux extrémités de la colonne) :

$$\begin{cases} (4.a)\quad \frac{\partial^2}{\partial x\partial y}\sigma(u) + \frac{\partial^2 v}{\partial y^2}=0\\ (4.b)\quad -\frac{\partial^2 v}{\partial x\partial y} + \frac{\partial^2 u}{\partial y^2}=0\end{cases}$$

La transformation du système du premier ordre en un "système équivalent" du second ordre a permis d'obtenir des résultats aussi bien pour le problème direct (géométrie imposée) que pour le problème inverse (pression ou vitesse imposée sur une paroi). Par "système équivalent", il faut comprendre - admettant la même solution. Il est facile de montrer que c'est en effet le cas. L'intégration des équations (4) par rapport à y fait apparaître une constante d'intégration qui est mise à zéro par application des conditions complémentaires (1).

La généralisation de cette approche consistant à élever l'ordre du système a pu être réalisée de façon systématique par l'introduction d'une méthode variationnelle de type moindres carrés. Les avantages d'une telle formulation sont multiples. D'une part, comme nous le verrons plus bas, l'utilisation de schémas centrés conduit à des matrices bien conditionnées. D'autre part la fonctionnelle ainsi introduite permet d'envisager de nouveaux types de traitements tels que pénalisation de la fonctionnelle et contrôle optimal. Enfin, cette formulation est bien adaptée à des applications de type éléments finis.

Considérons tout d'abord le problème de l'écoulement de fluide parfait dans une tuyère très élancée. L'utilisation des équations d'Euler apparaît comme un modèle physique beaucoup plus significatif que le modèle isentropique utilisant l'équation du potentiel, pour lequel la position du choc ne dépend pas de la pression aval. Les équations d'Euler quasi-monodimensionnelles s'écrivent :

(5.a) $$\frac{\partial \rho u g}{\partial x} = 0$$

$$x \in \mathcal{D} = [x_1, x_m]$$

(5.b) $$\frac{\partial}{\partial x}[(\rho u^2 + p)g] - p g' = 0$$

ρ est la masse volumique

u est la vitesse

$g(x)$ représente la section de la tuyère

p est la pression, calculée à partir de l'enthalpie d'arrêt h_0

(6) $$\frac{\gamma}{\gamma - 1}\frac{p}{\rho} + \frac{u^2}{2} = h_0$$

γ représente le rapport des chaleurs spécifiques supposées constantes.

Les conditions aux limites sont les conditions physiques de fonctionnement d'une tuyère :

(7) $$\begin{cases} s_1 = 0 & \text{entropie amont fixée,} \quad s = C_v \operatorname{Log} \dfrac{p}{p_\infty \rho^\gamma} \\ p_m & \text{pression de sortie fixée} \end{cases}$$

Le système (5)-(7) admet des solutions avec sauts. Pour assurer l'unicité de la solution, il faut ajouter l'inégalité sur l'entropie :

$$(8) \quad \frac{\partial}{\partial x} \rho u g s \geq 0$$

La formulation variationnelle introduite dans [17] s'écrit comme suit :

$$\begin{cases} (9.a) \quad \operatorname*{Inf}_{\rho} I \quad , \quad I = \int_{\mathcal{D}} \left\{ \frac{\partial}{\partial x} \rho u g \right\}^2 dx \\ (9.b) \quad \operatorname*{Inf}_{u} J \quad , \quad J = \int_{\mathcal{D}} \left\{ \frac{\partial}{\partial x} [(\rho u^2 + p) g] - p g' \right\}^2 dx \end{cases}$$

L'équation (9.a) est linéaire par rapport à ρ ; (9.b) en revanche est non-linéaire par rapport à u et admet deux solutions en chaque point. La sélection de la "bonne solution" doit tenir compte de la condition d'entropie. Pour cette raison, la fonctionnelle J est modifiée et la formulation complète s'apparente à un problème de contrôle optimal :

$$(10) \quad \operatorname*{Sup}_{h>0} \operatorname*{Inf}_{u} K \ , \ K = J + H \ , \ H = -\int_{\mathcal{D}} h \left\{ \frac{\partial}{\partial x} \rho u g s \right\} dx$$

L'extremum sur h , $\operatorname*{Sup}_{h>0}$, est recherché itérativement avec l'algorithme d'Uzawa [18] :

$$(11) \quad h^{n+1} = \operatorname{Max} \left\{ h^n - \omega \frac{\partial}{\partial x} \rho u g s , 0 \right\}$$

ω est un facteur de relaxation choisi égal au pas de discrétisation Δx.
n est un indice indiquant le niveau de l'itération. On peut remarquer que la résolution de l'équation (10) implique celle de (9.b) et de l'inégalité d'entropie (8), car le terme H est nul à convergence.

Les équations (9) sont discrétisées au moyen de schémas centrés . Les matrices correspondantes sont tridiagonales à diagonale dominante. La méthode itérative utilise un schéma implicite d'une extrémité à l'autre du domaine. Par contre le terme H dans l'équation (10) est discrétisé au moyen d'un schéma décentré vers l'amont et est traité explicitement. Des résultats obtenus dans le cas d'une tuyère de forme parabolique en régime non bloqué (écoulement subsonique) et bloqué (écoulement subsonique et supersonique avec choc) sont représentés sur les figures 1 et 2 (référence [17]). On peut noter en particulier pour le cas bloqué l'influence de la condition de pression imposée à l'aval sur la position de l'onde de choc, et le saut d'entropie qui lui est associé .

Cette même approche a été mise en oeuvre pour résoudre un problème modèle hyperbolique. Il s'agit d'une première étape en vue de la résolution des équations d' Euler stationnaires en écoulement bidimensionnel. L'équation de conservation de la masse, à champ de vitesse donné, correspond à un problème de convection. Le domaine physique Ω représente une tuyère bidimensionnelle de forme parabolique (fig. 3). Le problème du premier ordre s'écrit :

$$(12) \quad \frac{\partial \rho u}{\partial x} + \frac{\partial \rho v}{\partial y} = 0$$

avec la condition amont :

$$(13) \quad \rho(x,y) = \rho_1(x,y) \quad si \quad (x,y) \in \Sigma_1$$

($\vec{q}.\vec{n}<0$, $\vec{n}$ normale extérieure) Σ_1 est la section d'entrée de la tuyère.

Comme nous l'avons indiqué plus haut, la formulation de ce problème hyperbolique du premier ordre ne permet pas l'emploi de schémas centrés et d'une méthode de résolution itérative par relaxation. Par contre il est possible de résoudre (12) (13) par des méthodes d'intégration pas à pas dans la direction x [8 - 10] , en tenant compte du domaine de dépendance d'un point. Une application par éléments finis a été présentée dans [19] dans le cas d'un champ de vitesse uniforme.

La formulation moindres carrés introduite dans [20] autorise une plus grande souplesse dans le traitement numérique. Soit I la fonctionnelle des moindres carrés, le problème s'écrit :

$$(14) \quad \operatorname*{Inf}_{\rho} I \quad , \quad I = \int_{\Omega} \left\{ \frac{\partial \rho u}{\partial x} + \frac{\partial \rho v}{\partial y} \right\}^2 dx\,dy$$

La discrétisation de (13) et (14) a été réalisée, sur un même maillage, par une méthode de différences finies et d'éléments finis. Pour les différences finies des schémas centrés et précis au second ordre sont utilisés aux noeuds intérieurs, et des schémas décentrés vers l'amont, précis au premier ordre, sont utilisés aux points de la frontière Σ_2 de Ω où l'écoulement vérifie $\vec{q}.\vec{n} \geq 0$. Pour la méthode d'éléments finis les produits ρu et ρv sont approchés sur les quadrangles par des polynômes du premier degré en x et y (approximation de type Q_1). Le maillage est présenté sur la figure 3. Des résultats ont été obtenus avec les deux méthodes de discrétisation dans le cas d'un champ de vitesse régulier figures 4, 5 et d'un champ discontinu simulant la présence d'un choc dans le divergent, figures 6 - 7. Les deux méthodes donnent de bons résultats (conservation du débit à travers la tuyère). On peut noter à la fois une plus grande précision de la méthode d'éléments finis et une plus grande simplicité dans le traitement des points frontière. Pour ce dernier exemple un calcul a été également effectué avec une méthode instationnaire utilisant un maillage légèrement plus fin dans la direction Ox (réf. [21]). Une comparaison des valeurs de la masse volumique sur l'axe de symétrie de la tuyère ($y = 0$) obtenues par les trois méthodes est présentée sur la figure 8.

2 - PROBLEMES INSTATIONNAIRES -

Pour les problèmes instationnaires, la résolution des systèmes du premier ordre par différences finies utilisant des schémas explicites est bien établie. Par analogie avec les schémas semi-implicites ou implicites de relaxation qui contribuent à l'efficacité de ces méthodes, des schémas implicites ont été proposés pour les équations instationnaires. Nous avons signalé l'importance qu'il faut attacher au bon conditionnement de la matrice du système discrétisé. Récemment, Lerat [22] a mis en évidence une classe de schémas implicites conservatifs, précis au second ordre, stables et à dominance diagonale. Nous montrons ici que la formulation moindres carrés s'étend aux problèmes instationnaires et permet, au moyen de schémas centrés simples possédant essentiellement les mêmes propriétés que les schémas de Lerat, de résoudre une équation non linéaire de propagation d'une onde. La variable temps est traitée de façon implicite (comme la variable x du problème précédent). Des schémas "implicites en temps" n'ont, à notre connaissance, pas été utilisés dans les méthodes de discrétisation, la variable temps jouant toujours un rôle privilégié. Morchoisne [23] utilise un schéma implicite en temps pour la résolution des équations de Navier-Stokes par une méthode pseudo-spectrale. La fonctionnelle est pénalisée pour tenir compte de "l'inégalité d'entropie".

Le problème à résoudre est le suivant :

$$(15)\qquad \frac{\partial u}{\partial t} + \frac{\partial u^2/2}{\partial x} = 0 \qquad \text{dans } \Omega$$

Les conditions aux limites et initiale sont :

$$(16)\qquad \begin{cases} u(t,0) = u_0 > 0 & \text{sur } \Sigma_0 \\ u(t,1) = u_1 < 0 & \text{sur } \Sigma_1 \\ u(0,x) = u^0(x) & \text{sur } S^0 \end{cases}$$

Il n'y a pas de condition sur S^1 (figure 9).

L'équation (15) est non-linéaire et admet plusieurs solutions faibles. L'inégalité suivante permet de sélectionner la solution ayant un saut lorsque $u^0_g > u^0_d$ $(x_g \leq x_d)$

$$(17)\qquad \frac{\partial u^2/2}{\partial t} + \frac{\partial u^3/3}{\partial x} \leq 0$$

Sous forme variationnelle équivalente à (15) - (17) le problème s'écrit :

$$(18)\qquad \operatorname*{Inf}_{u} I \ , \quad I = \int_{\Omega} \left\{ \frac{\partial u}{\partial t} + \frac{\partial u^2/2}{\partial x} \right\}^2 dt\,dx + \alpha \int_{\Omega} \left\{ \left| \frac{\partial u^2/2}{\partial t} + \frac{\partial u^3/3}{\partial x} \right| + \frac{\partial u^2/2}{\partial t} + \frac{\partial u^3/3}{\partial x} \right\}^2 dt\,dx$$

$\alpha > 0$ est d'ordre unité.

La minimisation de I conduit à une équation de second ordre dans Ω , qui est discrétisée au moyen d'un schéma centré en espace et en temps :

$$(19) \quad \left\{\frac{\partial}{\partial t} + u\frac{\partial}{\partial x}\right\}\left\{\frac{\partial u}{\partial t} + \frac{\partial u^2/2}{\partial x}\right\} + \alpha\left\{u\frac{\partial}{\partial t} + u^2\frac{\partial}{\partial x}\right\}\left\{\left|\frac{\partial u^2/2}{\partial t} + \frac{\partial u^3/3}{\partial x}\right| + \frac{\partial u^2/2}{\partial t} + \frac{\partial u^3/3}{\partial x}\right\} = 0 \quad \text{dans } \Omega$$

Sur S^1 l'équation (18) conduit à une équation du premier ordre qui est discrétisée au moyen d'un schéma décentré en espace et en temps suivant la direction caractéristique locale:

$$(20) \quad \frac{\partial u}{\partial t} + \frac{\partial u^2/2}{\partial x} + \alpha u\left\{\left|\frac{\partial u^2/2}{\partial t} + \frac{\partial u^3/3}{\partial x}\right| + \frac{\partial u^2/2}{\partial t} + \frac{\partial u^3/3}{\partial x}\right\} = 0 \quad \text{sur } S^1$$

Les équations (19) et (20) sont résolues par une méthode de relaxation par point. Le champ u est initialisé en tout point du domaine $\Omega = (0,1) \times (0,1/2)$ avec une fonction qui n'est pas solution des équations (19) et (20).

$$\begin{cases} u(x,t) = 1 & x < \frac{1}{2} + t \\ u(x,t) = -1 & x \geq \frac{1}{2} + t \end{cases}$$

La solution finale, obtenue à convergence, correspond pour ce problème modèle à la solution exacte (figure 10).

$$\begin{cases} u(x,t) = 1 & x < \frac{1}{2} \\ u(x,t) = -1 & x \geq \frac{1}{2} \end{cases}$$

CONCLUSION -

La méthode de moindres carrés proposée pour le traitement des équations et systèmes du premier ordre a été appliquée à des problèmes stationnaires hyperboliques et mixtes. Les avantages de cette formulation se manifestent par la simplicité des schémas qu'elle autorise et le bon conditionnement des matrices qui en découle, permettant l'emploi de schémas de résolution semi-implicites et implicites contribuant à l'efficacité de la méthode.

Une application récente à une équation non-linéaire de propagation d'une onde indique qu'il est possible d'étendre cette approche aux problèmes instationnaires. L'intérêt d'une telle approche peut se justifier pour s'affranchir d'une condition de stabilité (critère de Courant-Friedrichs-Lewy) lors de la recherche de solutions périodiques dans le temps.

REFERENCES

[1] SINGLETON R.E. - "Lax-Wendroff difference scheme applied to the transonic airfoil problem", Transonics Aerodynamics, AGARD Conf. Proc. Paris, sept. 1968

[2] LAVAL P. - "Méthodes instationnaires de calcul de l'écoulement transsonique dans une tuyère", N.T. ONERA n° 173, déc. 1970.

[3] MURMAN E.M. et COLE J.D. - "Calculation of plane steady transonic flows", AIAA Journal, janv. 1971.

[4] JAMESON A. -" Iterative solution of transonic flows over airfoils and wings, including flows at Mach 1.", Comm. Pure and Applied Math., vol 27 (1974), p. 288-309.

[5] CHATTOT J.J., COULOMBEIX C. et TOME C. - "Ecoulements transsoniques autour d'ailes", La Recherche Aérospatiale 1978-4, p. 143-159

[6] HOLST T.L. - "A fast, conservative algorithm for solving the transonic full potential equation", AIAA Paper n° 79-1456

[7] BEAM R.M. et WARMING R.F. - "Numerical calculations of two-dimensional unsteady transonic flows with circulation", NASA TN D-7605, fev. 1974.

[8] Mac CORMACK R.W. et WARMING R.F. - "Survey of computational methods for three-dimensional supersonic inviscid flows with shocks", in Adv. in Num. Fl. Dyn., AGARD LS-64, 1973.

[9] CHUSHKIN P.I., Dok. Akad. Nauk SSSR 125, 748-751, 1959.

[10] EUVRARD D. et TOURNEMINE G. - "Méthode directe de calcul de l'écoulement sonique autour d'un profil d'aile donné. Cas d'un écoulement symétrique"., Journal de Mécanique, vol. 12 n° 3, sept. 1973.

[11] CHATTOT J.J. - "Symmetrical flow past a double wedge at high subsonic Mach numbers", J. Fluid Mech., vol 86, part 1, pp. 161-177, 1978.

[12] HOLT M., Numerical Methods in Fluid Dynamics, Springer-Series in Computational Physics, Springer-Verlag, 1977.

[13] MURMAN E.M. - "Analysis of embedded shock waves calculated by relaxation methods", AIAA Journal, vol 12, n° 5, p. 626-633, 1974

[14] COLLINS D.J. et KRUPP J.A. - "Experimental and theoretical investigations in two-dimensional transonic flow", AIAA Paper n° 73-659.

[15] STEGER J.L. et LOMAX H. - "Generalized relaxation methods applied to problems in transonic flow", Lecture Notes in Physics, vol. 8, Springer-Verlag, p. 193-198, 1971

[16] CHATTOT J.J. - "Une méthode de relaxation entièrement conservative pour les écoulements transsoniques", La Recherche Aérospatiale n° 1975-6, p. 339-346. English version available as T.P. 1975-93 E.

[17] CHATTOT J.J. - "Relaxation approach to the steady Euler equations in transonic flow", AIAA Paper n° 77-636

[18] CEA J., Optimisation, Théorie et Algorithmes, Méthodes Mathématiques de l'Informatique, Dunod, 1971.

[19] LESAINT P. et RAVIART P.A. - "On a Finite Element Method for Solving the Neutron Transport Equation", Math. Aspect of Finite El. Meth., Partial Diff. Equations, Academic Press, 1974

[20] CHATTOT J.J. - GUIU-ROUX J. et LAMINIE J. - "Résolution numérique d'une équation de conservation par une approche variationnelle", Proceedings VI int. Conf. Num. Meth. Fl. Dyn., Tbilisi, June 20-25, 1978, USSR, Springer-Verlag 1979

[21] VEUILLOT J.P. et VIVIAND H. - "A Pseudo-Unsteady method for the computation of Transonic Potential Flows", 11e Conférence sur la Dynamique des Fluides et des Plasmas, organisée par l'AIAA Seattle, 10-11 Juillet 1978, AIAA Paper n° 78-1150.

[22] LERAT A. - "Une classe de schémas aux différences implicites pour les systèmes hyperboliques de lois de conservation", C.R. Acad. Sc. Paris, t. 288, série A, 18 Juin 1979

[23] MORCHOISNE Y. - "Résolution des équations de Navier-Stokes par une méthode pseudo-spectrale en espace-temps", La Recherche Aérospatiale n° 1979-5, p. 293.306.

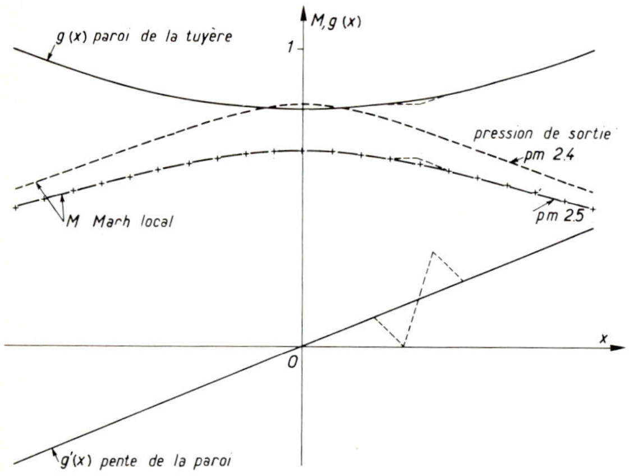

Fig. 1 - Ecoulement quasi-monodimensionnel dans une tuyère.

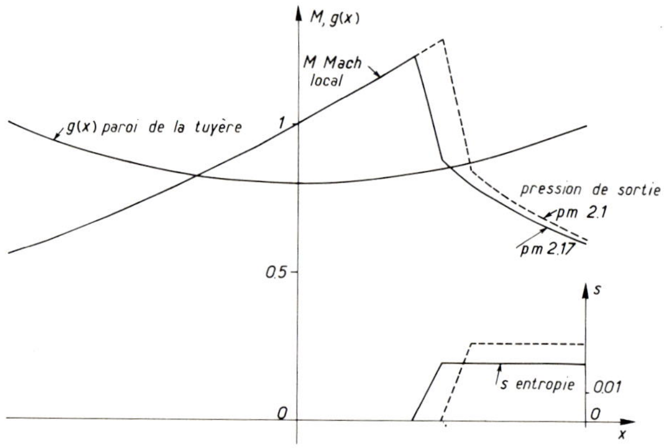

Fig. 2 - Ecoulement quasi-monodimensionnel dans une tuyère.

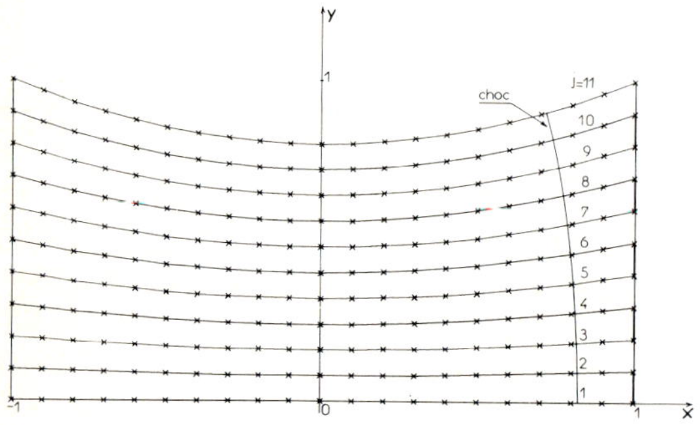

Fig. 3 - Maillage.

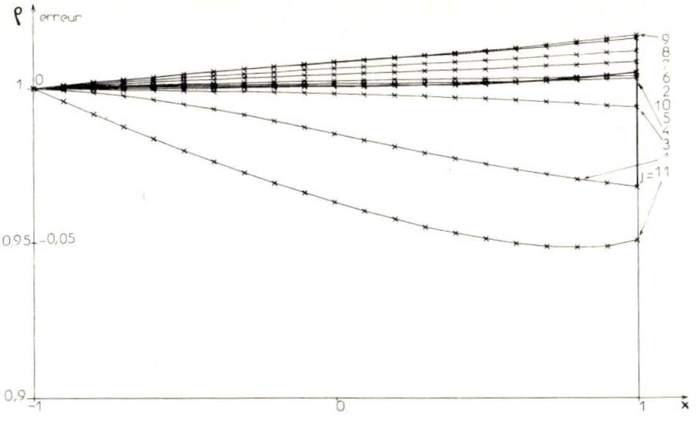

Fig. 4 - Champ de vitesse à divergence nulle. Méthode des différences finies.

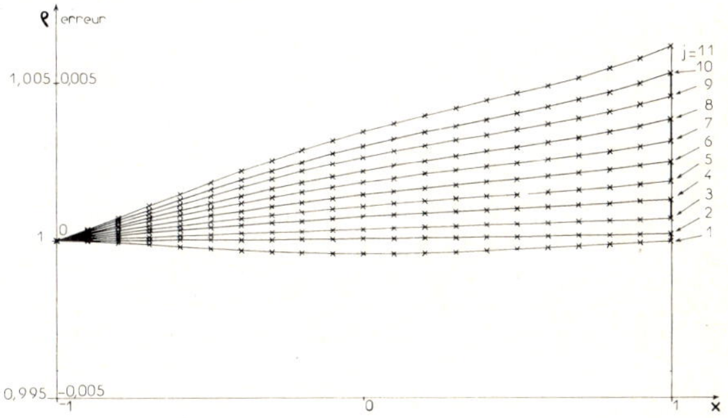

Fig. 5 - Champ de vitesse à divergence nulle. Méthodes des éléments finis.

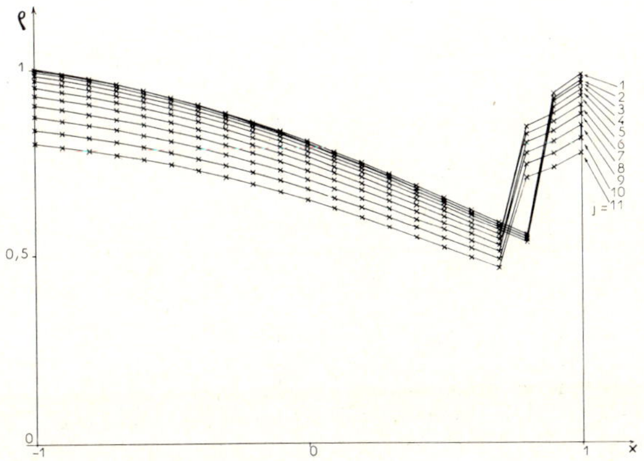

Fig. 6 - Simulation d'un choc. Méthode des différences finies.

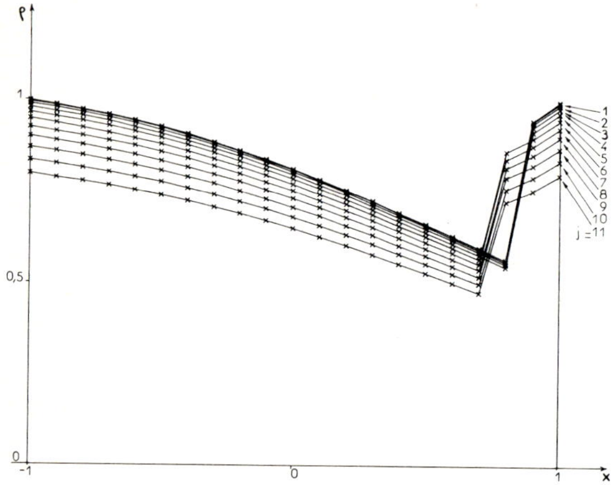

Fig. 7 - Simulation d'un choc. Méthode des éléments finis.

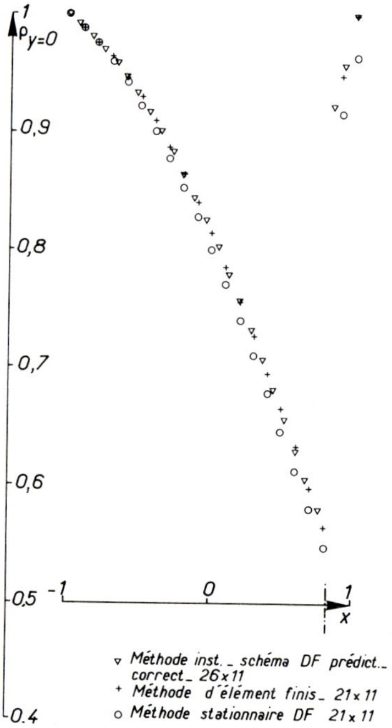

Fig. 8 - Masse volumique sur l'axe de la tuyère. Comparaison entre les méthodes des différences finies et des éléments finis.

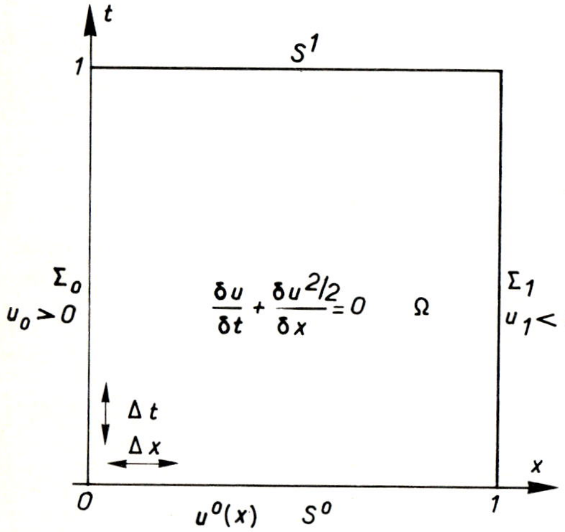

Fig. 9 Problème aux valeurs initiales et aux limites pour une équation non-linéaire de propagation d'une onde.

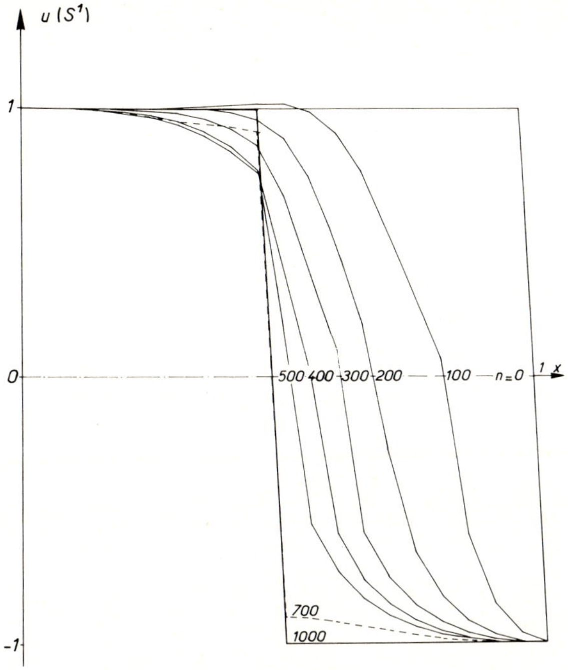

Fig. 10 - Convergence de la solution sur la frontière S^1 en fonction du nombre d'itérations.

COMPUTING METHODS IN APPLIED SCIENCES AND ENGINEERING
R. Glowinski, J.L. Lions (editors)
North-Holland Publishing Company

VORTEX METHOD WITH FINITE ELEMENT FOR THE EULER EQUATION

C. Bardos*, M. Bercovier†, O. Pironneau*

Introduction :

We propose to use finite elements approximation in the vortex method for the Euler equation. The vortex method has been investigated by several authors, Christiansen [6], Chorin [5], Baker [2] and Hald [10]. In these works finite differences and transport equation of discrete vortex plus regularization are generally used instead of the finite element method. We believe that the finite elements are more suitable for domain with boundaries and we will obtain an error bound which is valid for all time. The error bound turn out to be very weak for large time and this is related to the estimate of the so called "pair dispersion". The study of the "pair dispersion" namely the behaviour of the distance of two particles of fluid is a basic ingredient both in the theoretical study of the regularity and in the computation of the error bound. In fact the weakness of the error bound (for $t \to \infty$) is related to the fact that for $t \to \infty$ two particles can come very close one from the other (at a rate increasing like $s^{e^{kt}}$ $(0 < s < 1)$)

The paper is organized as follows. In section 1 we describe the exact and approximate problem. In section 2 we prove the error bound in section 3 we give some application to the exterior problem. Section 4 contains some numerical examples. Finally detailed proofs will appear elsewhere.

I. Exact and approximate problem.

In an open set Ω of $\mathbb{R}^2$, we will consider the Euler equation for an inviscid incompressible fluid.

(1) $$\frac{\partial u}{\partial t} + u\,\nabla u = -\nabla p \quad , \quad \nabla.u = 0 \ , \quad \text{in} \quad \Omega \times \mathbb{R}_+$$

(2) $$u(x,t).\nu(x) = 0 \quad \text{for} \quad x \in \partial\Omega \quad ; \quad u(x,0) = u_0(x) \ \text{in} \ \Omega .$$

(*) University of Paris 13.
(†) University of Jerusalem.

Unless specified otherwise, we will assume that Ω is a smooth bounded open and convex set in $\mathbb{R}^2$. $\nu(x)$ denotes the exterior normal to the boundary $\partial\Omega$ of Ω. It is known (Kato [11], Schaeffer [13], Wolibner [14]) that, for smooth initial data $(u(x,0) = u_0(x) \in C^{1,\alpha}(\Omega))$ the solution of (1), (2) is uniquely determined and smooth.

Taking the end of the equation (1) one obtains a transport equation for $\omega = \nabla \wedge u$:

$$(3) \qquad \frac{\partial\omega}{\partial t} + u\nabla\omega = 0 \ .$$

When Ω is simply connected u is uniquely determined by the relations :

$$(4) \qquad u = \nabla\wedge\psi = \left(\frac{\partial\psi}{\partial x_2}, -\frac{\partial\psi}{\partial x_1}\right) \ ; \ -\Delta\psi = \omega \ ; \ \psi\big|_{\partial\Omega} = 0$$

therefore we will denote A the operator $\omega \to u$ defined by (4), then the equation (3) reads

$$(5) \qquad \frac{\partial\omega}{\partial t} + A(\omega).\nabla\omega = 0 \ , \ \omega(x,0) = \omega_0(x) = (\nabla\wedge u_0)(x) \ .$$

For the approximate problem, we will introduce a convex polygonal approximation of Ω, Ω_h constructed with a regular triangulation, $\mathcal{T}_h$ we will assume that if a side of one triangle belongs to $\partial\Omega_h$, then the corresponding two vertices belong to $\partial\Omega$. V_h will denote the space of continuous function on Ω_h which are linear and affine on each triangle $\mathcal{T}_h$ and which vanishes on the boundary $\partial\Omega_h$. For any function $\omega_h \in L^\infty(\Omega_h)$ we define the operator A_h by the relations :

$$(6) \qquad u_h = A_h\,\omega_h = \nabla\wedge\,\psi_h \ .$$

$$(7) \qquad \psi_h \in V_h \text{ and } \int_{\Omega_h} \nabla\psi_h\,\nabla\theta_h\,dx = \int_{\Omega_h} w_h\,\theta_h\,dx \quad \forall\,\theta_h \in V_h \ .$$

Now the discrete approximation of (5) will be the problem (P_h) :

$$(8) \qquad \frac{\partial\omega_h}{\partial t} + A_h(\omega_h^k)\,\nabla\,\omega_h = 0 \ , \ \omega_h(x,0) = \omega_0(x)$$

where on each interval of time $[k\Delta t,(k+1)\Delta t[$ ω_h^k is defined by the relation :

$$\text{for } x \in T_i \quad \omega_h^k(x,t) = \omega_h(\xi_i^k, k\Delta t) \ ,$$

where ξ_h^k may be chosen arbitrarily in the triangle T_i. The function ω_h^k is constant on every triangle and for t in any intervalle $[k\Delta t,(k+1)\Delta t[$. Furthermore $A_h(\omega_h^k)$ is a divergence free (in the distribution sense) vector field which is constant on every region of the form $T_i \times [k\Delta t,(k+1)\Delta t[$.

Remark : The solution of (8) is completely determined by the value of ω_h at the points $(\xi_h^k, k\Delta t)$ therefore it is a problem of finite dimension. The solution depends on the choice of the mesh points $\xi^k \in T_h$, but we will prove that the error estimate is uniform and does not depends on this choice.

Numerical solution of (8)

Let u_h^k denotes $A_h(\omega_h^k)$. The characteristics of (8) are the solutions $X_h^{x,t}(\tau)$ of

$$\frac{dX_h}{dt} = u_h(X_h(\tau),\tau) \qquad X_h(t) = x \ .$$

Since u_h is constant over each prism $P_{ij} = T_i \times [k\Delta t\ (k+1)\Delta t]$, $\{(X_h(\tau),\tau) : \tau \leq t\} \cap P_{ik}$ is a straight segment and X_h is parallel to u_h(figure 1) ; therefore each characteristic can be computed in a finite number of steps : Once $X_h^{x,t}(0)$ is known then the solution of (8) is

$$\omega_h(x,t) = \omega_0(X_h^{x,t}(0)) \ .$$

To compute ω_h^{k+1} from ω_h^ℓ, $\ell \leq k$, we may proceed as follows :

1. Compute $\eta = X_h^{\xi_i^k, k\Delta t}((k+1)\Delta t)$ for all i and set $\xi_{j(i)}^{k+1} = \eta_i$ for all i,j(i) such that $\eta_i \in T_{j(i)}$. Set $\omega_h(\xi_{j(i)}^{k+1},(k+1)\Delta t) = \omega_h(\xi_i^k,k\Delta t)$.

2. Let $J = \{j : j \neq j(i)\ \forall i\}$ then for all $j \in J$ choose ξ_j^{k+1} compute $X^{\xi_j^{k+1},(k+1)\Delta t}(0)$ and set $\omega_h(\xi_j^{k+1}(k+1)\Delta t) = \omega_0(x_h^{\xi_j^{k+1}}$

For $\Delta t \simeq h/u_h$ the total number of multiplications of this method is of $O(h^{-3})$ over $[0,T]$. It is similar in cost to the solution of a Dirichlet problem.

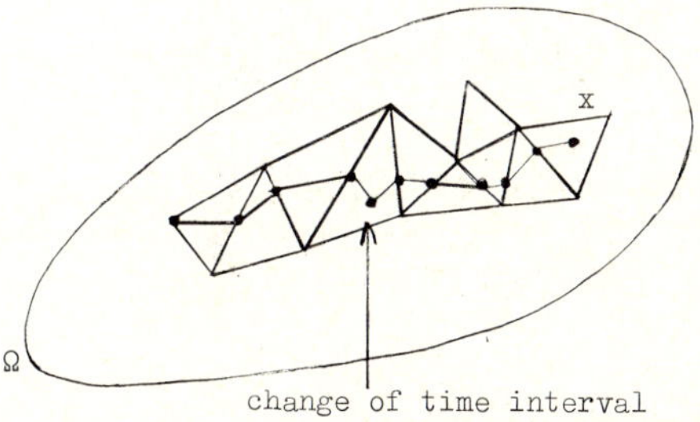

Figure 1.

II. Error Estimates.

$\omega_h(.,t)$ is the solution of the transport equation (8) therefore it is uniformly bounded in $L^\infty(\Omega)$ for all t

$$(|\omega_h(x,t)| \leq \sup_x |\omega_0(x)|) .$$

and of course it is also uniformly bounded in $L^p(\Omega)$ $(1 \leq p \leq \infty)$. To prove the regularity of the solution of (3) one usually start with this estimate, which turns out to be some how not sufficient for the computations. Therefore using the fact that ω is constant along the characteristics one will estimate the Holder norm $|\omega|_{0,\alpha}$ of the vorticity ω. All these methods are based on the two following facts : first it is not possible to obtain a uniform bound for the second derivatives $\nabla^2\Psi$ of the Dirichlet problem

$$(9) \qquad -\Delta \Psi = f \quad \text{in} \quad \Omega \quad , \quad \Psi|_{\partial\Omega} = 0$$

in term of the L^∞ norm of f. But on the other hand it is possible to obtain and "almost" lipschitzian bound for the gradient of Ψ. Namely, due to the singularity (as $\log|x-y|$) of the green of function of (9), one has

$$(10) \qquad |\nabla\Psi(x) - \nabla\Psi(y)| \leq C|x-y| \operatorname{Log} \frac{D}{|x-y|} \; |f|_\infty$$

Now for two particles of fluid x(t) and y(t), one has

$$(11) \qquad \dot{x}(t) = u(x(t),t) \;, \quad \dot{y}(t) = u(y(t),t)$$

therefore one obtains setting $\rho(t) = |x(t) - y(t)|$

$$(12) \qquad |\dot{\rho}(t)| \leq |u(x(t),t) - u(y(t),t)| \leq C\rho(t) \operatorname{Log} \frac{D}{\rho(t)} \quad (^1)$$

From (12) one finally obtains

$$(13) \qquad \left(\frac{\rho(0)}{D}\right)^{e^{c|\omega_0|_\infty t}} \leq \frac{\rho(t)}{D} \leq \left(\frac{\rho(0)}{D}\right)^{e^{-c|\omega_0|_\infty t}}$$

---

(1) D denotes the diameter of Ω.

This is the estimate for the "pair dispersion" which is used by Wolibner [14] Kato [11] and Schaeffer [13] for the regularity.

We will follow for the error bound the same method. It is convenient to define $\bar{\omega}_h$ and $\tilde{\omega}_h$ by the two following relations :

$$(14) \qquad \bar{\omega}_h(x,t) = \omega_h(\xi_i^k, \Delta t) \quad \text{if} \quad \xi_i^k \in T_i$$

$$\bar{\omega}_h(x,t) = 0 \quad \text{if} \quad x \in \Omega - \Omega_h$$

$$(15) \qquad \frac{\partial \tilde{\omega}_n}{\partial t} + A(\bar{\omega}_h).\nabla\tilde{\omega}_h = 0 \;,\; \tilde{\omega}_h(x,0) = \omega_0(x) \;.$$

$\tilde{\omega}_h$ will not be explicitely computed but will appear as a comparison term for the error.

Let $u_h(x,t)$ and $\tilde{u}_h(x,t)$ be the two vector fields.

$$(16) \qquad u_h(x,t) = A_h(\bar{\omega}_h(.,k\Delta t))(x) \;,$$

$$\tilde{u}_h(x,t) = A(\bar{\omega}_h(.,t))(x) \;.$$

We have, for the corresponding characteristics

$$(17) \qquad \dot{x}\,(s) = u_h(x(s),s), \quad \dot{\tilde{x}}(s) = u_h(\tilde{x}_h(s),s)$$

Therefore

$$(18) \qquad |\tilde{\omega}_h(x,t) - \bar{\omega}_h(x,t)| = |\omega_0(x(0)) - \omega_0(\tilde{x}(0))|$$

$$\leq |\nabla\omega_0|\; |x(0) - \tilde{x}(0)| \;.$$

where $x(s)$ and $\tilde{x}(s)$ denote the solutions of (17) which satisfies the condition :

$$(19) \qquad x(t) = \tilde{x}(t) = x.$$

Now, for $s \in [k\Delta t, (k+1)\Delta t[$ we compute

$$(20) \qquad |\dot{x}(s) - \dot{\tilde{x}}(s)| = |u_h(x(s),s) - \tilde{u}_h(\tilde{x}(s),s)|$$

$$\leq |u_h(x(s),k\Delta t) - \tilde{u}_h(\tilde{x}(s),k\Delta t)|$$

$$+ |\tilde{u}_h(\tilde{x}(s),s) - \tilde{u}_h(\tilde{x}(s),k\Delta t)|$$

$$\leq |\tilde{u}_h(\tilde{x}(s),k\Delta t) - \tilde{u}_h(x(s),k\Delta t)|$$

$$+ |\tilde{u}_h(x(s),k\Delta t) - u_h(x(s),k\Delta t)|$$

$$+ |\tilde{u}_h(\tilde{x}(s),s) - \tilde{u}_h(\tilde{x}(s),k\Delta t)|$$

The operator A is continuous from $L^\infty(\Omega)$ to $W^{1,p}(\Omega)$ for any p, $1 < p < \infty$. Therefore we have for the last term of the right hand side of (20), the relation

$$(21) \qquad |\tilde{u}_h(\tilde{x}(s),s) - \tilde{u}_h(\tilde{x}(s),k\Delta t)|$$

$$= |A(\bar{\omega}_h(.,s)(\tilde{x}) - A(\bar{\omega}_h(.,k\Delta t))(\tilde{x})|$$

$$\leq C|\bar{\omega}_h(.,s) - \bar{\omega}_h(.,k\Delta t)| \leq |u_h(.,t\Delta t)|_\infty \Delta t$$

Now if we define Ψ_h as the solution of $\Psi_h \in V_h$

$$\int_{\Omega_h} \nabla\Psi_h \nabla\Theta_h \, dx = \int_{\Omega_h} \bar{\omega}_h \Theta_h \, dx \qquad \forall \Theta_h \in V_h :$$

$$\forall \Theta_h \in V_h$$

and $\tilde{\Psi}_h$ as the solution of $-\Delta\tilde{\Psi}_h = \bar{\omega}_h$, $\tilde{\Psi}_h|_{\partial\Omega} = 0$, we obtain for the second term of the right hand side of (20) the estimate

$$(22) \qquad |\tilde{u}_h(x(s),k\Delta t) - u_h(x(s),k\Delta t)| \,. \leq |\Psi_h - \tilde{\Psi}_h|_{W^{1,\infty}(\Omega)}$$

Finally for the first term of (22)we use the estimate (10) and we obtain finally

$$(23) \qquad |x(t) - \tilde{x}(t)| \leq |\Psi_h - \tilde{\Psi}_h|_{W^{1,\infty}(\Omega)}$$

$$+ |u_h(.,k\Delta t)|_\infty \Delta t$$

$$+ \; C\,|x(t) - \tilde{x}(t)| \,\mathrm{Log}\, \frac{D}{|x(t)-\tilde{x}(t)|} \, |\omega_0|_\infty$$

It is clear (use Sobolev Theorem) that $\tilde{\Psi}_h$ is uniformly bounded in $W^{1,\infty}(\Omega)$, therefore if we obtain a uniform bound for $|\Psi_h-\tilde{\Psi}_h|$ in $W^{1,\infty}(\Omega)$, as consequence we will recover a uniform bound for $(u_h|_{L^\infty(\Omega)}$.

The bound for $|\Psi_h-\tilde{\Psi}_h|$ is related to the evaluation of the error in P_1 approximation it has been studied by Nitsche [12] and this is also described in the book of Ciarlet [9]. The only changes we have to consider are the following :

1. We want an estimate in term of the $L^\infty(\Omega)$ norm of the data, instead of the second derivative of the solution.

2. We consider a polygonal approximation of Ω : Ω_h.

These two difficulties can be easily overcome, this is described in a forth coming paper by the authors, and we obtain :

$$(24) \qquad |\Psi_h - \tilde{\Psi}_h|_{W^{1,\infty}(\Omega)} \leq C\, |\omega_h|_\infty\, h^{1-1/p}$$

From (23), (24) and the uniform estimate on ω_h we finally obtain, setting $\rho(t) = |x(t) - \tilde{x}(t)|$ the relation :

$$(25) \qquad |\dot{\rho}(s)| \leq C\, |\omega_0|_\infty [(\Delta t + h^{1-1/p}) + C\,\rho\, \mathrm{Log}\,\frac{D}{\rho}\,], \quad \rho(t) = 0.$$

The relation (25) gives a bound for $\rho(0)$ in term of

$$(26) \qquad \rho(0) \leq C(t, |\omega_0|_\infty)(\Delta t + h^{1-1/p})^{-e^{K|\omega_0|_\infty t}}$$

Using (18), we obtain

$$(27) \qquad |\bar{\omega}_h(.,t) - \tilde{\omega}_h(.,t)|_. \leq C\, (\Delta t + h^{1-1/p})^{\bar{e}^{K|\omega_0|_\infty t}}$$

The constant appearing in the right hand side of (27) depends on the C^2 norm of the initial data u_0, and on the interval of time $(0,T)$ where the computation is performed. It is independant of Δt and h.

Finally, we compare the curl of the exact solution $\omega = \nabla \Lambda u$ with $\tilde{\omega}_h$ and ω_h with ω_h, we write :

$$(28) \qquad 0 = \frac{\partial\omega}{\partial t} + A(\omega)\nabla\omega - \frac{\partial\tilde{\omega}_h}{\partial t} + A(\bar{\omega}_h)\,\nabla\tilde{\omega}_h =$$

$$\frac{\partial}{\partial t}(\omega - \tilde{\omega}_h) + A(\bar{\omega}_h)\nabla(\omega - \tilde{\omega}_h) - A(\bar{\omega}_h - \omega)\,\nabla\omega$$

$$= \frac{\partial}{\partial t}(\omega - \tilde{\omega}_h) + A(\bar{\omega}_h)\nabla(\omega - \tilde{\omega}_h) + A(\omega - \tilde{\omega}_h)\nabla\omega - A(\bar{\omega}_h - \tilde{\omega}_h)\,\nabla\omega\,.$$

From (28) we deduce that we have

$$(29) \qquad \left| \frac{\partial}{\partial t}(\omega-\tilde{\omega}_h) + A(\bar{\omega}_h)\nabla(\omega-\tilde{\omega}_h) \right| \leq C_1 |\omega-\tilde{\omega}_h| + C_2 |\bar{\omega}_h-\tilde{\omega}_h| .$$

Using (29), (27) and the Bellman-Gronwall lemma, one can finally conclude by the

<u>Theorem 1.</u> Assume that Ω is a smooth convex bounded set of $\mathbb{R}^2$; denote by ω the solution of (5) and by ω_h the solution of (8) : assume, that the corresponding initial data belongs to $C^2(\Omega)$, then one has the estimate

$$(30) \qquad |\omega - \tilde{\omega}_h| \leq C(\Delta t + h^{1-1/p})^{e^{-K|\omega_0|_\infty t}}$$

<u>Remark</u> : p can be arbitrary large ; on the other hand the exponent exp $- K|\omega_0|_\infty t$ will give, for t large a very bad error estimate. This seems related to the simplycity and the generality of the method. It may be possible to improve the error by chosing conveniently the point ξ_h^k ; however this will greatly increase the cost of the method. On the other hand an estimate of the type (30) appears often in fluid mechanic in two dimensions (cf. for instance Arnold [1]).

III. <u>The exterior problem and the Kutta Joukovski condition</u>.

In this section we show that the vortex with finite elements is convenient for the computation of the variations of the lift of an airfoil travelling through a region of turbulence. However we just give some general ideas, proofs and explicit computations will appear in some forthcoming paper. In the plane $\mathbb{R}^2 = (x,y)$ we consider the non simply connected domain Ω of figure 2.

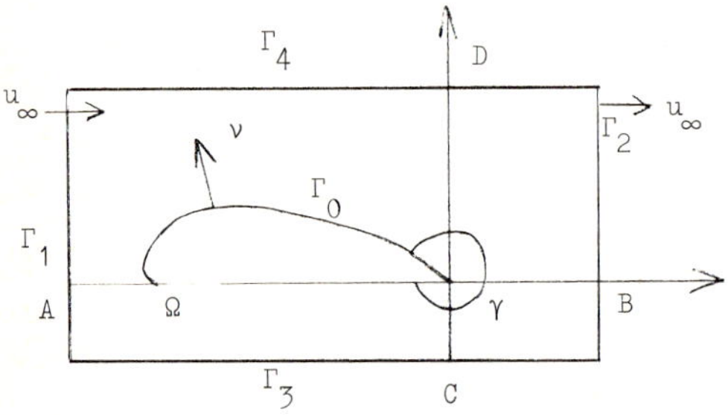

<u>Figure 2</u>

The Euler equation in this domain is

$$(31) \qquad \frac{\partial u}{\partial t} + u\nabla u = - \nabla p$$

$$(32) \qquad \nabla . u = 0$$

$$(33) \qquad u(x,0) = u_0(x)$$

and a boundary condition. We assume that the boundary condition is the following : $u.\nu|_{\Gamma_i} = 0$ for $i = 0,3,4$.

and $u.\nu|_{\Gamma_i} = u_\infty$ (given) for $i = 1,2$; $\nabla \times u|_{\Gamma_1} = \omega_1$ given.

The boundary condition satisfy the relation

$$\int_{\partial\Omega} u.\nu \, d\sigma = 0 \quad \text{(where } \nu \text{ denotes the outward normal to } \partial\Omega).$$

using the method of Bardos [3] one can prove the existence of solution of this problem

$$u(x,t) = (u_1(x,t),\ u_2(x,t)).$$

It is easy to show that $\omega(.,t) = \nabla\Lambda u$ is uniformly bounded in $L^\infty(\Omega)$, therefore we have $u \in L^\infty(\mathbb{R}_t; H^1(\Omega))$. However the existence of the "corner" (the angle $\gamma > M$) implies that u will not generally be $L^\infty(\Omega)$. Therefore we replace the problem (31) (32) (33) plus boundary condition, by the following problem.

(34) $$\frac{\partial\omega}{\partial t} + u\,\nabla\omega = 0\ ,$$

(35) $$-\Delta\Psi = \omega \quad \text{in} \quad \Omega\ .$$

(36) $$\Psi|_{\Gamma_\infty} = C$$

(37) $$\Psi|_{\Gamma_\infty} = u_\infty y \qquad \Gamma_\infty = \Gamma_1 \cup \Gamma_2 \cup \Gamma_3 \cup \Gamma_4$$

(38) $$\omega(t=0) = \omega_0 \quad \text{in} \quad \Omega \quad ; \quad \omega|_{\Gamma_1} = \omega_1(y,t)$$

(39) $$u = \nabla\Lambda\Psi\ .$$

In (36) the constant C has to be determined in such a way that $\nabla\Lambda\Psi$ belongs to $L^\infty(\Omega)$ (which means that the velocity is bounded).

In fact due to the angle γ any solution, behave, for $(x_1, x_2) \to 0$ like

$$\lambda_1\, r^{\pi/\gamma} \sin \frac{\pi\theta}{\gamma} + V$$

which belongs to the space $W^{2,p}(\Omega)$ $(p \leq 2/(1-\frac{\Pi}{\gamma}))$ (cf. Grisvard [8]).

Now λ_1 is an affine linear function of C which inturn can be uniquely determined by the relation

(40) $$\lambda_1 = 0.$$

Taking the curl of the expression $\frac{\partial u}{\partial t} + u\nabla u$ we obtain :

(41) $$\text{curl}\left(\frac{\partial u}{\partial t} + u\nabla u\right) = \frac{\partial \omega}{\partial t} + u\nabla\omega = 0$$

this imply that locally $\frac{\partial u}{\partial t} + u\nabla u$ is a gradient, however since the open set Ω is no more simply connected we have :

(42) $$\frac{\partial u}{\partial t} + u\nabla u = -\nabla p + \alpha(t)\left(\frac{-y}{x^2+y^2}, \frac{x}{x^2+y^2}\right)$$

$\alpha(t)$ which appears in the right and side of (42) due to the condition $u \in L^\infty(\Omega)$ (which determines C and then $\alpha(t)$ is an external force corresponding to the Kutta Joukowski condition.

The approximation of the problem (33), (39) will lead to the following problem

(43) $$\frac{\partial \omega_n}{\partial t} + A_h\,(\omega_h(\cdot, k\Delta t)).\,\nabla\omega_h = 0\ .$$

The main difference lies in the definition of the operator A_h. The problem is not homegenous, the open set is not simply connected and one must determine C, following the idea of Grisvard [8] or by some other method of fluid mechanics.

IV. Numerical experiments

Test 1 : we have solved problem (12-2) with

$$\Omega =]0,1[^2$$

$$u_0 = \begin{cases} 10 \text{ in } [.22, .33] \times [.11, .22] \\ 1 \text{ elsewhere.} \end{cases}$$

and we have triangulated Ω with 3 families of parallel lines $h = 1/9$, $\Delta t = 1$. At later times u can only be equal to 10 or 1. Therefore the initial spot, $u = 10$, is expected to turn with the flow and the area of $\{x : u(x,t) = 10\}$ remains constant.

Since our scheme is conservative in a statistical sens only the area mentionned above is not constant, but it fluctivates around the exact value. The spot can be seen to turn with the flow.

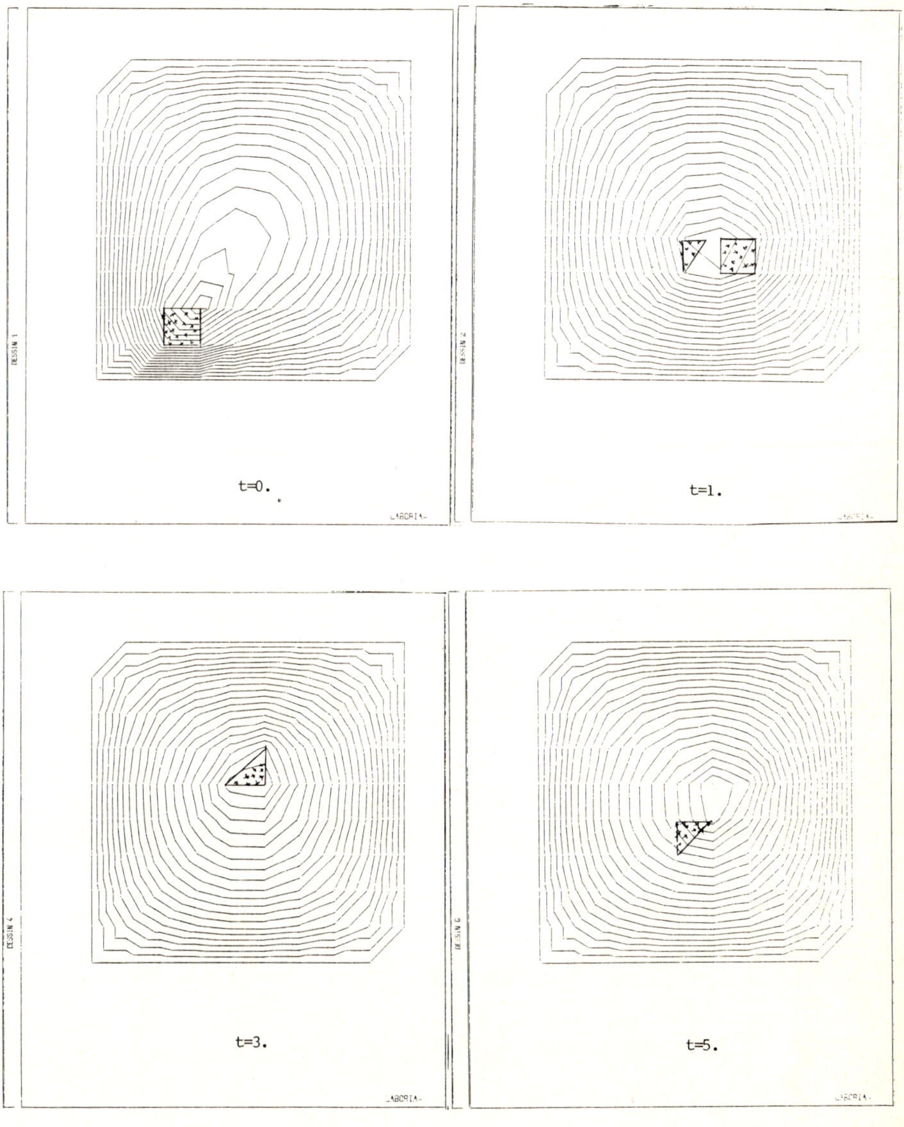
DESSIN 1
t=0.
LABORIA
DESSIN 2
t=1.
DESSIN 4
t=3.
LABORIA
DESSIN 6
t=5.
LABORIA

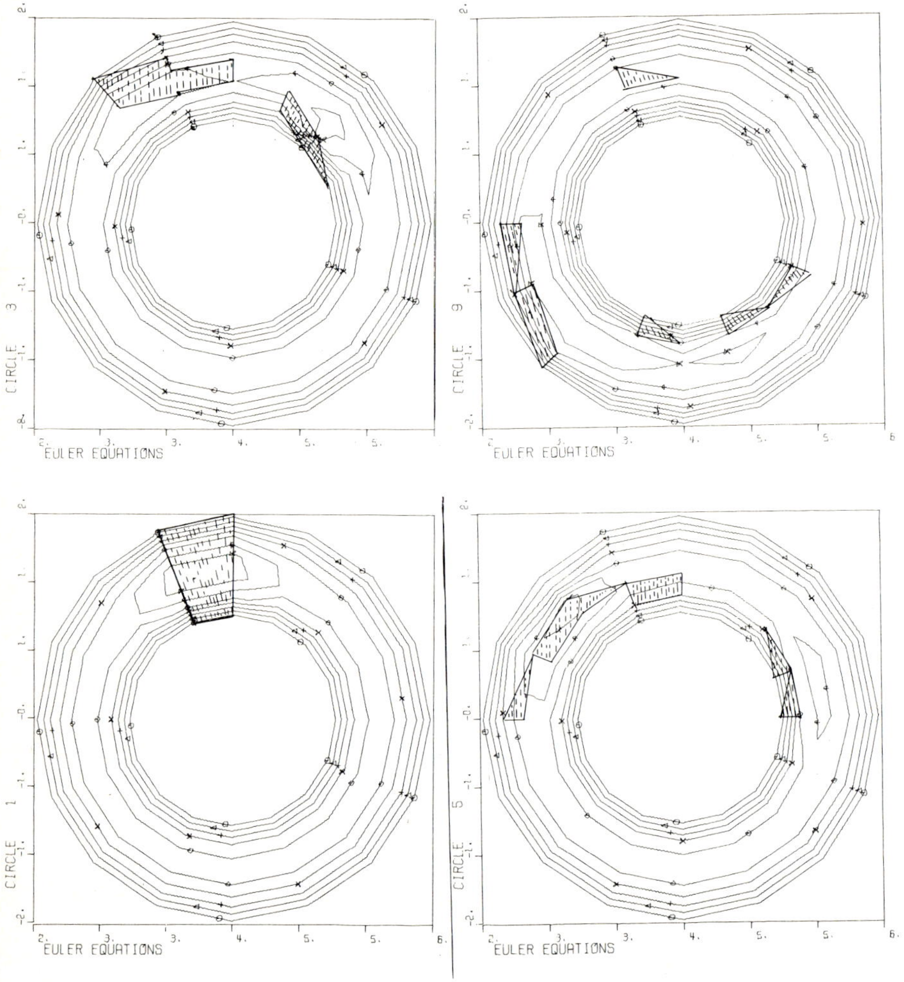
CIRCLE 3
EULER EQUATIONS
CIRCLE 9
EULER EQUATIONS
CIRCLE 1
EULER EQUATIONS
CIRCLE 5
EULER EQUATIONS

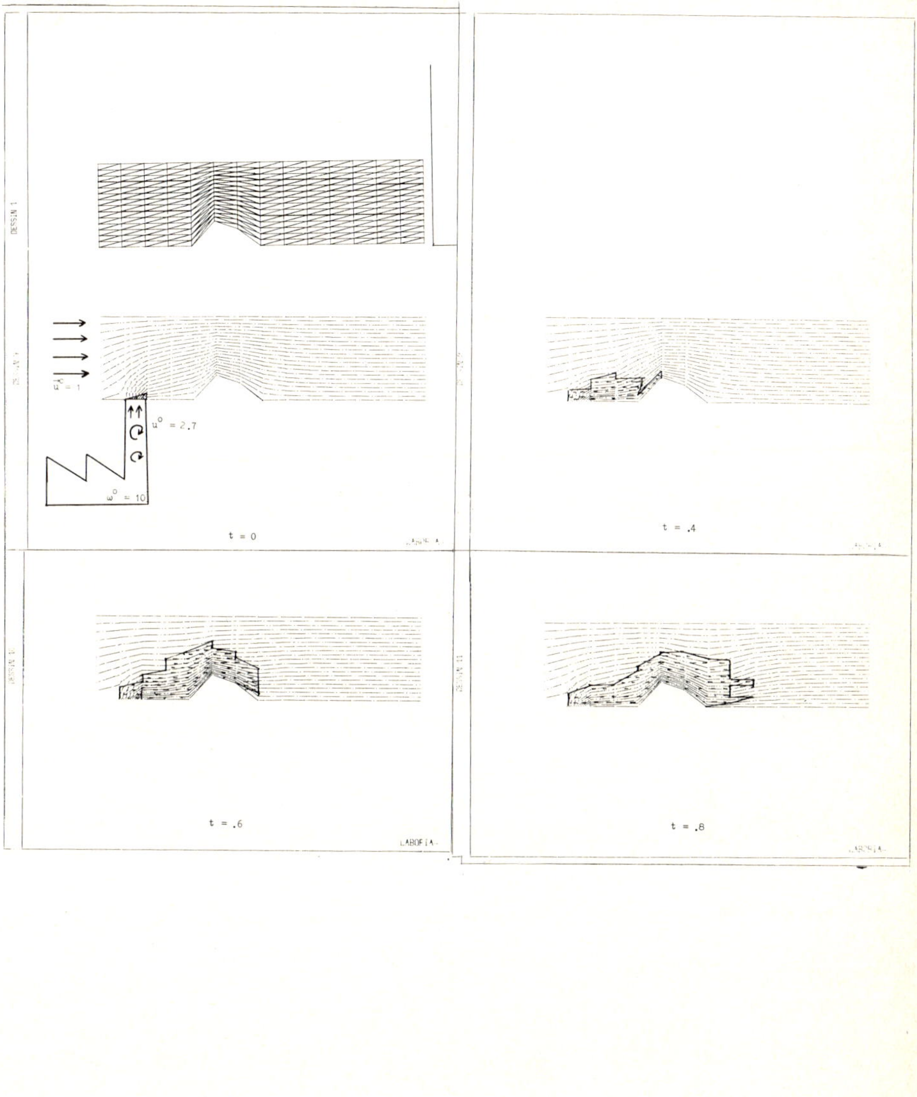
u° = 1
u° = 2.7
ω° = 10
t = 0
t = .4
t = .6
LABORIA
t = .8

Test 2 : Similar problem in a ring.

These problems are particularly difficult for dissipative numerical methods.

Test 3 : This test simulates the dispersion of a pollutant (cheminey smoke) by a wind. The pollutant comes out of Γ_1 a part of the boundary Γ with initial velocity 27 times the initial velocity of the wind and a vorticity of 10, while the wind has no initial vorticity ; Ω contains a hump (hill) to illustrate the feasibility of the finite element method.

Therefore we have solved

$$\frac{\partial\omega}{\partial t} + (\nabla\Lambda\Psi)\nabla\omega = 0$$

$$\omega(t = 0) = 0$$

$$\omega|_{\Gamma_1} = 10 \quad , \quad \omega|_{\Gamma_2} = 0 \ .$$

$$- \Delta\Psi = \omega$$

$$\frac{\partial\Psi}{\partial y}\Big|_{\Gamma_2 \cup \Gamma_3} = 1 \quad , \quad - \frac{\partial\Psi}{\partial x}\Big|_{\Gamma_1} = 2.7 \quad , \quad \Psi|_{\Gamma_4} = \text{constant}.$$

The triangulation shows 225 nodes and $\Delta t = .2$.

Computing times are 30" on an IBM 370. 168 for tests 1 and 2 and 1 mn for test 3. The plottings of the vorticity (shaded area on the figures) are done by hand because it is piecewise constant on the triangulation.

References

[1] Arnold V. Méthodes mathématiques en Mécanique classique.

[2] Baker G. The clouds and cells technique applied to the roll up of vortex sheets J. of Comp. phys. 75-95 (1979).

[3] Bardos C. Existence et Unicité de la solution de l'équation d'Euler en dimension deux. J. of. Math. Anal. and Appl. 40, 769-790 (1972).

[4] Ciarlet Ph. The finite element method for elliptic problems. Studied in Math. and its applications, North Holland (1978).

[5] Chouh A.J. Numerical study of slightly viscous flow, J. of Fluid. Mech. 57,

784-796 (1973).

[6] Christiansen J.P. Numerical solution of hydrodynamics by the method of point vortices J. of Comp. Phys. 13, 863-879 (1973).

[7] Covet B., Buneman O. and Leonard A. Three dimensional simulation of the free shear Layer using the vortex in cell method, Proceeding 2^{d} symposium on Turbulent shear flows, London (1979).

[8] Grisvard P. Behaviour of the solution of an elliptic boundary value problem in a polygonal of polyhedral domain. Numerical solutions of Differential Equation, 3 (1976). (Ed . B. A. Press, 267-274.

[9] Hald M.O. Convergence of vortex method for the Euler equation. Siam J. Numer. Anal. vol. 16, n° 5, October 1979, 726-755.

[10] Hald M.O. and Manceri del Prete V. Convergence of vortex method for the Euler equation. Math. Comp. 27 (1973) 719-728.

[11] Kato T. On the classical solution of the two dimensional non stationary Euler equation, Arch. Rat. Mech. Anal. 24 (1967). 302-324.

[12] Nitsche J. L_∞ convergence of finite element approximation "Mathematical aspects of finite element method' Rome (1975), Springer.

[13] Schaeffer A.C. Existence Theorem for the flow of an incompressible fluid in two-dimension, Trans. Am. Math. Soc. 42, 497 (1967).

[14] Wolibner W. Un théorème sur l'existence du mouvement plan d'un fluide parfait homogène et incompressible pendant un temps infiniment long, Math. Zeit. 37 (1935) 727-738.

[15] Zerner, M. Equations d'évolution quasi-linéaire du premier ordre : le cas lipschitzien (à paraître).

COMPUTING METHODS IN APPLIED SCIENCES AND ENGINEERING
R. Glowinski, J.L. Lions (editors)
North-Holland Publishing Company

NUMERICAL METHODS FOR USE IN COMBUSTION MODELING

Alexandre Joel Chorin*
Department of Mathematics
University of California
Berkeley, California

In recent years a number of methods have been developed for use in combustion modeling; some of these methods rely in an essential way on a random walk or a sampling procedure. The purpose of this talk is to explain in an elementary manner how these methods work and why they are useful.

Combustion problems have the following characteristics, each one of which presents a major challenge to computational modeling: they are time dependent phenomena occuring in a three dimensional space; the number of equations to be solved is often substantial; there are several distinct length and time scales to be resolved; the flow in which the combustion occurs is usually turbulent, and on the scales important to combustion, very intermittent. The solution is very sensitive to a numerically induced diffusion or conduction.

Stochastic methods provide a particularly promising way for overcoming these difficulties. They are typically insensitive to stiffness; they provide a way of controlling numerical diffusion (and thus of ensuring that the effects of viscosity or conduction are represented accurately), and they concentrate computing effort where it would do the most good. We shall explain the methods by means of simple examples.

A Reaction-Diffusion Equation

Consider the equation

$$v_t = \kappa v_{xx} + f(v) \tag{1}$$

where κ is a small parameter, and $f(v)$ is a function of v such that: $f(v) = 0$ for $0 \leq v \leq \varepsilon$; $f(v) > 0$ for $\varepsilon < v < 1$, $f(1) = 0$, $f'(1) < 0$. Equation (1) is a model for many equations which occur in combustion theory. Consider in particular the initial data

* Partially supported by the Engineering, Mathematical, and Geosciences Division of the U.S. Department of Energy under contract W-7405-ENG-48, and by the Office of Naval Research under contract N00014-76-0-0316.

$$v(x,0) = \begin{cases} 1 & x < 0, \\ 0 & x > 0. \end{cases} \qquad (2)$$

The condition $f(v) = 0$ for $0 \leq v \leq \varepsilon$ (ε is an "ignition temperature") makes the problem well conditioned and allows a simple discussion (see e.g. [4], [10]).

Equation (1) has a solution which tends to a traveling wave solution. Both the thickness and the velocity of the wave are $O(\sqrt{\kappa})$. Suppose one were to compute v by a difference method with a fixed grid size. Clearly, the amount of labor required to reach a fixed time T with a given accuracy would be $O(\kappa^{-\frac{1}{2}})$. What is needed is a procedure for refining the mesh in the region where v has large gradients, without disturbing the rest of the calculation (Equation (1) may be describing the flow in a small area only) and without incurring an instability. We wish to show that a random procedure will do the job perfectly, in the sense that the amount of labor for given accuracy and given integration time will be exactly independent of κ. The algorithm will be explicit and unconditionally stable.

Approximate $v(x,t)$ by a piecewise constant function of x, with jumps Δv_i, $i = 1, \ldots, N$ at the points x_i, $i = 1, \ldots, N$. Assume $|\Delta v_i| < \Delta_{max}$ for all i, Δ_{max} a small quantity. Assume for simplicity that $v(x = +\infty, t) = 0$. Let the time be descretized, $t = nk$, n = integer, $n \geq 0$, k = time step. Then

$$v(x,nk) = \sum_{x_i > x} \Delta v_i \qquad (3)$$

It is not required that the x_i be distinct. Thus, if $\Delta_{max} = 1/N$, the initial data (2) can be represented by

$$\Delta v_1 = \ldots = \Delta v_N = 1/N,$$

$$x_1 = \ldots = x_N = 0.$$

To progress from $t = nk$ to $t = (n+1)k$, let each of the x_i perform a gaussian jump,

$$x_i^{n+1} = x_i^n + \eta,$$

where η is drawn from a gaussian distribution with mean zero and variance $2\kappa k$. The expected value of (3) is then a solution of the heat equation $v_t = \kappa v_{xx}$. Note that what is diffused by a random walk is not v itself, but its x derivative. This is the key to success.

We have not yet taken $f(v)$ into account. Consider a region S_i

in which v is a constant, $S_i = \{x | x_i \leq x \leq x_{i+1}\}$. $v = v_i$. Solve the ordinary differential equation

$$\frac{dw}{dt} = f(w), \quad w(0) = v_i.$$

Let $\delta_i = w(k) - v_i$ be the change in v due to f in S_i. Then v_i should become $v_i + \delta_i$ in S_i, and thus Δv_{i+1} becomes $\Delta v_{i+1} + \delta_i$ and v_i becomes $\Delta v_i - \delta_i$. Do this for all S_i. In after this step one of the $|\Delta v_i|$ exceeds Δ_{max}, break it into smaller pieces which do satisfy $|\Delta v_i| \leq \Delta_{max}$. If a Δv_i becomes very small, delete it (some care is required to ensure that $v(x = -\infty, t) = 1$ with the initial data (2)). This last step ensures that no work is performed where the gradients of v are small. The overall algorithm is easy to program and use, and with appropriate variance reduction it is also very accurate. Furthermore, as $\kappa \to 0$, the solution scales exactly with $\sqrt{\kappa}$ and the amount of labor remains constant. For details see [4].

The Boundary Layer Equations

The boundary layer equations for an incompressible fluid are conceptually similar to a reaction diffusion equation. They describe the interaction between the creation of vorticity at a wall, its diffusion and its transport. The creation process is more interesting than in a reaction-diffusion equation.

Consider a flow over a heat plate in two dimensions. The Prandtl equations are

$$\partial_t \xi + (\underline{u} \cdot \underline{\nabla})\xi = \nu \frac{\partial^2 \xi}{\partial y^2}, \tag{4a}$$

$$\partial_x u + \partial_y v = 0, \tag{4b}$$

$$\frac{\partial u}{\partial y} = -\xi, \tag{4c}$$

where $\underline{u} = (u,v)$ is the velocity, ξ is the vorticity, x is a coordinate parallel to the wall, y is a coordinate normal to the wall, and ν is a (small) viscosity. The boundary conditions include: $u\ (x,y = +\infty, t) = U$, $\underline{u}(x,y = 0,t) = 0$. We shall solve the equations by diffusing ξ, which by (4c) is a velocity gradient.

Consider a collection of N segments S_i of vortex sheets. Each S_i, $i = 1, \ldots, N$, is a segment of a straight line, parallel to the wall, of length h, with center $\underline{x}_i = (x_i, y_i)$, and such that if $u+$ is u just above S_i and $u-$ is u just below

S_i, $u+ - u- = \xi_i$. Consider the motion of the S_i as equations (4) would impose it. Equations (4b) and (4c) yield

$$u(x,y) = U - \int_y \xi(x,z)\,dz, \tag{5a}$$

$$v(x,y) = -\partial_x \int_0^y u(x,z)\,dz\ , \tag{5b}$$

i.e., given $\xi(x,y)$, u and v can be determined. Equation (5a) states that the fluid element at (x,y) is slowed down from U by all the sheets above that element. Equation (5b) states that whatever fluid cannot leave a rectangular region through its sides must leave upwards. These observations lead quickly to rules which determine $\underline{u}_i = \underline{u}(x_i)$ given the array of $S_j, j = 1, \ldots, N$, and one can write approximately

$$u_i^{n+1} = u_i^n + ku_i \tag{6b}$$

$$y_i^{n+1} = y_i^n + kv_i \tag{6b}$$

where $t = nk$, n = integer, k = time step, as before. Again, the diffusive term in (4a) can be taken into account by writing, instead of (6b),

$$y_i^{n+1} = y_i^n + kv_i + \eta, \tag{7}$$

where η is a gaussian random variable with mean 0 and variance $2k\nu$. (5a) and (5b) clearly lead to rules which satisfy the condition $u = U$ at $y = +\infty$ at $v = 0$ at $y = 0$.

Suppose the boundary condition $u = 0$ fails to be satisfied at a point $(x,0)$ on the wall. Imagine the flow is continued from $y > 0$ to $y < 0$ by the rule $\underline{u}(x,-y) = -\underline{u}(x,y)$. As one crosses the wall downwards, both u and y change signs, and therefore

$$\xi(x,y) = -\frac{\partial u}{\partial y}(x,y) = -\frac{\partial u}{\partial y}(x,-y) = \xi(x,-y)\ .$$

Now let the flow described by equation (4a) take place. The diffusion term will guarantee that the boundary condition $u = 0$ is satisfied.

It u is not zero at the wall, the antisymmetry $u(x,-y) = -u(x,y)$ will introduce at the wall a line of discontinuity. This line of discontinuity can be broken into vortex sheets of the type described above, which then take part in the subsequent motion. This is the process of vorticity generation, which mimics the physical process of vorticity generation. For details, implementation, and variance reduction, see e.g. [3], [5]. The algorithm is unconditionally stable, and contains an automatic scaling which scales the thickness of the layer with $\sqrt{\nu}$.

Vortex Method for the Navier-Stokes Equations

The step from the Prandtl equations to the Navier-Stokes equation is short. In the plane, the latter equations can be written as

$$\partial_t \xi + (\underline{u}\cdot\underline{\nabla})\xi = \frac{1}{R}\Delta\xi, \tag{8a}$$

$$u = -\partial_x \psi, \quad v = \partial_y \psi, \tag{8b,c}$$

$$\Delta\psi = -\xi, \quad \underline{u} = (u,v) , \tag{8d}$$

where u, v, ξ have the same meaning as in (4). ψ is the stream function and R is the Reynolds number (which we assume is large). The main difference between these quations and the Prandtl equations lies in the fact that the relationship between $\underline{u}$ and ξ is more complex, and also in the fact that diffusion takes place in both spatial direction. Equation (8a) states, just as does equation (4a), that in the absence of diffusion vorticity is merely transported by the velocity field which itself induces.

What should be done is plain: consider ξ to be a sum of small elements. Move each according to the velocity field induced by all the others. Add a small random component to represent diffusion. Create vorticity at walls. For details, see [1],[5], [11].

Two words of caution are needed: at the wall, $\underline{u}(x,y) = -\underline{u}(x,-y)$ does not lead to $\xi(x,y) = \xi(x,-y)$ for the Navier-Stokes equations. The reason is that here,

$$\xi = -\frac{\partial u}{\partial y} + \frac{\partial v}{\partial x}$$

and though u, v and y change signs as one crosses the wall, x does not (we assume as before that the wall coincides with $y = 0$). This leads to some minor tehcnical difficulties. The easiest thing to do is use sheets near walls.

Finally, note that the cost of counting all vortex interactions can be substantial. One may be tempted to impose a grid on the domain of interaction, extrapolate the vorticity to the grid, compute the velocity from (8b,c,d) on the grid by some fast method, and interpolate to the vorticity elements (this is the "vortex in cell" or "vortex cloud" method). The idea is attractive but dangerous. Interpolation introduces a numerical viscosity, which, if one is careless, may destroy the purpose of the method, which is to represent the diffusive length scales correctly. In many problems, of course, the diffusion does not matter as long as it is small (e.g. in a periodic domain or in the absence of boundaries) but then one may just as well compute with more standard methods.

Glimm's Method for Reacting Gas Flow

We now turn to a somewhat different order of ideas, which goes back to Glimm's construction of solutions of hyperbolic problems [8].

Consider the simplicity the single equation

$$v_t = f_x, \quad f = f(v) \tag{9}$$

A Riemann problem for this equation is the initial value problems with data

$$v(x,0) = \begin{cases} v_R & x > 0 , \\ v_L & x < 0 . \end{cases}$$

v_R, v_L constants. Let the solution of this problem be $w(v_R, v_L, x,t)$. Consider now the initial value problem for (9) with data $v(x,0) = g(x)$. Let $x = ih$, i = integer, h = spatial increment (we are now for the first time introducing a spatial grid). Let $t = nk$. Given v_i^n, v_{i+1}^n, we set

$$v_{i+1/2}^{n+1/2} = w(v_{i+1}^n, v_i^n, \theta h, k/2),$$

where θ is a variable whose values are equidistributed on $[-\frac{1}{2}, \frac{1}{2}]$. This formula defines an algorithm for solving (9) on a staggered grid. The algorithm proceeds by constructing exact solutions of the Riemann problems, and then sampling them. In the simple case $f(v) = v$, the equation to be solved becomes $v_t = v_x$; the solution is $v(x,t) = g(x+t)$. The Glimm scheme reduces to

$$v_{i+1/2}^{n+1/2} = \begin{cases} v_{i+1}^n & \text{if} \quad \theta h \geq -k/2 \\ v_i^n & \text{if} \quad \theta h < -k/2. \end{cases}$$

A brief recursion argument shows that

$$v_i^n = g(x+\phi, t),$$

where ϕ is a variable whose mean is zero and whose variance tends to zero as $h \to 0$, $k/h \leq 1$. ϕ is independent of x. The Glimm scheme for this simple equation preserves the shape of the solution exactly, at the cost of a small random displacement.

We adopted the Glimm scheme in a combustion model because it had a small intrinsic diffusion. (In two dimensional problems it turns out that the diffusion is small but not zero). This is particularly significant in the following circumstances: consider a flow in which thin combustion layers are imbedded. On the natural outer scale one may often view the diffusion and conduction coefficients as being zero and the reaction rates as infinite. The equations of gas dynamics acquire then many of the properties of non convex hyperbolic systems (see e.g. [2], [5]). A simple non convex hyperbolic equation is an equation (9) in which f'' is allowed to change signs. In such a problem, the wave pattern which emerges depends on the global shape of the function f, and diffusion in a rather substantial amount is needed to spread the variation of the solution sufficiently for a difference scheme to discern how it should behave (see [9]). The Glimm scheme overcomes this problem entirely by building the right wave patterns into the Riemann problems, without a need for extra diffusion. For examples, see [6], [7], [11].

Conclusions

The real problems of combustion theory are of course much more complex than the simple examples sketched above, and the actual programs in use are hybrids of substantial complexity. All the methods described above have in common the fact that they rely heavily on known properties of the specific equations to be solved. In very complicated problems, this may well be the right path to choose.

BIBLIOGRAPHY

[1] A.J. Chorin, J. Fluid Mech., 57 785 (1973).

[2] A.J. Chorin, J. Comp. Phys., 25, 257 (1977).

[3] A.J. Chorin, J. Comp. Phys., 27 428 (1978).

[4] A.J. Chorin, Stochastic Solution of Reaction/Diffusion Equations, Berkeley (1979).

[5] A.J. Chorin and J.E. Marsden, A Mathematical Introduction to Fluid Mechanics, Springer (1979)

[6] P. Colella, Ph.D. Thesis, Dept. of Math, Berkeley (1979).

[7] P. Concus and W. Proskurowski, J. Comp. Phys., 30, 153 (1979).

[8] J. Glimm, Comm. Pure Appl. Math, 18, 697 (1965).

[9] A. Harten, J. Hyman and P.D. Lax, Comm. Pure Appl. Math, 29, 297 (1976).

[10] S.S. Lin, Ph.D. Thesis, Dept. of Math, Berkeley (1975).

[11] A. Majda, A Qualitative Model for Dynamic Combustion, Berkeley, Dept. of Math (1979).

[12] A. Sod, Numerical Methods in Fluid Mechanics, Academic Press (1980).

COMPUTING METHODS IN APPLIED SCIENCES AND ENGINEERING
R. Glowinski, J.L. Lions (editors)
North-Holland Publishing Company

NUMERICAL STUDY OF COASTAL UPWELLINGS BY A FINITE ELEMENT METHOD

Lien HUA
Laboratoire d'Océanographie Physique du
Muséum d'Histoire Naturelle de Paris
43 rue Cuvier
F-Paris (5ème)

François THOMASSET
IRIA-LABORIA
Domaine de Voluceau
B.P. 105
F-78150 Le Chesnay

-=-=-=-=-

1. - INTRODUCTION

In this paper, we are concerned with the physical process of *coastal upwelling* induced by winds in sea (or lake) waters : this is an ascending motion by which deep, cool water is brought to the surface.
This process is modelled by *shallow waters* equations in a *two layer* ocean ; this is a good representation of the stratification observed in summer on the continental shelf.
We have in mind the real case of the upwellings observed in the Gulf of Lions. From infra-red satellite data for surface temperature we get a synoptic view of the process which shows how important the shape of the shore-line is for the localization of upwelling "cells" (see figure 1).
This suggested us to use a *finite element method*, so as to be able to take into account a complex geometry. Furthermore we used *linear, non conforming* finite elements which appeared suiting better our problem.
As regards time discretization we adapted a powerful semi implicit scheme devised by Kwizak and Robert [7] in a finite difference context.
The method has been tested on an analytic solution ; then applied to the case of the Gulf of Lions. The results show a good agrement with the observations.

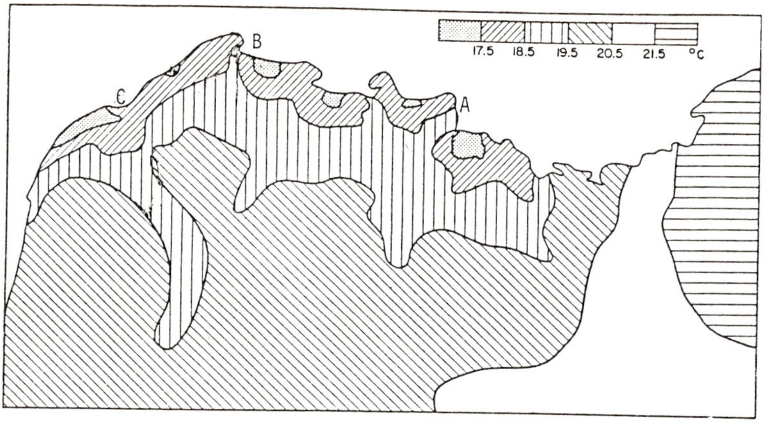

Figure 1

Isolines of surface temperature from satellite NOAA IV (averaged over several gusts of wind). The dotted regions are locations of intense upwellings.

A = Marseille B = Pointe de l'Espiguette C = Cap d'Agde

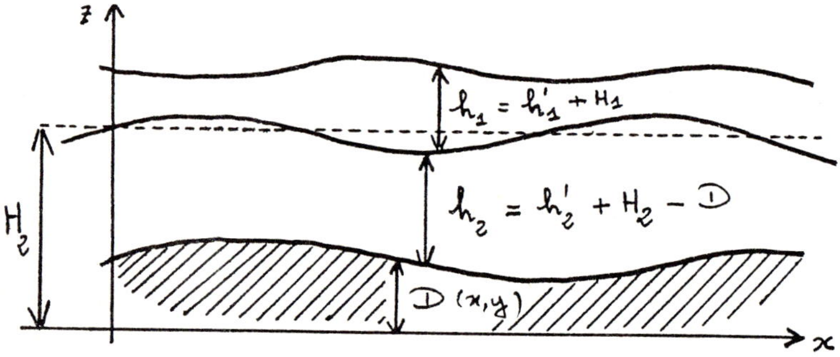

Figure 2

2. - EQUATIONS OF THE PROBLEM

We assume a two layer ocean, each layer having a constant density : ρ_1 in the upper layer, ρ_2 in the lower layer ($\rho_1 < \rho_2$).

We define $h_i(x,y,t)$ (i=1,2) as the height of water in layer i (i=1,2), and $\vec{u}_i$ as the mean velocity in layer i .

The equations are obtained by integration of Navier Stokes equations in Z direction, in each layer (see Hurlburt and O'Brien [6])

$$\begin{cases} \dfrac{\partial \vec{u}_i}{\partial t} + (\vec{u}_i \cdot \vec{\text{grad}})\vec{u}_i + f\,\vec{k}\times\vec{u}_i + g\,\vec{\text{grad}}\,(\alpha_i h_1 + h_2 + D) - \nu\Delta\vec{u}_i = (\vec{\tau}_{s_i} - \vec{\tau}_{B_i})/\rho_i h_i & (1) \\[2ex] \dfrac{\partial h_i}{\partial t} + \text{div}(h_i\,\vec{u}_i) = 0 \qquad i = 1,2. & (2) \end{cases}$$

- $\alpha_1 = 1$, $\alpha_2 = 1 - \varepsilon$,

  $\varepsilon = \dfrac{\rho_2 - \rho_1}{\rho_2} = 2.10^{-3}$

- surface wind stress : $\vec{\tau}_{s_1} = 1.6\ 10^{-3}\ |\vec{w}|\ \vec{w}$

  ($\vec{w}$ is the *given* wind velocity 10 meters above sea level)

- friction stress between the two layers :

  $\vec{\tau}_{B_1} = \vec{\tau}_{s_2} = \rho\, C_I\, |\vec{u}_1 - \vec{u}_2|\, (\vec{u}_1 - \vec{u}_2)$ with $C_I = 10^{-5}$ MKSA

- bottom stress :

  $\vec{\tau}_{B_2} = \rho_2\ C_B\ |\vec{u}_2|\ \vec{u}_2$

- Coriolis parameter :

  $f = 10^{-4}\ s^{-1}$

- Coriolis force : $f\,\vec{k}\times\vec{u}_i = \begin{cases} -f\,u_i^2 \\ +f\,u_i^1 \end{cases}$

- gravity acceleration :

  $g = 9.8\ m.s^{-2}$

- horizontal eddy viscosity : $\nu = 10^2\ m^2\ s^{-1}$.

INITIAL CONDITIONS

We integrated the equations from rest ; the wind was set up at time $t = 0$, and then kept constant (in time but not in space) in most of our computations ; (maximum wind velocity = 10 $m.s^{-1}$).

BOUNDARY CONDITIONS

We took *free slip* boundary conditions for $\vec{u}_i$ ($\vec{u}_i.\vec{n} = 0$ at the boundaries) ; thus the boundary layer has not any serious influence far inside the domain.
Other boundary conditions that can be considered are :

- general circulation ($\vec{u}_i.\vec{n}$ given $\neq 0$)
- "radiation" boundary conditions, allowing waves to go away, instead of being reflected against the oceanic boundaries.

GRAVITY WAVES

Among the many free oscillation modes of the system, the gravity waves have the fastest speed of propagation. These are solutions of the simplified and linearized equations (in the case of D=0)

$$\frac{\partial \vec{u}_i}{\partial t} + g\ \vec{\text{grad}}(\alpha_i h_1' + h_2') = 0$$

$$\frac{\partial h_i'}{\partial t} + \text{div}(H_i\ \vec{u}_i) = 0$$

$$h_i = H_i + h_i'(x,y,t)$$

which yields the wave equation :

$$\frac{\partial^2}{\partial t^2} h_i' - g\ H_i \Delta(\alpha_i h_1' + h_2') = 0\ .$$

This induces two modes of propagation, with celerities :

$$C_1 = (g(h_1+h_2))^{\frac{1}{2}} \quad \text{for the external (surface) gravity waves}$$

$$C_2 = \left(\varepsilon g\ \frac{h_1\ h_2}{h_1+h_2}\right)^{\frac{1}{2}} \quad \text{for the internal (interface) gravity waves.}$$

For typical values of the parameters, $C_1 \simeq 100$ m/s , $C_2 \simeq 1$ m/s.

In the physics of upwelling, the most important are internal gravity waves modified by the earth rotation.
Analytical results show that a characteristic spatial scale of such waves is given by the Rossby ratio $C_2/f \sim 10$ km (f = Coriolis parameter). Therefore the mesh size in the upwelling region cannot be larger than this value.
On the other hand any explicit time discretization is submitted to the C.F.L. condition :

$$\frac{\delta x}{\delta t} > C_1 \quad .$$

With $\delta u \simeq 3$ km, this yields $\delta t < 30$ s. Since the model is to be integrated over several days, this is a very severe constraint, which can be alleviated by an implicit treatment of gravity waves.

3. THE NUMERICAL METHOD

Owing to the geometry of the Gulf of Lions, and to the need for a variable grid, we were naturally let to use finite elements.
In a previous paper, Bégis, Crépon and Millot [3] used linear conforming elements (i.e. with degrees of freedom at the vertices) to solve the above equations.
This classical choice has the main following drawbacks :

- they need to make the mass lumping approximation, so as to make the mass matrix diagonal, and use an explicit time discretization.
- the treatment of free-slip boundary is questionable, since a normal direction to the boundary needs to be defined at the corner nodes.

The use of *non-conforming finite elements* for $\vec{u}_i$ and h_i can free us from such questions. In this approximation, the unknowns are :

. piecewise polynomials of degree 1 *per triangle* ;
. continuous at the *mid-side points*.

(See Crouzeix and Raviart [5] , Thomasset [10], [11]).
Therefore the boundary conditions are applied at the mid-side points, where the normal direction is well-defined. On the other hand the *mass matrices are diagonal* without the mass lumping approximation.
The time-marching scheme is the same as Kwizak and Robert [7] that is leap frog implicit with respect to gravity waves :

$$h_1^m = H_1 + h_1'^m \quad , \quad h_2^m = H_2 - D + h_2'^m \qquad \text{(see notations on fig. 2)}$$

(3) $$\frac{\vec{u}_i^{m+1} - \vec{u}_i^{m-1}}{2\delta t} + g\ \mathrm{grad}\left(\alpha_i \frac{h_1'^{m+1} + h_2'^{m-1}}{2} + \frac{h_2'^{m+1} + h_2'^{m-1}}{2}\right) + (\vec{u}_i^m \cdot \vec{\nabla})\vec{u}_i^m$$

$$+ f\ \vec{k}.\vec{u}_i^m - \nu \Delta \vec{u}_i^{m-1} = (\vec{\tau}_{s_i}^m - \vec{\tau}_{B_i}^m)/\rho_i h_i \qquad (i=1,2)$$

(4) $$\frac{h_1'^{m+1} - h'^{m-1}}{2\delta t} + H_1\ \mathrm{div}\ \frac{\vec{u}_1^{m+1} + \vec{u}_1^{m-1}}{2} + \mathrm{div}(h_1'^m \vec{u}_1^m) = 0$$

(5) $$\frac{h_2'^{m+1} - h_2'^{m-1}}{2\delta t} + \mathrm{div}(H_2 - D)\ \frac{\vec{u}_2^{m+1} + \vec{u}_2^{m-1}}{2} + \mathrm{div}(h_2'^m \vec{u}_2^m) = 0$$

(For the first step we took a Matsuno predictor-corrector scheme).

We set :

$$\mathcal{H}_i^m = \text{vector of degrees of freedom of } h_i'$$

$$\mathcal{U}_i^m = \text{vector of degrees of freedom of } \vec{u}_i \ .$$

The equations (3)(4)(5) take the following form (i=1,2), in the case D = constant :

(6) $$\mathcal{M}\mathcal{U}_i^{m+1} + g\ \delta t\ \mathcal{B}(\alpha_i\ \mathcal{H}_1^{m+1} + \mathcal{H}_2^{m+1}) = \mathcal{V}_i^m$$

$$\mathcal{N}\mathcal{H}_i^{m+1} - H_i\ \delta t\ \mathcal{C}\mathcal{U}_i^{m+1} = \mathcal{K}_i^m$$

with $\mathcal{M}$, $\mathcal{N}$ = *diagonal* mass matrices.
(See the Appendix, or [14] for the details).

Elimination of velocities yields :

(7) $$\mathcal{N}\mathcal{H}_i^{m+1} + g\ H_i\ \ \delta t^2 \mathcal{C}\mathcal{M}^{-1}\ \mathcal{B}(\alpha_2\ \mathcal{H}_1^{m+1} + \mathcal{H}_2^{m+1}) = \mathcal{K}_i^m + H_i \delta t\ \mathcal{C}\mathcal{M}^{-1}\mathcal{V}_i^m$$

Owing to the diagonal form of $\mathcal{M}$, the matrix $\mathcal{C}\mathcal{M}^{-1}\mathcal{B}$ is sparse : an element a_{qr} of the matrix (for any two nodes q and r) is non zero :

- either if q and r belong to a common triangle :
- or if there exists another node p such that q is neighbour of p , and p neighbour of r . (See Fig. 3)

The set of equations (7) can be solved using block-SOR. Note that the matrix is time invariant, so that the optimal relaxation parameter can be experimentally determined at the beginning of the computation.

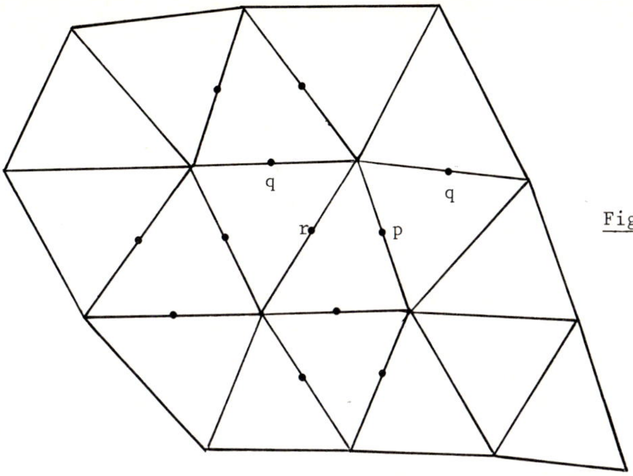

Figure 3

CORE REQUIREMENTS AND COMPUTING TIME

If NT is the number of triangles, the core requirement is around 100 NT four-bytes words.

This high figure is due to :

- the need for some book-keeping arrays, as usual in finite elements ;
- the use of double precision to store the values of h_i and $\vec{u}_i$:
- the choice of linear non conforming finite elements : indeed, for a given triangulation the number of mid-side nodes is of order $\frac{3}{2}$ NT, while the number of vertex nodes is around $\frac{1}{2}$ NT.

As to the computing time, the improvements brought in by the implicit scheme are put into light by the following table ; this shows the computing time (on IBM 370/168) necessary to advance from 0 to 5 hours, the academic model described in the next section (number of triangles = 680).

	explicit (leap-frog)	semi-implicit			
t	1'	5'	10'	30'	60'
(relaxation parameter)	--	1.2	1.4	1.63	1.68
CPU time (to integrate from 0 to 5 hours)	7'	2'50"	1'26"	46"	33"

4.- NUMERICAL EXPERIMENTS

Some tests with variable topography (D(x,y) are presently being run ; the results here below concern only constant D.

i) Academic model

To test the validity of the method, we applied it to an academic problem for which we have an *analytic solution* to the *linear* problem (that is, when h_i is taken constant in the terms $\vec{\tau}/\rho h$ and $\mathrm{div}(h_i\ \vec{u}_i)$).

The domain is shown on Figure 4.

Figure 4

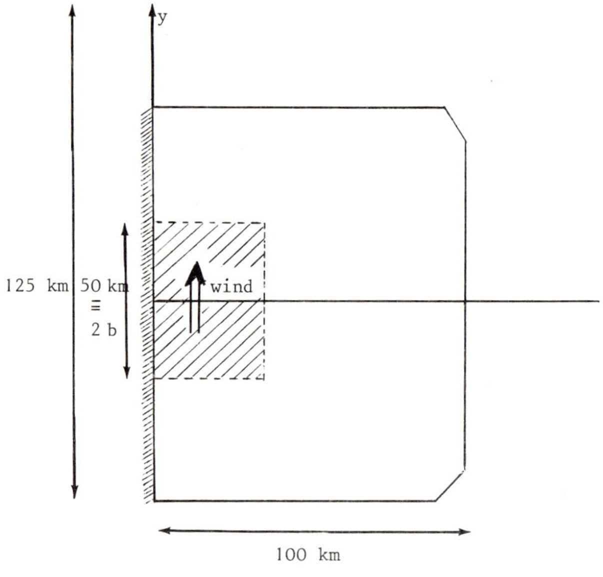

The wind blows *northwards* (i.e. parallel to the coast in y direction) in the shaded region ; it decays exponentially in x direction and the dependance in y is such that the wind stress $\tau_y = |w|\ w_y$ has a hat-like profile (Fig. 5 a)). The mesh is shown on Figure 6 (680 triangles). The analytical solution ([1]) from Crépon [4] is shown at different times on Figure 5 b) to 5 e) ; we can see the *southwards* propagation of a Kelvin wave the amplitude of which tends to an asymptotic value as time goes to infinity.

This process is well represented in the numerical results.

([1]) For an unbounded ocean, and the linear problem.

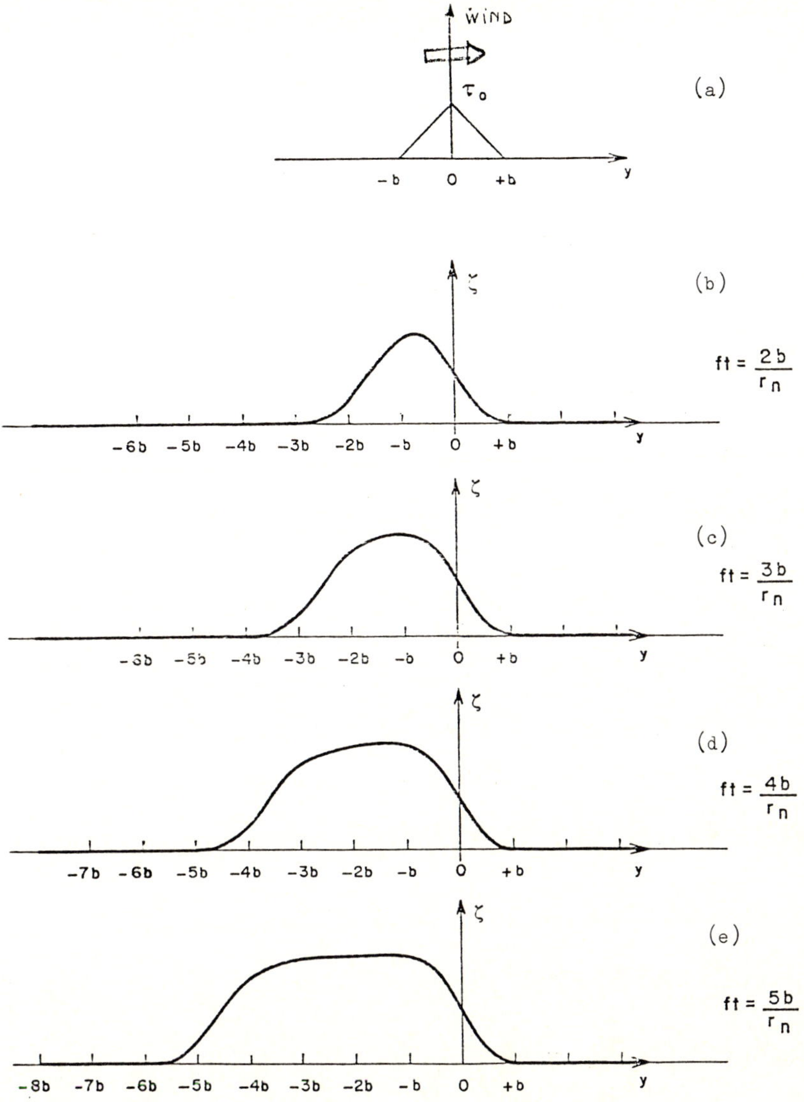

Figure 5 : (a) wind stress profile versus y

(b) ... (e) analytical solution from Crépon [4]: thermoline /y at different times.

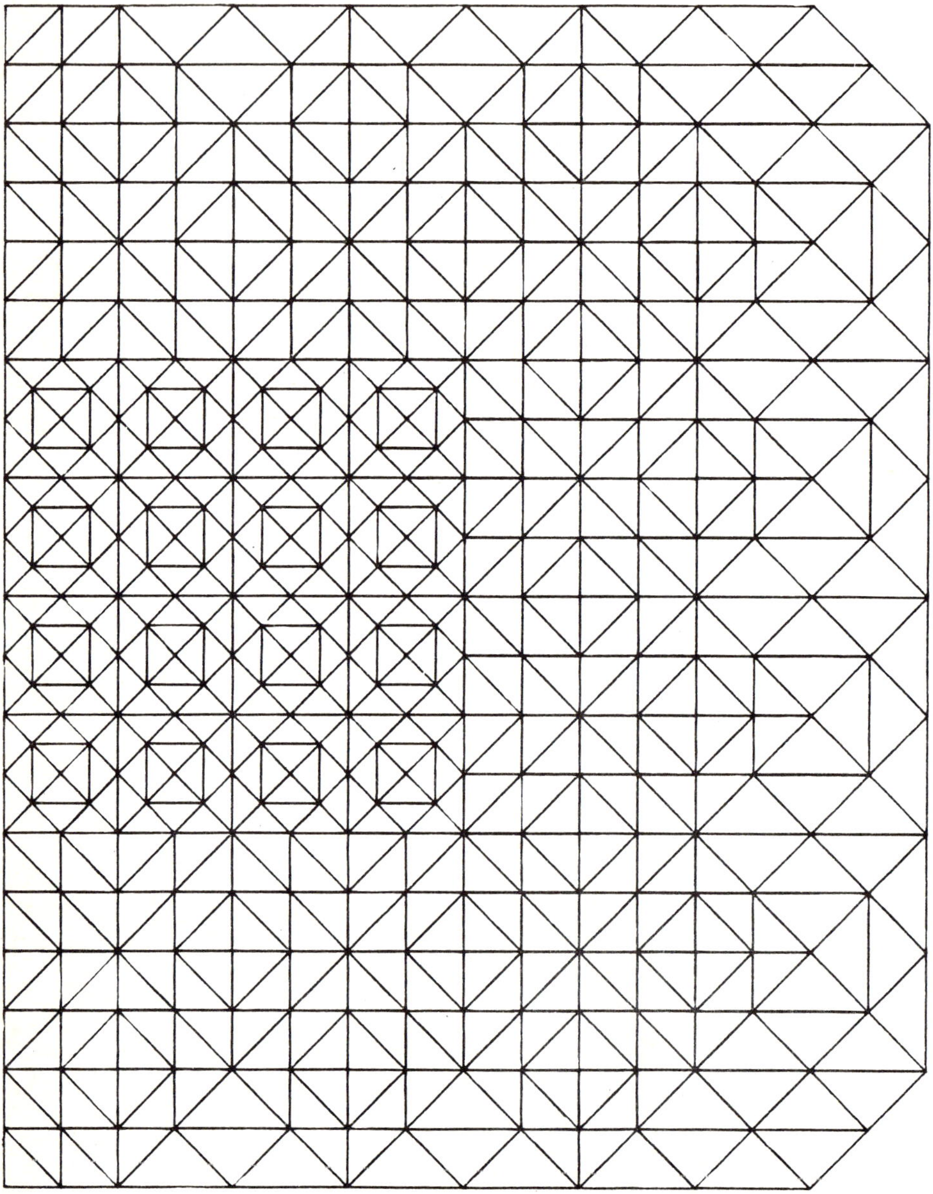

Figure 6

On Fig.7, we compare the analytical and computed[1] solutions for the *thermocline*[2] profile versus y at the coast, at time t=16h.

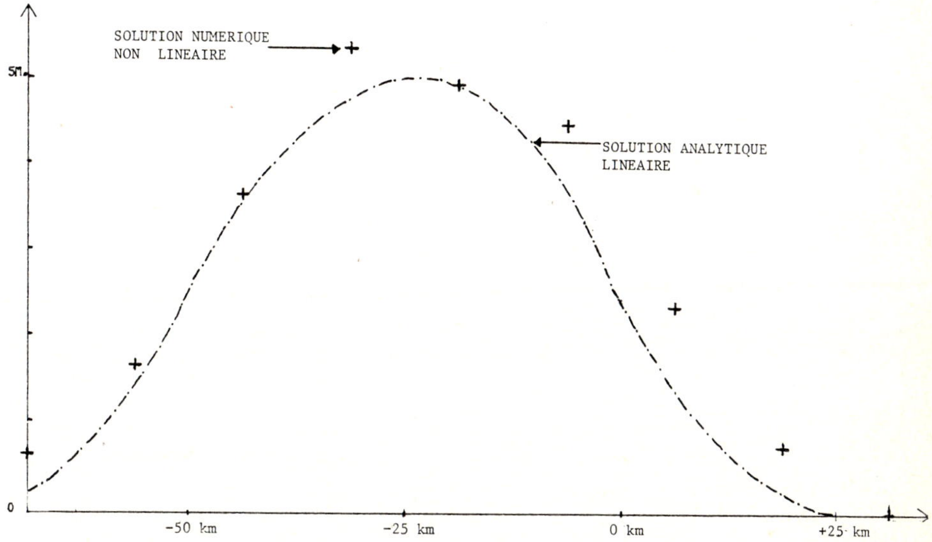

Figure 7

Figures 8, 9, show the isolines of thermocline h'_2 at times 16h. and 30h.

The results are in good agreement with the analytical solution, and the qualitative results on non-linear effects agree with Adamec and O'Brien [1 .

ii) Gulf of Lions

We show on Figures 10, 11, the meshes [3] used for the computation, resp. 524 triangles ($\delta x \simeq 5.6$ km) and 1733 triangles ($\delta x \simeq 3$ km).

---

(1) of the non linear problem ; computations with the linear problem are presently being run.

(2) interface height, that is h'_2

(3) these have been obtained from MODULEF subroutines.

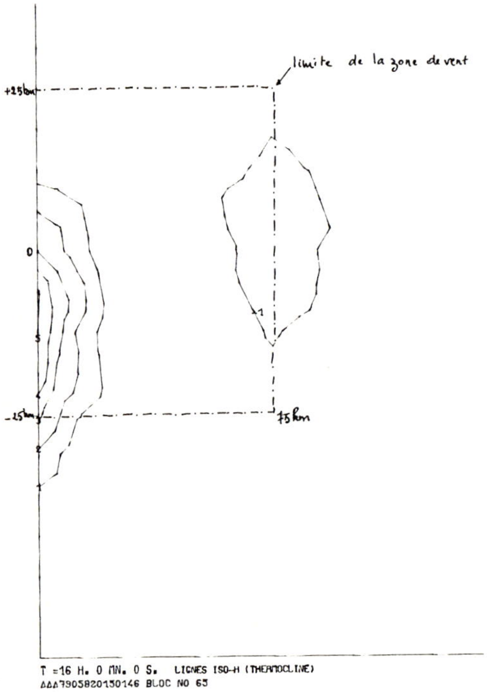

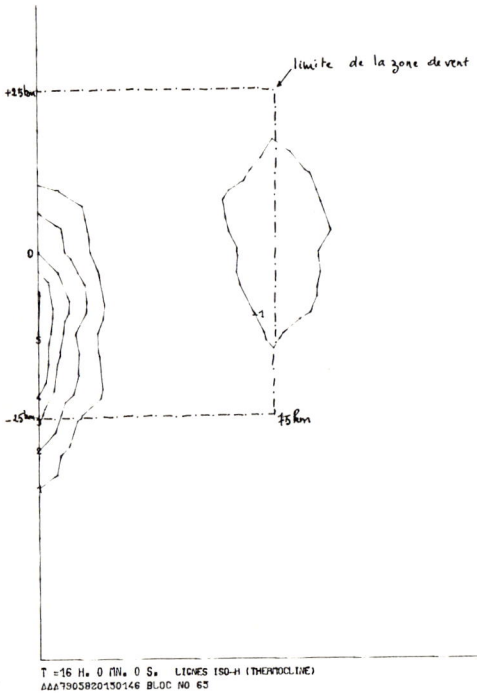

Figure 8 : time t=16h

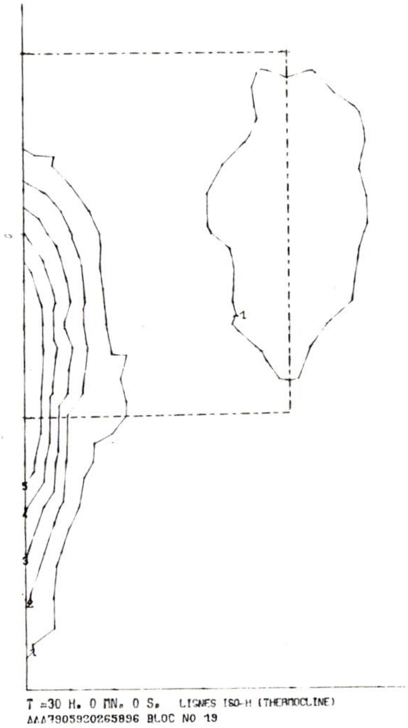

Figure 9 : time t=30h

Isolines of thermocline

The wind field is shown on Figure 12. (It vanishes beyond 200 km from Cap d'Agde). The boundaries have been pushed away as far as possible, to the Baleares in the South, and to the Corsica coast in the East.

For the coarse mesh (524 triangles), the isolines of the thermocline after 60 hours are shown on Figure 13, the velocity fields are shown on Figures 14 and 15.

Corresponding results for the refined mesh are shown on Figures 16, 17 and 18.

Figures 13 and 16 can be compared with surface temperature measurements on Fig. 1. Of course when an upwelling occurs, the thermocline comes to the surface and our equations are no longer valid ; thermodynamic processes are involved by which the surface layers are cooled. However the numerical results indicate that the model can predict fairly well the location of the upwelling cells.

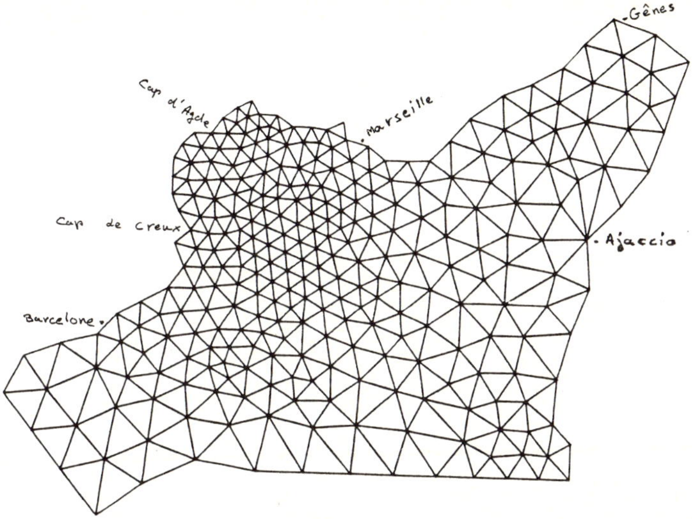

Figures 10 et 11 : Triangulations for the Gulf of Lions :
524 triangles (fig. 10) and 1733 triangles (fig.11).

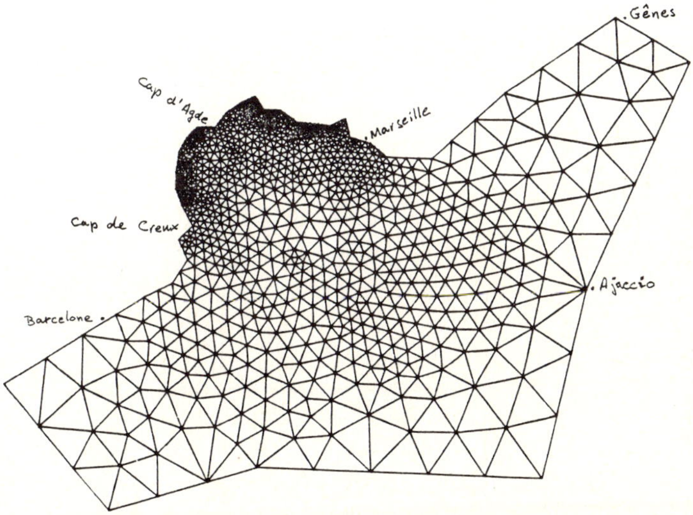

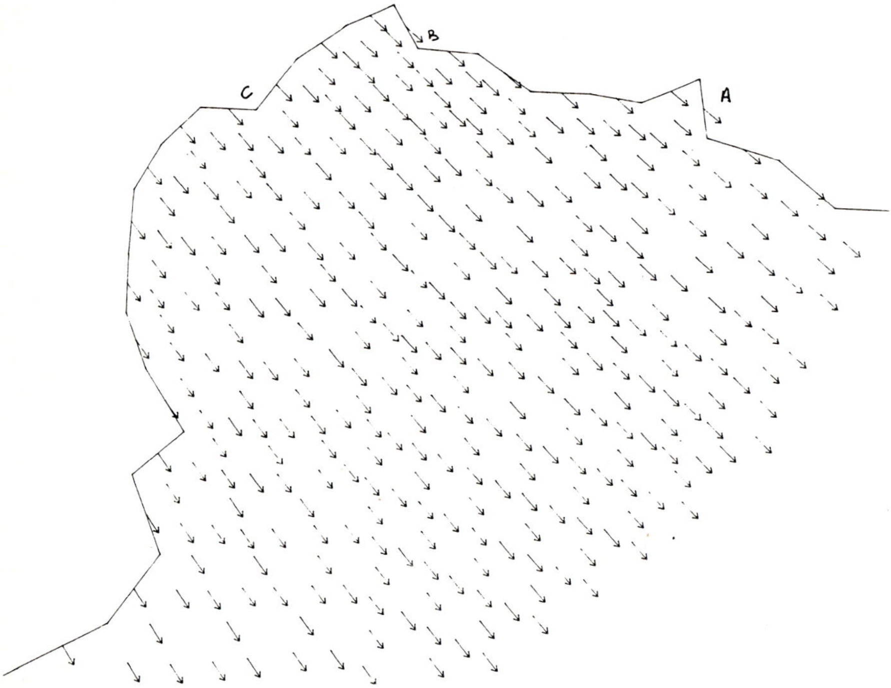

Figure 12 : Wind field.

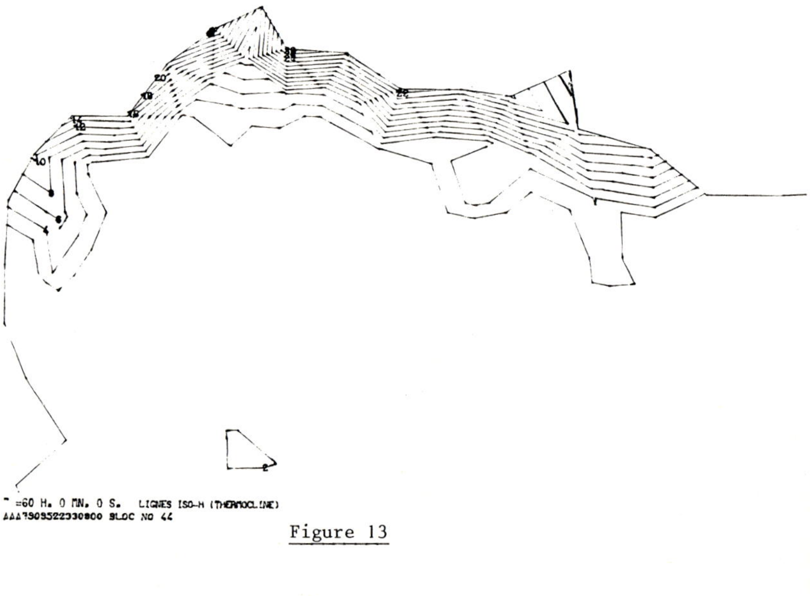

Figure 13

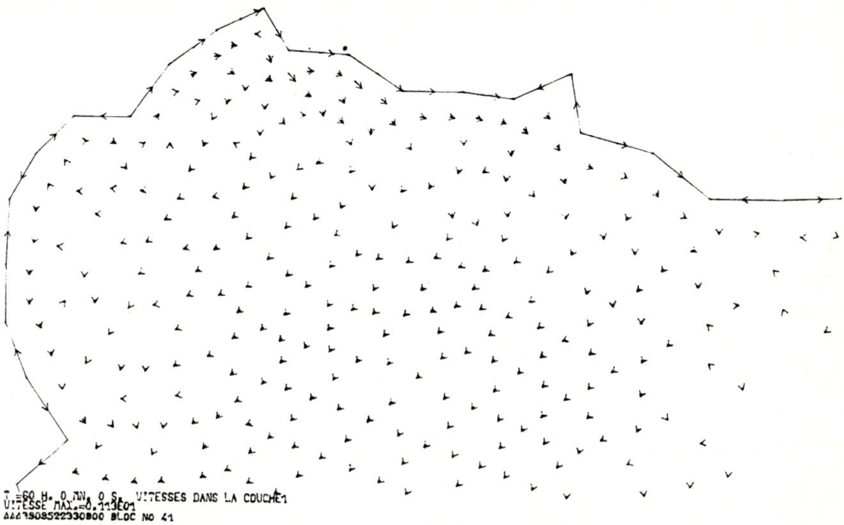

Figure 14

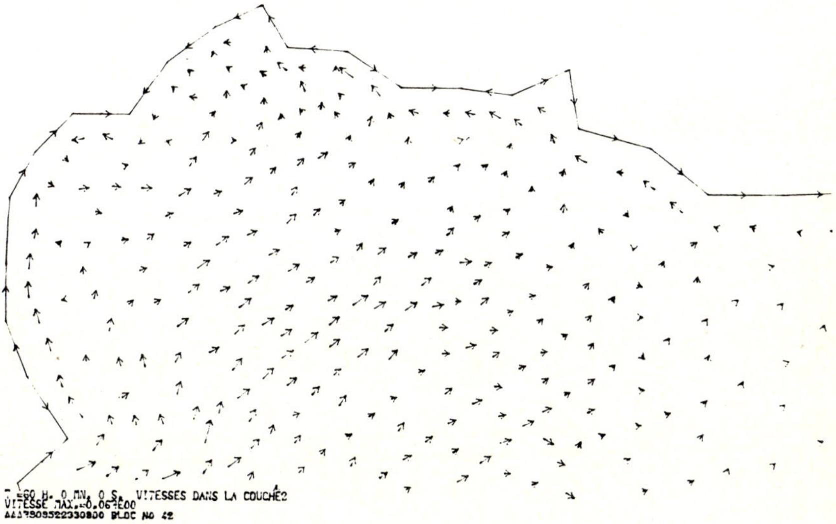

Figure 15

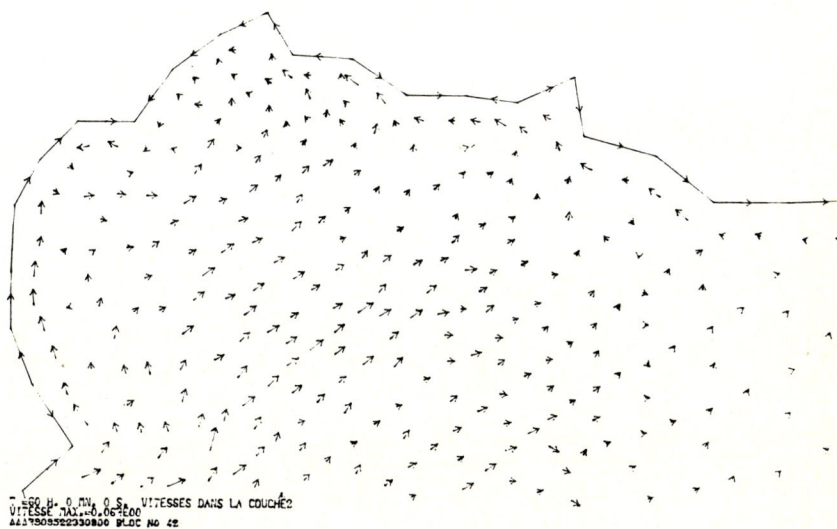

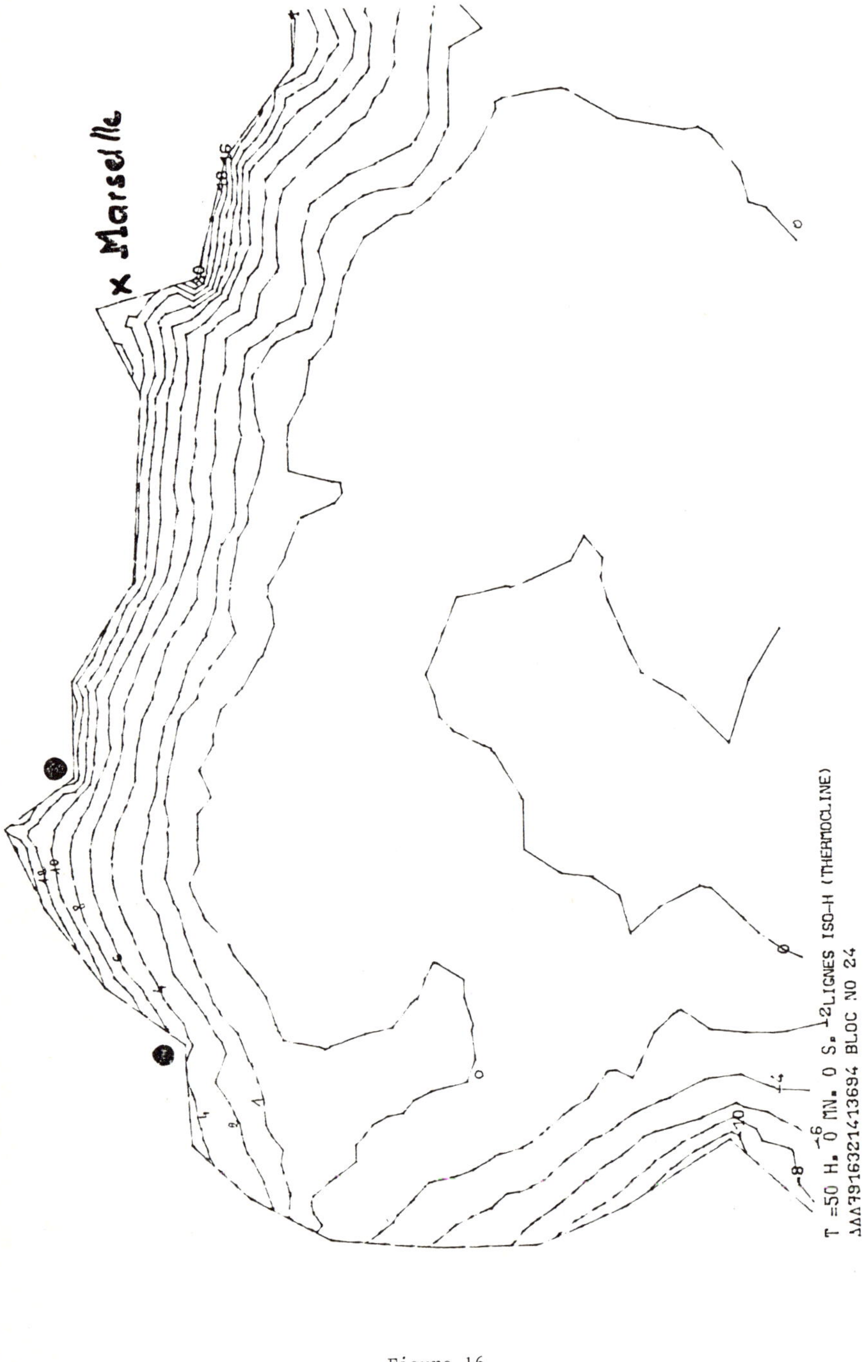

Figure 16

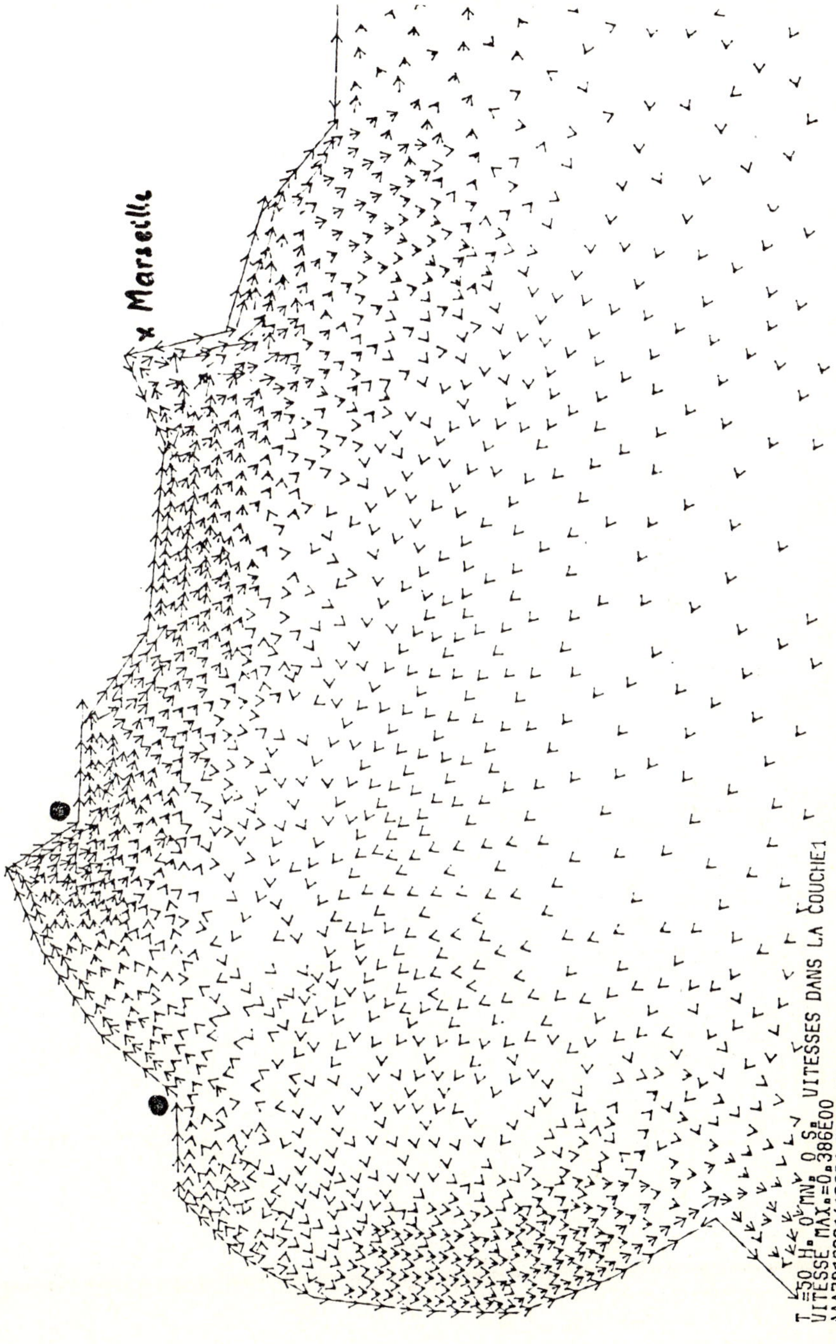
x Marseille
T =50 H. 0 MN. 0 S. VITESSES DANS LA COUCHE1
VITESSE MAX.=0.386E00
ΔΔΔ7916321413694 BLOC NO 21

Figure 17

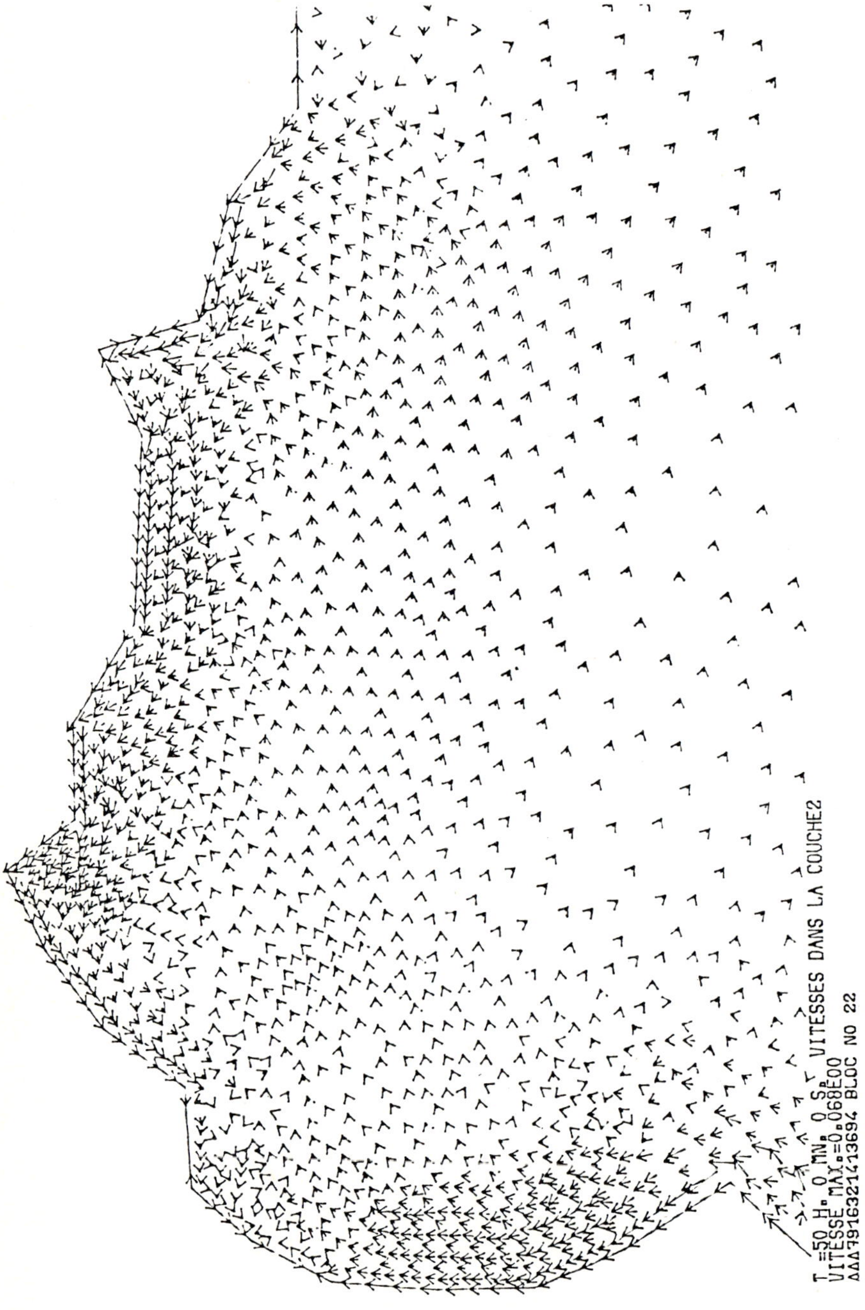
T.=50 H. 0 MN. 0 S. VITESSES DANS LA COUCHE2
VITESSE MAX.=0.068E00
ΔΔΔ7916321413694 BLOC NO 22

Figure 18

CONCLUSION

A finite element method has been developed which appears well suited to reproduce the influence of a complex shoreline.
Further work includes the analysis of the influence of topography, and taking into account thermodynamic effects.

BIBLIOGRAPHY

[1] ADAMEC and O'BRIEN "The seasonal upwelling in the Gulf of Guinea to remote forcing". J. of Physical Oceanography, vol. 8, n° 6, Nov. 1978, pp. 1050-1060.

[2] J.S.ALLEN "Some aspects of the forced-wave response of stratified coastal regions". J. of Physical Oceanography, vol. 6, n° 1, 1976, pp. 113-119.

[3] D. BEGIS, M. CREPON, C. MILLOT "Un problème d'environnement : modélisation de la dynamique océanique du plateau continental". IRIA, Rapport de Recherche n° 236, Mai 1977.

[4] M. CREPON "Transient upwelling generated by two-dimensional atmospheric forcing (1979), to appear.

[5] M. CROUZEIX, P.A. RAVIART "Conforming and non conforming finite element methods for solving the stationary Stokes equations". R.A.I.R.O. (R-3), Déc. 1973, pp. 33-75.

[6] H.E. HURLBURT and J.J. O'BRIEN "A numerical model of coastal upwelling". J. of Physical Oceanography. vol. 2, N° 1, January 1972, pp. 11-26.

[7] M. KWIZAK, A.J. ROBERT "A semi-implicit scheme for grid-point atmospheric models of the primitive equations". Mon. Weather Review., 99, pp. 32-36, 1971.

[8] C. MILLOT "Wind induced upwellings in the Gulf of Lions", Oceanologica Acta, 1979, vol. 2, n° 3.

[9] PEFFLEY, J.J. O'BRIEN "A three-dimensional simulation of coastal upwelling of Oregon. J. of Physical Oceanography, vol 6, n° 2, 1976, pp. 164-180.

[10] F. THOMASSET "Numerical solution of the Navier Stokes equations by finite elements methods. Von Karman Institure (Rhode St. Genèse, Belgium), Lecture Series 86 (1977).

[11] R. TEMAM and F. THOMASSET "Numerical solution of the Navier Stokes equations by a finite element method". Proc. of the IInd Int. Symposium on Finite Element Methods in Flow Problems". Santa Margherita Ligure (Italy) (1976) ed. Springer.

[12] W.C. THACKER "Comparison of Finite Element and Finite Difference Schemes" J. of Physical Oceanography, vol. 8 (1978).

[13] M. BENAHMED "Identification de non linéarités ou de paramètres répartis dans 2 équations auw dérivées partielles non linéaires modélisant un gisement d'hydrocarbures et un écoulement dans un canal ouvert". Thèse d'Etat, Paris IX, (1978).

[14] L. HUA and F. THOMASSET "Modélisation numérique de phénomènes d'upwelling côtiers par une méthode d'éléments finis non conformes".IRIA, Rapport de Recherche n° 366.

APPENDIX

WEAK FORMULATION AND FINITE ELEMENT EQUATIONS

The weak formulation of equation (1) (resp.(2)) is obtained in the standard way : multiply by a test function $\vec{v}$ (resp. φ), integrate over Ω (domain of the ocean) and apply Green's formula

$$(A_1) \qquad \left(\frac{\partial \vec{u}_i}{\partial t} + (\vec{u}_i.\vec{\nabla})\vec{u}_i + f\,\vec{k}\,\vec{u}_i + g \text{ grad } (\alpha_i h_1 + h_2 + D), \vec{v}\right) + \nu(\vec{\text{grad}}\,\vec{u}_i, \vec{\text{grad}}\,\vec{v}) = \left(\frac{\vec{\tau}_{s_i} - \vec{\tau}_{B_i}}{\rho_{h_i}}, \vec{v}\right)$$

$$(A_2) \qquad \left(\frac{\partial h_i}{\partial t}, \varphi\right) - (h_i\,\vec{u}_i, \vec{\text{grad}}\,\varphi) = 0$$

(this last form ensures that mass conservation will be respected in the discrete equation).
Here (.,.) stands for the usual L^2 scalar product over Ω .

Let us introduce some notations :

$\mathcal{S}$ = set of all mid-side nodes,
$\mathcal{S}_o$ = set of *interior* mid-side nodes,
$\mathcal{S}_1$ = set of *boundary* mid-side nodes.

We introduce the scalar shape function w_p , associated with any node $p \in \mathcal{S}$:

$$\left\{\begin{array}{l} .\ w_p \text{ is the polynomial of degree at most 1 on each triangle,} \\ w_p(q) = \begin{cases} 1 & \text{if } q = p \\ 0 & \text{if } q \neq p, \quad q \in \mathcal{S} . \end{cases} \end{array}\right.$$

Note that w_p is non zero only on the two triangles T, T' neighbour to node p , a very useful property for the implementation of the semi-implicit scheme.

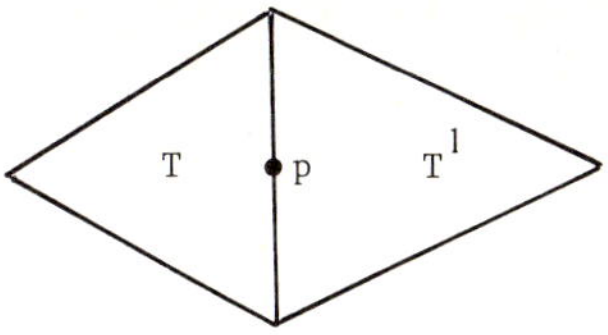

Set $S_p = \frac{\text{area (T) + area (T')}}{3}$

(see joint figure)

Then :

$$(A_3) \qquad (w_p, w_q) = \begin{cases} S_p & \text{if } p = q \\ 0 & \text{if } p \neq q \end{cases}$$

Next if $p \in \mathcal{S}_1$ (boundary node), we set :

$\vec{\tau}_p$ = unit vector tangent to the boundary at node p .

Then we write the discrete variables as :

$$\vec{u}_i(x,y,t) = \sum_{p \in \mathcal{S}_o} \begin{pmatrix} u_{x,i,p}(t) \\ u_{y,i,p}(t) \end{pmatrix} w_p(x,y) + \sum_{p \in \mathcal{S}_1} u_{i,p}(t)\, \vec{\tau}_p\, w_p(x,y)$$

$$h_i(x,y,t) = \sum_{p \in \mathcal{S}} h_{i,p}^{(t)}\, w_p(x,y)$$

Finally we require (A_1) and (A_2) to be true with $\vec{v} = (w_p, o)$ or $\vec{v}$ (o, w_p), for all $p \in \mathcal{S}_o$, $\vec{v} = \vec{\tau}_p\, w_p$ for all $p \in \mathcal{S}_1$, $\varphi = w_p$ for all $p \in \mathcal{S}$.

(the spatial derivatives being taken constant per triangle).

In view of (A_3) we get instead of the first equation of (A_1) :

$$(A_4) \qquad S_p \frac{\partial u_{x,i,p}}{\partial t} - S_p\, f\, u_{y,i,p} + ((\vec{u}_i . \vec{\nabla})\, u_{x,i}, w_p) + g\, (\frac{\partial}{\partial x}(\alpha_i\, h_1 + h_2 + D), w_p) + $$
$$+ \nu\, (\vec{\text{grad}}\, u_{x,i}, \text{ grad } w_p) = (\frac{\tau_{S_{x,i}} - \tau_{B_{x,i}}}{\rho_i h_i}, w_p) \quad (p \in \mathcal{S}_o).$$

The equations for $u_{y,i,p}$ and $u_{i,p}$ $(p \in \mathcal{S}_1)$ are obtained in a similar way ; the equation (A_2) yields :

$$(A_5) \qquad S_r \frac{\partial h_{i,r}}{\partial t} - (h_i, \vec{u}_i, \vec{\text{grad}}\, w_r) = 0 \qquad \forall\, r \in \mathcal{S} \text{ .}$$

Semi-implicit scheme :

From $(A_4)(A_5)$ it is straightforward to write the semi implicit scheme as follows $(\forall\, r \in \mathcal{S})$

$$(A_6) \qquad S_r\, h'^{m+1}_{1r} + g\delta t^2\, H_1 \sum_{q \in \mathcal{S}} a_{qr}(h'^{m+1}_{1q} + h'^{m+1}_{2q}) = \text{right handside}$$

with

$$a_{qr} = \sum_{p \in \mathcal{S}_0} S_p^{-1}\left((w_p, \frac{\partial w_r}{\partial x})(w_p, \frac{\partial w_q}{\partial x}) + (w_p, \frac{\partial w_r}{\partial y})(w_p, \frac{\partial w_q}{\partial y})\right)$$
$$+ \sum_{p \in \mathcal{S}_1} S_p^{-1}\left(\tau_{x,p}(w_p, \frac{\partial w_r}{\partial x}) + \tau_{y,p}(w_p, \frac{\partial w_r}{\partial y})\right) \times \left(\tau_{x,p}(w_p, \frac{\partial w_q}{\partial x}) + \tau_{y,p}(w_p, \frac{\partial w_q}{\partial y})\right).$$

If D is constant the equation in the second layer is similar to (A_6) :

$$S_r\, h'^{m+1}_2 + g\delta t^2\, H_2 \sum_{q \in \mathcal{S}} a_{qr}\left((1-\varepsilon)h'^{m+1}_{1,q} + h'^{m+1}_{2q}\right) = \text{r.h.s.}$$

Otherwise the equation for $h'_{2,r}$ is as follows :

$$S_r\, h'^{m+1}_{2,r} + g\delta t^2 \sum_{q \in \mathcal{S}} a^{(2)}_{qr}\left((1-\varepsilon)h'^{m+1}_{1,q} + h'^{m+1}_{2,q}\right) = \text{r.h.s.}$$

with

$$a^{(2)}_{qr} = \sum_{p \in \mathcal{S}_0} S_p^{-1}\left(((H_2-D)w_p, \frac{\partial w_r}{\partial x})(w_p, \frac{\partial w_q}{\partial x}) + ((H_2-D)w_p, \frac{\partial w_r}{\partial y})(w_p, \frac{\partial w_q}{\partial y})\right) +$$
$$+ \sum_{q \in \mathcal{S}} S_p^{-1}(t_{x,p}((H_2-D)w_p, \frac{\partial w_r}{\partial x}) + t_{y,p}((H_2-D)w_p, \frac{\partial w_r}{\partial y})) \times$$
$$\times \left(t_{x,p}(w_p, \frac{\partial w_q}{\partial x}) + t_{y,p}(w_p, \frac{\partial w_q}{\partial y})\right).$$

SESSION IV

PLASMA PHYSICS

COMPUTING METHODS IN APPLIED SCIENCES AND ENGINEERING
R. Glowinski, J.L. Lions (editors)
North-Holland Publishing Company

ALTERNATING DIMENSION PLASMA TRANSPORT IN THREE DIMENSIONS

Harold Grad
Courant Institute of Mathematical Sciences
New York University
New York, NY 10012/USA

ABSTRACT

The *alternating dimension* ($1\frac{1}{2}$ D) method of solving macroscopic adiabatic and transport problems is here generalized to arbitrary 3-D toroidal plasma confinement systems. The principal new result is the derivation of an evolution equation for the poloidal and toroidal fluxes in which second derivatives can be explicitly exhibited to show that the system is diffusive. This extends previous results in 2-D, axial symmetry and helical symmetry where the flux functions for the magnetic field are explicit consequences of an ignorable coordinate and the EBT closed magnetic line configuration. The eigenvalues (diffusion coefficients) are evaluated and are shown to represent one-dimensional *relative diffusion* among the adiabatic variables, independent of the representation (e.g. whether diffusion is measured relative to mass, or toroidal flux, or poloidal flux). The skin effect diffusion coefficient decouples from the other coefficients and represents diffusion of one magnetic field component relative to the other. Other transport coefficients such as those for mass and energy flow are intrinsically coupled. As in previously implemented alternating dimension codes, a 3-D code built to these specifications should be expected to be extremely accurate and efficient.

1. INTRODUCTION

The macroscopic evolution of a plasma is governed by some form of conservation of mass, momentum, energy and magnetic flux. In both ideal and dissipative dynamic models, each conservation law is an evolution equation with a time derivative of density, velocity, temperature, or magnetic field. A model for *slow* evolution [1,2,3] is obtained by dropping inertia, $\rho du/dt = \rho(\partial u/\partial t + u\cdot\nabla u)$, from the momentum equation, leaving

$$\nabla p = J \times B, \quad J = \text{curl } B \ . \tag{1.1}$$

Formally, this can be obtained by placing a small parameter in front of all time derivatives, velocity components, and transport coefficients; all equations are homogeneous in $\partial/\partial t$, u, and transport coefficients and are unchanged, except for the equation of momentum conservation which reduces to (1.1). The interpretation of the slow evolution model is that one is following a time sequence of equilibria, (1.1). Without transport (the adiabatic model), the equilibrium varies as a consequence of changing external constraints or boundary conditions. Transport causes an additional slow evolution through static equilibrium states. The crucial point is that the equilibrium equation (1.1) is incomplete (in a toroidal, closed magnetic surface configuration, which is all we shall consider) when only boundary conditions are specified. In two dimensions, for example (n is the unit vector in the ignorable z-direction),

$$B = n \times \nabla\psi + n\, B_z \,, \tag{1.2}$$

and equation (1.1) reduces to

$$\Delta\psi = - dp/d\psi - f\, df/d\psi \tag{1.3}$$

where specifying the two profiles $p = p(\psi)$ and $B_z = f(\psi)$ reduces the equilibrium problem to a nonlinear elliptic equation. The two profiles $p(\psi,t)$, $f(\psi,t)$ will vary with time as a consequence of external constraints or dissipation.

In the *adiabatic* problem, the evolution of $p(\psi,t)$ and $f(\psi,t)$ will be governed by specifying fixed adiabatic profiles, e.g. $\mu(\psi)$ and $\nu(\psi)$

$$\begin{cases} \mu = p/(\psi')^{\gamma} \\ \nu = f/\psi' \end{cases} \tag{1.4}$$

Here $\psi'(V)$ is the derivative with respect to the volume V (area in two dimensions), where $\psi(V)$ is the inverse function to $V(\psi)$, the volume within a surface ψ = const. Eliminating p and f in favor of μ and ν gives the generalized (or *queer*) differential equation (QDE)

$$\Delta\psi = F(V,\psi,\psi',\psi'') \tag{1.5}$$

where the right side is a second order ordinary differential operator in $\psi(V)$. A certain amount of analytic theory has been done [4,5,6], and there are very effective numerical methods [4,7-14] for solving the generalized differential equation (1.5). For the moment, assume that the problem of finding an equilibrium solution with given adiabatic profiles $\mu(\psi)$ and $\nu(\psi)$ is fully understood. We now show how to reduce the problem of slow transport to the adiabatic problem (up to now done only in configurations with an ignorable coordinate [15]).

The key to solving the transport problem is to *eliminate the velocity* (appearing in the mass, energy and flux equations) which no longer evolves in accordance with a given value of $\partial u/\partial t$ but is determined by the constraint that p, B, and J (which do satisfy evolution equations) must be in pressure balance at all times, $\nabla p = J \times B$. The velocity is eliminated by first taking in each transport equation a microcanonical volume average on a flux surface, and then introducing adiabatic independent and dependent variables. The result is a set of one-dimensional equations for the *mutual diffusion of all adiabatic variables*. We may take, for example, as a complete set of adiabatic variables the extensive variables, mass, $M = \int \rho dV$, entropy $S = \int \sigma dM$, and two fluxes, ψ (poloidal), and χ (toroidal) [note that μ and ν in (1.4) are given by $\mu = \sigma(dM/d\psi)^{\gamma}$, $\nu = d\chi/d\psi$ in terms of the present adiabatic set]. Taking M as the independent variable, S(M), $\psi(M)$, and $\chi(M)$ are invariant profiles in an adiabatic problem, and the mass densities $\partial S/\partial M$, $\partial\psi/\partial M$, $\partial\chi/\partial M$ satisfy a set of three coupled second order diffusion equations in M and t. The three eigenvalues, $\lambda_i^{(M)}$, of the matrix of coefficients of second derivatives ($\partial^2\sigma/\partial M^2$, etc.) are the *diffusion coefficients*, with dimensionality M^2/t. The values of these eigenvalues are essentially independent of the choice of independent variable (see Appendix). For example, taking ψ as the independent variable, $\partial M/\partial\psi$, ν, and σ satisfy a diffusion system with the eigenvalues (dimensionally ψ^2/t)

$$\lambda_i^{(\psi)} = (\partial\psi/\partial M)^2 \lambda_i^{(M)} .$$

To be more precise, the 1-D diffusion system with is obtained by averaging is not self-contained since it involves geometrical coefficients such as the inductance of a shell, and thermal or electrical resistivity of a shell. Also, the three diffusive equations contain a fourth dependent variable, viz., the Jacobian from volume to the independent variable, e.g. $M'(V) = \rho$ or $\chi'(V)$, etc. The equation count (after losing an equation by introducing an adiabatic independent variable) is maintained by supplementing the 1-D diffusive system by the average pressure balance.

The common key to solving either the adiabatic or the dissipative problem is to alternate between solution of the 1-D system and determination of the geometrical coefficients in the 1-D system from the 2-D (or 3-D) equilibrium (this is the *Alternating Dimension algorithm*, at one time called "$1\frac{1}{2}$-D" algorithm [3,4,13]). For the adiabatic problem the entire 1-D system is the average pressure balance which is a second order ordinary differential equation for $\psi(V)$ given $\mu(\psi)$ and $\nu(\psi)$ and the appropriate inductance coefficients $L_{ij}(V)$. Given an approximation to $L_{ij}(V)$ [$\mu(\psi)$ and $\nu(\psi)$ are fixed] the ordinary differential equation gives an approximation to $\psi(V)$ and to the current [$F(\psi)$ in (1.5)]; this is a inhomogeneous term for an elliptic equation for $\psi(x,y)$. From such an approximation to the current distribution in physical space (2-D or 3-D), one finds the flux surfaces and corresponding inductance coefficients and repeats the procedure.

This iterative procedure has been found to converge much more rapidly than standard elliptic iteration, e.g. than $\Delta\psi_{n+1} = F(\psi_n)$ for given $F(\psi)$. In the dissipative problem, the 2-D (or 3-D) geometry has to be recomputed only after hundreds of diffusion time steps. Thus the diffusion problem is almost entirely 1-D with occasional corrections to the geometric coefficients from 2-D or 3-D.

What is done in this paper is:

(1) find an evolution equation for the flux ($\partial\psi/\partial t$, $\partial\chi/\partial t$) in a general, 3-D configuration;

(2) identify this evolution equation as diffusive by explicitly identifying second derivatives of the adiabatic variables;

(3) calculate the eigenvalues (diffusion coefficients) in complete generality;

(4) Show that the diffusion coefficients are independent of the representation; in other words, the physical picture is 1-D diffusion of adiabatic variables relative to one another (note that volume is not a proper variable, nor is mass flow relative to a fixed frame or relative to V).

At least one 3-D code of this type is being constructed [16]. One important advantage of the Alternating Dimension algorithm, not considered in this paper, is the efficient treatment of complex and changing topology [17,4,8].

2. PRELIMINARIES

Toroidal Configuration

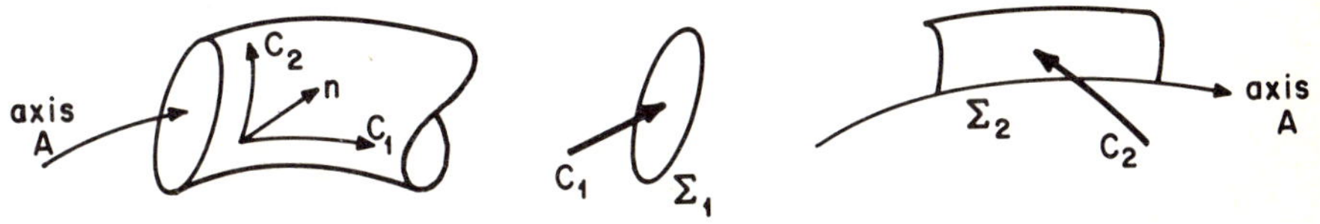

Figure 1

Consider a family of nested toroidal surfaces with a simple closed curve, A, as axis. Take C_1 and C_2 , as indicated in Fig. 1, as a basis for the homology of one-dimensional closed curves on a toroidal surface, oriented such that (C_1, C_2, n) is a right-handed system (n is the outward normal to the surface). The open surfaces Σ_1 and Σ_2 are taken as in the figure, with oriented boundaries

$$\partial\Sigma_1 = -C_2 , \qquad \partial\Sigma_2 = C_1 - A . \tag{2.1}$$

The intersection numbers are given by $\{C_i, \Sigma_j\} = \delta_{ij}$.

Surface Potentials and Periods [18]

Consider the family of nested toroidal surfaces to be parameterized by the values of a parameter p through the level surfaces of a given function $p(x,y,z) = \text{const.}$ Suppose X' is a single-valued two-dimensional vector field on a surface p. We use the notation $X' = \nabla'\alpha$ for the *surface gradient* of a *surface potential* α defined by the property

$$\oint_C X' \cdot dx = 0 \tag{2.2}$$

for all closed curves C homologous to zero on the surface. The surface periods

$$[\alpha]_i = \int_{C_i} \nabla'\alpha \cdot dx \tag{2.3}$$

are path independent on the given surface, but are functions of p. This is in contrast to a conventional (three-dimensional) gradient, $X = \nabla\phi$.

Its projection, $\nabla'\phi$, on a surface is a special case of a surface gradient with periods which are independent of p. Note that $\nabla\phi$ is single-valued (and $\int \nabla\phi\cdot dx$ path-independent) within the three dimensional domain, whereas $\nabla\alpha$ is single-valued only in a domain cut by Σ_1 and Σ_2, or when projected on each surface. For a three-dimensional vector field X, the property n·curl X = 0 on a p-surface implies that the projection X' is a surface gradient.

A very useful identity involving two arbitrary surface potentials α and β is

$$\oint_p \nabla\alpha \times \nabla\beta\cdot dS = [\alpha]_1[\beta]_2 - [\alpha]_2[\beta]_1 \tag{2.4}$$

where the integral is taken over a toroidal p-surface.

Next, suppose there is given a vector field B satisfying

$$\begin{cases} \text{div } B = 0 \\ B\cdot\nabla p = 0 \end{cases} \tag{2.5}$$

We define the magnetic flux periods

$$\psi_i = \int_{\Sigma_i} B\cdot dS \tag{2.6}$$

Clearly, $\psi_i = \psi_i(p)$ and, using $p = p(x,y,z)$, we can extend ψ_i as point functions $\psi_i(x,y,z)$. The derivative $d\psi_2/d\psi_1$ (or its reciprocal) is the *rotation number* of a B-line on a given toroidal surface (depending on whether the rotation number is measured by crossings of Σ_1 or Σ_2).

For each of the $\psi_i(x,y,z)$, surface potentials α_i can be defined (on each surface p) such that

$$B = \nabla\alpha_i \times \nabla\psi_i \ . \tag{2.7}$$

From the identity $B\cdot dS = d\alpha\, d\psi$, the following period relations are easily established:

$$\begin{cases} [\alpha_1]_1 = -d\psi_2/d\psi_1 \ , & [\alpha_1]_2 = 1, \\ [\alpha_2]_1 = -1, & [\alpha_2]_2 = d\psi_1/d\psi_2 \end{cases} \tag{2.8}$$

Furthermore, $\alpha_1 = (d\psi_2/d\psi_1)\alpha_2 + \alpha_0(p)$ pointwise (α_1 and α_2 are unique within an added constant on each surface). We shall sometimes use the notation (ψ,χ) for (ψ_1,ψ_2) or (ψ_2,ψ_1) in which case we shall write

$$B = \nabla\alpha \times \nabla\psi = \nabla\beta \times \nabla\chi \tag{2.9}$$

To continue, assume now that B and p satisfy

$$\left\{ \begin{array}{l} J \times B = \nabla p, \quad J = \text{curl } B \\ \text{div } B = 0 \end{array} \right. \tag{2.10}$$

[this is compatible with (2.5)]. Since n·curl B = 0, one can introduce the surface potential ϕ,

$$B = \nabla'\phi \tag{2.11}$$

Defining the currents

$$I_i = \int_{\Sigma_i} J \cdot dS \tag{2.12}$$

and mmf's (magneto-motive forces)

$$\Phi_A = \oint_A B \cdot dx \,, \qquad \Phi_i = \oint_{C_i} B \cdot dx = [\Phi]_i \tag{2.13}$$

(A is the axis) we verify

$$I_1 = -\Phi_2 \quad , \qquad I_2 = \Phi_1 - \Phi_A \,. \tag{2.14}$$

From div J = 0 and $J \cdot \nabla p = 0$ we can introduce

$$J = \nabla\zeta \times \nabla p \tag{2.15}$$

Using the identity $J \cdot dS = d\zeta \, dp$ and the notation $\zeta_i = [\zeta]_i$, we easily verify

$$\zeta_1 = -dI_2/dp, \qquad \zeta_2 = dI_1/dp \tag{2.16}$$

$$\zeta_i = -d\Phi_i/dp \tag{2.17}$$

Other useful identities are

$$J \cdot \nabla\alpha_i = -dp/d\psi_i \tag{2.18}$$

$$B \cdot \nabla\zeta = 1 \tag{2.19}$$

$$B = \zeta \, \nabla p + \nabla\Phi \tag{2.20}$$

and the Jacobian identities

(2.21) $$\nabla\alpha_i \times \nabla\psi_i \cdot \nabla\Phi = B^2, \quad dV = d\alpha\, d\psi\, d\Phi/B^2$$

(2.22) $$\nabla\alpha_i \times \nabla\psi_i \cdot \nabla\zeta = 1\ , \quad dV = d\alpha\, d\psi\, d\zeta$$

To interpret these as Jacobians in three dimensions, it is simplest to cut the interior of the torus at Σ_1 and Σ_2 , which leaves (Φ,α,ψ) or (ζ,α,ψ) as single-valued coordinates in the cut domain.

As an example, using (α,ψ,Φ) as variables we can evaluate

(2.23) $$\zeta_\Phi = 1/B^2\ , \qquad \zeta_\alpha = \frac{J\cdot B}{B^2}\frac{d\psi}{dp}\ , \qquad \zeta_\psi = (B\times\nabla\alpha\cdot\nabla\zeta)/B^2$$

and write

(2.24) $$J = \frac{dp}{d\psi}\,[\zeta_\alpha B + \frac{B\times\nabla\psi}{B^2}]$$

Surface Averages [18,2,4]

The torus ψ < const. has a volume $V(\psi)$ which can be inverted as $\psi(V)$. From $\psi(x,y,z)$ we can extend $V(\psi)$ to a point function $V(x,y,z)$. We recall the well known identities

(2.25) $$\oint_\psi \frac{dS}{|\nabla\psi|} = \frac{1}{\psi'(V)}\ , \qquad \oint \frac{dS}{|\nabla V|} = 1$$

leading to the microcanonical (volume weighted) surface averages,

(2.26) $$\begin{aligned} \langle\phi\rangle &\equiv \int \phi\,\frac{dS}{|\nabla V|} \\ &= \psi'\oint \phi\,\frac{dS}{|\nabla\psi|} \\ &= \frac{d}{dV}\int \phi\, dV \end{aligned}$$

where the volume integral is extended over the interior of the torus V < const.

We introduce the standard notation for derivatives of general point functions and functions which are constant on V = const.,

(2.27) $$\frac{\partial}{\partial t}\,\phi(x,y,z,t), \qquad \psi_t \equiv \frac{\partial}{\partial t}\,\psi(V,t), \qquad p' \equiv \frac{\partial}{\partial V}\,p(V,t)\ .$$

Some easily derived formulas are

(2.28) $$\langle \mathrm{div}\, X\rangle = \langle X\cdot\nabla V\rangle'$$

(2.29) $$<\phi>' = \left\langle \text{div}\left(\frac{\phi\ \nabla V}{|\nabla V|^2}\right)\right\rangle$$

(2.30) $$<\frac{\partial V}{\partial t}> = 0$$

(2.31) $$<\frac{\partial \phi}{\partial t}> = <\phi>_t + <\phi\ \frac{\partial V}{\partial t}>$$

(2.32) $$\frac{\partial V}{\partial t} = \frac{1}{\psi'}\left(\frac{\partial \psi}{\partial t} - \psi_t\right)$$

where $\phi(x,y,z,t)$ and $X(x,y,z,t)$ represent arbitrary scalar and vector functions respectively. In (2.32), ψ can be replaced by any function which is constant on V surfaces (e.g. p). For $f = f(V,t)$, we have the important special case of (2.31),

(2.33) $$<\frac{\partial f}{\partial t}> = f_t$$

This together with (2.28) is the reason why volume is the natural independent variable in averaged conservation equations.

The formula (2.4) for two surface potentials can now be written

(2.34) $$<\nabla\alpha \times \nabla\beta \cdot \nabla V> = [\alpha]_1[\beta]_2 - [\alpha]_2[\beta]_1 .$$

Applying this identity to (2.21), we obtain

(2.35) $$<B^2> = \Phi_1\psi_1' + \Phi_2\psi_2'$$

Similarly, from (2.18),

(2.36) $$-\ p' = \Phi_1'\psi_1' + \Phi_2'\psi_2'$$

and from (2.15),

(2.37) $$<J\cdot B> = \Phi_2\Phi_1' - \Phi_1\Phi_2' = \Phi_1 I_1' + \Phi_2 I_2'$$

For later purposes, it is important to note that there is no simple formula involving periods for $<J^2>$.

Surface Harmonics [18,19]

If two surface potentials f and g are related by

(2.38) $$\nabla' f = n \times \nabla' g$$

where n is the normal to the surface, then f and g are *conjugate harmonics* in the surface metric. Similarly, defining $\omega = \alpha_1 \psi_1'$, from

$$B = \nabla\omega \times \nabla V = \nabla'\Phi \tag{2.39}$$

we recognize that B is a generalized harmonic vector in the surface V = const. with ω and Φ as conjugate surface harmonics; ω and Φ satisfy an elliptic equation which differs from the Cauchy-Riemann equations (2.38) by the appearance of a weight function $|\nabla V|$ presumed given. If the surface and the weight $|\nabla V|$ are given, then B is uniquely determined by a pair of periods, either $[\omega]_i$ (ψ_1' and ψ_2') or Φ_i. From this follows the existence of 2×2 symmetric, positive definite surface inductance and susceptance (or capacitance) matrices $L = L_{ij}$ and $\Lambda = L^{-1} = \Lambda_{ij}$ respectively, relating the conjugate periods

$$\begin{aligned} \psi_i' &= L_{ij}\Phi_j \\ \Phi_i &= \Lambda_{ij}\psi_j' \end{aligned} \tag{2.40}$$

From (2.35), the magnetic energy density per volume is

$$\begin{aligned} \langle \tfrac{1}{2} B^2 \rangle &= \tfrac{1}{2} (\Phi_1\psi_1' + \Phi_2\psi_2') \\ &= \tfrac{1}{2} \Phi_i L_{ij} \Phi_j \\ &= \tfrac{1}{2} \psi_i' \Lambda_{ij} \psi_j' \end{aligned} \tag{2.41}$$

The inductance coefficients are purely geometrical, depending only on the function V(x,y,z) [$|\nabla V|$ on a given surface].

The average pressure balance (2.36) is most useful when written in terms of a mixed basis (ψ_1', Φ_2) or (ψ_2', Φ_1) (eliminating the other two periods by the inductance formulas (2.40)],

$$-\frac{dp}{d\psi_1} = (\psi_1'/L_{11})' + \Phi_2 \frac{d\Phi_2}{d\psi_1} / \Lambda_{22} - \Phi_2 \frac{d}{d\psi_1} (L_{12}/L_{11}) , \tag{2.42}$$

and the same formula interchanging indices 1 and 2. In configurations with an ignorable coordinate, (2.42) is the average of an elliptic equation for ψ_1. For example in two dimensions [cf. (1.3)], $(\psi_1'/L_{11})'$ is the average of $\Delta\psi_1$, $\Phi_2 = f(\psi_1)$ and $\Lambda_{22} = 1$, $\Lambda_{12} = 0$.

3. CONSERVATION OF MASS AND ENERGY

The conservation of mass, including sources, takes the form

$$\frac{\partial \rho}{\partial t} + \text{div}\ (\rho u) = \Sigma \tag{3.1}$$

We assume that ρ is constant on pressure surfaces (this will follow from the fact that it holds for pressure and temperature [14]). Defining

$$U = \oint u \cdot dS = \langle u \cdot \nabla V \rangle \tag{3.2}$$

We compute the averaged mass equation

$$\rho_t + (\rho U)' = \langle \Sigma \rangle \tag{3.3}$$

Note that the independent variable V is defined by p = const. which has not yet appeared in the present discussion. Defining total mass by

$$M = \int \rho dV, \qquad M' = \rho \tag{3.4}$$

we obtain

$$M_t + M'U = \int \Sigma dV. \tag{3.5}$$

Without sources mass is conserved, and the derivative holding mass fixed, d/dt, is

$$\frac{d}{dt} \equiv \frac{\partial}{\partial t} \phi(M,t) = \phi_t + U\phi' \tag{3.6}$$

We have

$$\frac{dM}{dt} = 0 \tag{3.7}$$

$$\frac{dV}{dt} = U \tag{3.8}$$

Note that, even without sources, (3.1) does not imply that the flow velocity, u, carries ρ = const. surfaces into ρ = const.; (nor does u carry ψ or p). M is defined as the mass within a V surface. This is not a Lagrangian variable in (x,y,z) since the fluid element within a V surface moving as in (3.8) [or in a fixed domain M(x,y,z,t)

$<$ const.] is not preserved. However (with $\Sigma = 0$) M *is* a Lagrangian variable in (V,t).

The pointwise conservation of energy takes the form

$$\frac{\partial p}{\partial t} + u \cdot \nabla p + \gamma p \operatorname{div} u = (\gamma-1) \operatorname{div}(\hat{\kappa} \nabla T) + (\gamma-1)\eta J^2 + (\gamma-1)Q \tag{3.9}$$

where $T = p/\rho$ is the temperature, $\hat{\kappa} = \kappa(T)/\omega^2\tau^2 = \kappa_1 \rho^2/T^{1/2}B^2$ is the transverse heat flow (κ_1 is an absolute constant), $\eta = \eta_1 T^{-3/2}$ is the resistivity, and Q is an energy source (per volume). Averaging, under the assumption that T = const. coincides with p = const., gives

$$p_t + Up' + \gamma pU' = (\gamma-1)(\kappa_* T')' + (\gamma-1) \langle \eta J^2 \rangle + (\gamma-1)\langle Q \rangle \tag{3.10}$$

where

$$\kappa_* \equiv \frac{\kappa_1 \rho^2}{T^{1/2}} \left\langle \frac{|\nabla V|^2}{B^2} \right\rangle = \frac{\kappa_1 \rho^2}{T^{1/2}(\psi')^2} \left\langle \frac{|\nabla \psi|^2}{B^2} \right\rangle \tag{3.11}$$

Introducing the entropy per unit mass $\sigma = C_V \log (p/\rho^\gamma)$, yields

$$\frac{p}{C_V(\gamma-1)} \frac{d\sigma}{dt} = (\kappa_* T')' + \langle \eta J^2 \rangle + \langle Q \rangle \tag{3.12}$$

Without dissipation or sources we have

$$\frac{dM}{dt} = 0, \quad \frac{d\sigma}{dt} = 0$$

M and σ are adiabatic variables and $\sigma(M)$ or $M(\sigma)$ can be taken as a fixed adiabatic profile.

If the motion is incompressible, div u = 0 [or incompressible on the average, $U(p) = 0$], then V is also an adiabatic variable. This is the reason why in some models (e.g. a low β, small poloidal field, straight Tokamak [2,8]), V has been used successfully as the independent 1-D variable.

4. FLUX CONSERVATION

To obtain appropriately averaged flux conservation equations, we use Maxwell's equations

$$\frac{\partial B}{\partial t} + \text{curl } E = 0 \ , \quad \text{div } B = 0 \tag{4.1}$$

and Ohm's law

$$E + u \times B = \eta J \ . \tag{4.2}$$

Letting $B = \nabla\alpha \times \nabla\psi$,

$$\frac{\partial B}{\partial t} = \nabla \frac{\partial \alpha}{\partial t} \times \nabla\psi + \nabla\alpha \times \nabla \frac{\partial \psi}{\partial t}$$

$$= \text{curl } \left(\frac{\partial \alpha}{\partial t} \nabla\psi - \frac{\partial \psi}{\partial t} \nabla\alpha\right)$$

and from (4.2),

$$E = \nabla\alpha \frac{\partial \psi}{\partial t} - \nabla\psi \frac{\partial \alpha}{\partial t} + \nabla f \ . \tag{4.3}$$

To begin, f must be restricted to the cut domain (Σ_1 and Σ_2); projecting (4.3) on a ψ surface shows that $\nabla' f$ is single-valued, thus f is a surface potential. Adding

$$u \times B = \nabla\alpha(u \cdot \nabla\psi) - \nabla\psi(u \cdot \nabla\alpha)$$

to (4.3) yields

$$\eta J = \nabla\alpha\left(\frac{\partial \psi}{\partial t} + u \cdot \nabla\psi\right) - \nabla\psi\left(\frac{\partial \alpha}{\partial t} + u \cdot \nabla\alpha\right) + \nabla f \ . \tag{4.4}$$

Equations (4.3) and (4.4) are valid with (β,χ) replacing (α,ψ) and a related potential g instead of f.

We remark that

$$\left\langle\frac{\partial \psi}{\partial t} + u \cdot \nabla\psi\right\rangle = \psi_t + U\psi' = \frac{d\psi}{dt} \ . \tag{4.5}$$

With this as motivation, take the scalar product of (4.4) first with J, then with B, and then average, to obtain

$$\begin{aligned} \langle\eta J^2\rangle &= - \frac{dp}{d\psi} \frac{d\psi}{dt} + \langle J \cdot \nabla f\rangle \\ &= - \frac{dp}{d\psi} \frac{d\psi}{dt} + p'\{f_1\zeta_2 - f_2\zeta_1\} \end{aligned} \tag{4.6}$$

$$\langle \eta J\cdot B\rangle = \langle B\cdot\nabla f\rangle$$
(4.7)
$$= \psi'\{f_1\alpha_2 - f_2\alpha_1\}$$

It remains to evaluate the periods f_1 and f_2. Taking the jump of (4.3) across either Σ_1 or Σ_2,

$$0 = [\nabla\alpha]\,\frac{\partial\psi}{\partial t} - \nabla\psi\,[\frac{\partial\alpha}{\partial t}] + [\nabla f]\ .$$

Note that only the projection of $[\nabla\alpha]$ is zero for a surface potential α. First take $\alpha = \alpha_1$ and the indicated jump across Σ_2 where $[\alpha_1]_2 = 1$ [cf. (2.8)]. This implies that $[\nabla\alpha_1]_2 = 0$ and $[\partial\alpha_1/\partial t]_2 = 0$, hence $[f]_2 = c_1(t)$. Similarly, for $\alpha = \alpha_2$, $[g]_1 = -c_2(t)$. The nonzero period of f or g can then be eliminated between (4.6) and (4.7) giving

$$\frac{d\psi_i}{dt} = -\frac{d\psi_i}{dp}\langle\eta J^2\rangle + \frac{dI_i}{dp}\langle\eta J\cdot B\rangle - c_i = -\frac{\psi_i'}{p'}\langle\eta J^2\rangle + \frac{I_i'}{p'}\langle\eta J\cdot B\rangle - c_i \tag{4.8}$$

These are the two averaged flux conservation equations. They reduce to the conventional equations in the case of an ignorable coordinate (two dimensions, axial symmetry, helical symmetry). These equations could be used, in principle, to advance ψ_i in time, since the right side is known in terms of a given, instantaneous equilibrium state, $\nabla p = J\times B$. There is, however, no resemblance to a diffusion equation, since second derivatives, ψ_i'', are not visible. From (2.14), (2.40), (2.37) we see that I_i', p', and $\langle J\cdot B\rangle$ are each linear expresions in ψ_1'' and ψ_2''. The second term in (4.8) is explicit in second derivatives, but it is not at all a conventional diffusion operator. Furthermore, $\langle J^2\rangle$ is not expressible in terms of periods. Examination of the special case of axial symmetry shows that there is cancellation between undesirable terms arising from $\langle J^2\rangle$ and $\langle J\cdot B\rangle$, also that geometrical expressions other than inductance arise.

Before treating the general case, note the cancellation in the following linear combination of the two equations (4.8)

$$\frac{1}{\psi_2'}\frac{d\psi_2}{dt} - \frac{1}{\psi_1'}\frac{d\psi_1}{dt} = -\frac{1}{\psi_1'\psi_2'}\langle\eta J\cdot B\rangle + \frac{c_1}{\psi_1'} - \frac{c_2}{\psi_2'} \tag{4.9}$$

Also, from (2.37) and (2.40)

$$\langle J\cdot B\rangle = \Delta(\psi_2'\psi_1'' - \psi_1'\psi_2'') - A \tag{4.10}$$

where

(4.11)
$$A = (\psi_1')^2(\Lambda_{11}\Lambda_{12}' - \Lambda_{12}\Lambda_{11}') + \psi_1'\psi_2'(\Lambda_{11}\Lambda_{22}' - \Lambda_{22}\Lambda_{11}') + (\psi_2')^2(\Lambda_{12}\Lambda_{22}' - \Lambda_{22}\Lambda_{12}')$$

and

(4.12)
$$\Delta = \Lambda_{22}\Lambda_{11} - \Lambda_{12}^2$$

Introducing ψ_1 as independent variable with the notation D/Dt holding ψ_1 fixed,

(4.13)
$$\partial\phi/\partial t = D\phi/Dt + (\partial\phi/\partial\psi_1)(\partial\psi_1/\partial t)$$

We obtain, for (4.9)

(4.14)
$$\frac{D\psi_2}{Dt} = \lambda_0 \frac{\partial^2\psi_2}{\partial\psi_1^2} + \frac{\eta}{\psi_1'} A + c_1 \frac{\psi_2'}{\psi_1'} - c_1$$

(4.15)
$$\lambda_0 = \eta(\psi_1')^2\Delta$$

Of course this equation is incomplete without the rest of the system, but it strongly suggests that ψ_2 diffuses relative to ψ_1 at a rate given by the diffusion coefficient λ_0 [similarly, ψ_1 diffuses relative to ψ_2 at a rate given by the diffusion coefficient $\eta(\psi_2')^2\Delta$]. Measured relative to V (this is only dimensional analysis, for there is no such diffusion equation), the diffusicn coefficient is simply $\eta\Delta$. It is interesting to note that Δ becomes large if the flux surface is rippled, and it may be unbounded at a separatrix.

Next we turn to the coefficient $<|\nabla\psi|^2/B^2>$ which appears in the average heat conductivity (3.11) [in axial symmetry $<|\nabla\psi|^2/B^2> = <r^2>$]. It also makes an appearance as part of $<J^2>$, viz. $<J_\perp^2>$ (perpendicular to B)

$$J_\perp = \frac{B\times\nabla p}{B^2}$$

$$<J_\perp^2> = <|\nabla p|^2/B^2> = \left(\frac{dp}{d\psi}\right)^2 <|\nabla\psi|^2/B^2>$$

(4.16)
$$\frac{d\psi}{dp} <J_\perp^2> = \frac{p'}{\psi'} <|\nabla\psi|^2/B^2>$$

Assuming that $<|\nabla\psi|^2/B^2>$ is geometrical, the expression (4.16), which is what appears in (4.8), is linear in second derivatives of ψ through the factor p'.

We must generalize the concept of *geometrical* coefficient (such as Λ_{ij}, or $<r^2>$ in axial symmetry) to *quasigeometrical* which involves the geometry alone and first derivatives, ψ_i', preferably as a homogeneous function of order zero in ψ_i' (i.e. a function of the rotation number ψ_2'/ψ_1'). Many properties of a second order parabolic system will be preserved with such dependence of the coefficients. At first glance, the energy equation (3.12) seems to give second derivatives via $(\kappa^*)' \sim <|\nabla\psi^2|/B^2>'$. However, in accordance with the adiabatic format described in the Introduction and in the Appendix, the proper dependent variables associated with flux are ψ_i' rather than ψ_i; thus quasigeometrical implies a coefficient which contains only the dependent variable itself, and no derivatives.

To show that $<|\nabla\psi|^2/B^2>$ is quasigeometrical we recall that B satisfies an elliptic equation in a surface. The manifold of solutions of this elliptic equation on a torus is a two-dimensional linear vector space. As a basis in this space, introduce

$$B^{(i)} = \nabla\theta^{(i)} \times \nabla V = \nabla\phi^{(i)}, \qquad i = (1,2)$$
$$[\theta^{(i)}]_j = \delta_{ij} \tag{4.17}$$

From the defining equations (4.17), the $B^{(i)}$ depend only on the geometry, viz., on the function V(x,y,z). Evidently,

$$B = \psi_1' B^{(1)} - \psi_2' B^{(2)} \tag{4.18}$$

(however, note that these are *not* Hamada coordinates). From (4.18) we conclude that $<|\nabla\psi_i|^2/B^2>$ depends only on ψ_i' as a homogeneous function of order zero [and, of course, on V(x,y,z)].

Next we turn to $<J_\parallel^2>$. On a given surface with a given vector field $B = \nabla\alpha \times \nabla\psi$, the current J and its potential ζ are almost determined by the ODE (2.19), $B \cdot \nabla\zeta = 1$. To be precise (Ref. [4], Appendix), if ζ_0 is any solution of this ODE, the general solution is given by

$$\zeta = \zeta_0 + c\zeta \tag{4.19}$$

where c is an arbitrary constant (i.e. function of ψ). Similarly,

$$J = J_0 + c\,\frac{dp}{d\psi}\,B \tag{4.20}$$

For a specific value of c, ζ in (4.19) and J in (4.20) are the originally given equilibrium quantities. Note that $J_\perp$ is independent of c, and $J\times B = J_0\times B = \nabla p$. There are two choices of c such that one of the periods ζ_i (or ϕ_i') is zero; such a choice has the property that the corresponding direction field, J, consists of closed lines. These two distinguished values of c lead to $\zeta_{(i)}$ and $J_{(i)}$,

$$(4.21)\qquad \begin{aligned} \zeta_{(1)} &= \zeta - [\zeta]_2\alpha_1 , & \zeta_{(2)} &= \zeta + [\zeta]_1\alpha_2 \\ [\zeta_{(1)}] &= (1/\psi_1',0) , & [\zeta_{(2)}] &= (0,\ 1/\psi_2') \end{aligned}$$

and

$$(4.22)\qquad \begin{aligned} J_{(1)} &= J + \frac{d\Phi_2}{d\psi_1} B , & J_{(2)} &= J - \frac{d\Phi_1}{d\psi_2} B \\ I'_{(1)} &= (0,\ -p'/\psi_1'), & I'_{(2)} &= (p'/\psi_2',0) \end{aligned}$$

(this representation *is* related to Hamada coordinates). From the construction of these solutions of the ODE, $B\cdot\nabla\zeta = 1$, with one period of ζ zero, it is clear that the potentials ζ depend only on the geometry and the periods ψ_i' of B. Consistent with this, the periods of $\zeta_{(1)}$ and $\zeta_{(2)}$ also depend only on ψ_i' (4.21), whereas the periods of ζ itself, I_i'/p' contain second derivatives. In other words, the second derivatives enter into solution of the ODE only as specified side conditions. Since

$$J_{(i)} = \nabla\zeta_{(i)} \times \nabla p = p'\ \nabla\zeta_{(i)} \times \nabla V,$$

$J_{(i)}/p'$ is quasigeometric and homogeneous of order -1 in ψ_i'.

This is sufficient to evaluate $\langle J^2\rangle$ in terms of $\langle J^2_{(i)}\rangle$ and periods; after some manipulation,

$$(4.23)\qquad \frac{1}{\eta}\frac{d\psi_i}{dt} = -\frac{d\psi_i}{dp}\langle J^2_{(i)}\rangle - \phi_j\phi_j'/\psi_i' - \frac{c_i}{\eta}\quad (j \neq i)$$

$$(4.24)\qquad = -\frac{dp}{d\psi_i} R_i - \phi_j \frac{d\phi_j}{d\psi_i} - \frac{c_i}{\eta}\quad (j \neq i)$$

where R_i is quasigeometric and can be written as

$$(4.25)\qquad \left\{ \begin{aligned} R_i &= S_i + T_i \\ S_i &= \langle |\nabla\psi_i|^2/B^2\rangle \\ T_1 &= \langle(\frac{\partial\zeta}{\partial\alpha_1} - \zeta_2)^2B^2\rangle , \quad T_2 = \langle(\frac{\partial\zeta}{\partial\alpha_2} + \zeta_1)^2B^2\rangle \end{aligned}\right.$$

The second derivatives in (4.24) are now explicitly visible in $d\phi/d\psi_i$ and $d\phi_j/d\psi_i$. For later use we introduce

(4.26)
$$\hat{R}_i = R_i - \phi_j^2/\langle B^2\rangle \qquad (j \neq i)$$
$$\hat{T}_i = T_i - \phi_j^2/\langle B^2\rangle \qquad (j \neq i)$$

and note that

(4.27)
$$\hat{R}_i > \hat{T}_i > 0$$

(4.28)
$$(\psi_2')^2\hat{T}_1 = (\psi_1')^2\hat{T}_2$$

The inequality $\hat{T}_i > 0$ follows from

$$\langle f^2B^2\rangle\langle B^2\rangle > \langle f\ B^2\rangle^2$$

and the identity is merely tedious.

5. COMPLETE ONE-DIMENSIONAL SYSTEM AND DIFFUSION COEFFICIENTS

In terms of V as independent variable, the complete system of averaged equations, collected from (2.3), (3.12), (4.23), and (2.36) is (dropping Σ and Q)

(5.1)
$$\rho_t + (\rho U)' = 0, \qquad M_t + UM' = 0$$

(5.2)
$$\frac{p}{C_V(\gamma-1)}(\sigma_t + U\sigma') = (\kappa_* T')' + \eta\langle J^2\rangle$$

(5.3)
$$\frac{d\psi_i}{dt} = -\frac{\eta}{\psi_i'}(p'R_i + \phi_j\phi_j') - c_i \qquad (j \neq i)$$

(5.4)
$$-p' = \phi_1'\psi_1' + \phi_2'\psi_2'$$

Since ρ is not an adiabatic variable, introduce

(5.5)
$$\xi_i = \rho/\psi_i' = \partial M/\partial\psi_i$$

and combine (5.1) with (5.3) to obtain

(5.6)
$$\frac{1}{\xi_i}\frac{d\xi_i}{dt} = \frac{1}{\psi_i'}\left\{\frac{\eta}{\psi_i'}(R_i p' + \phi_j\phi_j')\right\}'$$

Only one of the two equations (5.6) is to be used to replace (5.1) if ψ_i or $\int \sigma dM$ is taken as the independent variable; both are used [replacing (5.3)] if M is the independent variable.

To be specific, choose M as the independent variable. Using the notation

$$\dot{\phi} \equiv \partial\phi/\partial M = \phi'/\rho \tag{5.7}$$

$$\dot{\psi}_i \equiv f_i = 1/\xi_i \tag{5.8}$$

we obtain for the adiabatic variables (σ, f_i) and the auxiliary variable ρ,

$$\frac{p}{C_V(\gamma-1)}\frac{d\sigma}{dt} = \rho(\kappa_* \rho \dot{T})^{\bullet} + \eta\langle J^2\rangle \tag{5.9}$$

$$\frac{df_i}{dt} = -(\eta R_i \dot{p}/f_i)^{\bullet} - (\eta\phi_j\dot{\phi}_j/f_i)^{\bullet} \qquad (j \neq i) \tag{5.10}$$

$$-\dot{p} = \rho(f_1\dot{\phi}_1 + f_2\dot{\phi}_2) \tag{5.11}$$

with the supplementary relations

$$p = \rho^{\gamma} \exp(\sigma/C_V)$$

$$T = \rho^{\gamma-1} \exp(\sigma/c_V)$$

$$\phi_i = \rho \sum \Lambda_{ij} f_j$$

$$\langle B^2\rangle = \rho(f_1\phi_1 + f_2\phi_2)$$

$$\langle J \cdot B\rangle = \rho(\phi_2\dot{\phi}_1 - \phi_1\dot{\phi}_2)$$

$$\langle J^2\rangle = (\dot{p}/f_i)^2 R_i - 2(\dot{\phi}_j/f_i)\langle J\cdot B\rangle - (\dot{\phi}_j/f_i)^2\langle B^2\rangle \qquad (j\neq i)$$

There is a choice of two expressions for $\langle J^2\rangle$. Noting

$$\dot{p}/p = \gamma\,\dot{\rho}/\rho + \dot{\sigma}/c_V \tag{5.12}$$

we see that (5.11) expresses $\dot{\rho}$ as a linear combination of $(\dot{\sigma}, \dot{f}_i)$ which allows $\ddot{\rho}$ to be eliminated from (5.9) and (5.10) in favor of $(\ddot{\sigma}, \ddot{f}_i)$ which are now explicit.

A tedious calculation yields the eigenvalues of the 3×3 matrix of coefficients of second derivatives, λ_0 and $\lambda_\pm$,

$$\lambda_0 = \eta\rho^2\Delta \tag{5.13}$$

$$\lambda_\pm = \frac{1}{2}(\lambda_1 + \lambda_2) \pm \frac{1}{2}[\lambda_1^2 + \lambda_2^2 - 2\delta\lambda_1\lambda_2]^{1/2} \tag{5.14}$$

where

$$\delta = \frac{p+(2/\gamma - 1)<B^2>}{p+<B^2>}, \qquad 0 < \delta < 1 \tag{5.15}$$

$$\lambda_1 = \eta(\hat{R}_i/g_i^2)\frac{\gamma p<B^2>}{\gamma p+<B^2>} \tag{5.16}$$

$$\lambda_2 = (\gamma-1)\kappa_1(S_i/g_i^2)\frac{\rho}{T^{1/2}}\frac{p+<B^2>}{\gamma p+<B^2>} \tag{5.17}$$

The geometric factors $\hat{R}_i$ and S_i are both positive [cf. (4.27)]; the combinations $\hat{R}_i/g_i^2$, S_i/g_i^2 are independent of i [cf. (4.28)]. These eigenvalues agree with previously calculated special cases in axial symmetry [15] and for EBT [20]. There is a more sophisticated transport model (tensor resistivity and heat conductivity) which has been carried out in axial symmetry and which involves additional geometrical coefficients [14]; so far we have not extended this analysis to the general 3-D geometry.

The expressions λ_1 and λ_2 are the contributions from resistivity and heat flow respectively, but they are not the eigenvalues (unless $\delta = 1$, which is approached only at high β, $<B^2>/p \to 0$). The geometric factors in λ_1 and λ_2 approach their "Pfirsch-Schlüter" values in the limit of a large aspect circular cross-section Tokamak. We see that these factors have been recovered in the eigenvalues, although they were lost in the mass flow and energy flow (which are nonlocal) in the transition from Pfirsch-Schlüter to Grad-Hogan.

6. DISCUSSION

The claim that we have given a general 3-D version of classical transport requires some interpretation. It was pointed out in 1967 [21,22] that the equation $\nabla p = J \times B$ is ill-posed in any toroidal configuration other than 2-D, axial symmetry, helical symmetry, and reflection symmetry (EBT). The (now pervasive) appearance of island structures was predicted, and, for example, the so-called *major disruption* in a Tokamak is evidently a disappearance of equilibrium rather than an instability. The problem of islation and complex topologies has been extensively studied using alternating dimension codes, but so far in configurations where existence of solutions is not in doubt.

The practical use of the present formulation lies in the fact that the failure of equilibrium to exist is sometimes unimportant and sometimes invisible. The latter is a property of some analytic expansions, for example, those in which the rotation number per period is small (all resonances are invisible to such expansions) [23,24]. Insofar as such expansions may approximate reality, the equilibrium calculation can be extended to include transport.

A second application is to numerical equilibrium calculations in which the existence of simple flux surfaces is forced [25]. Refining the mesh will, of course, lead to difficulty; but these difficulties may be hard to identify in a 3-D computation! Again, insofar as a numerical code is a black box which seems to give numerical equilibria, it can be extended to include transport.

From experience in 2-D, we should expect great accuracy and efficiency for these transport techniques. As one example, a computation simulating the resistive transfer from a Belt Pinch (simple topology) to a fully developed Doublet (figure eight magnetic surfaces), with greater accuracy than is physically meaningful, was run in approximately fifteen seconds of 7600 time [8]. In another example, the power output, $\int \eta J^2$ (in this case J is a second derivative since ψ is the dependent variable), was calculated for an oscillating separatrix with an accuracy of one part in 10^4 using a 32×64 2-D mesh [26].

One early operating version of the algorithm ([17], 1974) used piecewise constant (in time) geometrical coefficients. Later versions use linear (in time) interpolations of geometrical coefficients between successive calculations of equilibria, together with iteration of the diffusion step until the advance geometry converges. To optimize a 3-D transport calculation, one should be prepared to minimize the number of geometrical computations by using more sophisticated time dependence. However, in the above-quoted transfer from Belt Pinch to Doublet, using only five geometry time steps was sufficient to give adequate accuracy. In any less radical change of geometry, two or three equilibrium calculations (with the remainder of the computation in 1-D) should suffice for practical purposes (but much greater accuracy is required for mathematical purposes such as the study of singularities and boundary layers).

APPENDIX: INVARIANCE OF DIFFUSION COEFFICIENTS

Consider a set of dependent variables $u = (u_1,\ldots,u_n)$ which satisfy the diffusion system

$$\frac{\partial u_i}{\partial t} = \frac{\partial}{\partial x}\left(A_{ij}\frac{\partial u_j}{\partial x}\right) + R_i = B_{ij}\frac{\partial^2 u_j}{\partial x} + R_i' \tag{A.1}$$

A and R are functions of $(x,u,\partial u/\partial x)$; more generally, R can also include "lower order" terms such as $\int f(x,u,\partial u/\partial x)\,dx$. The principal part (i.e. excluding R) is in conservation form; the diffusion coefficients are defined to be the eigenvalues $\lambda_i(x,u,\partial u/\partial x)$ of the matrix B and are assumed to be negative (B may incorporate second derivatives from the expansion of $\partial A/\partial x$, but in the actual model in the text, A does not depend on $\partial u/\partial x$).

Theorem. If x is replaced by the new independent variable $\xi = \int u_1 dx$, then $u(\xi,t)$ satisfies a system similar to (A.1); specifically the new principal part is $A' = u_1^2 A$; the eigenvalues of B' are $\lambda_i' = u_1^2\lambda_i$.

We use the single expression "R" to represent any combination of lower order terms. With the notation $\partial\phi(\xi,t)/\partial t = d\phi/dt$,

$$\frac{d\phi}{dt} = \frac{\partial\phi}{\partial t} - \frac{\partial\phi}{\partial\xi}\frac{\partial\xi}{\partial t}$$

$$= \frac{\partial\phi}{\partial t} + R$$

by use of

$$\frac{\partial\xi}{\partial t} = \int\frac{\partial u_1}{\partial t}dx = A_{1j}\frac{\partial u_j}{\partial x} + \int R_1\,dx = R.$$

Also

$$\frac{\partial u}{\partial x} = u_1\frac{\partial u}{\partial\xi}$$

therefore

$$\frac{\partial}{\partial x}\left(A\frac{\partial u}{\partial x}\right) = u_1\frac{\partial}{\partial \xi}\left(u_1 A\frac{\partial u}{\partial \xi}\right) = \frac{\partial}{\partial \xi}\left(u_1^2 A\frac{\partial u}{\partial \xi}\right) + R$$

and

$$\frac{du}{dt} = \frac{\partial}{\partial \xi}\left(u_1^2 A\frac{\partial u}{\partial \xi}\right) + R .$$

A second transformation, replacing u_1 by $1/u_1$, leaves the eigenvalues unchanged (this result is classical for any transformation of dependent variables). We use the two transformations on a set of extensive adiabatic variables $(\sigma_0, \sigma_1, \ldots, \sigma_n)$ which are assumed to satisfy (A.1) where

$$x = \sigma_0 , \quad u = (\partial\sigma_1/\partial x, \ldots, \partial\sigma_n/\partial x)$$

The twice-transformed system is equivalent to interchanging σ_0 and σ_1,

$$\xi = \sigma_1 = \int u_1 \, dx$$

$$u_1' = \frac{\partial\sigma_0}{\partial\xi} = \frac{1}{u_1} .$$

In other words, choosing any σ_i as independent variable and $\partial\sigma_j/\partial\sigma_i$, $j \neq i$, as the set of dependent variables gives a system with the same eigenvalues, independent of i, except for a uniform factor.

ACKNOWLEDGMENT

This work was supported by the U. S. Department of Energy under Contract Number EY-76-C-02-3077.

REFERENCES

[1] Grad, H., and Hogan, J., Phys. Rev. Lett. 24, 1377, 1970.

[2] Grad, H., in "Proc. Congress of Mathematicians", Nice, Vol. 3, p. 105, 1970.

[3] Grad, H., unpublished memoranda "A Skeleton Key to Classical Resistive Diffusion in Tokamak," July 6, 1970, "Transient Diffusion," April 22, 1970.

[4] Grad, H., Hu, P. N., and Stevens, D. C., Proc. Nat. Acad. Sci. U.S.A. 72, 3789, 1975.

[5] Vigufsson, G., Thesis, New York Univ., 1977; Bull. Amer. Math. Soc., 85, 778, 1979.

[6] Teman, R., in Proc. Intl. Meeting on Recent Methods in Non-linear Analysis, Rome, May, 1978.

[7] Grad, H., Hu, P.N., Stevens, D. C., and Turkel, E., Proc. 2nd Eur. Conf. Comp. Phys., Garching, April 1976.

[8] Grad, H., Hu, P. N., Stevens, D. C., and Turkel, E., in Proc. 6th IAEA Conf., Berchtesgaden, Oct. 1976.

[9] Nelson, D. B., in Proc. of Varenna Conf., Sept. 1977.

[10] Stevens, D. C., in Proc. 8th Conf. Num. Simulation of Plasmas, Monterey, June 1978.

[11] Barnes, D. C., et al., in Proc. 7th IAEA Conf., Innsbruck, 1978.

[12] Hogan, J., Nucl. Fusion 19, 753, 1979.

[13] Grad, Harold, "Survey of $1\frac{1}{2}$ D Transport Codes," Report No. MF-93, Courant Inst., Oct. 1978.

[14] Nelson, D. B., and Grad, H., "Heating and Transport in Tokamaks of Arbitrary Shape and β," to appear.

[15] Grad, Harold, and Hu, P. N., "Classical Diffusion: Theory and Simulation Codes," in: Proc. Workshop on High β Plasmas, Varenna, Sept. 1977.

[16] Winsor, N. K., Miner, W. H., and Grad, H., "3-D EBT Equilibrium and Transport," Bull. Am. Phys. Soc. 24, 1050, 1979.

[17] Hu, P. N., Grad, H., and Stevens, D. C., Bull. Amer. Phys. Soc. 19, 865, 1974.

[18] Grad, H., and Rubin, H., in Proc. 2nd Geneva Conf. 31, 190, 1958.

[19] Bateman, G., Nucl. Fusion 13, 227, 1973.

[20] Grad, H., "Variable Dimensional Transport." in Proc. of EBT Workshop, Aug. 1979, U.S. Dept. of Energy.

[21] Grad, H., Phys. Fl. 10, 137, 1967.

[22] Grad, H., "Problems in Magnetostatic Equilibrium," Lille Colloq. on MHD, June 1969; also, Cour. Inst. Report MF-62, April 1970.

[23] Spies, G. O., and Lortz, D., Plasma Phys. 13, 799, 1971.

[24] Grad, H., in Proc. 5th IAEA Conf., Tokyo, 1974, Vol. II, 153, 1975.

[25] Bauer, F., Betancourt, O., Garabedian, P., "A Computational Method in Plasma Physics," Springer-Verlag, New York, 1978

[26] Grad, H., Jensen, T. H., Kress, M., "Low Frequency Ohmic Heating of Doublet," Bull. Amer. Phys. Soc. 24, 976, 1979.

COMPUTING METHODS IN APPLIED SCIENCES AND ENGINEERING
R. Glowinski, J.L. Lions (editors)
North-Holland Publishing Company

NUMERICAL SIMULATION IN PLASMA PHYSICS

A.A. Samarskii
The Keldysh Institute of Applied
Mathematics USSR Acad. of Sci.
Moscow, USSR

INTRODUCTION

Plasma physics is not only a field for development of physical theories and mathematical models but also an object of application of the computational experiment /I/ comprising analytical and numerical methods adapted for computers. In this paper we consider only MHD plasma physics problems.

Three main aspects of the numerical simulation should be pointed out:

I) the choice of physical approach and mathematical model, and its analytical study;

2) mathematical and physical accessories of computational experiment for the chosen model;

3) execution of computational experiment (computing runs in the frame of chosen model, correcting this model, analysing the computational results and comparing them with physical experiment data).

The mathematical accessories imply the numerical method and the coding technique. The physical accessories include the medium characteristics - the state equation, viscosity, heatconductivity, conductivity, radiation absorption path, etc. Specifying these parameters is a separate important problem involving a considerable amount of quantum-mechanical computations.

It is important to stress that the computational experiment involves not only the improvement of numerical methods but also the development of analytical methods for nonlinear equations to provide self-similarity and asymptotic solutions which describe the characteristic features of the process under investigation and may be also used for testing numerical methods.

Although many numerical methods appear in connection with particular problems, they are often of general importance. The results of investigations carried out under guidance and in cooperation with the author at the Keldysh Institute of Applied Mathematics, USSR Acad. of Sci., and at Lomonosov Moscow State University are presented here.

§ I DISSIPATIVE STRUCTURES IN PLASMA. METASTABLE LOCALIZATION PHENOMENON

Instability studies play an important role in plasma physics investigations. In many cases instabilities may cause formation of selfsupporting structures in the medium. A class of dissipative structures is of great interest because they also exist in many problems in astrophysics, meteorological and marine physics, nuclear physics, unified field theory, biology, etc. A review of stationary dissipative structure investigations may be found in the books by Prigogine et al. (for example, /2/).

Transient dissipative structures arising and evolving in a medium under aggravation regime are a subject of intensive study for the last years. These regimes imply an infinite growth of solutions till the finite time moments. They may be caused by aggravated boundary conditions or by nonlinear sources /3-10/. The T-layer effect /4/, /10/ is an example of nonstationary dissipative MHD structures predicted by numerical simulation and confirmed later in a number of physical experiments. In many cases the process of structure formation is that of threshold nature. The aggravation regimes exist in Z and Θ-pinches, laser fusion, implosions, Langmuir's turbulence, nonlinear optics, astrophysics, etc.

The investigation of aggravation regimes necessitates formulation of new physical concepts and development of new mathematical methods for their treatment. The new type solutions with paradoxical physical meaning have been obtained even from the simplest model of quasilinear diffusion in a medium with nonlinear sources. Consider an example of the Cauchy problem for the equation

$$\frac{\partial T}{\partial r} = \frac{\partial}{\partial r}\left(K(T)\frac{\partial T}{\partial r}\right) + Q(T),$$

$$t \geqslant 0, \quad -\infty \leq r \leq \infty,$$

$$T(r, 0) = T_0(r),$$

where $T_0(r) = 0$ only for $-0,5a < r < 0,5a$, $a > 0$; $K(T) = K_0 T^{\sigma}$, $K_0 > 0$, $\sigma > 0$; $Q(T) = q_0 T^{\beta}$, $q_0 > 0$, $\beta > 0$. This equation describes, for example, electron heatconduction and burning in a medium with $Q(T) > 0$. Depending on the values of parameters β and σ there exist /3/ - /5/ different medium burning regimes: if $\beta > I$ aggravation burning regime may appear when for the finite time t_f the temperature becomes infinite either in the whole space (if $I < \beta < \sigma + I$ – HS regime), or in a finite volume (if $\beta = \sigma + I$ – S regime), or at a single point only (if $\beta > \sigma + I$ – LS regime).

The paradoxical phenomenon of heat localization within certain spatial scales (fundamental lengths L_T) has been found under S and LS regimes. Under the S regime, $L_T = 2\pi\sqrt{K_0\ (\sigma+I)/q_0}/\sigma$ and depends on the medium specifications only. Under the LS regime, L_T depends also on the initial disturbance amplitude $T_{om} = \max T_0(r)$ and span a.

If $a \geqslant L_T^*$, where L_T^* is a threshold value ("resonant length"), then the burning is localized within the region of diameter $L_T \leqslant L_T^* = dT_{om}^{-[\beta-(\sigma+1)]/2}$, $d = \pi\sqrt{2K_0\ (\beta+\sigma+I)/(q_0\sigma(\beta-I))}$ till the aggravation time moment $t_f = (\beta+\sigma+I)\ /\ ((\sigma+2)\ (\beta-I)\ q_0\ T_{om}^{\beta-1})$. If $a < L_T^*$ there exist 2 cases: I) $\sigma + I < \beta < \sigma + 3$ localization appears after the finite time t_* within the fundamental length $L_T \lessapprox d^{2/(\sigma+\beta-3)} \times W_0^{-[\beta-(\sigma+1)]/(\sigma+3-\beta)}$ depending on the initial disturbance energy W_0; 2) $\beta > \sigma + 3$ the disturbance attenuates and approaches the selfsimilarity regime for point thermal source of finite energy.

The existence of characteristic space and time scales (L_T, t_f) in this nonlinear problem shows the importance of such analysis for the proper choice of steps in space and time in numerical simulations.

Because of the localization effect there may coexist, without interaction, a few nonstationary dissipative structures. If they have different aggravation moments, then the structure assembly degenerates into a single structure corresponding to the earliest aggravation moment. The localization regions may overlap one another. Then they interact. The conditions providing the concurrent aggravation of the structure assembly without the degeneration are given in /5, 9/. The most interesting results are obtained in the case of burning under LS regime ($\beta > \sigma + I$). The developed stage of burning corresponds to a self-similar problem with nonunique solution. The eigenfunctions of this problem are obtained by numerical methods for various values of the parameters β and σ /5, 9/. The linearization of self-similar equation provides an estimation of the number of eigenfunctions $N = [k - [[k]\ k^{-1}]] + I$, $k = (\beta - I)/(\beta - \sigma - I)$, where $[k]$ is an integer part of k. The proof of burning localization under LS regime is based on the comparison theorem, the analytical solution under S regime being used for majorant estimations of localization range under LS regime /5, 9, 3/. The first eigenfunction corresponds to the medium burning in a form of a simple structure with a single temperature maximum in the localization region. The subsequent eigenfunctions describe the burning with a few maxima.

Initiation of burning structures is produced by the proper choice of initial temperature profiles which coincide with the self-similar solutions in the localization range. The localization results are extended for the case of more general dependency K(T) and Q(T) /II/. The localization takes place also in the case of constant K The burning localization is studied in a moveless medium with K = K (r, T) and Q = Q (r, T) /7/ and in the case of plasma compression under S regime in presence of many dissipative processes /4/.
tive processes /4/.

The self-similarity problems are of great importance in localization studies. Invariant solutions are deduced in that connection by means of group theory /I2/including the aggravation regimes. In the particular case considered here 6I invariant solution types are found. For $K = K_0$, $Q(T) = 0$ or $K = K_0 T^{\sigma}$, $Q = 0$, the invariant solutions were earlier studied by S.Lie and L.Ovsyannikov.

In the last years the theory of nonlinear system organization was discussed at three International conferences on synergetics. But the laws of complication of nonlinear medium organization obtained in /3/ - /IO/ are essentially connected with the heat and burning inertia which causes the nonstationary dissipative structure development, and by this the above works are different from others including those of Prigogine. It is paradoxical that the world of such structures may be described by localized processes of different complexity. The structure architecture and evolution are determined by the eigenfunctions of nonlinear medium which, in distinction from linear problems, are independent of boundary conditions and boundary location and depend only on the nonlinear medium parameters (because of localization effect). The multidimensional computations confirm metastable existence of this structures /8/. These investigations establish not only a new principle of heat and burning localization in laser fusion but represent also an important advance in the theory of nonlinear system organization.

§ 2 MHD MODEL OF SOLAR DYNAMO

The numerical simulation is of great importance in cosmophysics because in many cases the actual experiments or measurements are impossible. Consider an interesting problem of solar activity cycle with period of 22 years. Longlasting observations show the correlation between solar magnetic fields and solar activity phenomena and draw attention to the explanation of largescale alternations of this field consisting of three main components: a dipole oriented along the solar axis plus azimuthal field (in solar spots) plus a dipole oriented in the equator plane (sectorial field), the axisymmetrical component variations having the period of 22 years.

The mathematical model of this periodical field (solar dynamo) was proposed in 1969 /13/ and generalized later in USA, FRG and USSR. I dwell briefly on the results obtained at our Institute /14/ - /16/ on the basis of equation taking into account the magnetic field generation, diffusion and advection and turbulent diamagnetism

$$\frac{\partial \vec{B}}{\partial t} = \mathrm{rot}\left([\vec{v} \times \vec{B}] + \alpha \vec{B} - \beta\, \mathrm{rot} \frac{\vec{B}}{\mu}\right) \qquad \text{(I)}$$

where $\vec{B}$ is the magnetic induction, $\vec{v} = [\vec{\omega} \times \vec{r}]$ is the solar rotation velocity, β-is the turbulent magnetic diffusitivity, μ-is the turbulent permeability (essentially less than 1), α- is the characteristic of turbulent velocity field spiralness ($\alpha = \langle(\vec{V}\,\mathrm{rot}\,\vec{V})\rangle$) being of great importance for magnetic field generation. Equation (I) is solved within a sphere of solar radius R, coefficients β, α being nonzero (and μ being nonunity) only in a turbulent layer with thickness of about 0,3 R. In the axisymmetrical case the problem is reduced to finding $B_\varphi = B$ and a component of vector potential $A_\varphi = A$ from the following equations

$$\frac{\partial A}{\partial t} = \alpha B + \frac{\beta}{r}\left[\frac{\partial}{\partial r}\left(\frac{1}{\mu}\frac{\partial}{\partial r} rA\right) + \frac{1}{r}\frac{\partial}{\partial \theta}\left(\frac{1}{\mu \cdot \sin\theta}\frac{\partial}{\partial \theta} A \sin\theta\right)\right]$$
$$\frac{\partial B}{\partial t} = \frac{1}{r}\left(\frac{\partial \omega}{\partial r}\frac{\partial}{\partial \theta} - \frac{\partial \omega}{\partial \theta}\frac{\partial}{\partial r}\right) rA \cdot \sin\theta + \frac{1}{r}\left[\frac{\partial}{\partial r}\left(\beta \frac{\partial}{\partial r} r \frac{B}{\mu}\right) + \frac{1}{r}\frac{\partial}{\partial \theta}\left(\frac{\beta}{\sin\theta}\frac{\partial}{\partial \theta}\frac{B}{\mu}\sin\theta\right)\right] \qquad (2)$$

Around the Sun ($r > R$) the field is approximately potential ($\beta \rightarrow \infty$), so

$$\frac{\partial^2}{\partial r^2}(rA) + \frac{1}{r}\frac{\partial}{\partial\theta}\left(\frac{1}{\sin\theta}\frac{\partial}{\partial\theta}A\sin\theta\right) = 0, \quad A \to 0, r \to \infty \qquad (3)$$

$$B = 0$$

There are matching conditions at the boundary r = R

$$[A] = 0, \quad \left[\frac{1}{\mu}\frac{\partial}{\partial r}(rA)\right] = 0, \quad \left[\frac{B}{\mu}\right] = 0 \qquad (4)$$

The square brackets designate a step of parameter value at the boundary. At the initial moment A, B are given. The functions α, ω, β, μ are considered known. In the dimensionless form the problem involves only one parameter — dynamo-number $D = \frac{\alpha_0 \omega_0}{\beta_0^2} R^3$ which determines threshold conditions for the existence of oscilating decayless solution.

The substitution of solution of problem (3) expressed in the form of Poisson's integral $A(r, \theta) = \int_0^{\pi} K(r, \theta', \theta)\, A(R, \theta') \sin\theta' d\theta'$ in relations (4) yields a nonlocal boundary condition at r = R. An algebraic equation system with respect to an unknown grid function A at r = R is obtained as a result of Poisson's integral discretization and subsequent elimination (from the centre to the boundary) applied to Eq. (2) approximated by Peaceman-Rachford scheme. The coefficients of this system are time independent.

Thus the system matrix is once inverted and stored and then repeatedly used during the solution computation.

The numerical simulation shows the generation of magnetic field oscillations without attenuation when $D = D_{thr}(\sim 10^4)$ which are stabilized in the nonlinear case of $\alpha = \alpha(B)$. Their period approaches the value of about 22 years influenced mainly by the permeability profile. This fact was unexpectable and discovered by the computational experiment. The previous models supposed $\mu = 1$ that resulted in too short cycle period.

A 3-dimensional solar dynamo model without axial symmetry is now under construction /16/ which is expected to explain such interesting phenomena as sectorial magnetic field and the so called coronal holes discovered during the Skylab mission.

§ 3 SUPERNOVA EXPLOSION SIMULATION

Another example of numerical experiment in astrophysics is an investigation /I9/ of the magnetorotational mechanism of supernova explosion carried out on the basis of model /I7/ involving the conversion of neutron star rotational energy.

A star is considered as a rotating cylinder of infinite length in the presence of gravitation and magnetic fields. Its core corresponding to a stable neutron star is supposed incompressible and surrounded by the corona with ideal conductivity. The problem concerns but the star corona allowing to introduce a radial magnetic field component. Heat conduction is absent and energy losses are created by neutrino emission. The reference dimensional parameters of the problem are the following: the radius of neutron star core $R_o \approx 10^6$ cm, the constant radial magnetic field component $H_r \sim 10^{18}$ Gauss and the mass M of corona per unit azimuthal angle and unit length.

The problem is described by dimensionless MHD equations in Lagrangian mass coordinates

$$\frac{\partial}{\partial t}\left(\frac{1}{\rho}\right) = \frac{\partial}{\partial s}(r v_r), \quad \frac{\partial r}{\partial t} = v_r, \quad r\frac{\partial \varphi}{\partial t} = v_\varphi = \omega r,$$

$$\frac{\partial v_r}{\partial t} - \frac{v_\varphi^2}{r} = -r\frac{\partial p}{\partial s} - \frac{\alpha}{2r}\frac{\partial H^2}{\partial s} + g, \quad g = -\frac{1 + s/\beta}{r},$$

$$\frac{\partial}{\partial t}(r v_\varphi) = \alpha \frac{\partial H}{\partial s}, \quad H = r H_\varphi, \quad \alpha = \text{const}, \quad \beta = \text{const}, \tag{5}$$

$$\frac{\partial}{\partial t}\left(\frac{H}{\rho r^2}\right) = \frac{\partial}{\partial s}\left(\frac{v_\varphi}{r}\right),$$

$$\frac{\partial \varepsilon}{\partial t} = -p\frac{\partial}{\partial s}(r v_r) - f(\rho, T),$$

$$p = p(\rho, T), \quad \varepsilon = \varepsilon(\rho, T).$$

where t is the time, r is the radius, s ($ds = \rho r dr$) is the mass coordinate, ρ is the density, p is the pressure, g is the gravity acceleration, H_φ is the azimuthal magnetic field component, ε is the specific internal energy, T is the temperature, f (ρ, T) is the energy losses due to neutrino emission and (v_r, v_φ) is the velocity vector. The parameter α specifies the initial ratio of magnetic field and kinetic energies and β is the mass ratio of the star core and atmosphere (let $\beta = 1$). The state equations are taken from /I8/ and neutrino energy losses are obtained by means of interpolation of well known tables. Equations (5) are solved in the region $0 < s < 1$ ($1 < r < R(t)$) for $\alpha = 10^{-2}, 10^{-4}, 10^{-8}$. The initial and boundary conditions are as follows:

$$v_r = T = H = 0 \quad \text{at } t = 0$$

$$\rho\big|_{t=0} = a\exp(-b(r-1)^2),\ a = \text{const} > 0,\ b = \text{const} > 0,$$

$$\left(\frac{v_\varphi^2}{r} - r\frac{\partial p}{\partial s} + g\right)\Big|_{t=0} = 0 \tag{6}$$

$$p(1,t) = H(1,t) = 0,\ v_r(0,t) = 0,\ \left(\frac{\beta}{2}\frac{\partial H}{\partial s} - H\right)_{s=0} = 0.$$

The condition for H(s, t) at internal boundary expresses the conservation of rotational impulse.

Consider some features of this problem. There exist two time scales in the problem with the ratio $\sqrt{\alpha}$ ($\alpha \sim 10^{-8}$, 10^{-10}), therefore the minimization of computational time is of great importance. The problem is approximated by the implicit complete conservative difference scheme (CCDS) /20/, the artificial viscosity depending on the density ρ and magnetic field H. The implicit equation system is solved by the separated successive elimination /20/ involving Newtonian iterations.

The scaling $t_\alpha = \sqrt{\alpha}\, t$, $v_{r\alpha} = v_r/\sqrt{\alpha}$, $H_\alpha = \sqrt{\alpha}\, H$, $f_\alpha = f/\sqrt{\alpha}$ makes easier the analysis of computational experiment, the respective curves being matched at small α as can be seen from Fig.1,3. That is one of the main results. Consider the basic characteristic features of the solution. At the initial stage the corona is divided into two parts: the one falling onto the core and the other expanding in vacuum. The rotation of the internal part is retarted, a slow shock MHD wave is generated moving outward (see Fig. 1, 2). The heating at the wave front is followed by cooling due to the neutrino emission.

The simulation results show that the rotation energy is mainly converted into the magnetic energy E_M and the thermal energy E_ε (Fig.3), the great deal of which being emitted through neutrinos. The kinetic energy of radial movement is negligible because of scaling $v_r = \sqrt{\alpha}\, v_{r\alpha}$.

Interpretating the results one has to keep in mind that gravitational potential of a real star (spherical) is zeroed out at infinity allowing some matter to escape. The estimations based on this computational experiment show that independently of α (if it is small) the escaping matter makes I3% of the corona mass and carries 3,5% of the total energy. The released energy ($3{,}5\cdot IO^{44}$ J) is great enough to explain the supernova explosion phenomenon.

§ 4 TWO-DIMENSIONAL MHD PROBLEMS. PLASMA COMPRESSION BY A LINER

Consider a two-dimensional problem about toroidal plasma compression by a quasispherical metal liner containing frozen closed magnetic field, which was initiated by R.Kurtmullaev et al (Kurchatov Institute of Atomic energy) in I974. Plasma heating is produced as a result of squeezing the initially cylindrical liner. Because of the prevailing velocity at the ends, the liner takes quasispherical configuration (Fig. 4) /2I/. The main drawback of this approach isthe formation of cumulative jets destructing the configuration. The first runs revealed such jets (Fig. 6). Thus the problem arises concerning the optimal liner shape which provides the proper cumulation. The numerical experiment was aimed at searching the liner form which results in almost simultaneous approach of it to the core. In this case, the succession of jets is generated which drastically slacken each other and consequently the resulting jet /2I/ (Fig.7).

§ 5 NUMERICAL METHODS

Consider two questions of major importance for numerical simulation in MHD plasma physics:

I) the construction of grid approximations, and

2) the solution of linear and nonlinear difference equations.

While elaborating difference schemes it is necessary to follow strictly the principle of complete conservativeness (PCC). There exists a simple euristic way of obtaining CCDSs, which was applied to gasdynamics, MHD and the Landau equation. If a few equivalent differential equation systems are given, each of them being converted into others by means of the proper identity transformations, then the PCC implies that a difference scheme approximating any of these systems may be converted in a similar way into schemes corresponding to other systems /20/.

In my report at the preceding conference in I977 the idea was discussed concerning the usage of variational principles for the straightforward construction of CCDSs in the one- and two-dimensional cases, the conservation laws being inevitably satisfied. At present the variational principle in the Hamilton-Ostrogradskii form is intensively studied theoretically /22, 23, 24/, CCDSs being thus obtained are used widely for solving many problems in plasma physics and continuum mechanics.

It is necessary to account for diffusive processes of different kind in plasma dynamics studies. In many cases it is important to know not only functions themselves but also the respective fluxes. The significant difficulties are encountered while constructing conservative schemes with self-adjoint space operator for arbitrary irregular grids, especially in the case of considerable variations of coefficient values being typical in plasma physics.

The particular variational approach /25/ may be the effective means for deriving the appropriate schemes. Consider it in connection with the heat conductance equation taken in the form

$$\frac{\partial u}{\partial t} + \operatorname{div} \vec{W} = 0 \tag{7}$$

$$\vec{W} + K \operatorname{grad} u = 0 \tag{8}$$

where u - is the unknown function (temperature), $\vec{W}$ is its flux vector, $K > 0$ is the conductivity. Consider the plane geometry with Cartesian coordinates (x, y). Let the condition $W_n = 0$ is given at the boundary Γ of the region Ω in which the problem is to be solved, and when $t = 0$ the initial values of u are known in the region $\Omega + \Gamma$. At each fixed time moment the heat fluxes field minimize the functional

$$F(\vec{W}) = \int_\Omega \frac{|\vec{W}|^2}{K}\, d\Omega + \frac{\partial}{\partial t} \int_\Omega u^2 d\Omega$$

in the class of continuous functions satisfying the above boundary condition. While varying the functional the functions u and k are considered given and the variation of the derivative $\partial u/\partial t$ is obtained from equation (7) which represents a coupling equation. Eliminating $\partial u/\partial t$ via (7) transform F(W) into the form

$$F(\vec{W}) = \int_\Omega \left(\frac{|\vec{W}|^2}{K} - 2u \operatorname{div} \vec{W}\right) d\Omega$$

Thus the fluxes W are deduced through minimization of F(W) and u is found from balance equation (7). When the boundary condition is taken in other forms the expression for F(W) involves pertinent integrals along the boundary /26/.

For constructing a variational-difference scheme replace the region Ω with a discrete point array. Let ξ (x, y), η (x, y) produce smooth one-to-one mapping of Ω into a square $0 \leq \xi, \eta \leq$ I. Taking ξ and η as curvilinear coordinates, cover the region Ω with a quadrangle grid which corresponds to the regular rectangular grid in the plane ξ, η . The grid function u^h and coefficient k^h values are supposed to be constant within each grid cell. The flux field is described by two grid functions W_ξ^h and W_η^h which are projections of the vector $\vec{W}^h$ on the perpendiculars erected from the middle of cell sides. By using the integro-interpolation method the balance equation for a cell is obtained (see Fig. 5)

$$|\Omega^h| u_t^h + \nabla_\xi (S_\xi \cdot W_\xi^h) + \nabla_\eta (S_\eta \cdot W_\eta^h) = 0 ,$$

which approximates differential equation (7) to the second order in space and to the first order in time. Here ∇_ξ and ∇_η are the forward difference operators, S_ξ and S_η are the approximate lengths of cell sides, $|\Omega^h|$ is the cell area. In the operator form the above equation becomes

$$Du_t^h = R\,\vec{W}^h ; \quad u_t^h = (u^h (t + \tau) - u^h(t))/\tau, \ \tau = \Delta t,$$

where $\vec{W}^h = (W_\xi^h, W_\eta^h)$ is the grid vector function, R is the block operator mapping an element from the space of fluxes $\vec{W}^h$ into the space of grid functions u^h, D is the diagonal operator. The functional F(W) is discretized as

$$F^h(\vec{W}^h) = \sum_{\Omega^h} \left\{ \frac{|\vec{W}^h|^2}{K^h} |\Omega^h| - 2\,u^h [\nabla_\xi (S_\xi \cdot W_\xi^h) + \nabla_\eta (S_\eta \cdot W_\eta^h)] \right\}$$

The value $\vec{W}^h$ at the center of a cell Ω_{ij}^h is calculated from the formula

$$|\vec{W}^h|_{ij}^2 = 0.25 \sum_{\alpha,\beta=0}^{1} \frac{(W_{\xi\, i+\alpha,j})^2 + (W_{\eta\, i,j+\beta})^2 + (-1)^{\alpha+\beta+1} \cdot \cos\varphi_{ij}^{i+\alpha,j+\beta} \cdot W_{\xi\, i+\alpha,j} \cdot W_{\eta\, i,j+\beta}}{\sin^2 \varphi_{ij}^{i+\alpha,j+\beta}}$$

where $\varphi_{ij}^{i+\alpha,j+\beta}$ is the angle between the cell sides intersected at the node ($i+\alpha$, $j + \beta$). From the functional minimum condition the difference equation for fluxes is derived

$$L\,\vec{W}^h = G\,u^h$$

where the operator $G = -R^*$, $D = D^* \geq \delta_1 \cdot E$, $\delta_1 > 0$, $L = L^* \geq \delta_2 \cdot E$, $\delta_2 > 0$, E is a unit operator.

Eliminating u^h from the equations $D u_t^h = R\vec{W}^h$, $L\vec{W}^h = G u^h$ results in the divergent equation for $\vec{W}^h$

$$L\vec{W}_t^h + A\vec{W}^h = 0, \quad A = -GD^{-1}R = A^* \geq \delta_3 \cdot E, \ \delta_3 > 0,$$

of the second order in space. This scheme is absolutly stable.

When MHD problems are to be solved in complicated regions in the presence of considerable sound velocity variations it is necessary to use implicit CCDSs avoiding time step reduction. But implicit schemes involve solving nonlinear systems of high order. The Newton iterations and their modifications acounting for the peculiarities of MHD equations are very effective for this purpose. Instead of ordinary procedure of velocity iterations /27/ the pressure iterations /28/ may be used providing the compactness, symmetry and positive definitness of linearized system matrix which allow to apply fast iteration techniques of single structure among which the alternative triangular method (ATM) and its modification /29/are usefull. Elliptic difference problems with variable coefficients are solved now by different techniques involving a factorized operator B at the upper time layer (at the new iteration). If operators $A = A^* > 0$ and $B = B^* > 0$ the solution of the equation

$$Au = f$$

may be found by means of the implicit method of conjugate gradients (CG)

$$B\, y_{k+I} = \alpha_{k+I}(B - \tau_{k+I} A)\, y_k + (I - \alpha_{k+I})\, B\, y_{k-I} + \alpha_{k+I}\tau_{k+I}, \ k \geq I,$$

$$B\, y_I = (B - \tau_I A)\, y_o + \tau_I f$$

where $\tau_{k+I} = (w_k, r_k)/(A w_k, w_k)$, $k \ \ 0$, $r_k = A y_k - f$, $w_k = B^{-I} r_k$,

$$\alpha_{k+I} = (I - \tau_{k+I}/\tau_k \cdot (w_k, r_k)/(w_{k-I}, r_{k-I})/\alpha_k), \ k \geq I, \ \alpha_1 = I$$

The number of iterations $n \geq n(\varepsilon) = \frac{I}{2\sqrt{\xi}} \ln \frac{2}{\varepsilon}$ providing the accuracy $\varepsilon > 0$ (here $\xi = \gamma_1/\gamma_2$, $\gamma_1 B \leq A \leq \gamma_2 B$, $\gamma_1 > 0$) is of the same order as in the case of two-layer scheme $B(y_{k+I} - y_k)/\tau_{k+1} + Ay_k = f$, $k = 0, I, \ldots$ with the Chebyshev set of parameters $\{\tau_k\}$ /I/. The proper choice of operator B is of major importance. Consider two methods:

I. The Cholesky method of incomplete decomposition of the operator A /30/ (ICCG (0)): B = LDL, where L is a triangular matrix of the same sparse structure as that of A. There is no theoretical estimations for this method.

2. The alternative triangular method (ATM-CG) /I, 29/

$$B = (D + \omega R_I) D^{-1} (D + \omega R_2),\ R_I = R_2^*,\ R_I + R_2 = A,$$
$$D = D^* > 0,$$

which has the proper theoretical estimations.

These methods were compared /3I/ through an example of difference Dirichlet problem approximating div (k grad u) = -f(x), $u|_\Gamma = g(x)$ in a rectangle. The operators A, D, R_I and R_2 are taken in accordance with /29,32/. The computations show that the average iteration number for ICCG(0) is proportional to the number of grid points N and for ATM - CG is to $\sqrt{N}$. In the case of a particular problem, ATM-CG efficiency exceeds that of ICCG (0) by a factor of I,I-I,6, both of them being superior to SOR method.

In some cases of multidimensional heat transfer (for example, in the problems involving phase transitions) it is beneficial to use ordinary implicite schemes instead of any additive one (alternative direction schemes, local onedimensional schemes, etc.) if the accuracy is a matter of concern. For example, the implicit scheme for the heat conduction equation $\partial u/\partial t = Lu + F$ leads to the difference equation Ay = f, where y is the grid function at the upper time layer, $A = 1/\tau\, E - \sigma\Lambda$, Λ is the difference approximation to the elliptic operator L, σ is the weight factor. The equation Ay = f can be solved by ATM with the Chebyshev set of parameters or ATM - CG. If $L = \Delta$ (Δ - is the Laplace operator), $\sigma = 0{,}5$ and $\tau = O(h)$ then the iteration number is proportional to $1/\sqrt{h}$. ATM-CG is usefull if the evaluation of γ_1, γ_2 is complicated (e.g. for arbitrary grids).

In some cases one can construct such an operator B and choose such an initial approach that the error $y - y_k$ (y is the exact solution, y_k is the k-th iteration) belongs to some subspace of substantially minor dimension in which the operators A and B preserving all their properties that allows to apply all the above methods, particularly, the methods of conjugate directions. The inequality

$\gamma_1 B \leq A \leq \gamma_2 B$ is now considered in the subspace and thus the ration $\xi = \gamma_1/\gamma_2$ increases diminishing the number of iterations $N(\varepsilon)$ (E.Nicolaev).

REFERENCES

I. Samarskii, A.A. The theory of finite-difference schemes, M., "Nauka", I977.
2. Glansdorf, P., Prigogine J. Thermodynamic theory of structure, stability and fluctuations M., "Mir", I973.
3. Samarskii, A.A., Zmitrenko, N.V., Kurdjumov, S.P., Mikhailov,A.P. Dokl. Akad. Nauk SSSR, 223, N 6, I344 (I975); 227, N 2, 32I (I976); Pis'ma v JETF, 26, 9, 620 (I977); Technical document IAEA-200, p.I85 (Vienna, I977).
4. Kurdjumov, S.P., Zmitrenko, N.V. PMTF, N I, 3 (I977).
5. Samarskii, A.A., Yelenin, G.G., Kurdjumov, S.P., Zmitrenko, N.V., Mikhailov, A.P. Dokl. Akad. Nauk SSSR, 237, N 6, I330 (I970).
6. Samarskii, A.A., Galaktionov, W.A., Kurdjumov, S.P., Mikhailov, A.P. Dokl. Akad. Nauk SSSR, 247, N 2, 349 (I979).
7. Kurdjumov, S.P., Kurkina, E.S., Malinetskii, G.G., Samarskii,A.A. Preprint Inst.Appl.Math. USSR Acad.Sci., I979, N I6; Dokl.Acad. Nauk SSSR (to appear).
8. Kurdjumov, S.P., Malinetskii, G.G., Poveshchenko, Yu.A., Popov,Yu.P. Samarskii, A.A. Preprint Inst.Appl.Math. USSR Acad.Sci., I978, N 77; Dokl.Akad.Nauk SSSR (to appear).
9. Kurdjumov, S.P. Preprint Inst.Appl.Math.Acad.Sci., I979, N 29.
I0. Samarskii, A.A. Vestnik Akad. Nauk SSSR, N 5, (I979); "Plasma theory problems" (Proceed. of Second Internat. Conference on Plasma Theory, SSSR, Kiev, I974), p.262, Kiev, "Naukova Dumka", I976.
II. Galaktionov, W.A., Kurdjumov, S.P., Mikhailov, A.P., Samarskii, A.A. Preprint Inst.Appl.Math. USSR Acad.Sci., I979, N 2I; Dokl. Akad.Nauk SSSR (to appear), Zh.Vychisl.Mat. i Mat.Fiz (to appear).
I2. Dorodnitsyn, W.A. Preprint Inst.Appl.Math.USSR Acad.Sci., I976, N I43; I979, N 57.
I3. Steenbeck, M., Krause, F. Astronom.Nachr., 29I, 49-84, I969.
I4. Ivanova, T.S., Ruzmaikin, M.Astronom. Zh., 53, 398 (I976).
I5. Ivanova, T.S., Ruzmaikin, M. Astronom.Zh., 54, 846 (I977).
I6. Ivanova, T.S., Ruzmaikin, M. Preprint Inst.Appl.Math. USSR Acad. Sci., I979, N I05.
I7. Bisnovatyi-Kogan, G.S. Astronom. Zh., 47, 8I3 (I970).
I8. Bisnovatyi-Kogan, G.S., Popov, Yu.P., Samokhin A.A. Preprint Inst.Appl.Math. USSR Acad.Sci., I975, N I6.
I9. Ardeljan, N.W., Popov, Yu.P. Preprint Inst.Appl.Math. USSR Acad. Sci., I979, N I07.
20. Samarskii, A.A., Popov, Yu.P. The finite-difference schemes of gas dynamics, M., "Nauka", I975.
2I. Gasilov, W.A., Goloviznin, W.M., Kurtmullaev, R.H., Semenov, W.N., Sosnin, N.W., Tishkin, W.F., Favorskii, A.P., Shashkov, M.Yu. Proceed. of Second Internat. Conference on Megagauss Magnetic Field Generation and Related Topics, 29 May - I June, I979, Washington.
22. Tishkin, W.F., Tjurina, N.N., Favorskii, A.P. Preprint Inst.Appl. Math. USSR Acad. Sci., I978, N 23.
23. Goloviznin, W.M., Samarskii, A.A., Favorskii, A.P. Dokl.Akad. Nauk SSSR, I979, 246, N 5.
24. Goloviznin, W.M., Korshija, T.K., Samarskii, A.A., Favorskii,A.P. Proceed. of Sixth Internat. Conference on Numerical Methods in Fluid Dynamics, Tbilisi, I978, Springer-Verlag, p.248-252.

25. Tishkin, W.F., Favorskii, A.P., Shashkov, M.Yu. Dokl.Akad. Nauk SSSR, I979, 246, N 6.
26. Korshija, T.K., Tishkin, W.F., Favorskii, A.P., Shashkov, M.Yu. Preprint Inst.Appl.Math. USSR Acad.Sci. № I, I979.
27. Mazhorova, O.S., Popov, Yu.P. Preprint Inst.Appl.Math. USSR Acad. Sci., I977, N 55.
28. Gasilov, W.A., Goloviznin, W.M., Taran, M.D., Tishkin, W.F., Tjurina, N.N., Shashkov, M.Yu. Preprint Inst.Appl.Math. USSR Acad. Sci. № I00, I978.
29. Samarskii, A.A., Nikolaev, E.S. Solution methods for Grid equations, Moscow, "Nauka", I978.
30. D.Kershaw. J.Comput.Phys., v.26, N I (I978).
3I. Bogdanova, M.S., Kucherov, A.B., et al. Preprint Inst. Appl.Math. USSR Acad. Sci. № II5, I978.
32. Kucherov, A.B., Nikolaev, E.S. Zh.vychisl.Mat. i Mat.Fiz., I976, I6, N 5; Zh.vychisl.Mat. i Mat.Fiz., I977, N 3, I7.

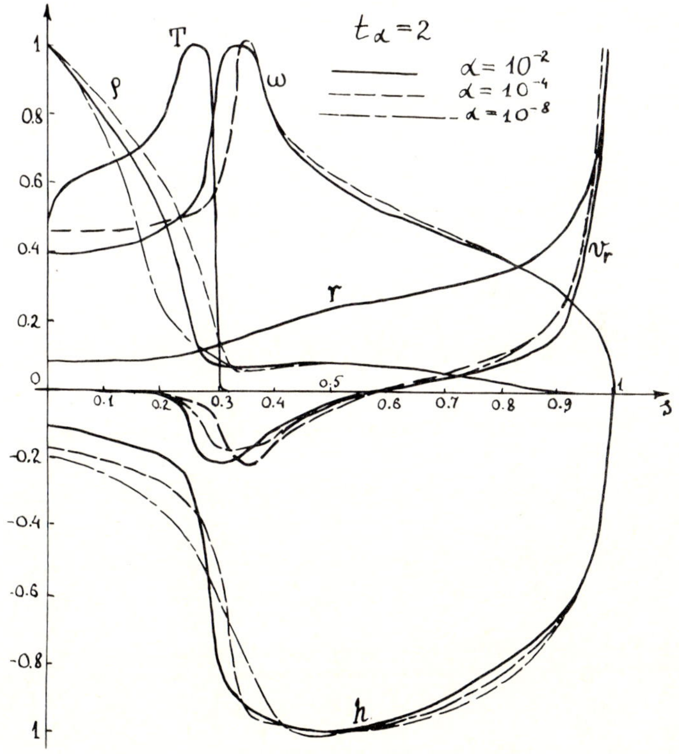

Fig. I

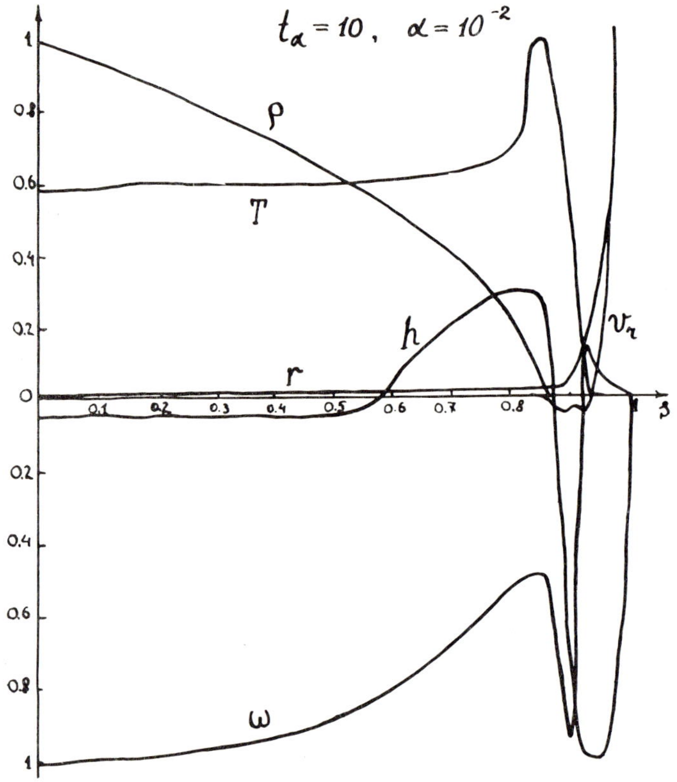
$t_\alpha = 10$, $\alpha = 10^{-2}$
1
0.8
0.6
0.4
0.2
0
0.2
0.4
0.6
0.8
1
ρ
T
h
r
v_r
ω
0.1
0.2
0.3
0.4
0.5
0.6
0.7
0.8
1
ξ

Fig. 2

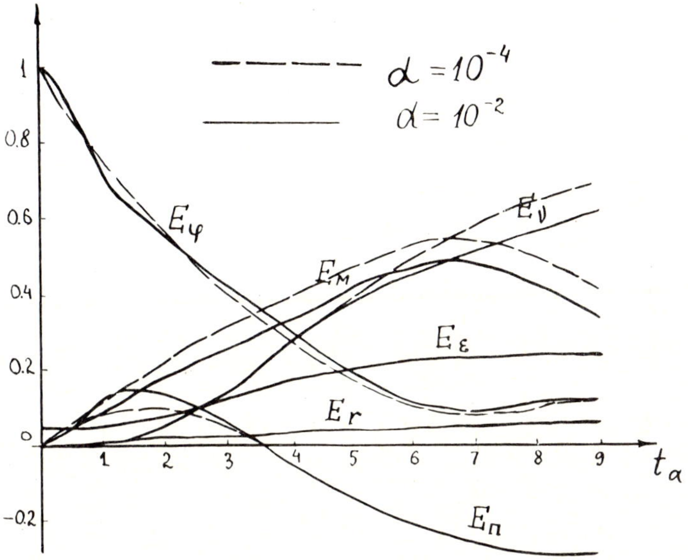
$\alpha = 10^{-4}$
$\alpha = 10^{-2}$
1
0.8
0.6
0.4
0.2
0
-0.2
1
2
3
4
5
6
7
8
9
t_α
E_φ
E_ν
E_M
E_ε
E_r
E_Π

Fig. 3

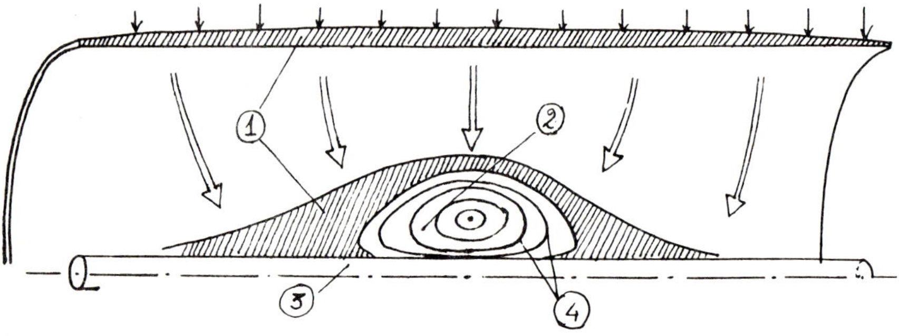

I - liner; 2 - plasma; 3 - core; 4 - magnetic surface

Fig. 4

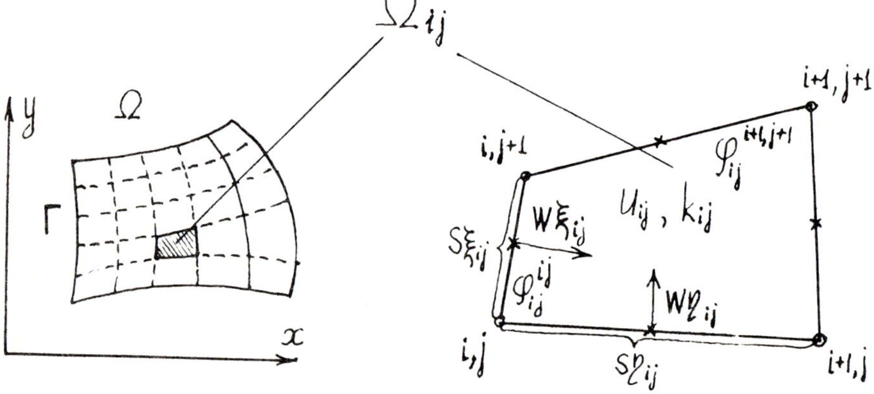

Fig. 5

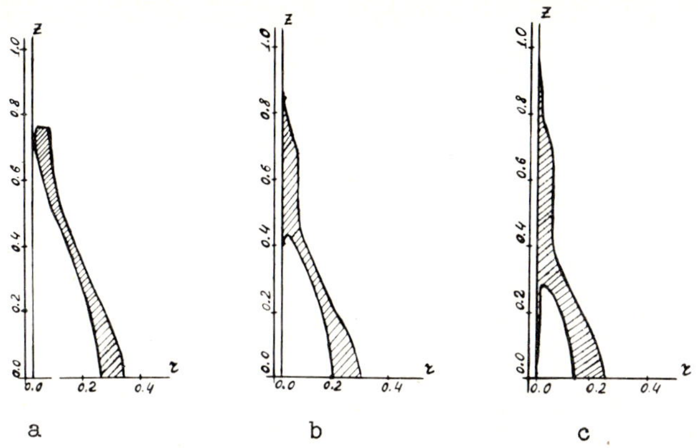

Fig. 6. The compression history

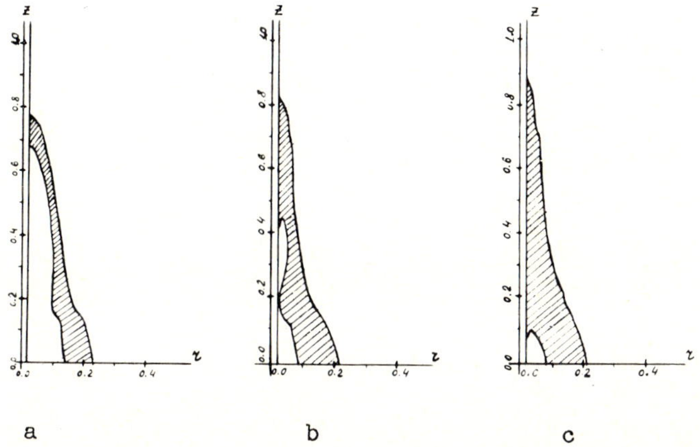

Fig. 7. The multijet compression history

COMPUTING METHODS IN APPLIED SCIENCES AND ENGINEERING
R. Glowinski, J.L. Lions (editors)
North-Holland Publishing Company

LINEAR AND NON-LINEAR CALCULATION OF RESISTIVE MAGNETOHYDRODYNAMIC INSTABILITIES*

J. Killeen, D. D. Schnack and A. I. Shestakov
Magnetic Fusion Energy Computer Center
Lawrence Livermore Laboratory
University of California
Livermore, California 94550

Abstract

The time-dependent, linear and non-linear, resistive magnetohydrodynamic, numerical models that have been developed at MFECC are reviewed. The purpose of these codes is to compute growth rates, mode structure and saturation of tearing, rippling, and interchange modes in fusion experiments. Cartesian, cylindrical, helical and toroidal geometries are used in the applications. The numerical methods are described and applications to reversed field configurations are presented.

1. INTRODUCTION

In order to achieve the high densities and temperatures required for a successful fusion reactor, a plasma must be confined by a magnetic field for a sufficiently long time. In the attempts to achieve this confinement, the problem of stability has emerged as one of the most important. The most dangerous type of instabilities are the magnetohydrodynamic (MHD) instabilities in which the plasma is assumed to behave as a conducting fluid and the instabilities involve displacement of macroscopic portions of the plasma. It is a particular MHD instability, the resistive instability which is considered in this paper.

Resistivity can destroy the stabilization achieved by the shearing of the lines of force. In the case of a magnetic field which has shear or which changes direction, the magnetic energy can be reduced by allowing the fields to mix and annihilate. This is prevented by a perfectly conducting plasma, but with finite conductivity an instability can develop in which the magnetic lines of force are torn into "islands". This type of resistive instability is known as a resistive tearing mode. [1]

There are three types of resistive modes: (1) the rippling mode, which is driven by a gradient in the resistivity and is usually not important when large temperature gradients are unlikely; (2) the gravitational mode (g-mode) which is the resistive equivalent of the interchange instability and is important in sheared systems; and (3) the tearing mode, which is the resistive equivalent of the kink mode and involves displacement of the whole plasma.

*Work performed under the auspices of the U.S.D.O.E. by Lawrence Livermore Laboratory under contract W-7405-ENG-48.

The modes grow on a time scale intermediate between the resistive diffusion time $\tau_R = 4\pi a^2/\eta c^2$ and the hydromagnetic transit time $\tau_H = a(4\pi\rho)^{1/2}B^{-1}$ where 'a' is a characteristic dimension of the plasma layer, η is the resistivity, ρ is the mass density of the plasma, B is the magnetic field, and c is the speed of light.

The resistive instability of an incompressible plasma was first systematically investigated by Furth, Killeen, and Rosenbluth [1]. They used the plane slab model, in which the equilibrium depends only on y, the magnetic field is $\hat{x}\, B_{xo} + \hat{z}\, B_{zo}$, and $\vec{v}_0 = 0$. In that paper perturbations of the form $f_1(y)\exp[i(k_x x + k_z z) + \omega t]$ are assumed, and the problem is to solve an eigenvalue problem for ω, the growth rate of the instability. In order to solve the problem the plasma is divided into two regions, a narrow inner region about the plane for which the wave vector is perpendicular to the zero-order magnetic field ($\vec{k} \cdot \vec{B}_0 = 0$) and an outer region where the infinite conductivity equations hold. By matching the solutions within the resistive layer to the outer ideal MHD solutions, FKR found resistive tearing modes with growth rates, $p = \omega\tau_R$, proportional to $S^{2/5}$, where $S = \tau_R/\tau_H$.

Due to the many possible equilibrium configurations and the many approximations necessary to make the problem analytically tractable, it is usually not possible to analytically describe the general parameter dependence of the growth rates. In order to obtain results for specific and wide choices of equilibrium magnetic fields and boundary conditions, numerical models have been developed to study these resistive instabilities.

In this paper we assume an arbitrary time-dependence and the problem becomes an initial-value problem. Two regions are not used, i.e., the same equations hold throughout the plasma. The initial-value problem is then solved numerically. This method of solution was developed [2] for the linear model simultaneously with the analytic technique and is described in Ref. [3]. The initial-value codes, RIPPLE, use the same basic equations and assumptions as FKR and are capable of finding tearing, rippling, and gravitational modes as well as mixed modes.

In order to consider more realistic equilibrium magnetic fields, a cylindrical model, RIPPLE IV, was developed [3], which also used the equations of incompressible magnetohydrodynamics. This model has been extensively applied to the study of tearing modes in reversed field pinch [4-8], and tokamak [8] equilibria. Dibiase developed a new cylindrical model [8,9] which includes the effects of compressibility, viscosity, and thermal conductivity along with finite resistivity. This model has also been applied to the RFP [5-10].

Recently there has been interest in the effect of equilibrium flow on the tearing mode [11-13]. We have developed a new linear initial-value code, RIPPLE V, [12] to study this problem. For this work we have gone back to the plane slab model using the incompressible MHD equations.

We have also applied the linear slab model to the double tearing mode [14]. In this case there are two neighboring singular surfaces, i.e., surfaces for which $\vec{k} \cdot \vec{B}_0 = 0$. If these surfaces lie close to one another, the modes at each singular surface may interact leading to an enhanced growth rate.

In all of the above linear models the initial-value problems solved are one-dimensional, i.e., the zero-order fields are given by $\vec{B}_0(y)$ or $\vec{B}_0(r)$ and the perturbed variables take the form $f_1(y,t) \exp [i(k_x x + k_z z)]$ or $f_1(r,t) \exp [i(m\theta + k_z z)]$.

In many toroidal confinement devices it is not possible to specify the equilibrium fields as functions of one variable. In tokamaks and other axisymmetric toroidal devices the zero-order field can be specified by $\vec{B}_0(r,z)$. To study tearing modes in such configurations we have developed a new two-dimensional, linear code, RIPPLE VI, in which the perturbations are of the form $f_1(r,z,t) \exp [in\phi]$, $n \geq 1$. We use the incompressible MHD equations to derive a set of eight coupled linear partial differential equations. For the case, $n = 0$, we have developed a 2D axisymmetric linear code (ALIMO), which makes use of field and velocity stream functions, resulting in a system of four equations.

The above linear models are discussed in more detail in Section 2 of this paper. These codes are used for extensive parameter studies of prospective equilibria. Stable and unstable regions of wave number space, growth rates of exponentially growing modes and their mode structure are calculated. In order to study the long-time, large amplitude behavior of these modes, and to simulate experimental devices in controlled fusion research, the non-linear fluid equations must be solved. In general, such a calculation requires the simultaneous advancement in time of eight non-linear partial differential equations in several spatial dimensions.

As discussed above, one of the effects of non-vanishing resistivity is the occurance of unstable modes which have no counterpart in ideal MHD theory [1]. These resistive instabilities grow on time scales which can be long relative to the fastest time scales of the system, leading to severe computational problems. Recently, the evolution and interaction of these modes in tokamaks has been successfully and extensively studied [15-19] by assuming an ordering that eliminates the fastest time scales from the problem, and results in a reduced set of equations for the scalar flux and stream functions [20]. This allows the calculation to proceed rapidly for the large values of S, the magnetic Reynolds number, typical of tokamak discharges. However, certain effects, such as those due to finite plasma pressure, are excluded. These effects can be important in controlled fusion devices, such as Reversed Field Pinches and High Beta Tokamaks.

In Section 3 we describe a two-dimensional, non-linear, resistive MHD model [21,22] which retains all the normal modes of the system. It is general in that the equations are cast in orthogonal curvilinear coordinates, making calculations in a variety of coordinate systems possible. For certain cases, the model employs a transformation to helical coordinates which allows the solution of the three-dimensional equations under the assumption that helical symmetry is preserved. We present the mathematical model in which the relevant equations are expressed as a set of conservation laws. Computational techniques for the solution of these equations are discussed. The boundary conditions, including the important case of singular boundaries, are considered. Examples of the non-linear evolution of tearing modes [23] and resistive g-modes for the Reversed Field Pinch are presented.

2. LINEAR CALCULATIONS

2.1 Basic equations and assumptions

We assume that the hydromagnetic approximation is valid, and the ion pressure and inertia terms are neglected in Ohm's law. An isotropic resistivity is assumed, the fluid is assumed to be incompressible, and perturbations in resistivity result only from convection. The basic equations are:

$$\frac{\partial \vec{B}}{\partial t} = \text{curl}\ (\vec{v} \times \vec{B}) - \text{curl}\ (\frac{\eta}{4\pi}\ \text{curl}\ \vec{B}) \quad , \tag{2.1}$$

$$\text{div}\ \vec{B} = 0 \quad , \qquad \text{div}\ \vec{v} = 0 \quad , \tag{2.2}$$

$$\text{curl}\ (\rho \frac{d\vec{v}}{dt}) = \text{curl}\ (\frac{1}{4\pi}\ \text{curl}\ \vec{B} \times \vec{B}) \ , \tag{2.3}$$

$$\frac{\partial \eta}{\partial t} + \vec{v} \cdot \nabla \eta = 0 \ . \tag{2.4}$$

In Eqs. (2.1) - (2.3) we consider $\vec{B} = \vec{B}_0 + \vec{B}_1$ and $\vec{v} = \vec{v}_0 + \vec{v}_1$, where $\vec{B}_0$ and $\vec{v}_0$ are given and the subscript 1 denotes perturbed quantities. We obtain, to first order, the following set of linearized equations:

$$\frac{\partial \vec{B}_1}{\partial t} = \text{curl}\ (\vec{v}_0 \times \vec{B}_1 + \vec{v}_1 \times \vec{B}_0) - \frac{1}{4\pi}\ \text{curl}\ (\eta_0\ \text{curl}\ \vec{B}_1 + \eta_1\ \text{curl}\ \vec{B}_0) \tag{2.5}$$

$$\text{div}\ \vec{B}_1 = 0 \ , \ \text{div}\ \vec{v}_1 = 0 \quad , \tag{2.6}$$

$$\rho_0\ \text{curl} \left[\frac{\partial \vec{v}_1}{\partial t} + (\vec{v}_0 \cdot \nabla)\vec{v}_1 \right] = \frac{1}{4\pi}\ \text{curl} \left[(\vec{B}_0 \cdot \nabla)\vec{B}_1 + (\vec{B}_1 \cdot \nabla)\vec{B}_0 \right] \ . \tag{2.7}$$

2.2 Effect of equilibrium flow on the tearing mode

The resistive tearing instability of an incompressible plasma is investigated for the plane sheet pinch in which the equilibrium magnetic field, $\hat{x}\, B_{xo} + \hat{z}\, B_{zo}$, depends only on y. The effect of a non-zero v_0 is studied. For a symmetric magnetic equilibrium and modes $\alpha = a(k_x^2 + k_z^2)^{\frac{1}{2}} < 1$ an exponential growth develops. The growth rate, $p = \omega\tau_R$, is computed as a function of α and $S = \tau_R/\tau_H$, for several values of v_0. The effect is to reduce p for all α, and to reduce the marginal α for instability for values of v_0 of the order of the resistive diffusion velocity. For asymmetric tearing, the effect of the diffusion velocity depends on its sign. The velocity may have either a stabilizing or destabilizing influence on both the growth rates and the critical α for instability [12,13].

From Eqs. (2.5) - (2.7), a pair of equations can be separated which involve only B_{y1} and v_{y1}. The remaining quantities are not needed for the analysis of tearing modes. We define the following parameters $\alpha = ka$, $k = (k_x^2 + k_z^2)^{\frac{1}{2}}$, $\nu = v_0\tau_R/a$ where $\vec{v}_0 = \hat{y}\, v_0$. We now define the dimensionless variables, $\mu = y/a$, $\tau = t/\tau_R$, $\psi = B_{y1}/B$, $w = -iv_{y1}k\tau_R$, $\eta = \frac{\eta_o}{\langle\eta\rangle}$, $\theta = -\frac{i\eta_1}{\langle\eta\rangle}$ and define $F = (1/kB)(k_x B_{xo} + k_z B_{zo})$. We have

$$\frac{\partial\psi}{\partial\tau} = \eta\left[\frac{\partial^2\psi}{\partial\mu^2} - \alpha^2\psi\right] - Fw - \nu\,\frac{\partial\psi}{\partial\mu} + \alpha F'\theta \quad , \tag{2.8}$$

$$\frac{1}{\alpha^2 S^2}\,\frac{\partial}{\partial\tau}\left[\frac{\partial^2 w}{\partial\mu^2} - \alpha^2 w\right] = -\,\frac{\nu}{\alpha^2 S^2}\,\frac{\partial}{\partial\mu}\left[\frac{\partial^2 w}{\partial\mu^2} - \alpha^2 w\right]$$

$$+ F\left[\frac{\partial^2\psi}{\partial\mu^2} - \alpha^2\psi\right] - F''\psi \,. \tag{2.9}$$

A difference equation corresponding to (2.8) is obtained by a Crank-Nicholson scheme [24]. Eq. (2.9) is differenced as follows:

$$\frac{1}{\alpha^2 S^2 \Delta\tau}\left[(\delta^2 w)_j^{n+1} - \alpha^2 w_j^{n+1} - (\delta^2 w)_j^n + \alpha^2 w_j^n\right] = -\,\frac{\nu}{2\alpha^2 S^2}\left[(\delta^3 w)_j^{n+1} + (\delta^3 w)_j^n\right]$$

$$+ \frac{\upsilon}{4S^2\Delta\mu}\,(w_{j+1}^{n+1} - w_{j-1}^{n+1} + w_{j+1}^n - w_{j-1}^n) - \frac{1}{2}\,F_j''\,(\psi_j^{n+1} + \psi_j^n) \tag{2.10}$$

$$+ \frac{1}{2}\,F_j\left[(\delta^2\psi)_j^{n+1} - \alpha^2\,\psi_j^{n+1} + (\delta^2\psi)_j^n - \alpha^2\,\psi_j^n\right]$$

The symbol $(\delta^2\psi)^n_j$ denotes the usual second divided difference of ψ at the point $(j\Delta\mu,\ n\Delta\tau)$, while $(\delta^3 w)^n_j$ denotes the third divided difference of w which is correct to second order and is spread over four mesh widths. We can use either free space or conducting wall boundary conditions on ψ and w. [12]

The implicit nature of the difference scheme involves solving a large linear system of equations at each time step. This system can be expressed as $M\vec{u}^{n+1} = \vec{z}^n$, where $\vec{u}^{n+1} = (\ldots, \psi^{n+1}_j, w^{n+1}_j, \ldots)$ and z^n contains the known values ψ^n_j, w^n_j. Since none of the coefficients in (2.8) and (2.9) vary in time, neither does M. At the start of each run after M is generated, it is decomposed via Gaussian elimination into a product of a lower and an upper triangular matrix, M=LU. The matrices LU are stored over M, and at each time step we need only solve two triangular systems by back substitution.

We consider the effect of variation of the parameters S, α, and ν on the growth rates, p. The background equilibrium magnetic field F and the resistivity η must also be specified. We use $F(\mu) = \tanh \mu$ for symmetric tearing and $\eta = 1$ i.e. $\theta = 0$ in this section. At the start of each run we specify an initial arbitrary perturbation in the function w. The program is then allowed to run until an exponentially growing mode dominates.

For symmetric tearing the general shape of the eigenfunctions is exhibited in Fig. 1. The upper function, ψ, exhibits the FKR[1] criteria for instability: $\Delta' > 0$ where Δ' is the change in logarithmic derivative on either side of the resistive layer about y = 0. The behavior of the eigenfunctions for various α and S is very much the same for zero or non-zero ν. For fixed S, as $\alpha \to 0$, the peak in ψ about y = 0 grows and Δ' becomes larger. As $\alpha \to 1$, the peak in ψ begins to flatten out and $\Delta' \to 0$. If instead α is kept fixed and S allowed to increase, the height of the peak in ψ does not vary appreciably, but does come to a

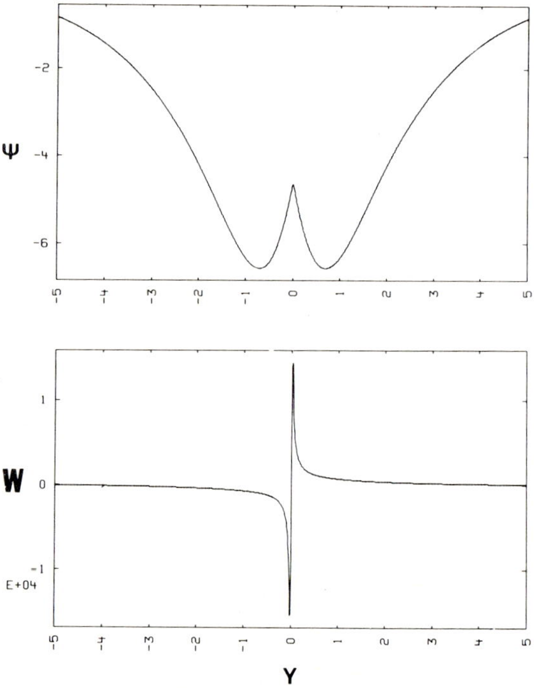

Fig. 1 Perturbed eigenmodes ψ and w, α=0.55, S=10^5, ν=1, p=83.6.

sharper point, and the locations of the spikes in w tend to come closer in to zero indicating a shrinking resistive layer (see Fig. 1).

A non-zero ν does effect the size of the growth rates. The behavior is shown in Fig. 2 which plots the growth rates as functions of α for several values of S, and for two values of ν: 0 and 1.

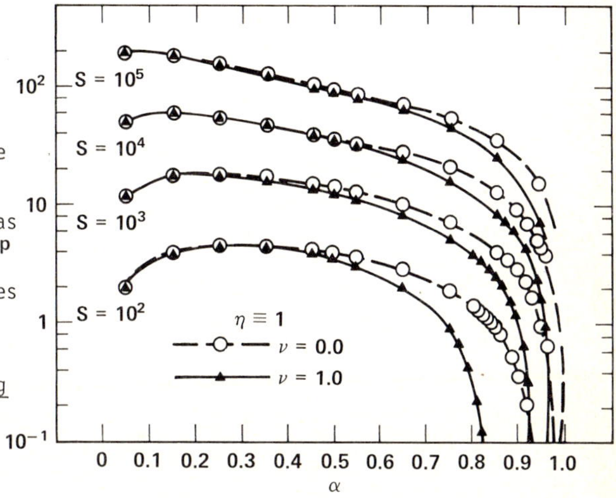

Fig. 2 Symmetric tearing, F=tanh μ. Growth rates, $p=\omega\tau_R$, as a function of α; $S=10^2$, 10^3, 10^4, 10^5 for $\nu=0$ and $\nu=1$.

2.3 Transition between tearing and rippling modes

The tearing modes considered in the preceding section are long wave-length modes, i.e. an exponentially growing mode develops for $\alpha<1$. If we include zero-order resistivity gradients, i.e. $d\eta_0/dy \neq 0$, short wave-length or "rippling" modes can develop for $\alpha>1$ when the fields are sheared ($F=\vec{k}\cdot\vec{B}_0=0$ at $y\neq 0$). In this section we assume $v_0=0$. From Eq. (2.4) we have

$$\frac{\partial\theta}{\partial\tau} + \frac{w}{\alpha}\frac{d\eta}{d\mu} = 0 \quad , \qquad (2.11)$$

which determines the perturbed resistivity θ in Eq. (2.8). Fig 3 is a plot of p vs α which displays the regions of pure tearing ($\theta=0$), pure rippling ($\alpha>>1$), as well as the transition region between them. These results were obtained using the code, RIPPLE V, with

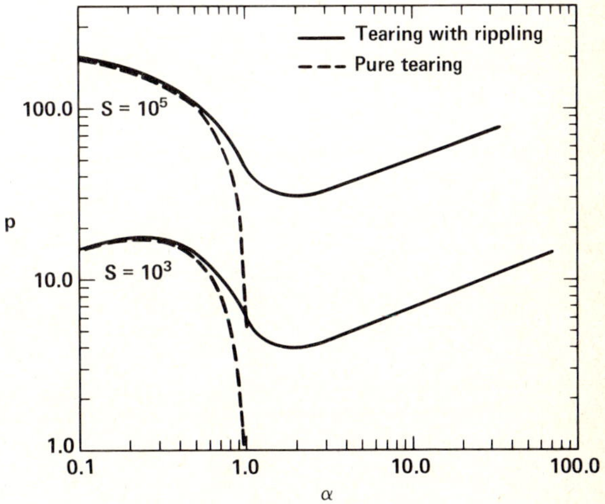

Fig. 3 Growth rate, p, versus wave number, α. Comparison between growth rates obtained for pure tearing and tearing with rippling (i.e. $d\eta_0/dy\neq 0$).

$F = \tanh\mu - \tanh\mu_o$ and $\eta = \cosh^2\mu/\cosh^2\mu_o$. The results of Fig. 3 show $p \sim \alpha^{2/5}$ for $\alpha >> 1$ in agreement with FKR [1].

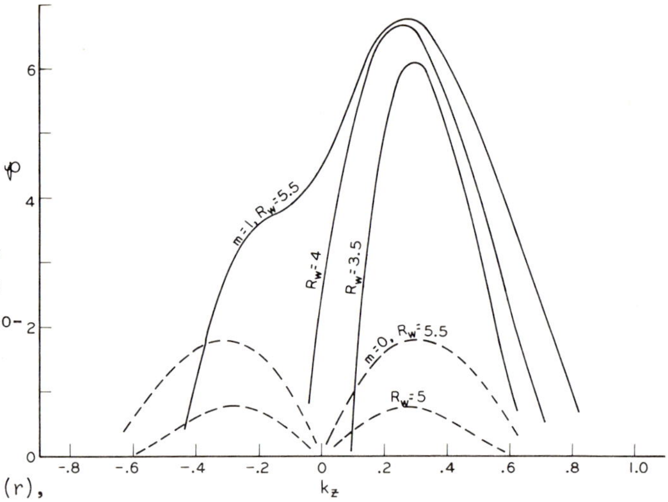

Fig. 4 Variation of the growth rate, $p=\omega\tau_R$, with k_z for the incompressible BFM with m=0,1 at various R_W for S=100.

2.4 Cylindrical models

The linear model given by Eqs. (2.5) - (2.7) can be applied in cylindrical geometry in order to study specific diffuse pinch configurations. The equilibrium is given by $\vec{B}_0(r) = \hat{\theta}\, B_{\theta 0}(r) + \hat{z}\, B_{z0}(r)$, $\vec{v}_0 = 0$, and $\eta_0 = \eta_0(r)$. These functions are chosen to describe a particular experiment, and the stabilizing effect of the location of the conducting walls (R_W) with reference to the singular surface can be determined.

We assume perturbations of the form $f_1(r,t)\exp\left[i(m\theta + k_z z)\right]$. We can find a consistent system of four equations involving the components B_{r_1}, B_{θ_1}, v_{r_1}, v_{θ_1}, and an equation for η_1 [3]. These equations are solved by an initial-value code, RIPPLE IV, using numerical methods analogous to those of the previous sections.

This code has been applied to the study of tearing modes in reversed field pinch equilibria [4-8]. An analytic example of such an equilibrium is the Bessel Function Model (BFM) given by $B_{\theta 0} = J_1(r)$ and $B_{z0} = J_0(r)$. Typical results are shown in Fig. 4, which illustrate the dependence of the growth rate on m, k_z and R_W. The effects of viscosity and compressibility on these modes can also be studied with the cylindrical code, RESTAB [9].

2.5 Toroidal model, n≠0

In axisymmetric toroidal devices, such as tokamaks with non-circular minor cross-section, the zero-order fields can be specified by

$$\vec{B}_0 = \hat{r}\, B_{r0}(r,z) + \hat{\phi}\, B_{\phi 0}(r,z) + \hat{z}\, B_{z0}(r,z) \quad .$$

To study tearing modes in such configurations we have developed a linear code, RIPPLE VI, in which the perturbations are of the form $f_1(r,z,t)\exp[in\phi]$.

In Eqs. (2.5) - (2.7), we consider the case where $\vec{v}_o = 0$, $\eta_1 = 0$, and η_o = constant. Eqs. (2.5) and (2.6) then yield four equations for $B^R_{r_1}$, $B^I_{r_1}$, $B^R_{z_1}$, $B^I_{z_1}$ where the superscripts denote the real and imaginary parts of the complex perturbations. The equation for $B^R_{z_1}$ is typical

$$\frac{\partial B^R_{z_1}}{\partial t} = \frac{\eta_o}{4\pi} L\,(B^R_{z_1}) + B_{ro}\frac{\partial v^R_{z_1}}{\partial r} - B_{\phi o}\frac{n}{r} v^I_{z_1} + B_{zo}\frac{\partial v^R_{z_1}}{\partial z}$$

$$- \frac{\partial B_{zo}}{\partial r} v^R_{r_1} - \frac{\partial B_{zo}}{\partial z} v^R_{z_1} \tag{2.12}$$

$$\text{where} \quad L(u) = \frac{1}{r}\frac{\partial}{\partial r}\left(r\frac{\partial u}{\partial r}\right) - \frac{n^2}{r^2} u + \frac{\partial^2 u}{\partial z^2} \tag{2.13}$$

As in previous sections we take the curl of Eq. (2.7) and use Eq. (2.6) to obtain

$$- \rho_o \frac{\partial}{\partial t}(\nabla^2 \vec{v}_1) = \frac{1}{4\pi}\,\text{curl curl}\left[(\vec{B}_o\cdot\nabla)\vec{B}_1 + (\vec{B}_1\cdot\nabla)\vec{B}_o\right] \tag{2.14}$$

Eq. (2.14) yields four equations for $v^R_{r_1}$, $v^I_{r_1}$, $v^R_{z_1}$, $v^I_{z_1}$. The eight equations can be summarized in the following

$$\frac{\partial B^R_{r_1}}{\partial t} = f_1(B^R_{r_1}, B^R_{z_1}, v^R_{r_1}, v^R_{z_1}, v^I_{r_1}) \tag{2.15a}$$

$$\frac{\partial B^R_{z_1}}{\partial t} = f_2(B^R_{z_1}, v^R_{r_1}, v^R_{z_1}, v^I_{z_1}) \tag{2.16a}$$

$$\frac{\partial}{\partial t}\left[L(v^R_{r_1}) + \frac{2}{r}\frac{\partial v^R_{r_1}}{\partial r} + \frac{v^R_{r_1}}{r^2} - \frac{2}{r}\frac{\partial v^R_{z_1}}{\partial z}\right]$$

$$= f_3(B^R_{r_1}, B^I_{r_1}, B^R_{z_1}, B^I_{z_1}) \tag{2.17a}$$

$$\frac{\partial}{\partial t}\left[L(v^R_{z_1})\right] = f_4(B^R_{r_1}, B^I_{r_1}, B^R_{z_1}, B^I_{z_1}) \tag{2.18a}$$

The operator L is defined in Eq. (2.13) and the functions f_i are functions of the first-order variables indicated and their derivatives, as well as functions of the zero-order variables (not shown). Additional Eqs. (2.15b) - (2.18b) for the imaginary components are similar to (2.15a) - (2.18a). Eqs. (2.17) and (2.18) are non-standard in form and f_3, f_4 involve derivatives up to third order in r and z.

The initial-value problem given by Eqs. (2.15) - (2.18) is solved by implicit finite-difference methods as in previous sections. In the (η_0 = constant) case that we are considering, Eq. (2.16ab) does not contain B_{r_1}, so we first solve for $B^R_{z_1}$, $B^I_{z_1}$ implicitly using $v^R_{r_1}$, $v^R_{z_1}$, $v^I_{z_1}$ from the preceding time step. We then solve Eq. (2.15ab) in a similar manner but using the new B_{z_1}. The velocity equations (2.17ab) and (2.18ab) are then solved separately implicitly using time-centered difference equations, analogous to Eq. (2.10), making use of the predicted B_{r_1}, B_{z_1} already computed from (2.15ab) and (2.16ab). Since the eight equations are solved separately, we provide for a correction cycle using the new velocities in Eqs. (2.15ab) and (2.16ab) and then the corrected fields in Eqs. (2.17ab) and (2.18ab). The correction cycle can then be iterated to satisfy a convergence criterion. In practice the correction cycles have not been needed.

The code, RIPPLE VI, has been tested using the one-dimensional equilibria, $\vec{B}_0(r)$, described in the preceding section. The equilibria tested were the Bessel Function Model, the Pitch and Pressure Model [25], and the Peaked Tokamak Model [26]. Tearing mode parameter studies of these models have been run [4,8] with the RIPPLE IV code and good agreement was obtained with earlier results. In particular the growth rates shown in Fig. 4, for $m = 1$ and $R_W = 5.5$, were also obtained with RIPPLE VI.

2.6 Toroidal model, n=0

The axisymmetric linear MHD code, ALIMO, treats the special case n=0, i.e. where the perturbations are of the form $f_1(r,z,t)$. It is convenient to express the zero-order fields as

$$\vec{B}_0(r,z) = \hat{r}\left(-\frac{1}{r}\frac{\partial\psi_0}{\partial z}\right) + \hat{\phi}\, B_{\phi 0} + \hat{z}\left(\frac{1}{r}\frac{\partial\psi_0}{\partial r}\right) \qquad (2.19)$$

where $B_{\phi 0}$ and ψ_0 are given, and ψ_0 is the usual poloidal flux function. In Eqs. (2.5)- (2.7) we consider the case where $\vec{v}_0 = 0$, $\eta_1 = 0$, but η_0 is not necessarily constant. The field and velocity perturbations are expressed by

$$\vec{B}_1(r,z,t) = \hat{r}\left(-\frac{1}{r}\frac{\partial\psi}{\partial z}\right) + \hat{\phi}\, B_{\phi_1} + \hat{z}\left(\frac{1}{r}\frac{\partial\psi}{\partial r}\right) \qquad (2.20)$$

$$\vec{v}_1(r,z,t) = \hat{r}\left(-\frac{1}{r}\frac{\partial\phi}{\partial z}\right) + \hat{\phi}\, v_{\phi_1} + \hat{z}\left(\frac{1}{r}\frac{\partial\phi}{\partial z}\right) \qquad (2.21)$$

where ψ, ϕ are the perturbed poloidal flux function and velocity stream function, respectively. Eqs. (2.5 - (2.7) yield the following

$$\frac{\partial\psi}{\partial t} = \frac{\eta_o}{4\pi}\Delta^*\psi - \frac{1}{r}\left(\frac{\partial\psi_o}{\partial z}\frac{\partial\phi}{\partial r} - \frac{\partial\psi_o}{\partial r}\frac{\partial\phi}{\partial z}\right) \tag{2.22}$$

$$\frac{\partial B_{\phi 1}}{\partial t} = \frac{1}{4\pi}\left[\frac{\partial}{\partial r}\left(\frac{\eta_o}{r}\frac{\partial(rB_{\phi 1})}{\partial r}\right) + \frac{\partial}{\partial z}\left(\eta_o\frac{\partial B_{\phi 1}}{\partial z}\right)\right]$$

$$- \frac{1}{r}\left[\frac{\partial\psi_o}{\partial z}\frac{\partial v_{\phi 1}}{\partial r} - \frac{\partial\psi_o}{\partial r}\frac{\partial v_{\phi 1}}{\partial z} - \frac{\partial\psi_o}{\partial z}\frac{v_{\phi 1}}{r}\right. \tag{2.23}$$

$$\left. + \frac{\partial B_{\phi o}}{\partial z}\frac{\partial\phi}{\partial r} - \left(\frac{\partial B_{\phi o}}{\partial r} - \frac{B_{\phi o}}{r}\right)\frac{\partial\phi}{\partial z}\right]$$

$$\frac{\partial v_{\phi 1}}{\partial t} = \frac{1}{4\pi\rho_o r}\left[-\frac{\partial\psi_o}{\partial z}\frac{\partial B_{\phi 1}}{\partial r} + \frac{\partial\psi_o}{\partial r}\frac{\partial B_{\phi 1}}{\partial z} - \frac{\partial\psi_o}{\partial z}\frac{B_{\phi 1}}{r}\right.$$

$$\tag{2.24}$$

$$\left. + \frac{\partial B_{\phi o}}{\partial z}\frac{\partial\psi}{\partial r} - \left(\frac{\partial B_{\phi o}}{\partial r} + \frac{B_{\phi o}}{r}\right)\frac{\partial\psi}{\partial z}\right]$$

$$\frac{\partial(\Delta^*\phi)}{\partial t} = \frac{1}{4\pi\rho_o r}\left[-\frac{\partial\psi_o}{\partial z}\frac{\partial(\Delta^*\psi)}{\partial r} + \frac{\partial\psi_o}{\partial r}\frac{\partial(\Delta^*\psi)}{\partial z} + \frac{2}{r}\frac{\partial\psi_o}{\partial z}\Delta^*\psi\right.$$

$$- \frac{\partial(\Delta^*\psi_o)}{\partial r}\frac{\partial\psi}{\partial z} + \frac{\partial(\Delta^*\psi_o)}{\partial z}\frac{\partial\psi}{\partial r} + \frac{2}{r}\Delta^*\psi_o\frac{\partial\psi}{\partial z} \tag{2.25}$$

$$\left. + 2r\left(B_{\phi o}\frac{\partial B_{\phi 1}}{\partial z} + \frac{\partial B_{\phi o}}{\partial z}B_{\phi 1}\right)\right]$$

where

$$\Delta^*\psi = r\frac{\partial}{\partial r}\left(\frac{1}{r}\frac{\partial\psi}{\partial r}\right) + \frac{\partial^2\psi}{\partial z^2} \tag{2.26}$$

The numerical scheme used to solve Eqs. (2.22) - (2.25) is analogous to that of the preceding section. Eqs. (2.22) - and (2.23) are solved implicitly for ψ^{n+1}, $B_{\phi 1}^{n+1}$ using ϕ^n and $v_{\phi 1}^n$. These predicted values of ψ^{n+1} and $B_{\phi 1}^{n+1}$ are then used in time-centered implicit difference equations for Eqs. (2.24) - (2.25), in order to obtain ϕ^{n+1} and $v_{\phi 1}^{n+1}$. Eqs. (2.22) - (2.23) can then be solved again using a time-centered scheme to obtain corrected values of ψ^{n+1}, $B_{\phi 1}^{n+1}$, which are then used to solve Eqs. (2.24 - (2.25) once more.

ALIMO has also been applied to the BFM equilibrium discussed in Sect. 2.4. The growth rates agree with those of Fig. 4 for m=0.

3.0 NONLINEAR CALCULATIONS

3.1 Mathematical Model

As described in section 2, the resistive MHD equations relate the electromagnetic fields $\vec{E}$ and $\vec{B}$ to the fluid velocity $\vec{v}$ and the thermodynamic variables (the pressure p, the mass density ρ, and the specific internal energy e). They may be combined into a set of conservation laws for the magnetic flux density $\vec{B}$, the fluid momentum density $\rho\vec{v}$, the mass density ρ, and the total energy density $u = \rho v^2 + B^2 + \rho e$. In order to make our model applicable to a variety of coordinate systems, we assume a metric of the form $ds^2 = h_1^2 dx_1^2 + h_2^2 dx_2^2 + h_3^2 dx_3^2$, where x_1, x_2, and x_3 are orthogonal curvilinear coordinates with scale factors $h_i = h_i\ (x_1, x_2)$. Then the relevant equations may be written as

$$\frac{\partial B^i}{\partial t} = \frac{1}{h_1 h_2 h_3} \frac{\partial}{\partial x^k} (h_1 h_2 h_3\ \sigma^{ik}) \tag{3.1}$$

$$\frac{\partial(\rho v^i)}{\partial t} = \frac{-1}{h_1 h_2 h_3} \frac{\partial}{\partial x^k} (h_1 h_2 h_3\ \tau^{ik}) + \tau^{nk} \begin{Bmatrix} i \\ nk \end{Bmatrix} \tag{3.2}$$

$$\frac{\partial \rho}{\partial t} = \frac{-1}{h_1 h_2 h_3} \frac{\partial}{\partial x^k} (h_1 h_2 h_3\ \rho v^k) \tag{3.3}$$

$$\frac{\partial u}{\partial t} = \frac{-1}{h_1 h_2 h_3} \frac{\partial}{\partial x^k} (h_1 h_2 h_3\ f^k) \tag{3.4}$$

where σ^{iK} are the contravariant components of the antisymmetric tensor $\underline{\underline{S}} = \vec{B}\,\vec{v} - \vec{v}\,\vec{B} + \frac{\eta}{S} (\nabla\vec{B} - \nabla\vec{B}^{\dagger})$, τ^{ik} are the contravariant components of the symmetric tensor $\underline{\underline{T}} = \rho\vec{v}\,\vec{v} - \vec{B}\,\vec{B} + \frac{1}{2} (p+B^2)\,\underline{\underline{I}}$, ρv^k are the contravariant components of the momentum, f^k are the contravariant components of the energy flux $\vec{F} = (u+p)\,\vec{v} + (B^2\underline{\underline{I}} - 2\,\vec{B}\,\vec{B})\cdot\vec{v} + \frac{2\eta}{S} (\vec{B}\cdot\nabla\vec{B} - \nabla\vec{B}\cdot\vec{B})$, $S = \tau_R/\tau_H$ and we have invoked the summation convention. The Christoffel symbols appearing in (3.2) arise because of the dependence of the unit vectors on the coordinates, and are defined as in Morse and Feshbach [27].

As discussed in Sect. 1, the unstable eigenmodes of a cylindrical plasma are in general functions of (r,θ,z). To describe these perturbations in our non-linear code, we must reduce the dimensionality of the problem from three to two. This is accomplished by applying the coordinate transformation $\phi = m\theta + k_z z$ to the fully three-dimensional equations. The equations in this coordinate system have been detailed elsewhere [23].

3.2 Computational techniques

Our basic equations form a set of Eulerian conservation laws in two spatial dimensions. When finite difference approximations are introduced it proves convenient to do so in such a way as to maintain the conservation properties of the original differential equation. Appropriate spatial difference approximations for scalar (3.3, 3.4) and vector (3.2) conservation laws are obtained in a straightforward manner by the application of Gauss' law to a computational cell. To obtain a conservative scheme for a pseudovector conservation law, $\partial\vec{w}/\partial t = \nabla\times\vec{G}$ (e.g., Eq. 3.1), we must apply Stokes' theorem. We introduce temporal differencing by means of the Alternating Direction Implicit (ADI) method. The final set of difference equations may then be written as

$$[1 - \frac{\Delta t}{2}(D_1 + S)]\underline{U}^{n+\frac{1}{2}} = [1 + \frac{\Delta t}{2}(D_2 + D_{12} + S)]\underline{U}^n \tag{3.5}$$

$$[1 - \frac{\Delta t}{2}(D_2 + S)]\underline{U}^{n+1} = [1 + \frac{\Delta t}{2}(D_1 + D_{12} + S)]\underline{U}^{n+\frac{1}{2}} \tag{3.6}$$

where $\underline{U}^n$ represents the state of the system at time t_n, D_i is a finite difference operator in the ith coordinate direction, D_{12} represents the mixed derivatives, and S represents the Christoffel symbols.

Since the operators discussed above are in general non-linear, each step of the ADI algorithm represents a set of M x J (where M is the length of U and J is the number of mesh points in one direction) non-linear algebraic equations to be solved on each row of the mesh. These equations must be linearized and solved iteratively. We write $(fg)^{(\ell)} = \frac{1}{2}\, f^{(\ell-1)}g^{(\ell)} + g^{(\ell-1)}f^{(\ell)}$ where $f^{(\ell-1)}$ and $g^{(\ell-1)}$ are considered as coefficients in the first and second terms, respectively. When Eqs. (3.5)-(3.6) are linearized in this manner, the resulting system is block tridiagonal, and may be written in the form

$$-\underline{\underline{A}}^{(\ell-1)}_{i,j}\cdot\underline{U}^{(\ell)}_{i+1,j} + \underline{\underline{B}}^{(\ell-1)}_{i,j}\cdot\underline{U}^{(\ell)}_{i,j} - \underline{\underline{C}}^{(\ell-1)}_{i,j}\cdot\underline{U}^{(\ell)}_{i-1,j} = \underline{D}_{i,j} \tag{3.7}$$

$$i = 2,3,\ldots,I-1$$

subject to the boundary conditions

$$\underline{\underline{G}}_1 \cdot \underline{U}^{(\ell)}_{1,j} = \underline{\underline{H}}_1 \cdot \underline{U}^{(\ell)}_{2,j} + \underline{J}_1 \quad \text{and} \quad \underline{\underline{G}}_I \cdot \underline{U}^{(\ell)}_{I,j} = H_I \cdot \underline{U}^{(\ell)}_{I-1,j} + \underline{J}_I \tag{3.8}$$

at the left hand boundary, and at the right hand boundary respectively. Eq. (3.7) may be solved by the well-known algorithm $\underline{U}^{(\ell)}_{i,j} = \underline{\underline{E}}_i \cdot \underline{U}^{(\ell)}_{i+1,j} + \underline{F}_i$ where $\underline{\underline{E}}_i$ and $\underline{F}_i$ are defined recursively in terms of the boundary conditions. A similar solution can be defined when periodic boundary conditions are imposed. In that case $\underline{U}^{(\ell)}_{i,j} = \underline{\underline{E}}_i \cdot \underline{U}^{(\ell)}_{i+1,j} + \underline{\underline{S}}_i \cdot \underline{U}^{(\ell)}_{I-1} + \underline{F}_i$ where $\underline{\underline{E}}$, $\underline{\underline{S}}$, and $\underline{F}$ are also determined recursively. Further details may be found in ref. [22].

Once the solution has been advanced to a new time level, the coefficients appearing in (3.7) are updated and the procedure is repeated until the solution converges to within a given tolerance. The time step is adjusted according to the number of iterations required for convergence. If convergence cannot be achieved within a specified number of iterations, the time step is decreased. Conversely, if the solution converges rapidly, the time step is increased. Thus the code always uses the largest possible time step to maintain the desired accuracy.

3.3 Boundary Conditions

The boundary conditions on field ($\vec{B}$) and flow ($\vec{v}$) variables at a perfectly conducting boundary are well known. For the thermodynamic variables ρ and u, we have found it convenient in some cases to impose boundary conditions which require that mass and energy be conserved. For example, if x_2 is taken as the coordinate normal to the wall, a scalar density is advanced according to

$$\frac{\partial \rho_J}{\partial t} = \frac{2}{(h_1 h_2 h_3)_J \Delta_- x_{2_J}} (h_1 h_3 F_2)^{n+1}_{J-\frac{1}{2}} \;.$$

For example, the finite difference approximation to the continuity equation (where $F_2 = \rho v_2$) is $-\, C\, (\rho v_2)^{n+1}_J + \rho^{n+1}_J = C\, (\rho v_2)^{n+1}_{J-1} + \rho^n_j$, with $C = \Delta t (h_1 h_3)_{J-\frac{1}{2}} \,/\, (h_1 h_2 h_3)_J (\Delta_- x_2)_J$. The coefficients appearing in this equation become elements in the boundary condition matrices $\underline{\underline{G}}$ and $\underline{\underline{H}}$.

The problem of numerically advancing the solution at the origin of coordinates (r = 0) is a difficult one, for unless symmetry conditions exist there is no natural boundary condition to be imposed at this point. For instabilities characterized by azimuthal mode number m = 0, we can pose the problem in (r,z) cylindrical coordinates. In this case we can apply a modification of the conservative boundary conditions discussed above, where axial fluxes are included in a straightforward manner. When $m^2 > 0$

no axial symmetry exists; one of the independent variables is an angle (θ), and the origin represents the same point in space for all values of this coordinate.

Scalar quantities, such as ρ and u, must have unique values on axis for any direction of approach, i.e., as we near $r = 0$ on any ray θ = constant, these quantities must approach the same limiting value. In polar and helical coordinates this is also the case for B_z and ρv_z. Evolutionary equations for such quantities may be found by integrating the appropriate equation over a small cylindrical cell of radius $\Delta r/2$, yielding the equation

$$\frac{\partial u_o}{\partial t} = \frac{-2}{\pi \Delta r} \sum_i \Delta\theta_i F_{i,3/2} \quad .$$

The ADI method decouples the spatial mesh by proceeding in partial time steps, so that on the partial step in which radial terms are treated explicitly, this equation can be used directly. On the other partial step, F must be treated implicitly, thus coupling the solution for all rays (lines for which θ-constant) at $r = 0$. This difficulty is avoided by noticing that F will in general be non-linear and will contain derivatives of u. After linearization and discretization, we can write

$$F^{n+1}_{i,3/2} \doteq P_i u^{n+1}_{i,2} + Q_i u^{n+1}_o + R_i \quad .$$

The coupling is now removed by reordering the recursive solution on the interior of the mesh as

$$u^{n+1}_{i,j} = e_{i,j} \; u^{n+1}_{i,j-1} + f_{i,j} \; , \; j = 2,3,\ldots\ldots,J \qquad (3.9)$$

i.e., we define the solution from "left to right", instead of "right to left". We then arrive at

$$u^{n+1}_o = \left\{1 + \frac{\Delta t}{\pi \Delta r} \sum_i \Delta\theta_i (P_i e_{i,2} + Q_i)\right\}^{-1} \cdot \left\{u^n_o - \frac{\Delta t}{\pi \Delta r} \sum_i \Delta\theta_i (P_i f_{i,2} + R_i)\right\} \quad (3.10)$$

as the expression for the scalar u at the origin. The complete solution is thus obtained by sweeping all rays from the outer boundary to the origin to determine $e_{i,j}$ and $f_{i,j}$ recursively for j=J-1, j-2,...,2, applying (3.10) and then using (3.9) to obtain the interior solution.

When $m^2>1$, we know that components of vector quantities must vanish at the origin. However, modes with m=1 are characterized by gross motion across the origin, and we use the uniqueness of the cartesian components to advance vector quantities at r=0. The general form for the equation describing the evolution of a vector quantity is

$$\frac{\partial\vec{V}}{\partial t} = -\frac{1}{r}\frac{\partial}{\partial r}(r\vec{F}) - \frac{1}{r}\frac{\partial\vec{G}}{\partial\theta} - \frac{\partial\vec{H}}{\partial z} + \vec{C} - \nabla f$$

where $\vec{F}$, $\vec{G}$, and $\vec{H}$ are the vector fluxes of $\vec{V}$, and $\vec{C}$ represents possible Christoffel terms. We assume that the cartesian representation of $\vec{V}$ can be obtained by the transformation $\vec{V}_c = \underline{\underline{\alpha}} \cdot \vec{V}$. Then the cartesian components of $\vec{V}$ evolve according to the equation

$$\frac{\partial\vec{V}_c}{\partial t} = -\frac{1}{r}\frac{\partial}{\partial r}(r\underline{\underline{\alpha}}\cdot\vec{F}) - \frac{1}{r}\frac{\partial}{\partial\theta}(\underline{\underline{\alpha}}\cdot\vec{G}) - \frac{\partial}{\partial z}(\underline{\underline{\alpha}}\cdot\vec{H}) + \frac{\partial\underline{\underline{\alpha}}}{\partial z}\cdot\vec{H} - \underline{\underline{\alpha}}\cdot\nabla f \tag{3.11}$$

where we have assumed $\underline{\underline{\alpha}}$ to be independent of r, and have used $\underline{\underline{\alpha}}\cdot\vec{C} = -(1/r)\partial\underline{\underline{\alpha}}/\partial\theta\cdot\vec{G}$ which, indeed, serves as a definition of $\vec{C}$. When transformation to helical coordinates is performed, equation (3.11) becomes

$$\frac{\partial\vec{V}_c}{\partial t} = -\frac{1}{r}\frac{\partial}{\partial r}(r\underline{\underline{\alpha}}\cdot\vec{F}) - \frac{1}{r}\frac{\partial}{\partial\phi}\left[\underline{\underline{\alpha}}\cdot(\vec{G}+rk_z\vec{H})\right] + k_z\frac{\partial\underline{\underline{\alpha}}}{\partial\theta}\cdot\vec{H} - \underline{\underline{\alpha}}\cdot\nabla f \tag{3.12}$$

which points out the reason for allowing $\underline{\underline{\alpha}}$ to depend explicitly on the z coordinate. Proceeding in a manner similar to that used for scalar quantities we arrive at the equation

$$\frac{\partial\vec{V}_c}{\partial t} = -\frac{2}{\pi\Delta r}\sum_i \Delta\theta_i\underline{\underline{\alpha}}_i\cdot\vec{W}_{i,3/2}$$

where $\vec{W} = \vec{F} - k_z\Delta r\,\underline{\underline{A}}\cdot\vec{H}/4 + (\Delta r/4)\nabla f$, and we have used the fact that, for polar coordinates, $\partial\underline{\underline{\alpha}}/\partial\theta = \underline{\underline{\alpha}}\cdot\underline{\underline{A}}$ where $A_{rs} = (-1)^r(1-\delta_{rs})$, r,s = 1,2. A self-consistent solution can now be defined in a manner analogous to that previously described. When we use the linearization $\vec{W}^{n+1}_{i,3/2} = \underline{\underline{P}}_i\cdot\vec{V}^{n+1}_{i,2} + \underline{\underline{Q}}_i\cdot\vec{V}^{n+1}_{c_0} + \vec{R}_i$ we arrive at $\vec{V}^{n+1}_{c_0} = (\underline{\underline{I}} + \underline{\underline{S}})^{-1}\cdot(\vec{V}^n_{c_0} - \underline{\underline{T}})$ where $\underline{\underline{I}}$ is the identity matrix, and

$$\underline{\underline{S}} = \frac{\Delta t}{\pi\Delta r}\sum_i \Delta\theta_i\underline{\underline{\alpha}}_i\cdot\left[\underline{\underline{P}}_i\cdot\underline{\underline{E}}_{i,2}\cdot\underline{\underline{\alpha}}_i^{-1} + \underline{\underline{Q}}_i\right]$$

$$\underline{\underline{T}} = \frac{\Delta t}{\pi\Delta r}\sum_i \Delta\theta_i\underline{\underline{\alpha}}_i\cdot\left[\underline{\underline{P}}_i\cdot\vec{F}_{i,2} + \vec{R}_i\right] \quad .$$

The polar components are then obtained by inverting the transformation for each ray, and the interior solution is then found as described previously.

3.4 Applications

As discussed in Section 1, one of the features of the resistive MHD equations is the occurrence of phenomona which may evolve on widely separated time scales. Requirements of accuracy and stability on the numerical solutions of these equations require that such simulations evolve on the fast time scale. In that case, computational studies of events which evolve on the slow time scale become difficult whenever S>0(1). Such is the case with resistive instabilities which grow on time scales which can approach the resistive diffusion time. We present sample calculations of resistive instabilities which demonstrate the performance of our model. For these examples we have taken $S \approx 10^{2-3}$, so that the time scales are well separated. We perform calculations in two different coordinate systems: axisymmetric cylindrical (r,z), and helical (r,ϕ). The first case is an example of the nonlinear evolution of a mode which appears only when the plasma possesses a finite pressure gradient.

As stated in section 3.3, when m=0 the problem can be posed in (r,z) coordinates. For computational purposes we take $x_1 = z$, $x_2=r$, $x_3=\theta$. The corresponding scale factors are $h_1=2\pi/\alpha, h_2=1, h_3=x_2$, where $\alpha=k_z a$ is the nondimensional axial wave number. We consider an equilibrium in which the axial magnetic field B_z changes sign in the outer regions of the pinch. Such fields are characteristic of devices known as Reversed Field Pinches. This specific equilibrium is stable against tearing modes, but is unstable to pressure driven resistive interchange modes (g-modes). [7]

In Fig. (5) we show the flux surfaces resulting from the nonlinear evolution of this mode. Note the extreme distortion of the flux surfaces between the two large magnetic islands. The large radial extent of the islands makes this mode particularly dangerous to plasma confinement.

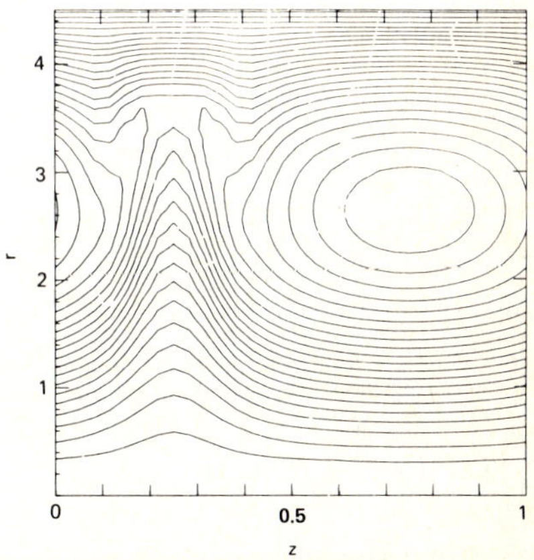

Fig. 5 Magnetic flux surfaces for resistive interchange mode, m=0, $t=595.5t_H$

Modes for which both m and k_z are non-vanishing possess helical symmetry. The natural coordinate system for these calculations consists of a radial (r) and an angular (ϕ) variable. For such cases we take $x_1=\phi$, $x_2=r$, $x_3=z$, $h_1=x_2$, $h_2=1$, $h_3=1/\alpha$. These cases also require the use of the boundary conditions described in section 3.3. We consider the Bessel Function model (BFM); this equilibrium has the property of being force-free, i.e., the Lorentz force JxB vanishes identically. Thus no equilibrium pressure gradient exists, and we expect tearing modes to constitute the primary resistive instabilities.

In Fig. (6) we plot the reconnected helical flux (which is proportional to the square of the island size) as a function of time for the case m = 1. Note the extended period of exponential growth followed by complete nonlinear saturation. The evolution of the flux surfaces is shown in Fig. (7). For this mode we see that the magnetic island can grow to large size and, in the non-linear phase, can occupy a considerable portion of the plasma volume. A more complete discussion of this behavior, which is qualitatively similar to that found for tokamak plasmas, can be found in Ref. [23]. Here we simply note that we have successfully simulated transport of both scalar and vector quantities across the origin by applying the methods described in Section 3.3.

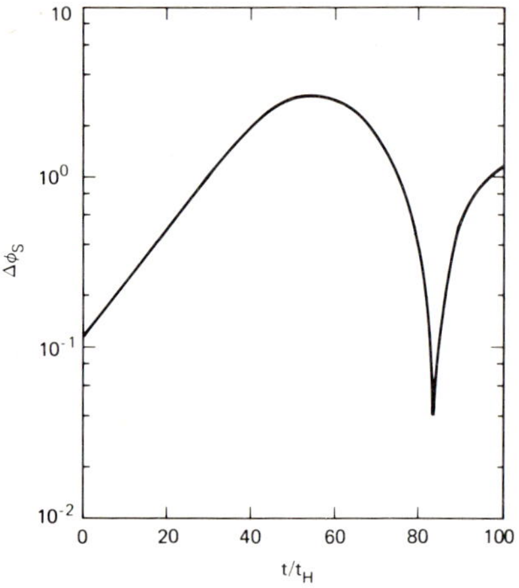

Fig. 6 Reconnected flux versus time for m=1.

REFERENCES

1. H. P. Furth, J. Killeen, and M. N. Rosenbluth, Phys. Fluids 6, 459 (1963).
2. J. Killeen and H. P. Furth, Bull. Am. Phys. Soc. 6, 193 (1961).
3. J. Killeen in Physics of Hot Plasmas edited by B. J. Rye and J. C. Taylor, Plenum, New York, 1970, p. 202.
4. J. E. Crow, J. Killeen and D. C. Robinson, Sixth European Conference on Controlled Fusion and Plasma Physics, Moscow (1973) 269.
5. D. C. Robinson, J. E. Crow, J. A. Dibiase, A. S. Furzer, and J. Killeen, "The Growth of Resistive Instabilities in a Diffuse Pinch" (1979) to be published.
6. A. J. L. Verhage, A. S. Furzer, and D. C. Robinson, Nuclear Fusion 18, 457 (1978).
7. D. C. Robinson, Nuclear Fusion 18, 939 (1978).
8. J. Dibiase, Ph.D. Thesis, Univ. of Calif., Davis, UCRL-51591 (1974).
9. J. Dibiase and J. Killeen, J. Comp. Phys. 24, 158 (1977).
10. J. A. Dibiase, J. Killeen, D. C. Robinson, D. Schnack, in Third Topical Conf. on Pulsed High Beta Plasmas, (Culham 1975) Pergamon Press, Oxford 1976 p. 283-289.
11. D. Dobrott, S. C. Prager, and J. B. Taylor, Phys. Fluids 20, 1850 (1977).
12. J. Killeen and A. I. Shestakov, Phys. Fluids 21, 1746, (1978).
13. R. K. Pollard and J. B. Taylor, Phys. Fluids 22, 126 (1979).
14. D. Schnack and J. Killeen, in Theoretical and Computational Plasma Physics, (Trieste, 1977), IAEA Vienna 1978, 337-360.
15. B. V. Waddell, M. N. Rosenbluth, D. A. Monticello, and R. B. White, Nucl. Fusion 16, 528 (1976).
16. H. R. Hicks, B. Carreras, J. A. Holmes, and B. V. Waddell, ORNL/TM-60966 (1977).

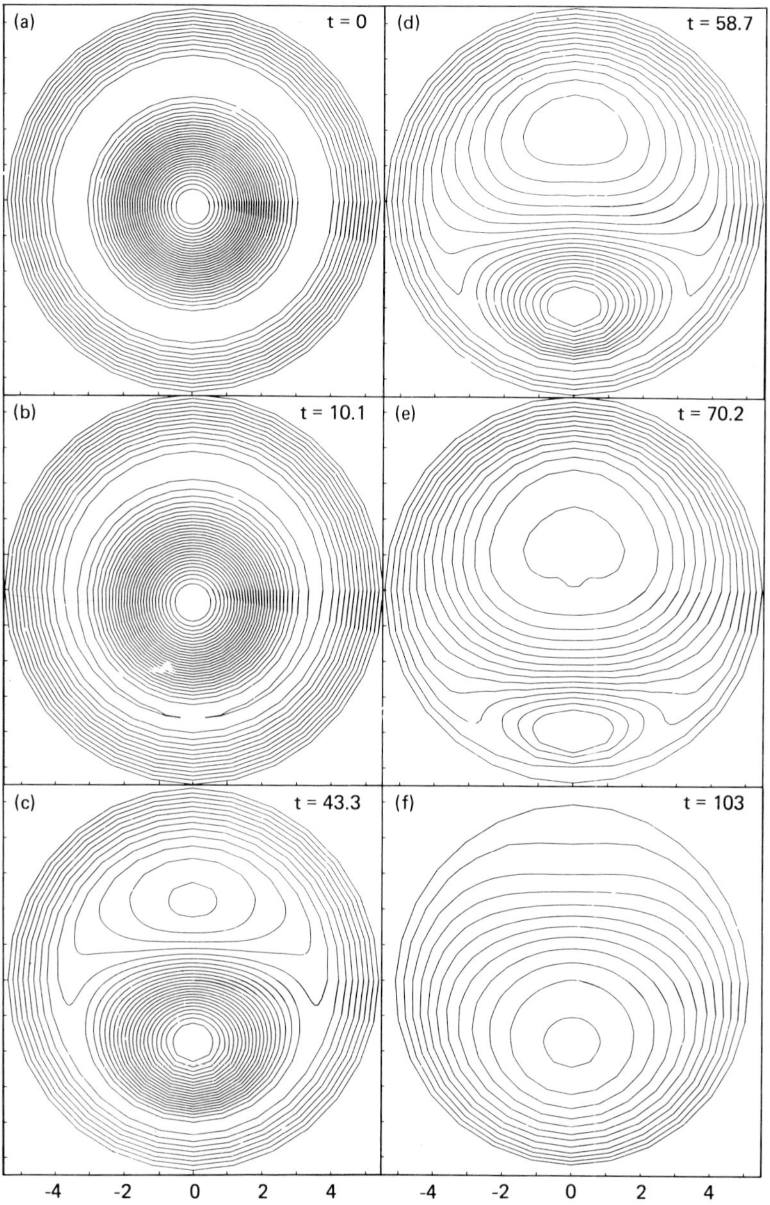

Fig. 7 Magnetic flux surfaces for m=1

17. B. V. Waddell, B. Carreras, H. R. Hicks, J. A. Holmes, and D. K. Lee, ORNL/TM-6213 (1978).
18. B. Carreras, H. R. Hicks, and B. V. Waddell, ORNL/TM-6570 (1978).
19. G. L. Jahns, M. Soler, B. V. Waddell, J. D. Callen, and H. R. Hicks, Nucl. Fusion 18, 609 (1978).
20. M. N. Rosenbluth, D. A. Monticello, H. R. Strauss, and R. B. White, Phys. Fluids 19, 1988 (1976).
21. D. Schnack, Ph.D. Thesis, Univ. of Calif., Davis, UCRL-52399 (1978).
22. D. Schnack and J. Killeen, Nonlinear, Two-Dimensional Magnetohydrodynamic Calculations (1979) accepted for publication in J. Comp. Phys.
23. D. Schnack and J. Killeen, Nuclear Fusion 19, 877 (1979).
24. R. D. Richtmyer and K. W. Morton, Difference Methods for Initial-Value Problems, Interscience, N.Y. 1967.
25. D. C. Robinson, Plasma Phys. 13, 439 (1971).
26. H. P. Furth, P. H. Rutherford, and H. Selberg Phys. Fluids 16, 1054, (1973).
27. P. M. Morse and H. Feshbach, Methods of Theoretical Physics, McGraw-Hill, New York (1953).

COMPUTING METHODS IN APPLIED SCIENCES AND ENGINEERING
R. Glowinski, J.L. Lions (editors)
North-Holland Publishing Company

NUMERICAL TREATMENT OF LINEARIZED EQUATIONS DESCRIBING INHOMOGENEOUS COLLISIONLESS PLASMAS

H. Ralph Lewis
Los Alamos Scientific Laboratory, P. O. Box 1663
Los Alamos, New Mexico 87545, U. S. A.

I. INTRODUCTION

There is considerable current interest in the initial-value problem for the linearized equations which describe small departures from equilibrium of a fully ionized plasma in which one or more of the particle species can be treated as collisionless. Present-day research in controlled thermonuclear fusion requires information about the stability of such systems and the effects of phase mixing in them; this is also true in other fields, such as space physics, in which the physics of collisionless plasmas plays a role. During any specified period of time, the collisionless description for a particular particle species in an experimental plasma applies if the temperature of the species is sufficiently high. The linearized equations for spatially inhomogeneous plasmas in which there is a collisionless species are difficult to solve, even computationally, because all three velocity compponents and at least one spatial coordinate must be considered as independent variables in the analysis. This means that the equations are a system of coupled integrodifferential equations in which there are at least five independent variables--time, three velocity components, and at least one spatial coordinate. Recently, progress has been made in the formulation of the problem in terms of a dispersion matrix, and applications of the fomulation to interesting equilibria with one nonignorable coordinate have been made. When there is one nonignorable coordinate in the equilibrium, only that spatial coordinate appears as an independent variable in the system of integrodifferential equations. The formulation is in terms of a description of the three-dimensional equilibrium motion of particles which is obtained by using an equivalent one-dimensional potential. Integrals with respect to time arise which extend over the times appropriate for the equivalent one-dimensional problem; for orbits which are trapped in the equivalent one-dimensional potential, the integrals extend over the bounce periods in the one-dimensional potential and not over the infinite time history of the equilibrium three-dimensional orbits. This approach has been described in the context of a general discussion of the initial-value problem for linearized equations which describe plasma systems in which there is a collisionless species.[1] Applications of the general formalism have been made to the stability of a plasma column within the framework of the Vlasov-fluid model[2] and to the stability of large-amplitude Bernstein-Greene-Kruskal equilibria.[3] The basic approach has also been used independently in the context of the Vlasov-fluid model to study the stability of a rotating theta pinch,[4] and to investigate the effects of resonant particles on kinetic stabilization in screw pinches.[5] Additional work is in progress.

In Section II, the basic linearized equations are presented in a general context, specialization for one nonignorable coordinate is indicated, and a formulation for numerical work is introduced. Numerical aspects of the problem are discussed in Section III, including choice of matrix representation and methods of solution. Some concluding remarks are given in Section IV.

II. GENERAL THEORETICAL FRAMEWORK

We consider a plasma which consists of one or more collisionless particle species which are governed by a linearized Boltzmann equation for each collisionless species s,

$$\left(\frac{\partial}{\partial t} + iL_s\right)f_s^{(1)} = U_s\phi^{(1)} \; ; \tag{1}$$

and we assume that the plasma can be described by these equations and a set of field equations of the form

$$K\phi^{(1)} = \sum_s \int d^3\underset{\sim}{v}\; J_s(\underset{\sim}{r},\underset{\sim}{v})f_s^{(1)}(\underset{\sim}{r},\underset{\sim}{v},t) \; . \tag{2}$$

The quantities L_s, U_s, and K are linear operators, and $f_s^{(1)}$ is the perturbation of a single-particle distribution function f_s about an equilibrium distribution function $f_s^{(0)}$:

$$f_s(\underset{\sim}{r},\underset{\sim}{v},t) = f_s^{(0)}(\underset{\sim}{r},\underset{\sim}{v}) + f_s^{(1)}(\underset{\sim}{r},\underset{\sim}{v},t) \; . \tag{3}$$

The quantity $\phi^{(1)}$ is the perturbation of an array ϕ of potential functions about an equilibrium array $\phi^{(0)}$:

$$\phi(\underset{\sim}{r},t) = \phi^{(0)}(\underset{\sim}{r}) + \phi^{(1)}(\underset{\sim}{r},t) \; . \tag{4}$$

The symbols $\underset{\sim}{r}$ and $\underset{\sim}{v}$ denote position and velocity vectors.

A simple example of equations of this form is the set of linearized equations for a one-dimensional electron gas in a background of immobile ions of number density $n_0(x)$:

$$\left(\frac{\partial}{\partial t} + v\frac{\partial}{\partial x} + \frac{e}{m}\frac{d\phi^{(0)}}{dx}\frac{\partial}{\partial v}\right)f^{(1)} = -\frac{e}{m}\frac{\partial f^{(0)}}{\partial v}\frac{\partial \phi^{(1)}}{\partial x} \; , \tag{5}$$

$$\frac{\partial^2\phi^{(1)}}{\partial x^2} = 4\pi e\int dv\, f^{(1)}(x,v,t) \; . \tag{6}$$

The symbols x and v denote the one-dimensional position and velocity variables. The quantity ϕ represents a single potential function, the scalar potential for the electric field; $\phi^{(0)}$ is a function of x only which is related to $f^{(0)}$ by

$$\frac{\partial^2\phi^{(0)}}{\partial x^2} = 4\pi e\left[\int dv\, f^{(0)}(x.v) - n_0(x)\right] \; . \tag{7}$$

The electron mass is m, the electron charge is -e, and the ion charge is e.

An example of equations of the form of (1)-(2) which are useful for describing a plasma in a magnetic field is given by the Vlasov fluid model.[6] This model is a low-frequency model for an ion-electron plasma in which the ions are treated as collisionless and the electrons are treated as a massless, pressureless fluid. The linearized equations are

$$\left[\frac{\partial}{\partial t} + \underset{\sim}{v}\cdot\underset{\sim}{\nabla}_{\underset{\sim}{r}} + \frac{e}{M}\left(\underset{\sim}{E}^{(0)} + \frac{1}{c}\,\underset{\sim}{v}\times\underset{\sim}{B}^{(0)}\right)\cdot\underset{\sim}{\nabla}_{\underset{\sim}{v}}\right]f^{(1)} = -\,e\,\frac{df^{(0)}(\varepsilon)}{d\varepsilon}\,\underset{\sim}{v}\cdot\underset{\sim}{E}^{(1)} \; , \tag{8}$$

$$\frac{1}{4\pi}[(\nabla\times B^{(0)})\times B^{(1)} + (\nabla\times B^{(1)})\times B^{(0)}] - en^{(0)}E^{(1)}$$

$$= e\int d^3v \left(E^{(0)} + \frac{1}{c}\, v\times B^{(0)}\right)f^{(1)} \ . \tag{9}$$

Here, e and M are the ion charge and mass, respectively; c is the speed of light; E and B are the electric and magnetic fields, respectively; f is the ion single-particle distribution function; the equilibrium distribution function is assumed to be a function of the energy ε only; and ε and n_0 are defined by

$$\varepsilon = \frac{1}{2}\,Mv^2 + e\phi^{(0)}(r) \ , \tag{10}$$

$$n^{(0)}(r) = \int d^3v\ f^{(0)}(\varepsilon) \ . \tag{11}$$

The array of perturbation potentials in (8)-(9) is the set of components of a displacement vector $\xi(r,t)$ from which $E^{(1)}$ and $B^{(1)}$ are derived:

$$E^{(1)} = -\frac{1}{c}\frac{\partial\xi}{\partial t}\times B^{(0)} - \nabla(E^{(0)}\cdot\xi) \ , \tag{12}$$

$$B^{(1)} = \nabla\times(\xi\times B^{(0)}) \ , \tag{13}$$

where

$$\xi\cdot B^{(0)} = 0 \ . \tag{14}$$

For some examples of equations (1)-(2), it can be useful to replace the perturbation distribution function $f_s^{(1)}$ by an auxiliary function g_s which is a linear functional of the perturbation potentials:

$$g_s = f_s^{(1)} - P_s\phi^{(1)} \ , \tag{15}$$

where P_s is a linear operator. The equations for the auxiliary functions and the perturbation potentials are of the same form as (1)-(2):

$$\left(\frac{\partial}{\partial t} + iL_s\right)g_s = W_s\phi^{(1)} , \tag{16}$$

$$\Lambda\phi^{(1)} = \sum_s \int d^3\underset{\sim}{v}\ J_s(\underset{\sim}{r},\underset{\sim}{v})g_s(\underset{\sim}{r},\underset{\sim}{v},t) , \tag{17}$$

where

$$\Lambda = K - \sum_s \int d^3\underset{\sim}{v}\ J_s(\underset{\sim}{r},\underset{\sim}{v})P_s . \tag{18}$$

The utility of such a transformation of dependent variables is that the field operator Λ which appears in (17) can chosen arbitrarily while preserving the form of the equations. In particular, the operator on the left hand side of (16) is unchanged by this transformation. The introduction of auxiliary functions in this way can be advantageous in numerical approximation schemes, a point to which we return later. Of course, the original equations, (1)-(2), are an example of (16)-(17). The original field operator, K, involves time differentiation for some physical systems of interest; an example is the case of the Vlasov-fluid model as indicated in (9). For these systems it is possible to introduce auxiliary functions such that Λ does not involve t or the operator $\partial/\partial t$. Henceforth, we consider (16)-(17) as the basic equations and assume that Λ does not involve t or $\partial/\partial t$.

It is convenient to consider the solution of (16)-(17) for the evolution of g_s and the perturbation potentials in terms of Laplace transforms. We denote the Laplace transform of a function h(t) by $\hat{h}(\omega)$, where

$$h(t) = \frac{1}{2\pi}\int_C d\omega\ e^{-i\omega t}\hat{h}(\omega) , \quad \hat{h}(\omega) = \int_0^\infty dt\ e^{i\omega t}h(t) , \tag{19}$$

and C is a suitable Bromwich contour. For the physical applications which we envision, the linear operator W_s may involve the time differentiation operator $\partial/\partial t$, but time does not occur in W_s in any other way. Therefore, it is appropriate to define $\hat{W}_s(\omega)$ as the result of substituting $-i\omega$ for $\partial/\partial t$ in W_s. The solution of (16)-(17) for the Laplace transforms $\hat{g}_s(\omega)$ and $\hat{\phi}^{(1)}(\omega)$ is

$$\hat{\phi}^{(1)}(\omega) = - iD^{-1}(\omega)\sum_s\int d^3\underset{\sim}{v}\; J_s[L_s - \omega]^{-1}[\Phi_s(\omega) + g_s(0)] \;, \tag{20}$$

$$\hat{g}_s(\omega) = - [L_s - \omega]^{-1}\hat{W}_s(\omega)D^{-1}(\omega)\sum_{s'}\int d^3\underset{\sim}{v}\; J_{s'}[L_{s'} - \omega]^{-1}[\Phi_{s'}(\omega) + g_{s'}(0)]$$

$$- i[L_s - \omega]^{-1}[\Phi_s(\omega) + g_s(0)] \;, \tag{21}$$

where

$$D(\omega) = \Lambda + i \sum_s\int d^3\underset{\sim}{v}\; J_s[L_s - \omega]^{-1}\hat{W}_s(\omega) \;, \tag{22}$$

and where $g_s(0)$ is the value of $g_s(t)$ at t=0 and $\Phi_s(\omega)$ can be constructed from the values at t=0 of $\phi^{(1)}(t)$ and its time derivatives. The contour C for the Laplace transforms must be above all singularities of $\hat{\phi}^{(1)}(\omega)$ and $\hat{g}_s(\omega)$. The inverse of the operator $[L_s - \omega]$ is nonsingular except on the real ω axis because L is a Hermitian operator.[1] Singularities off the real axis and not associated with initial conditions are singularities of $D^{-1}(\omega)$. The operator $D(\omega)$ is called the <u>dispersion operator</u>. It plays a crucial role in determining the stability properties of the system, and it will be important in our discussion of numerical approximation schemes. Note that the dispersion operator does not depend on the operators P_s in terms of which auxiliary functions g_s may be defined.

It can be shown[1] that the singularities of $\hat{\phi}^{(1)}(\omega)$ and $\hat{g}_s(\omega)$ are located at the roots of the equation

$$\left[\prod_s \det(L_s - \omega\underset{\sim}{I})\right]\det D(\omega) = 0 \;, \tag{23}$$

where $\underset{\sim}{I}$ is the unit operator in $(\underset{\sim}{r},\underset{\sim}{v})$ space. Because L_s is Hermitian, any root of this eqaution which has an imaginary part, corresponding to exponential growth or decay, must be a zero of $\det D(\omega)$. The stability of the system (16)-(17) can be studied numerically by finding a suitable finite-dimensional approximation to $D(\omega)$. A numerical approximation to the solution of the initial-value problem for a fixed length of time can be obtained by finding a suitable approximation of the roots of (23) in terms of a finite number of points in the ω plane. Since the eigenvalue spectrum of L_s is a set of real continua, approximating the roots of (23) with a finite number of points means approximating continuous spectra with discrete spectra of finite size. We now turn to the question of constructing numerical approximations.

III. NUMERICAL ASPECTS OF THE PROBLEM

For numerical purposes it is useful to introduce eigenfunctions of L_s as a basis for the $(\underset{\sim}{r},\underset{\sim}{v})$ space for species s. This is also useful for some analytical calculations. The eigenfunctions of L_s are a complete set of functions in $(\underset{\sim}{r},\underset{\sim}{v})$ space which we take to be orthonormal:[1]

$$L_s w_{sr} = \mu_{sr} w_{sr} \, , \tag{24}$$

$$(w_{sr}, w_{sr'}) = \delta_{rr'} \, , \tag{25}$$

where the index r stands for whatever set of labels is needed to specify w_{sr}, the parenthesis notation denotes an inner product, and $\delta_{rr'}$ stands for a product of Kronecker deltas and Dirac delta functions--one Kronecker delta for each pair of discrete labels, and one Dirac delta function for each pair of continuous labels. If at most one coordinate is nonignorable in the equilibrium configuration, then the eigenfunctions w_{sr} and eigenvalues μ_{sr} can be found explicitly in terms of definite integrals.[1] The eigenvalues μ_{sr} play an important role in resonance denominators arising from the inverse of the operator $[L - \omega]$ which appears in the Laplace transforms (20)-(21).

In order to have a matrix representation of $D(\omega)$, we also introduce a basis for configuration space with basis functions $\eta_n(\underset{\sim}{r})$. In some cases, the basic equations can be formulated such that one of the eigenfunctions of the operator Λ which appears in (17) is a good approximation to $\hat{\phi}^{(1)}(\omega_0)$ for a complex frequency ω_0 of interest. Then, choosing the basis functions η_n to be eigenfunctions of Λ is advantageous,

$$\Lambda \eta_n = \lambda_n \eta_n \, . \tag{26}$$

The operator Λ can be chosen to be Hermitian for most problems. However, it may not be Hermitian, in which case it can be useful to define a dual set of functions $\zeta_n(\underset{\sim}{r})$,

$$(\zeta_n, \eta_{n'}) = \delta_{nn'} \, , \tag{27}$$

where again $\delta_{nn'}$ stands for a product of Kronecker deltas and Dirac delta functions.

Even if one of the eigenfunctions of Λ is not a good approximation to $\hat{\phi}^{(1)}(\omega_0)$, a linear combination of a few of the eigenfunctions may be good; in this case it would still be a good idea to let the basis functions η_n be eigenfunctions of Λ because a small truncated matrix representation of $D(\omega_0)$ could be a good approximation. Of course, the optimal choice of basis would be one with which the representation of $D(\omega)$ would be diagonal; then, $\det D(\omega) = 0$ would be satisfied by setting any diagonal element to zero. If the equilibrium is spatially homogeneous, $D(\omega)$ can be made diagonal by choosing basis functions proportional to $\exp(i\underset{\sim}{\kappa}\cdot\underset{\sim}{r})$, a familiar situation in the stability theory for an infinite homogeneous equilibrium in plasma physics. However, when the equilibrium is spatially inhomogeneous, the basis functions which diagonalize $D(\omega)$ generally will depend on ω, and are usually not known for any given ω.

There is a systematic way of determining a Λ whose eigenfunctions are good basis functions. Sometimes the procedure can be carried out. The idea is to try to diagonalize $D(\omega)$ for a mode of interest whose frequency is near $\omega = \omega_0$. If the basis functions are eigenfunctions of Λ, then the operators $\hat{P}_s$ must satisfy

$$\hat{W}_s(\omega_0) \approx 0 ,$$

which means

$$i(L_s - \omega_0)\hat{P}_s(\omega_0)\hat{\phi}^{(1)}(\omega_0) \approx \hat{U}_s(\omega_0)\hat{\phi}^{(1)}(\omega_0) . \tag{28}$$

This corresponds to an approximate solution of the original linearized Boltzmann equation. The parameter ω_0 is determined by solving the dispersion relation obtained from the approximate solution of (28). The $\hat{P}_s$ determined in this way depends parametrically on ω_0 and it determines, through (18), a Λ whose eigenfunctions form a suitable basis for working with a severely truncated dispersion matrix in the neighborhood of $\omega = \omega_0$.

The procedure just outlined for determining the basis functions for configuration space is good if it can be carried out. For example, it was effective in studying the stability of a large-amplitude Bernstein-Greene-Kruskal equilibrium.[3] However, it does occur for some problems of physical interest that Λ has to be chosen rather carefully. For example, in the case of a magnetized plasma, there are modes for which spatial variations of short scale length, on the order of the ion gyroradius, are present in addition to variations of much longer scale length.[4,7] For the procedure of using only a few of the eigenfunctions of Λ to be effective, they must be able to represent all of the important spatial variations with sufficient accuracy, the short scale length variations as well as those with long scale length. For a specific problem it may not be possible to solve (28) to sufficient accuracy, or it may be inconvenient to do so.

When it is not feasible to use a very small truncation of $D(\omega)$ by taking the eigenfunctions of a suitable Λ as the basis for configuration space, it is necessary to choose a basis with which arbitrary variations of the appropriate scale lengths can be adquately represented. This can easily lead to a matrix representation of $D(\omega)$ whose dimension is large enough that very serious computer storage problems are encountered. If there is more than one nonignorable coordinate in the equilibrium, there is at present no general numerical procedure for finding the eigenfrequencies of the system. Even in the case of one nonignorable coordinate, there has not been a generally applicable numerical procedure so far. Recently, however, an approach has been found which should render it feasible to find the eigenfrequencies numerically for a general equilibrium with one nonignorable coordinate.[8] The approach makes detailed use of the explicit form of the eigenfunctions and eigenvalues of the operator L_s,[1] organizes the computational work in a way which minimizes computer storage problems, and relies on some empirical simplicity of the dependence on ω of elements of the dispersion matrix. A computer code based on this approach is being constructed.

An approximation to the solution of the initial-value problem for (16)-(17) can be obtained by approximating the entire spectrum of solutions of (23), including the continuous branches. The continuous part of the spectrum is responsible for phase mixing in the evolution of the system, an example of which is a decay in time of electrostatic perturbations which is known as Landau damping. If a finite set of basis functions w_{sr} and η_n is used, as will be the case in computational work, then (23) is in fact always a polynomial equation.[1] Despite the fact that the polynomial can be of very large degree, it can be evaluated because (23) represents the polynomial as the product of the determinant of a diagonal matrix times the determinant of the dispersion matrix. The determinant of the diagonal matrix is trivial to evaluate, and the determinant of the dispersion matrix can be evaluated if the number of basis functions η_n is not too large. All of the roots of such a polynomial equation of large degree can be found simultaneously by means of a quadratically convergent iteration method proposed by Aberth,[9] Kerner,[10], and Durand.[11] The method also converges when there are multiple roots.

IV. CONCLUSION

The equations governing the small-signal response of spatially inhomogeneous collisionless plasmas have practical significance in physics, for example in controlled thermonuclear fusion research. Although the solutions are very complicated and the equations are difficult to solve numerically, effective methods for them are being developed which are applicable when the equilibrium involves only one nonignorable coordinate. The general theoretical framework probably will provide a basis for progress when there are two or three nonignorable coordinates.

REFERENCES

1. H. R. Lewis and K. R. Symon, J. Math. Phys. 20, 413 (1979); also Errata, to be published in J. Math. Phys.
2. H. R. Lewis and L. Turner, Nucl. Fusion 16, 993 (1976).
3. J. L. Schwarzmeier, H. R. Lewis, B. Abraham-Shrauner, and K. R. Symon, Phys. Fluids 22, 1747 (1979).
4. C. E. Seyler, to be published in Phys. Fluids.
5. C. E. Seyler and J. P. Freidberg, to be published in Phys. Fluids.
6. J. P. Freidberg, Phys. Fluids 15, 1102 (1972).
7. T. E. Cayton and H. R. Lewis, to be published in Phys. Fluids.
8. C. E. Seyler and H. R. Lewis, work in progress.
9. O. Aberth, Math. Comp. 27, 339 (1973).
10. I. O. Kerner, Numer. Math. 8, 290 (1966).
11. E. Durand, Solutions Numériques des Équations Algébriques, Tome I (Masson, Paris, 1960), pp. 277-281.

COMPUTING METHODS IN APPLIED SCIENCES AND ENGINEERING
R. Glowinski, J.L. Lions (editors)
North-Holland Publishing Company

A FLUID MODEL NUMERICAL CODE SYSTEM FOR TOKAMAK FUSION RESEARCH

M. AZUMI, G. KURITA, T. MATSUURA*, T. TAKEDA,
Y. TANAKA*, and T. TSUNEMATSU

Japan Atomic Energy Research Institute
Tokai, Ibaraki, Japan
(presented by T. TAKEDA)

Abstract

In this article we describe the project TRITON, the main objective of which is development of a fluid model numerical code system (TRITON) for the tokamak fusion research. Numerical methods and applications of constituent codes of the TRITON system are presented illustratively. Basic concept and concrete form of the system are also demonstrated with explanation of codes in supporting subsystems.

1. Introduction

In the tokamak fusion research role of computation becomes more and more important. Larger-scale and more complicated computer codes are required as large tokamaks are designed and more precise analyses of experiments are carried out. Moreover, because of the recent progress of computer science and technology efficient use of computer codes in this field is being prompted.

As necessary numerical codes become larger and more complicated, systematic development of the codes, manageability of them, and efficient way to utilize them are much more required. The reason is that duplication of efforts in development of codes, incorrect utilization of them should be avoided and revision of the codes should be carried out as easily as possible even by a different user from the original author.

Needs for the inclusive code system of this kind are understood more concretely from different viewpoint by the following description. If one would like to analyze the optimum beta value of a certain tokamak device comprehensively, an MHD equilibrium, linearized MHD stability, and transport codes are necessary. In the past it was rather difficult to treat consistently the problems of this kind, because the most convenient MHD equilibrium from the viewpoint of solution methods is, for example, assumed by neglecting self-consistency. It is also the case when one analyze time-evolutional nonlinear MHD phenomena or energy balances of a tokamak plasma.

Among various kinds of fusion codes fluid model numerical codes and relating ones are most easily consolidated into a code system and are expected to be utilized

---

* On loan from Fujitsu Co. Ltd., Shinkamata, Ohta-Ku, Tokyo, Japan

most efficiently because utilization technique of the codes are developed very well. On the basis of experience obtained during the last decade, therefore, we mapped out the project TRITON, which is composed of three main items, that is, (1) development of computer codes which analyze plasma behaviour on the basis of fluid model, (2) development of supporting codes for the development and efficient use of large-scale fusion codes, and (3) utilization of numerical codes in the TRITON system for practical problems.

The more important items among them are the first two, that is, developments of the fluid model numerical code system TRITON and supporting system for the development of numerical codes PEGASUS (Fig.1). The TRITON system is an inclusive fluid model numerical code system for the analyses of a tokamak plasma and the main part of the TRITON system is a subsystem ASTRAEA which consists of numerical codes for MHD stability analyses, multi-dimensional transport calculation and MHD equilibrium calculations. The TRITON system also contains subsystem PARIS (a subroutine package for large-scale numerical analyses) and ARGUS (an assembly of graphic I/O subroutines). The PEGASUS system is composed of assembly of supporting numerical codes, such as, automatic code generators written in a formula manipulation language "REDUCE-2", preprocessors for large-scale matrix handling and so on. Details of the PEGASUS system will be described elsewhere.

Section 2 describes the ASTRAEA subsystem in some detail with examples of application of the codes in this subsystem. Explanations on the supporting subsystems PARIS and ARGUS are presented in sections 3 and 4, respectively. The concept of the code system, supervisor code, and some utility packages are summarized in section 5.

2. Numerical codes for specific physical problems and their application

2-1. ASTRAEA

The ASTRAEA subsystem is the most essential part of the TRITON system. All the physics codes belong to this subsystem. The present constituent codes of this subsystem are summarized in Table 1.

In the following subsections explanation of the codes and their applications are demonstrated concerning three important problems of the fluid codes, i.e., the beta optimization, establishment of high beta states and analyses of the nonlinear MHD phenomena.

2-2. Equilibrium calculation and beta optimization in relation with linear ideal MHD stability

From the economical and technological point of view it is required for a fusion reactor to achieve as high beta value as possible. One of the most important phenomena which determine the limit of the beta value is the MHD property of the plasma. The simplest limit is obtained by the ideal and linear stability theory. The value may be too pessimistic because the nonlinear saturation of unstable modes or the kinetic stabilizing effects are not taken into account. On the other hand, the beta limit is reduced by the resistive mode. However, the analyses of ideal MHD equilibria and linear stabilities provide the good guidline to the achievable beta value of a fusion reactor.

We have been developing and implementing the following computer codes for the analyses of the ideal MHD equilibria and linear stabilities. In this section we make brief remarks on SELENE code (equilibrium code), JAERI version ERATO code (two-dimentional stability code) and BOREAS code (high mode number ballooning stability code).

<< SELENE >>

It is necessary to calculate equilibrium with high accuracy near a magnetic axis and in the region where magnetic surfaces are closely located, because conventional numerical codes for equilibrium calculation are formulated on the basis of square meshes in a r-z poloidal plane but stability calculation is carried out on square meshes in a ψ-θ plane[1]. To avoid this discrepancy we developed a new equilibrium code in which the structure of meshes are corrected during iterations so that the ψ value of a certain mesh is always same and the magnetic axis is always on a mesh point. The algorithm to solve the Grad-Shafranov equation is, in principle, the usual solution method of a partial differential equation by the finite element method. Up to now two kinds of fixed boundary versions which use above scheme are completed. One (SELENE20) solves the equation as a nonlinear eigenvalue problem[2] and the other (SELENE30) solve it under the flux-conserving (FCT) condition. A free boundary version which solves equation on a conventional mesh structure (SELENE40) is also completed and used.

Numerical Method

In our problem the functional L for the equilibrium equation is

$$L(\psi^{n+1}) = -\int_V rdrdz \, \{(\frac{1}{r}\frac{\partial\psi^{n+1}}{\partial r})^2 + (\frac{1}{r}\frac{\partial\psi^{n+1}}{\partial z})^2 - \frac{2}{r} f^n \psi^{n+1}\} \quad , \tag{1}$$

and the boundary condition is

$$\psi^{n+1} = 0 \qquad \text{on the plasma surface} \quad , \tag{2}$$

where the index n means the iteration step and f^n is the toroidal current distribution calculated by using $\psi^n(r,z)$. By using an appropriate set of basis functions, the functional L (eq.(1)) is represented by N parameters (ψ_1^{n+1},--------,ψ_N^{n+1}) where N is the number of the node points. Simultaneous linear equations with respect to ψ_i^{n+1} is immediately derived from eq.(1), that is,

$$A\psi^{n+1} = b^n \quad , \tag{3}$$

where ${}^{t}\psi^{n+1} = (\psi_1^{n+1}, \text{--------}, \psi_N^{n+1})$ and the matrix A is the finite element representation of the operator $\Delta^* \; (\equiv r \frac{\partial}{\partial r}\frac{1}{r}\frac{\partial}{\partial r} + \frac{\partial^2}{\partial z^2})$.

In SELENE20 f^n is determined so that ψ^n may be $-1 \le \psi^n \le 0$, that is,

$$f^n = -\lambda^n \mu_0 \; (r \frac{dp}{d\psi}(\psi^n) + \frac{1}{\mu_0 r}\frac{1}{2}\frac{d}{d\psi} T^2(\psi^n)) \quad , \tag{4}$$

and

$$\lambda^n = - \frac{1}{\psi_0{}^n} \lambda^{n-1} \quad . \tag{5}$$

The parameter λ^n corresponds to the eigenvalue of the Euler-Lagrange equation of the functional (1). The function $p(\psi)$ and $T(\psi)$ are given a priori or by the results of transport codes. In SELENE30 f^n is determined under the flux-conserving condition where the profile of $p(\psi)$ and the safety factor $(q(\psi))$ are given, that is,

$$f^n = -\mu_0 \; (r \frac{dp}{d\psi}(\psi^n) + \frac{1}{\mu_0 r}\frac{1}{2}\frac{dT^2}{d\psi}(\psi^n)) \quad , \tag{6}$$

and

$$T^n = 4\pi^2 q \; (\frac{d\psi}{dV})^n \frac{1}{A^n} \quad , \tag{7}$$

where $\frac{d\psi}{dv}$ is obtained by solving the surface-averaged Grad-Shafranov equation as usual FCT calculation[3] and $A^n = (\oint \frac{1}{r^2} \frac{d\ell}{B_p{}^n} / \oint \frac{d\ell}{B_p{}^n})_{\psi=\psi^n}$.

The overall iteration procedure is summarized as follows,

Box 1 : Prepare intinal meshes.

Box 2 : Calculate f^n.

Box 3 : Solve eq.(3).

Box 4 : Move mesh points on constant ψ lines.

Box 5 : If $|r_{axis}^{n+1} - r_{axis}^{n}| > \varepsilon$ or $|\lambda^{n+1} - \lambda^n| > \varepsilon'$ (SELENE20), or $|r_{axis}^{n+1} - r_{axis}^{n}| > \varepsilon$ or $|\frac{d\psi^{n+1}}{dV} \;\; \frac{d\psi^n}{dV}| > \varepsilon'$ (SELENE30), then return to Box 2.

Box 6 : End of iterations.

Numerical Results

For an example of calculations we show the result by using the following profile[4] and elliptic shape of the cross-section,

$$p = \beta_J p_0 \{(\psi-\psi_b) - \frac{\alpha}{2}[(\psi-\psi_m)^2 - (\psi_b-\psi_m)^2] + \frac{\gamma}{L+1}[(\psi-\psi_m)^{L+1} - (\psi_b-\psi_m)^{L+1}]\} \quad , \quad L \ge 4 \tag{8}$$

$$\gamma = \frac{\alpha(\psi_m-\psi_b) - 1}{(\psi_b-\psi_m)^L} \quad , \tag{9}$$

$$T \frac{dT}{d\psi} = R_0^2 \, (1/\beta_J - 1) \frac{dp}{d\psi} \quad , \tag{10}$$

where ψ_b and ψ_m are the ψ values at the plasma boundary and the magnetic axis, and α and β_J are the parameters which determine the width of the current column and the poloidal beta value, respectively. Figure 2 shows the result for $\alpha = 1.03$, $\beta_J = 2$, E = 1.7 and A = 3.3, where E and A are the ellipticity and the aspect ratio. Figure 3 shows the convergence of the eigenvalue (λ) and the safety factor at the magnetic axis (q_0). From this figure we find the quadratic convergence of λ versus 1/N, where $N = N_\psi = N_\theta$, and N_ψ and N_θ are the number of the division in the radial and poloidal direction. For the check of the accuracy, we compared the magnetic shear $S \equiv -\frac{d}{d\psi}[T\oint(1/r^2 B_p)d\ell]/(\frac{dV}{d\psi})^3$ between the analytic ($A = \infty$, E = 1) and the computed (A = 40, E = 1) cases for the parabolic current distribution (Fig.4). Both results coincide in very high accurracy.

<< ERATO >>

Fully two-dimensional linear and ideal MHD stability code such as ERATO[1)] developed at Lausanne and PEST[5)] developed at Princeton is the indispensable tool for the analyses of the MHD stability of an axisymmetric toroidal plasma. We implemented the ERATO code in the JAERI computer system as a code in the TRITON system and have been investigating MHD stability properties of various kinds of equilibria using the ERATO code. During the implementation we adapted the high speed I/O procedure to our FACOM 230-75 computer system by using assembly language and changing the file-organization. We also implemented the ERATO code in FACOM 230-75 APU (Array processor unit of FACOM 230-75) successfully and obtained the results by more than 3 times faster (see also section 3).

Numerical Results

For an expamle of the analyses by using ERATO code we show the dependence of stabilizing effect of a conducting shell on a poloidal beta value for the equilibrium given by eqs.(8)-(10) with E = 1 (circular cross-section) and A = 3.3. Figure 5 shows the critical beta values (β_c) versus the toroidal mode number n for various positions of the shell ($\Lambda \equiv a_w/a_p$, a_w and a_p are the minor radii of a shell and a plasma surface, respectively). The n = 1 mode is completely stabilized by a shell which is located at a practically possible position (Λ = 1.1~1.2) and the critical position of the shell becomes closer to the plasma surface with increasing β_p. The stabilizing effect on the n = 2 mode is remarkable for higher β_p when the shell is placed sufficiently close to the plasma but the shell far from the plasma surface has hardly an effect on the stability property of a higher β_p plasma.

<< BOREAS >>

ERATO code is useful for the analysis of the low mode number stabilities (n=1-5 n: the toroidal mode number). On the other hand, the code for the analysis of the high n mode is necessary. The criteria for the stability of the high n mode are the Mericier criterion[5] and one for the high-n ballooning mode[6]. The Mercier criterion is satisfied by choosing the appropriate value of safety factor at the magnetic axis (q_0). The criterion for the high-n ballooning mode gives the sufficient condition for the stability (Fig.5).

Numerical Method

We use the energy integral given by Connor et al[7]. By Fourier-transformation of the normal displacement $X \equiv rB_p\xi_\psi$ the potential energy is written as follows.

$$W = \pi \int d\psi W(\psi) \quad , \tag{11}$$

$$W(\psi) = \int_{-\infty}^{\infty} d\eta \{ f(\psi,\eta) \left| \frac{dX_0}{d\eta} \right|^2 + g(\psi,\eta) |X_0|^2 \} \quad , \tag{12}$$

$$f(\psi,\eta) = \frac{1}{Jr^2B_p^2} + \frac{r^2B_p^2}{JB^2} \left(\int_{-\infty}^{\eta} \frac{\partial \nu}{\partial \psi} d\eta \right)^2 \quad , \tag{13}$$

$$g(\psi,\eta) = -\frac{2J}{B^2}\frac{dp}{d\psi} \left[\frac{\partial}{\partial \psi} \left(\mu_0 p + \frac{B^2}{2} \right) - \frac{T}{B^2}\frac{1}{J}\frac{\partial}{\partial \eta} \left(\frac{B^2}{2} \right) \int_{-\infty}^{\eta} \frac{\partial \nu}{\partial \psi} d\eta \right] \quad , \tag{14}$$

$$B^2 = B_p^2 + \frac{T^2}{r^2} \quad , \tag{15}$$

$$\nu = \frac{TJ}{r^2} \quad , \tag{16}$$

where J is the Jacobian and X_0 is the Fourier coefficient of X. All quantities except $\int_{-\infty}^{\eta} \frac{\partial \nu}{\partial \psi} d\eta$ are periodic in 2π with respect to η.

By using the finite hybrid element method[1], the functional $W(\psi)$ (eq.(12)) becomes the quadratic form

$$\hat{W} = {}^tX\,A\,X \quad , \tag{17}$$

$${}^tX = (X_1, \text{--------}, X_N) \quad , \tag{18}$$

where X_i is the nodal value at a mesh point on a magnetic surface and N is the number of mesh points. The minimum value of $\hat{W}$ is given by the minimum eigenvalue (λ_{min}) of the matrix A. If λ_{min} is negative, the plasma is ballooning-unstable. The matrix A is a tridiagonal one so that the λ_{min} is obtained very fast by using the usual bisection method.

Numerical Results

The computed λ_{min} gives the pessimistic value as ERATO, but it remarkably does not seem to change the sign as $N \to \infty$[8]. The convergence property is shown in Fig.6.

Though this property has not been theoretically proved yet, we can determine whether the plasma is stable or unstable without extrapolation, if our conjecture is correct. As an application of BOREAS code, we tested several classes of equilibria,

Class 1 : given by eqs.(8)-(10),

$$\text{Class 2 : } p(\psi) = \left(\frac{\psi-\psi_b}{\psi_m-\psi_b}\right)^{\alpha} ,$$
$$T\frac{dT}{d\psi} = R_o^2\,(1/\beta_J-1)\,\frac{dp}{d\psi} , \tag{19}$$

$$\text{Class 3 : } p(\psi) = \frac{\alpha_1}{2}\psi^2 - \frac{\alpha_2}{3}\psi^3 ,$$
$$T\frac{dT}{d\psi} = \alpha_2 R_o^2 \psi^2 + \alpha_4 \psi^4 . \tag{20}$$

Results are shown in Fig.7.

2-3. Establishment of high beta equilibria

The 1-dimensional tokamak code is the powerful tool to study the temporal evalution of plasma parameters during the discharge, and various theoretical and numerical models have been developed to give better understandings of the tokamak discharge[9]. The basic assumption of this code is that the geometry of the magnetic field is constant in time. However, since the plasma equilibrium depends on the plasma parameters, it evolves with the plasma, and this evolution of the magnetic geometry affects the plasma transport. This process is especially important in a tokamak with finite β value and/or non-circular cross section. To solve this problem, the consistent treatment of the plasma transport and the MHD equilibrium is required[10]. The APOLLO is the 2(or 1 1/2)-dimensional tokamak code, which has been developed to study this reciprocal relation between the plasma transport and the MHD equilibrium. The other aspect of this code is to give the realistic (or non-intuitive) equilibrium configuration for the MHD stability analysis.

The basic structure of the APOLLO is the combination of the transport step and the equilibrium step and these two steps are solved alternately. In the transport step, plasma parameters are advanced in time for the fixed magnetic geometry. Using the fact that all transport processes along the magnetic field are dominant processes, the problem is reduced to the 1-dimensional problem. Choosing $\rho \equiv \sqrt{\chi/\pi B_o}$ (χ is the toroidal flux and B_o is constant) as the radial co-ordinate, the transport equations to be solved are

$$\frac{\partial}{\partial t}\left[ne\,\frac{\partial V}{\partial \rho}\right] + \frac{\partial}{\partial \rho}\left[\frac{\partial V}{\partial \rho}\,\Gamma_e\right] = S , \tag{21}$$

$$\frac{3}{2}\frac{\partial}{\partial t}[p_e(\frac{\partial V}{\partial \rho})^{5/3}] + (\frac{\partial V}{\partial \rho})^{2/3}\frac{\partial}{\partial \rho}\{\frac{\partial V}{\partial \rho}[q_e + \frac{5}{2}T_e\Gamma_e]\}$$

$$= (\frac{\partial V}{\partial \rho})^{5/3}\{Q^{ohm} + Q_e^{inj} - \frac{3m_e}{m_i}\frac{n_e}{\tau_e}(T_e - T_i)$$

$$- \frac{\Gamma_e}{n_e}\frac{dp_e}{d\rho} - \mu_i T_i A_{1i} - Q^{rad}\} \quad , \qquad (22)$$

$$\frac{3}{2}\frac{\partial}{\partial t}[p_i(\frac{\partial V}{\partial \rho})^{5/3}] + (\frac{\partial V}{\partial \rho})^{2/3}\frac{\partial}{\partial \rho}\{\frac{\partial V}{\partial \rho}[q_i + \frac{5}{2}T_i\Gamma_e]\}$$

$$= (\frac{\partial V}{\partial \rho})^{5/3}\{Q_i^{inj} + \frac{3m_e}{m_i}\frac{n_e}{\tau_e}(T_e - T_i)$$

$$+ \frac{\Gamma_e}{n_e}\frac{dp_e}{d\rho} + \mu_i T_i A_{1i} - Q^{cx}\} \quad , \qquad (23)$$

$$\frac{3}{2}\frac{\partial}{\partial t}B_{po} = \frac{\partial}{\partial \rho}\{\frac{\eta\, T}{\langle R^{-2}\rangle \frac{\partial V}{\partial \rho}}\frac{\partial}{\partial \rho}[\frac{\langle|\nabla V/R|^2\rangle}{\frac{\partial V}{\partial \rho}}B_{po}]\} \quad , \qquad (24)$$

where $V(\rho)$ is the volume contained by the ρ(or χ) surface, and the angular bracket $\langle\ \rangle$ denotes the surface averaged quantity. S is the particle source and the Q's represent energy source or loss due to ohmic heating, injection heating, radiation and charge exchange processes. The particle flux Γ_e and the energy flux $q_{e,i}$ depend on the transport model and on the geometric factors.

Numerical Method

This system of transport equations is solved by the finite element method and the implicit time-integrating scheme. The former makes it easy to introduce the variable mesh without reducing the accuracy, while the latter guarantees the numerical stability.

In the equilibrium step, the Grad-Shafranov equation

$$R^2\nabla \cdot \frac{\nabla\psi}{R^2} = -R^2\frac{dp}{d\psi} - T\frac{dT}{d\psi} \qquad (25)$$

is solved under the adiabatic constraints, which requires that following quantities must be conserved: $\mu(\psi) = p(\psi)(\frac{\partial V}{\partial \psi})^{5/3}$, $q(\psi) \equiv \rho B_o/R_o B_{po}(\psi) = T(\psi)\frac{\partial V}{\partial \psi}\frac{\langle R^{-2}\rangle}{(2\pi)^2}$ and ψ values at the magnetic axis and on the plasma surface. These constraints can be insured by solving the surface averaged equilibrium equation

$$\frac{d}{dV}[\langle|\frac{\nabla V}{R}|^2\rangle\frac{d\psi}{dV}] = -\frac{d}{d\psi}\mu(\frac{\partial \psi}{\partial V})^{5/3} - q\frac{\partial}{\partial V}[\frac{q}{\langle R^{-2}\rangle}\frac{\partial \psi}{\partial V}] \quad , \qquad (26)$$

for given boundary values of ψ. The combined system of eqs.(25) and (26) gives the solution by the iteration. Since we have to calculate several tens of equilibria in one transport simulation, the equilibrium solver with high speed and high accuracy is required. For this purpose, we developed two kinds of equilibrium modules. The one, SELENE30, solves the fixed boundary equilibrium. The basic procedure of this module is the same as SELENE20, which employ the mesh structure based on the constant ψ surface and solve Eqs.(25) and (26) by the finite element method. This module gives us the equilibrium with sufficient accuracy near the magnetic axis. The other module SELENE40, solves the free boundary equilibrium. The basic procedure of this module is the combination of the Buneman solver of the Poisson type equation with the method of the surface Green function, which gives the magnetic flux on the artificial boundary of the Buneman solver. Both equilibrium modules include the ballooning code and automatically checks the ballooning stability.

Numerical Results

Figure 8 shows an example of results of analyses of ballooning-stable equilibria. Results are obtained by using the APOLLO30 code which includes anomalous thermal diffusion due to the ballooning instability. Figure 9 shows the sequence of equilibria during the major radius compression. For simplicity, we have neglected all kinds of dissipations. This numerical result is in good agreement with the theoretical prediction.

2-4. Analyses of nonlinear MHD phenomena

The analyses of nonlinear MHD phenomena require the calculations in the fully tree-dimensional space even in the axisymmetric toroidal case, if we intend to solve the full sets of MHD equations. The three-dimensional calculation calls for huge amount of computational times. At the first step, therefore, we are developing the following reduced codes, that is,

AEOLUS-R : Code for the analyses of nonlinear resistive MHD phenomena by using the reduced set of equations in a fixed boundary[11].

AEOLUS-P : Code for the analyses of nonlinear phenomena of ideal positional (two-dimensional) instability by using the full set of equations in a free boundary.

<< AEOLUS-R >>

This code solves a system of reduced equations as an initial value problem. In the usual explicit time integration, the time increment (Δt) is limited by the relation $\Delta t \leq C(\Delta x)^2/D (C \lesssim 0.5)$, where Δx is the mesh size in space and D is the magnitude of the diffusion. We used the leap-frog method (explicit time integration) for MHD convective motion and Crank-Nicolson implicit method for resistive diffusion term to remove the restriction on Δt by the diffusion. Here we illustrate our method by using the simplified equation

$$\frac{\partial \psi}{\partial t} = D \frac{\partial^2 \psi}{\partial x^2} - v \frac{\partial \psi}{\partial x} \quad . \tag{27}$$

In our method eq.(27) is written in the following finite difference form

$$\frac{\psi_i^{n+1} - \psi_i^{n-1}}{2\Delta t} = D\left[\frac{\psi_{i+1}^{n+1} - 2\psi_i^{n+1} + \psi_{i-1}^{n+1}}{2\Delta x^2} + \frac{\psi_{i+1}^{n-1} - 2\psi_i^{n-1} + \psi_{i-1}^{n-1}}{2\Delta x^2}\right] - \frac{\psi_{i+1}^{n} - \psi_{i-1}^{n}}{2\Delta x} \tag{28}$$

The amplification factor (G) of eq.(28) is always less than unity for all values of D and $|v\Delta t/\Delta x| \leq 1$. Figure 1C shows the stable region where $G < 1$ for our method and the usual explicit method. Our method is similar to the DuFort-Frankel leap frog method[12]. The latter one, however, has the error of $O(\Delta t,(\Delta x)^2)$. On the other hand, the error in our method is one of $O((\Delta t)^2,(\Delta x)^2)$, though the implicit integration requires the inversion of a tridiagonal matrix. As a test case of the above scheme, resistive modes with single-helicity are obtained with a longer time step by comparison with a fully explicit code and CPU time reduced to 1/5 for S = 1000. Further development of multi-helicity codes including toroidal geometry is in progress by using above technique.

<< AEOLUS-P >>

Before a fully three-dimensional nonlinear code we are developing a two-dimensional axisymmetric one to investigate the nonlinear behavior of the tokamak with a poloidal divertor such as JT-60. The Arbitrary Lagrangian-Eulerian grid[13] is favorable for the free-boundary problem. We construct the meshes under the constraint where a Jacobian is spatially constant[14]. We represent the grid point in the polar coordinate (ρ,θ) in a poloidal plane, where ρ is the distance between the magnetic axis and the grid point. For convenience we introduce the parameters r and χ, that is,

$$\rho = \rho_o(\theta) f(r) \quad , \tag{29}$$

$$\theta = \theta(r,\chi) \quad , \tag{30}$$

$$0 \leq r \leq 1 \quad \text{and} \quad 0 \leq \theta \leq 2\pi \quad , \tag{31}$$

where f(r) is the monotone increasing function with f(0) = 0 and f(1) = 1. The positions $\rho = \rho_o(\theta)$ give a plasma surface which is deformed by the perturbation and the function $\rho_o(\theta)$ is obtained by dividing the poloidal cross-section in equal area. Using the relation (29)-(31), the constraint that the Jacobian is spatially constant gives the differential equation

$$\rho_o(\theta)\,\frac{d\theta}{d\chi} = C/\left(\frac{df}{dr}\right) \quad , \tag{32}$$

where

$$C = \frac{1}{2\pi} \int_0^{2\pi} \rho_o(\theta) d\theta \left.\frac{df}{dr}\right|_{r=1} \quad . \tag{33}$$

Eqs.(29) and (32) give the mesh points (ρ_j, θ_j) for the equi-intervals of Δr and $\Delta\chi$. The simplest choice of f(r) is f(r) = r where θ depends only χ. We show a several grid configurations in Fig.11 determined by the above procedure.

The ideal MHD equations are expressed in a weak form by the finite element method and the resultant equations are in conservative form. All equations are solved in explicit scheme except the equation of a total pressure to remove the fast wave phenomena which limits the maximum time step for the explicit scheme[14).

3. Supporting subsystem PARIS

When one develops and uses such large-scale computer codes as in the ASTRAEA subsystem, which also includes a new kind of numerical problems, speed-up of computation and efficient usage of main memory of a computer should be, first, considered. Consequently, development of new algorithms and substantial improvement of programming technique are required. Usual SSL's (Scientific Subroutine Library) which are provided in a standard configuration of a computer system could not meet our requirements.

From this point of view we are developing the PARIS subsystem for development of numerical codes in the ASTRAEA. Requirements for the PARIS are summarized in Table 2.

3-1. ATLAS

We are often required to handle such a large-scale matrix that all the matrix elements cannot be stored on the main memory of a computer even if it is a band matrix, and, consequently, efficient I/O algorithms should be adopted. In the field of the structural analyses various routines are developed which could handle such a large matrix by using external memories efficiently. Usually, however, it is rather difficult to extract such matrix handling routines from complete codes of the structural analyses and use them separately for the MHD problems. Moreover, matrices appearing in the MHD problems are usually more dense by compared with those in the structural analyses, which is the consequence of the fact that couplings of different degrees of freedom are stronger in the MHD problem than the structural analyses.

From this viewpoint the first version of a subroutine package ATLAS[15)] (Assembly of The Large-scale Array-handling Subroutines) was developed to handle large-scale matrices which appear in developing and using the fusion codes in the ASTRAEA. The package is composed of three kinds of subroutine groups, i.e., basic arithmetic subroutines (ATLAS-A), subroutines for solving linear simultaneous equations (ATLAS-L), and subroutines for solving general eigenvalue problems (ATLAS-E).

3-2. Optimization of codes with respect to I/O operations and for an array processor

In the case of a code with a large-scale matrix handling optimization of the code with respect to I/O operations are very important. We are improving computer codes by rewriting the basic I/O routines to make the USE time of the computer code sufficiently short. For the ERATO code we optimized the I/O operation and the speed-up of about factor 4 in the I/O time was attained.

To process a job which deals a large-scale linear algebra, vector computers (array processors) begin to be used. In order to demonstrate the effectiveness of the vector computers enough, however, it is required to optimize codes for vector operations. We are begining to optimize some fusion codes for an array processor (FACOM 230-75 APU). The following items are important for efficient use of vector computers:

i) Rewriting of Fortran source program in vector-operation-oriented Fortran language.

ii) Separation of arithmetic operation blocks and I/O operation blocks.

iii) Separation of scalar operation blocks and vector operation blocks.

iv) Efficient use of "vector temporaries" (vector registors) for temporarily obtained results.

Among the above items, ii)~iv) are important in the case of FACOM 230-75 APU because of its hardware architecture. For the ERATO calculation 3-4 times faster speed has been already attained by compared with calculations by the conventional FACOM 230-75 CPU.

4. Supporting subsystem ARGUS

In the process of development and utilization of large-scale computer codes it is very effective to use graphic I/O functions of a computer system. When one analyze a nonlinear MHD phenomenon in a 3-dimensional space, for example, a good graphic I/O system is very helpful and even indispensable for initial value setting and display of results. Because of recent progress of the computer technology it becomes possible for users to write a computer code including a graphic output being insensible what kind of output device is used.

The ARGUS subsystem of the TRITON is developed to offer a mean to write easily fusion codes which include both graphic input and output processes. This subsystem is subdivided into three groups, i.e., ARGUS-V (a basic graphic output subsystem), ARGUS-W (an advanced graphic output subsystem which includes an option to produce a motion picture), and ARGUS-I (a graphic input subsystem). Presently, the first version of ARGUS-V (ARGUS-V1) was developed and used efficiently for development and utilization of the TRITON codes.

The basic idea of the ARGUS-V is to define attributes of input data clearly and classify them definitely into two groups, i.e., a graphic data set (GDS) and control data set (CDS). When one plot data on a sheet it is, at least, necessary to convert the data into actual lengths on the sheet. This process is carried out in the GDS subroutines. Users are not required to take care how the GDS data are stored in a memory of a computer. Allocation of the memory location and disk files are automatically carried out by the ARGUS subsystem. Usually, a frame of the figure, scales of axes, comments, symbols on curves and so on are, hopefully, added to complete the figure. Data necessary to plot these informations are included in the CDS. Therefore, if one prepares a set of default values for the CDS (actually these are prepared in the ARGUS-V1), one can plot a simplified figure very easily during the course of the code development, only by providing a GDS by calling a GDS subroutine. If one would like to plot a more refined figure after the development of the code is finished, one should only add some appropriate CDS. Actually if one is satisfied by a figure which contains minimum informations one is required to call only 2 subroutines, that is, a GDS subroutine and Display subroutine, even when one would like to plot a very complicated figure such as a projection of a three dimensional object on a sheet.

In order to demonstrate the easiness of plotting an example is shown in Fig.12 with corresponding Fortran source program.

5. Supervisor of the code system and utility program packages

The TRITON system is not a conventional closed modular code system but a growing open system which has no definite boundaries. It is a natural consequence of the fact that production of new fusion codes are progressing and they are being improved rather constantly. It is very difficult to maintain such codes so that improvement and utilization of them may be always manageable. In order to overcome this difficulty we applied the following concept of a code system to the TRITON. By this concept rules which define relations between constituent codes are first prepared, and actual codes and relations between them are put into the system according to the rule. This is the marked difference between this system and conventional modular code systems in which a set of constituent codes are first prepared, and relations between the elements of the system are looked for by considering the structure of original codes.

To realize this kind of code systems we are developing three types of supporting codes for manageable maintenance and easy development of numerical codes.

5-1. HARMONIA

HARMONIA is the supervisor code of the TRITON system. This code manages a set of complete programs, subprograms, job control programs, series of input data, and series of output data. Each element of the set is related, doubly, to other elements by a longitudinal tree structure (a family line of the codes, see Fig.13) and by horizontal functional relations (Fig.14).

When one would like to analyze a certain phenomenon by using some computer codes in the TRITON, one can find most appropriate codes by sorting codes in the longitudinal tree structure and after that retrieve a group of elements (programs, job control programs, input data, and so on) which are connected each other by the horizontal functional relation. Relations between the elements and attribute of the elements are stored in INDEX FILE defined on a magnetic disk and the elements of the set are stored in magnetic tapes serially (Fig.15).

5-2. PLUTO

To facilitate maintenance, improvement, and adaptation of codes PLUTO is being developed. The main functions of the PLUTO are semi-automatic documentation of a code, and extraction of necessary parts of a large-scale computer code. By using the former function of the PLUTO exchange of computer codes with other researchers becomes very easy and by using the latter function one can write a new code easily by reconstructing parts obtained from other codes.

5-3. EOS

EOS is the code which facilitate development of a new code. Originally the EOS was designed to be similar to the Olympus System[16]. Present version of the EOS is, however, different from the Olympus System in that the EOS has no rule which defines a program structure. The EOS code has functions of insertion and extraction of COMMON blocks and it has extended Olympus utility programs which mainly facilitate programming of line-printer outputs (the BELLEROPHON package).

Acknowledgments

The authors would like to thank Dr. Masatoshi Tanaka for valuable discussions and continuing encouragements. They are also grateful to Drs. Sigeru Mori and Yukio Obata for their encouragements.

References

1) R. Gruber, "ERATO stability code", submitted to Comput. Phys. Commun.
2) T. Takeda and T. Tsunematsu, "A numerical code SELENE to calculate axisymmetric toroidal MHD equilibria", JAERI-M 8042 (1979).
3) J.F. Clarke, "High beta flux-conserving tokamaks", ORNL/TM-5429 (1976).
4) Y-K. M. Peng, R.A. Dory, D.B. Nelson, and R.O. Sayer, Phys. Fluids 21 (1978) 467.
5) R.C. Grimm, J.M. Greene, and J.L. Johnson, "Computation of the magnetohydrodynamic spectrum in axisymmetric toroidal confinement system", in Methods in

Computational Physics, Vol.16, (Academic Press N.Y., 1976).
6) C. Mercier, Nucl. Fusion 1 (1960) 47.
7) J.M. Connor, R.J. Hastie, and J.B. Taylor, Phys. Rev. Lett. 6 (1978) 396.
8) T. Takeda and T. Tsunematsu, "The high-n ballooning stability code BOREAS", to appear in JAERI-M.
9) J.T. Hogan, "Multi-fluid tokamak transport model", in Methods in Computational Physics, Vol.16, (Academic Press, N.Y., 1976).
10) H. Grad, "Survey of 1-1/2 D transport codes", COO-3077-154, MF-93 (1978).
11) H.R. Hicks, B. Carreras, J.A. Holmes, and B.V. Waddel, "Interaction of tearing modes of different pitch in cylindrical geometry", ORNL/TM-6096 (1977).
12) P.J. Roache, Computational Fluid Dynamics, (Hermosa Publishers, New Mexico, 1976).
13) J.V. Brackbill and W.E. Pracht, J. Comp. Phys. 13 (1973) 455.
14) S.C. Jardin, J.L. Johnson, J.M. Greene, and R.C. Grimm, J. Comp. Phys. 29 (1978) 101.
15) T. Tsunematsu, T. Takeda, K. Fujita, T. Matsuura, and N. Tahara, "Large-scale matrix-handling subroutines ATLAS", JAERI-M 7573 (1978) (in Japanese).
16) K.V. Roberts, Comput. Phys. Commun. 7 (1974) 237.

Table 1: The present constituent codes of the ASTRAEA subsystem.

SELENE	MHD equilibrium analysis
NOTUS ERATO	Linear ideal MHD stability analysis for the low wave number modes
BOREAS	Ballooning mode analysis for the high wave number modes
AEOLUS	Nonlinear MHD behavior analysis
APOLLO	2-dimensional tokamak transport analysis

Table 2: Requirements for the PARIS.

Problems	Requirements
Eigenvalue Problems of Large-Scale Matrices	High speed (CPU time, I/O time) Efficient use of memories
Large-Scale Simultaneous Linear Equations	High speed (CPU time, I/O time) Efficient use of memories
FFT	High speed (CPU time) Possibility to choose sample points arbitrarily
Interpolation and Smoothing	2D and 3D codes
Usage of Vector Processor	Development of technique to use efficiently a vector processor

Project TRITON

\# TRITON

(The Reliable system of Inclusive TOkamak Numerical codes)

* ASTRAEA

  (Assembly of numerical codes on Stability, TRAnsport, and Equilibrium of Axisymmetric toroidal plasmas)

* PARIS

  (Package of ARIthmetic Subroutines)

* ARGUS

  (All-Round Graphic Utility Subroutines)

\# PEGASUS

(Programming languages for formula Expansion and code Generation: A SUpporting System)

Fig. 1: TRITON and a supporting system.

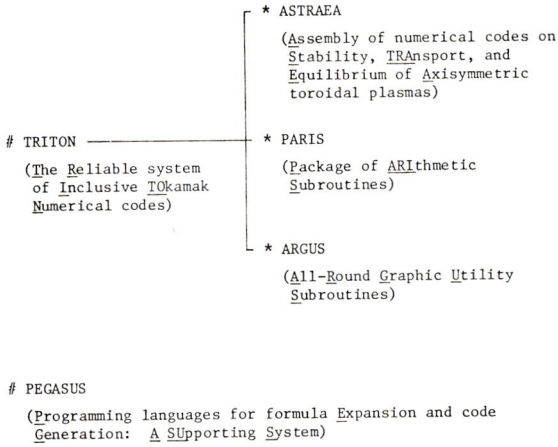

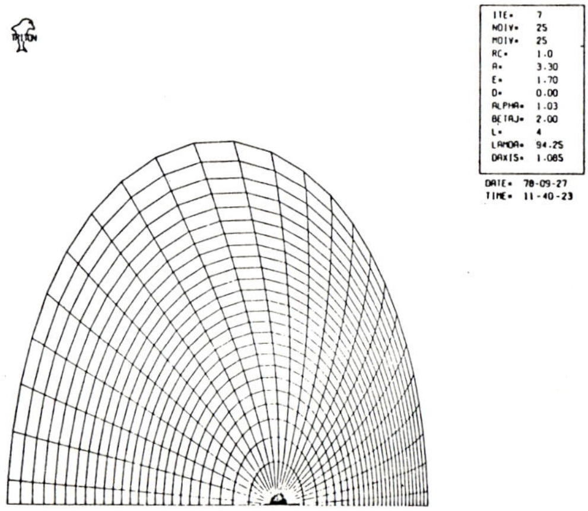

Fig. 2: Computed magnetic surfaces for α=1.03, β_J=1.7, and A=3.3. Final mesh points construct the magnetic surfaces (SELENE20).

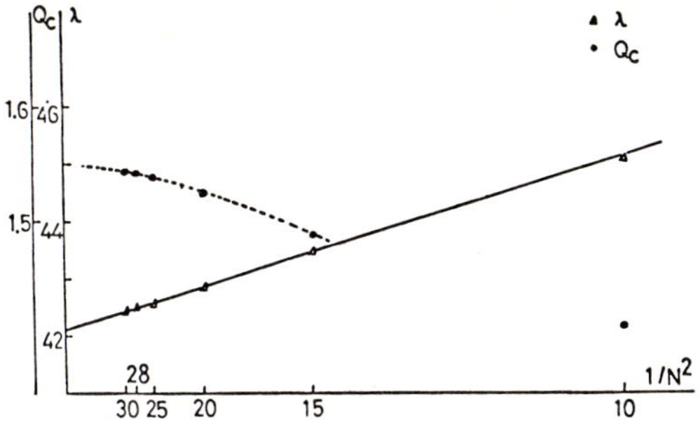

Fig. 3: Convergence of the eigenvalue λ and safety factor at the magnetic axis (q_o), (SELENE20).

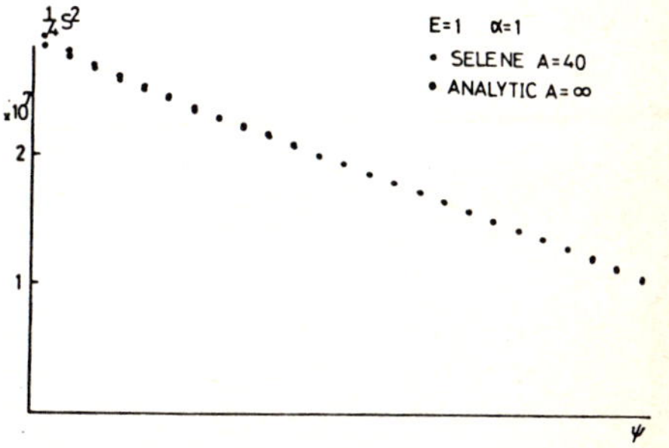

Fig. 4: Magnetic shears of a large aspect ratio plasma (numerical) and a cylindrical plasma (analytical), (SELENE20).

Λ	∞	1.25	1.20	1.0
$\beta_J = 1.0$	o	□		◇
$\beta_J = 1.65$			▼	◆

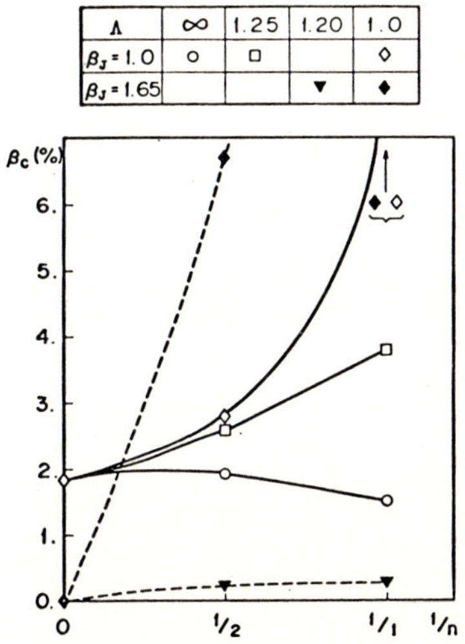

Fig. 5: Critical beta values versus 1/n (ERATO and BOREAS20).

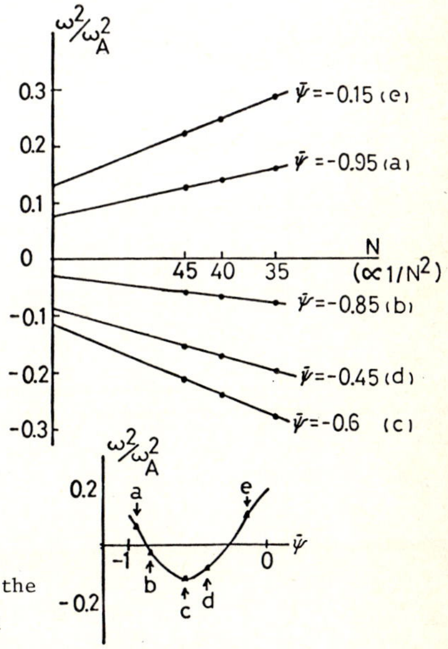

Fig. 6: Convergence property of the minimum eigenvalues λ_{min} (BOREAS20).

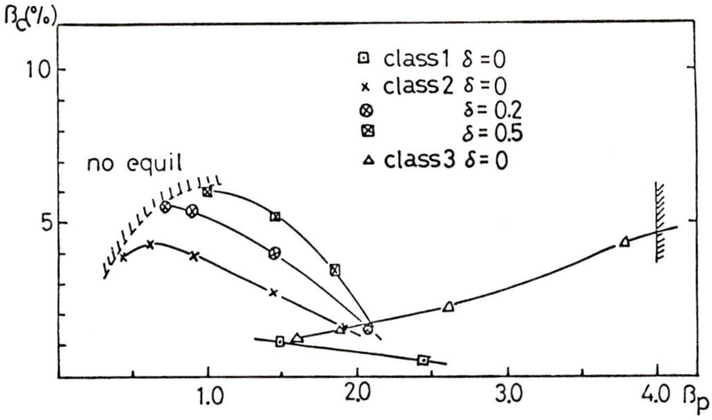

Fig. 7: Critical beta values versus β_p and A=4.2. δ denotes the triangularity (BOREAS20).

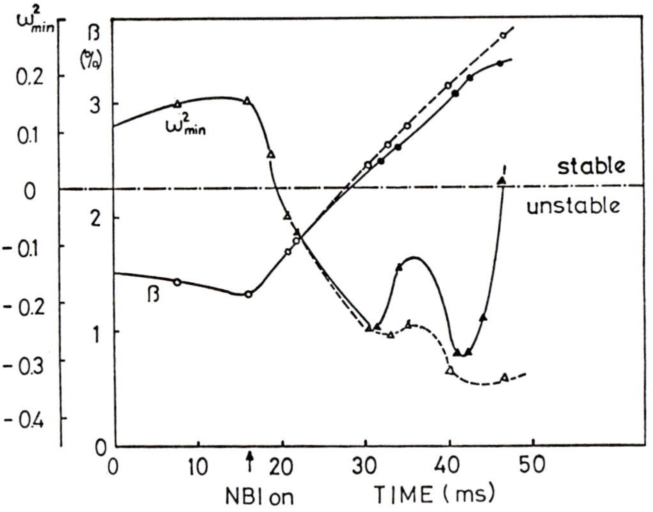

Fig. 8: Examples of equilibria obtained with and without enhanced themal diffusion due to ballooning mode (APOLLO30).

(a) Time evolution of beta value and ω^2_{min} for a bollooning mode.

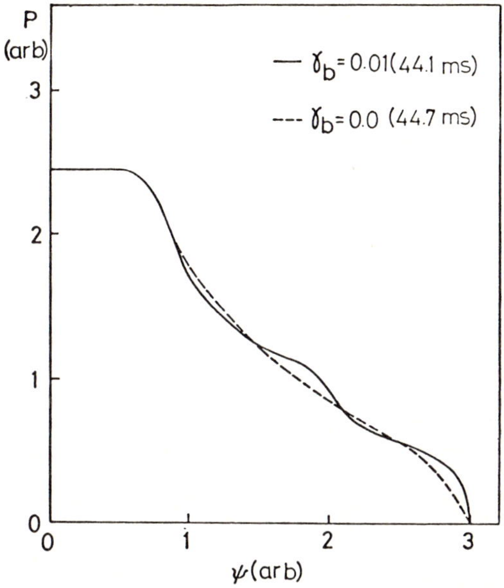

Fig.8 (b) Pressure profiles.
Solid line: a case with diffusion due to ballooning mode.
Broken line: a case without diffusion due to ballooning mode.

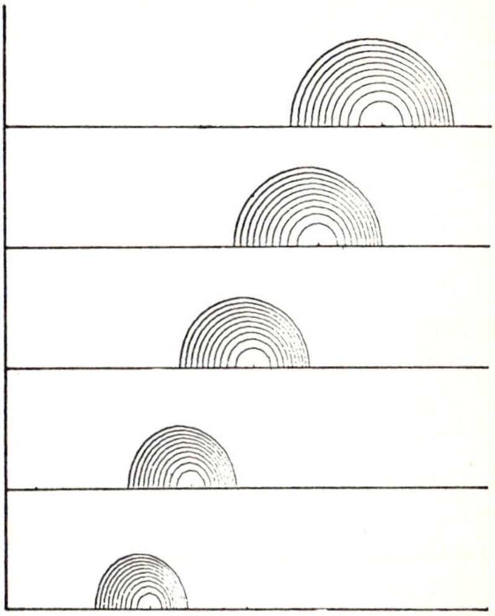

Fig. 9: Sequence of equilibria during the major radius compression (APOLLO40).

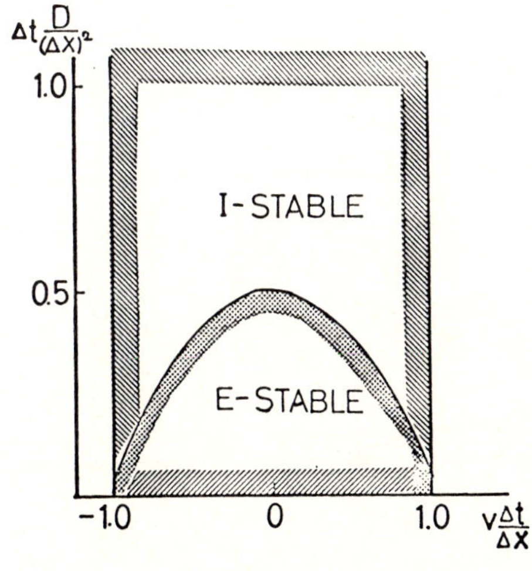

Fig.10: The region of G<1 (AEOLUS-R).

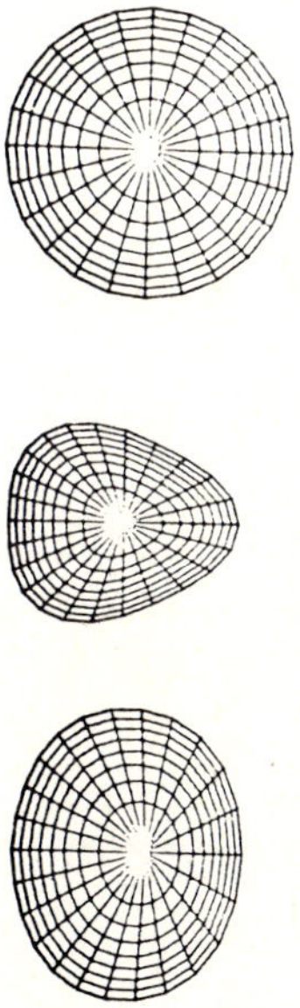

Fig.11: Grid configurations (AEOLUS-P).

Fig.12 : An example of ARGUS output.

(a) : [1]+[5]

(b) : [1]+[2]+[5]

(c) : [1]+[2]+[3]+[5]

(d) : [1]+[2]+[3]+[4]+[5]

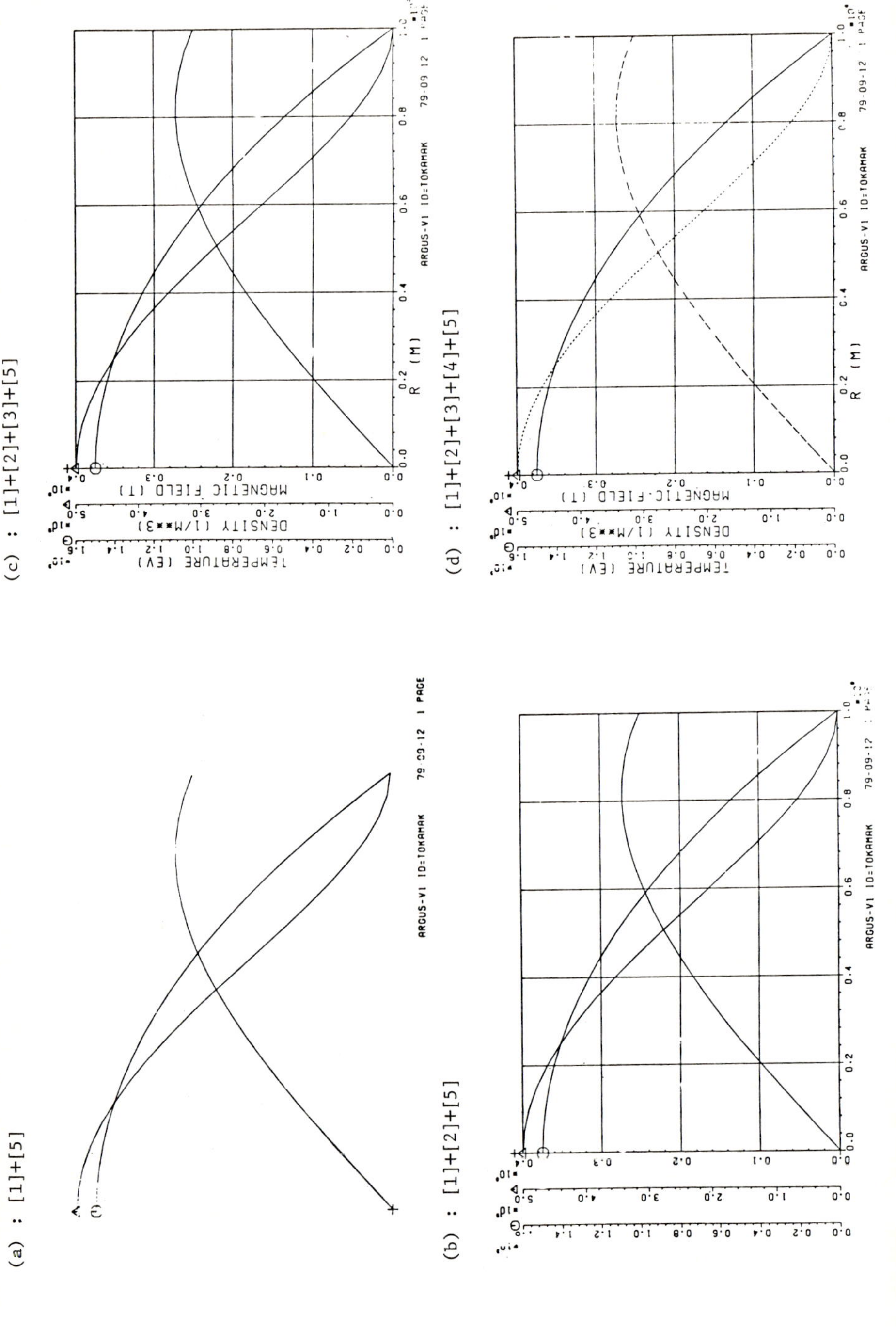

```
COMMENT     ARGUS
COMMENT
COMMENT
COMMENT     *** INITIALIZATION ***
      CALL ARGUS1(0,IDUM)
COMMENT
COMMENT     *** GDS SUBROUTINE ***                                   :
      CALL STOR1D('TOKAMAK ',N1,-3,R,0.0,T,N1)                       : [1]
      CALL STOR1D('TOKAMAK ',N1,-3,R,0.0,D,N1)                       :
      CALL STOR1D('TOKAMAK ',N1,-3,R,0.0,B,N1)                       :
COMMENT
COMMENT     *** CDS SUBROUTINES (1) ***                              :
      CALL FRAME('TOKAMAK ',1,1,1)                                   : [2]
      CALL GRID ('TOKAMAK ',1,5,5)                                   :
COMMENT
COMMENT     *** CDS SUBROUTINES (2) ***                              :
      CALL COMENT('TOKAMAK ','X    ',1,'R  (M)',6,0.0,0.0,0.0,0.0)    : [3]
      CALL COMENT('TOKAMAK ','Y1   ',2,                              :
     1     'TEMPERATURE (EV)',16,0.0,0.0,0.0,0.0)                    :
      CALL COMENT('TOKAMAK ','Y2   ',3,                              :
     1     'DENSITY (1/M**3)',16,0.0,0.0,0.0,0.0)                    :
      CALL COMENT('TOKAMAK ','Y3   ',4,                              :
     1     'MAGNETIC FIELD (T)',18,0.0,0.0,0.0,0.0)                  :
COMMENT
COMMENT     *** CDS SUBROUTINES (3) ***                              :
      IL(1)=1                                                        : [4]
      IL(2)=3                                                        :
      IL(3)=5                                                        :
      CALL CURVE('TOKAMAK ',IL)                                      :
COMMENT
COMMENT     *** DISPLAY SUBROUTINE ***                               :
      CALL ONEDIM('TOKAMAK ')                                        : [5]
      STOP                                                           :
      END
```

Fig.12 (e) Fortran program for Fig.12 (a)-(d).

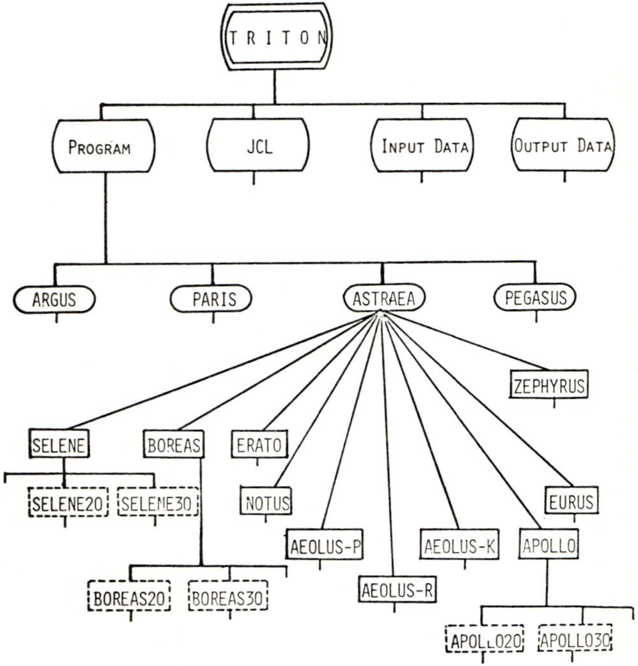

Fig.13: An example of a tree structure in the HARMONIA.

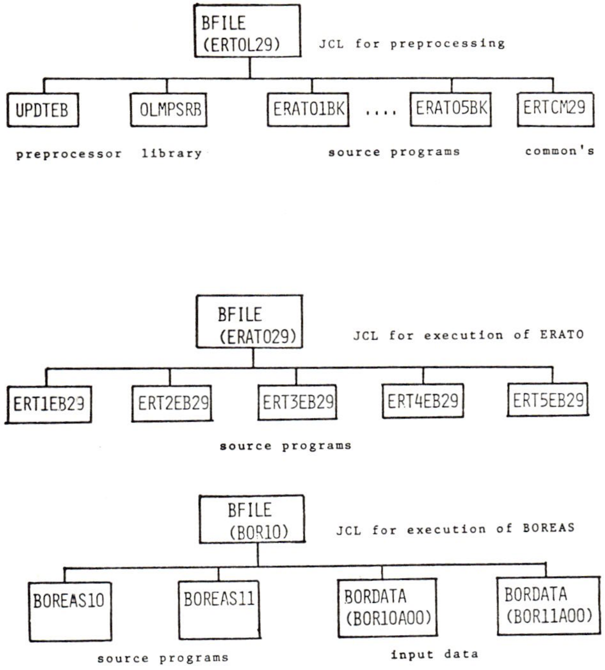

Fig.14: An example of a horizontal functional relation in the HARMONIA.

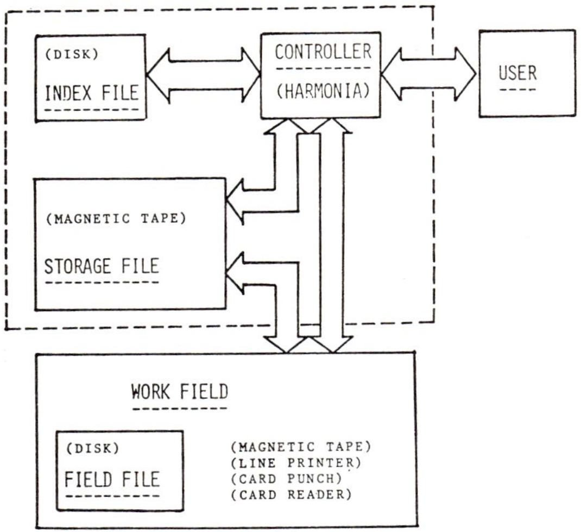

Fig.15: The schematic diagram of the HARMONIA.

SESSION V

FREE BOUNDARY PROBLEMS

COMPUTING METHODS IN APPLIED SCIENCES AND ENGINEERING
R. Glowinski, J.L. Lions (editors)
North-Holland Publishing Company

FINITE ELEMENT METHOD
FOR THE WATER FLOODING PROBLEM

-=-=-=-

Guy Chavent
IRIA
Domaine de Voluceau
F-78150 Le Chesnay

-=-=-

INTRODUCTION

The purpose of this paper is to present the research work made at IRIA on the numerical resolution, by finite elements, of the non linear diffusion convection equation governing the displacement of oil by water in porous media (water flooding techniques for the secondary recovery of oil).

A very early attempt to use variational methods for the resolution of these equations was made by Douglas and al [1] . Since, various authors (Vermvelen [2], Verner and al [3], Sanchez[4] , Douglas and al [5], Douglas [7], for instance) have tried several finite element methods (one could also mention all the unpublished work made by the Intercomp Company for the oil companies). However the superiority of the Finite Element Method (F.E.M.) against finite differences has not yet been established for the simulation of water flooding problems.

The method presented here was introduced first by Chavent and al. [7] for the 1-D case, and developed for the 2-D case by J. Jaffre [8][9]. Though we do not claim such a general superiority against the finite differences for our F.E.M., it seems to present some specific advantages which can make it useful :

- it is based on a mathematical formulation where each physical effect (capillary diffusion, gravity, capillarity pressure, heterogeneity for instance), appear in separate terms and hence can be discritized accordingly to their mathematical nature (diffusion, transport, etc..).
- the use of discontinuous finite elements for the first order term (according to the work of Lesaint and Raviart [10] simultaneously with mixed finite elements for the approximation of the diffusion term (according to the work of Raviart-Thomas [11], gives approximations of all physically important quantities (water

saturation, oil and water flows, pressure) which satisfy exactly the mass balance.

- the use of the work of Leroux [12] for the discretization and of Leroux-Bardos-Nedelec [13] for the boundary conditions of the *non-monotonous* transport equation makes it possible to give a comprehensive discretization of the transport term and of the flux boundary conditions in the cases where gravity and/or capillarity pressure heterogeneity, are present (counter-flow problems).
- for one-dimensional problems, the implementation of the method is easy and the computation time, for a given accuracy, less than for finite difference.
- for two-dimensional problems (no three-dimensional test has yet been performed), the above advantages are counter-balanced by a heavier implementation and larger(*) computation times than for finite differences, but no definitive conclusion can be given, as the 2-D experiment program is not yet completed.

1. THE GENERAL EQUATIONS OF TWO-PHASE FLOWS IN POROUS MEDIA

We refer to Chavent [14] [15] for the establishment and the mathematical study of the following equations :

1 = index for water

2 = index for oil

$\Omega \subset \mathbb{R}^n$ = porous medium

$\Gamma = \partial\Omega$ = boundary of porous medium

P = global pressure (joint function of P_1, P_2 and S)

S = reduced water saturation

$\vec{\varphi}_1$ = water-flow vector

$\vec{\varphi}_2$ = oil-flow vector

$\vec{r}$ = part of $\frac{\vec{\varphi}_1 - \vec{\varphi}_2}{2}$ due to capillarity diffusion

$\vec{q}_o = \frac{1}{2}(\vec{\varphi}_1 + \vec{\varphi}_2)$ = mean water + oil flow vector

$\Gamma = \Gamma_D \cup \Gamma_N$ with $\Gamma_D \cap \Gamma_N = \emptyset$ where (cf. figure) :

Γ_D= part of Γ where the <u>pressure is specified</u> (D stands for Dirichlet)

Γ_N= part of Γ where the <u>overall flow is specified</u> (N stands for Neumann)

$\Gamma = \Gamma_- \cup \Gamma_+$ with $\Gamma_- \cap \Gamma_+ = \emptyset$ where (cf. figure) :

$\Gamma_- = \{s \in \Gamma \mid Q = \vec{q}_o \cdot \vec{\nu} \leq 0\}$ - <u>overall injection</u> boundary

$\Gamma_+ = \{s \in \Gamma \mid Q = \vec{q}_o \cdot \vec{\nu} > 0\}$ = <u>overall production</u> boundary

(*) on the basis of the same number of unknowns for the saturation u.

Equations governing the global pressure P for every $t \in [o,T]$:

. inside Ω:

(1) $\operatorname{div} \vec{q}_o = 0$

(2) $\vec{q}_o = -\psi d(S) \operatorname{grad} P + d(S) \sum_{j=1}^{2} \gamma_j(S)\, \vec{q}_j$

. on the boundary Γ: we use the partition $\Gamma = \Gamma_D \cup \Gamma_N$:

(3) $P = P_e$ on Γ_D (Dirichlet)

(4) $\vec{q}_o \cdot \vec{\nu} = Q$ on Γ_N (Neuman)

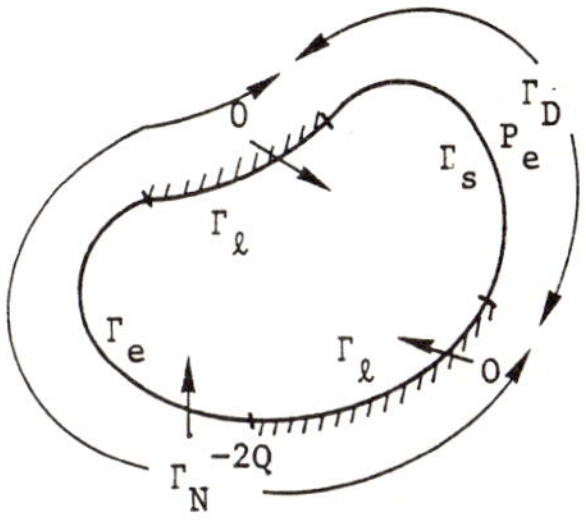

Example of pressure boundary conditions

Equations governing the saturation S :

. inside $Q = \Omega \times]0T[$:

(5) $\varphi \frac{\partial S}{\partial t} + \operatorname{div} \{\vec{r} + \sum_{j=0}^{2} b_j(S)\, \vec{q}_j\} = 0$

(6) $\vec{r} = -\psi P_{cM} \operatorname{grad} \alpha(S)$

. on the boundary Γ: we use the partition $\Gamma = \Gamma_- \cup \Gamma_+$:

- on the overall injection boundary Γ_- one can take :

(7) $S = S_e$ (Dirichlet)

or

(8) $\vec{\varphi}_1 \cdot \vec{\nu} = Q_1$ (Neuman)

- on the overall production boundary Γ_+ one can take :

(9) $S = S_e$ (Dirichlet)

or

(10) $S \leq 1 \;,\; \vec{\varphi}_1 \cdot \vec{\nu} \geq 0 \;,(1-S)\vec{\varphi}_1 \cdot \vec{\nu} = 0$ (unilateral)

. at the initial time $t = 0$

(11) $S = S_o(x)$ on Ω

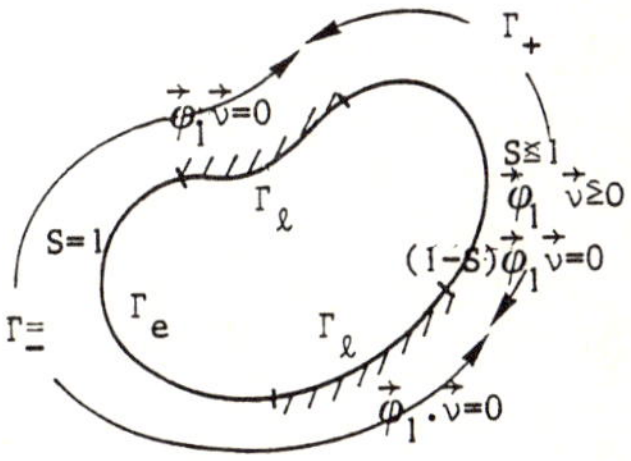

Example of saturation boundary conditions compatible with the pressure conditions of fig. 19 when P_e is constant over Γ_D

where :

$\psi(x)$, $\varphi(x)$ are given functions of space related to permeability and porosity of rock;

$P_{CM}(x)$ is the given maximum capillarity pressure at point x ;

$\vec{q}_1(x)$, $\vec{q}_2(x)$ are given vector field related to capillarity pressure heterogeneity and to gravity ;

$d(S)$ $b_j(s)$, $j=0,1,2$, $\alpha(S)$ are given functions of saturation of the forms :

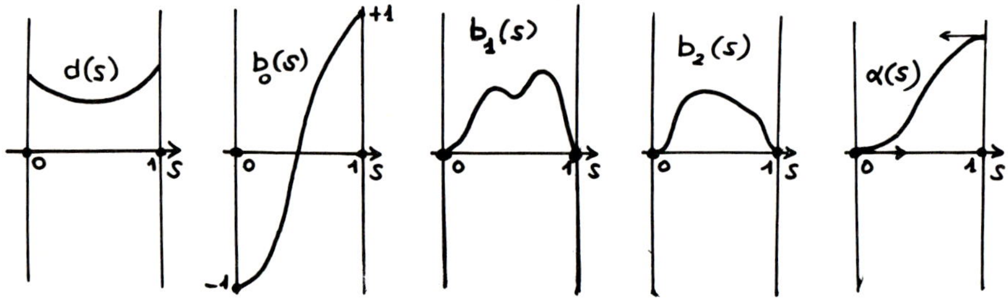

P_e, S_e, Q, Q_i are given boundary data

$S_o(x)$ is the given initial saturation

The equations for separate phase pressure and flows are then :

Equations for separate phase pressures and flows :

$$P_1 = P - [\gamma(S) - \frac{1}{2} p_c(S)] P_{CM}$$

$$P_2 = P - [\gamma(S) + \frac{1}{2} p_c(S)] P_{CM}$$

$$\vec{\varphi}_1 = \vec{q}_o + \vec{r} + \sum_{j=0}^{2} b_j(S)\, \vec{q}_j = (1 + b_o(S))\vec{q}_o + \vec{r} + \sum_{j=1}^{2} b_j(S)\, \vec{q}_j$$

$$\vec{\varphi}_2 = \vec{q}_o - \vec{r} - \sum_{j=0}^{2} b_j(S)\, \vec{q}_j = (1 - b_o(S))\vec{q}_o - \vec{r} - \sum_{j=1}^{2} b_j(S)\, \vec{q}_j$$

One sees that the transport term in (5) can be non-monotonous when gravity or capillarity pressure heterogeneity are present.

2. THE ONE-DIMENSIONAL CASE WITH GRAVITY

In this case, the above system reduces to

(12) $$\varphi \frac{\partial S}{\partial t} + \frac{\partial r}{\partial x} + \frac{\partial}{\partial x} f(S) = 0 \qquad \text{on }]0L[\times]0T[$$

(13) $$r = -\psi P_{CM} \operatorname{grad} \alpha(S) \qquad \text{on }]0L[\times]0T[$$

where f(S) is given by

$$f(S) = q_o\, b_o(S) + q_2\, b_2(S) \qquad q_o , q_2 \in \mathbb{R}$$

and is not necessarily monotonous.

This case has been studied by G. Salzano (cf.[16]) with various boundary and initial conditions.

The main feature of this approximation is that it uses the generalization to non-monotonous function f of the "upward differentiating" of the first order term $\frac{\partial}{\partial x} f(S)$ introduced by Godunov (cf.[17]) and studied by Leroux [12] in the case of finite differences.

For the transport equation ($\alpha \equiv 0$ hence $r \equiv 0$) the scheme uses a discretization of the Dirichlet boundary conditions for non monotonous transport equations introduced by Leroux, Bardos and Nedelec [13] :

$$S_{gn}(\gamma(S)-k)\ [f(\gamma(S)-f(k)] \leq 0 \qquad \forall\ k \in I\ [(S),\ S_e]$$

where

$\gamma(S)$ is the trace of S on boundary Γ of Ω

$$\begin{cases} I(a,b) = \begin{cases} [a,b] & \text{if } a \leq b \\ [b,a] & \text{if } b \leq a \end{cases} \\ S_{gn}(a) = \begin{cases} +1 & \text{if } a > 0 \\ 0 & \text{if } a = 0 \\ -1 & \text{if } a < 0 \end{cases} \end{cases}$$

These conditions do not necessarily require the equality of $\gamma(S)$ *and* S_e *!*

For a simple function f the exact solution of (12) may be calculated as reference. We show in Figures 3 and 4 some numerical results for the effect of gravity on oil and water in a vertical core sample under hydrostatic pressure equilibrium ($q_o = 0$) for a two-case of boundary conditions.

When diffusion is present ($\alpha \equiv 0$) the scheme can take into account either the usual Dirichlet boundary conditions or oil or water flow conditions. The discretization of these latter conditions is not obvious, but a careful study shows that it resumes numerically to the inversion of a monotonous function, and so, is always well posed. The influence of the capillarity diffusion on the equilibrium saturation profiles in a vertical slab under hydrostatic pressure equilibrium ($q_o=0$) with closed extremities, is shown on Figure 5. We refer for more details to the report[16] by Salzano.

3. THE ONE-DIMENSIONAL CASE : SPACE-TIME FINITE ELEMENTS

In order to reduce the computation time, we tried to use some implicit schemes for the resolution of equation (12). As the first order term was predeterminant, space and time played very similar roles, and we used space-time rectangular discontinuous finite elements, using the same guidelines as mentioned in the Introduction.

This study was made for the monotonous transport equation (no diffusion). In that case the resolution of the system resumes to a sequence of resolution of systems of four non linear equations with four unknowns, which was done using Newton types algorithms.

This is done by F. Forges(*), who has significantly reduced the computation times required with explicit schemes. We show for instance on Figure 6, the deterioration of the water-oil front when the time step increases from 10s up to 120s (the maximum admissible time step for the explicit scheme being around 5s). The computation time is already divided by 3 for a time step of 50s.

This study also showed the so-called "implicit-linearized" schemes, which are very favoured by oil engineers, and which consist in one simple Newton step, lead to satisfactory shape of the front and to reduction of computation time, but also give a propagation speed for the front which is an increasing function of the time step, and so, can lead to erroneous determination of water break-through times. This phenomenon disappears when freely implicit schemes are used, as it appears clearly on Figure 6.

4. THE TWO-DIMENSIONAL CASE

The difficulty of the problem is considerably increased : the whole system of equations (1) to (4) for the pressure and (5) to (11) for the saturation has now to be solved. Even in the case (for instance if d(S) is constant, where the two systems of equations are decoupled, the matrix giving the approximation r_h of r in equation (6) is never diagonal (as it was in the 1-D case when an adequate numerical quadrature formula was used), so that one has at least to solve one linear system at each time-step. So the computation times are rather important and the implementation

(*) with the help of a subsidiary of I.F.P. (Institut Français du Pétrole).

of the two-dimensional mixed elements is not easy.

Numerical programmes using these techniques are developed at IRIA by J. Jaffré and Gary Cohen in the frame of research contracts with the I.F.P. (Institut Français du Pétrole) and the French National Oil Company ELF-Aquitaine.

The first results (for monotonous transport terms, $u\vec{q}_1 = \vec{q}_2 = 0$), have been presented at TICOM in 1979 (cf. Jaffre [8]) and used triangular finite elements. They are briefly recalled on Figure 7, where their main default is apparent : they present a numerical diffusivity which is very much depending on the orientation of the grid with respect to the field $\vec{q}_0$ governing the transport term. However, in the "good" directions those elements give a very good approximation of the sharp fronts (cf. Figure 8).

Research is currently underway to try to overcome the difficulty encountered with triangular elements (use of equilateral triangles and of quadrilateral finite elements) and to take into account in the 2-D programs the non-monotonous terms coming from the gravity and from the capillarity pressure heterogeneity.

For more details, we refer to the papers [8][9] by Jaffre.

REFERENCES

[1] J. Douglas Jr. T. Dupont, H.H. Rachford Jr. "The application of variational methods to water flooding problems", in The Journal of Canadian Petroleum Technology, July/Sept. 1969.

[2] J.L. Vermvelen "Numerical simulation of Edge Water Drive with Well effect by Galerkin method". Paper SPE 4634, 48th Fall Meeting of Society of Petroleum Engineers of AIME, Las Vegas, 30 Sept.-3 Oct. 1973.

[3] E.A. Verner, R.W. Lewis, O.C. Zienkiewicz "Finite elements for two-phase flow in porous media". In "Flow in porous media I", 1975.

[4] J.M. Sanchez "Résolution par éléments finis d'un problème de déplacement eau-huile en dimension 2". Thèse de 3ème cycle, Université de Paris, 1978.

[5] J. Douglas Jr., B.L. Darlow, M. Wheeler, R.P.Kendall. "Self-adaptative Galerkin methods for one-dimensional, two-phase immiscible flow". Paper SPE, 7679, Symposium on Reservoir Simulation, Denver, 1979.

[6] J. Douglas Jr. "Interior Penalty Galerkin methods for transient problems", US-Japan Seminar on Interdisciplinary Finite Elements Anlysis, Cornell University, August 7-11, 1978.

[7] G. Chavent, G. Cohen "Numerical approximation and indentification in a 1-D parabolic degenerated non-linear diffusion and transport equation", in Lecture Notes in Control and Information Sciences, Vol. 6, pp. 283-293, Springer, 1977.

[8] J. Jaffrè "Approximation of a diffusion convection equation by a mixed finite element method ; application to the water-flooding problem", to appear in Computers and Fluids, 1979.

[9] J. Jaffre "Approximation par une méthode d'éléments finis mixtes d'une équation du type diffusion-convection linéaire stationnaire", LABORIA Report, to appear.

[10] P. Lesaint and P.A. Raviart " On a finite element method for solving the neutron transport equation", in "Mathematical aspects for finite elements in partial differential equations", Ed. Carl de Boor, Academic Press, 1974.

[11] P.A. Raviart, J.M. Thomas, "A mixed finite element method for second order elliptic problems", in "Mathematical aspects of finite elements methods, Rome, 1975, Ed. Galligani and Magenes, Lecture Notes in Mathematics, Springer, vol. 606, 1977.

[12] A.Y. Leroux "Approximation de quelques problèmes hyperboliques non linéaires". Thèse de Doctorat ès Sciences Mathématiques, Université de Rennes 1979.

[13] C. Bardos, A.Y. Leroux, J.C. Nedelec, "First order quasilinear equations with boundary conditions". Report Nr. 38, Centre de Mathématiques Appliquées, Ecole Polytechnique, 91128, Palaiseau Cedex, France.

[14] G. Chavent "A new formulation of diphasic incompressible flows in porous media in "Application of Methods of Functional Analysis to Problems of Mechanics". Lecture Notes in Mathematics, Vol. 503, Springer 1976.

[15] G. Chavent "About the identification and modelling of miscible or immiscible displacement in porous media", in "Modelling and Identification of distributed parameter systems". Lecture Notes in Control and Information Sciences, Vol. 1, Springer, 1977.

[16] G. Salzano "Résolution numérique d'un déplacement diphasique en milieu poreux en présence de gravité et de forces capillaires", to appear as a LABORIA Report, 1979.

[17] S.K. Godunov "Finite Difference Method for Numerical Computation of discontinuous Solutions of the Equations of Fluid Dynamics". Math. Sb.47(3) 1959, p. 27. USSR.

--=-=-=-=-=-=---

Figure 3 : Counter flow due to gravity in a vertical slab.

Boundary conditions : $S_e=1$ at x=0 and $S_e=0$ at x=1. Then the exact and computed oil and water flows through x=0 and x=1 are non zero. There is permanently water coming down and oil going up through the slab.

Saturation profiles at :

- ········ t=0
- – – – – t=0,7 (exact)
- ——— t=0,7 (computed)
- –+–+– t=∞ (exact)

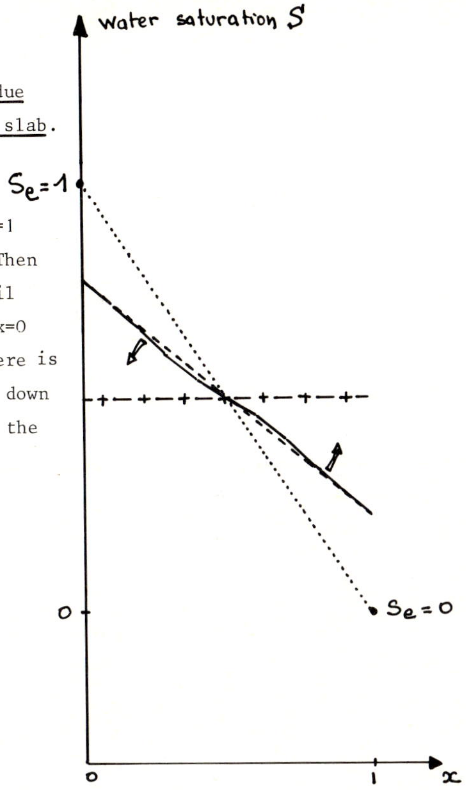

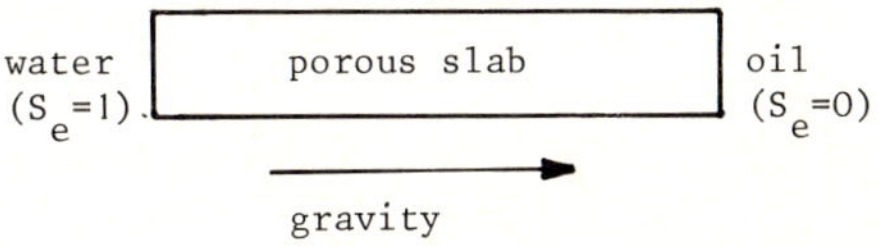

Figure 4 : Separation by gravity of oil and water in absence of capillary diffusion.

Boundary conditions :

$S_e=0$ at $x=0$ and $S_e=1$ at $x=1$. Then it turns out that the oil and water flows through $x=0$ and $x=1$ are equal to zero.

Saturation profiles at

········· $t=0$

- - - - - $t=0,7$ (exact)

——— $t=0,7$ (computed)

+—+— $t\geq 1,5$ (exact and computed)

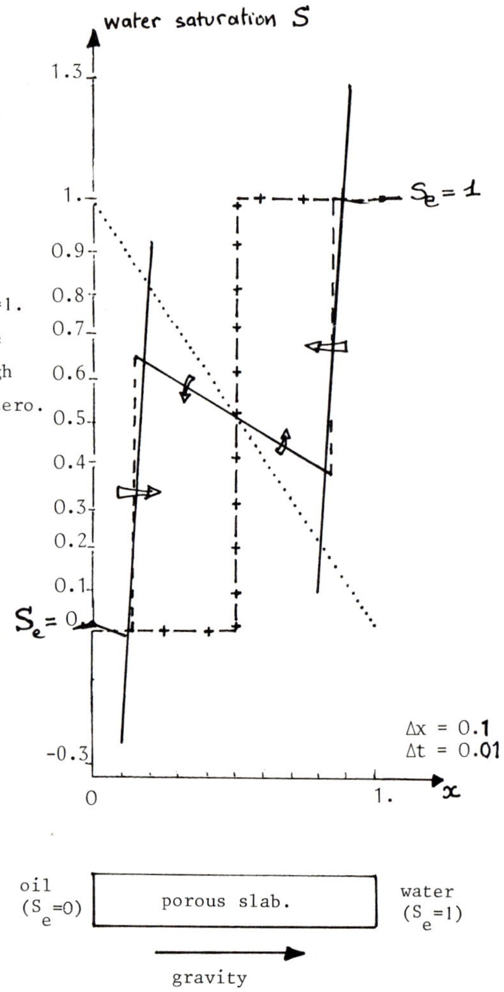

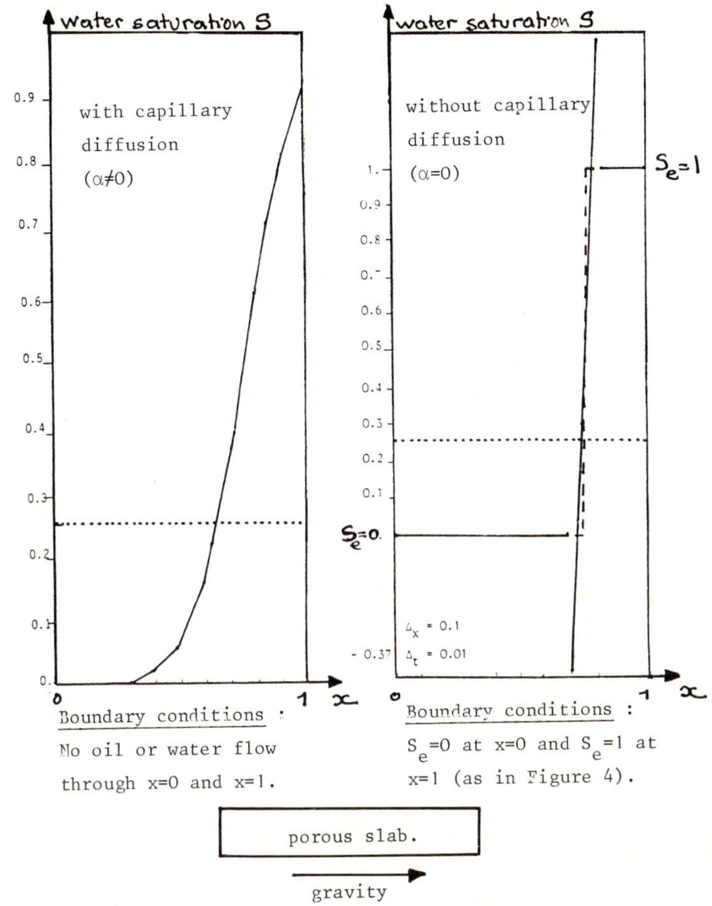

Saturation profiles shown at : t=0 ....

t=infinity (exact) - - - - -

t=infinity (computed) ————

Figure 5 : Influence of the capillary diffusion on the stationary saturation profile in a vertical insulated slab

Figure 6 - Influence of the time step on the shape of the water-oil front for a space time FEM : the front smears as the time step increases, but its position remains unchanged.
(maximum admisible time step with explicit schemes : ca 4 s)

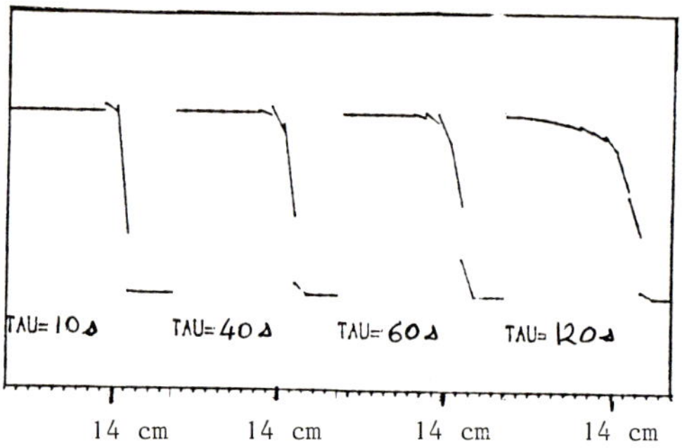

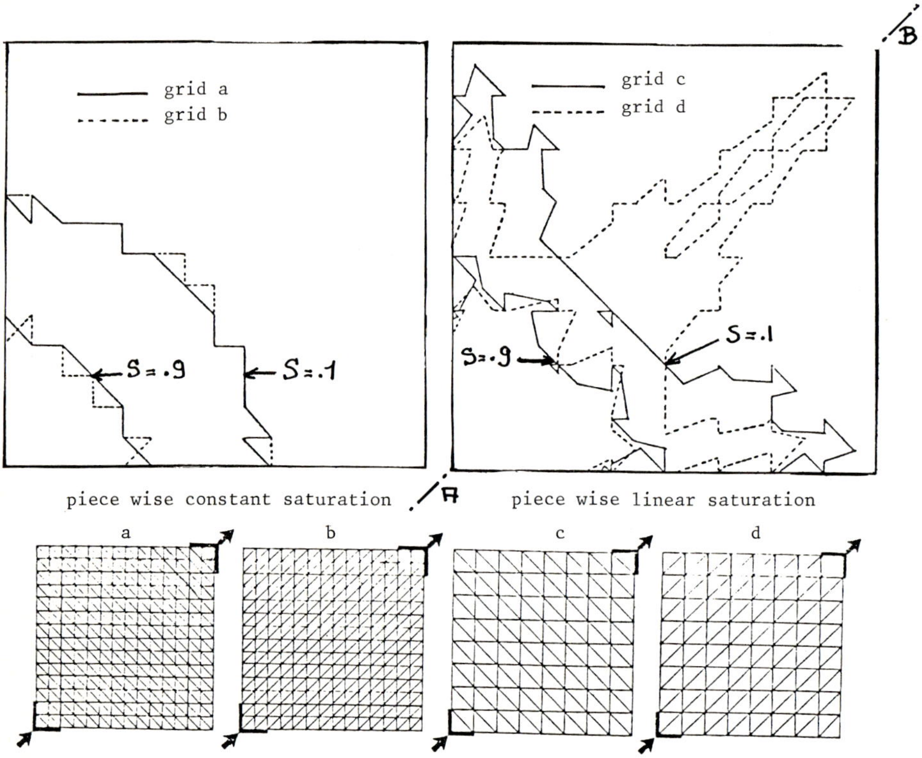

bottom left : water injection well
top rigt : production well.

Figure 7 - Water flooding of one quarter of five spot by mixed triangular F.E.M.

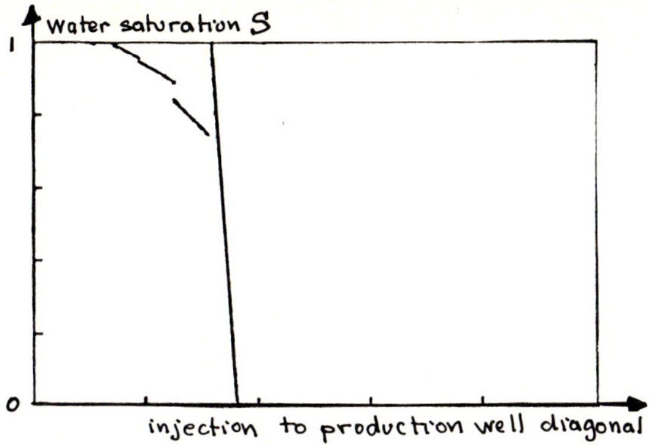

Figure 8 - Cross section of the saturation on the line A-B with the grid C of figure 7.

COMPUTING METHODS IN APPLIED SCIENCES AND ENGINEERING
R. Glowinski, J.L. Lions (editors)
North-Holland Publishing Company

NUMERICAL SOLUTION OF FREE BOUNDARY PROBLEM FOR IDEAL FLUID

Hideo KAWARADA
Department of Applied Mathematics
University of Tokyo
Hongo, Tokyo 113/JAPAN

Introduction

From a view-point of optimum design, the free boundary problems for elliptic equations have been successfully studied by means of the extented Hadamard's formula and others. In this paper, we present a simple method by means of the penalty method and solve numerically the free boundary problem(P) with Laplace equation $\Delta u = 0$ defined in the annulus in R^2. On the free boundary γ, $u = 1$ and $|\text{grad } u| = \frac{C}{\ell(\gamma)}$ ($\ell(\gamma)$ is the length of γ) are prescribed. This problem occurs in cryogenics experiments and is connected with the Stefan problem ; that kind of problem is also introduced to modelize plasma configulations in tokamak machines when the plasma is subject to spin effect (see for example A.S. DEMIDOV [1]). In section 1 we state the position of problem. Section 2 is devoted to analytical study of (P). This is partial results obtained by A. DERVIEUX and myself during my stay in IRIA-LABORIA (1977/12-1978/12). In section 3, we construct the numerical solution of (P). If we approximate the boundary value problem of an elliptic type in a bounded domain Ω by the penalized problem defined in B ($\supset \overline{\Omega}$), we found that the value of the approximate solution on $\Gamma = \partial\Omega$ is closely related with the flux of the unpenalized solution on Γ. We transform the original problem into some optimization problem by applying the fact mentioned above.

1. Position of the problem

Let D be a doubly connected region in R^2 limited by the point at infinity and a convex boundary component. Convex boundary denotes the boundary of convex body.

Let us consider the Free Boundary Problem (P) in D ;

(1.1) $\Delta u = 0$ in $\Omega \subset D$,

(1.2) $u|_\Gamma = 0$,

(1.3) $u|_\gamma = 1$,

(1.4) $$|\text{grad } u|_\gamma = \frac{C}{\ell(\gamma)} \ .$$

Ω is an annulus in D having Γ as one boundary component and another boundary component γ, the Free Boundary. C is a given positive number and $\ell(\gamma)$ is the length of γ.

Remark 1.1 Let $\left.\frac{\partial u}{\partial n}\right|_\gamma$ be the outward normal derivative of u on γ. By means of the maximum principle and (1.3), we have

(1.5) $$|\text{grad } u|_\gamma = \left.\frac{\partial u}{\partial n}\right|_\gamma \ .$$

Remark 1.2 By (1.4) and (1.5),

(1.6) $$C = \int_\gamma |\text{grad } u| d\sigma = \int_\gamma \frac{\partial u}{\partial n} d\sigma \ .$$

If u is the velocity potential, the number C denotes the Circulation of this quantity along γ .

On the other hand, from Green's formula

(1.7) $$C = \int_\Omega |\text{grad } u|^2 dx\, dy \ .$$

C of (1.7) represents the total energy of the system (1.1) $\sim$ (1.4).

The physical meaning of (P) is ; "Find the domain $\Omega \subset D$ and the velocity potential u of an ideal fluid defined in Ω under the circulation C (total energy) prescribed."

2. Existence and uniqueness of solution of (P)

2.1 Existence and uniqueness theorem and approximating process.

In this subsection we state some analytical results (A.DERVIEUX - H. KAWARADA [1]);

Theorem 2.1 (Existence and uniqueness) (P) has a unique solution $(\bar{u},\bar{\gamma})$ for any C (>0). The free boundary $\bar{\gamma}$ is analytically convex closed.

Definition 2.2 The polar parametrization of γ, Γ is possible; the origin of the axis is chosen inside Γ;

$$\gamma \in C^\infty(0,2\pi) \ , \text{ periodic with period } 2\pi \ ,$$

the boundary $\gamma = \{(r \cos\theta,\ r \sin\theta),\ \theta \in [0,2\pi],\ r = \gamma(\theta)\}$.

Similarly Γ is defined.

In order to obtain an approximate sequence for solution of (P), we consider another F.B. problem in which (1.4) is replaced by

$$(2.1) \quad |\mathrm{grad}|_\gamma = k \, .$$

k is a given positive number. This problem was studied by A. BEURLING [1] and D.E. TEPPER [1,2]. Hereafter, we denote (1.1) $\sim$ (1.3), (2.1) by (P′). Let us take the following iterative process;

Step 1 : Choose k_o such that

$$C_o = k_o \cdot \ell(\gamma_o) > C \, .$$

Step 2 : Define $k_{n+1} = \dfrac{C}{\ell(\gamma_n)}$,

$$C_{n+1} = k_{n+1} \cdot \ell(\gamma_{n+1}) \quad ,$$

where $\gamma_n (n=0,1,2,\cdots)$ is the F.B. solution of (P′) corresponding to the datum k_n .

Theorem 2.3 (Approximating process)
The above algorighm converges monotonically to the unique solution of (P) for the datum C :

$$\gamma_n \nearrow \gamma \qquad \text{in } C^\circ(0,2\pi) \text{ (i.e., uniformly)},$$

$$\ell(\gamma_n) \nearrow \ell(\gamma),$$

$$C_n \searrow C \, ,$$

$$\tilde{u}_n \searrow \tilde{u} \qquad \text{in } C^\circ(D) \text{ and } H^1(D) \, ,$$

where $\tilde{u}_n$, $\tilde{u}$ are 1-exstension of u_n , u in D .

2.2 Proofs

2.2.1 Some preliminaries

Let us recall the following classical result (see for example M. BERGER [1])

Lemma 2.4 Let Ω_i (i=1,2) be two closed convex regions of $\mathbb{R}^2$ with smooth boundaries $\gamma_i = \partial\Omega_i$ such that

$$\Omega_2 \subset \Omega_1 \, , \; \Omega_2 \neq \Omega_1$$

then

$\ell(\gamma_1) > \ell(\gamma_2)$

<u>Lemma 2.5</u> Let (k_n) $(n=1,2,3,\cdots)$ be a sequence of positive numbers such that

$$k_1 > k_2 > \cdots > k_n > k_{n+1} \longrightarrow k_o \in]0,\infty[\tag{2.4}$$
$$\text{when } n \longrightarrow +\infty \ ;$$

Let us denote by γ_n (resp. ℓ_n, u_n) the corresponding Free Boundary (resp. its length, and the distributed solution) for problem (P')(n= 0,1,2,...).

Then, when n tends to infinity we have

(i) $\gamma_n \leq \gamma_{n+1} \longrightarrow \gamma_o$ in $C^\circ(0,2\pi)$ (see (24))

(ii) $\ell_n \to \ell_o$.

Moreover if $\tilde{u}_n$ denotes the 1-extension of u_n in the open set Ω whose boundaries are Γ and γ_o $(n=0,1,2,\cdots)$, then

(iii) $\tilde{u}_n \longrightarrow u_o$ in $C^\circ(\Omega)$,

(iv) $\tilde{u} \longrightarrow u_o$ in $H^1(\Omega)$.

The following result is trivially deduced from Lemma 2.5 :

<u>Lemma 2.6</u> : The mapping $k \to C_k = k\cdot\ell(\gamma_k)$ is a continuous function for $0 < k < \infty$.

<u>Lemma 2.7</u> <u>$(k \to 0)$</u> : Let k_n be a sequence of positive numbers such that

$$k_1 > k_2 > \cdots > k_n > k_{n+1} \longrightarrow 0 \quad \text{as} \quad n \to \infty$$

then

$$\gamma_1 < \gamma_2 < \cdots < \gamma_n < \gamma_{n+1} \longrightarrow +\infty \text{ every where,} \tag{2.6}$$

$$u_1 \geq u_2 \geq \cdots \geq u_{n+1} \longrightarrow 0 \ , \tag{2.7}$$

weakly in $H^1(\Omega)$ for every open bounded set Ω included in D and uniformly in every compact subset of D .

Moreover, for every compact subset K' of Ω_{n+1}, we have

$$u_n > u_{n+1} \quad \text{on} \quad K' . \tag{2.8}$$

Lemma 2.8 $(k \to \infty)$: Let k_n be a sequence of positive numbers such that

$$k_1 < k_2 < \cdots < k_n < k_{n+1} \longrightarrow \infty \quad \text{as} \quad n \to \infty ,$$

then

(2.9) $\gamma_n \to \Gamma$ uniformly, i.e. in $C^\circ(0,2\pi)$

and

(2.10) $\tilde{u}_n \to 1$ in $L^2(D)$.

Lemma 2.9 : The mapping $k \to C_k$ defined in Lemma 2.6 have the following properties

(i) If $k_1 < k_2$ then $C_{k_1} < C_{k_2}$

(ii) $C_k \to 0$ when $k \to 0$,

(iii) $C_k \to \infty$ when $k \to \infty$.

2.2.2 Proof of Theorem 2.2

Taking into account the above Lemmas, we need only to note that $k \to C_k$ is a continuous, monotone and strictly increasing function defined on $[0, +\infty[$, with range $[0, +\infty[$:
thus for any given C in $[0, +\infty[$ there exists a unique corresponding k in $[0, +\infty[$ such that $C = C_k$, and Theorem 2.2 is proved.

2.2.3 Proof of Theorem 2.3

Let us note first that Step 1 is always possible monotonicity properties are explicited in the following lemmas.

Lemma 2.10 : The sequence $\{k_n, \Omega_n, C_n\}$ verify

(i) $k_o > k_1 > k_2 > \cdots > k_n > k_{n+1} > \cdots$,

(ii) $\Omega_o \subset \Omega_1 \subset \Omega_2 \subset \cdots \subset \Omega_n \subset \Omega_{n+1} \subset \cdots$,

(iii) $C_o > C_1 > C_2 > \cdots > C_n > C_{n+1} > \cdots > C > 0$.

Lemma 2.11 : The sequence (k_n) converges to a positive constant k_∞.

To complete the proof of Theorem 2.3, we use Lemma 2.4 :

$$\gamma_n \longrightarrow \gamma = \gamma(k_\infty) \quad \text{uniformly}$$

$$\tilde{u}_n \longrightarrow u = u(k_\infty) \quad \text{in} \quad H^1(D) \ ,$$

$$\ell_n \longrightarrow \ell(\gamma) \ ,$$

and indeed

$$C_n = k_n \cdot \ell_n \longrightarrow k \cdot \ell(\gamma) = \lim_{n \to +\infty} k_{n+1} \cdot \ell(\gamma_n) = C \ .$$

3. Computational method

3.1 Transformation of (P) into an optimization problem

Our plan to obtain the numerical solution of (P) is ;
(i) to transform (P) into some optimization problem (P_ε) by means of penalty method ;
(ii) to discretize (P_ε) by means of finite difference method and to solve (P_ε) by means of symplex method.

3.2 Penalty method

Let us consider the boundary value problem defined in a bounded domain $\Omega_o \subset R^2$ with smooth boundary $\partial\Omega_o = \Gamma_o$:

(3.2) $\quad -\Delta u = f \quad \text{in} \quad \Omega_o \ ,$

(3.2) $\quad u|_{\Gamma_o} = 0 \ .$

We approximate (3.1) $\sim$ (3.2) by the following penalized problem: for $\varepsilon > 0$,

(3.3) $\quad -\Delta u_\varepsilon + \frac{1}{\varepsilon}\chi u_\varepsilon = \tilde{f} \quad \text{in} \quad B \ ,$

(3.4) $\quad u_\varepsilon|_{\Gamma_1} = 0 \quad (\Gamma_1 = \partial B) \ .$

B is a bounded domain in R^2 which contains $\overline{\Omega}$. χ is the characteristic function of $\Omega_1 = B\backslash\overline{\Omega}_o$ in B .

$$\tilde{f} = \begin{cases} f & \text{in} \quad \Omega_o \ , \\ 0 & \text{in} \quad \overline{\Omega}_1 \ . \end{cases}$$

<u>Remark 3.1</u> For any $f \in L^2(\Omega_o)$, problems (3.1) $\sim$ (3.2), (3.3) $\sim$ (3.4) have the solutions $u_o \in H^2(\Omega_o) \cap H^1_o(\Omega_o)$ and $u_\varepsilon \in H^2(B) \cap H^1_o(B)$ for any $\varepsilon > 0$, respectively.
Then we have

<u>Theorem 3.2</u> Let ε be sufficiently small.

Then

$$\left.\frac{u_\varepsilon}{\sqrt{\varepsilon}}\right|_{\Gamma_o} = -\left.\frac{\partial u_o}{\partial n}\right|_{\Gamma_o} + 0(\sqrt{\varepsilon}) \quad \text{in} \quad H^{\frac{1}{2}}(\Gamma_o) \ . \tag{3.5}$$

$\frac{\partial u_o}{\partial n}$ denotes the outward normal derivative of u_o on Γ_o with respective to Ω_o .

Proof Our plan to prove (3.5) is to expand u_ε by singular perterbation method. First we prepare two lemmas.

Let u_ε be the solution of the following boundary value problem;

$$-\varepsilon\Delta u_\varepsilon + u_\varepsilon = 0 \quad \text{in} \quad \Omega_1 \ , \tag{3.6}$$

$$u_\varepsilon|_{\Gamma_o} = -h(s)\cdot\sqrt{\varepsilon} \quad (s \in \Gamma_o) \tag{3.7}$$

$$u_\varepsilon|_{\Gamma_1} = 0 \ . \tag{3.8}$$

$h(s)$ is sufficiently smooth on Γ_o .

Then we have

Lemma 3.3

$$\left.\frac{\partial u_\varepsilon}{\partial n}\right|_{\Gamma_o} = h(s) + \sigma_o(s)\cdot\sqrt{\varepsilon} \quad \text{in} \quad H^{-\frac{1}{2}}(\Gamma_o) \ , \tag{3.9}$$

where $\|\sigma_o(s)\|_{H^{-\frac{1}{2}}(\Gamma_o)} \leq C_o$ (C_o is a constant dependent of h).

Proof. This may be proved by refering to Lions [1].

Let u_ε be the solution of the following boundary value problem;

$$\Delta u_\varepsilon = 0 \quad \text{in} \quad \Omega_o \ , \tag{3.10}$$

$$u_\varepsilon|_{\Gamma_o} = -h(s)\cdot\sqrt{\varepsilon} \ . \tag{3.11}$$

Then we have

Lemma 3.4 (S. AGMON - A. DOUGRIS - L. NIRENBERG [1])

Let $h \in H^{m+\alpha}(\Gamma_o)$ $(m=1,2,\cdots,0 < \alpha < 1)$. Then there exists a linear continuous mapping T ; $H^{m+\alpha}(\Gamma_o) \longrightarrow H^{m+\alpha-1}(\Gamma_o)$ such that

$$\left.\frac{\partial u_\varepsilon}{\partial n}\right|_{\Gamma_o} = T(-h(s)\sqrt{\varepsilon}) = -\sqrt{\varepsilon}\cdot(Th)(s) \quad (s \in \Gamma_o). \tag{3.12}$$

Define

$$a_1(u,v) = \int_{\Omega_1} u\cdot v \, dx \, dy \ ,$$

$$a(u,v) = \int_B \text{grad } u \cdot \text{grad } v \, dx \, dy ,$$

$$(u,v) = \int_B u \cdot v \, dx \, dy .$$

(3.3) and (3.4) may be rewritten in the following form :

$$\varepsilon a(u_\varepsilon, v) + a_1(u_\varepsilon, v) = \varepsilon(\tilde{f}, v) \qquad (\forall v \in H_0^1(B)) . \tag{3.13}$$

As 0-th approximation of the solution u_ε of (3.13), we take

$$\tilde{u}_0 = \begin{cases} u_0 & \text{in } \overline{\Omega}_0 , \\ 0 & \text{in } \Omega_1 . \end{cases} \tag{3.14}$$

Let $h(s) = \left.\frac{\partial u_0}{\partial n}\right|_{\Gamma_0}$ $(s \in \Gamma_0)$. Then $\tilde{u}_0$ satisfies

$$\varepsilon a(\tilde{u}_0 , v) = \varepsilon \int_{\Gamma_0} h(s) \; v(s) \; d\sigma \; + \varepsilon(\tilde{f}, v), \quad (\forall v \in H_0^1(B)). \tag{3.15}$$

The 1st approximation is chosen as follows. Let $\varphi^1_{i\varepsilon}$ (i=0,1) be the solutions of the boundary value problems defined in Ω_i :

$$\Delta\varphi^1_{0\varepsilon} = 0 \quad \text{in } \Omega_0 , \tag{3.16}$$

$$\left.\varphi^1_{0\varepsilon}\right|_{\Gamma_0} = -h(s)\cdot\sqrt{\varepsilon} \qquad (s \in \Gamma_0) . \tag{3.17}$$

$$-\varepsilon\Delta\varphi^1_{1\varepsilon} + \varphi^1_{1\varepsilon} = 0 \quad \text{in } \Omega_1 , \tag{3.18}$$

$$\left.\varphi^1_{1\varepsilon}\right|_{\Gamma_0} = -h(s)\cdot\sqrt{\varepsilon} \qquad (s \in \Gamma_0) , \tag{3.19}$$

$$\left.\varphi^1_{1\varepsilon}\right|_{\Gamma_1} = 0 . \tag{3.20}$$

Put $\theta^1_\varepsilon = \begin{cases} \varphi^1_{0\varepsilon} & \text{in } \Omega_0 , \\ \varphi^1_{1\varepsilon} & \text{in } \Omega_1 . \end{cases}$

By means of Lemmas (3.3), (3.4), θ^1_ε satisfies

$$\varepsilon a(\theta^1_\varepsilon , v) + a_1(\theta^1_\varepsilon , v) = \varepsilon\int_{\Gamma_0} [-h-(\sigma_0 + Th)\sqrt{\varepsilon}] \; v \; d\sigma \qquad (\forall v \in H_0^1(B)) . \tag{3.21}$$

Similarly we may construct 2nd approximation. Let $\varphi^2_{i\varepsilon}$ (i=0,1) be the solutions of the boundary value problems (3.16) ~ (3.20) in which the boundary conditions (3.17) and (3.19) on Γ_0 are replaced by

(3.22) $$\varphi^2_{0\varepsilon}\Big|_{\Gamma_o} = (\sigma_o + Th)\cdot\varepsilon \quad ,$$

(3.23) $$\varphi^2_{1\varepsilon}\Big|_{\Gamma_o} = (\sigma_o + Th)\cdot\varepsilon \quad ,$$

respectively.

Put $$\theta^2_\varepsilon = \begin{cases} \varphi^2_{0\varepsilon} & \text{in } \Omega_o \ , \\ \varphi^2_{1\varepsilon} & \text{in } \Omega_1 \ . \end{cases}$$

Then θ^2_ε satisfies

(3.24) $$\varepsilon a(\theta^2_\varepsilon , v) + a_1(\theta^2_\varepsilon , v) = \varepsilon\int_{\Gamma_o} [\varepsilon T(\sigma_o + Th) + \sqrt{\varepsilon}(\sigma_o + Th) + \varepsilon\sigma_1] v d\sigma \ , \qquad (\forall v \in H^1_o(B)) \ .$$

Combining (3.15), (3.21) and (3.24), we have

(3.25) $$\varepsilon a(\tilde{u}_o + \theta^1_\varepsilon + \theta^2_\varepsilon , v) + a_1(\tilde{u}_o + \theta^1_\varepsilon + \theta^2_\varepsilon , v) = \varepsilon^2\int_{\Gamma_o} [T(\sigma_o + Th) + \sigma_1] v d\sigma + \varepsilon(\tilde{f}, v) \quad , \qquad (\forall v \in H^1_o(B)) \quad ,$$

Let $w_\varepsilon = u_\varepsilon - (\tilde{u}_o + \theta^1_\varepsilon + \theta^2_\varepsilon)$.

Then w_ε satisfies

(3.26) $$\varepsilon a(w_\varepsilon , v) + a_1(w_\varepsilon , v) = \varepsilon^2\int_{\Gamma_o} g v d\sigma \quad , \qquad (\forall v \in H^1_o(B)) \quad ,$$

where $g = T(\sigma_o + Th) + \sigma_1$.

Let $v = w_\varepsilon$ in (3.26). We have

(3.27) $$\varepsilon||w_\varepsilon||^2_{H^1(B)} + ||w_\varepsilon||^2_{L^2(\Omega_1)} \leq \varepsilon^2||g||_{H^{-\frac{1}{2}}(\Gamma_o)} \cdot ||w_\varepsilon||_{H^{\frac{1}{2}}(\Gamma_o)} \ .$$

Using (3.27) and trace theorem (J. L. LIONS - E. MAGENES [1]), we have

(3.28) $$\frac{1}{\sqrt{\varepsilon}}||w_\varepsilon||_{H^{\frac{1}{2}}(\Gamma_o)} \leq \sqrt{\varepsilon}||g||_{H^{-\frac{1}{2}}(\Gamma_o)} \ .$$

On Γ_o ,

(3.29) $$\frac{w_\varepsilon}{\sqrt{\varepsilon}}\Big|_{\Gamma_o} = \frac{u_\varepsilon}{\sqrt{\varepsilon}}\Big|_{\Gamma_o} + h - \sqrt{\varepsilon}\cdot(\sigma_o + Th) \quad \text{in } H^{\frac{1}{2}}(\Gamma_o) \quad ,$$

From (3.29) and (3.28), we obtain (3.5).

3.3 Optimization problem

Let us remind that the F.B. γ of (P) is convex closed. We denote K by the set of $\eta(\theta)$ ($\theta \in [0,2\pi]$) which satisfies (i) $\eta(\theta)$ is convex boundary; (ii) $L \geqslant \eta(\theta) \geqslant \Gamma(\theta)$ ($\theta \in [0,2\pi]$), where L is a sufficiently large positive number. It is well known that convex boundary is Jordan curve and rectifiable.

Let B be a bounded open domain which includes the disk of radius L. Let u_ε be the solution of the boundary value problem defined in an annulus D having Γ and ∂B as boundary components; for $\varepsilon > 0$,

$$-\Delta u_\varepsilon + \frac{1}{\varepsilon}\chi(u_\varepsilon - 1) = 0 \quad \text{in } D , \tag{3.30}$$

$$u_\varepsilon\big|_\Gamma = 0 , \tag{3.31}$$

$$u_\varepsilon\big|_{\partial B} = 1 . \tag{3.32}$$

χ is the characteristic function of the annulus having η and ∂B as boundary components.

Remark 3.5 There exists a unique solution $u_\varepsilon \in H^2(D)$ for (3.1) - (3.3). u_ε satisfies (i) $u_\varepsilon \in C^\circ(\bar{D})$ by Sobolev; (ii) the mapping : $\eta \in K \longrightarrow u_\varepsilon \in C^\circ(\bar{D})$ is continuous.

Define

$$J(\eta) = \int_0^{2\pi} \left| \frac{u_\varepsilon}{\sqrt{\varepsilon}}(\eta) - \frac{C}{\ell(\eta)} \right|^2 d\sigma_\eta(\theta) , \qquad (\forall \eta \in k) , \tag{3.33}$$

where $d\sigma_\eta(\theta)$ is bounded variation of length of $\eta(\theta)$. We should note that $J(\eta)$ is Stieltjes integral.

Remark 3.6 From (3.5) in Theorem 3.2 we see $\frac{u_\varepsilon}{\sqrt{\varepsilon}}(\eta)$ in (3.33) approximates $\frac{\partial u_0}{\sigma n}(\eta)$.

Now let us consider the optimization problem:

Find $\eta^\circ \in k$ such that

$$J(\eta^\circ) = \inf_{\eta \in k} J(\eta) . \tag{3.34}$$

Theorem 3.7 (P_ε) has at least one solution $\eta^\circ \in k$.

Proof : Let $\tilde{K}$ be the set of convex body having $\eta \in K$ as boundary. Introduce Hausdorff metric into $\tilde{K}$. Then $\tilde{K}$ is compact by Blaschke's selection rule.

Using (ii) in Remark 3.5 and the limiting theorem of Stieltjes integral we see the continuity of $J(\eta)$ to $\eta \in K$.

Thus $J(\eta)$ has a minimum in K.

REFERENCES

(1) S.AGMON-A.DOUGRIS-L.NIERNBERG

[1] Estimates near the boundary for solutions of elliptic partial differential equations satisfying general boundary conditions.I, Comm. Pure Appl. Math. 12 (1959), 623-727.

(2) M.BERGER

[1] Géométric, Cedic, Fernand Nathau, Paris (1977).

(3) A.BEURLING

[1] Free Boundary Problems for the Laplace equation. Institute for advanced study, Seminar, Princeton, N.J., I (1957).

(4) A.S.DEMIDOV

[1] The form of a steady plasma subject to the spin effect in a tokamak with non-circular cross-section. Nuclear Fusion 15 (1975) 765-768.

(5) A.DERVIEUX-H.KAWARADA

[1] Free Boundary Problems for the Laplace equation. LABORIA Report (to appear).

(6) J.L.LIONS

[1] Perturbations Singulières dans les Problèms aux Limites et en Contrôl. Springer-Verlag 1973.

(7) J.L.LIONS-E.MAGENES

[1] Problemes aux limites non homogènes et applications. Paris: Dunod 1968 (Vol.1 and 2).

(8) D.E.TEPPER

[1] Free Boundary Problem, SIAM J. Math. Anal., 5 (1974) 841-846.

[2] On a Free Boundary Problem, the Starlike Case, SIAM J. Math. Anal. 6 (1975) 503-505.

COMPUTING METHODS IN APPLIED SCIENCES AND ENGINEERING
R. Glowinski, J.L. Lions (editors)
North-Holland Publishing Company

CONTROLE DE SYSTEMES A FRONTIERE LIBRE

APPLICATION A LA COULEE CONTINUE D'ACIER

C. SAGUEZ* - M. LARRECQ**

Résumé : A partir de l'exemple de la coulée continue d'acier, on présente une méthode numérique de résolution des problèmes de contrôle optimal de systèmes à frontière libre. Dans la formulation du problème, on utilise l'enthalpie du système qui intervient comme un opérateur maximal monotone. A partir de l'approximé de Yosida de ce dernier, on obtient une méthode de simulation et des conditions nécessaires d'optimalité. Numériquement on calcule le contrôle optimal par un algorithme de type gradient. Des résultats numériques dans le cas d'une installation complète de coulée continue d'acier sont donnés.

Introduction : Des problèmes à frontière libre apparaissent dans de nombreux domaines de la physique, mécanique des milieux continus [9], physique des plasmas [19], hydrodynamique [2], ... Dans cet article, nous considérons le problème de la solidification de l'acier en coulée continue. Sur le plan physique une coulée continue est un échangeur de chaleur où est réalisé le changement de phase liquide-solide. Ce phénomène est modélisé en utilisant l'enthalpie du système, qui peut-être une fonction multivoque de la température.

Le problème de l'optimisation considéré ici pour une telle installation consiste à déterminer le système optimal de refroidissement de l'acier, permettant d'obtenir la meilleure productivité (vitesse maximale de coulée), compte tenu des contraintes physiques et métallurgiques existantes. Mathématiquement il s'agit d'un problème de contrôle optimal d'un système à frontière libre, le contrôle apparaissant dans les conditions frontières.

Après avoir, dans une première partie, brièvement présenté le procédé de coulée continue, nous donnons dans une deuxième partie la formulation mathématique du problème. Dans une troisième partie nous proposons une méthode numérique, d'une part pour la simulation du système, d'autre part pour la détermination du contrôle optimal. Enfin, dans une dernière partie, nous donnons des résultats numériques pour des installations réelles.

---

* IRIA - Domaine de Voluceau - 78150 LE CHESNAY

** IRSID - Station d'Essais - 57210 MAIZIERES-Les-METZ

CHAPITRE I

PRESENTATION DU PROCEDE DE COULEE CONTINUE

Le procédé de coulée continue occupe une place de plus en plus importante dans la chaine d'élaboration des produits sidérurgiques. Il permet à la fois des gains qualitatifs sur les propriétés du métal coulé et d'importantes réductions du prix de revient [16].

Le principe de la coulée continue est de couler l'acier dans un moule dont le fond est formé par la partie solide du lingot que l'on extrait en continu; l'alimentation du moule en acier liquide étant elle-même continue. Si ce principe est connu depuis plus d'un siècle (Bessemer 1857-1891), les difficultés de mise au point ont demandé de très nombreuses années d'effort. En 1933, furent déposés par Junghams les premiers brevets et les premiers essais datent des années 1948-1950. C'est à partir de 1962-1964 que le nombre de machines a augmenté rapidement. En 1977, l'acier coulé en continue représentait en France 24 % de la production totale.

Le schéma de principe d'une installation de coulée continue est représenté sur la Figure 1.

L'acier est amené au-dessus de l'installation dans une poche. Un répartiteur assure la distribution de l'acier dans une ou plusieurs lignes de coulée et permet la coulée de plusieurs poches en séquence sans arrêt du fonctionnement de la machine.

L'acier est coulé dans une lingotière en cuivre,refroidie par circulation d'eau. La lingotière assure la solidification d'une croûte solide capable de résister aux différents efforts qui lui seront imposés par la suite.

En sortant de la lingotière, le produit pénètre dans le refroidissement secondaire qui est formé de rouleaux de soutien et d'un système de pulvérisation d'eau permettant d'achever la solidification de l'acier. Les rampes de pulvérisation sont regroupées en zones d' arrosage, chacune de ces zones ayant son propre système de réglage du débit d'eau pulvérisée.

La machine comporte enfin des rouleaux extracteurs permettant le défilement en continu du produit et un système de coupe des barres coulées (chalumeau ou cisaille).

Les produits coulés peuvent être de formes différentes : produits ronds, produits carrés (Billettes - Blooms), produits plats (Brames).

FIGURE 1

Schéma de Principe d'une Machine de Coulée Continue

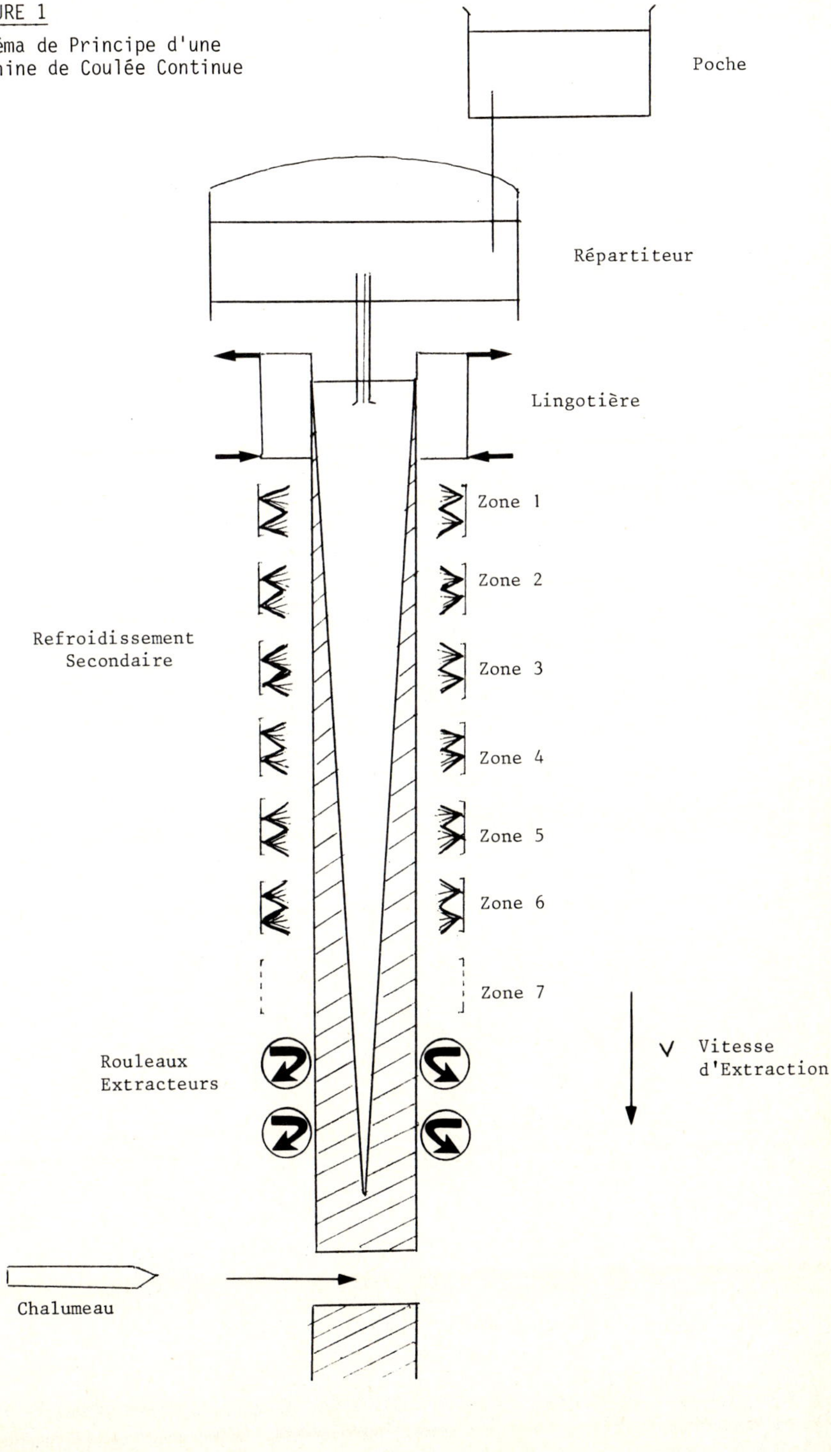

CHAPITRE II

LE PROBLEME ETUDIE

Le problème, dans son ensemble, consiste à déterminer le meilleur fonctionnement des zones d'arrosage du refroidissement secondaire afin de pouvoir atteindre une vitesse d'extraction maximale. Nous présentons successivement la formulation du système d'état (modèle thermique de l'acier) et du problème de contrôle optimal associé.

1) Modèle mathématique de la coulée continue :

On note z la hauteur dans la machine de coulée continue et Ω le domaine déterminé par une coupe transversale du produit. Pour $x \in \Omega$, on désigne la température de l'acier (qu'il soit liquide ou solide) par T(x,z). T vérifie le système :

$$\begin{cases} \rho V \dfrac{\partial v}{\partial z} - \operatorname{div}(\lambda(T) \operatorname{grad} T) = 0 \quad x \in \Omega, \quad z \in]0, Z[& (2.1) \\ v \in H(T) & (2.2) \end{cases}$$

avec ρ la densité de l'acier

λ sa conductibilité thermique

V la vitesse d'extraction

H(T) désigne l'enthalpie du système, H est un opérateur maximal monotone de graphe :

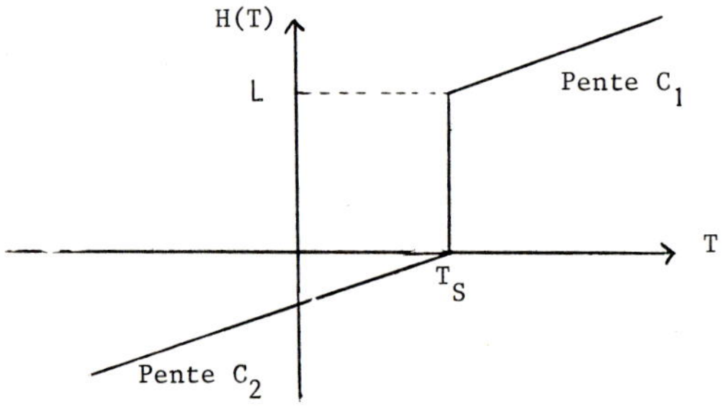

FIGURE 2

avec C_1(resp. C_2) la capacité calorifique de l'acier liquide (resp. solide)

L la chaleur latente de solidification

T_S la température de solidification

Les conditions aux bords sont données par : ($\vec{n}$ désignant la normale extérieure) :

- au niveau de la lingotière, on connait, grâce à des mesures, le flux extrait :

$$\lambda(T)\ \vec{\text{grad}}\ T.\ \vec{n} = g \tag{2.3}$$

- au niveau du refroidissement secondaire, chaque zone i est caractérisée par un coefficient d'échange h_i :

$$\lambda(T)\ \vec{\text{grad}}\ T.\ \vec{n}\Big|_{\Gamma_i} = -hi(T - Te)\Big|_{\Gamma_i} \tag{2.4}$$

avec Γ_i frontière correspondante à la zone i
T_e température de l'eau de refroidissement.

On a enfin les conditions initiales :

$$T(x,o) = T_o(x)\ ;\ v(x,o) = v_o(x) \in H(T_o) \tag{2.5}$$

2) Le problème de contrôle optimal

Les coefficients d'échange h_i(i=1,...,Nz ; Nz nombre de zones) sont les variables de contrôle.

Problème 1 : Optimisation à vitesse de coulée donnée

On considère le problème de contrôle optimal suivant : déterminer les coefficients h_i de sorte que l'acier soit totalement solide au niveau z_1 de l'installation, ceci pour une vitesse de coulée donnée. Pour cela on introduit la fonctionnelle :

$$J(h) = \int_o^{z_1} \int_\Omega [(v(x,z) - L)^+]^2\, d\Omega\, dz \tag{2.6}$$

J représente la quantité de chaleur restant à extraire pour avoir la solidification complète.

Les contrôles h_i doivent vérifier les contraintes suivantes :

$$o < h_{im} \le h_i \le h_{iM} \qquad i = 1,\dots,Nz \tag{2.7}$$

$$\sum_{i=1}^{Nz} \alpha_i\, h_i \le D \tag{2.8}$$

où α_i désigne "l'efficacité de l'eau" dans la zone i et D le débit maximal d'eau dont on dispose.

La température T doit d'autre part vérifier au point de redressage z_2, à cause de considérations métallurgiques, la contrainte :

(2.9) $T(x,z_2)|_{\partial\Omega} \not\subset]T_1, T_2[$

Pour traiter cette contrainte, on étudie successivement les problèmes avec la contrainte $T(x,z_2)|_{\partial\Omega} \leq T_1$ et $T(x,z_2)|_{\partial\Omega} \geq T_2$.

D'autres contraintes pourraient être considérées, en particulier le gradient de température sur le bord doit rester inférieur à une valeur donnée. Nous ne tenons pas compte ici de ce type de contraintes. Le problème de contrôle se formule ainsi :

(2.10) $\begin{cases} \text{Trouver } \bar{h} \in \mathcal{U}ad = \{h_i \in \mathbb{R},\ i=1,\dots,Nz \mid \text{vérifiant } (2.7),(2.8),(2.9)\} \\ \text{tel que } J(\bar{h}) \leq J(h)\ \forall h \in \mathcal{U}ad \end{cases}$

Problème 2 : Détermination de la vitesse optimale de coulée

Trouver $(\tilde{V},\tilde{h})$ tel que :

(2.11) $\tilde{V} = \text{Max}\ \{ V \mid J(\bar{h}) = o\}$

On se ramène, dans la résolution, à une succession de problème du type problème 1.

CHAPITRE III

METHODES NUMERIQUES

Afin de simplifier l'exposé, nous nous plaçons dans le cas $\lambda(T) = \gamma$ (constante donnée) et avec, comme unique condition aux bords, la condition (2.4). Si on effectue le changement de variable :

(3.1) $\theta = T_S - T$

on obtient le système :

$$\begin{cases} \rho V \frac{\partial u}{\partial z} - \gamma \Delta \theta = o \ ; \quad u \in G(\theta) & (3.2) \\ \frac{\partial \theta}{\partial n}\Big|_\Gamma + h(\theta-\theta_e)\Big|_\Gamma = o & (3.3) \\ \theta(x,o) = \theta_o(x) \ ; \quad u(x,o) = u_o(x) \in G(\theta_o) & (3.4) \end{cases}$$

avec $\theta_e = T_S - T_e$ et G de graphe :

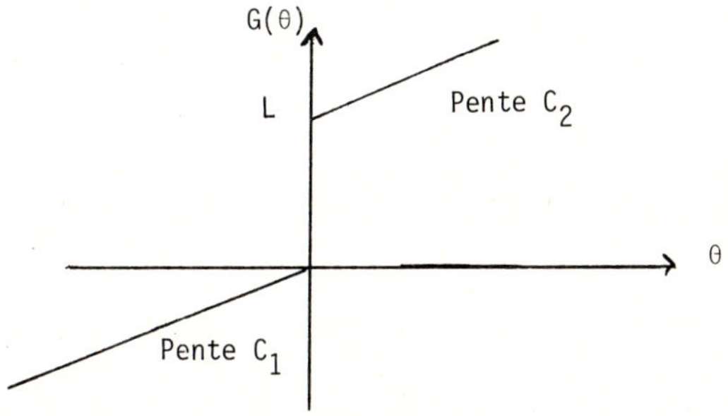

FIGURE 3

Nous ne nous intéressons pas ici aux problèmes d'existence. Pour ces points, nous renvoyons à A. Damlamian [7], O. Grange, F. Mignot [11], C. Saguez [17]. Dans le cas présent, la méthode consiste à étudier l'existence sur un problème semi-discrétisé en z et à passer à la limite.

On considère le problème semi-discrétisé en z suivant :

$$
\begin{cases}
(3.5) \quad \rho V\left(\frac{W^{n+1}-W^n}{k}, \psi\right) + \frac{\omega}{k}(\theta^{n+1} - \theta^n, \psi) + \gamma(\text{grad }\theta^{n+1}, \text{grad }\psi) + \\
+ \int_\Gamma h^{n+1}(\theta^{n+1} - \theta_e^{n+1})\psi d\Gamma = 0 \qquad \forall \psi \in H^1(\Omega) \\
(3.6) \quad W^{n+1} \in G_\omega(\theta^{n+1}) = G(\theta^{n+1}) - \frac{\omega}{\rho V}\theta^{n+1} \qquad n = 0,\dots, N-1 \\
(3.7) \quad \theta^0 = \theta_0(x), \quad W^0 = u_0 - \frac{\omega}{\rho V}\theta_0
\end{cases}
$$

avec $0 < \omega < \rho V \text{ Min } (C_1, C_2)$.

1) Résolution de l'équation d'état :

A chaque pas en z, on doit résoudre le problème (3.5), (3.6), (3.7). Utilisant un résultat de A. Bermudez, C. Moreno [3], on constate que :

$$(3.8) \qquad W \in G_\omega(\theta) \Longleftrightarrow W = G_\omega^\mu(\theta + \mu W) \qquad \forall \mu > 0$$

où G_ω^μ est l'approximé de Yosida de $G_\omega(\theta)$.

On en déduit l'algorithme, à l'étape (n+1) (W^n, θ^n sont donnés)

$$(3.9) \qquad \frac{\omega}{k}(\theta_j^{n+1},\psi) + \gamma(\text{grad }\theta_j^{n+1}, \text{grad }\psi) + \int_\Gamma h^{n+1}(\theta_j^{n+1} - \theta_e^{n+1})\psi\, d\Gamma =$$

$$= -\frac{\rho V}{k}(W_j^{n+1} - W^n,\psi) + \frac{\omega}{k}(\theta^n,\psi) \qquad \forall \psi \in H^1(\Omega)$$

$$(3.10) \qquad W_{j+1}^{n+1} = G_\omega^\mu(\theta_j^{n+1} + \mu W_j^{n+1})$$

On a le résultat de convergence :

Proposition 1 : Pour $\mu > \mu_0 = \frac{\rho V}{k} \times \frac{1}{2 \text{ Min}(\gamma, \frac{\omega}{k})}$, on a la convergence:

$$W_j^{n+1} \longrightarrow W^{n+1} \quad \text{dans } L^2(\Omega) \text{ faible}$$

$$\theta_j^{n+1} \longrightarrow \theta^{n+1} \quad \text{dans } H^1(\Omega) \text{ fort}$$

Démonstration : D'après la lipschitzianité de G_ω^μ on obtient :

$$(3.11) \qquad ||W^{n+1} - W_{j+1}^{n+1}||_{L^2}^2 \leq ||W^{n+1}-W_j^{n+1}||_{L^2}^2 + \frac{2}{\mu}(\theta^{n+1}-\theta_j^{n+1}, W^{n+1}-W_j^{n+1}) + \frac{1}{\mu^2}||\theta^{n+1}-\theta_j^{n+1}||_{L^2}^2$$

De plus à partir de (3.8) on déduit

$$(3.12) \quad \gamma||\mathrm{grad}(\theta^{n+1}-\theta_j^{n+1})||^2_{L^2} + \frac{\omega}{k}||\theta^{n+1}-\theta_j^{n+1}||^2_{L^2} + \frac{\rho V}{k}(W^{n+1}-W_j^{n+1},\theta^{n+1}-\theta_j^{n+1}) + \int_\Gamma h(\theta^{n+1}-\theta_j^{n+1})^2\, d\Gamma = o$$

De (3.10), (3.11) on obtient

$$||W^{n+1}-W_{j+1}^{n+1}||^2_{L^2} \leq ||W^{n+1}-W_j^{n+1}||^2_{L^2} + \left(\frac{1}{\mu^2} - \frac{2}{\mu}k\frac{\mathrm{Min}(\gamma,\frac{\omega}{k})}{\rho V}\right)||\theta^{n+1}-\theta_j^{n+1}||^2_{H^1}$$

Donc, avec $\mu > \mu_o$, on en déduit que $||W^{n+1}-W_{j+1}^{n+1}||^2_{L^2}$ converge et que $||\theta^{n+1}-\theta_j^{n+1}||_{H^1} \to o$ et grâce à (3.8) on a $W_j^{n+1} \longrightarrow W^{n+1}$ dans $L^2(\Omega)$ faible.

Remarque 1 : Il faut signaler, dans un certain nombre de cas, la possibilité par un changement de variable convenable, de transformer le problème initial en inéquation variationnelle ([9], [10], [13], [14] , [18]).

2) Résolution du problème de contrôle optimal

On considère le problème de contrôle régularisé où (3.6) est remplacé par :

(3.13) $W_\varepsilon^{n+1} \in G_\omega^\varepsilon(\theta_\varepsilon^{n+1})$ où G_ω^ε est le sous-différentiel d'une fonction ψ_ω^ε, strictement convexe, s.c.i., propre tel que $G_\omega^{\varepsilon,\mu}$ soit une fonction très régulière et $|\psi_\omega^\varepsilon(x) - \psi_\omega(x)| \leq k(\varepsilon)x$ ($k(\varepsilon) \to o$ quand $\varepsilon \to o$, et $\partial\psi_\omega = G_\omega$)

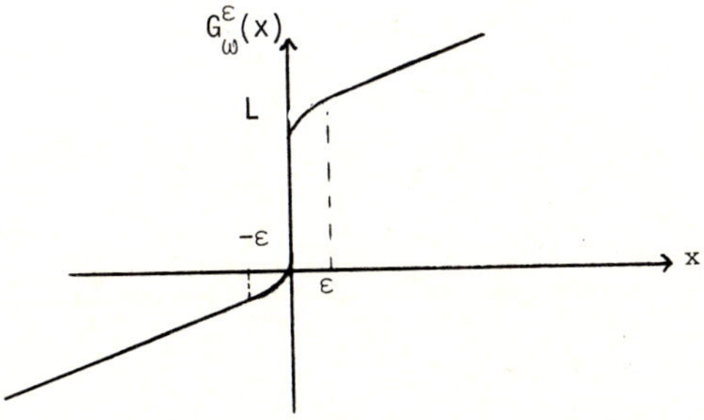

FIGURE 4

La fonctionnelle J est approchée par:

$$(3.14) \qquad J_k(h) = \sum_{n=1}^{NT} ||(u_\varepsilon^n - L)^-||^2_{L^2(\Omega)}$$

On a donc, comme pour le problème initial, l'existence d'une solution $\{h_\varepsilon^n\}$. De plus on la convergence :

Proposition 2 : Toute valeur d'adhérence de la suite $\{h_\varepsilon^n\}$ des solutions du problème régularisé est solution du problème de contrôle optimal initial.

On obtient, d'autre part, les conditions nécessaires d'optimalité :

Proposition 3 : Toute solution $\{(W_\varepsilon^n, \theta_\varepsilon^n, h_\varepsilon^n\})$ du problème régularisé vérifie les conditions nécessaires d'optimalité :

Etat du système :

$$(3.15) \qquad \frac{\rho V}{k}(W_\varepsilon^{n+1} - W_\varepsilon^n, \psi) + \gamma(\mathrm{grad}\,\theta_\varepsilon^{n+1}, \mathrm{grad}\,\psi) + \frac{\omega}{k}(\theta_\varepsilon^{n+1} - \theta_\varepsilon^n, \psi) + \int_\Gamma h_\varepsilon^{n+1}(\theta_\varepsilon^{n+1} - \theta_e^{n+1})\psi\, d\Gamma = o$$

$$\forall \psi \in H^1(\Omega)$$

$$(3.16) \qquad W^{n+1} = G_\omega^{\varepsilon,\mu}(\theta_\varepsilon^{n+1} + \mu W_\varepsilon^{n+1}) \qquad n = o, \ldots, NT-1$$

$$(3.17) \qquad \theta_\varepsilon^o = \theta_o \quad ; \quad W_\varepsilon^o = W_o^\varepsilon \in G_\omega^\varepsilon(\theta_o)$$

Etat Ajoint :

$$(3.18) \qquad \frac{\rho V}{k}(q_\varepsilon^n G_\omega'^{\varepsilon\mu}(\theta_\varepsilon^{n+1} + \mu W_\varepsilon^{n+1}), \psi) + \gamma(\mathrm{grad}\,p_\varepsilon^n, \mathrm{grad}\,\psi) + \frac{\omega}{k}(p_\varepsilon^n - p_\varepsilon^{n+1}, \psi) + \int_\Gamma h_\varepsilon^{n+1} p_\varepsilon^n \psi\, d\Gamma = -2((u_\varepsilon^{n+1} - L)^-, \psi) \qquad \forall \psi \in H^1(\Omega)$$

$$(3.19) \qquad p_\varepsilon^n - p_\varepsilon^{n+1} = q_\varepsilon^n(1 - \mu G_\omega'^{\varepsilon\mu}(\theta_\varepsilon^{n+1} + \mu W_\varepsilon^{n+1})) \qquad n = o, \ldots, NT-1$$

$$(3.20) \qquad p_\varepsilon^{NT} = o$$

Condition nécessaire d'optimalité :

$$(3.21) \qquad \sum_{n=1}^{NT} \int_\Gamma (\theta_\varepsilon^n - \theta_e^n)\, p_\varepsilon^{n-1}(s^n - h_\varepsilon^n)\, d\Gamma \le o \qquad \forall \{s^n\} \in \mathcal{U}_{ad}$$

Principe de la démonstration :

i) On étudie la G-différentiabilité de θ_ε^{n+1} et W_ε^{n+1} par rapport à h^{n+1}

- si on note $y_\alpha = \dfrac{\theta_\varepsilon^{n+1}(h^{n+1}+\alpha s) - \theta_\varepsilon^{n+1}(h^{n+1})}{\alpha}$; $r_\alpha \doteq \dfrac{W_\varepsilon^{n+1}(h^{n+1}+\alpha s) - W_\varepsilon^{n+1}(h^{n+1})}{\alpha}$,

en utilisant (3.15) et la lipschitzianité de $G_\omega^{\varepsilon,\mu}$ on montre que $||y_\alpha||^2$ et (r_α, y_α) sont bornés indépendamment de α

- en utilisant la stricte convexité de ψ_ω^ε, on prouve que r_α est borné, indépendamment de α dans $L^2(\Omega)$

- on peut donc extraire des sous-suites telles que

$y_\alpha \longrightarrow y$ dans $H^1(\Omega)$ faible
$r_\alpha \longrightarrow r$ dans $L^2(\Omega)$ faible

et on en déduit que y et r vérifient :

$$\frac{\rho V}{k}(r, \psi) + \gamma(\text{grad}y, \text{grad}\psi) + \frac{\omega}{k}(y,\psi) + \int_\Gamma hy\psi + \int_\Gamma s(\theta_\varepsilon^{n+1}(h^{n+1}) - \theta_e^{n+1})\psi d\Gamma \tag{3.22}$$

$\forall \psi \in H^1(\Omega)$

$$r = G_\omega'^{\varepsilon,\mu}(\theta_\varepsilon^{n+1}(h^{n+1}) + \mu W_\varepsilon^{n+1}(h^{n+1}))(y+\mu r) \tag{3.23}$$

ii) en raisonnant par récurrence, on généralise au cas $\{(\theta_\varepsilon^n, W_\varepsilon^n) | n=1,\dots,NT\}$ par rapport à $\{h^n \; | n = 1,\dots,NT\}$.

iii) on introduit l'état adjoint (3.15), (3.17). Si on note (y_i^n, r_i^n) la G-différentielle de $(\theta_\varepsilon^n, W_\varepsilon^n)$ par rapport à h^i dans la direction s_i, on obtient :

$$\sum_{n=o}^{NT-1} \{\frac{\rho V}{k}(q^n G_\omega'^{\varepsilon,\mu}(\theta_\varepsilon^{n+1} + \mu W_\varepsilon^{n+1}), y_i^{n+1}) + \gamma(\text{grad } p^n, \text{grad } y_i^{n+1}) + \frac{\omega}{k}(p^n - p^{n+1}, y_i^{n+1}) + \int_\Gamma h^{n+1} p^n y_i^{n+1} = - \sum_{n=o}^{NT-1} (2(u_\varepsilon^{n+1} - L)^-, y_i^{n+1}) \tag{3.24}$$

et

$$\sum_{n=o}^{NT-1} \frac{\rho V}{k}(r_i^{n+1} - r_i^n, p^n) + \gamma(\text{grad } y_i^{n+1}, \text{grad } p^n) + \frac{\omega}{k}(y_i^{n+1} - y_i^n, p^n) + \int_\Gamma h^{n+1} y_i^{n+1} p^n \, d\Gamma + \int_\Gamma s^i \, \delta_i^{n+1}(\theta_\varepsilon^{n+1} - \theta_e^{n+1}) p^n \, d\Gamma = o \tag{3.25}$$

(δ_i^{n+1} symbole de Kronecher).

Soustrayant (3.24) de (3.25) on déduit, grâce à (3.16) et (3.19) que

$$\int_\Gamma s^{n+1}(\theta_\varepsilon^{n+1} - \theta_e^{n+1})\, p^n d\Gamma = \sum_{n=o}^{NT-1} 2((u_\varepsilon^{n+1} - L)^-, y_i^{n+1})$$

et donc, on a la condition d'optimalité :

$$\sum_{n=1}^{NT-1} \int_\Gamma (\theta_\varepsilon^n - \theta_e^n)\, p_\varepsilon^{n-1}(s^n - h_\varepsilon^n) d\Gamma \le o \qquad \forall \ \{s_n\} \in \mathcal{U}\ ad$$

L'étude de ce problème ainsi que les démonstrations détaillées sont faites ans C. Saguez [17].

Pour résoudre le problème de contrôle optimal (2.10) on utilise un algorithme du type gradient. Pour cela on distingue deux types de contraintes :

Cas 1 $\mathcal{U}\ ad = \{h_{im} \le h_i \le h_{iM} \mid \Sigma\ \alpha_i\ h_i = D\}$

On utilise l'algorithme suivant, développé dans J. Henry [12]. A chaque itération, on note $\{h_i^l\}$ la valeur précédente et $\{G_i^l\}$ le gradient correspondant, on obtient $\{h_i^{l+1}\}$par les itérations :

$$(3.26) \quad \begin{cases} h_i^{l+1,j} = \mathrm{Max}(h_{im}, \mathrm{Min}(h_{iM}, h_i^l - \rho_1^j\ (G_i^l - \lambda^j \alpha_i))) \\ \lambda^{j+1} = \lambda^j - \rho_2^j (\sum_i \alpha_i\ h_i^{l+1,j} - D) \end{cases}$$

Cas 2 $\mathcal{U} ad = \{h_{im} \le h_i \le h_{iM} \mid \theta\ (x, z_2)_{|\partial\Omega} \le \theta_2\}$

On utilise une méthode de pénalisation pour la contrainte sur l'état. Les contraintes de borne sur le contrôle sont traitées par projection.

CHAPITRE IV

RESULTATS NUMERIQUES

Nous présentons des résultats obtenus sur des données réelles d'installation de coulée continue.

1) Simulation de l'état du système

Le problème (3.9) est discrétisé avec des éléments finis P_1. Le problème discret, non linéaire (λ dépend de T) est résolu par une méthode de Newton. Nous présentons les résultats sur deux exemples en dimension 1 et en dimension 2. La conductibilité thermique est donnée par :

T(oC)	λ(cal/cm/s/oC)	T(oC)	λ(cal/cm/s/oC)
550	0,091	1100	0,068
600	0,086	1150	0,070
650	0,081	1200	0,071
700	0,076	1250	0,072
750	0,071	1300	0,073
800	0,068	1350	0,075
850	0,065	1400	0,076
900	0,064	1450	0,077
950	0,065	1500	0,078
1000	0,066	1550	0,079
1050	0,067	1600	0,080

i) Exemple en Dimension 1 : Solidification d'une brame

Nous avons les données suivantes :

- Caractéristiques de l'installation :
  - Hauteur lingotière 70 cm
  - Refroidissement secondaire :
    - zone 1 60 cm
    - zone 2 200 cm
    - zone 3 150 cm
    - zone 4 350 cm
    - zone 5 390 cm
    - zone 6 430 cm
    - zone 7 1250 cm
  - Longueur totale de l'installation 29 m
  - Demi-épaisseur de brame 12,5 cm
- Caractéristiques thermiques de l'acier :

  L'enthalpie est donnée par la figure 1 avec :

  C_1=0,2 cal/g/oC ; C_2=0,16 cal/g/oC ; L=59,09 cal/g ; T_S=1502 oC

  Densité de l'acier 7 g/cm^3

  Température initiale de l'acier 1537 oC

Vitesse de coulée 1,23 m/mm

Flux extrait en lingotière :

hauteur (cm)	flux (cal/cm^2/s)
10	20,70
20	48,98
30	45,46
40	37,59
50	32,36
60	26,28
70	26,89

- Le pas de discrétisation en z est de 10 cm, en x 12,5 /40 cm.

Sur chaque zone, les valeurs des coefficients d'échange sont :

zone	1	2	3	4	5	6	7
h_i cal/cm^{2}s°c	0,015	0,010	0,010	0,006	0,003	0,003	0

Le paramètre μ est égal à 2,5.

Nous présentons figures 5 et 6, les résultats obtenus pour 200 pas de z pour la température en bord de brame et l'évolution du front de solidification. Le temps de calcul sur IBM 370-168 est de 35 s.

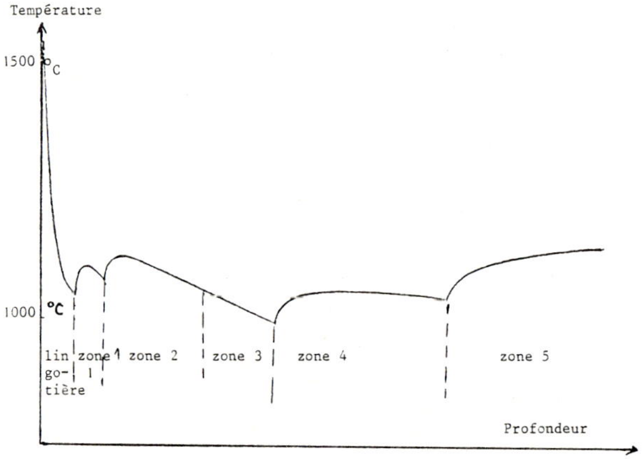

FIGURE 5 : Température en bord de brame

On constate le très fort refroidissement au niveau de la lingotière et l'influence des différents coefficients d'échange.

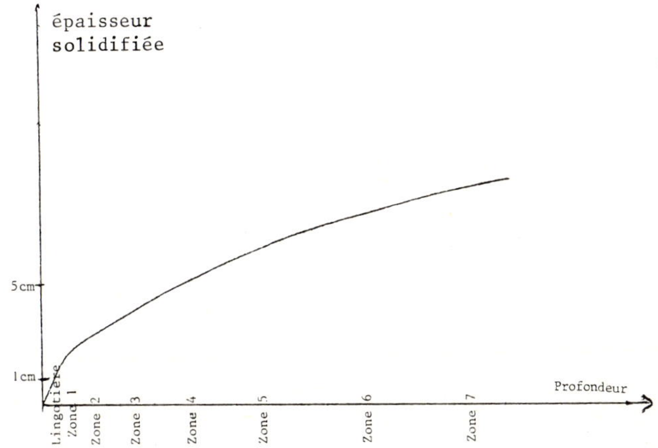

FIGURE 6 : Front de solidification

Exemple en dimension 2 : Solidification d'une billette

- les caractéristiques de l'installation sont les mêmes qu'en dimension 1. La billette est un carré de 10,5 cm de côté.On considère ici 1/4 de billette par raison de symétrie.
- caractéristiques thermiques de l'acier

On considère le cas d'un problème à deux fronts liquidus et solidus. L'enthalpie est alors de la forme :

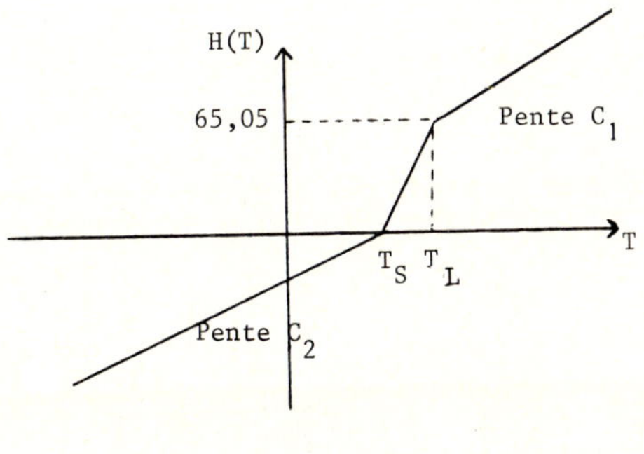

FIGURE 7

avec C_1 = 0,2 cal/g/°c ; C_2 = 0,16 cal/g/°c ; T_S = 1476°c ; T_L = 1508°c
densité de l'acier 7g/cm^3 ; Température initiale 1546°c ; Vitesse de coulée 1,23 m/mn
Le flux extrait en lingotière a les mêmes valeurs qu'en dimension 1.

Nous présentons sur les figures ci-dessous les résultats obtenus : pour les températures et l'évolution du front de solidification. Des résultats complets sont présentés dans M. Thirion [20].

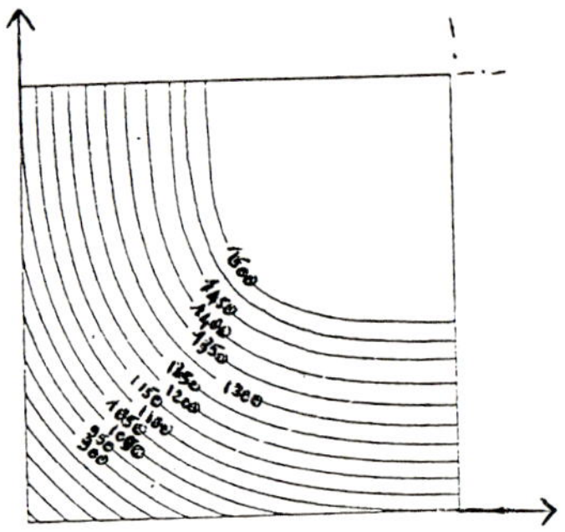

FIGURE 8 : Isothermes au niveau 120 cm

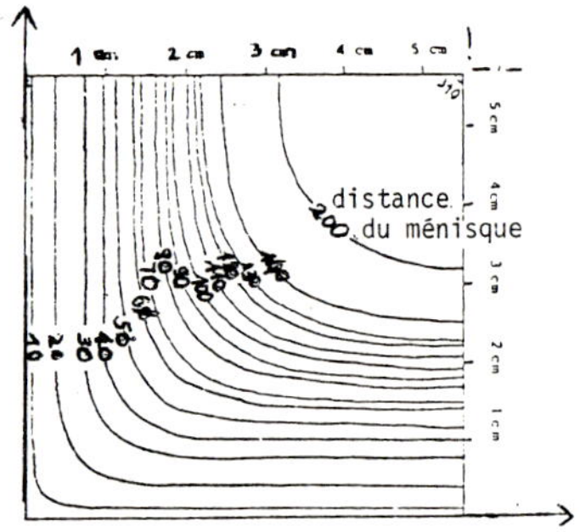

FIGURE 9 : évolution du front du solides

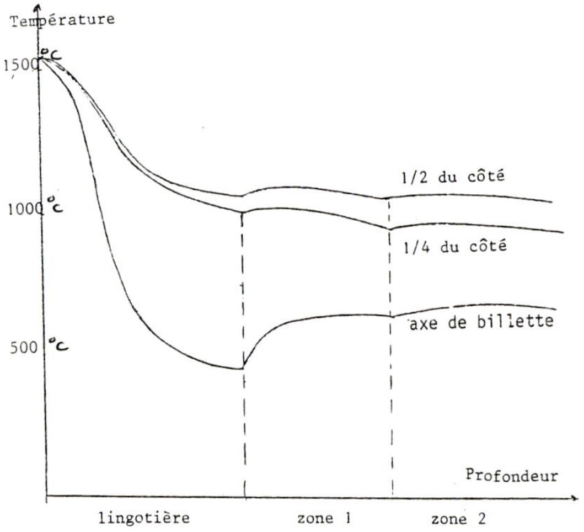

FIGURE 10 : Température au bord de la billette

2) Résolution du problème de contrôle

Nous présentons des résultats uniquement sur le problème 1 (optimisation à vitesse de coulée donnée) dans le cas de la dimension 1. Pour obtenir la vitesse optimale, on opère par dichotomie sur V. Nous considérons successivement les deux types de contraintes :

cas i) Contrainte sur le contrôle

Les caractéristiques de l'installation et les données thermiques de l'acier sont identiques au cas de la simulation dans l'exemple en dimension 1.

La vitesse de coulée est 1,50 m/mn

Les contraintes sont les suivantes :

zone	h_{im}	h_{iM}	α_i
1	0,015	0,020	0,40
2	0,010	0,016	0,5790
3	0,010	0,018	0,4902
4	0,0060	0,0157	0,8285
5	0,0030	0,0144	1,0492
6	0,0030	0,0066	0,9125
7	0,0025	0,0025	0

D = 160/4180 (débit maximum)

On obtient la solution suivante :

zone	1	2	3	4	5	6
h_i cal/cm^{2}s°c	0,015	0,0128	0,010	0,0132	0,00439	0,00483

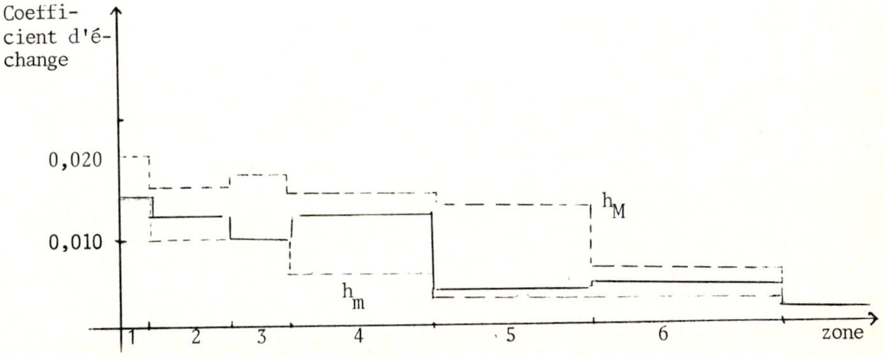

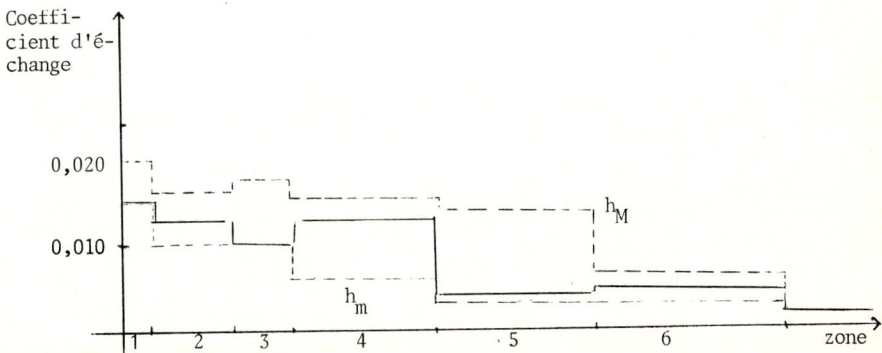

FIGURE 11 : Contrôle optimal et contraintes de borne

cas ii) Contrainte sur l'état

Les caractéristiques de l'installation et les données thermiques de l'acier sont toujours les mêmes, mais on considère seulement une installation de 16,50 m. La vitesse de coulée est de 0,9 m/mn. Les contraintes de borne sur le contrôle sont les mêmes qu'au cas précédent.

La contrainte sur l'état est $\theta \leq 502°c$ au niveau 16 m. On prend un coefficient de pénalisation $\varepsilon = 2 \times 10^{-1}$.

On obtient la solution :

zone	1	2	3	4	5	6
h_i $cal/cm^2 s°c$	0,01858	0,01544	0,01576	0,01146	0,00313	0,003

Pour ces valeurs, tout l'acier est solidifié et on a $\theta = 495°c$ au niveau 16 m.

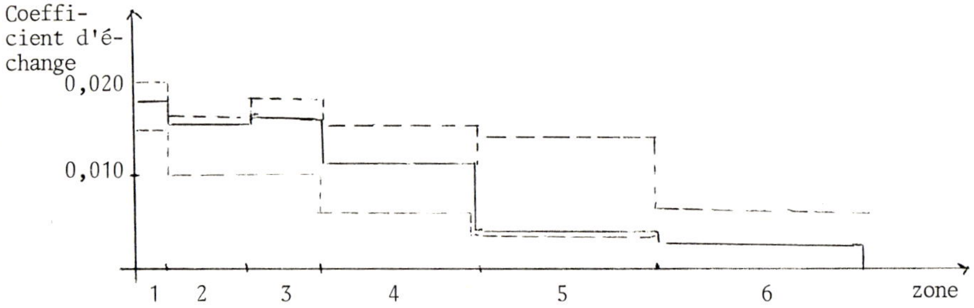

FIGURE 12 : Contrôle optimal et contraintes de borne

On remarque la structure très différente du contrôle optimal pour les deux cas ci-dessus.

La méthode présentée ici s'est avérée efficace pour résoudre le problème de l'optimisation du refroidissement secondaire en coulée continue d'acier. Il s'agit d'une méthode générale pour les problèmes associés à des phénomènes de changement de phase qui peuvent être modélisés avec l'enthalpie. Ce travail doit permettre, à terme, d'étudier le problème de l'automatisation complète d'une coulée continue.

REFERENCES

[1] R. ALBERNY Présentation de la coulée continue (Rapport Interne IRSID, Nov. 78)

[2] C. BAIOCCHI
Free Boundary problems in the theory of fluid flow through porous media (Int. Congress of mathematicians - Vancouver, Canada Vol.II (1974)

[3] A. BERMUDEZ, C. MORENO
Duality methods for solving variational inequalities (1978)

[4] F. BONNANS Contrôle optimal d'un problème à frontière libre. (Stage IRIA 1979)

[5] H. BREZIS Opérateurs maximaux monotones et semi-groupes de contraction dans les espaces de Hilbert (North Holland Publishing Company 1973)

[6] J.F. CIAVALDINI
Analyse numérique d'un problème de Stefan à deux phases par une méthode d'éléments finis. (SIAM J. of Num. Anal. Vol. 2, n° 3, Juin 1975).

[7] A. DAMLAMIAN
Problèmes aux limites non linéaires du type du problème de Stefan et inéquations variationnelles d'évolution (Thèse Paris 1976)

[8] A. DEGUEIL Résolution par une méthode d'éléments finis d'un problème de Stefan en termes de température et teneur en matériau non gelé (Thèse 1977)

[9] G. DUVAUT Problèmes à frontière libre en théorie des milieux continus (Rapport Laboria n° 185, 1976).

[10] R. GLOWINSKI, J.L. LIONS, R. TREMOLIERES
Analyse numérique des inéquations variationnelles (Dunod 1972)

[11] O. GRANGE, F. MIGNOT
Sur la résolution d'une équation et d'une inéquation parabolique d'évolution (J. of Functional Analysis 1972).

[12] J. HENRY Contrôle d'un réacteur enzymatique à l'aide de modèles à paramètres distribués. Quelques problèmes de controlabilité de systèmes paraboliques (Thèse 1978).

[13] J.M. LARRECQ, C. SAGUEZ, M. WANIN
Modèle mathématique de la solidification en coulée continue en tenant compte de la convection à l'interface solide-liquide (Revue de métallurgie Juin 1978).

[14] J.L. LIONS Introduction to some aspects of free boundary problems (Colloque Synspade Maryland 1975).

[15] R. ALBERNY, A. PERROY, D. AMORY, M. LAHOSSE
Optimisation du refroidissement secondaire en coulée continue de brames d'acier extra-doux (Revue de métallurgie Juin 1978).

[16] F. PICHON Aspects économiques de la coulée continue (rapport CESSID 76-125, Janvier 1977).

[17] C. SAGUEZ Contrôle optimal d'un système à frontière libre, application à l'optimisation du refroidissement secondaire en coulée continue d'acier (Rapport Laboria à paraître).

[18] C. SAGUEZ Contrôle optimal d'inéquations variationnelles avec observation de domaines (Rapport Laboria n° 286, Mars 1978).

[19] R. TEMAM A non linear eigenvalue problem. The shape at equilibrium of a confined plasma (A.R.M.A. Tome 60; 1975).

[20] M. THIRION Programmes de calcul de température dans une brame de coulée continue par la méthode des éléments finis (Rapport de Stage Ecole des Mines, 1979).

SESSION VI

NUMERICAL LINEAR ALGEBRA AND RELATED TOPICS

COMPUTING METHODS IN APPLIED SCIENCES AND ENGINEERING
R. Glowinski, J.L. Lions (editors)
North-Holland Publishing Company

RECENT DEVELOPMENTS IN THE SOLUTION OF LARGE SPARSE LINEAR EQUATIONS

Iain S. Duff
A.E.R.E. Harwell,
Didcot, Oxon OX11 ORA, G.B.

Abstract

We examine some recent techniques used in the solution of sparse linear equations with particular emphasis on their relative merits when used to solve practical problems. We consider both the solution of symmetric and unsymmetric systems and look at methods which use auxiliary storage devices in addition to those which work entirely in main memory. We also comment on some of the implications of the advent of parallel and pipeline processors. Finally, although not strictly covered by the title of this talk, we present a brief report on some recent work applicable to the in-core solution of problems in sparse non-linear optimization.

Contents

1. Introduction

This paper is not intended to be a comprehensive survey of techniques used in the solution of sparse problems. Space does not permit that. We have contented ourselves with a review of some of the areas in which there is much active research and in which we believe significant contributions will shortly be made to mathematical software. The references are by no means exhaustive. In each area we give either the basic reference or a very recent one and so we must apologise in advance to any researcher who feels neglected.

Throughout most of this paper we will be concerned with techniques for the solution of the set of equations

$$A\underline{x} = \underline{b} \tag{1.1}$$

where A is a large sparse matrix. This problem naturally divides into two distinct cases according to whether A is or is not positive definite. For symmetric positive definite A no pivoting is required to ensure stability of direct methods and a wide class of iterative techniques are effective; on the other hand, direct methods for unsymmetric A usually require some form of numerical pivoting. We discuss these two cases in sections 2 and 3 respectively. Even though the primary storage available on some of today's large mainframes is very large, the use of direct methods on very big problems will often require the use of auxiliary storage devices. The explosion in the usage of minicomputers also implies a need for out-of-core methods. We discuss methods using secondary storage in section 4. The fourth generation computers which employ forms of parallelism and/or pipeline processing are capable of performing arithmetic about 100 times faster than the best machines of the previous generation. In section 5, we comment on the effect of such architectures on sparse matrix software. Finally, because of the great interest voiced in private discussions at the GAMNI Congress in Paris last December (Duff (1979b)) we discuss, in section 6, some recent developments in the solution of sparse non-linear optimization problems.

Unless otherwise stated, the times quoted in the tables are in milliseconds on an IBM 370/168 and the codes were written in Fortran and compiled using the H extended compiler at an optimization level of 2.

2. Symmetric systems

Many of the symmetric matrices arising in practice, for example from the use of energy principles or discretizations of simple self-adjoint elliptic operators, are positive definite. For systems which are indefinite, we refer the reader to Duff (1979b) where we discuss the use of 2x2 block pivoting and techniques based on the Lanczos recurrence. If a system is nearly positive definite we still might be able to use some of the techniques described in this section (Golub and van Loan (1980)) or we could follow Concus et al (1975) by using the matrix splitting

$$A = M-N,$$

where M is positive definite, and solving (1.1) by the iteration

$$M\underline{x}^{(r+1)} = N\underline{x}^{(r)} + \underline{b} \quad .$$

We further discuss unsymmetric and indefinite problems in sections 3 and 4 but, for the rest of this section, we will concentrate on cases where A is symmetric and positive definite.

Perhaps the most significant trend in direct methods for the solution of symmetric problems is towards utilization of the underlying structure of the system. Thus, for example, when we consider the discretization of a partial differential equation over a region, we work with the geometry of the region and the mesh itself rather than the resulting sparse matrix. There are two distinct benefits in using this approach. On the one hand, the routines for determining pivot order in a Cholesky-based decomposition scheme can take proper advantage of any peculiarities in the underlying structure and can execute orders of magnitude faster than the actual numerical factorization. This is in marked contrast to the codes used at the beginning of the 70's when the ordering phase was more time-consuming than the actual factorization. On the other hand, when we work with the underlying structure, we can often avoid generating even the original matrix in its entirety and thus save on core requirements. This we discuss in section 4.

In this section, and later in this paper, we will use the term "model problem". We mean, by this, the matrix obtained from the 5-point discretization of Poisson's equation with Dirichlet boundary conditions on a uniform square grid with n internal grid nodes in each direction. Thus, the order of the matrix involved is n^2.

The best general purpose pivoting technique is to select as pivot the diagonal element in the row with fewest non-zeros (minimum degree criterion). Traditionally, this has been implemented as a local technique where, at each stage, this same criterion is used to select the next pivot from the current reduced or active matrix. It is difficult for such a local strategy to take account of the global connectivity of a problem. This is particularly true of regular grid problems where there are many ties (that is, many rows with the same minimum number of non-zeros). For example, on a matrix arising from the 9-point discretization of Laplace's equation on a 64x64 grid two different tie-breaking strategies yield 111328 and 184565 non-zeros in their respective factors (Duff et al (1976)). This, of course, makes the local minimum degree algorithm difficult to study theoretically. Additionally, it is difficult to tell in advance how many non-zeros will be in the resulting decomposition, so that we will often allocate far more storage than is necessary to ensure successful pivot selection and subsequent decomposition. These problems are addressed by more globally oriented methods which additionally can substantially reduce the time spent in pivot selection and can be organised so that the storage required to implement the pivot selection is known a priori and is dependent on the number of non-zeros in the original matrix rather than the factored form.

A class of such methods are the dissection methods (for example, George (1973)) whose performance on regular grids can be easily determined and which achieve the proven lower asymptotic bound for all direct methods based on Gaussian elimination of $O(n^3)$ floating point operations* and $O(n^2 \log n)$ storage for the model problem. The time to select pivots is negligible and the constants multiplying these highest order terms are dependent on the implementation; 20 and 8 might be typical figures. Much recent work has examined the automation of these dissection methods and their extension to more general regions (George and Liu (1978)). These techniques are particularly useful on grid-based problems such as arise in the discretization of elliptic operators but are not so effective on problems in networks where the connectivity is not so regular.

A fairly new approach (for example, George and Liu (1979), or Eisenstat et al (1979)) which is particularly useful on matrices arising from finite element discretizations involves the use of "generalized elements". We illustrate this in Figure 2.1 below.

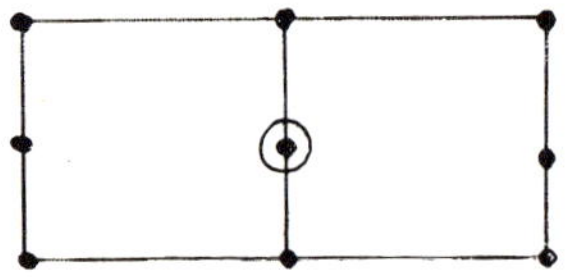

Figure 2.1
Illustration of generalized element approach

In this figure there are two elements each with 6 nodes. If variable(s) corresponding to the circled node are chosen as pivots in Gaussian elimination, then all the remaining nodes in the two elements become connected and we have, in effect, a new element containing 8 nodes. This illustrates the innermost loop of the generalized element approach. New "generalized elements" formed in

---

We use the term floating point operation to mean a single floating point operation of the type +,-,,÷.

this way together with any original unchanged elements will again have a node or nodes in common with each other (like the circled node above). If these common nodes were eliminated then the elements sharing these nodes would become amalgamated to form a single new element. This process of node elimination and element amalgamation continues until only one element is left whence its nodes are eliminated to complete the ordering. The principal importance of these techniques is that each individual connection between nodes is not explicitly recorded but rather a note of inter-element connectivities and which variables are in each element is kept. Thus, at the innermost loop in the generalized element model, storage is linear in the number of variables in each element as is the work involved in node elimination and element amalgamation. In numeric factorization and traditional symbolic factorization (as employed by the local strategy mentioned earlier), work and storage at this innermost level is quadratic in the number of variables. Recently, Reid (personal communication) has been experimenting with the use of such techniques to implement a minimum degree algorithm on both element and non-element problems and we show the results of some runs using his code in Table 2.1. A prepass through the non-element problems artificially converts them to element form. The time for this prepass is included in the times given in Table 2.1.

	Finite Element Problems		Non-Element Problems	
Order of matrix	1270	3466	199	130
Non-zeros in lower ◺ (incl.diag.)	4969	13681	536	713
Local Strategy	1551	6048	86	83
Generalized Elements	730	2047	90	173

Table 2.1 Times for pivot selection using two implementations of minimum degree algorithm

The times in this talk are rather flattering to the local strategy since it takes some advantage of the structure by effectively eliminating many nodes simultaneously. The 1975 Harwell minimum degree subroutine MA17E, which can be much faster than the original 1972 subroutine (MA17A) when there are many non-zeros in the factors, takes 62 seconds on the example of order 3466, that is, over an order of magnitude worse than the local strategy in the table. We will see more evidence of the efficacy of such a generalized element approach in section 4.

Recently, there has been much effort expended to increase the scope and power of a class of methods often termed "fast methods". They are so called because the asymptotic behaviour of the work required to solve the model problem is $O(n^2 \log n)$ or better as opposed to the optimal behaviour of $O(n^3)$ for methods based on Gaussian elimination. The storage for these fast methods is proportional to n^2 and the constants are, in general, less than those for the nested dissection methods. Once some of the stability problems had been overcome (for example, Buneman (1969)) these methods have become increasingly popular and most recent work has gone into extending these techniques to more general equations (Swarztrauber (1974)), regions (Buzbee et al (1971), Temperton (1977a)) grids and boundary conditions (Schumann and Sweet (1975)).

The two techniques which such methods use are the fast Fourier transform (FFT) and block cyclic (or odd-even) reduction (see, for example, Buzbee et al (1970)).

If our basic 5-point equation is of the form

$$x_{i-1,j} + x_{i+1,j} + x_{i,j-1} + x_{i,j+1} - 4\, x_{i,j} = b_{i,j}$$

at the $(i,j)^{th}$ grid point $(0<i<n,\ 0<j<m)$, where $x_{i,j}$ is the value of the unknown at that point, then, if

$$\sum_{k=1}^{n-1} \hat{x}_{k,j} \sin({}^{ik\pi}/n)$$

is the Fourier sine expansion for $x_{i,j}$, the equations for the Fourier coefficients $\hat{x}$ are of the form

$$\hat{x}_{k,j-1} + \lambda_k \hat{x}_{k,j} + \hat{x}_{k,j+1} = \hat{b}_{k,j}$$

where $\lambda_k = 2\cos({}^{k\pi}/n) - 4$ and

$$\hat{b}_{k,j} = ({}^2/n) \sum_{i=1}^{n-1} b_{i,j} \sin({}^{ik\pi}/n)$$

We can thus solve our problem by

(i) Using an FFT to obtain $\hat{b}$.

(ii) Solving a tridiagonal system.

(iii) Performing the inverse transform using an FFT to obtain our solution x.

The other technique is that of block cyclic reduction. If $\underline{x}_j$ is our vector of solution values along grid line j in our grid (we assume variables are ordered in the i direction first), then our equations are of the form

$$\underline{x}_{j-1} + T\,\underline{x}_j + \underline{x}_{j+1} = \underline{b}_j \qquad (2.1)$$

where T is a tridiagonal matrix whose diagonal entries are -4 and whose off-diagonal entries are 1. If we multiply equation (2.1) by (-T) and add it to the equations for the $j-1^{th}$ and $j+1^{th}$ lines we get equations of the form

$$\underline{x}_{j-2} + (2I-T^2)\,\underline{x}_j + \underline{x}_{j+2} = \underline{b}_{j-1} - T\,\underline{b}_j + \underline{b}_{j+1} .$$

We can thus obtain equations for variables on every other line of the grid and so have halved the dimension of the problem we have to solve. This can be repeated on these resulting equations to reduce our problem even further and in principle we can continue for μ steps where $2^{\mu} \leq n < 2^{\mu+1}$. This basic reduction is unstable but can easily be made stable by expressing the right hand side differently (Buneman (1969)).

By optimally combining these techniques very fast "fast methods" can be obtained. For example, in FACR(ℓ) of Hockney (Hockney (1965), Temperton (1977b)), he first performs ℓ steps of cyclic reduction after which he uses the FFT technique using cyclic reduction to solve the tridiagonal system at step (ii). By optimally choosing ℓ, Hockney produces a method with $O(n^2 \log\log n)$ behaviour.

It is important to realize that these methods are so far only applicable on a restricted (although commonly occurring) set of problems and, even within this class, are noticeably degraded by more irregular grids or more general equations. We put these techniques in perspective by the results in Table 2.2 where the results using the fast solvers were obtained by Thompson at Harwell (personal communication). The methods in Table 2.2 are listed in order of increasing applicability. We indicate the range of problems on which each works below:

POT1 : Poisson solver on regular mesh.
POIS : Poisson in one direction only.
BLKTRI : General separable equation.
MA31 : Symmetric solver, here using minimum degree criterion.
MA28 : General purpose unsymmetric solver.

Method \ Grid	32x32	64x64	128x128
POT1 (Hockney)	19.5	66.4	273.4
POIS (NCAR)	73.3	356.2	1710
BLKTRI (NCAR)	107	572	2818
[Start up time]	[51]	[189]	[686]
MA31 (Harwell)	58	299	794
[Start up time]	[1470]	[13400]	[132000]
MA28 (Harwell)	60	-	-
[Start up time]	[8830]		

Table 2.2 A comparison of different on methods on regular model problems. (Times in milliseconds on an IBM 3033)

The start up times for the two Harwell subroutines (MA31 and MA28) are the time taken to choose pivots and factor the matrix. Although we see that their solution times are competitive, this high start up time and the much larger storage overheads make such techniques increasingly unattractive for very large regular grids. We see, from the results in this table, that, as the fast solver becomes more general in its applicability it also becomes slower on the simple model problem.

A popular current area of research is that of preconditioning to accelerate simple iterative techniques. We give some examples to illustrate the flavour of this approach in this paper and draw attention to the papers by Manteuffel and Glowinski et al at this conference which deal in more detail with some preconditioning techniques. This class of methods shows perhaps the best promise for solving problems arising from the discretization of very large two dimensional or three dimensional structures.

Since one of the most popular such methods is that of preconditioned conjugate gradients (see for example, Duff (1979c)) we describe it briefly here.

The conjugate gradient algorithm (Hestenes and Steifel (1952)) only requires that the user can form the vector $A\underline{x}$ given the vector $\underline{x}$ and is thus an ideal algorithm for utilising sparsity. With exact arithmetic this method requires at most n iterations and even with roundoff good approximations to the solution can often be found in far less than n steps. Indeed since the conjugate gradient algorithm minimizes the quantity $\underline{r}^T A^{-1} \underline{r}$ (where $\underline{r}$ is the residual vector) over all vectors of the form $\underline{x}^{(0)}+P_{i-1}(A)\underline{r}^{(0)}$ where P_{i-1} is any polynomial in degree i-1 and $\underline{x}^{(0)}$ and $\underline{r}^{(0)}$ are the initial guess and residual respectively, we have that the residual at the ith step, $\underline{r}^{(i)}$ is given by

$$\begin{aligned} \underline{r}^{(i)} &= \underline{b}-A\underline{x}^{(0)} - AP_{i-1}(A)\, \underline{r}^{(0)} \\ &= (I-AP_{i-1}(A))\, \underline{r}^{(0)} \\ &= Q_i(A)\, \underline{r}^{(0)} \end{aligned}$$

and we see that, if we expand $\underline{r}^{(0)}$ in terms of the eigenvectors of A, then, if a polynomial of degree i exists whose values at every eigenvalue is small and whose value at the origin is unity, vectors of the form $Q_i(A)\, \underline{r}^{(0)}$ with small norm exist. Because of its minimization property conjugate gradients will find such a vector. Indeed, if the spectrum of A has its eigenvalues in, say, k tight clusters then only k iterations would be required. In the extreme case of the identity matrix only one iteration would be needed. The idea of preconditioning is to alter the spectrum of the iteration matrix so that it is favourable for the conjugate gradient algorithm in the sense that k, above, is small. One method of generating such a preconditioning matrix is to obtain factors of a perturbation of the original matrix. That is to obtain $\tilde{L}$ such that

$$A+E = \tilde{L}\tilde{L}^T$$

whence the preconditioned matrix

$$\tilde{L}^{-1} A \tilde{L}^{-T}$$

will hopefully have most of its eigenvalues near one. The ICCG(0) method (Meijerink and van der Vorst (1977)) forces $\tilde{L}$ to have the same sparsity as A and has been shown by Kershaw (1978) to work well on matrices arising from discretizations of variable coefficient diffusion equations. Some additional benefits may be obtained if we allow $\tilde{L}$ to also have non-zeros in one or more diagonal bands near the main diagonal or immediately inside the outermost non-zero band of A (for example, Meijerink and van der Vorst (1978)). A more general preconditioning matrix is employed by Harwell subroutine MA31 (Munksgaard (1979)). This allows $\tilde{L}$ to have a pattern between that of A and that of the true factors of A, non zeros being dropped from the factors if they fall below a tolerance level, as suggested by Jennings and Malik (1978). By changing the tolerance level this code can be adjusted between the extremes of no fill (ICCG(0)) and all fill (i.e. a direct method based on Gaussian elimination).

It is important to note that the effectiveness of these techniques lies in their ability to cluster the eigenvalues as much as in their reduction of the condition number of A. Indeed many of these preconditionings do not affect the condition number greatly but do cause a significant drop in conjugate gradient iterations because of eigenclustering (see, for example, the talk by Manteuffel at this conference). This makes the analysis of such methods and the theoretical prediction of good preconditioning methods difficult (for example, Greenbaum (1979)).

One can also look at these methods the other way round, as it were. Assume we have an approximation $\underline{x}'$ to the solution so that the residual $\underline{r}$ is given by

$$\underline{r} = \underline{b} - A\underline{x}'$$

and assume that the matrix M is some approximation to A^{-1}. Then we use the iteration

$$\underline{\delta x}' = M\underline{r}$$

to hopefully get $\underline{x}'+\underline{\delta x}'$ as a better approximation to the solution. The convergence of the iteration can be very slow and so techniques like Chebyshev acceleration or conjugate gradients can be used to accelerate it. Additionally various methods can be used for obtaining M, for example SSOR splitting, incomplete Cholesky decomposition, or more general splittings (see section 3). The various combinations of preconditionings and iterative methods give rise to a whole battery of techniques from which no front-runner has yet clearly emerged. For regular problems, however, incomplete Cholesky decomposition accelerated by conjugate gradients appears to do well.

3. Unsymmetric systems

When using direct methods to solve general unsymmetric systems, we need to perform numerical pivoting. Because we need to monitor the size of the non-zero elements as the elimination progresses, we must, even during the ordering phase, retain all (or at least, part) of the partially decomposed system. This militates against the rapid pivot selection procedures discussed in section 2 although, as we shall see in the next section, it may still be possible to use the underlying structure of the problem. Fortunately, the freedom to pivot off the diagonal can give some compensating gains in speed and storage. Thus codes for the solution of such systems (for example, Duff (1977)) can be very efficient. We discuss the use of such general purpose techniques further in Duff (1979a) and Duff (1979b).

Another area of very active current research lies in the extension of the preconditioned iterative techniques of section 2 to the unsymmetric case. It is, of course, possible to work with the normal equations

$$A^T A\underline{x} = A^T\underline{b} \tag{3.1}$$

which will be symmetric positive definite but then the amount of work at each step is increased and convergence can be poor because of the bad conditioning of system (3.1). Most of the recent work avoids the formation of these equations.

If we can write A as

$$A = M-N$$

where $M=(A+A^T)/2$ is positive definite and $N = -(A-A^T)/2$ is skew-symmetric then the equations (1.1) can be written in iterative form

$$M\underline{x}^{(r)} = N\underline{x}^{(r-1)} + \underline{b}$$

and various techniques have been proposed for accelerating this iteration. One of the original methods, that of generalized conjugate gradients, was described at a previous IRIA meeting (Concus and Golub (1975)), while more recently Chebyshev acceleration techniques have been proposed by Manteuffel (1977) and acceleration by Lanczos recurrences by Widlund (1978).

Recently, van der Vorst and van Kats (1979) have had great success with a shifted version of Manteuffel's algorithm where incomplete Cholesky decompositions are used to precondition M.

4. Techniques using auxiliary storage

Even using asymptotically optimal ordering techniques, the storage required to hold the factors of the decomposed matrix increases as $O(n^2 \log n)$ for our model problem consequently restricting the size of n permissible for an in-core solution. Thus, one is forced to consider either iterative or semi-iterative methods, or to employ auxiliary storage devices (I include all storage available through paging systems as primary storage).

One technique which lends itself very readily to out-of-core implementation is that of the variable band scheme (for example, Jennings (1977)).

Irons (1970) describes a frontal method scheme for the solution of symmetric systems and this was generalized by Hood (1976) to handle unsymmetric systems. We illustrate this solution procedure in Figure 4.1 where we have shown the situation part-way through the assembly and elimination process. The whole of this figure corresponds to the reduced or active matrix at an intermediate stage, the rows and columns which have already been pivotal are not included and in practice could have been written to a buffer or auxiliary store. When more elements are assembled the front may grow and more rows and columns may become fully assembled. To maintain the integrity of our diagram these would be permuted to augment A_1 and the corresponding rows and columns of the complete reduced matrix. The only part of the reduced matrix which must be held in core is the front matrix of Figure 4.1. Usually it will be much smaller relative to the size of the complete reduced matrix than our figure suggests.

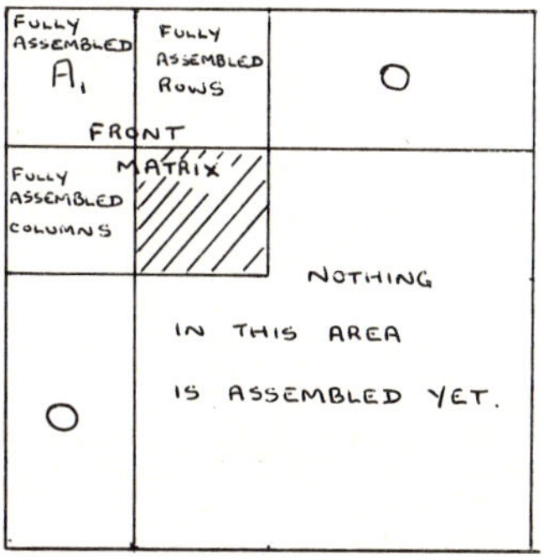

Figure 4.1
Illustration of frontal scheme

Although the term element assembly refers to finite element discretizations such a technique can also be applied to equations generated using finite differences where a variable is fully summed whenever the equation in which it appears for the last time is reached. Clearly the equations can be generated as required or held on backing store and read as needed.

The block A_1 in Figure 4.1 contains all the non-zeros which are eligible as pivot candidates at that particular stage in the assembly/elimination. Since we can choose any diagonal as pivot in the symmetric positive definite case without dangers of numerical instability, the entire block A_1 can be eliminated before requesting further assemblies. In the symmetric indefinite case, we will often be able to similarly choose the non-zero diagonal elements as pivots or to pivot on a 2x2 symmetric block in A_1. Only in fairly perverse cases will we need to do further assemblies before eliminating all the variables in A_1 and so again we should be able to keep the size of the fully assembled but uneliminated block low. However, in the unsymmetric case, the more stringent numerical criterion (usually requiring that the pivot is large relative to other non-zeros in its column and/or row) often means that we are unable to do many eliminations before further assemblies are required. Thus the size of the fully assembled but uneliminated block will be higher as will the core requirements of the unsymmetric solver on the same underlying geometry.

It might be thought that the overheads of transferring information to and from backing store could make such schemes costly in computing time but some theoretical physicists at Harwell (Cliffe et al (1978)) have been solving large (order over 7000) problems in fluid flow with a much improved version of Hood's 1976 program and have found that, even with these I/O overheads, the time per inner-loop operation was only about 1.5 microseconds on the IBM 370/168. We reported on the results of their runs at greater length in Duff (1979b). Since then we have greatly modified their subroutines, particularly the user interface, and have incorporated them in the Harwell Subroutine Library under the name MA32AD. Our experience with this routine has been that, while it is slower on our model problem than an efficient minimum degree algorithm working entirely in core it is still competitive (particularly if the number of grid points in one direction is small) and, of course, will normally require considerbly less core storage.

At any stage during the frontal method we have a partially assembled active matrix and we note that this is rather like a single element in structure with fully assembled variables corresponding to internal nodes and partially assembled ones to boundary nodes (the shaded region in Figure 4.1). Therefore there is no reason why we cannot store this "superelement" and start assembling and eliminating on another part of the structure. This gives rise to a multi-frontal approach (for example, Speelpenning (1978)) which, in addition to handling more complex geometries than the uni-front approach, can give significant gains when several "limbs" have identical structure. Although engineers have used methods akin to these for some time under the term "substructuring", it is only recently that numerical analysts have begun developing good mathematical software along these lines.

This is very similar to the work on generalizing nested dissection (the generalized element approach) already mentioned in section 2 and although codes based on such methods are still in their infancy some really dramatic results are being obtained.

Reid has run an out-of-core multi-frontal scheme code on some medium-to-large (order about 3000) finite element problems and has only required 1.7 seconds to find the pivot order (actually a globally oriented minimum degree scheme) and 71 seconds, at about .8 Megaflops (million floating point operations per second), to subsequently obtain the numerical factors. This supports a remark we made at the beginning of section 2 where we said that with modern codes the selection of pivots can be insignificant compared with the actual factorization time.

5. Impact of fourth generation computing

There has been much excitement generated recently by the advent of computers using parallel or pipeline processing. Jordan (1979) gives a very nice description of several of these new generation machines and their individual attributes. In pipeline processing each task (for example, multiplication) is broken up into a sequence of subtasks and the separate subtasks can be performed simultaneously on different operands. Thus, for example, when we are adding two vectors the various components at an intermediate stage may be fully or partially summed depending on how far they have progressed through the pipeline. Using these techniques some pipelined machines (for example, the Cray-1) have been able to achieve operating speeds in excess of 100 Megaflops, that is about 100 times faster than most of the large third generation computers. In Table 5.1 below we show timings obtained by Reid (private communication, 1978) for Gaussian elimination on full matrices on the IBM 370/168 and the Cray-1. We must stress that, for these runs, no changes were made to the Fortran code to make it vectorize more easily

Randomly generated values. Order n.	Cray-1	168	Ratio
50	7	.84	8.3
100	11.3	.92	12.3
200	16.8	.96	17.5
400	22.6	.98	23.1

Table 5.1 A comparison of Megaflop rates for a full matrix code run on the Cray-1 and the 168

and that it is almost certainly possible to achieve even higher Megaflop rates for the Cray-1 were this so done. Naturally such changes would have little effect on the IBM 370 times. Yet further gains might be obtainable if Cray assembly coding were used.

However, to take advantage of this ability to process different subtasks simultaneously the pipeline must be full and there is often a significant start up time required to fill it before full operating speeds can be attained. Thus, we might expect such machines to work well on calculations with long vectors where the start up times will be comparatively small. We see this trend in the results shown in Table 5.1 above, where as our average vector length increases so does our Megaflop rate. However, in using typical techniques for general sparse systems, most vectors will be short. Additionally, most storage schemes store these short vectors in packed form where the position of each non-zero in the matrix is indicated by an integer in an auxiliary array. Both these features militate against efficient vectorization and even using many different techniques to code the inner loop of a typical sparse matrix calculation (shown in Figure 5.1 below), Dembart and Neves (1977) have been unable to achieve the high speeds of Table 5.1.

```
      DO 100 JJ=J1,J2
      J=ICN(JJ)
      W(J)=W(J)-AMULT*A(JJ)
  100 CONTINUE
```

Figure 5.1

Typical innermost-loop code in sparse matrix calculation. W is a work vector of dimension the order of the system and the packed form is held in arrays A (real values) and ICN (column indices)

For his problems, also arising principally from network calculations, Pottle (1979) has argued that the use of large vector machines is not cost effective even if one can afford the very high price tag of a machine like the Cray.

We illustrate this poor performance, in Table 5.2, where we show runs of a general purpose sparse matrix code (MA28A) on the 168 and the Cray. The difference in times can be completely explained by the faster scalar speed of the Cray so that no vectorization of our code is apparent. We should add that this is a particularly extreme example of non-vectorization.

Model problem grid-size	Cray-1	IBM 168	Ratio
10x10	62	140	2.3
20x20	688	1487	2.2
30x30	5824	10940	1.9

Table 5.2 Times (in milliseconds) for a general purpose sparse matrix code (MA28A) on the IBM 370/168 and the Cray-1

In this section, we have presented the bad news first principally to counteract the euphoria with which many people view these new machines. Their advent is certainly not the panacea to all ills in the sparse matrix world although there are areas where their impact is having a profound influence on numerical computation. This is particularly true where large problems must be solved in real time. It is no coincidence that weather prediction centres in Europe and the United States have been among the first customers for these powerful machines. The large elliptic partial differential equations which are commonplace in such an environment give rise to sparse problems of a far more regular pattern than those from network applications; we follow Dembart and Neves in calling these grid problems.

For these grid problems much software has been developed which takes advantage of the architecture of these new machines. Many efficient methods have been developed for handling tridiagonal and band systems (for example, Sameh and Kuck (1978)) and efficient implementations of the fast Fourier transfom (Temperton (1979)) have been studied.

Another approach has been considered by Pottle (1979) where he creates a parallel machine by linking together inexpensive mini or micro computers connected through a common data bus. This technique is still in its early development as is the work at Carnegie-Mellon (for example, Kung and Leiserson (1979)) but I suspect we shall see a burgeoning of special purpose machines for particular classes of problems in the near future. Both Pottle and Dembart and Neves have been very impressed at the cost performance ratio of the AP-120B pipelined floating point unit from Floating Point Systems, Inc. which they claim to give 1/10th the performance of the Cray at 1/100th of the cost. It still is not the universal panacea since it needs to be front-ended by another machine and requires specifically designed software to operate at peak efficiency.

The implications of these new architectures on numerical algorithm design are significant. Certainly, general purpose pivoting techniques such as the minimum degree algorithm are difficult to implement efficiently and methods which either preorder the system so that it can be decomposed into subsystems to be solved in parallelor which use longer vectors are to be preferred. In the first category, preordering to block triangular form (see Duff (1979b), for example) or to bordered block diagonal form (for example, Pottle (1979)) are well established and further research into orderings to other suitable forms is well-advanced. In the second category, where we would like to achieve the performance in Table 5.1 as opposed to the performance in Table 5.2, techniques like the frontal and multi-frontal approach discussed in the previous section look promising since the inner-loop is full vector code. It has even been suggested (Dembart and Neves (1977)) that one can improve performance by adding explicitly held zeros to the packed form in order to create artificially longer vectors! It is possible that the extra computation involved is more than compensated for by the relative reduction in start up times for vector processing.

6. Sparse non-linear optimization

We report on some recent work in this area even though it is not strictly covered by the title of this paper. This is largely because of the interest shown when I discussed this work informally during the GAMNI Congress in December 1978 (see reference, Duff (1979b)).

Although gradient methods have historically been popular for solving sparse non-linear problems, recently much work has been done on adapting more powerful second order techniques to take sparsity into account. Dennis, Marwil, Powell, Schnabel, Shanno and Toint have all contributed much to this area and for our discussion in this section we have drawn heavily on the work of Powell and Toint (for example, Toint (1978) and Powell and Toint (1979)) and are indebted to comments made by Powell while giving a recent seminar at Harwell.

The basis of these second order techniques is to minimize over vectors $\underline{d}$, at each stage, the quadratic approximation

$$F(\underline{x}_k+\underline{d}) = F(\underline{x}_k) + \underline{d}^T\underline{g}_k + \tfrac{1}{2}\underline{d}^T B_k\underline{d}$$

to F in the neighbourhood of $\underline{x}_k$. In quasi-Newton methods, which we now discuss, B_k is an approximation to the Hessian, or second derivative matrix, of F at point $\underline{x}_k$ and we obtain the correction $\underline{d}$ to $\underline{x}_k$ (which we may scale) by solving the linear system

$$B_k\underline{d} = -\underline{g}_k .$$

If the actual Hessian is sparse we would like B_k to be also.

Recently, two approaches have been developed to generate a sparse approximation to the sparse second derivative matrix. The first uses finite difference schemes utilizing the sparsity structure of the Hessian to reduce the number of gradient evaluations required. The second technique uses an updating formula which preserves sparsity. In the rest of this section, we discuss these two approaches.

At each stage, we wish to obtain an estimate for B_k without using analytic derivatives (it is difficult for the user to program them correctly even when it is possible to calculate them, and they are often expensive to evaluate) but we do not require high accuracy in our approximation. We will consider the use of finite differences and updating formulae in obtaining the B_k.

If

$$\underline{\delta} = (\underline{x}_{k+h} - \underline{x}_k)$$

and

$$\underline{\gamma} = (\underline{g}_{k+h} - \underline{g}_k) \qquad \text{where } \underline{g}_k = \underline{g}(\underline{x}_k)$$

then we can calculate B_k by requiring it to satisfy the equation

$$B_k\underline{\delta} = \underline{\gamma}$$

for a carefully chosen set of displacements $\underline{\delta}$. In general, we would expect to require n+1 gradient evaluations (where n is the dimension of our problem) to calculate all the elements of B_k. However, as in the case of sparse Jacobians (Curtis et al (1974)), we can use the sparsity structure of B_k, assuming it is known a priori, to reduce this number. For example, if B_k is tridiagonal, then a displacement of the form (h,0,0,h,0,0,h,...) enables the 1st,4th,7th,10th... columns of B_k to be calculated simultaneously and the remaining columns can be obtained using the successive displacements (0,h,0,0,h,...) and (0,0,h,0,0,h,0,0,h,...). Thus, irrespective of the order of B_k, we can calculate all of its non-zero entries with only four gradient evaluations. Powell and Toint (1979) have shown how the symmetry of the Hessian can be used to reduce this work even further. For example, the evaluation of a matrix whose pattern has non-zeros only on the diagonal and in the last row and column would require n+1 gradient evaluations but, if symmetry is exploited successive displacements of (h,h,h,h,...h,0) and (0,0,0,...0,h) can be used to find diagonal entries and non-zeros in the last column respectively, whereupon the symmetry immediately gives us the elements in the last row.

If we use a conventional updating formula to get our estimate at step k+1, B_{k+1}, from B_k then the traditional formulae of DFP or BFGS do not take account of sparsity and so B_{k+1} will be full even if B_k is sparse. If we express our update in variational form we have the following. If

$$E = B_{k+1} - B_k, \qquad \underline{\delta}_k = \underline{x}_{k+1} - \underline{x}_k, \qquad \underline{\gamma}_k = \underline{g}_{k+1} - \underline{g}_k$$

then we require that

$$||E||_F \text{ is minimized subject to}$$

(i) The quasi-Newton condition

$$B_{k+1}\underline{\delta}_k = \underline{\gamma}_k$$

is satisfied.

(ii) E is symmetric.

(iii) $E_{ij} = 0$ for all (i,j) for which $\nabla^2 F_{ij} \equiv 0$, $(i,j) \in S$, say

and we can express this as a constrained minimization problem with Langrangian

$$\tfrac{1}{2}\Sigma E_{ij}^2 + \underline{\lambda}^T [\underline{r}_k - \tfrac{1}{2}(E + E^T)\,\underline{\delta}_k] + \sum_S \mu_{ij} E_{ij}$$

where

$$\underline{r}_k = \underline{\gamma}_k - B_k\underline{\delta}_k \,.$$

This gives us a solution of the form

$$E_{ij} = \tfrac{1}{2}[\lambda_i\delta_j + \delta_j\lambda_i] \qquad (i,j) \notin S$$

which, together with equation

$$E\,\underline{\delta}_k = \underline{r}_k$$

gives a set of linear equations for $\underline{\lambda}$ whose coefficient matrix is positive definite and of the same structure as $\nabla^2 F$.

In practice, Powell and Toint currently recommend that a good technique is to use finite difference approximations every 5 or 6 iterations and to use the updating scheme at the intermediate steps. Work is currently being done for weighted problems and for variable metric methods where the B_k are also required to be positive definite (for example, Toint (1979)).

Acknowledgements

I would like to thank Al Erisman, Gene Golub and Tom Manteuffel for their constructive criticism and comments on an extended abstract of this paper and John Reid for his useful comments on the first draft of this manuscript. Additionally, I am indebted to Ben Dembart and Christopher Pottle for comments on their experience with fourth generation machines.

References

Buneman, O. (1969). A compact non-iterative Poisson solver. Inst. for Plasma Res. Stanford. Report 294.

Buzbee, B.L., Dorr, T.W., George, J.A. and Golub, G.H. (1971). The direct solution of the discrete Poisson equation on irregular regions. SIAM J. Numer. Anal. 8, pp. 722-736.

Buzbee, B.L., Golub, G.H. and Nielson, C.W. (1970). On direct methods for solving Poisson's equations. SIAM J. Numer. Anal. 7, pp. 627-656.

Cliffe, K.A., Jackson, C.P., Rae, J. and Winters, K.H. (1978). Finite element flow modelling using velocity and pressure variables. Harwell Report AERE-R.9202.

Concus, P. and Golub, G.H. (1975). A generalized conjugate gradient method for non-symmetric systems of linear equations. Comput.Sci.Dept. Stanford Report STAN-CS-75-535.

Concus, P., Golub, G.H. and O-Leary, D.P. (1975). A generalized conjugate gradient method for the numeric solution of elliptic partial differential equations. Comput.Sci.Dept. Stanford Report STAN-CS-75-533.

Curtis, A.R., Powell, M.J.D. and Reid, J.K. (1974). On the estimation of sparse Jacobian matrices. J. Inst. Math. Appl. 13, pp.117-120.

Dembart, B. and Neves, K.W. (1977). Sparse triangular factorization on vector computers in Special Report EL-566-SR. Electric Power Research Institute.

Duff, I.S. (1979a). Practical comparisons of codes for the solution of sparse linear systems. pp.107-134. In Duff and Stewart (1979).

Duff, I.S. (1979b). Some current approaches to the solution of large sparse systems of linear equations. Harwell Report CSS 65.

Duff, I.S. (Ed.)(1979c). Conjugate gradient methods and similar techniques. Harwell Report AERE-R.9636.

Duff, I.S. (1977). MA28 - a set of Fortran subroutines for sparse unsymmetric linear equations. Harwell Report AERE-R.8730.

Duff, I.S., Erisman, A.M. and Reid, J.K. (1976). On George's nested dissection method. SIAM J. Numer. Anal. 13, pp.686-695.

Duff, I.S. and Stewart, G.W. (Ed.) (1979). Sparse matrix proceedings 1978. SIAM Press.

Eisenstat, S.C., Schultz, M.H. and Sherman, A.H. (1979). Software for sparse Gaussian elimination with limited core storage. pp.135-153 in Duff and Stewart (1979).

George, J.A. (1973). Nested dissection of a regular finite element mesh. SIAM J. Numer. Anal. 10, pp.345-363.

George, J.A. and Liu, J.W.H. (1978). An automatic nested dissection algorithm for irregular finite element problems. SIAM J. Numer. Anal. 15, pp.1053-1069.

George, J.A. and Liu, J.W.H. (1979). A quotient graph model for symmetric factorization. pp.154-175 in Duff and Stewart (1979).

Golub, G.H. and van Loan, C. (1980). Unsymmetric positive definite linear systems. Lin. Alg. and its Applics. To appear.

Hestenes, M.R. and Stiefel, E. (1952). Methods of conjugate gradients for solving linear systems. J. Res. Nat. Bur. Standards, 49, pp.409-436.

Greenbaum, A. (1979). Comparison of splittings with the conjugate gradient algorithm. Numer. Math. 33, pp.181-193.

Hockney, R.W. (1965). A fast direct solution of Poisson's equation using Fourier analysis. J. Assoc. Comput. Mach. 12, pp.95-113.

Hood, P. (1976). Frontal solution program for unsymmetric matrices. Int. J. Numer. Meth. Engng. 10, pp.379-399.

Irons, B.M. (1970). A frontal solution program for finite element analysis. Int. J. Numer. Meth. Engng. 2, pp.5-32.

Jennings, A. (1977). Matrix computation for engineers and scientists. Wiley.

Jennings, A. and Malik, G.M. (1978). The solution of sparse linear equations by the conjugate gradient method. Int. J. Numer. Methods Eng. 12, pp.141-158.

Jordan, T.L. (1979). A performance evaluation of linear algebra software in parallel architectures. pp.59-76 in Performance Evaluation of Numerical Software. L. Fosdick (Ed). North Holland.

Kershaw, D.S. (1978). The incomplete Cholesky conjugate gradient method for the iterative solution of systems of linear equations. J. Comput. Phys. 26, pp.43-65.

Kung, H.T. and Leiserson, C.E. (1979). Systolic arrays (for VSLI), pp.256-282, in Duff and Stewart (1979).

Manteuffel, T.A. (1977). The Tchebychev iteration for non-symmetric linear systems. Numer. Math. 28, pp.307-327.

Meijerink, J.A. and van der Vorst, H.A. (1977). An iterative solution method for linear systems of which the coefficient matrix is a symmetric M-matrix. Math. Comp. 31, pp.148-162.

Meijerink, J.A. and van der Vorst, H.A. (1978). Guide lines for the usage of incomplete decompositions in solving sets of linear equations as occur in practical problems. Report TR-9, ACCU, Utrecht.

Munksgaard, N. (1979). Solving sparse symmetric sets of linear equations by pre-conditioned conjugate gradients. Harwell Report CSS 67.

Pottle, C. (1979). Solution of sparse linear equations arising from power system simulation on vector and parallel processors. ISA Transactions 18, pp.81-88.

Powell, M.J.D. and Toint, Ph. L. (1979). On the estimation of sparse Hessian matrices. To appear in SIAM J. Numer. Anal.

Sameh, A.H. and Kuck, D.J. (1978). On stable parallel linear system solvers. J. Assoc. Comput. Mach. 25, pp.81-91.

Schumann, U. and Sweet, R.A. (1976). A direct method for the solution of Poisson's equation with Neumann boundary conditions on a staggered grid of arbitrary size. J. Comp. Phys. 20, pp.171-182.

Speelpenning, B. (1978). The generalized element method. Report UIUCDCS-R-78-946. Dept. of Computer Science. University of Illinois at Urbana-Champaign.

Swarztrauber, P.N. (1974). A direct method for the discrete solution of separable elliptic equations. SIAM J. Numer. Anal. 11, pp.1136-1150.

Temperton, C. (1977a). An improved algorithm for the direct solution of Poisson's equation over irregular regions. ECMWF Research Dept. Internal Report No.5.

Temperton, C. (1977b). On the FACR(ℓ) algorithm for the discrete Poisson equation. ECMWF Research Dept. Internal Report No.14.

Temperton, C. (1979). Fast Fourier Transforms on CRAY-1. ECMWF Research Dept. Internal Report No. 21.

Toint, Ph. L. (1978). Some numerical results using a sparse matrix updating formula in unconstrained minimization. Math. Comp. 32, pp.839-851.

Toint, Ph. L. (1979). A note about sparsity exploiting quasi-Newton updates. Report 79/5. Dept. of Maths. FUN. Namur. Presented at Math. Prog. Meeting in Montreal, Aug. 1979.

van der Vorst, H.A. and van Kats, J.M. (1979). Manteuffel's algorithm with preconditioning for the iterative solution of certain sparse linear systems with a non-symmetric matrix. Report TR-11, ACCU, Utrecht.

Widlund, O. (1978). A Lanczos method for a class of non-symmetric systems of linear equations. SIAM J. Numer. Anal. 15, pp.801-812.

COMPUTING METHODS IN APPLIED SCIENCES AND ENGINEERING
R. Glowinski, J.L. Lions (editors)
North-Holland Publishing Company

SOLVING STRUCTURES PROBLEMS ITERATIVELY WITH A SHIFTED INCOMPLETE CHOLESKY PRECONDITIONING

T. A. MANTEUFFEL

SANDIA LABORATORIES

ALBUQUERQUE, NEW MEXICO 87185

ABSTRACT

This paper describes a technique for solving the large sparse symmetric linear systems that arise from the application of finite element methods. The technique combines an incomplete factorization method called the shifted incomplete Cholesky factorization with the method of generalized conjugate gradients. The shifted incomplete Cholesky factorization produces a splitting of the matrix A that is dependent upon a parameter α. It is shown that if A is positive definite, then there is some α for which this splitting is possible and that this splitting is at least as good as the Jacobi splitting. The method is shown to be more efficient on a set of test problems than either direct methods or explicit iteration schemes.

1. INTRODUCTION

High order finite element models of three dimensional structures problems yield large sparse positive definite linear systems with some unique characteristics. They are generally more dense than their finite element counterparts. The zero pattern is more complex. It reflects the complexity of the domain of the model and the irregularity of the mesh. These matrices do not possess Young's property A (Young [10]) nor are they M-matrices, H-matrices nor diagonally dominant. Finally, the condition of the system is due more to the condition of the structure itself than the size of the system. For structures of interest, the condition is usually very poor.

In this paper we describe an iterative technique for solving such systems. The iterative technique is a variant of the incomplete Cholesky factorization - generalized conjugate gradient method (ICCG) described by Meijerink and van der Vorst [7]. In their paper it is shown that if A is an M-matrix it can always be split into

$$A = M - N$$

where

$$M = L \Sigma U$$

and L and U are sparse unit lower and upper triangular matrices and Σ a positive diagonal matrix. The splitting is a regular splitting. If A is also symmetric, then M will be symmetric and this splitting can be used in conjunction with the generalized conjugate gradient algorithm. If A is not an M-matrix, this splitting will certainly not be a regular splitting, and in fact, the factorization of M may not always yield a positive Σ (see Section 3). This paper will describe the implementation of a similar procedure called the shifted incomplete Cholesky factorization (SIC) on general symmetric positive definite matrices and discuss its efficiency.

In Section 2 we will review the iteration technique and discuss the qualities of a good preconditioning. In Section 3 we will review incomplete factorization and discuss the class of matrices for which the factorization is positive. In Section 4 we will introduce the shifted incomplete factorization and present results that show this splitting is at least as good as the Jacobi splitting. Finally, Section 5 will include experimental results that demonstrate the strengths of this method.

This paper is mainly concerned with symmetric positive definite matrices. However, most of the results can be generalized to nonsymmetric positive definite systems. A more detailed account may be found in MANTEUFFEL [5, 6].

2. ACCELERATION OF MATRIX SPLITTINGS

Given the linear system

$$A\underline{x} = \underline{b},$$

A is split into two parts

$$A = M - N$$

where M is easily invertible. By this we mean that M is nonsingular and it is easy to solve systems of the type

$$M\underline{x} = \underline{y}\ .$$

We can implicitly form the preconditioned system

$$M^{-1}A\underline{x} = M^{-1}\underline{b}.$$

If M is a "good" approximation of A, then $M^{-1}A$ will approximate the identity matrix and the preconditioned system will be easier to solve by iterative techniques than the original system.

If both M and A are positive definite, the preconditioned system may be solved by a variant of the conjugate gradient algorithm (Concus, Golub, and O'Leary, [1]). Otherwise, methods based upon the Tchebychev polynomials in the complex plane may be employed (Manteuffel, [4]). These methods are polynomial methods in that at each step the approximate solution is improved by adding a vector that is a linear combination of members of a Krylov sequence in $M^{-1}A$. Given an initial guess $\underline{x}_0$ let the residual be

$$\underline{r}_0 = \underline{b} - A\underline{x}_0\ .$$

Let

$$\underline{h}_0 = M^{-1}\underline{r}_0$$

be the generalized residual and let

$$V_i = \{h_o, M^{-1}A\underline{h}_o, (M^{-1}A)^2\underline{h}_o, \ldots, (M^{-1}A)^i\underline{h}_o\}$$

be the subspace spanned by the first i+1 terms of the Krylov sequence of $\underline{h}_o$ and $M^{-1}A$. After i steps of iteration, a polynomial method will yield

$$\underline{x}_i = \underline{x}_o + \underline{p}_{i-1}$$

for some $\underline{p}_{i-1} \varepsilon V_{i-1}$. If $\underline{x}$ is the exact solution and

$$\underline{e}_i = \underline{x} - \underline{x}_i$$

is the error at step i, then

$$\underline{e}_i = \underline{e}_o - \underline{p}_{i-1} .$$

Since $\underline{p}_{i-1} \varepsilon V_{i-1}$, there is some polynomial $P_{i-1}(z)$ of degree i-1 such that

$$\underline{p}_{i-1} = P_{i-1}(M^{-1}A)\underline{h}_o = P_{i-1}(M^{-1}A)M^{-1}A\underline{e}_o.$$

Thus, the error at step i may be expressed as

$$\underline{e}_i = \left(I - P_{i-1}(M^{-1}A)\, M^{-1}A\right) \underline{e}_o .$$

The conjugate gradient iteration will, in theory, automatically choose the polynomial that minimizes $\|\underline{e}\|_A = < A\underline{e}, \underline{e} >$. The polynomial $P_{i-1}(M^{-1}A)$ is an approximation to $(M^{-1}A)^{-1}$. The rate at which the iteration converges depends upon how well a polynomial of degree i-1 can approximate $(M^{-1}A)^{-1}$. This depends mainly upon the condition of $M^{-1}A$; that is, the ratio of the largest eigenvalue of $M^{-1}A$ to the smallest eigenvalue of $M^{-1}A$:

$$C(M^{-1}A) = \frac{\lambda_{max}}{\lambda_{min}} .$$

More precisely, the rate at which the conjugate gradient iteration will converge depends upon all of the ratios

$$\frac{\lambda_i}{\lambda_j} , \; i \geq j .$$

The best matrix splitting is the splitting that yields the best spectrum of $M^{-1}A$ (Greenbaum, [3]).

3. INCOMPLETE FACTORIZATION

If A is positive definite, a triangular decomposition of A yields

$$A = L \Lambda U, \quad \Lambda = \text{diag}(\rho_1,\dots,\rho_N), \ \rho_i > 0 ,$$

where L and U are unit lower and unit upper triangular. Incomplete factorization is a method by which the positive definite matrix A is split into

$$A = M-R$$

where

$$M = \hat{L} \Sigma \hat{U} , \quad \Sigma = \text{diag}(\sigma_1,\dots,\sigma_n)$$

and $\hat{L}$ and $\hat{U}$ are unit lower and unit upper triangular and sparse. If A is symmetric, then $C = \hat{L} \Sigma^{1/2}$ yields $M = CC^T$, which is known as incomplete Cholesky factorization.

The incomplete Cholesky factorization was first described by Varga [8] as a method of constructing a regular splitting of certain finite difference operators. Meijerink and van der Vorst [7] showed how it could be applied to M-matrices with arbitrary zero-structure, and that $\hat{L}$ and $\hat{U}$ could be constructed to have a predetermined zero-structure. Given G, a set of ordered pairs of integers (i,j), $1 \le i, j \le N$, we construct $\hat{L}$ and $\hat{U}$ so that $\hat{\ell}_{ij} \neq 0$ (i>j), $\hat{u}_{ij} \neq 0$ (i<j) only if $(i,j) \in G$. We will refer to G as the nonzero set of the factorization.

Once G has been chosen the factorization is defined recursively by

$$\sigma_i = a_{ii} - \sum_{k=1}^{i-1} \hat{\ell}_{ik} \hat{u}_{ki} \sigma_k ,$$

and for $j = i+1,\dots,N$

$$\sigma_i \hat{\ell}_{ji} = \begin{cases} a_{ji} - \sum_{k=1}^{i-1} \hat{\ell}_{jk} \hat{u}_{ki} \sigma_k & (j,i) \in G \\ 0 & (j,i) \notin G \end{cases}$$

$$\sigma_i \hat{u}_{ij} = \begin{cases} a_{ij} - \sum_{k=1}^{i-1} \hat{\ell}_{ik} \hat{u}_{ki} \sigma_k & (i,j) \in G \\ 0 & (i,j) \notin G \end{cases}$$

If we write

$$M = \hat{L}\,\Sigma\,\hat{U} = A + R$$

we see we have the exact factorization of the matrix A + R. A quick calculation shows that A and M match on the diagonal and the nonzero set G.

When implementing this procedure the elements of $\hat{L}$ and $\hat{U}$ are computed only if they are not to be set to zero later. Thus, the entire computation of $\hat{L}$, $\hat{U}$, and Σ can be carried out in the storage space that $\hat{L}$ and $\hat{U}$ will eventually occupy. The diagonal matrix Σ can be stored over the unit diagonal of L. If A is symmetric, then $\hat{L} = \hat{U}^T$ and only $\hat{L}$ need be stored.

The decomposition defined above will be stable as long as $\sigma_i \neq 0$ at each step. In addition, we would like $\sigma_i > 0$, for every i. Otherwise, $M^{-1}A$ would be indefinite. In general, indefinite systems are much harder to solve using iterative techniques than definite systems. If A is symmetric and the conjugate gradient iteration is to be used to accelerate the splitting, then the iteration may break down unless M is positive definite. We make the following definition.

<u>Definition</u> The incomplete factorization of the matrix A using nonzero set G is said to be <u>positive</u> if

$$\sigma_i > 0 \qquad i = 1, \ldots, N. \tag{3.6}$$

When will the factorization be positive? It is well known that if A is positive definite, then complete factorization will yield pivots $\rho_i > 0$. Let us examine one step of complete factorization. Write

$$A_1 = \left(\begin{array}{c|c} a_{11} & \underline{b}_1^T \\ \hline \underline{a}_1 & B_1 \end{array}\right) = \left(\begin{array}{c|c} 1 & \underline{0}^T \\ \hline \underline{\ell}_1 & I \end{array}\right) \left(\begin{array}{c|c} \rho_1 & \underline{0}^T \\ \hline \underline{0} & A_2 \end{array}\right) \left(\begin{array}{c|c} 1 & \underline{u}_1^T \\ \hline \underline{0} & I \end{array}\right)$$

where

$$A_2 = B_1 - \rho_1\,\underline{\ell}_1\,\underline{u}_1^T .$$

It is easy to show that if A_1 is positive definite, then A_2 will be also.

The modification of B_1 by the rank one matrix is the source of fill-in. In incomplete factorization we discard those elements of $\rho_1\,\underline{\ell}_1\,\underline{u}_1^T$ that do not fall on the predetermined nonzero set G. This modification may cause A_2 to no longer be positive definite. We would like to know sufficient conditions which guarantee the factorization will be globally positive, that is, positive for any nonzero set. We have the following results.

Definition A matrix $A = (a_{ij})$ is an M-matrix if $a_{ij} \leq 0$ for $i \neq j$, A is nonsingular and $A^{-1} \geq 0$.

Theorem (Meijerink and van der Vorst [7]) If A is an M-matrix, then incomplete factorization with any nonzero set will be positive.

This result can be extended.

Definition A matrix $A = (a_{ij})$ is an H-matrix if the matrix $B = (b_{ij})$ with $b_{ii} = |a_{ii}|$ and $b_{ij} = -|a_{ij}|$ for $i \neq j$ is an M-matrix. (A diagonally dominant matrix is an H-matrix.)

Theorem If A is an H-matrix with positive diagonal elements, then incomplete factorization with any nonzero set will be positive.

This result is not necessary, however. One can construct matrices that are not H-matrices but are globally positive.

We would also like to know if allowing more fill-in imprcves the condition. We have the following result.

Theorem If A is an M-matrix,then choosing a larger nonzero set will improve the condition of the splitting (up to a factor of 2).

This result is not known for general matrices. Moreover, it is not clear whether the improved condition outweighs the additional work per iteration caused by more nonzeros in the factors. This seems to be problem dependent.

If one of the pivots becomes negative,then it is clear that the factorization is not positive. However, even if all the pivots stay positive it is important that they not become too small. A small pivot would yield a nearly singular M and thus $M^{-1}A$ would most likely have poor condition. We may ask: How small is too small? We need a way to compare the size of the pivot to the size of the elements of A.

If we use the diagonal scaling matrix

$$D = \text{diag}\ (\ldots a_{ii}^{-1/2} \ldots),$$

then the matrix $\tilde{A} = DAD$ has unit diagonal. If we perform an incomplete factorization on $\tilde{A}$ to get pivots $\Sigma = \text{diag}\ (\sigma_1, \ldots \sigma_N)$, then it is easy to see that

$$\sigma_1 = 1 \;,\; \sigma_i \leq 1, \; i = 2,\ldots,N.$$

Hence,

$$S = 1/\min_i (\sigma_i)$$

yields a good measure of the positivity.

The matrix splitting is invariant of diagonal scaling. We have the following result.

Theorem Incomplete factorization of the scaled matrix $\tilde{A}$ yields a splitting with the same eigenvalues as would incomplete factorization (with the same nonzero set) of the matrix A.

Let us assume that A has been diagonally scaled. If we knew the spectrum of $M^{-1}A$, we could predict the convergence of the conjugate gradient iteration. Unfortunately, this information is computationally unavailable at the time of the splitting. In fact, even C(A) may not be known accurately. We do know that for the optimal splitting, the complete factorization, we have $M^{-1} = A^{-1}$, $C(M) = C(A)$. It is also clear that the condition of M should not be allowed to be significantly larger than the condition of A. If an estimate of C(A) is available, then a reasonable computational strategy is to accept the factorization only if

$$S \leq C(A).$$

This bound has worked well in practice. In fact, in the examples to be described later the factorizations which performed best corresponded to values of S many orders of magnitude smaller than C(A).

4. THE SHIFTED INCOMPLETE CHOLESKY FACTORIZATION.

The symmetric linear systems that arise from structural mechanics are positive definite, but are not M-matrices nor are they H-matrices. It was found through experiment that the incomplete factorization with the nonzero set of L equal to the nonzero set of A was not positive for many systems. To overcome this problem one might increase the nonzero set and allow more fill-in. This proved to be inefficient. As an alternative, consider the incomplete factorization of a matrix close to A but more nearly diagonally dominant. Suppose we have the diagonally scaled symmetric positive definite matrix

$$A = I - B .$$

Consider the pencil of matrices

$$A(\alpha) = I - \frac{1}{1+\alpha} B.$$

Clearly, for α sufficiently large $A(\alpha)$ is diagonally dominant. Thus, there is some $\alpha \geq 0$ for which the incomplete factorization of $A(\alpha)$ with a given nonzero set is positive. Suppose we have

$$A(\alpha) = L \, \Sigma \, L^T - R(\alpha) = M(\alpha) - R(\alpha) .$$

(Here L and Σ also depend upon α). Then, we may write

$$I - \frac{1}{1+\alpha} B = M(\alpha) - R(\alpha)$$

and

$$I - B = M(\alpha) - (R(\alpha) + \frac{\alpha}{1+\alpha} B)$$

or

$$A = M(\alpha) - N(\alpha) \quad .$$

For each α for which the incomplete factorization of $A(\alpha)$ is positive we have a splitting of A.

Although we are interested in the factorization for small α, examining the asymptotic effect for large α is instructive. Notice that as α gets large $A(\alpha)$ approaches I. One would expect that $M(\alpha)$ also approaches I. In fact, if the nonzero set of L includes the nonzero set of A we have

$$M(\alpha) = I - \frac{1}{1+\alpha} B + \mathcal{O}\left(\frac{1}{(1+\alpha)^2}\right)$$

$$N(\alpha) = \frac{\alpha}{1+\alpha} B + \mathcal{O}\left(\frac{1}{(1+\alpha)^2}\right)$$

for large α. A quick calculation yields

$$M^{-1}(\alpha) = I + \frac{1}{1+\alpha} B + \mathcal{O}\left(\frac{1}{(1+\alpha)^2}\right) .$$

This yields

$$M^{-1}(\alpha)A = I - B + \frac{1}{1+\alpha} (B-B^2) + \mathcal{O}\left(\frac{1}{(1+\alpha)^2}\right) .$$

Asymptotically, this is the Jacobi splitting. Suppose μ_i is an eigenvalue of $A = I - B$ and $\mu_i(\alpha)$ is the corresponding eigenvalue of $M^{-1}(\alpha)A$. Then, for α sufficiently large we have

$$1 < \mu_i(\alpha) < \mu_i, \text{ for } \mu_i > 1 ,$$

and

$$\mu_i < \mu_i(\alpha) < 1, \text{ for } \mu_i < 1 .$$

Thus the condition of $M^{-1}(\alpha)A$ is less than the condition of the Jacobi splitting. In fact, one can show that for sufficiently large α we have the following

$$\frac{\mu_i(\alpha)}{\mu_j(\alpha)} < \frac{\mu_i}{\mu_j} \text{ for } \mu_i \geq \mu_j.$$

This gives the following result:

Theorem For α sufficiently large, the shifted incomplete Cholesky factorization is better than the Jacobi splitting for acceleration by the generalized conjugate gradient iteration.

The behavior of the splitting for small values of α is of greater interest, but not well understood. Suppose we are given the matrix A and the nonzero set G. Let α_a be such that the incomplete factorization is stable for every $\alpha > \alpha_a$. In general α_a is not known. Bounds on α_a may sometimes be found by examining the problem from which the linear system arose. We do have the following results.

Corollary If A is an H-matrix, then $\alpha_a \leq 0$.

Let α_c be the value of α that yields the minimum condition for $M^{-1}(\alpha)A$. Again, this is not known in general. We have the following.

<u>Theorem</u> If A is an M-matrix and if the nonzero set includes the nonzero set of A, then if $0 \leq \alpha_1 < \alpha_2$ the condition of $M^{-1}(\alpha_1)A$ is smaller than the condition of $M^{-1}(\alpha_2)A$ (up to a factor of 2).

<u>Corollary</u> If A is an M-matrix, then $\alpha_c \leq 0$ (up to a factor of 2).

Let us consider the result of allowing more fill-in. The most convenient nonzero set is the same as the nonzero set of A. This allows the factorization to be stored in the same data structure as A. Let us call this nonzero set G_1. Using nonzero set G_1 may be thought of as allowing fill-in in location (i,j) of L whenever unknown i and unknown j are neighbors in the graph of A. Suppose we extend this association and allow fill-in in location (i,j) of L whenever unknown i and unknown j have a common neighbor in the graph of A. Not all such locations will actually fill in. We need only have $(i,j) \in G$ whenever there exists $k \leq j < i$ such that (i,k) and (j,k) are in the nonzero set of A.

We will refer to this first level of extension as G_2. Clearly, the nonzero set may be extended any number of levels until all possible fill-in is accounted for. Let G_i be the nonzero set which includes (j,k) whenever there is a path in the graph of A of at most i edges connecting unknown j to unknown k. Let α_a^i , α_c^i , be the minimum α and α of best condition respectively for the splitting with nonzero set G_i. In practice, one finds

$$\alpha_a^{i+1} \leq \alpha_a^i$$

and

$$\alpha_c^{i+1} \leq \alpha_c^i$$

However, this has not been proven. Let

$$c^i = C(M^{-1}(\alpha_c^i)A)$$

be the condition of the splitting using graph G_i and α_c^i. In practice one finds

$$c^{i+1} \leq c^i .$$

Again this has not been proven except in the case of M-matrices. A more important question is whether the improved condition justifies the additional work and storage required to perform the factorization with a larger nonzero set and thus more fill-in.

5. NUMERICAL RESULTS

This work was motivated by a three dimensional model of the structural deformation of a cylinder with varying thickness and holes (see Figure 1). Assuming linear deformation and using isoparametric 20-node brick finite elements, the model required the solution of a positive definite linear system with approximately 18,000 unknowns, a half-bandwidth of 1,300 and 1,040,000 nonzero elements in the upper triangular part. The number of unknowns varied slightly according to the boundary conditions imposed. The condition of the matrix was estimated to be on the order of 10^8.

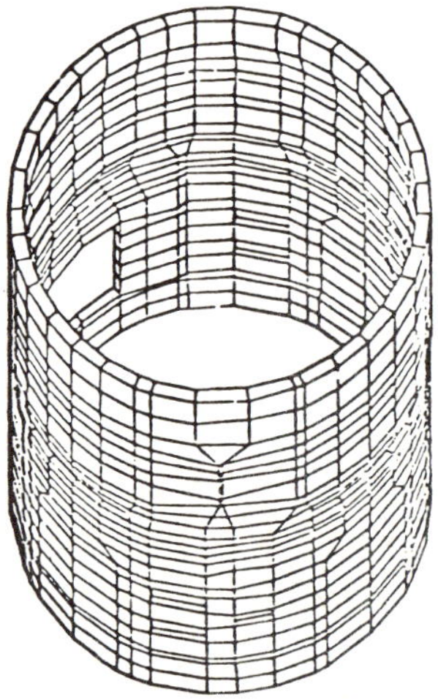

Figure 1. Motivating Problem: Cylinder

A direct solution to this system was sought but never achieved due to the large amount of storage required for the upper triangular band (23,000,000). It was estimated from partial runs to require 9,000 CP seconds on the CDC-7600 at Sandia Laboratories in Albuquerque, or 50,000-70,000 CP seconds on the CDC-6600 at Sandia Laboratories in Livermore.

Conjugate gradients with diagonal scaling, that is, acceleration of the Jacobi splitting, was only moderately successful. This implementation required 2,000,000 words of storage for the matrix. A series of runs were made in which the algorithm was restarted at each run using the solution from the previous run as an initial guess. Some of the runs were made on the CDC-7600 and some on the CDC-6600. The Euclidean norm of the residual was reduced by a factor of 10^{-6} after a total of about 4,000 iterations which is estimated to cost 7,000 CP seconds on the CDC-7600 or 40,000-45,000 CP seconds on the CDC-6600 (see Table 1).

The shifted incomplete Cholesky factorization was more successful. It required 3,000,000 words of storage. Using nonzero set G equal to the nonzero set of A, the factorization required 700 CP seconds on the CDC-6600. The problem was solved for several sets of boundary conditions on the CDC-6600. Using ad hoc values of α ranging from .05 to .10 an acceptable solution was achieved after approximately 200 iterations or 6,000 CP seconds. A solution of much higher resolution was achieved in approximately 700 iterations or 20,000 CP seconds (see Table 1).

	Time (6600)	Error Reduction	Storage
DIRECT	50,000 - 70,000	---------	23,000,000
CG	40,000 - 45,000	10^{-6}	2,000,000
SICCG	6,000 - 10,000	10^{-5}	3,000,000
	20,000 - 25,000	10^{-10}	

Table 1

It is clear that the savings in both time and storage were significant. Also notice that the time required to perform the factorization was small compared to the overall effort. In such a problem it is feasible to spend time searching for a good value of the parameter α.

The SICCG procedure was tested extensively on much smaller problems, where the advantage over direct methods is no longer clear. The algorithm behaved similarly in each case. The remainder of this section will be devoted to exploring in depth the results from one test problem.

The test problem was a three dimensional model of structural deformation in a tapered slab (see Figure 2). Again linear deformation was assumed and isoparametric 20-node brick finite elements were used. The boundary conditions corresponded to pressing the thin edge against a wall like a wedge. This gave a system with 3090 unknowns, a half-bandwidth of 286 and 170,000 nonzeros in the upper triangular part. The matrix was scaled symmetrically to have unit diagonal. The condition of the system was estimated to be on the order of 10^{+7}.

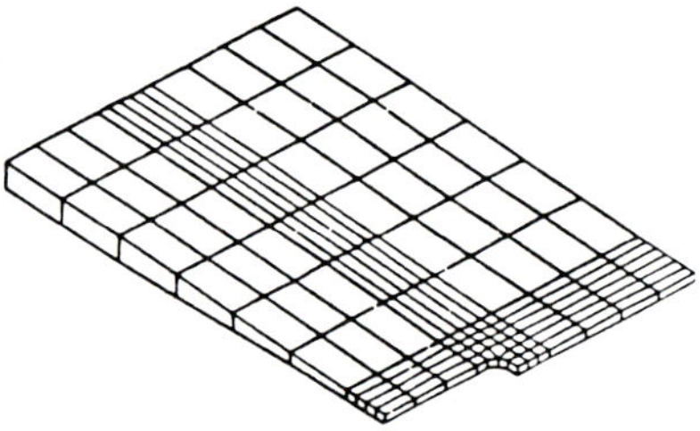

Figure 2. Test Problem: Tapered Slab.

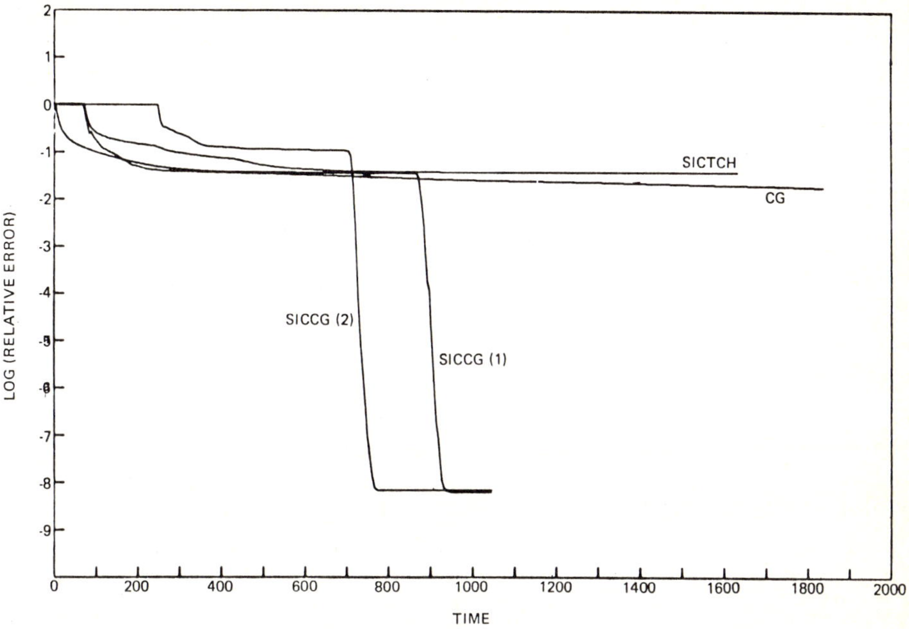

Figure 3. Method Comparison

Figure 3 shows a comparison of the log of the Euclidean norm of the error versus cp time in seconds on a CDC 6600 for several algorithms. The competing methods were a conjugate gradient acceleration (CG) of the Jacobi splitting, a Tchebychev acceleration of the shifted incomplete Cholesky factorization with graph G_1 (SICTCH), and conjugate gradient acceleration of the shifted incomplete Cholesky factorization with graph G_1 and G_2 (SIC(1), SIC(2)). The values of α used were α = .005 for SIC(1) and α = .0015 for SIC(2). The initial plateau corresponds to the time required to perform the factorization. Notice that although the factorization time was much longer for SIC(2) than SIC(1), total time to convergence was smaller for SIC(2). The steep cliff is characteristic of the conjugate gradient iteration and may be due to a bunching of the eigenvalues of $M^{-1}(\alpha)A$.

The number of iterative steps required to reach the cliff was dependent upon the parameter α. Figure 4 shows

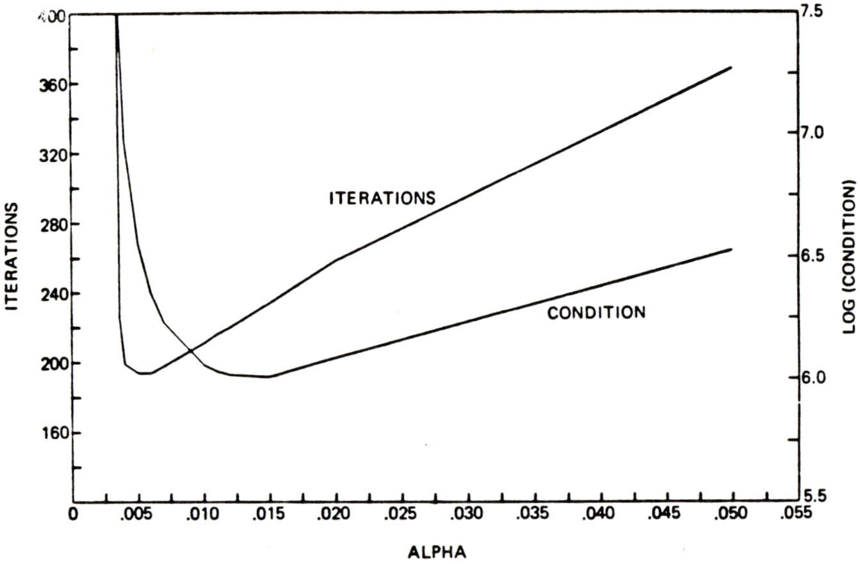

Figure 4. Iterations and Condition vs. α

the number of iterations required to reduce the error by a factor of 10^{-6} for various values of α using SIC(1). Notice the unimodal shape of the graph. It was found experimentally that the factorization was unstable for $\alpha < .0036$ and that convergence was fastest for $\alpha_b = .0055$. Notice that for $\alpha = .05$, an order of magnitude larger, convergence still occurred within a reasonable number of iterations.

The condition of the splitting was estimated by taking advantage of the relationship between the Lanczos algorithm for finding eigenvalues and the conjugate gradient algorithm. The log of the condition for various values of α is also shown in Figure 4. Notice that the best condition occurs at $\alpha_c = .015$ and that

$$0 < \alpha_a < \alpha_b < \alpha_c \ .$$

This was the case in every test. Perhaps the value α_b produced a greater bunching of the eigenvalues which outweighed the enhanced condition.

6. CONCLUSION

The incomplete Cholesky factorization provides a bridge that spans the gap between direct and iterative methods. It appears that a factorization which retains only a fraction of the fill-in yields an approximation that is sufficiently accurate on a large set of vectors to effectively reduce the degree of the minimal polynomial of the splitting. The conjugate gradient iteration exploits this situation.

The shifted factorization provides a means by which the incomplete Cholesky factorization can be extended to any positive definite linear system. The asymptotic behavior of the shift gives confidence that the factorization does yield improved condition. It does not, however, explain the remarkable performance for smaller values of α. Hopefully, theory will be developed that will allow the a priori estimation of acceptable values for the shift parameter α.

References

1. P. Concus, G. H. Golub, D. P. O'Leary, A Generalized Conjugate Gradient Method for the Numerical Solution of Elliptic Partial Differential Equations, Lawrence Berkeley Laboratory Pub. LBL-4604, Berkeley, CA.
2. Ky Fan, Note on M-matrices, Quart. J. Math., Oxford (2nd Series) 11 (1960), pp. 43-49.
3. A. Greenbaum, Comparison of Splittings Used With The Conjugate Gradient Method, Lawrence Livermore Laboratories Report, UCRL-80800, March 1978.
4. T. A. Manteuffel, The Tchebychev Iteration for Nonsymmetric Linear Systems, Numer. Math., 28, 307-327 (1977).
5. T. A. Manteuffel, The Shifted Incomplete Cholesky Factorization, Sandia Report, SAND78-8226, Livermore, CA (1978).
6. T. A. Manteuffel, An Incomplete Factorization Technique for Positive Definite Linear Systems, Math. Comp., to appear.
7. J. A. Meijerink and H. A. van der Vorst, An Iterative Solution Method for Linear Systems of Which the Coefficient Matrix is a Symmetric M-matrix, Math. Comp., vol. 31, 137 (1977), pg. 148-162.
8. R. S. Varga, Factorization and Normalized Iterative Methods, Boundary Problems in Differential Equations, R. E. Langer, Ed., University of Wisconsin Press, Madison (1960).

9. R. S. Varga, Matrix Iterative Analysis, Prentice Hall, Englewood Cliffs, New Jersey (1962).

10. D. Young, Iterative Solution of Large Linear Systems, Academic Press, New York (1972).

COMPUTING METHODS IN APPLIED SCIENCES AND ENGINEERING
R. Glowinski, J.L. Lions (editors)
North-Holland Publishing Company

AN EFFICIENT PRECONDITIONNED CONJUGATE GRADIENT METHOD APPLIED TO NONLINEAR PROBLEMS IN FLUID DYNAMICS VIA LEAST SQUARE FORMULATIONS

R. GLOWINSKI, B. MANTEL**, J. PERIAUX**, O. PIRONNEAU***, G. POIRIER***

ABSTRACT : We present in this paper a new technique for the incomplete Cholesky factorization of large, sparse, symmetric, positive definite matrices. This technique has been applied to scale conjugate gradient algorithms, and appears therefore as a very useful tool to solve, via nonlinear least square formulations, important problems in Fluid Dynamics, such as transonic potential flows for compressible, inviscid fluids ant Navier-Stokes equations for viscous, incompressible fluids.

1. INTRODUCTION.

Conjugate gradient methods with *scaling* (one also says *preconditionning)* have been very popular these very last years for solving nonlinear boundary value problems in Mechanics and other fields (let us mention among others BARTELS-DANIEL [1], DOUGLAS-DUPONT [2], CONCUS-GOLUB-O'LEARY [3]) ; in the specific field of Fluid Dynamics they have been used - via a convenient *nonlinear least square formulation* - to solve complicated nonlinear problems like *transonic potential flows for compressible inviscid fluids* and also the *Navier-Stokes equations for incompressible viscous fluids* ; we refer for more details to [4],[5],[6],[7]. Actually these methods are successfully used at the moment by L. Reinhart and the first author for the numerical solution of the (nonlinear) Von Karman equations for plates.

However as the size and complexity of the problem under consideration increase, the scaling operation (usually the solution of a linear system whose matrix is symmetric

* Université Pierre et Marie Curie, L.A.N. 189, Tour 55.65, 4 Place Jussieu, 75230 Paris Cedex 05, and INRIA, B.P. 105, 78150 Le Chesnay, France

** AMD/BA, 78 Quai Carnot, B.P. 300, 92214 St-Cloud, France

*** Université Paris-Nord, Département de Mathématiques, 93200 St-Denis and INRIA, B.P. 105, 78150 Le Chesnay, France and positive definite) may become a formidable task by itself ; this will be a standard situation if one wishes to solve problems associated to industrial applications ; it is to overcome that crucial difficulty that several authors have introduced in the last five years those methods generically called "*incomplete factorization methods*" ; we shall mention among several others AXELSSON [8], MEIJERINK-VAN DE VORST [9], MANTEUFEL [10], KERSHAW [11].

Our goal in this paper is to introduce a new incomplete Cholesky factorization technique and apply it to the numerical solution of nonlinear problems in Fluid Dynamics, namely those mentioned above, i.e. transonic potential flows for *compressible inviscid fluids* and Navier-Stokes equations for *incompressible viscous fluids*.

Numerical experiments will show the efficiency of this new method in comparison to more classical incomplete factorization methods (like those discussed in [9]). One may find in GLOWINSKI-PERIAUX-PIRONNEAU [12] a preliminary discussion of the methods discussed in this paper.

2. - GENERALITIES ON THE PRECONDITIONED SOLUTION OF BOUNDARY VALUE PROBLEMS.

We shall discuss in this section the numerical solution of boundary value problems, by iterative methods with preconditioning ; to avoid abstraction we have chosen to discuss the solution of two families of test problems.

2.1. A first test problem. The linear heat equation.

Let Ω be a *bounded* domain of $\mathbb{R}^N$ (N=1,2,3 in practice) and let Γ be its boundary ; using a convenient system of physical units, heat conduction phenomenon are described ($]0,T[$ is a *time* interval) by

$$\frac{\partial\phi}{\partial t} - \Delta\phi = f \text{ in } \Omega \times]0,T[, \tag{2.1}$$

$$\phi = g \text{ on } \Gamma \times]0,T[\ , \tag{2.2}$$

$$\phi(x,0) = \phi_o(x) \text{ in } \Omega. \tag{2.3}$$

Using the standard backward (implicit) *time discretization* scheme we derive from (2.1)-(2.3) the following semi-discretization scheme, where $\phi^n(x) \simeq \phi(x,n\Delta t)$,

$$\phi^o = \phi_o \ , \tag{2.4}$$

and for $n \geq 0$,

$$\left(\frac{I}{\Delta t} - \Delta\right)\phi^{n+1} = f^{n+1} + \frac{\phi^n}{\Delta t} \text{ on } \Omega, \tag{2.5}$$

(2.6) $\phi^{n+1} = g^{n+1} = g(\cdot,(n+1)\Delta t)$ on Γ

(with $f^n(x) = f(x,n\Delta t)$, $g^n(x) = g(x,n\Delta t)$).

Then using a convenient *space discretization* (by finite elements or finite differences) we obtain a *fully discrete* approximation of (2.1)-(2.3) by

(2.7) $\phi_h^o = \phi_{oh}$

and for $n \geq 0$

(2.8) $A_h\phi_h^{n+1} = F_h^{n+1}$

where (2.8) approximates simultaneously (2.5),(2.6).

The above matrix A_h is *symmetric, positive definite, sparse,* and in the present case *independent of* n ; in practice this $N \times N$ matrix A_h may be such that $N > 10^4$; from these properties it is therefore very tempting to use a *Cholesky factorization* $A_h = L_hL_h^t$, done once and for all, to solve the sequence of problems (2.8). A most important reason to use Cholesky factorization is that it preserves the bandwidth. In fact even if L_h has the same bandwidth it contains in general much more non zero elements than A_h. We have represented on Figure 2.1 the matrix A_h and the matrix L_h^t with a visualization of the zero and non zero elements ; this figure illustrates very well the structure of A_h and L_h^t (therefore L_h) concerning the distribution of zero and non zero elements.

From the above considerations it follows that there are situations in which it is possible to store *in core* A_h (in fact its lower triangular part, for example, if A_h is symmetric) but not L_h.

Therefore solving the two triangular systems associated with the solution of (2.8), namely

(2.9) $L_hz_h^{n+1} = F_h^{n+1}$

(2.10) $L_h^t\,\phi_h^{n+1} = z_h^{n+1}$,

will require tape or disk transfers, resulting in a possibly prohibitively execution time for very large problems. A classical remedy to the above difficulty is to introduce a matrix $\tilde{L}_h$, *sparser than* L_h, which is such that $\tilde{A}_h$, defined by

CHOLESKY FACTORIZATION

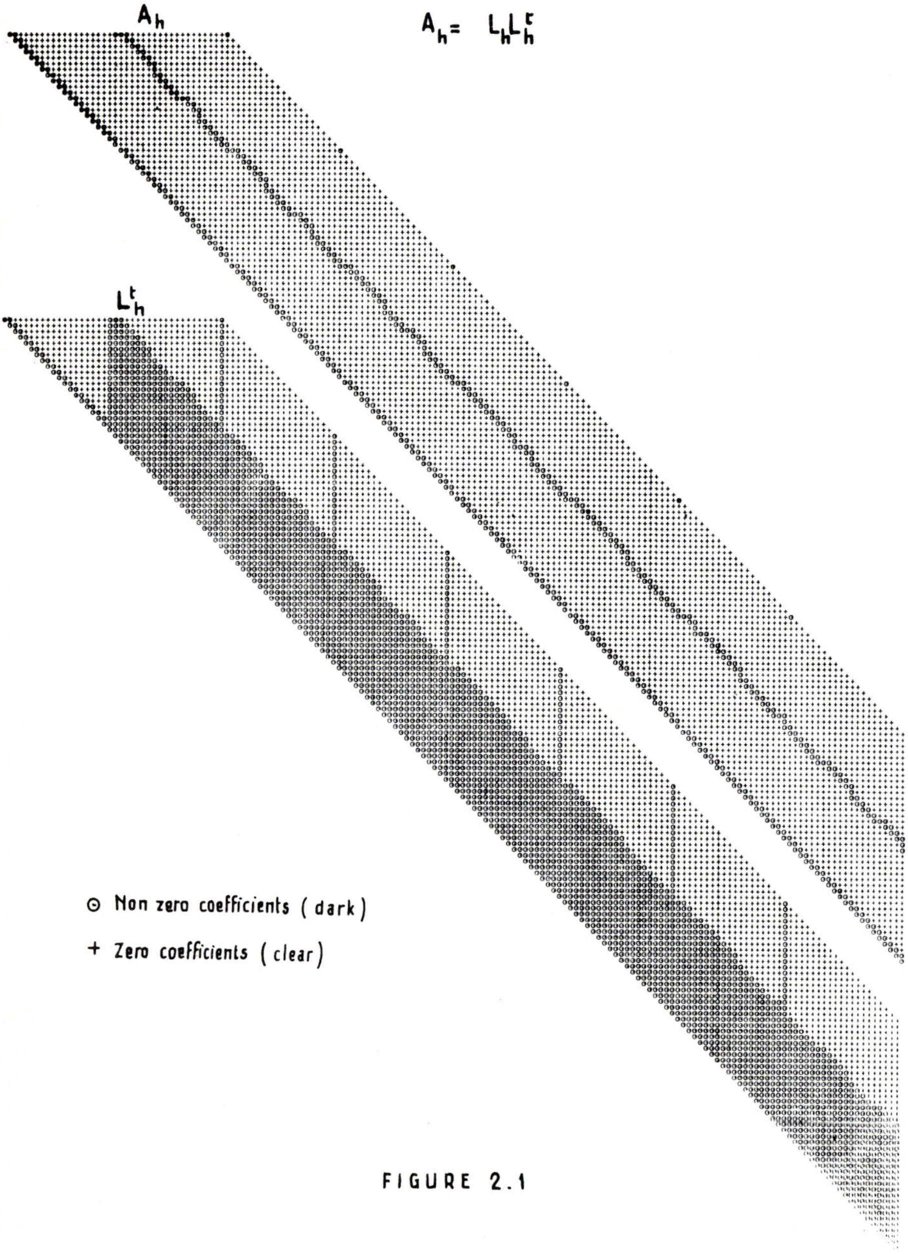

FIGURE 2.1

(2.11) $\tilde{A}_h = \tilde{L}_h \tilde{L}_h^t$,

is spectrally as close as possible to A_h ; then (omitting indices n and h) one may solve (2.8) by

(2.12) ϕ_o given,

then for $m \geq 0$,

(2.13) $\tilde{A}\phi_{m+1} = (\tilde{A}-A)\phi_m + F$

or by conjugate gradient variants of (2.12),(2.13).

In order to construct $\tilde{L}_h$ we would like to discuss two types of situations :

(i) The matrix A_h being given the number of problems (2.8) that we have to solve (differing one to each other by their right hand sides) *is not very large* ; in that case we highly recommend *approximate factorization methods* like those discussed in, e.g., [9],[10],[11].

(ii) The number of problems that we have to solve for the same A_h is very large ; in that latter case it may be interesting to construct a "better" $\tilde{L}_h$ via a more costly process, involving for example as a preliminary step the construction of L_h. That second point of view will be detailed in Sec. 3.

2.2. Description of a preconditioned conjugate gradient algorithm for solving linear equations.

Before discussing a second family of examples, related to nonlinear problems we shall describe a conjugate gradient algorithm with preconditioning to solve *linear systems whose matrix is symmetric and positive definite*. Suppose that this system is

(2.14) $\underset{\sim}{A}\underset{\sim}{\phi} = \underset{\sim}{F}$

where $\underset{\sim}{A}$ is a $N \times N$ symmetric positive definite matrix and where $\underset{\sim}{\phi}, \underset{\sim}{F} \in \mathbb{R}^N$; we denote by $(\cdot,\cdot)$ the usual inner product of $\mathbb{R}^N$. Then (2.14) is equivalent to the minimization problem

(2.15) $$\begin{cases} \text{Find } \phi \in \mathbb{R}^N \text{ such that} \\ J(\phi) \leq J(\psi) \quad \forall \psi \in \mathbb{R}^N \end{cases}$$

where $J : \mathbb{R}^N \to \mathbb{R}$ is defined by

$$J(\underset{\sim}{\psi}) = \frac{1}{2}(\underset{\sim}{A}\underset{\sim}{\psi},\underset{\sim}{\psi}) - (\underset{\sim}{b},\underset{\sim}{\psi}). \tag{2.16}$$

We introduce now a second matrix, $\underset{\sim}{S}$, which is also *symmetric* and *positive definite* ; using $\underset{\sim}{S}$ as a preconditioner we can solve (2.14),(2.15) using the following conjugate gradient algorithm

Step 0 : Initialization

$$\underset{\sim}{\phi}^o \text{ arbitrarily given in } \mathbb{R}^N, \tag{2.17}$$

$$\underset{\sim}{g}^o = \underset{\sim}{S}^{-1}(\underset{\sim}{A}\underset{\sim}{\phi}^o - \underset{\sim}{F}), \tag{2.18}$$

$$\underset{\sim}{w}^o = \underset{\sim}{g}^o, \tag{2.19}$$

then for $n \geq 0$, *assuming that* $\underset{\sim}{\phi}^n, \underset{\sim}{g}^n, \underset{\sim}{w}^n$ *are known we compute* $\underset{\sim}{\phi}^{n+1}$, $\underset{\sim}{g}^{n+1}$, $\underset{\sim}{w}^{n+1}$ *using*

Step 1 : Descent

$$\underset{\sim}{\phi}^{n+1} = \underset{\sim}{\phi}^n - \lambda_n \underset{\sim}{w}^n \tag{2.20}$$

where

$$\lambda_n = \operatorname*{Arg\,Min}_{\lambda \in \mathbb{R}} J(\underset{\sim}{\phi}^n - \lambda \underset{\sim}{w}^n)$$

i.e.

$$\lambda_n = \frac{(\underset{\sim}{S}\underset{\sim}{g}^n, \underset{\sim}{w}^n)}{(\underset{\sim}{A}\underset{\sim}{w}^n, \underset{\sim}{w}^n)} \quad \left(= \frac{(\underset{\sim}{S}\underset{\sim}{g}^n, \underset{\sim}{g}^n)}{(\underset{\sim}{A}\underset{\sim}{w}^n, \underset{\sim}{w}^n)} \right) \tag{2.21}$$

Step 2 : Calculation of the new descent direction

$$\underset{\sim}{g}^{n+1} = \underset{\sim}{g}^n - \lambda_n \underset{\sim}{S}^{-1} \underset{\sim}{A}\underset{\sim}{w}^n, \tag{2.22}$$

$$\gamma_{n+1} = \frac{(\underset{\sim}{S}\underset{\sim}{g}^{n+1}, \underset{\sim}{g}^{n+1})}{(\underset{\sim}{S}\underset{\sim}{g}^n, \underset{\sim}{g}^n)}, \tag{2.23}$$

$$\underset{\sim}{w}^{n+1} = \underset{\sim}{g}^{n+1} + \gamma_{n+1}\underset{\sim}{w}^n, \tag{2.24}$$

$n = n+1$, go to (2.20).

The convergence of (2.17)-(2.24) is a well-known result.

Remark 2.1 : The closer is $\tilde{S}$ to $\tilde{A}$ the faster is the convergence with regard to the number of iterations (if $\tilde{S}=\tilde{A}$ we need only *one* iteration). Actually, in practice, $\tilde{S}$ has to be much easier to factorize than $\tilde{A}$ but has to be such that the spectrum of $\tilde{S}^{-1}\tilde{A}$ is a narrow one in order to give good convergence properties to algorithm (2.17)-(2.24).

Remark 2.2 : The above preconditioned conjugate gradient algorithm appears as a compromise between a purely iterative method and a direct method, for solving the linear problem (2.14).

2.3. A second family of test problems. Least square solution of nonlinear Poisson problems.

2.3.1. Formulation of the test problems.

We follow [5, Sec. 2] ; let $\Omega \subset \mathbb{R}^N$ be a bounded domain with a smooth boundary. We consider the *nonlinear Dirichlet problem*

$$(2.25) \qquad \begin{cases} -\Delta\phi - T(\phi) = 0 \text{ in } \Omega, \\ \phi = 0 \text{ on } \Gamma. \end{cases}$$

We do not discuss here the *existence* and *uniqueness* properties of the solution of (2.25) since we do not want to be very specific about the operator T.

Since various iterative methods for solving (2.25) are described in [5, Sec. 2] we shall concentrate on *conjugate gradient methods with scaling* via convenient *least square formulations*.

2.3.2. A H^{-1} least square method for solving (2.25).

Let introduce first some *Sobolev's functional spaces* which seem optimally suited for the following study

$$(2.26) \qquad H^1(\Omega) = \{v \in L^2(\Omega), \frac{\partial v}{\partial x_i} \in L^2(\Omega) \quad \forall i=1,\dots N\},$$

$$(2.27) \qquad H^1_0(\Omega) = \{v \in H^1(\Omega), v=0 \text{ on } \Gamma\}.$$

Then $H^1(\Omega)$ is an *Hilbert space* with the scalar product

$$(v,w)_{H^1(\Omega)} = \int_\Omega (vw+\nabla v\cdot\nabla w)\,dx$$

and the corresponding norm. Moreover $H^1_0(\Omega)$ is a *closed subspace of* $H^1(\Omega)$. Since Ω is bounded, then $H^1_0(\Omega)$ is a Hilbert space with the scalar product

(2.28) $$(v,w)_{H_0^1(\Omega)} = \int_\Omega \nabla v \cdot \nabla w \, dx ,$$

and the corresponding norm $\|v\|_{H_0^1(\Omega)} = (\int_\Omega |\nabla v|^2 dx)^{1/2}$ is equivalent to the norm induced by $H^1(\Omega)$. Let $H^{-1}(\Omega) = (H_0^1(\Omega))'$ be the dual topological space of $H_0^1(\Omega)$. If $L^2(\Omega)$ has been identified to its dual space, then

$$H_0^1(\Omega) \subset L^2(\Omega) \subset H^{-1}(\Omega) ,$$

moreover $\Delta = \nabla^2$ is an *isomorphism* from $H_0^1(\Omega)$ onto $H^{-1}(\Omega)$. In the following the duality pairing between $H^{-1}(\Omega)$ and $H_0^1(\Omega)$ will be denoted by $<\cdot,\cdot>$ where $<\cdot,\cdot>$ is such that

(2.29) $$<f,v> = \int_\Omega fv \, dx \quad \forall \, f \in L^2(\Omega), \quad \forall v \in H_0^1(\Omega) .$$

The topology of $H^{-1}(\Omega)$ is defined by $\|\cdot\|_{-1}$, where

(2.30) $$\|f\|_{-1} = \sup_{v \in H_0^1(\Omega)-\{0\}} \frac{|<f,v>|}{\|v\|_{H_0^1(\Omega)}} .$$

From the above considerations a natural *least square formulation* for solving the model problem (2.25), is

(2.31) $$\operatorname{Min}_{v \in H_0^1(\Omega)} \|\Delta v + T(v)\|_{-1} .$$

It is clear that if (2.25) has a solution, then this solution will be a solution of (2.31) for which the cost function will vanish. Let us introduce $\xi \in H_0^1(\Omega)$ by

(2.32) $$\begin{cases} \Delta\xi = \Delta v + T(v) \; \textit{in} \; \Omega, \\ \xi = 0 \; \textit{on} \; \Gamma, \end{cases}$$

then (2.31) reduces to

(2.33) $$\operatorname{Min}_{v \in H_0^1(\Omega)} \|\Delta\xi\|_{-1}$$

where, in (2.33), ξ is a function of v via (2.32) ; actually it can be proved that if $\|\cdot\|_{-1}$ is defined by (2.30) with $<\cdot,\cdot>$ obeying (2.29), then

(2.34) $$\|\Delta v\|_{-1} = \|v\|_{H_0^1(\Omega)} \quad \forall \, v \in H_0^1(\Omega) .$$

It follows then from (2.34) that (2.33) may be formulated also by

$$\min_{v \in H_0^1(\Omega)} \int_\Omega |\nabla \xi|^2 dx \tag{2.35}$$

where ξ is still a function of v through (2.32).

Remark 2.3 : Nonlinear boundary value problems have been treated by CEA-GEYMONAT [13] and LOZI [14] using formulations very close to (2.32),(2.35).

2.3.3. Conjugate gradient solution of the least square problem (2.32),(2.35).
Let us define $J : H_0^1(\Omega) \to \mathbb{R}$ by

$$J(v) = \frac{1}{2} \int_\Omega |\nabla \xi|^2 dx \tag{2.36}$$

where ξ is a function of v in accordance with (2.32) ; then (2.35) may also be written as

$$\min_{v \in H_0^1(\Omega)} J(v). \tag{2.37}$$

To solve (2.37) we shall use a *conjugate gradient* algorithm. Among the possible conjugate gradient algorithms we have selected the *Polak-Ribière* version (cf. [15]), since this algorithm produced the best performances (compared to other variants) in the preliminary numerical tests we made (the good performances of the Polak-Ribière algorithm are discussed in [16]). Let us denote by $J'(\cdot)$ the differential of $J(\cdot)$; then the Polak-Ribière version of the conjugate gradient method, applied to the solution of (2.37) is

Step 0 : Initialization

$$\phi^o \in H_0^1(\Omega) \text{ given,} \tag{2.38}$$

then compute $g^o \in H_0^1(\Omega)$ *from*

$$\begin{cases} -\Delta g^o = J'(\phi^o) \text{ in } \Omega, \\ g^o = 0 \text{ on } \Gamma, \end{cases} \tag{2.39}$$

and set

$$w^o = g^o. \tag{2.40}$$

Then for $n \geq 0$, *assuming* ϕ^n, g^n, w^n *known, compute* $\phi^{n+1}, g^{n+1}, w^{n+1}$ *by*

Step 1 : Descent

(2.41) $$\lambda^n = \operatorname{Arg}\min_{\lambda \in \mathbb{R}} J(\phi^n - \lambda w^n),$$

(2.42) $$\phi^{n+1} = \phi^n - \lambda^n w^n.$$

Step 2 : Construction of the new descent direction

Define $g^{n+1} \in H_0^1(\Omega)$ *by*

(2.43) $$-\Delta g^{n+1} = J'(\phi^{n+1}) \text{ in } \Omega, \quad g^{n+1} = 0 \text{ on } \Gamma,$$

then

(2.44) $$\gamma^{n+1} = \frac{\int_\Omega \nabla g^{n+1} \cdot \nabla (g^{n+1} - g^n)\, dx}{\int_\Omega |\nabla g^n|^2\, dx}$$

(2.45) $$g^{n+1} = g^n + \gamma^{n+1} w^{n+1},$$

$n = n+1$, *go to* (2.41).

The two non trivial steps of algorithm (2.38)-(2.45) are

(i) The solution of the *single variable* minimization problem (2.41) ; the corresponding *line search* can be achieved by *dichotomy* of *Fibonacci* methods (see for example [15], [17],[18]). We observe that each evaluation of $J(v)$, for a given argument v, requires the solution of the linear Dirichlet problem (2.32) to obtain the corresponding ξ.

(ii) The calculation of g^{n+1} from ϕ^{n+1} which requires the solution of two linear Dirichlet problems (namely (2.32) with $v = \phi^{n+1}$ and (2.43)).

Calculation of $J'(\phi^n)$ and g^n : We refer to [5, Sec. 2.4] where we prove that $\forall\, v,w \in H_0^1(\Omega)$ we have

(2.46) $$\langle J'(v), w\rangle = \int_\Omega \nabla\xi \cdot \nabla w\, dx - \langle T'(v)\cdot w, \xi\rangle$$

where, in (2.46), T' is the differential of T at v and ξ the solution of (2.32) corresponding to v. Therefore $J'(v) \in H^{-1}(\Omega)$ may be identified with the linear functional defined on $H_0^1(\Omega)$ by

(2.47) $$w \to \int_\Omega \nabla\xi\cdot\nabla w\,dx - \langle T'(v)\cdot w, \xi\rangle .$$

It follows then from (2.43),(2.46),(2.47) that g^n is the solution of the following *linear variational problem*

(2.48) $$\begin{cases} \text{Find } g^n \in H_0^1(\Omega) \text{ such that } \forall\, w \in H_0^1(\Omega) \\ \int_\Omega \nabla g^n\cdot\nabla w\,dx = \int_\Omega \nabla\xi^n\cdot\nabla w\,dx - \langle T'(\phi^n)\cdot w, \xi^n\rangle \end{cases}$$

where ξ^n is the solution of (2.32) corresponding to $v = \phi^n$.

Remark 2.4 : The fact that $J'(v)$ is known through (2.46) is not at all a drawback if a *Galerkin* or a *finite element* method is used to approximate (2.25). Indeed we only need to know the value of $\langle J'(v), w\rangle$ for w belonging to a *basis* of the *finite dimensional subspace* of $H_0^1(\Omega)$ corresponding to the approximation under consideration.

We conclude this subsection 2.3.3 by observing that, from the points (i) and (ii) of the above discussion, an *efficient Poisson solver* will be a basic tool in the solution of (2.25) (in fact of *an approximation* of it) by the conjugate gradient algorithm (2.38)-(2.45) ; this very important point will be discussed with more details in Sec. 2.3.5.

2.3.4. A nonlinear least square approach to arc length continuation methods.

Despite the fact that it is not exactly the main object of this paper we would like to show that the above nonlinear least square methodology can be (slightly) modified in order to be used to solve nonlinear problems by an *arc length continuation method*, directly inspired from H.B. KELLER [19], [20] (where the basic iterative methods are *Newton's* and *Quasi-Newton's*, instead of *conjugate gradient)*.

We have chosen as test problem a variant of the nonlinear Poisson problem (2.25). Let us consider the following family of nonlinear Poisson problems, parametrized by $\lambda \in \mathbb{R}$,

(2.49) $$\begin{cases} -\Delta\phi = \lambda T(\phi) \text{ in } \Omega \\ \phi = 0 \text{ on } \Gamma; \end{cases}$$

(2.25) corresponds to $\lambda=1$.

Following [19], [20] (for which we refer to justification) we associate to (2.49) a *"continuation"* equation ; we have chosen[1]

(2.50) $$\int_\Omega |\nabla\dot{\phi}|^2 dx + \dot{\lambda}^2 = 1,$$

where $\dot{\phi} = \frac{\partial\phi}{\partial s}$, $\dot{\lambda} = \frac{d\lambda}{ds}$, and where the *curvilinear abcissa* s is defined by

(2.51) $$\delta s = \dot{\lambda}\delta\lambda + \int_\Omega \nabla\dot{\phi}\cdot\nabla\delta\phi \, dx$$

or equivalently by

(2.52) $$(\delta s)^2 = (\delta\lambda)^2 + \int_\Omega \nabla\delta\phi\cdot\nabla\delta\phi \, dx ;$$

in fact we are considering a path in $H_0^1(\Omega)\times\mathbb{R}$ whose arc length if defined by (2.50)-(2.52).

Then in order to solve (2.49) we consider the family of nonlinear systems (2.49), (2.50). In practice we shall approximate it by the following *discrete family* of nonlinear problems where Δs is an *arc length step, positive or negative* (possibly varying with n) and where $\phi^n \simeq \phi(n\Delta s)$:

(2.53) $$\phi^o = 0, \ \lambda^o = 0 \ ; \ \dot{\phi}(0), \ \dot{\lambda}(0) \ \textit{given}$$

(since $\phi=0$ is the unique solution of (2.49) if $\lambda=0$), *then for* $n \geq 0$ *assuming* ϕ^{n-1}, λ^{n-1}, ϕ^n, λ^n *known we obtain* $\{\phi^{n+1},\lambda^{n+1}\} \in H_0^1(\Omega)\times\mathbb{R}$ *by*

(2.54) $$\begin{cases} -\Delta\phi^{n+1} = \lambda^{n+1}\, T(\phi^{n+1}) \ \textit{in}\ \Omega, \\ \phi^{n+1} = 0 \ \textit{on}\ \Gamma, \end{cases}$$

and

(2.55) $$\begin{cases} \int_\Omega \nabla(\phi^1-\phi^o)\cdot\nabla\dot{\phi}(0)\,dx + (\lambda^1-\lambda^o)\dot{\lambda}(0) = \Delta s \\ \textit{if}\ n=0, \end{cases}$$

(2.56) $$\begin{cases} \int_\Omega \nabla(\phi^{n+1}-\phi^n)\cdot\nabla\left(\frac{\phi^n-\phi^{n-1}}{\Delta s}\right)dx + (\lambda^{n+1}-\lambda^n)\left(\frac{\lambda^{n+j}-\lambda^{n+j-1}}{\Delta s}\right) = \Delta s \\ \textit{with}\ j=0\ \textit{or}\ 1,\ \textit{if}\ n \geq 1 \ ; \end{cases}$$

obtaining $\dot{\phi}(0)$ and $\dot{\lambda}(0)$ is an easy task since from (2.49) we have

---

(1) Other choices are possible.

(2.57)
$$\begin{cases} -\Delta\dot{\phi}(0) = \dot{\lambda}(0)T(0) \ \textit{in}\ \Omega, \\ \dot{\phi}(0) = 0 \ \textit{on}\ \Gamma, \end{cases}$$

and therefore

(2.58)
$$\dot{\lambda}^2(0)\left(1+\int_\Omega |\nabla\hat{\phi}|^2 dx\right) = 1$$

where $\hat{\phi} \in H_o^1(\Omega)$ is the solution of

$$\begin{cases} -\Delta\hat{\phi} = T(0) \ \textit{in}\ \Omega, \\ \hat{\phi} = 0 \ \textit{on}\ \Gamma, \end{cases}$$

(then we clearly have $\dot{\phi}(0) = \dot{\lambda}(0)\hat{\phi}$).

Relations (2.53)-(2.56) suggest clearly a discretization scheme for solving the Cauchy problem for *first order Ordinary Differential Equations* ; from this analogy we can derive many other discretization schemes for the approximation of (2.49), (2.50) (more implicit, multistep, Runge-Kutta, etc...).

Nonlinear least square and conjugate gradient solution of (2.54)-(2.56) : Without going into details we can solve (2.54)-(2.56) by a variant of (2.38)-(2.45) defined on the Hilbert space $H_o^1(\Omega)\times\mathbb{R}$ equipped with the metric and inner-product corresponding to

(2.59)
$$\{v,\mu\} \to \int_\Omega |\nabla v|^2\, dx + \mu^2 \ ;$$

it is clear that many other choices than (2.59) are possible, however in all cases the scaling of the conjugate gradient algorithm using a discrete variant of[1]

(2.60)
$$\begin{pmatrix} -\Delta & 0 \\ 0 & 1 \end{pmatrix} \ \textit{(or similar operators)}$$

will require an efficient solver and the conclusions of Sec. 2.1 and Sec. 2.3.3 still hold.

---

[1] In (2.60), Δ corresponds to the homogeneous Dirichlet boundary conditions.

2.3.5. On the practical implementation of the preconditioning operation in algorithm (2.38)-(2.45).

In practice, we shall use a *finite dimensional* variant of (2.38)-(2.45) to solve a *finite dimensional approximation of* (2.25) ; from this observation we shall discuss the numerical solution by a preconditioned conjugate gradient algorithm, via a nonlinear least square formulation, of the following problem in $\mathbb{R}^N$

(2.61) $$\underset{\sim}{F}(\underset{\sim}{\phi}) = \underset{\sim}{0}$$

where $\underset{\sim}{F} : \mathbb{R}^N \to \mathbb{R}^N$, $\phi \in \mathbb{R}^N$.

Let $\underset{\sim}{B}$ be a *symmetric, positive definite* N×N matrix ; to $\underset{\sim}{v} \in \mathbb{R}^N$ we associate $\underset{\sim}{\xi}(\underset{\sim}{v}) = \underset{\sim}{\xi} \in \mathbb{R}^N$ by

(2.62) $$\underset{\sim}{B}\underset{\sim}{\xi} = \underset{\sim}{F}(\underset{\sim}{v}).$$

A nonlinear least square formulation of (2.61) is then

(2.63) $$\underset{\underset{\sim}{v}\in\mathbb{R}^N}{\mathrm{Min}} \frac{1}{2}(\underset{\sim}{B}\underset{\sim}{\xi},\underset{\sim}{\xi})$$

where, in (2.63), ξ is a function of v via (2.62) and where $(\cdot,\cdot)$ denotes the usual inner product of $\mathbb{R}^N$. We define $J : \mathbb{R}^N \to \mathbb{R}$ by

$$J(\underset{\sim}{v}) = \frac{1}{2}(\underset{\sim}{B}\underset{\sim}{\xi}(\underset{\sim}{v}),\underset{\sim}{\xi}(\underset{\sim}{v})) ;$$

let us detail now the discrete analogue of (2.38)-(2.45) for solving (2.61) via (2.62),(2.63) :

Step 0 : Initialisation

(2.64) $$\underset{\sim}{\phi}^o \in \mathbb{R}^N, \textit{ given},$$

(2.65) $$\underset{\sim}{g}^o = \underset{\sim}{B}^{-1}J'(\underset{\sim}{\phi}^o)$$

(2.66) $$\underset{\sim}{w}^o = \underset{\sim}{g}^o .$$

Then for $n \geq 0$, *assuming that* $\underset{\sim}{\phi}^n, \underset{\sim}{g}^n, \underset{\sim}{w}^n$ *are known we define* $\underset{\sim}{\phi}^{n+1}, \underset{\sim}{g}^{n+1}, \underset{\sim}{w}^{n+1}$ *by*

Step 1 : Descent

(2.67) $$\lambda^n = \underset{\lambda\in\mathbb{R}}{\mathrm{Arg\ min}}\ J(\phi^n - \lambda w^n) ,$$

$$\phi^{n+1} = \phi^n - \lambda^n w^n \tag{2.68}$$

<u>Step 2</u> : <u>Construction of the new descent direction</u>

$$g^{n+1} = B^{-1} J'(\phi^{n+1}) \ , \tag{2.69}$$

$$\gamma^{n+1} = \frac{(Bg^{n+1}, g^{n+1} - g^n)}{(Bg^n, g^n)} \ , \tag{2.70}$$

$$w^{n+1} = g^{n+1} + \gamma^{n+1} w^n \tag{2.71}$$

$n = n+1$, *go to* (2.67).

Concerning $J'(v)$ we have

$$J'(v) = (F'(v))^t \xi = (F'(v))^t B^{-1} F(v) \quad \forall\, v \in \mathbb{R}^N . \tag{2.72}$$

From the above relations a most important tool for solving (2.61) via (2.62),(2.63) and algorithm (2.64)-(2.71) will be an efficient technique for solving linear systems associated to matrix B ; let us consider a Cholesky factorization of B such as

$$B = LL^t \ ; \tag{2.73}$$

if L is not sparse enough compared to B, it may be necessary to solve also these linear systems by an *iterative method* with preconditioning.
We may use in particular to solve these systems

$$Bx = c \quad (x, c \in \mathbb{R}^N) \tag{2.74}$$

conjugate gradient method with a preconditioning associate to the matrix

$$\tilde{B} = \tilde{L}\tilde{L}^t \ , \tag{2.75}$$

where $\tilde{L}$ is precisely obtained from B via an incomplete Cholesky factorization process. One may find the above approach a bit complicated since it involves the *conjugate gradient outer loop* (2.64)-(2.71) and *conjugate gradient inner loops* for solving various internal linear systems like (2.74) ; at that point it is very tempting to use $\tilde{B}$ instead of B in (2.62),(2.63) and also in (2.64)-(2.74). If $\tilde{B}$ is well chosen, *using* $\tilde{B}$ *instead of* B *will increase the number of outer iterations*, but since the internal linear systems (2.74) are replaced by linear systems like

$$\tilde{B}x \ (= \tilde{L}\tilde{L}^t x) = c \quad (x, c \in \mathbb{R}^N), \tag{2.76}$$

much easier to solve, the resulting global algorithm may be faster.

3. - ON A NEW INCOMPLETE CHOLESKY FACTORIZATION METHOD.

Let $\underset{\sim}{A} = (a_{ij})_{1\le i,j\le N}$ be a N×N *symmetric, positive definite, banded* matrix of band-width 2m+1 (one says that $\underset{\sim}{A}$ is 2m+1-*diagonal*) ; we clearly have

(3.1) $$a_{ij} = 0 \textit{ if } |i-j| > m.$$

Consider the *linear system*

(3.2) $$\underset{\sim}{A}\underset{\sim}{x} = \underset{\sim}{b} \ ;$$

we may use, to solve (3.2), a *Cholesky's factorization*

(3.3) $$\underset{\sim}{A} = \underset{\sim}{L}\underset{\sim}{L}^t$$

of $\underset{\sim}{A}$, with $\underset{\sim}{L}$ lower triangular. It is well known that (3.1) implies

(3.4) $$\ell_{ij} = 0 \textit{ if } i-j > m,$$

but as mentioned above (see Fig. 2.1), some *zero elements* in the band of $\tilde{A}$ may be such that the corresponding elements of $\underset{\sim}{L}$ are $\neq 0$.

We shall use $\underset{\sim}{L}$ to construct an *incomplete Cholesky's factorization* of $\underset{\sim}{A}$; to define and discuss the properties of the following methods it is convenient to introduce

(3.5) $$K = \{\{i,j\} \mid 0 \le i-j \le m,\ a_{ij} = 0\}$$

(3.6) $$K^* = \{\{i,j\} \mid 0 \le i-j \le m,\ \ell_{ij} \neq 0\}\ ,$$

(3.7) $$n(K) = \text{Card } (K).$$

We give now a *positive constant* C, and from $\underset{\sim}{L}$ and C we define the two following *lower triangular matrices*

(3.8) $$\begin{cases} \tilde{\underset{\sim}{L}}_c = (\tilde{\ell}_{ij}), \textit{ with} \\ \tilde{\ell}_{ij} = 0 \textit{ if } \{i,j\} \in K \textit{ and } |\ell_{ij}| < C, \\ \tilde{\ell}_{ij} = \ell_{ij} \textit{ if not,} \end{cases}$$

$$(3.9)\quad \begin{cases} \tilde{L}'_c = (\tilde{\ell}'_{ij}), \textit{ with} \\ \tilde{\ell}'_{ij} = 0 \textit{ if } \{i,j\} \in K \textit{ and } |\ell_{ij}| < C \min\limits_{i,j} (\ell_{ii},\ell_{jj}) \\ \tilde{\ell}'_{ij} = \ell_{ij} \textit{ if not.} \end{cases}$$

<u>Remark 3.1</u> : We observe that

(i) *If* $C < \min\limits_{\{i,j\}\in K} |\ell_{ij}|$ *then* $\tilde{L}_c = L$,

(ii) *If* $C < \min\limits_{\{i,j\}\in K} \{\frac{|\ell_{ij}|}{\min\limits_{i,j}(|\ell_{ii}|,|\ell_{jj}|)}\}$ *then* $\tilde{L}'_c = L$

(iii) *If* $C > \max\limits_{\{i,j\}} |\ell_{ij}|$ *then* $\{\{i,j\}|\tilde{\ell}_{ij} = 0\} = K$,

(iv) *If* $C > \max\limits_{\{i,j\}} \{\frac{|\ell_{ij}|}{\min\limits_{i,j}(\ell_{ii},\ell_{jj})}\}$ *then* $\{\{i,j\}|\tilde{\ell}'_{ij} = 0\} = K$.

If (iii) and (iv) hold, then $\tilde{L}_c$ and $\tilde{L}'_c$ have their nonzero elements in the same position then those of A, and are therefore close to the incomplete Cholesky operators discussed in [9],[11]. However the "approximate" Cholesky's factors in [9],[11] are constructed *during* the Cholesky's factorization of A, and not *after* like in our methods ; from this fundamental difference it follows that

- The approximate factors in [9],[11] are more economical to construct than those defined by (3.8),(3.9), and therefore may be interesting to use if the number of linear systems associated to A is small.
- The matrices $\tilde{A}_c$ and $\tilde{A}'_c$, defined from $\tilde{L}_c$ and $\tilde{L}'_c$ by

$$(3.10)\qquad \tilde{A}_c = \tilde{L}_c\, \tilde{L}_c^t ,$$

$$(3.11)\qquad \tilde{A}'_c = \tilde{L}'_c\, \tilde{L}'^t_c$$

are *spectrally closer* to A than the corresponding matrices in [9],[11] ; moreover, *once constructed*, their factors have basically the same storage requirement than those in [9], [11]. Therefore in industrial environment where the number of linear systems associated to a given A may be very large, our approach appears to lead to more efficient algorithms than those in [9], [11], even if the construction of L requires an *out of core* process.

<u>Practical construction of $\tilde{L}_c$ and $\tilde{L}'_c$</u> :
For practical applications it is feasible to choose C such that a given percentage of

non zero elements of $\tilde{\underset{\sim}{L}}_c$ and $\tilde{\underset{\sim}{L}}'_c$ is kept. Thus, giving d, $0 < d < 100$, we may define $\tilde{\underset{\sim}{L}}_{(d/100)}$ and $\tilde{\underset{\sim}{L}}'_{(d/100)}$ as follows :

We define from C *and* $\tilde{\underset{\sim}{L}}_c$, $\tilde{\underset{\sim}{L}}'_c$

(3.12) $$\tilde{K}_c = \{\{i,j\} \mid \{i,j\} \in K \,,\; \tilde{\ell}_{ij} \neq 0\} \,,$$

(3.13) $$\tilde{K}'_c = \{\{i,j\} \mid \{i,j\} \in K,\; \tilde{\ell}'_{ij} \neq 0\} \,,$$

(3.14) $$n(\tilde{K}_c) = \text{Card}\,(\tilde{K}_c),\; n(\tilde{K}'_c) = \text{Card}\,(\tilde{K}'_c)$$

and then we adjust C *in order to obtain either*

(3.15) $$\tilde{\underset{\sim}{L}}_{(d/100)} = \tilde{L}_c \text{ with C such that } n(\tilde{K}_c)/n(K) = d/100,$$

or

(3.16) $$\tilde{\underset{\sim}{L}}'_{(d/100)} = \tilde{\underset{\sim}{L}}'_c \text{ with C such that } n(\tilde{K}'_c)/n(K) = d/100.$$

We observe that d=0 (resp. d=100) leads to the same $\tilde{L}_c$ and $\tilde{L}'_c$ than (iii) and (iv) in Remark 3.1 (resp. leads to matrix $\underset{\sim}{L}$ of (3.3)).

Remark 3.2 : If $\underset{\sim}{A}$ comes from a *finite element approximation* one may also proceed as follows to obtain an approximate Cholesky's factorization of $\underset{\sim}{A}$:
Let $\mathcal{C}_h$ be a finite element partition of a given domain Ω[1] ; we have

(3.17) $$\bar{\Omega} = \bigcup_{T \in \mathcal{C}_h} T$$

where, in (3.17), the finite elements T are adjacent polyedras ; let

(3.18) $$\{P_i\}_{i=1}^{N}$$

be the set of the nodes of $\mathcal{C}_h$ associated to the finite element method under consideration. The *complementary* set $\bar{K}$ of K (see (3.5) for the definition of K) is defined by

(3.19) $$\bar{K} = \{\{i,j\} \mid 0 \le i-j \le m,\; \exists T \in \mathcal{C}_h \text{ such that } P_i, P_j \in T\} \quad ;$$

we introduce also

---

(1) For simplicity we suppose that Ω is a polyhedra and that we only deal with *Lagrangian* finite elements.

(3.20) $\quad \overline{\overline{K}} = \{\{i,j\} \mid 0 \le i-j \le m, \exists k \text{ such that } P_i, P_k \in T,\ P_j, P_k \in T' \text{ with } T, T' \in \mathcal{T}_h\}$.

We define finally, from $\overline{\overline{K}}$ and $\underset{\sim}{L}$, an approximate Cholesky's factorization of $\underset{\sim}{A}$ by

(3.21)
$$\begin{cases} \overline{\overline{\underset{\sim}{L}}} = (\overline{\overline{\ell}}_{ij}) , \\ \overline{\overline{\ell}}_{ij} = \ell_{ij} \textit{ if } \{i,j\} \in \overline{\overline{K}},\ \overline{\overline{\ell}}_{ij} = 0 \textit{ otherwise.} \end{cases}$$

Using such a construction $\overline{\overline{\underset{\sim}{L}}}$ is independent of the numbering of $\{P_i\}_{i=1}^N$; in two-dimension the number of non zero elements in the band of $\overline{\overline{\underset{\sim}{L}}}$ is approximately 20%, but in three-dimension it is around 50% which is quite large in view of the problems that we want to solve.

Remark 3.3 : Actually the introduction of $\tilde{\underset{\sim}{L}}'_c$ defined by (3.9) was also motivated by finite element method. Indeed a quick and easy analysis shows that if $\Omega \subset \mathbb{R}^3$ then $\ell_{ij} = O(h)$; on the other hand $\ell_{ij} = O(1)$ if $\Omega \subset \mathbb{R}^2$. Thus it is necessary to discard the small elements by a test on their relative magnitude instead of their absolute value ; observe however that this problem does not arises for $\tilde{\underset{\sim}{L}}_{(d/100)}$ and $\tilde{\underset{\sim}{L}}'_{(d/100)}$.

4. - NUMERICAL TESTS.

More details about the numerical tests may be found in [6].

4.1. Solution of linear problems.

We consider in this Sec. 4.1 the application of the above algorithms to the solution of *linear problems in finite dimension*

(4.1)
$$\underset{\sim}{A}\underset{\sim}{\phi} = \underset{\sim}{F}.$$

In our tests, (4.1) comes from a *finite element* approximation of the *elliptic boundary value problem*

(4.2)
$$\begin{cases} -\Delta\phi = f \textit{ in } \Omega, \\ \phi = g_1 \textit{ on } \Gamma_1,\ \dfrac{\partial\phi}{\partial n} = g_2 \textit{ on } \Gamma_2 \end{cases}$$

where, in (4.2), $\Gamma_1 \cap \Gamma_2 = \emptyset$, $\Gamma_1 \cup \Gamma_2 = \partial\Omega$;
the following results correspond to *piecewise linear, globally continuous approximations*.

We have used algorithm (2.17)-(2.24) of Sec. 2.2, with various preconditioning matrices $\underset{\sim}{S}$, and also with different numbering of the nodes.

The first example is the discretization of a two-dimensional problem (4.2) ; the corresponding number of unknowns is 600 and the Cholesky's factor $\underset{\sim}{L}$ in

(4.3) $$\underset{\sim}{A} = \underset{\sim}{L}\underset{\sim}{L}^t$$

has about 12000 non zero elements ; we have represented on Fig. 4.1 the number of iterations necessary to solve (4.1) by algorithm (2.17)-(2.24) using various incomplete Cholesky's factorization of $\underset{\sim}{A}$ to construct the preconditioning matrix $\underset{\sim}{S}$; the stopping test is

(4.4) $$\|\underset{\sim}{A}\underset{\sim}{\phi}^n - \underset{\sim}{F}\|^2 \leq \varepsilon$$

(here $\varepsilon = 10^{-7}$) ; the norm (4.4) is the usual $\mathbb{R}^N$-norm.

We observe the very good performances obtained with $d/100 \simeq 5\%$. We also have indicated the results corresponding to the methods in [9]. Other tests have shown that up to 5% the number of iterations with $\underset{\sim}{L}_{d/100}$ is not very sensitive to mesh refinement (of course the computer time increases).

We have shown on Fig. 4.2 the results of similar experiments related to a 3-dimensional problem (4.2) ; the number of unknowns is 5328 and the Cholesky's factor $\underset{\sim}{L}$ in (4.3) has about 1.5×10^6 non zero elements.

4.2. Application to transonic potential flows for compressible inviscid fluids.
For more details on this important problem we refer to [4]-[6].

4.2.1. Mathematical formulation.
Under simplifying assumptions for which we refer to, e.g., [4]-[6] the flows under consideration are modelled by

(4.5) $$\nabla\cdot\rho(\phi)\nabla\phi = 0 \ \textit{in}\ \Omega,$$

where

(4.6) $$\rho(\phi) = \rho_o\left(1 - \frac{|\nabla\phi|^2}{\frac{\gamma+1}{\gamma-1}C_*^2}\right)^{\frac{1}{\gamma-1}} .$$

In (4.5),(4.6) :

Ω is the domain of the flow, ϕ a velocity potential(1), ρ the fluid density, γ (1.4 in air) is the ratio of specific heats, C_* is the critical velocity.

(1) i.e. $\underset{\sim}{v} = \nabla\phi$ where $\underset{\sim}{v}$ is the flow velocity.

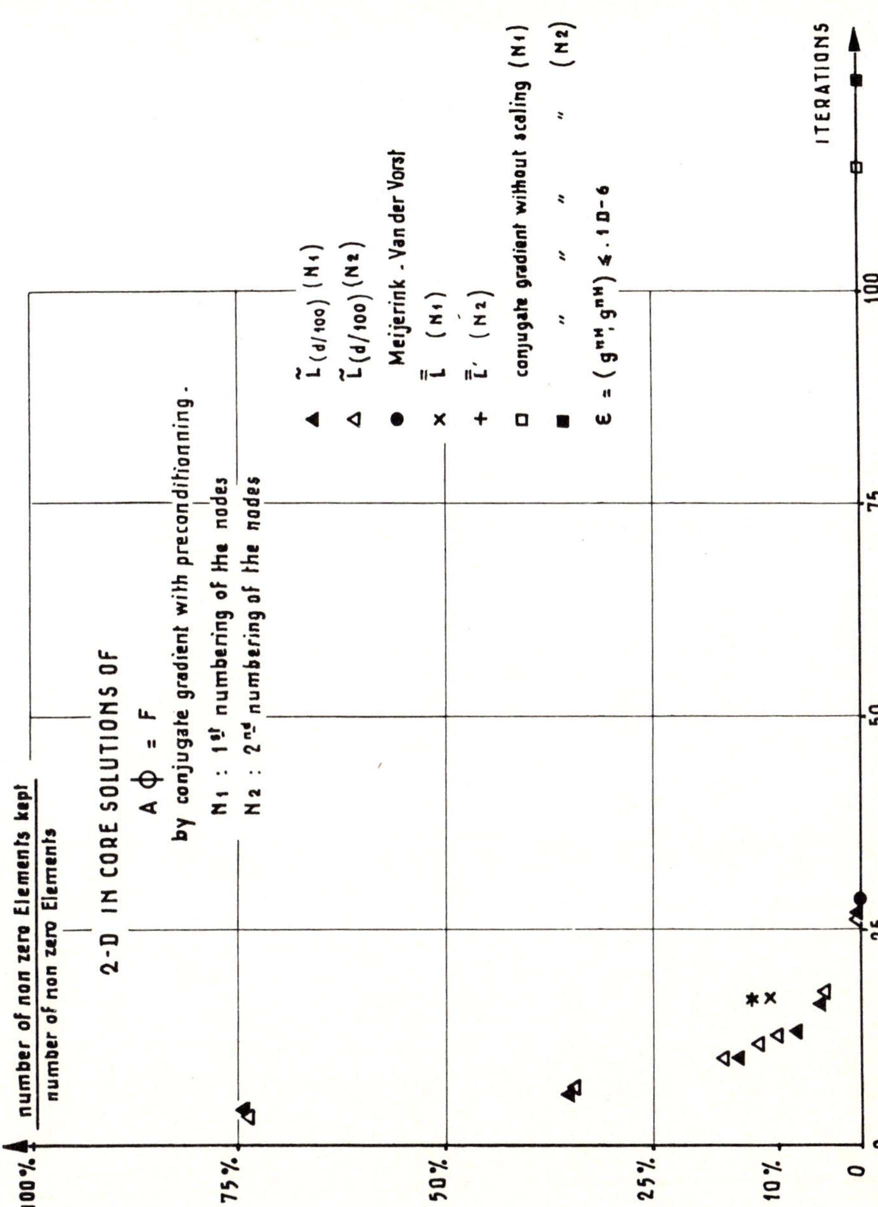
number of non zero Elements kept / number of non zero Elements
2-D IN CORE SOLUTIONS OF
$A \Phi = F$
by conjugate gradient with preconditionning.
N_1 : 1st numbering of the nodes
N_2 : 2nd numbering of the nodes
$\tilde{L}_{(d/100)}$ (N_1)
$\tilde{L}(d/100)$ (N_2)
Meijerink - Van der Vorst
$\bar{\bar{L}}$ (N_1)
$\bar{\bar{L}}'$ (N_2)
conjugate gradient without scaling (N_1)
" " " " (N_2)
$\epsilon = (g^{nM}, g^{nM}) \leq .1D-6$
100%
75%
50%
25%
10%
0
0
25
50
75
100
ITERATIONS

FIGURE 4.1

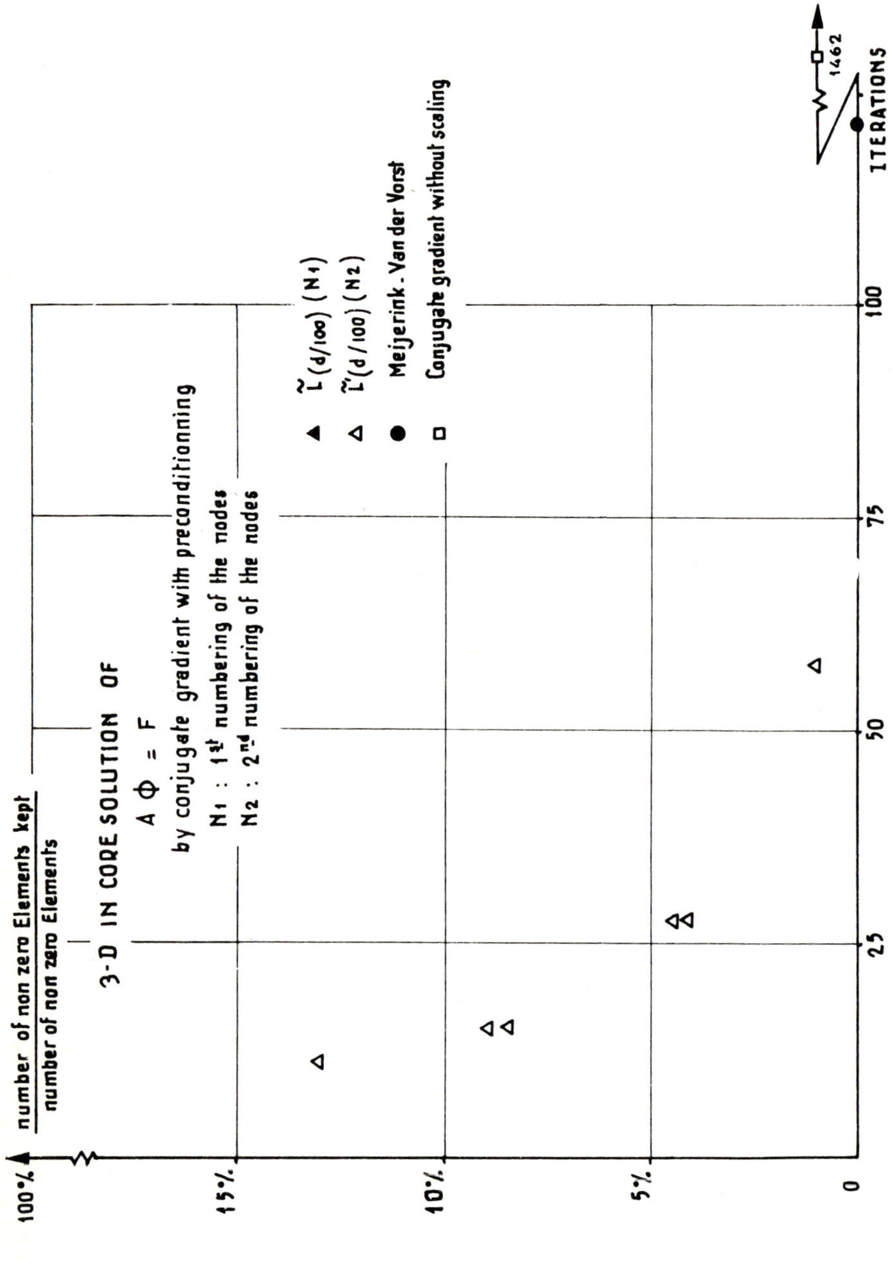

100%
15%
10%
5%
0
25
50
75
100
1462
ITERATIONS
number of non zero Elements kept / number of non zero Elements
3-D IN CORE SOLUTION OF
A Φ = F
by conjugate gradient with preconditionning
N1 : 1st numbering of the nodes
N2 : 2nd numbering of the nodes
L̃(d/100) (N1)
L̃(d/100) (N2)
Meijerink . Van der Vorst
Conjugate gradient without scaling

FIGURE 4-2

In addition to (4.5),(4.6) we have to prescribe *boundary conditions* like

$$(4.7)\quad \begin{cases} \phi = g_1 \ \text{on}\ \Gamma_1,\ \rho\dfrac{\partial\phi}{\partial n} = g_2 \ \text{on}\ \Gamma_2\ , \\ \text{where}\ \partial\Omega = \Gamma_1 \cup \Gamma_2,\ \Gamma_1 \cap \Gamma_2 = \emptyset\ , \end{cases}$$

and also *Kutta-Joukowsky conditions* if Ω is a *multi-connected domain* ; we have finally to include an *entropy condition* to *avoid rarefaction shocks* since such shocks are *unphysical*.

4.2.2. Finite element approximation and iterative solution.

The numerical solution of the above transonic problem using finite elements and non-linear least square - conjugate gradient methods with scaling is discussed with many details in [4]-[6],[21].

Suppose that Ω has been approximated by a polyhedral domain - still denoted by Ω - and denote by $\mathcal{T}_h$ a finite element partition of Ω, with $h = \max_{T\in\mathcal{T}_h} d(T)$, with $d(T)$ = diameter of T. Assuming that the T are triangles (resp. tetrahedrons) in 2-dimension (resp. in 3-dimension), we approximate $H^1(\Omega)$ by

$$H_h^1 = \{v_h \in C^0(\bar{\Omega}),\ v_h|_T \in P_1 \quad \forall\, T \in \mathcal{T}_h\}\ ,$$

with P_1 = space of polynomials of degree ≤ 1 ; we look for approximate solutions in

$$V_h = \{v_h \in H_h^1\ ,\ v_h|_{\Gamma_1} = g_{1h}\}$$

where g_{1h} is an approximation of g_1 ; we associate to V_h

$$V_{oh} = \{v_h \in H_h^1,\ v_h|_{\Gamma_1} = 0\}\ .$$

For simplicity we do not discuss Kutta-Joukowsky conditions ; then an approximate transonic flow problem can be formulated as follows

$$(4.8)\quad \underset{\phi_h\in V_h}{\text{Min}}\ \{\ \frac{1}{2}\int_\Omega |\nabla\xi_h|^2 dx + \mathcal{S}_h(\phi_h)\}\ ,$$

where, in (4.8), ξ_h is a function of ϕ_h via the *state problem*

$$(4.9)\quad \begin{cases} \displaystyle\int_\Omega \nabla\xi_h\cdot\nabla v_h\, dx = \int_\Omega \rho(\phi_h)\nabla\phi_h\cdot\nabla v_h dx - \int_{\Gamma_2} g_{2h}\, v_h\, d\Gamma \quad \forall v_h \in V_{oh}\ , \\ \xi_h \in V_{oh}\ , \end{cases}$$

and where $\mathcal{E}_h$ is a *discrete entropy functional,* taking *very large values* if the velocity field $\nabla\phi_h$ contains rarefaction shocks.

The discrete problem (4.8),(4.9), which is very close to (2.62),(2.63), can be solved by conjugate gradient methods with scaling very close to (2.64)-(2.71). It is very convenient to define $\underset{\sim}{B}$, in this variant of (2.64)-(2.71), from the bilinear form

$$\{v_h, w_h\} \to \int_\Omega \nabla v_h \cdot \nabla w_h dx\ dx\ ,$$

thus $\underset{\sim}{B}$ can be viewed as a discrete Laplace operator and therefore each iteration of the above conjugate gradient algorithm will require the solution of several discrete Poisson problems and the various comments done in Sec. 2.3.5 hold for this discrete transonic flow problem.

4.2.3. Numerical experiments.

Solution of a two-dimensional transonic flow problem : The problem that we consider is the flow around a NACA 0012 airfoil with $M_\infty = .8$ and a zero angle of attack[1] ; from these values there exists a *supersonic* pocket in the flow and a supersonic-subsonic transition with *shock* (see Fig. 4.5 for the mach distribution on the skin of the airfoil ; we observe a very sharp shock).

Following Sec. 4.2.2 the above flow problem has been approximated by a finite element method using about 1400 triangles and 600 nodes ; the Cholesky's factor associated to the corresponding discrete Laplace operator has about 1.1×10^4 non zero elements.

Following Sec. 4.2.2 the discrete transonic flow problem (4.8),(4.9) is then solved by a conjugate gradient method with a preconditionning operator $\underset{\sim}{B}$, which is a discrete variant of $-\Delta$ with appropriate *boundary conditions* ; using the above preconditioning method implies the solution at each iteration of several discrete Poisson problems ; these problems can also be solved by conjugate gradient methods with preconditioning, namely the methods discussed in Sec. 2.2, Sec. 3 and also Sec. 4.1.

We have indicated on Fig. 4.3 the C.P.U. time necessary to solve to a given accuracy the above flow problem, according to the incomplete Cholesky methods used to solve at each iteration the discrete Poisson problems required by the nonlinear least squares - conjugate gradient algorithm ; we observe again the very good performances of the method using $\underset{\sim}{\tilde{L}}_{(d/100)}$ even with rather small values of d ($d \simeq 10$).

If $\underset{\sim}{\tilde{L}}$ is an incomplete Cholesky's factor of $\underset{\sim}{B}$, it is very tempting, as mentioned in

---

(1) Thus the Kutta-Joukowsky condition is automatically satisfied

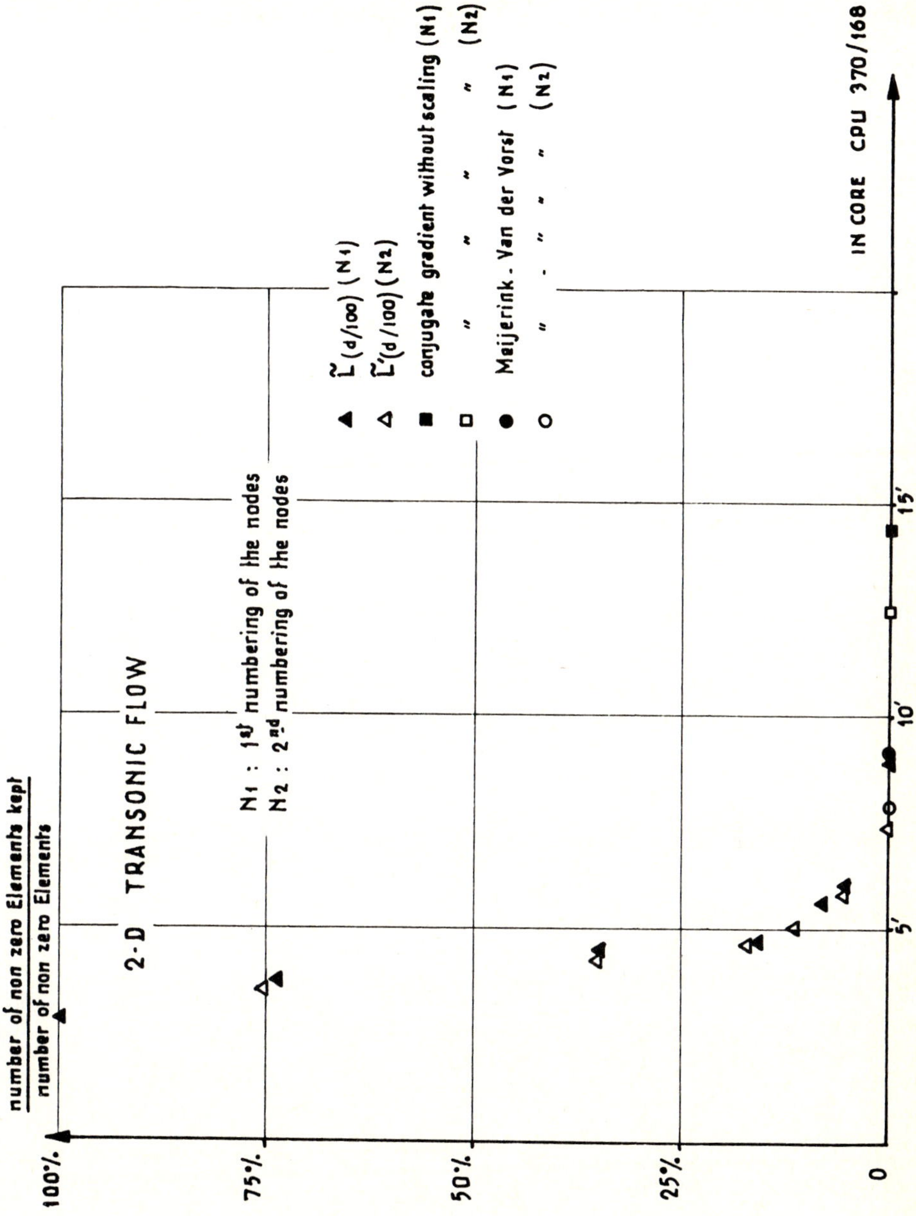
number of non zero Elements kept / number of non zero Elements
100%
75%
50%
25%
0
2-D TRANSONIC FLOW
N_1 : 1st numbering of the nodes
N_2 : 2nd numbering of the nodes
$\tilde{L}(d/100)$ (N_1)
$\tilde{L}'(d/100)$ (N_2)
conjugate gradient without scaling (N_1)
" " " " (N_2)
Meijerink - Van der Vorst (N_1)
" - " " (N_2)
5'
10'
15'
IN CORE CPU 370/168

FIGURE 4.3

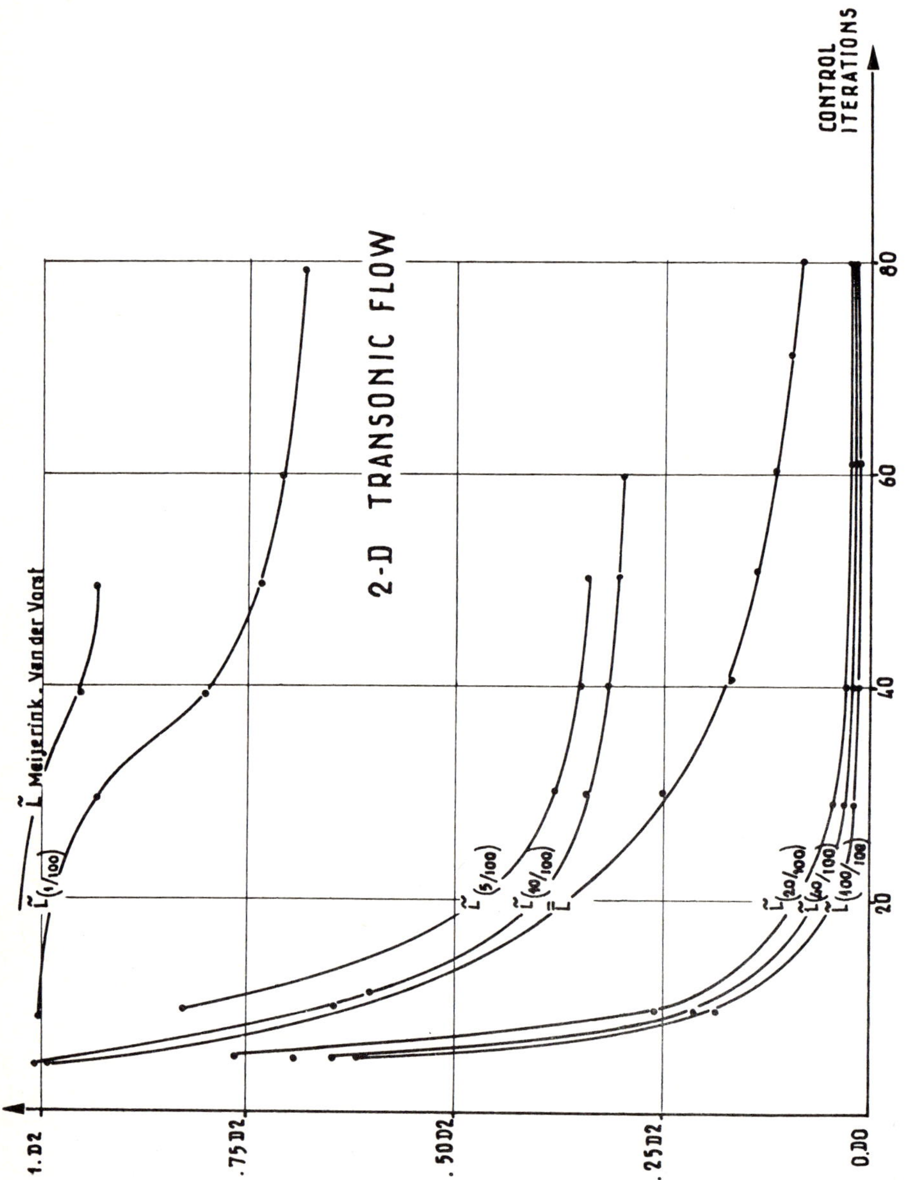
2-D TRANSONIC FLOW
CONTROL ITERATIONS
L̃ Meijerink, Van der Vorst
1.D2
.75D2
.50D2
.25D2
0.D0
20
40
60
80

FIGURE 4-4

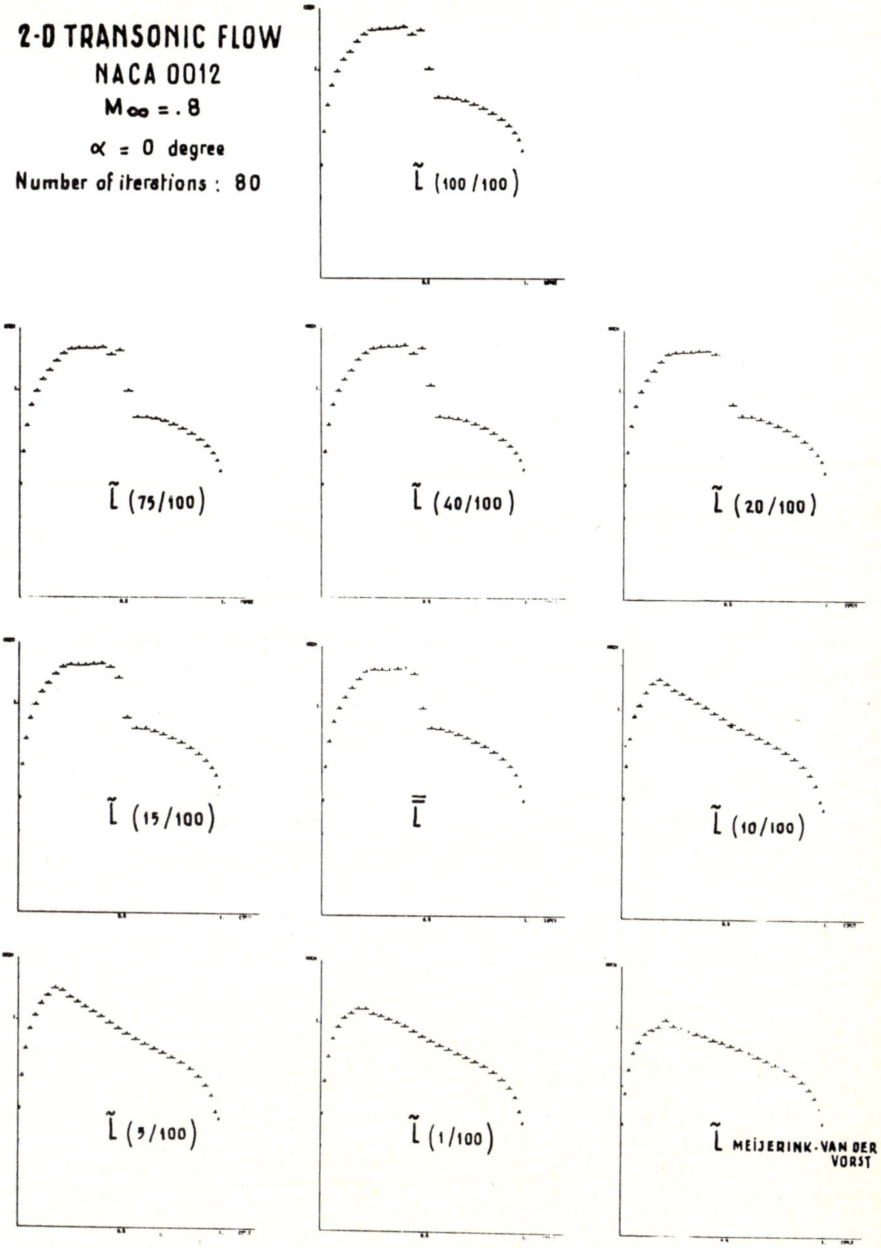
2-D TRANSONIC FLOW
NACA 0012
M∞ = .8
α = 0 degree
Number of iterations : 80
L̃ (100/100)
L̃ (75/100)
L̃ (40/100)
L̃ (20/100)
L̃ (15/100)
L̿
L̃ (10/100)
L̃ (5/100)
L̃ (1/100)
L̃ MEIJERINK-VAN DER VORST

FIGURE 4.5

Sec. 2.3.5, to use $\tilde{B} = \tilde{L}\tilde{L}^t$ to replace B introduced in Sec. 2.3.5 and 4.2.2 ; thus instead of the state equation (4.9) we shall introduce a state equation whose matrix in the left hand side will be $\tilde{B}$; similarly, instead of $\int_\Omega |\nabla\xi_h|^2 dx$ we shall use in (4.8) the quadratic form associate to $\tilde{B}$. Such a choice leads to some simplification of the iterative process[1], at the possible expense however of a greater number[1] of iterations. We have used the above strategy to solve the above transonic flow problem using $\tilde{B}$ $(=\tilde{L}\tilde{L}^t)$. The corresponding results are summarized on Figure 4.4 which shows the behavior of a *reference residual* as a function of the number of iterations ; as residual we have taken, using the notation of Sec. 2.3.5, $(B\hat{\xi}^n, \hat{\xi})_{\mathbb{R}^N}$ where $\hat{\xi}^n$ is defined from ϕ^n by

$$B\hat{\xi}^n = F(\phi^n) ,$$

where $\{\phi^n\}_n$ is the sequence of approximate solutions generated by the nonlinear least squares-conjugate gradient method[2] ; in addition to this residual behavior we have shown on Fig. 4.5 the velocity distribution on the airfoil at iteration 80 for different choices of the incomplete Cholesky's factors. As we can see the shock restitution is rather poor if the degree of incompleteness is too large ; this is not surprising since discontinuities mean large spectrum and clearly for small d or for the Meijerink-Van de Vorst factorization process the high frequencies have been too much damped.

<u>A 3-dimensional transonic flow problem</u> : The above two-dimensional transonic flow problem was in fact of sufficiently small size to be treated in core without difficulty. Now we would like to discuss the solution of a much more complicated problem whose standard solution needs out of core operations (mostly disk transfers). This three dimensional problem is the flow in and around the air intake of Fig. 4.6 ; in fact we want to simulate the effect of an oblique wind ; additional details concerning the problem are shown on Fig. 4.6.

Details of the finite element mesh which has been used are shown on Fig. 4.6, 4.7 ; we have about 6×10^3 nodes, 2.5×10^4 tetrahedrons, 7×10^4 non zero elements in the half band of the matrix approximating $-\Delta$; the corresponding Cholesky's factor has about 1.8×10^6 non zero elements ; these large numbers imply that an in core solution by the methods of Sec. 4.2.2 is impossible. This fact justifies incomplete Cholesky's procedures for our conjugate gradient algorithms.

Figure 4.8 show CPU times according to the type of preconditioning used to solve at each iteration (by conjugate gradient) the various discrete Poisson problems as-

---

(1) depending upon the degree of incompleteness

(2) Using $\tilde{B} = \tilde{L}\tilde{L}^t$ instead of B.

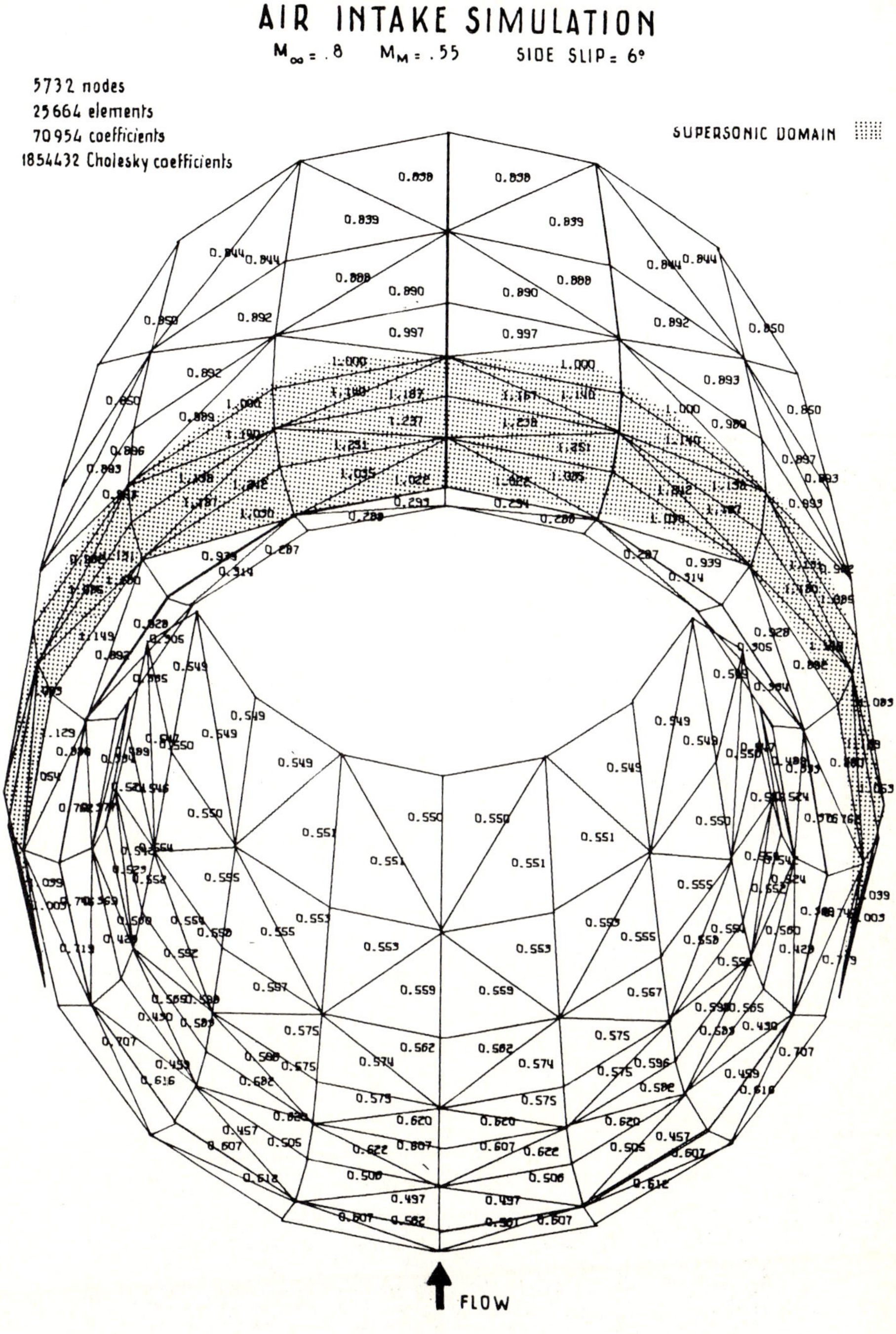
AIR INTAKE SIMULATION
M_∞ = .8 M_M = .55 SIDE SLIP = 6°
5732 nodes
25664 elements
70954 coefficients
1854432 Cholesky coefficients
SUPERSONIC DOMAIN
FLOW

FIGURE 4-6

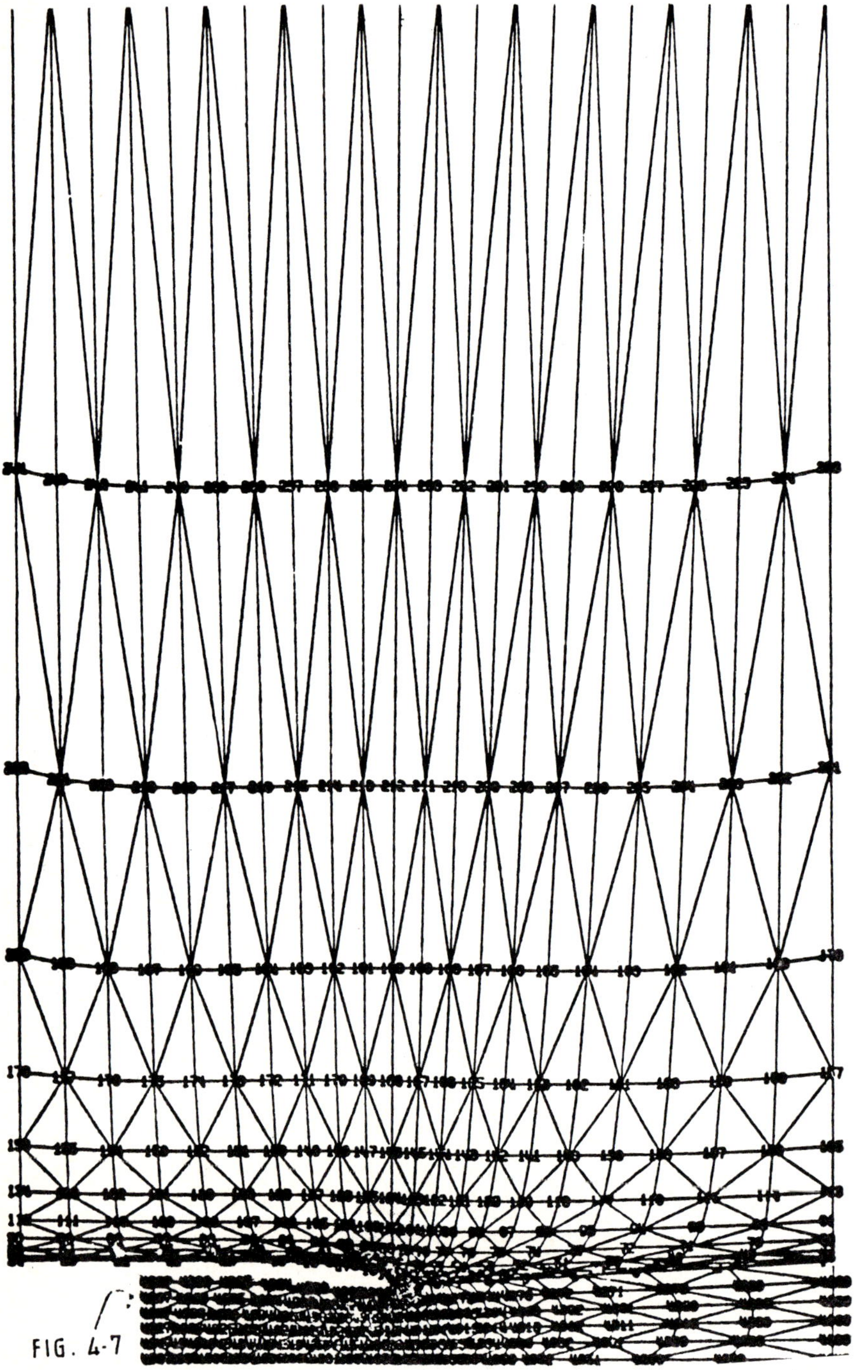

FIG. 4-7

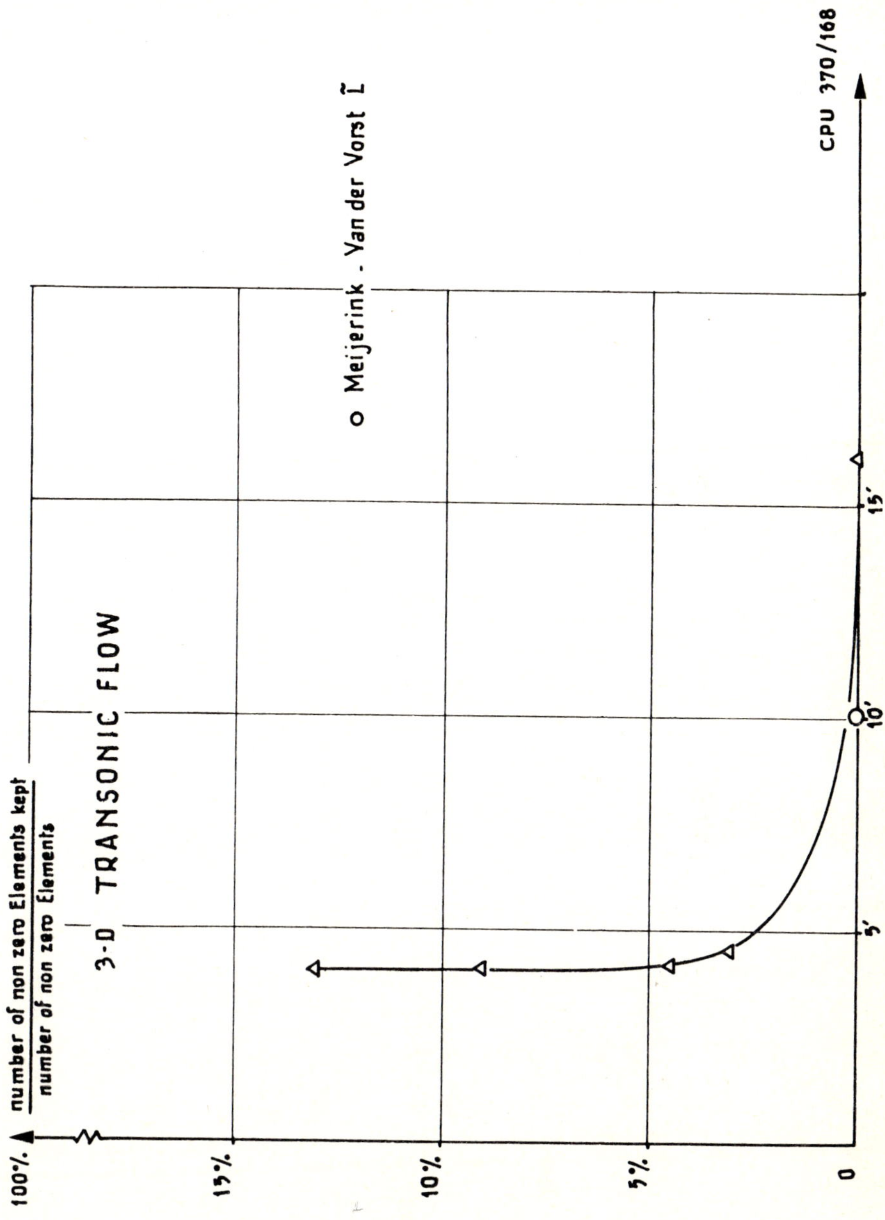
number of non zero Elements kept / number of non zero Elements
3-D TRANSONIC FLOW
o Meijerink - Van der Vorst Ĩ
CPU 370/168
100%
15%
10%
5%
0
5'
10'
15'

FIGURE 4.8

sociate to the nonlinear least squares-conjugate gradient methods discussed in Sec. 4.2.2 ; Figure 4.9 shows the variation of a *reference residual* (defined as in two-dimension) as a function of the iteration number, and according to the type of matrice $\tilde{B} = \tilde{L}\tilde{L}^t$ replacing the discrete variant $\tilde{B}$ of $-\Delta$ in the cost function (4.8) and the state equation (4.9).

The *supersonic* region on the surface of the air intake is indicated on Figure 4.6.

4.3. Application to the Navier-Stokes equations for incompressible viscous flows. We refer for more details on the methodology to follow to [4],[5],[6],[7],[22].

4.3.1. Mathematical formulation.

Let us consider a *Newtonian viscous, incompressible fluid.* If Ω and Γ denote respectively the region of the flow and its boundary, then this flow is governed by the *Navier Stokes equations*

$$\frac{\partial \tilde{u}}{\partial t} - \nu\Delta\tilde{u} + (\tilde{u}\cdot\tilde{\nabla})\tilde{u} + \tilde{\nabla}p = \tilde{f} \text{ in } \Omega, \tag{4.10}$$

$$\tilde{\nabla}\cdot\tilde{u} = 0 \text{ in } \Omega \text{ (incompressibility condition)}, \tag{4.11}$$

which in the steady case reduce to

$$-\nu\Delta\tilde{u} + (\tilde{u}\cdot\tilde{\nabla})\tilde{u} + \tilde{\nabla}p = \tilde{f} \text{ in } \Omega, \tag{4.12}$$

$$\tilde{\nabla}\cdot\tilde{u} = 0. \tag{4.13}$$

In (4.10)-(4.13) :

- $\tilde{u}$ is the *flow velocity*,
- p is the *pressure*,
- ν is the *viscosity* of the fluid ($\nu = 1/Re$, Re : Reynold's number),
- $\tilde{f}$ is a density of *external* forces.

Boundary conditions have to be added ; typical boundary conditions are for example

$$\tilde{u} = \tilde{g} \text{ on } \Gamma . \tag{4.14}$$

From (4.11),(4.13), $\tilde{g}$ has to obey $\int_\Gamma \tilde{g}\cdot\tilde{n}\, d\Gamma = 0$ where $\tilde{n}$ is the unit vector of the outward normal at Γ. Finally, for the time dependent problem (4.10),(4.11) an *initial condition* such as

$$\tilde{u}(x,0) = \tilde{u}_o(x) \text{ a.e. in } \Omega, \tag{4.15}$$

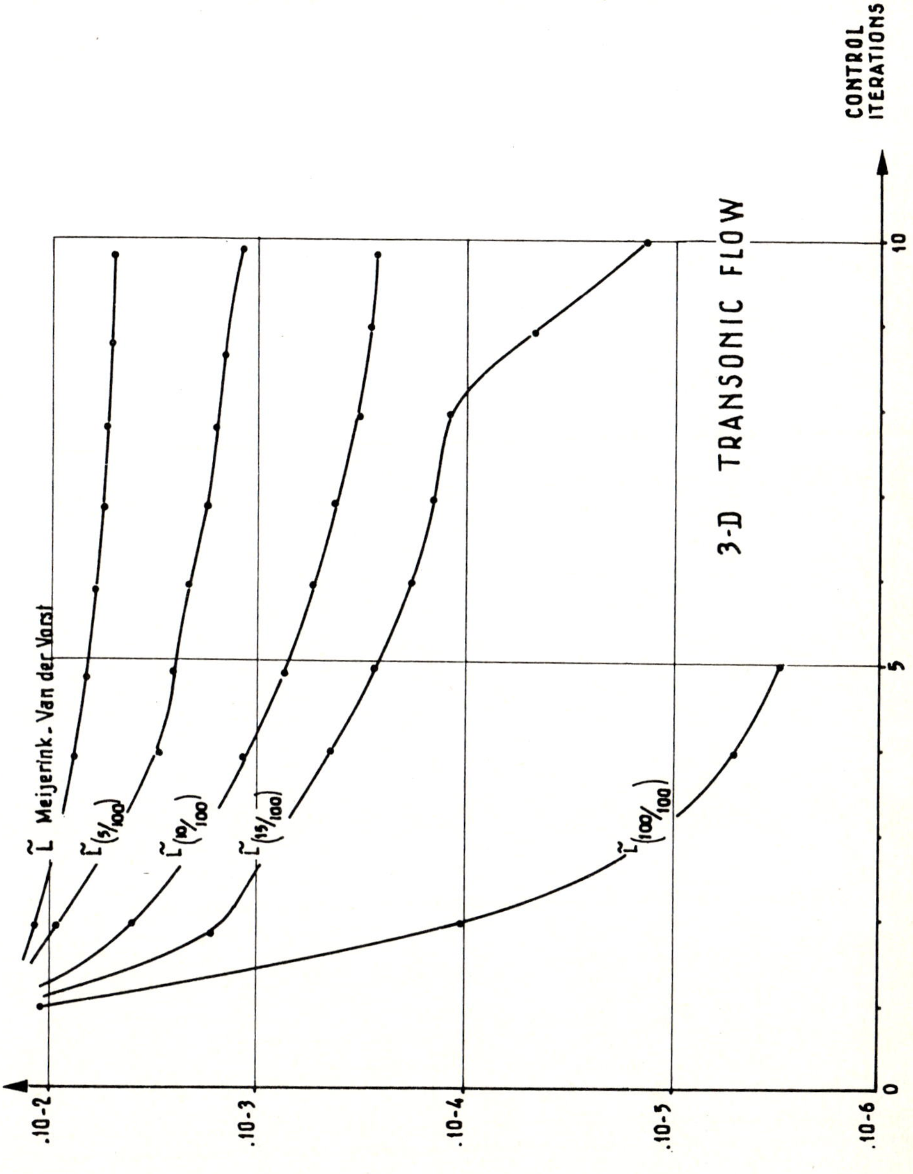

3-D TRANSONIC FLOW
CONTROL ITERATIONS
.10-2
.10-3
.10-4
.10-5
.10-6
0
5
10
L̃ Meijerink - Van der Varst
L̃(5/100)
L̃(10/100)
L̃(15/100)
L̃(100/100)

FIGURE 4·9

where $\underset{\sim}{u}_o$ is given, is prescribed.

A *linear* problem of particular interest, namely the *Stokes problem* is obtained by *cancelling the nonlinear terms* in (4.10),(4.12).

4.3.2. Finite element approximation and iterative solution.
Using *implicit time discretization schemes,* like those discussed in [4],[5],[6],[7],[22], we reduce the solution of the time dependent problem (4.10),(4.11),(4.14),(4.15) to the solution of a sequence of time independent problems very close to the steady Navier-Stokes problem (4.12)-(4.14).

Then using a convenient nonlinear least square-conjugate gradient method (see [4] for more details) we reduce the solution of the Navier-Stokes problem (4.12)-(4.14) to a sequence of Stokes problems. To summarize we can say that *solving the steady and unsteady Navier-Stokes equations can be reduced to the solution of a sequence of Stokes problems*. Using in turn the results of e.g. [4]-[7],[22],[23] it appears that *the solution of these Stokes problems can be reduced to the solution of a sequence of Poisson problems.*

If we suppose that the *finite element approximations* to be used are those described in [4],[6],[22]-[26], then the solution of the steady and unsteady Navier-Stokes equations is eventually reduced to the solution of a sequence of discrete Poisson problems. From these reduction properties it is clear that a most important tool in order to solve (4.10),(4.11),(4.14),(4.15) or (4.12),(4.13),(4.14) is again *efficient discrete Poisson solvers*. Thus the methods for solving linear problems discussed in Sections 3 and 4.1 are still the basic tools for solving the above Navier-Stokes problems. Moreover most of the comments of Sec. 4.2, related to the solution of another nonlinear fluid dynamics problem still hold for the present problem.

4.3.3. Numerical experiments.
Solution of a two dimensional unsteady problem : The problem that we consider is a fairly complicated one since it is the simulation at Re = 750 of a two-dimensional unsteady incompressible Navier-Stokes flow *in* and *around* an idealized *air intake* (the front part of which is shown on Fig. 4.10, 4.11), at a very large angle of attack (40 degrees) ; more details on this problem are given in Ref. [27] (and also in [6],[22] for a similar problem at Re = 250). Using the *finite element method* in [4],[6],[7],[22],[23],[25]-[27], we have introduced two finite element triangulations, one called $\mathcal{T}_h$ - to approximate the *pressure* (the coarser one, a part of which is shown on Fig. 4.10)-and one called $\tilde{\mathcal{T}}_h$ - to approximate the *velocity* (see Fig. 4.11). Using these grids we have defined globally continuous, piecewise linear approximations for both pressure and velocity. The corresponding number of nodes and triangles are respectively

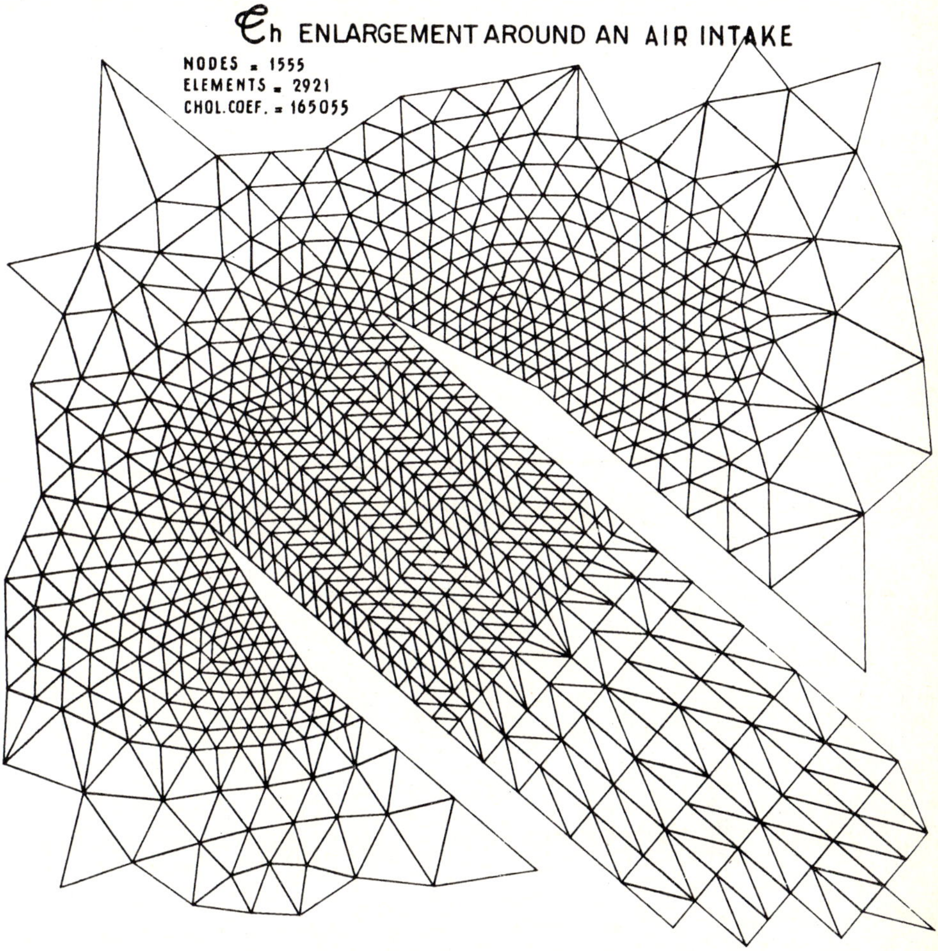

FIGURE 4.10 - (Pressure grid)

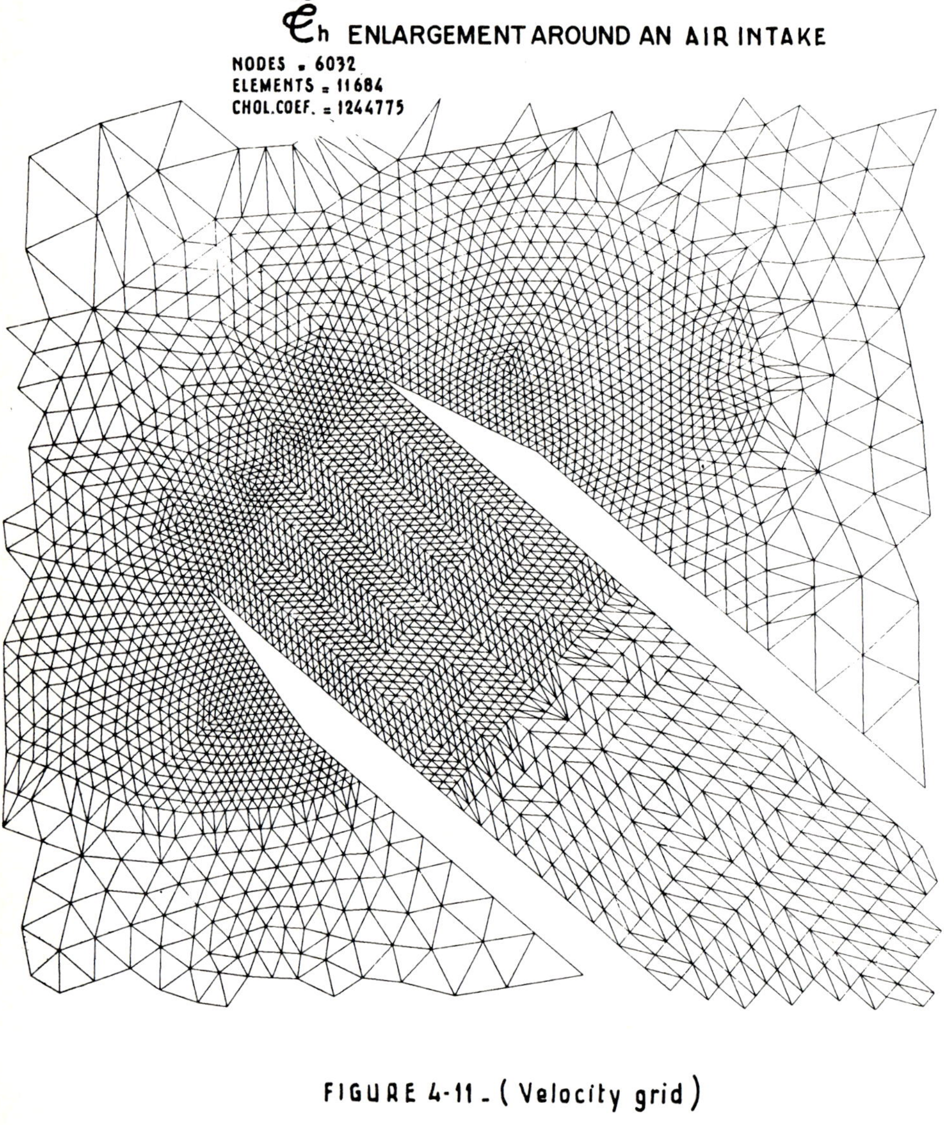

FIGURE 4-11 - (Velocity grid)

for the *pressure*, 1555 *nodes*, 2921 *triangles*,
for the *velocity*, 6032 *nodes*, 11684 *triangles*.

The Cholesky's factors of the discrete analogue of $-\Delta$ associate to $\mathcal{T}_h$ (resp. $\tilde{\mathcal{T}}_h$) contain about 1.65×10^5 (resp. 1.25×10^6) non zero elements ; thus from these large numbers a standard solution will require an *out of core* procedure.

Using incomplete Cholesky's factorization methods for solving by preconditioned conjugate gradient the discrete Poisson problems, we have been able, *once* the approximate Cholesky's factor $\tilde{L}_h$ is constructed, to solve *in core* the approximate Navier-Stokes problem : actually since the solution of each discrete Stokes problem is reduced - using the methods in [4],[6],[7],[22],[23],[25]-[27]- to the solution of a finite number of discrete Poisson problems (7 in two-dimensions, 9 in three-dimensions) *plus* a discrete boundary integral problem, whose matrix is *symmetric, positive definite*, and *full*, we have experimented the iterative solution of that latter problem by preconditioned conjugate gradient also. Let us denote by S_h the above boundary matrix ; since it is symmetric and positive definite we have

$$S_h = R_h R_h^t \tag{4.16}$$

where R_h is a regular, lower triangular full matrix.

We can define from R_h various incomplete factors of S_h, using for example the methods described in Sec. 3 ; doing so we obtain an incomplete Cholesky's factor $\tilde{R}_h$ from which we define $\tilde{S}_h = \tilde{R}_h\tilde{R}_h^t$ which is still symmetric and positive definite and which, therefore, can be used as a preconditioning matrix.

We have represented on Fig. 4.12 the cost of 10 iterations according to the preconditioning strategy for the discrete Poisson problems and the discrete above boundary integral problem ; from Fig. 4.12 it is clear that an important saving of computer resources can be obtained using $\tilde{L}_{(d/100)}$ and $\tilde{R}_{(d'/100)}$ with reasonably small values of d and d' (particularly of d).

We have indicated on Fig. 4.13 the velocity distribution at $t = 6$; we observe a rather complicated vortex configuration. For more details on that 2-dimensional simulation see [27].

Solution of a three-dimensional unsteady problem : Using three dimensional variants of the above methods we have also tested our incomplete factorization method on a three dimensional Navier-Stokes problem. We have taken as a test problem a flow around an idealized wing. We have indicated on Fig. 4.14 the CPU time corresponding to 10 itera-

STOKES SOLVER 2-D GLOWINSKI-PIRONNEAU

CONJUGATE GRADIENT WITH PRECONDITIONNING

$\tilde{L}\tilde{L}^t$ and $\tilde{R}\tilde{R}^t$

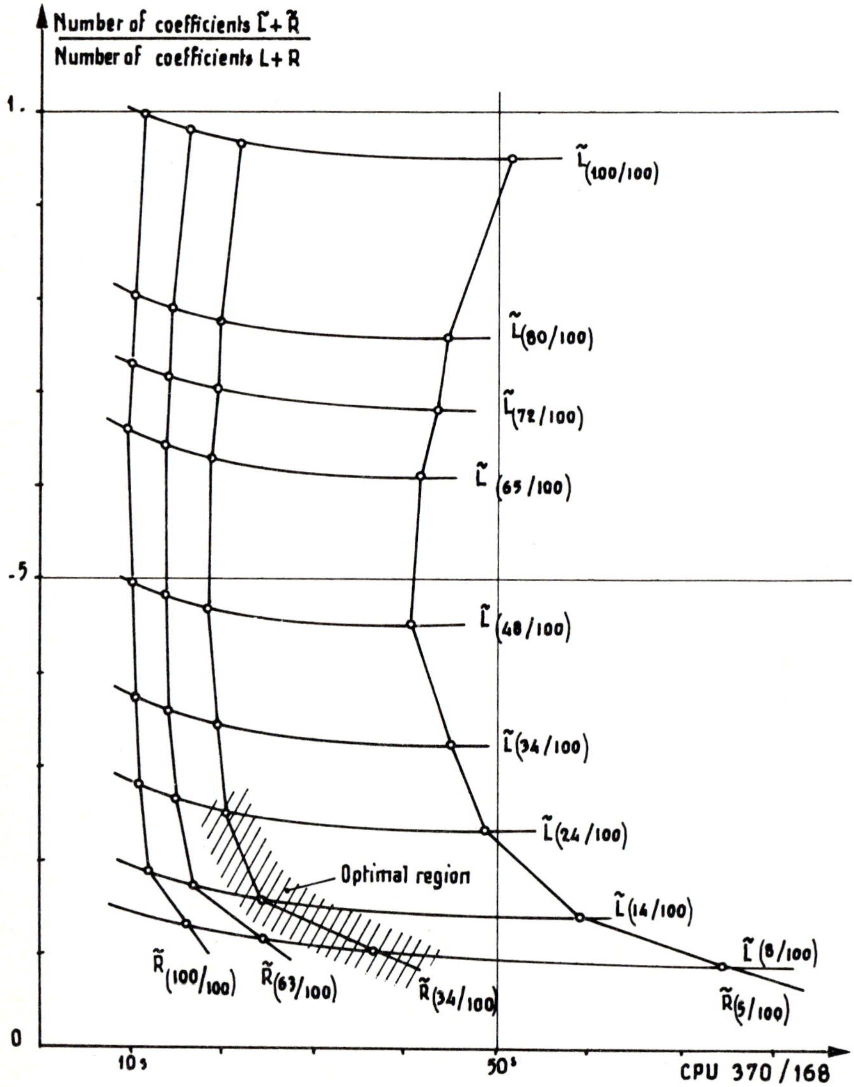

FIGURE 4.12

INCIDENCE 40.00

REYNOLDS 750.0

CYCLE ITER 120

TIME STEP 0.05

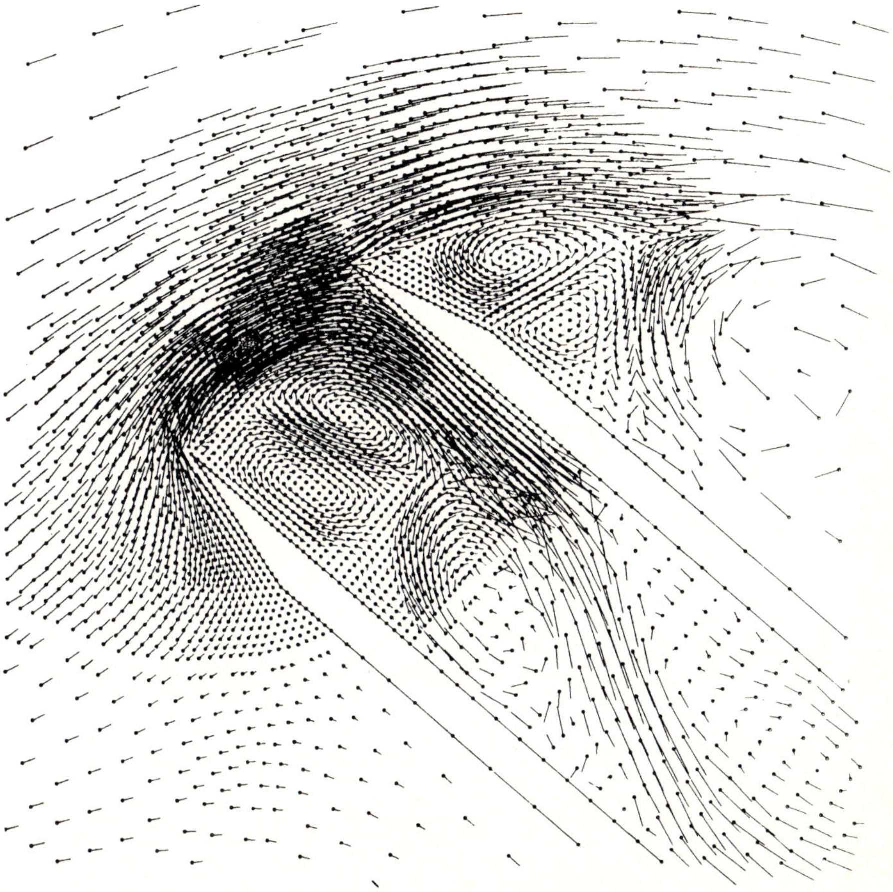

FIGURE 4-13

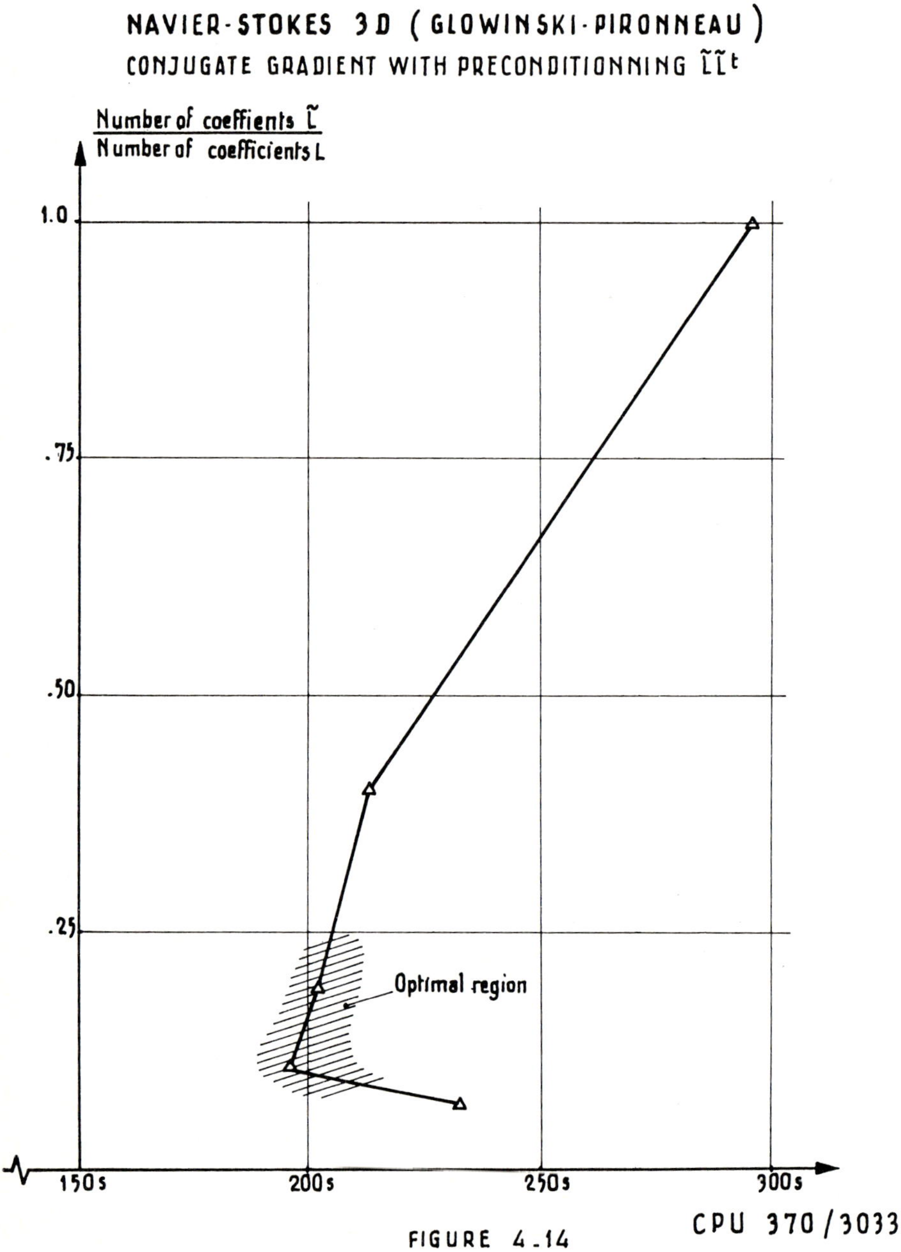
NAVIER-STOKES 3D (GLOWINSKI-PIRONNEAU)
CONJUGATE GRADIENT WITH PRECONDITIONNING L̃L̃ᵗ
Number of coeffients L̃ / Number of coefficients L
1.0
.75
.50
.25
Optimal region
150s
200s
250s
300s
CPU 370/3033

FIGURE 4.14

tions, according to the values of d in $\tilde{L}_{(d/100)}$; we see again that reasonably small values of d may imply an interesting saving of computer resources.

5. CONCLUSION.

From the above study it appears that conjugate gradient methods with preconditioning are well suited to the solution of large problems like those coming from industrial applications ; indeed up to 95% of memory storage can be saved when compared with a direct method and a factor of 100 can be gained on the computer time when compared with a *non-preconditioned* conjugate gradient.

In conclusion preconditioned conjugate gradient is no more time consuming than direct methods but if a convenient preconditioning is done it can use only 5% of the memory required by a direct method. From such properties this method is a very valuable tool for solving complicated Fluid Dynamics problems like those considered in the above sections.

REFERENCES

[1] BARTELS R., DANIEL J.W., A conjugate gradient approach to nonlinear boundary value problems in irregular region, in : G.A. Watson (ed.), *Conference on the Numerical Solution of Differential Equations, Dundee, 1973*, Lecture Notes in Math., 363, Springer, Berlin, 1973.

[2] DOUGLAS J., DUPONT T., Preconditioned conjugate gradient iteration applied to Galerkin methods for mildly nonlinear Dirichlet problem, in : J.R. Bunch and D.J. Rose (ed.), *Sparse Matrix Computations*, Acad. Press, New-York, 1976, pp. 333-348.

[3] CONCUS P., GOLUB G.H., O'LEARY D.P., Numerical solution of nonlinear partial differential equations by a generalized conjugate gradient method, *Computing*, 19, (1977), pp. 321-340.

[4] BRISTEAU M.O., GLOWINSKI R., PERIAUX J., PERRIER P., PIRONNEAU O., POIRIER G., Application of Optimal Control and Finite Element Methods to the calculation of Transonic Flows and Incompressible viscous Flows, in : B. Hunt (ed.) *Numerical Methods in Applied Fluid Dynamics*, Acad. Press, London, 1979/80 (and *LABORIA report* 294 (1978)).

[5] M.O. BRISTEAU, R. GLOWINSKI, J. PERIAUX, P. PERRIER, O. PIRONNEAU, On the numerical solution of nonlinear problems in Fluid Dynamics by least squares and finite element methods (I) Least square formulations and conjugate gradient solution of the continuous problems, *Comp. Meth. Appl. Mech. Eng.* 17/18, (1979), pp. 619-657.

[6] PERIAUX J., *Résolution de quelques problèmes non linéaires en aérodynamique par des méthodes d'éléments finis et de moindres carrés fonctionnels*, Thèse de 3ème Cycle, Université Pierre et Marie Curie, Juin 1979.

[7] LE TALLEC P., *Simulation numérique d'écoulements visqueux incompressibles par des méthodes d'éléments finis mixtes*, Thèse de 3ème cycle, Université Pierre et Marie Curie, 1978.

[8] AXELSSON O., A class of iterative methods for finite element equations, *Comp. Meth. Appl. Mech. Eng.*, 9, (1976), 2, pp. 123-138.

[9] MEIJERINK J.A., VAN DER VORST H.A., An iterative solution method for linear systems of which the coefficients matrix is a symmetric M-matrix, *Math. of Comp.*, 31, (1977), pp. 148-162.

[10] MANTEUFFEL T.A., The shifted incomplete Cholesky factorization, *SAND 78-8226, May 1978*, Sandia Laboratories, Livermore, California 94550.

[11] KERSHAW D.S., The incomplete Cholesky-conjugate gradient method for the iterative solution of systems of linear equations, *J. of Comp. Physics*, 26, (1978), pp. 43-65.

[12] GLOWINSKI R., PERIAUX J., PIRONNEAU O., An efficient preconditioning scheme for iterative numerical solutions of partial differential equations, *Applied Math. Modelling* (to appear) (and also Fascicule 8, Université Paris-Nord, Dept. Math., 1978-1979).

[13] CEA J., GEYMONAT G., Une méthode de linéarisation via l'optimisation, *Bol. Istituto Nazionale Alta Mat., Symp. Math.*, 10, (Bologna, 1972), pp. 431-451.

[14] LOZI R., *Analyse Numérique de certains problèmes de bifurcation*, Thèse de 3ème cycle, Université de Nice, 1975.

[15] POLAK E., *Computational Methods in Optimization*, Acad. Press, New-York, 1971.

[16] POWELL M.J.D., Restart procedure for the conjugate gradient method, *Math. Programming*, 12, (1977), pp. 148-162.

[17] WILDE D.J., BEIGHTLER C.S., *Foundations of Optimization*, Prentice Hall, Englewood Cliffs, N.J., 1967.

[18] BRENT R., *Algorithms for minimization without derivatives*, Prentice Hall, Englewook Cliffs, N.J., 1973.

[19] KELLER H.B., Constructive methods for bifurcation and nonlinear eigenvalue problems, in : *Computing Methods in Aoplied Sciences and Engineering, 1977, (I), Proceedings, IRIA, Paris*, R. Glowinski, J.L. Lions (Ed.), Lecture Notes in Math., 704, Springer, Berlin, 1979, pp. 241-251.

[20] KELLER H.B., Calculation of flows between rotating disks (these proceedings).

[21] BRISTEAU M.O., Application of optimal control theory to transonic flow computations by finite element methods, in : *Computing Methods in Applied Sciences and Engineering, 1977, (II), Proceedings, IRIA, Paris*, R. Glowinski, J.L. Lions (Ed.), Lecture Notes in Physics, 91, Springer, Berlin, 1979, pp. 103-124.

[22] BRISTEAU M.O., GLOWINSKI R., MANTEL B., PERIAUX J., PERRIER P., PIRONNEAU O., A finite element approximation of Navier-Stodes equations for incompressible viscous fluids. Iterative methods of solution, to appear in the *Proceedings of the IUTAM Conf. on Navier-Stokes equations*, Paderborn (R.F.A.), Sep. 1979, to be published by Springer in the Lecture Notes in Math. series.

[23] GLOWINSKI R., PIRONNEAU O., On numerical methods for the Stokes problem, in : *Energy Methods in Finite Element Analysis*, R. Glowinski, E.Y. Rodin, O.C. Zienkiew (Ed.), J. Wiley and Sons, Chichester, 1979, Chap. 13, pp. 243-264.

[24] BERCOVIER M., PIRONNEAU O., Error estimates for finite element method solution of the Stokes problem in the primitive variables, *Numerische Math.*, 33, (1979), pp. 211-224.

[25] GLOWINSKI R., PIRONNEAU O., On a mixed finite element approximation of the Stokes problem (I). Convergence of the approximate solutions, *Numerische Math.*, 33, (1979), pp. 397-424.

[26] GLOWINSKI R., PIRONNEAU O., Approximation par éléments finis mixtes du problème de Stokes en formulation vitesse-pression. Résolution des systèmes approchés. *C.R. Acad. Sc. Paris,* T. 286 A, (1978), pp. 181-183.

[27] GLOWINSKI R., MANTEL B., PERIAUX J., PIRONNEAU O., A finite element approximation of Navier-Stokes equations for incompressible viscous fluids. Functional least-square methods of solution,in : *Computer Methods in Fluids,* C. Taylor, K. Morgan, C.A. Brebbia (eds.), Pentech Press, 1980, pp. 84-133.

COMPUTING METHODS IN APPLIED SCIENCES AND ENGINEERING
R. Glowinski, J.L. Lions (editors)
North-Holland Publishing Company

SPARSE LEAST SQUARES PROBLEMS

Gene H. Golub*
Computer Science Department
Stanford University
Stanford, CA 94305

Robert J. Plemmons**
Departments of Computer Science and Mathematics
University of Tennessee
Knoxville, TN 37916

ABSTRACT

Two matrix decompositions are given for efficiently solving structured, sparse least squares problems. One decomposition is an orthogonalization procedure and the other is based on the Peters-Wilkinson method of solving linear least squares problems.

Invited paper presented at the Fourth International Symposium on Computing Methods in Applied Sciences and Engineeering, December 10-14, 1979, I R I A , Versailles, France.

* Research supported in part by U.S. Army Grant DAAG29-77-G0179 and by National Science Foundation Grant MCS78-11985.

** Research supported in part by U. S. Army Grant DAAG29-77-G0166.

1. INTRODUCTION

Large, sparse linear least squares problems occur in many scientific applications Amongst these are (1) econometric models, (2) analysis of seismological data, (3) photogrammetry [4] and (4) geodetic adjustments [6]. In addition, the natural factor formulation for the finite element method applied to the analysis of structures (cf.[1]) leads to the equivalent of a linear least squares problem. For a large class of problems the data matrix is not only sparse but is structured. That is, after a suitable permutation the non-zero elements occur in a very regular fashion. In our discussion, we shall discuss a particular structure which occurs frequently, especially in the applications listed above.

Let A be a known real $m \times n$ matrix with $m \geq n$ and throughout this paper assume rank A is n. We are given a vector b and we wish to determine $\hat{x}$ so that

$$\|b - A\hat{x}\|_2 = \min. \tag{1.1}$$

It is well known that $\hat{x}$ satisfies the normal equations

$$A^T A \hat{x} = A^T b . \tag{1.2}$$

It is not difficult to show that numerical inaccuracies may be introduced into the solution $\hat{x}$ by the formation of the normal equations. It is often useful to factor the matrix A in the form

$$A = BC$$

where B is an $m \times n$ matrix and C an $n \times n$ matrix. It is easy to verify that

$$A^+ = C^{-1} B^+$$

where A^+ indicates the unique pseudo-inverse of A. The matrix B is chosen to have some special property. In section 2, B will be an orthogonal matrix and in section 3, B will be lower trapezoidal.

In this paper, we shall assume that after a suitable permutation A has the following form:

A =

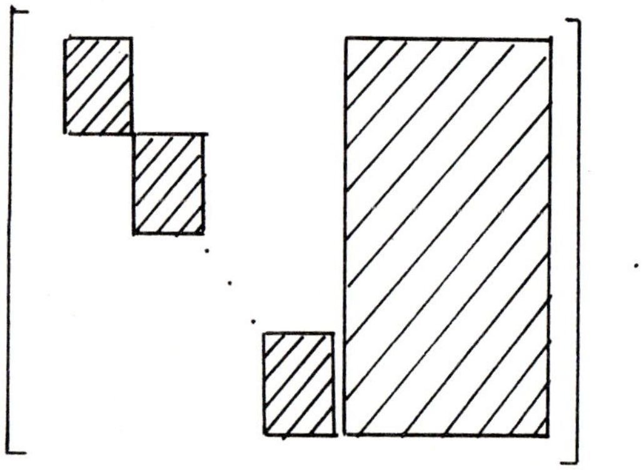

(1.3)

In [6], a procedure is described for determining such a structure in a geodetic least squares adjustment problem. We shall present two factorizations which take into account this form of A .

2. ORTHOGONAL DECOMPOSITION

One method of solving the normal equations (1.2) is by computing the Cholesky decomposition

$$A^TA = F^TF \quad \text{where} \quad F = \tag{2.1}$$

and then solving $F^Tw = A^Tb$ by forward substitution and $F\hat{x} = w$ by back substitution. The upper triangular matrix F is called the Cholesky factor of A . Another procedure for solving the least squares problem is to compute the orthogonal decomposition of A ; viz

$$A = Q\begin{bmatrix} R \\ 0 \end{bmatrix} \quad \text{where} \quad Q^TQ = I$$

and R is upper triangular. The orthogonal matrix Q is constructed via a sequence of Householder reflections or Givens rotations (cf. [7]). It is easy to verify that

$$F = D\,R$$

where D is a diagonal matrix with ± 1 on the diagonal. To solve the least squares problem, it is necessary to compute

$$\begin{bmatrix} c \\ d \end{bmatrix} = Q^Tb$$

and then solve $R\hat{x} = c$.

Now write

$$A = \begin{bmatrix} A_1 & & & & A_{1,t+1} & \cdot & \cdot & \cdot & A_{1,2t-1} \\ & A_2 & & & A_{2,t+1} & \cdot & \cdot & \cdot & A_{2,2t-1} \\ & & \cdot & & \cdot & & & & \cdot \\ & & & \cdot & \cdot & & & & \cdot \\ & & & \cdot & \cdot & & & & \cdot \\ & & & A_t & A_{t,t+1} & \cdot & \cdot & \cdot & A_{t,2t-1} \end{bmatrix} \tag{2.2}$$

We shall assume henceforth that $t = 2^k$. For a certain flexibility of the algorithm and also for simplicity in the notation, we do not altogether distinguish here between zero and nonzero blocks A_{ij} .

Algorithm 1. This algorithm computes R and the least squares solution $\hat{x}$ to $Ax \approx b$.

Step 1. Reduce each diagonal block A_i of A to upper triangular form by orthogonal transformations and merge the reduced blocks.

1) Do for $i = 1,2,\ldots,t$.

   1) Determine Q_i^T so that $Q_i^T A_i = \begin{bmatrix} R_i \\ 0 \end{bmatrix}$, $R_i =$ 

   (Note that Q_i^T need not be formed explicitly).

   2) Compute

$$Q_i^T[A_i, A_{i,t+1}, \ldots, A_{i,2t-1}]$$

$$\equiv \begin{bmatrix} R_i & A^0_{i,t+1} \cdots & A^0_{i,2t-1} \\ 0 & A^1_{i,t+1} \cdots & A^1_{i,2t-1} \end{bmatrix} .$$

2) Merge the reduced row blocks by row permutations so that the resulting matrix has the form

$$\begin{bmatrix}
R_1 & & & A^0_{1,t+1} & A^0_{1,t+2} & & A^0_{1,3t/2} & A^0_{1,(3t/2)+1} & \cdots & A^0_{1,2t-1} \\
& R_2 & & A^0_{2,t+1} & A^0_{2,t+2} & & A^0_{2,3t/2} & A^0_{2,(3t/2)+1} & \cdots & A^0_{2,2t-1} \\
& & \ddots & \vdots & \vdots & & \vdots & \vdots & & \vdots \\
& & R_t & A^0_{t,t+1} & A^0_{t,t+2} & \cdots & A^0_{t,3t/2} & A^0_{t,(3t/2)+1} & \cdots & A^0_{t,2t-1} \\
& & & A^1_{1,t+1} & & & & A^1_{1,(3t/2)+1} & \cdots & A^1_{1,2t-1} \\
& & & A^1_{2,t+1} & & & & A^1_{2,(3t/2)+1} & \cdots & A^1_{2,2t-1} \\
& & & & A^1_{3,t+2} & & & A^1_{3,(3t/2)+1} & \cdots & A^1_{3,2t-1} \\
& & & & A^1_{4,t+2} & & & A^1_{4,(3t/2)+1} & \cdots & A^1_{4,2t-1} \\
& & & & & \ddots & & \vdots & & \vdots \\
& & & & & & A^1_{t-1,3t/2} & A^1_{t-1,(3t/2)+1} & & A^1_{t-1,2t-1} \\
& & & & & & A^1_{t,3t/2} & A^1_{t,(3t/2)+1} & \cdots & A^1_{t,2t-1}
\end{bmatrix}$$

<u>Step 2.</u> Reduce and merge the intermediate-stage blocks.

1) Do for $u = t, t/2, \ldots, t/2^{k-1} = 2$.

1) Do for $v = 1,3,\ldots,u-1$

   1) Reduce each pair of row diagonal blocks

$$\begin{bmatrix} A^1_{v,t+v} \\ A^1_{v+1,t+v} \end{bmatrix}$$

   to upper triangular form by orthogonal transformation, as in Step 1.

   2) Merge the resulting reduced row blocks by row permutations so that the upper triangular blocks R_i appear first, as in Step 1 .

At the end of Step 2, A has been reduced by orthogonal transformations to the following form, where each R_i is upper triangular and where certain of the blocks A^0_{ij} are zero.

$$R = \begin{bmatrix} R_1 & & & & A^0_{1,t+1} & \cdots & A^0_{1,2t-1} \\ & R_2 & & & A^0_{2,t+1} & \cdots & A^0_{2,2t-1} \\ & & \ddots & & \vdots & & \vdots \\ & & & R_t & A^0_{t,t+1} & \cdots & A^0_{t,2t-1} \\ & & & & R_{t+1} & & \vdots \\ & & & & & \ddots & A^0_{2t-2,2t-1} \\ & & & & & & R_{2t-1} \end{bmatrix}$$

<u>Step 3</u>. Back Substitution. Let n_i denote the order of R_i , for $i = 1,\ldots,2t-1$. Let $c^T = (c_1,\ldots,c_{2t-1})^T$ denote the result of applying the same sequence of orthogonal transformations to b and let $\hat{x}^T = (\hat{x}_1,\ldots,\hat{x}_{2t-1})^T$ denote the least squares solution to $Ax \approx b$, where c_i and $\hat{x}_i$ are n_i-vectors, $i = 1,\ldots,2t-1$. Solve each of the following upper-triangular systems by back-substitution

1) $R_{2t-1}\,\hat{x}_{2t-1} = c_{2t-1}$

2) $R_i\hat{x}_i = c_i - \sum_{j=i+1}^{2t-1} A^0_{ij}\hat{x}_j, \qquad i = 2t-2,\ 2t-1,\ \ldots,\ t$

3) $R_i \hat{x}_i = c_i - \sum_{j=t+1}^{2t-1} A^0_{ij} \hat{x}_j \quad , \qquad i = t, t-1, \ldots, 1 .$

3. PETERS-WILKINSON DECOMPOSITION

As printed out earlier, we can solve the least squares problem (1.1) by writing

$$A = BC$$

and then computing $A^+ = C^{-1}B^+$. Since B is of full rank, it is easy to show that

$$B^+ = (B^TB)^{-1}B^T$$

so that

$$A^+ = C^{-1}(B^TB)^{-1}B^T . \tag{3.1}$$

Peters and Wilkinson [8] have devised an algorithm for which B is a lower trapezoidal matrix ($b_{ij} = 0$ when $j > i$) and C is upper triangular. This method has been used successfully in solving a variety of large sparse least squares problems (cf. [2]). We describe the algorithm for matrices of the form (1.3).

Algorithm 2.

Step 1. Perform an LU decomposition of A where row and column permutations are permitted on A only within blocks. For each diagonal block A_{ii} ,

$P_1 A_{ii} P_2 = L_i U_i \quad , \quad L_i =$  $\quad , \quad U_i =$ 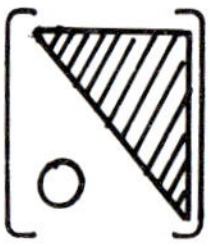

where P_1 and P_2 reflect row and column permutations within that block for numerical stability.

Note that in order to complete the LU decomposition of A , merging and shifting of blocks is necessary. In the end, we have

$$PAQ = LU$$

where P and Q are permutation matrices, and

$U =$ 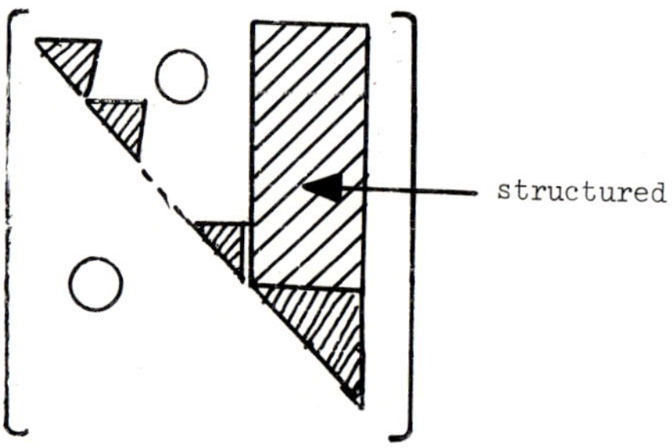

but where L is a block matrix in column permuted lower trapezoidal form. However, L^TL has the block diagonal form:

$L^TL =$ 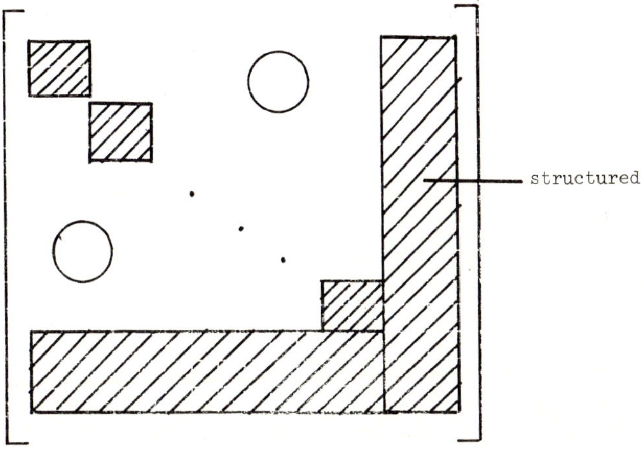

Step 2. Form L^TL in the form as above and solve

$$L^TLw = L^TP^Tb$$

by block Cholesky.

Step 3. Solve

$$Uz = w$$

by back substitution.

Step 4. Set $\hat{x} = Q^Tz$. Then $\hat{x}$ is the least squares solution.

4. FINAL REMARKS

We have described two decompositions for solving least squares problems with a particular structure. The first method is an orthogonalization procedure which will produce an accurate solution when the condition number is relatively small. In addition, it has the virtue that it is not difficult to add and delete variables and observations (cf.[3]). The second procedure is based on the Peters-Wilkinson decomposition [8]. This method may not be as accurate as orthogonalization procedure but it has been observed that it produces more accurate solutions than the normal equations and generally there is less fill in than in an orthogonalization procedure (see [2]). We will be reporting on numerical experiments in a future paper [5].

REFERENCES

[1] J. H. Argyris and O.E. Brönlund [1975], "The natural factor formulation of the stiffness matrix displacement method", Computer Methods in Applied Mechanics and Engineering, Vol. 5, 97-119.

[2] I. S. Duff and J. K. Reid [1976], "A comparison of some methods for the solution

of sparse and overdetermined systems of linear equations", J. Inst. Math. Applics. 17, 267-280.

[3] P. E. Gill, G. H. Golub, W. Murray and M. A. Saunders [1974], "Methods for updating matrix factorization", Math. Comp. 28, 505-535.

[4] G. H. Golub, F. T. Luk and M. Pagano [1979], "A sparse least squares problem in photogrammetry", Proc. of Computer Science and Statistics: Twelfth Annual Conference on the Interface, Waterloo, Canada.

[5] G. H. Golub and R. J. Plemmons [1980], "Dissection schemes for large sparse least squares and other rectangular systems", in preparation.

[6] G. H. Golub and R. J. Plemmons [1979], "Large scale geodetic least squares adjustment by dissection and orthogonal decomposition", Stanford University Report STAN-CS-79-774.

[7] C. L. Lawson and R. J. Hanson [1974], Solving Least Squares Problems, Prentice Hall, Englewood Cliffs, NJ.

[8] G. Peters and J. H. Wilkinson [1970], "The least squares problem and pseudo-inverses", The Computer Journal, 13, 309-316.

SESSION VII

NON LINEAR PROGRAMMING AND APPLICATIONS

COMPUTING METHODS IN APPLIED SCIENCES AND ENGINEERING
R. Glowinski, J.L. Lions (editors)
North-Holland Publishing Company

AN IMPLEMENTABLE ALGORITHM FOR THE OPTIMAL DESIGN CENTERING, TOLERANCING AND TUNING PROBLEM

E. Polak
Department of Electrical Engineering and Computer Sciences
and the Electronics Research Laboratory
University of California, Berkeley, California 94720

ABSTRACT

An implementable master algorithm for solving optimal DCCT problems is presented. This master algorithm decomposes the original nondifferentiable optimization problem into a sequence of ordinary nonlinear programming problems. The master algorithm generates sequences with accumulation points that are feasible and satisfy a new optimality condition which is shown to be stronger than the one previously used for these problems.

1. INTRODUCTION

Quite commonly, the engineering designer has to take into account the fact that the parameter values of the actual system, structure, or device, will be different from the nominal values in the design. In control system design, this discrepancy is largely due to identification errors; in steel structures and in electronic circuit design it is due to production tolerances. Recently, optimization algorithms have been proposed [1,2,14] which enable the designer to ensure satisfaction of specifications not only by the nominal design but also by all possible system or device realizations within a prescribed tolerance range. When using such algorithms, one can make the tolerance range a design parameter and maximize it while minimizing some other cost function of the other parameters by constructing an aggregate cost by means of weighted combinations. It has been known for some time that the requirement of 100% yield, i.e. that all realizations within tolerance range satisfy specifications, may result in very tight tolerances or very high precision of identification requirements [3,4]. To overcome this difficulty, it has become common to tune a control system or to trim, by laser beam, electronic devices after manufacture [5,6, 7,8,9]. Empirically, it has been found that tuning and trimming permits the relaxation of the error or tolerance range to acceptable levels and hence results in considerably increased yield. Quite recently, two conceptual† algorithms have been proposed for solving design problems with tolerances and tuning or trimming [3,10]. In the electronics literature such problems are referred to as design centering, tolerancing and tuning problems (DCTT).

Typically, a specification on the nominal value of the design parameter $x \in \mathbb{R}^n$ takes

†We say that an algorithm is conceptual if it contains infinite operations which cannot be easily approximated. We say that an algorithm is implementable if specific truncation rules are given for all such infinite operations.

the form

$$f^i(x) \leq 0, \tag{1}$$

with $f^i : \mathbb{R}^n \to \mathbb{R}^n$ continuously differentiable. When identification errors, or tolerances, $\varepsilon \in \mathbb{R}^m$, ranging over a compact set $E \subset \mathbb{R}^m$, need to be taken into account, the constraint (1) becomes modified to

$$\max_{\varepsilon \in E} \phi^i(x,\varepsilon) \leq 0, \tag{2}$$

with $\phi^i : \mathbb{R}^n \times \mathbb{R}^{n_\varepsilon} \to \mathbb{R}^1$. Finally, assuming that tuning or trimming will be performed by means of a parameter $\tau \in \mathbb{R}^\ell$, ranging over a compact set $T \subset \mathbb{R}^{n_\tau}$, after the process picks an ε, we modify (2) to

$$\psi_{E,T}(x) = \max_{\varepsilon \in E} \min_{\tau \in T} \max_{k \in K} \zeta^k(x,\varepsilon,\tau) \leq 0, \tag{3}$$

since τ must work for all the specifications $k \in K$, with $K = \{1,2,\ldots,m\}$. For a more detailed exposition of the design centering, tolerancing and trimming problem formulation as an optimization problem, the reader is referred to [3,4].

Although Lipschitz continuous [3], the function $\psi_{E,T}(\cdot)$ is, in general, not differentiable; in fact, it may even fail to have directional derivatives. In [3] we find a conceptual algorithm for optimization problems with constraints of the form (3), based on the concept of generalized gradients [11,12,13]. The algorithm in [3] consists of two parts: a master outer approximations algorithm, which replaces the set E in (3) with a discrete subset $E_i \subset E$ (as in [14]), and an inner, non-differentiable optimization sub algorithm which solves the resulting simpler problems. It is shown that the solutions of the simpler problems converge to a solution of the original problem. The inner, nondifferentiable optimization subalgorithm in [3] has two serious drawbacks. The first, common to many nondifferentiable optimization algorithms, is that it utilizes a very expensive bisection procedure to get adequate approximations to required bundles of subgradients (see, for example, [18,15,16]). The second, because of a requirement of semi-smoothness [13], is that it is applicable only to the case where there is only <u>one</u> constraint function ζ^k (i.e., $K = \{k\}$).

In this paper we present a new, implementable algorithm for solving optimization problems with constraints of the form (3). Just as the algorithm in [3,14], it makes use of outer approximations to the feasible set to decompose the original problem into an infinite sequence of simpler problems, by replacing the set E with discrete subsets E_i. However, because of the use of certain transformations, the resulting simpler problems are ordinary, <u>differentiable</u>, constrained optimization problems, solvable by a large number of existing, efficient algorithms. Thus, as a consequence of these simple, but not immediately obvious transformations, all the computational difficulties caused by the need for a nondifferentiable optimization subalgorithm in [3] have been removed. Truncation rules are given for all the major infinite operations in the new algorithm, making it implementable and hence easily programmable. Finally, it is shown in the Appendix that the optimality conditions, on which the present algorithm is based, are sharper than the ones used in [3].

It is our hope that this new algorithm will become a valuable tool in the arsenal of the engineering designer.

2. PROBLEM DECOMPOSITION VIA OUTER APPROXIMATIONS TO THE FEASIBLE SET

The most general form of an optimization problem arising in engineering design, that we shall consider in this paper, is as follows:

$$P_g : \min_x \{ f^0(x) \mid f^i(x) \leq 0,\ i \in I;$$

$$\max_{\omega^j \in \Omega^j} \phi^j(x,\omega^j) \leq 0,\ j \in J;$$

$$\max_{\varepsilon \in E} \min_{\tau \in T} \max_{k \in K} \zeta^k(x,\varepsilon,\tau) \leq 0\} \tag{4}$$

where I,J,K are sets of integers (e.g., K = (1,2,...,m)) $f^0 : \mathbb{R}^n \to \mathbb{R}^1$, $f^i : \mathbb{R}^n \to \mathbb{R}^1$, $i \in I$, $\phi^j : \mathbb{R}^n \times \mathbb{R}^{(n_\omega)j} \to \mathbb{R}^1$, $j \in J$, and $\zeta^k : \mathbb{R}^n \times \mathbb{R}^{n_\varepsilon} \times \mathbb{R}^{n_\tau} \to \mathbb{R}^j$, $k \in K$, are all continuously differentiable. In this context, x is the nominal design vector; the ω^j are tolerances, errors, or variables, such as frequency, or time, or temperature, which must be considered over a continuum of values; ε is an error or tolerance to be overcome by tuning, and τ is the tuning parameter. We shall assume that the sets Ω^j, E and T are <u>compact</u> and specified by differentiable inequalities, which we shall introduce, as needed, later.

There is no essential loss of generality, as far as the exposition of our method is concerned, in assuming that I and J contain only one index each, i.e., in considering only one constraint of each kind. On the other hand, there is a considerable simplification in notation when the indices i and j are eliminated. We shall therefore restrict ourselves to the simpler problem

$$P : \min_x \{ f^0(x) \mid f(x) \leq 0,\ \max_{\omega \in \Omega} \phi(x,\omega) \leq 0,$$

$$\max_{\varepsilon \in E} \min_{\tau \in T} \max_{k \in K} \zeta^k(x,\varepsilon,\tau) \leq 0\} \tag{5}$$

where $f : \mathbb{R}^n \to \mathbb{R}^1$, $\phi : \mathbb{R}^n \times \mathbb{R}^{n_\omega} \to \mathbb{R}^1$, are continuously differentiable, $\Omega \subset \mathbb{R}^{n_\omega}$, and $E \subset \mathbb{R}^{n}$, and $T \subset \mathbb{R}^{n_\tau}$ are compact and the remaining quantities are as in P_g.

Now, let $\Omega_i \subset \Omega$ and $E_i \subset E$ be discrete sets and consider the problem

$$P_i : \min_x \{ f^0(x) \mid f(x) \leq 0,\ \max_{\omega \in \Omega_i} \phi(x,\omega) \leq 0,$$

$$\max_{\varepsilon \in E_i} \min_{\tau \in T} \max_{k \in K} \zeta^k(x,\varepsilon,\tau) \leq 0\}. \tag{6}$$

We shall show that unlike the problem P, P_i is an ordinary nonlinear programming problem. Thus, suppose that

$$\Omega_i = \{\omega_j \in \Omega \mid j \in J_i\},\ E_i = \{\varepsilon_\ell \in E \mid \ell \in L_i\}, \tag{6a}$$

with J_i, L_i finite sets of integers and let

$$\phi^j(x) \triangleq \phi(x,\omega_j),\ j \in J_i \tag{6b}$$

$$\zeta^{k\ell}(x,\tau) \triangleq \zeta^k(x,\varepsilon_\ell,\tau),\ \ell \in L_i,\ k \in K. \tag{6c}$$

Now suppose that

$$T \triangleq \{\tau \mid g^s(\tau) \leq 0,\ s \in S\} \tag{6d}$$

with $S \triangleq \{1,2,\ldots,\sigma\}$, $g^s : \mathbb{R}^{n_\tau} \to \mathbb{R}^1$ continuously differentiable, and consider the problem

$$\hat{P}_i : \min_{(x,\tau_\ell)} \{f^0(x) \mid f(x) \leq 0;\ \phi^i(x) \leq 0,\ j \in J_i;$$

$$\zeta^{k\ell}(x,\tau_\ell) \leq 0,\ k \in K,\ \ell \in L_i;\ g^s(\tau_\ell) \leq 0,\ s \in S,\ \ell \in L_i\}, \tag{6e}$$

Proposition 1: $\hat{x}$ solves P_i if and only if $\{\hat{x};\hat{\tau}_\ell, \ell \in L_i\}$ solves $\hat{P}_i$, for some $\hat{\tau}_\ell \in T$, $\ell \in L_i$.

Proof: $\Rightarrow$ Suppose $\hat{x}$ solves P_i. Then $\phi^j(\hat{x}) \leq 0$ for $j \in J_i$ and there exist $\hat{\tau}_\ell \in T$, $\ell \in L_i$ such that $\max_{\ell \in L_i} \max_{k \in K} \zeta^{k\ell}(\hat{x},\hat{\tau}_\ell) \leq 0$, i.e., $\{\hat{x},\hat{\tau}_\ell, \ell \in L_i\}$ are feasible for $\hat{P}_i$. Now suppose that this triplet is not optimal for $\hat{P}_i$. Then there exist $\{\tilde{x};\tilde{\tau}_\ell, \ell \in L_i\}$ feasible for $\hat{P}_i$, and such that $f^0(\tilde{x}) < f^0(\hat{x})$. Now $\tilde{x}$ satisfies $f(\tilde{x}) \leq 0$, $\phi^j(\tilde{x}) \leq 0$, $j \in J_i$ and

$$\max_{\varepsilon \in E_i} \min_{\tau \in T} \max_{k \in K} \zeta^k(\tilde{x},\varepsilon,\tau)$$

$$= \max_{\ell \in L_i} \min_{\tau \in T} \max_{k \in K} \zeta^{k\ell}(\tilde{x},\tau)$$

$$\leq \max_{\ell \in L_i} \max_{k \in K} \zeta^{k\ell}(\tilde{x},\tilde{\tau}_\ell) \leq 0 \tag{6d}$$

i.e., $\tilde{x}$ is feasible for P_i. But then $f(\tilde{x}) < f(\hat{x})$ contradicts the optimality of $\hat{x}$.

$\Leftarrow$ Now suppose that $\{\hat{x};\hat{\tau}_\ell, \ell \in L_i\}$ solves $\hat{P}_i$. Then, because of (6d), $\hat{x}$ is feasible for P_i. Suppose that $\hat{x}$ is not optimal for P_i. Then there exists an $\tilde{x}$ and corresponding $\tilde{\tau}_\ell, \ell \in L_i$ such that $\tilde{x}$ is feasible for P_i, $f(\tilde{x}) < f(\hat{x})$ and $\max_{\ell \in L_i} \min_{\tau \in T} \max_{k \in K} \zeta^{k\ell}(\tilde{x},\tau) = \max_{\ell \in L_i} \max_{k \in K} \zeta^{k\ell}(\tilde{x},\tilde{\tau}_\ell) \leq 0$, so that $\{\tilde{x};\tilde{\tau}_\ell, \ell \in L_i\}$ is feasible also for $\hat{P}_i$. But this contradicts the optimality of $\hat{x}$ and we are done. ⌑

Next, let $\chi : \mathbb{R}^n \times \mathbb{R}^{n_\varepsilon} \to \mathbb{R}^1$ be defined by

$$\chi(x,\varepsilon) \triangleq \min_{\tau\in T} \max_{k\in K} \zeta^k(x,\varepsilon,\tau) \tag{7}$$

It was shown in [3] that $\chi(\cdot,\cdot)$ is Lipschitz continuous. Since $\Omega_i \subset \Omega$ and $E_i \subset E$,

$$\{x|\max_{\omega\in\Omega} \phi(x,\omega) \leq 0\} \subset \{x|\max_{\omega\in\Omega_i} \phi(x,\omega) \leq 0\} \tag{8}$$

and

$$\{x|\max_{\varepsilon\in E} \chi(x,\varepsilon) \leq 0\} \subset \{x|\max_{\varepsilon\in E_i} \chi(x,\varepsilon) \leq 0\}. \tag{9}$$

Consequently, the feasible set in (6) contains the feasible set in (5), and therefore, if $\hat{x}$ solves P and $\hat{x}_i$ solves P_i, we must have

$$f(\hat{x}_i) \leq f(\hat{x}). \tag{10}$$

Now suppose that we construct an infinite sequence of problems P_i, with solutions $\hat{x}_i$ such that $\hat{x}_i \to \tilde{x}$ as $i \to \infty$ and $\tilde{x}$ is feasible for P, i.e.,

$$\tilde{x} \in F \triangleq \{x|f(x) \leq 0,\ \max_{\omega\in\Omega} \phi(x,\omega) \leq 0,\ \max_{\varepsilon\in E} \chi(x,\varepsilon) \leq 0\}. \tag{11}$$

Then, by continuity of f^0 and (10)

$$f(\tilde{x}) \leq f(\hat{x}) \tag{12}$$

and hence $\tilde{x}$ is optimal for P. Thus, since we can solve problems of the form $\hat{P}_i$, the solution of problem P can be assured by constructing discrete sets Ω_i, E_i such that any accumulation point, $\hat{x}$, of a solution sequence $\{\hat{x}_i\}$ satisfies $\hat{x} \in F$. In [14] we find two specific schemes for achieving this result. We shall now summarize the most relevant results from [14].

<u>Proposition 2</u>: Let $\{\delta_i\}_{i=0}^{\infty}$ and $\{\hat{\delta}_i\}_{i=0}^{\infty}$ be a positive, monotonically decreasing sequences, with $\delta_i \to 0$, $\hat{\delta}_i \to 0$ as $i \to \infty$. Let $E_0 \subset E$ and $\Omega_0 \subset \Omega$ be discrete sets. For $i = 0,1,2,\ldots$ let $x_i \in \mathbb{R}^n$, $\Omega_i \subset \Omega$ and $E_i \subset E$ be defined recursively as follows: let x_i be such that

$$f(x_i) \leq \delta_i, \tag{13a}$$

$$\max_{\omega\in\Omega_i} \phi(x_i,\omega) \leq \delta_i, \tag{13b}$$

$$\max_{\varepsilon\in E_i} \chi(x_i,\varepsilon) \leq \delta_i. \tag{13c}$$

Let ω_i, ε_i be such that

$$\phi(x_i,\omega_i) = \max_{\omega\in\Omega} \phi(x_i,\omega); \tag{14a}$$

$$\chi(x_i,\varepsilon_i) = \max_{\varepsilon \in E} \chi(x_i,\varepsilon), \tag{14b}$$

and let

$$\Omega_{i+1} = \Omega_i \cup \{\omega_i\} \quad \text{if} \quad \phi(x_i,\omega_i) > \hat{\delta}_i$$

$$= \Omega_i \text{ otherwise,} \tag{15a}$$

$$E_{i+1} = E_i \cup \{\varepsilon_i\} \quad \text{if} \quad \chi(x_i,\varepsilon_i) > \hat{\delta}_i$$

$$= E_i \text{ otherwise.} \tag{15b}$$

Then any accumulation point $\hat{x}$ of $\{x_i\}_{i=0}^{\infty}$ is in F.

<u>Proof</u>: Suppose that $x_i \xrightarrow{I} \hat{x}$, with $I \subset \{1,2,3,\ldots\}$. Then, since $f(\cdot)$ is continuous and $\delta_i \to 0$,

$$f(\hat{x}) \leq 0 \tag{16}$$

Next, since the functions

$$\psi_\Omega(x) \triangleq \max_{\omega \in \Omega} \phi(x,\omega) \tag{17a}$$

and

$$\psi_{E,T}(x) \triangleq \max_{\varepsilon \in E} \chi(x,\omega) \tag{17b}$$

are both continuous, $\psi_\Omega(x_i) \xrightarrow{I} \psi_\Omega(x)$ and $\psi_{E,T}(x_i) \xrightarrow{I} \psi_{E,T}(\hat{x})$. For the sake of contradiction, suppose that $\psi_\Omega(\hat{x}) > 0$. Then, since $x_i \xrightarrow{I} \hat{x}$ and $\hat{\delta}_i \to 0$, there exists an $i_0 \in I$ such that $\psi_\Omega(x_i) > \delta_i$ for all $i \in I$, $i \geq i_0$ and hence for any $i_2 > i_1 > i_0$ in I,

$$\Omega_{i_2} \supset \{\omega_i \mid i \in I,\ i_0 \leq i \leq i_1\} \tag{18}$$

Because of (13b), we then have

$$\phi(x_{i_2},\omega_i) \leq \delta_{i_2} \quad \text{for all } i \in I,\ i_0 \leq i \leq i_1 \tag{19}$$

and therefore, in particular, $\overline{\lim}\, \phi(x_{i_2},\omega_{i_1}) \leq 0$ as $i_2 > i_1 \to \infty$, $i_2,i_1 \in I$. But, because Ω is compact and $x_i \xrightarrow{I} \hat{x}$,

$$|\phi(x_{i_2},\omega_{i_1}) - \phi(x_{i_1},\omega_{i_1})| \to 0 \tag{20}$$

as $i_2 > i_1 \to \infty$, $i_2,i_1 \in I$. Because of (19) and (20) we now obtain that

$$\psi_\Omega(\hat{x}) = \lim_{i_1 \in I} \phi(x_{i_1},\omega_{i_1}) \leq 0 \tag{21}$$

and we have a contradiction. In exactly the same way it can be shown that $\psi_{E,T}(\hat{x}) \leq 0$ and hence the proof is complete. ⌑

The simplest way of incorporating the construction of the Ω_i, E_i, described in Proposition 2, into a master algorithm for solving P is as follows (c.f. [14]).

Master Algorithm 1

Parameters: $\{\hat{\delta}_i\}_{i=0}^{\infty}$, $\hat{\delta}_i > 0$, $\hat{\delta}_i \to 0$ as $i \to \infty$.

Data: $\Omega_0 \subset \Omega$, $E_0 \subset E$, discrete sets.

Step 0: Set $i = 0$.

Step 1: Solve $\hat{P}_i$ for $\{\hat{x}_i;\hat{\tau}_\ell,\ell \in L_i\}$.

Step 2: Compute $\psi_\Omega(\hat{x}_i)$, $\psi_{E,T}(\hat{x}_i)$ and corresponding $\omega_i \in \Omega$ and $\varepsilon_i \in E$, satisfying (14a,b) for $x_i = \hat{x}_i$.

Step 3: a) If $\psi_\Omega(\hat{x}_i) > \hat{\delta}_i$, set $\Omega_{i+1} = \{\omega_i\} \cup \Omega_i$. Else set $\Omega_{i+1} = \Omega_i$.

b) If $\psi_{E,T}(\hat{x}_i) > \hat{\delta}_i$, set $E_{i+1} = \{\varepsilon_i\} \cup E_i$. Else set $E_{i+1} = E_i$.

Step 4: If $\psi_\Omega(\hat{x}_i) \leq 0$ and $\psi_{E,T}(\hat{x}_i) \leq 0$, stop; else proceed.

Step 5: Set $i = i+1$ and go to Step 1. ⌑

The following result is obvious in view of Proposition 2:

Theorem 1: If Master Algorithm 1 stops in Step 4, then $\hat{x}_i$ is optimal for P. If Master Algorithm 1 constructs an infinite sequence $\{\hat{x}_i\}$, then any accumulation point $\hat{x}$ of $\{\hat{x}_i\}$ is optimal for P. ⌑

Master Algorithm 1 has the disadvantage that the sets Ω_i and E_i grow without bound. To see this, suppose that $\psi_\Omega(x_i) \leq \hat{\delta}_i$ and $\psi_{E,T}(\hat{x}_i) \leq \hat{\delta}_i$. Then the algorithm simply increases the index i to i+1, sets $\hat{x}_{i+1} = \hat{x}_i$ and continues doing so until $\hat{\delta}_i$ declines enough for either $\psi_\Omega(\hat{x}_i) > \hat{\delta}_i$ or $\psi_\Omega(x_i) > \hat{\delta}_i$ to take place. In [14] we find a way of circumventing this undesirable phenomenon. Let $\{\delta_{ij}\}$ be a sequence such that $\delta_{ii} = 0$, $\delta_{ij} > 0$ for $i > j$ and let $\delta_{ij} \to \hat{\delta}_i$, as $i \to \infty$, e.g., $\delta_{ij} = 10 \max \{0,1/(i+j)-1/(1+i)\}$) and suppose that we include ω_j in E_i and ε_j in E_i for all $i \geq j+1$ such that $\psi_\Omega(\hat{x}_j) > \delta_{i-1,j}$, or $\psi_{E,T}(x_j) > \delta_{i-1,j}$, respectively. Now, suppose that $\psi_\Omega(x_j) \leq \hat{\delta}_j$. Then ω_j would not be included in Ω_{j+1} in Master Algorithm 1, but under the new scheme, ω_j would be included in Ω_{j+1}, if $\psi_\Omega(x_j) > 0$ and, furthermore, it would be retained for a certain number of iterations in Ω_i, $i = j+1,J=2,\ldots$, until $\psi_\Omega(x_j) \leq \delta_{i-1,j}$ took place. It would then be dropped, never to be used again. As a result, we get a scheme which tends to keep the cardinality of the Ω_i and E_i small. To make the cardinality of the Ω_i and E_i as small as possible, the elements of the sequence $\{\hat{\delta}_i\}$ should be large and decay to zero as slowly as possible. To formalize this discussion we summarize it in the form of

Master Algorithm 2 [14]

Parameters: $\{\delta_{ij}\}_{i,j=0}^{\infty}$ such that

a) $\delta_{ij} = 0$ for all $i \leq j$, and $\delta_{ij} > 0$ otherwise.

b) $\delta_{ij} \to \hat{\delta}_j$ as $i \to \infty$.

c) $\hat{\delta}_j \to 0$ as $j \to \infty$.

Data: $\Omega_0 \subset \Omega$, $E_0 \subset E$, discrete sets.

Step 0: Set $i = 0$.

Step 1: Solve $\hat{P}_i$ for $(\hat{x}_i, \hat{\tau}_{i_1}, \ldots, \hat{\tau}_{i_{t_i}})$.

Step 2: Compute $\psi_\Omega(\hat{x}_i)$, $\psi_{E,T}(\hat{x}_i)$ and corresponding $\omega_i \in \Omega$, $\varepsilon_i \in E$ satisfying (14a,b) for $x_i = \hat{x}_i$.

Step 3: Include ω_j in Ω_{i+1} for all $0 \leq j \leq i$ such that $\psi_\Omega(\hat{x}_j) > \delta_{ij}$, and include ε_j in E_{i+1}, for all $0 \leq j \leq i$ such that $\psi_{E,T}(\hat{x}_j) > \delta_{ij}$.

Step 4: If $\psi_\Omega(\hat{x}_i) \leq 0$ and $\psi_{E,T}(\hat{x}_i) \leq 0$ stop. Else proceed.

Step 5: Set $i = i+1$ and go to Step 1. ⌑

Since $\delta_{ij} \leq \hat{\delta}_j$ for all $i \geq j$, the following theorem is a trivial consequence of Theorem 1.

Theorem 2: If Master Algorithm 2 stops in Step 4, then $\hat{x}_i$ is optimal for P. If Master Algorithm 2 constructs an infinite sequence $\{\hat{x}_i\}$, then any accumulation point $\hat{x}$ of $\{\hat{x}_i\}$ is optimal for P. ⌑

So far we have assumed that we can solve the problems $\hat{P}_i$ exactly and also evaluate the functions $\psi_\Omega(\cdot)$ and $\psi_{E,T}(\cdot)$ exactly. In the next section we shall consider what happens when one solves the problems $\hat{P}_i$ only to the extent of finding approximate stationary points and when one evaluates the functions $\psi_\Omega(\cdot)$ and $\psi_{E,T}(\cdot)$ approximately, as will be the case in practice. We shall then summarize our findings in the form of an implementable algorithm.

3. THE IMPLEMENTABLE ALGORITHM

First we shall discuss the evaluation of the functions $\psi_\Omega(\cdot)$ and $\psi_{E,T}(\cdot)$, defined in (17a,b). We shall assume that

$$\Omega \triangleq \{\omega \mid h^j(\omega) \leq 0,\ j \in H_\omega\} \tag{22}$$

with $h^j : \mathbb{R}^{n_\omega} \to \mathbb{R}^1$ continuously differentiable and H_ω a finite set of integers. We then see that

$$\psi_\Omega(x) = \max\{\phi(x,\omega) \mid h^j(\omega) \leq 0,\ j \in H_\omega\} \tag{23}$$

is defined by an ordinary nonlinear programming problem, the difficulty of which depends entirely on the nature of the functions $\phi(x,\cdot)$ and $h^j(\cdot)$. In engineering design situations, the constraints $h^j(\omega) \leq 0$, $j \in H_\omega$, tend to be very simple and (23) is not too difficult to solve, at least approximately.

Next, we consider $\psi_{E,T}(x)$. First, we note that

$$\psi_{E,T}(x) = \max_{\varepsilon \in E} \min_{\tau \in T} \max_{k \in K} \zeta^k(x,\varepsilon,\tau)\}$$

$$= \max_{(\varepsilon^0,\varepsilon)} \{\varepsilon^0 | \min_{\tau \in T} \max_{k \in K} \zeta^k(x,\varepsilon,\tau) - \varepsilon^0 \geq 0, \ \varepsilon \in E\} \tag{24}$$

If we substitute a discrete set $T_i \subset T$ for T in (24), we get a larger feasible set and hence (24) can be evaluated by a straightforward modification of Master Algorithm 2, such as the one to be presented shortly. The use of this subalgorithm requires the development of the following details. Suppose that

$$E \triangleq \{\varepsilon | p^j(\varepsilon) \leq 0, \ j \in H_\varepsilon\} \tag{25}$$

with $p^j : \mathbb{R}^{n_\tau} \to \mathbb{R}^1$ all continuously differentiable and H_ε a finite set of integers, and that $T_i = \{\tau_j \in T | j \in G_i\}$, with G_i a finite set of integers. Then the discretized problem which the outer approximations Subalgorithm 1, below, requires to be solved at each iteration is

$$\max_{(\varepsilon^0,\varepsilon)} \{\varepsilon^0 | \min_{\tau \in T_i} \max_{k \in K} \zeta^k(x,\varepsilon,\tau) - \varepsilon^0 \geq 0; \ p^j(\varepsilon) \leq 0, \ j \in H_\varepsilon\}$$

$$= \max_{k \in K} \max_{(\varepsilon^0,\varepsilon)} \{\varepsilon^0 | \zeta^k(x,\varepsilon,\tau_\ell) - \varepsilon^0 \geq 0, \ell \in G_i; p^j(\varepsilon) \leq 0, \ j \in H_\varepsilon\} \tag{26}$$

which is seen to be a finite set of ordinary nonlinear programming problems. However, since one has to solve a number of such problems to get a reasonable approximation to $\psi_{E,T}(x)$, it is clear that the computation of approximations to $\psi_{E,T}(x)$ will be the most time consuming operation at each iteration of our outer approximations algorithm for solving P. For the sake of clarity, it seems preferable to state the subalgorithm for computing $\psi_{E,T}(x)$ in conceptual form.

<u>Subalgorithm 1</u>: Evaluates $\psi_{E,T}(x)$.

<u>Parameters</u>: $\{\delta_{ij}\}$ as in Master Algorithm 2.

<u>Data</u>: $x \in \mathbb{R}^n$, $T_0 \subset T$, a discrete set.

<u>Step 0</u>: Set $i = 0$.

<u>Step 1</u>: Solve (26) for $(\varepsilon_i^0, \varepsilon_i)$.

Step 2: Compute $\chi(x,\varepsilon_i)$ and τ_i by solving

$$\chi(x,\varepsilon_i) = \min_{\tau\in T}\max_{k\in K} \zeta^k(x,\varepsilon_i,\tau)$$

$$= \min_{(\tau^0,\tau)} \{\tau^0 | \zeta^k(x,\varepsilon_i,\tau) - \tau^0 \leq 0,\ k \in K,\ g^s(\tau) \leq 0,\ s \in S\} \tag{27}$$

Step 3: Include τ_j in T_{i+1} for all $0 \leq j \leq i$ such that $\varepsilon_j^0 - \chi(x,\varepsilon_j) > \delta_{ij}$.

Step 4: If $(\chi(x,\varepsilon_j) - \varepsilon_j^0) \geq 0$, stop; else proceed.

Step 5: Set i = i+1 and go to Step 1. ⌑

Since Subalgorithm 1 is merely a transcription of the Master Algorithm 2 to the solution of problem (24), the following result is a direct consequence of Theorem 2.

Theorem 3: If Subalgorithm 1 stops in Step 4, then $\psi_{E,T}(x) = \max_{k\in K} \zeta^k(x,\varepsilon_i,\tau_i)$. If Subalgorithm 1 constructs an infinite sequence $\{(\varepsilon_i,\tau_i)\}_{i=0}^{\infty}$ then any accumulation point $(\hat{\varepsilon},\hat{\tau})$ of this sequence satisfies $\psi_{E,T}(x) = \max_{k\in K} \zeta^k(x,\hat{\varepsilon},\hat{\tau})$. ⌑

Next we turn towards constructing an implementation for Master Algorithm 2 which will, by the same token, yield an implementation for Subalgorithm 1. For this purpose we must establish appropriate optimality conditions for P and the subproblems $\hat{P}_i$. Returning to problem $\hat{P}_i$ in (6e), it is clear that it can be rewritten in the following more convenient form:

$$\hat{P}_i : \min_{(x,y_{E_i})} \{f^0(x) | f(x) \leq 0;\ \phi(x,\omega) \leq 0,\ \omega \in \Omega_i;$$

$$\zeta^k(x,\varepsilon,\tau_\varepsilon) \leq 0,\ k \in K,\ \varepsilon \in E_i;\ g^s(\tau_\varepsilon) \leq 0,\ s \in S,\ \varepsilon \in E_i\}, \tag{28}$$

where

$$y_{E_i} \triangleq \{\tau_\varepsilon | \varepsilon \in E_i\}. \tag{28a}$$

Next, we define

$$\psi_{\Omega_i}(x) \triangleq \max_{\omega\in\Omega_i} \phi(x,\omega),$$

$$\psi_{E_i}(x,y_{E_i}) \triangleq \max_{\substack{\bar{\varepsilon}\in E_i \\ k\in K}} \zeta^k(x,\bar{\varepsilon},\tau_\varepsilon), \tag{28b}$$

$$\psi_{T,E_i}(y_{E_i}) \triangleq \max_{\substack{\varepsilon\in E_i \\ s\in S}} g^s(\tau_\varepsilon), \tag{28c}$$

and finally we define

$$\psi_{\hat{P}_i}(x,y_{E_i}) \triangleq \max\{0, f(x), \psi_{\Omega_i}(x), \psi_{E_i}(x,y_{E_i}), \psi_{T,E_i}(y_{E_i})\}. \tag{28d}$$

Proposition 3: Let $E_i = \{\varepsilon_\ell \in E | \ell \in L_i\}$ and let

$$\theta_{\Omega_i E_i}(x,y_{E_i}) \triangleq \min_{\substack{\|h\|_\infty \le 1 \\ \|\eta_\ell\|_\infty \le 1}} \{\max \langle \nabla f^0(x), h\rangle;$$

$$f(x) + \langle \nabla f(x), h\rangle\ ;\ \phi(x,\omega) + \langle \nabla_x \phi(x,\omega), h\rangle,\ \omega \in \Omega_i;\ \zeta^k(x,\varepsilon_\ell,\tau_\ell)$$

$$+ \langle \nabla_x \zeta^k(x,\varepsilon_\ell,\tau_\ell), h\rangle + \langle \nabla_\tau \zeta^k(x,\varepsilon_\ell,\tau_\ell), \eta_\ell\rangle\ \ell \in L_i;\ g^s(\tau_\ell)$$

$$+ \langle \nabla_\tau g^s(\tau_\ell), \eta_\ell\rangle\ ,\ s \in S, \ell \in L_i\} - \psi_{\hat{P}_i}(x,y_{E_i}). \tag{28e}$$

If $(\hat{x},\hat{y}_{E_i})$ is optimal for $\hat{P}_i$ then $\theta_{\Omega_i,E_i}(\hat{x},\hat{y}_{E_i}) = 0$.

Proof: $\theta_{\Omega_i,E_i}(\hat{x},\hat{y}_{E_i}) = 0$ is the Topkis-Veinott necessary optimality condition which was shown in [17] to be equivalent to the F. John optimality condition. ⌑

To extend this optimality condition to problem P, we proceed as follows. For every $\varepsilon \in E$, let τ_ε be some vector in $\mathbb{R}^{n_\tau}$. Then we see that P is equivalent to the following problem

$$\hat{P}:\ \min_{(x,y_E)} \{f^0(x) | f(x) \le 0;\ \phi(x,\omega) \le 0\ \ \forall\omega \in \Omega;$$

$$\zeta^k(x,\varepsilon,\tau_\varepsilon) \le 0,\ \ \forall k \in K,\ \ \forall \varepsilon \in E;$$

$$g^s(\tau_\varepsilon) \le 0,\ \text{for } s \in S,\ \ \forall \varepsilon \in E\} \tag{29}$$

with

$$y_E \triangleq \{\tau_\varepsilon | \varepsilon \in E\}. \tag{29a}$$

Next, let $\psi_{\hat{P}}(x,y_E)$ be defined as in (28d), with Ω, E replacing Ω_i, E_i, respectively, and let

$$\theta_{\Omega,E}(x,y_E) \triangleq \min_{\substack{\|h\|_\infty \le 1 \\ \|\eta_\varepsilon\|_\infty \le 1}} \max\{\langle \nabla f^0(x), h\rangle;$$

$$\zeta^k(x,\varepsilon,\tau_\varepsilon) + \langle \nabla_x \zeta^k(x,\varepsilon,\tau_\varepsilon) h\rangle + \langle \nabla_\tau \zeta^k(x,\varepsilon,\tau_\varepsilon), \eta_\varepsilon\rangle\ ,\ k \in K,\ \varepsilon \in E;$$

$$g^s(\tau_\varepsilon) + \langle \nabla_\tau g^s(\tau_\varepsilon), \eta_\varepsilon\rangle\ ,\ s \in S,\ \varepsilon \in E\} - \psi_{\hat{P}}(x,y_E). \tag{29b}$$

Theorem 3: Suppose that $(\hat{x},\hat{y}_E)$ is optimal for $\hat{P}$ (28), then $\theta_\Omega(\hat{x},\hat{y}_E) = 0$.

Proof: By construction, $\theta_{\Omega,E}(x,y_E) \le 0$ for any (x,y_E), and since $(\hat{x},\hat{y}_E)$ is feasible for $\hat{P}$, $\psi_{\hat{P}}(\hat{x},\hat{y}_E) = 0$. To obtain a contradiction, suppose that $\theta_{\Omega,E}(\hat{x},\hat{y}_E) = -\hat{\delta} < 0$. Then there exists a vector h, with $\|h\|_\infty \leq 1$, such that

$$\langle \nabla f^0(\hat{x}),\hat{h}\rangle \leq -\hat{\delta} \tag{30}$$

and hence there exists a $\lambda_0 > 0$ such that

$$f^0(\hat{x}+\lambda\hat{h}) - f^0(\hat{x}) \leq -\lambda\hat{\delta}/2 \tag{31}$$

for all $\lambda \in (0,\lambda_0]$. Next, we have

$$f(\hat{x}) + \langle \nabla f(\hat{x}),h\rangle \leq -\hat{\delta} \tag{32}$$

and hence, for $\lambda \in [0,1]$,

$$f(\hat{x}) + \lambda\langle \nabla f(\hat{x}),h\rangle \leq -\lambda\hat{\delta} + (1-\lambda)f(\hat{x}) \leq -\lambda\hat{\delta}, \tag{33}$$

since $f(\hat{x}) \leq 0$. Consequently, there exists a $\lambda_1 \in (0,\lambda_0]$ such that

$$f(\hat{x}+\lambda\hat{h}) \leq 0, \text{ for all } \lambda \in (0,\lambda_1]. \tag{34}$$

Similarly,

$$\phi(\hat{x},\omega) + \langle \nabla_x\phi(\hat{x},\omega),\hat{h}\rangle \leq -\hat{\delta} \tag{35}$$

implies that for $\lambda \in [0,1]$, since $\phi(\hat{x},\omega) \leq 0$ for all $\omega \in \Omega$,

$$\begin{aligned}\phi(\hat{x},\omega) + \lambda\langle \nabla_x\phi(\hat{x},\omega),h\rangle &\leq -\lambda\hat{\delta} + (1-\lambda)\phi(\hat{x},\omega)\\ &\leq -\lambda\hat{\delta},\end{aligned} \tag{36}$$

and hence, since Ω is compact, there exists a $\lambda_2 \in (0,\lambda_1]$ such that

$$\max_{\omega\in\Omega} \phi(\hat{x}+\lambda\hat{h},\omega) \leq 0, \text{ for all } \lambda \in [0,\lambda_2]. \tag{37}$$

Next, for each $\varepsilon \in E$, $k \in K$, there exist $\hat{\eta}_\varepsilon$, with $\|\hat{\eta}_\varepsilon\|_\infty \leq 1$, such that

$$\zeta^k(\hat{x},\varepsilon,\hat{\tau}_\varepsilon) + \langle \nabla_x\zeta^k(\hat{x},\varepsilon,\hat{\tau}_\varepsilon),\hat{h}\rangle + \langle \nabla_\tau\zeta^k(\hat{x},\varepsilon,\hat{\tau}_\varepsilon),\eta_\varepsilon\rangle \leq -\hat{\delta}, \tag{38}$$

and hence for $\lambda \in [0,1]$,

$$\begin{aligned}\zeta^k(\hat{x},\varepsilon,\hat{\tau}_\varepsilon) + \lambda\{\langle \nabla_x\zeta^k(\hat{x},\varepsilon,\hat{\tau}_\varepsilon),\hat{h}\rangle + \langle \nabla_\tau\zeta^k(\hat{x},\varepsilon,\hat{\tau}_\varepsilon),\hat{\eta}_\varepsilon\rangle\} &\leq -\lambda\hat{\delta} + (1-\lambda)\zeta^k(\hat{x},\varepsilon,\hat{\tau}_\varepsilon)\\ &\leq -\lambda\hat{\delta}, \text{ for all } k \in K, \text{ and } \varepsilon \in E,\end{aligned} \tag{39}$$

because $\zeta^k(\hat{x},\varepsilon,\hat{\tau}_\varepsilon) \leq 0$ for all $k \in K$ and $\varepsilon \in E$. Hence, since E is compact, we conclude that there exists a $\lambda_3 \in (0,\lambda_2]$ such that

$$\max_{\varepsilon\in E}\min_{\tau\in T}\max_{k\in K} \zeta^k(\hat{x}+\lambda\hat{h},\varepsilon,\tau) \leq \max_{\varepsilon\in E}\max_{k\in K} \zeta^k(\hat{x}+\lambda\hat{h},\varepsilon,\hat{\tau}_\varepsilon+\lambda\hat{\eta}_\varepsilon) \leq 0, \text{ for all } \lambda \in [0,\lambda_3]. \quad (40)$$

Finally,

$$g^s(\hat{\tau}_\varepsilon) + \langle \nabla g^s(\hat{\tau}_\varepsilon),\hat{\eta}_\varepsilon \rangle \leq -\hat{\delta}, \text{ for all } \varepsilon \in E \text{ and } s \in S, \quad (41)$$

and hence

$$g^s(\hat{\tau}_\varepsilon) + \lambda\langle \nabla g^s(\hat{\tau}_\varepsilon),\hat{\eta}_\varepsilon \rangle \leq -\lambda\hat{\delta}, \text{ for all } \varepsilon \in E,\ s \in S. \quad (42)$$

Since E is compact, we conclude that there exist a $\lambda_4 \in (0,\lambda_3]$ such that

$$\max_{\substack{\varepsilon\in E\\ s\in S}} g^s(\hat{\tau}_\varepsilon+\lambda\hat{\eta}_\varepsilon) \leq 0, \text{ for all } \lambda \in [0,\lambda_4]. \quad (43)$$

We thus have shown that $\{(\hat{x}+\lambda_4\hat{h}),(\hat{\tau}_\varepsilon+\lambda_4\hat{\eta}_\varepsilon,\varepsilon \in E)\}$ is feasible for $\hat{P}$ and results in a lower cost, which contradicts the optimality of $\hat{x}$. Hence the theorem must be true. ¤

We now establish an important relationship between the stationary points for the problems $\hat{P}_i$ and those of problem $\hat{P}$.

Theorem 4: Let $\Omega_i \subset \Omega$, $E_i \subset E$, $i = 0,1,2,\ldots$, be infinite sequences of discrete subsets and let $x_i \in \mathbb{R}^n$, $y_{E_i} = \{\tau_{\varepsilon,i} | \tau_{\varepsilon,i} \in E_i\}$, $i = 0,1,2,\ldots$, be such that

$$\psi_{\hat{P}_i}(x_i,y_{E_i}) \to 0 \text{ as } i \to \infty, \quad (44a)$$

$$\theta_{\Omega_i,E_i}(x_i,y_{E_i}) \to 0 \text{ as } i \to \infty, \quad (44b)$$

and $x_i \to \hat{x}$ as $i \to \infty$, with

$$\max\{f(\hat{x}),\psi_\Omega(\hat{x}),\psi_{E,T}(\hat{x})\} \leq 0. \quad (44c)$$

Then there exists a $\hat{y}_E = \{\tau_\varepsilon | \varepsilon \in E\}$ such that

$$\psi_{\hat{P}}(\hat{x},\hat{y}_E) = 0 \quad (44d)$$

and

$$\theta_{\Omega,E}(\hat{x},\hat{y}_E) = 0.$$

Proof: For every $\varepsilon \in E$ and $i = 0,1,2,\ldots$, let

$$y_i \triangleq \{\tau_{\varepsilon,i} | \tau_{\varepsilon,i} \in y_{E_i}\ \forall\varepsilon \in E_i,\ \tau_{\varepsilon,i} \in \arg\min_{\tau\in T}\max_{k\in K} \zeta^k(x_i,\varepsilon,\tau),\ \forall\varepsilon \in E \sim E_i\}. \quad (45)$$

(the definition of y_i is not necessarily unique). Then, because $\psi_{E,T}(\cdot)$ is continuous, and because $\psi_{E,T}(\hat{x}) \leq 0$, we have

$$\lim_{i\to\infty} \min_{\tau\in T} \max_{k\in K} \zeta^k(x_i,\varepsilon,\tau) \leq 0, \forall\varepsilon \in E, \tag{46}$$

and hence (with E replacing E_i in (28b))

$$\lim_{i\to\infty} \psi_E(x_i,y_i) \triangleq \lim_{i\to\infty} \max_{\substack{\varepsilon\in E\\ k\in K}} \zeta^k(x_i,\varepsilon,\tau_{\varepsilon,i})$$

$$= \lim_{i\to\infty} \max \{\max_{\substack{\varepsilon\in E \sim E_i\\ k\in K}} \zeta^k(x_i,\varepsilon,\tau_{\varepsilon,i}), \max_{\substack{\varepsilon\in E_i\\ k\in K}} \zeta^k(x_i,\varepsilon,\tau_{\varepsilon,i})\} \leq 0. \tag{47}$$

For every $\varepsilon \in E$, $\{\tau_{\varepsilon,i}\}_{i=0}^{\infty}$ is compact and hence has at least one accumulation point $\hat{\tau}_\varepsilon$. Let

$$\hat{y}_E \triangleq \{\hat{\tau}_\varepsilon | \varepsilon \in E\}, \tag{48}$$

with $\hat{\tau}_\varepsilon$ an accumulation point of $\{\tau_{\varepsilon,i}\}_{i=0}^{\infty}$. Then, because of (47),

$$\psi_E(\hat{x},\hat{y}_E) \leq 0, \tag{49}$$

and hence, because of (44c), we see that (44d) holds.

Next, by construction of y_i,

$$\theta_{\Omega_i,E_i}(x_i,y_{E_i}) \leq \theta_{\Omega,E}(x_i,y_i) - \psi_{\hat{P}_i}(x_i,y_{E_i}) + \psi_{\hat{P}}(x_i,y_i). \tag{50}$$

Since $\theta_{\Omega,E}(x_i,y_i) \leq 0$ for all i, and since $\theta_{\Omega_i,E_i}(x_i,y_{E_i}) \to 0$, $\psi_{\hat{P}_i}(x_i,y_{E_i}) \to 0$ and $\psi_{\hat{P}}(x_i,y_i) \to 0$ as $i \to \infty$, it now follows that (44e) is true, which completes the proof. ⌑

We can now state our implementable algorithm.

Master Algorithm 3

Parameters: $\{\alpha_i\}_{i=0}^{\infty}$ such that $\alpha_i \searrow 0$ as $i \to \infty$, and $\{\delta_{ij}\}_{i,j=0}^{\infty}$ such that

a) $\delta_{ij} = 0$ for all $i \leq j$, and $\delta_{ij} > 0$ otherwise.

b) $\delta_{ij} \nearrow \hat{\delta}_j$ as $i \to \infty$.

c) $\hat{\delta}_j \searrow 0$ as $j \to \infty$.

Data: $\Omega_0 \subset \Omega$, $E_0 \subset E$ discrete sets.

Step 1: Apply iterations of a nonlinear programming algorithm to $\hat{P}_i$ until an (x_i, y_{E_i}) pair is computed satisfying

$$-\alpha_i \leq \theta_{\Omega_i E_i}(x_i, y_{E_i}), \tag{51a}$$

$$\psi_{\hat{P}_i}(x_i, y_{E_i}) \leq \alpha_i. \tag{51b}$$

Step 2: Apply iterations of a nonlinear programming algorithm to (23) and iterations of Subalgorithm 1 to (24) to obtain an $\omega_i \in \Omega$ and an $\varepsilon_i \in E$, with the property that[†]

$$|\psi_\Omega(x_i) - \phi(x_i, \omega_i)| \to 0 \text{ as } i \to \infty, \tag{51c}$$

$$\psi_E(x_i) - \chi(x_i, \varepsilon_i) \to 0 \text{ as } i \to \infty. \tag{51d}$$

Step 3: Include ω_j in Ω_{i+1} for all $0 \leq j \leq i$ such that $\phi(x_j, \omega_j) > \delta_{ij}$, and include ε_j in E_{i+1} for all $0 \leq j \leq i$ such that $\chi(x_j, \varepsilon_j) > \delta_{ij}$.

Step 4: Set $i = i+1$ and go to Step 1. ⌑

Since we want δ_{ij} to decay very slowly, a good choice for δ_{ij} seems to be

$\delta_{ij} = M \max \{0, \dfrac{1}{(1+j)^{1/L}} - \dfrac{1}{(1+i)^{1/L}}\}$, with $M >> 1$ and $L \geq 10$ say.

Master Algorithm 3 has properties which are quite analogous to those of Master Algorithm 2, as we see from:

Theorem 5: Consider the sequence $\{x_i\}_{i=0}^{\infty}$ constructed by Master Algorithm 3. If $x_i \overset{I}{\to} \hat{x}$ as $i \to \infty$, $I \subset \{0,1,2,\ldots\}$, then there exists a $\hat{y}_E = \{\hat{\tau}_\varepsilon | \varepsilon \in E\}$ such that $\psi_{\hat{P}}(\hat{x}, \hat{y}_E) = 0$ and $\theta_{\Omega,E}(\hat{x}, \hat{y}_E) = 0$.

Proof: Because of (51b-d) and because $\psi_{E_i,T}(x_i) \leq \psi_{E_i}(x_i, y_{E_i}) \leq \psi_{\hat{P}_i}(x_i, y_{E_i})$, it follows by a trivial extension of Proposition 2 that $\max\{f(\hat{x}), \psi_\Omega(\hat{x}), \psi_E(\hat{x})\} \leq 0$ (i.e., $\hat{x} \in F$ (see (11)). The theorem now follows from (51a) and Theorem 4. ⌑

[†]This property is achieved by applying the appropriate algorithm for a progressively larger and larger number of iterations as $i \to \infty$.

4. CONCLUSION

The main contribution of this paper is to show that design centering, tolerancing and tuning (DCTT) problems can be treated outside of the framework of nondifferentiable optimization algorithms. As a result, a number of major obstacles to obtaining an implementable algorithm have been overcome and the first implementable algorithm for solving DCTT problems has been constructed. Without doubt this algorithm will have practical impact.

Finally, in the Appendix the optimality conditions used in this paper are compared to the ones based on generalized gradients, used in [3]. It seems that the optimality conditions in this paper are somewhat sharper than the ones used in [3].

APPENDIX: A COMPARISON OF OPTIMALITY CONDITIONS

In [3], problems such as (6) were treated as nondifferentiable optimization problems. Mifflin [13] has developed a necessary condition of optimality for such problems based on the theory of subgradients (see [12]). Unfortunately, Mifflin's condition is not verifiable for problems such as (6), with K = {1}, and hence his conditions were somewhat relaxed in [3], as follows:

Theorem A1 [3]: Suppose that x_i is optimal for P_i in (6), with K = {1}, then $f(x_i) \leq 0$, $\psi_{\Omega_i}(x_i \leq 0$, $\psi_{E_i,T}(x_i) \leq 0$ and

$$0 \in M(x_i),$$

where

$$M(x_i) = \mathrm{Co}\{\nabla f^0(x_i); \delta_f \nabla f(x_i);\ \delta_{\Omega_i} \nabla_x \phi(x_i,\omega), \omega \in \Omega_i(x_i);\ \delta_{E_i} \nabla_x \zeta^1(x_i,\varepsilon,\tau),$$

$$\varepsilon \in E_i(x_i), \tau \in T(x_i,\varepsilon)\} \tag{A2}$$

where $\delta_f(\delta_{\Omega_i},\delta_{E_i})$ is zero if $f(x_i) < 0$ $(\psi_{\Omega_i}(x_i) < 0, \psi_{E_i,T}(x_i) < 0)$ and is equal to one otherwise,

$$\Omega_i(x_i) \triangleq \arg\max_{\omega \in \Omega_i} \phi(x_i,\omega) \tag{A3}$$

$$E_i(x) \triangleq \arg\max_{\varepsilon \in E_i} \chi(x_i,\varepsilon) \tag{A4}$$

$$T(x_i,\varepsilon) \triangleq \arg\min_{\tau \in T} \zeta^1(x_i,\varepsilon,\tau) \tag{A5}$$

Now, by Proposition 3, if (x_i, y_{E_i}) are optimal for problem $\hat{P}_i$, (6e), then $x_i \in F$ and $\theta_{\Omega_i,E_i}(x_i,y_{E_i}) = 0$. Since y_{E_i} is not necessarily unique, to obtain a comparison with Theorem A1 we shall assume that $\tau_\varepsilon \in y_{E_i} \Rightarrow \tau_\varepsilon \in T(x_i,\varepsilon)$ for all $\varepsilon \in E_i$. Now, it is shown in Sections 1.2 and 4.4 of [17], that for $x_i \in F$, $\theta_{\Omega_i,E_i}(x_i,y_{E_i}) = 0$ if and only if the F. John condition [17] is satisfied at (x_i,y_{E_i}), i.e., there exist multipliers μ^0, μ_f, μ_ω, $\mu_{\varepsilon,\tau}$, $\mu^s \geq 0$, not all zero such that

$$\mu^0\nabla f^0(x_i) + \mu_f\nabla f(x_i) + \sum_{\omega\in\Omega(x_i)} \mu_\omega\nabla_x\phi(x_i,\omega) + \sum_{\substack{\varepsilon\in E(x_i)\\ \tau\in T(x_i,\varepsilon)}} \mu_{\varepsilon,\tau}\nabla\zeta^1(x_i,\varepsilon,\tau) = 0, \tag{A6}$$

$$\mu_{\varepsilon,\tau}\nabla_\tau\zeta^1(x_i,\varepsilon,\tau) + \sum_{s\in S}{}'\mu_\varepsilon^s\nabla g^s(\tau) = 0, \quad \forall\ \varepsilon \in E(x_i),\ \tau \in T(x_i,\varepsilon) \tag{A7}$$

and

$$\mu_f f(x_i) = 0, \tag{A8}$$

$$\mu_\omega\phi(x_i,\omega) = 0, \quad \forall\ \omega \in \Omega(x_i), \tag{A9}$$

$$\mu_{\varepsilon,\tau}\zeta^1(x_i,\varepsilon,\tau) = 0 \quad \forall\ \varepsilon \in E(x_i), \quad \forall\ \tau \in T(x_i,\varepsilon), \tag{A10}$$

$$\mu_\varepsilon^s g^s(\tau) = 0, \quad \forall\ \varepsilon \in E(x_i), \quad \forall\ \tau \in T(x_i,\varepsilon).$$

We note that (A7) together with (A11) is merely the F. John condition for the problems

$$\min_{\tau\in T} \zeta(x_i,\varepsilon,\tau),\ \varepsilon \in E(x_i). \tag{A12}$$

Next, (A6) sums a smaller number of vectors to zero than (A1) and hence, together with (A8-A10), is a sharper condition of optimality then Theorem A1. By the same token, the optimality condition $\theta(\hat{x},\hat{y}_E) = 0$ used in this paper for $\hat{P}$ is sharper than the optimality condition for P used in [3].

ACKNOWLEDGEMENT

Research sponsored by the National Science Foundation (RANN) Grant ENV76-04264 and the Joint Services Electronics Program Contract F44620-76-C-0100.

REFERENCES

[1] D. Q. Mayne, E. Polak, and R. Trahan, "An Outer Approximations Algorithm for for Computer Aided Design Problems," JOTA, in press.

[2] E. Polak, "On a Class of Computer Adied Design Problems," Proc. of the 7th IFAC World Congress, Helsinki, Finland, June 1978.

[3] E. Polak and A. Sangiovanni-Vincentelli, "Theoretical and Computational Aspects of the Optimal Design Centering, Tolerancing and Tuning Problem," IEEE CAS Trans., special issue on Computer Adided, Design, in press.

[4] J. W. Bandler, P. C. Liu and H. Tromp, "A Nonlinear Programming Approach to Optimal Design Centering, Tolerancing and Tuning," IEEE Trans. on Circuits and Systems, vol. CAS-23, no. 3, March 1976.

[5] P. V. Lopresti, "Optimum Design of Linear Tuning Algorithms," IEEE Trans. on Circuits and Systems, vol. CAS-24, no. 3, March 1977.

[6] J. F. Pinel, "Computer-Aided Network Tuning," IEEE Trans. on Circuit Theory, vol. CT-17, no. 1, January 1971, pp. 192-194.

[7] G. Müller, "On Computer-Aided Tuning of Microwave Filters," IEEE Proc. ISCAS 1976, Münich, May 1976, pp. 209-211.

[8] E. Lüder and B. Kaiser, "Precision Tuning of Miniaturized Circuits," IEEE Proc. ISCAS 1976, Münich, May 1976, pp. 722-725.

[9] R. L. Adams and V. K. Manaktala, "An Optimization Algorithm Suitable for Computer-Assisted Network Tuning," IEEE Proc. 1975 ISCAS, Boston 1975, pp. 210-212.

[10] E. Polak and A. Sangiovanni-Vincentelli, "On Optimization Algorithms for Engineering Design Problems with Distributed Constraints, Tolerances and Tuning," Proc. 1978 JACC, Philadelphia, October 1978.

[11] F. H. Clarke, "Generalized Gradients and Applications," Trans. Amer. Math. Soc., vol. 205, 1975, pp. 247-262.

[12] F. H. Clarke, "A New Approach to Lagrange Multipliers," Math. of Oper. Research, vol. 1, 1976, pp. 165-174.

[13] R. Mifflin, "Semismooth and Semiconvex Functions in Constrained Optimization," SIAM J. on Control and Optimization, November 1977, pp. 959-972.

[14] C. Gonzaga and E. Polak, "On Constraint Dropping Schemes and Optimality Functions for a Class of Outer Approximations Algorithms," Memorandum UCB ERL M77/68, Electronics Research Laboratory, University of California, Berkeley, November 1977, also SIAM J. on Control and Optimization, vol. 14, no. 4, 1979, pp. 477-494.

[15] C. Lemarechal, "An Extension of Davidon Methods to Nondifferentiable Problems," in Nondifferentiable Optimization, M. L. Balinski and P. Wolfe, eds., Mathematical Programming Study 3, North Holland, Amsterdam, 1975, pp. 95-109.

[16] P. Wolfe, "A Method of Conjugate Subgradients for Minimizing Nondifferentiable Functions," ibid, pp. 145-173.

[17] E. Polak, Computational Methods in Optimization, Academic Press, 1971.

[18] R. Mifflin, "An Algorithm for Constrained Optimization with Semismooth Functions," Math. of Oper. Research, vol. 2, no. 2, May 1977, pp. 181-207.

COMPUTING METHODS IN APPLIED SCIENCES AND ENGINEERING
R. Glowinski, J.L. Lions (editors)
North-Holland Publishing Company

NUMERICAL COMPUTATIONS IN NONLINEAR MECHANICS

Gilbert Strang
M.I.T.
Cambridge, USA

Hermann Matthies
Germanischer Lloyd,
Hamburg, Germany

ABSTRACT

We propose the application of quasi-Newton methods to the discrete equations of nonlinear mechanics. The paper discusses the properties of one particular algorithm, the BFGS update, from the viewpoint of numerical analysis and also of computational convenience. Its implementation for inelastic problems (incremental plasticity and nonlinear elasticity) is briefly described.

The support of the National Science Foundation (MCS 78-12363) is gratefully acknowledged.

1. Description of the method

Our goal is a more effective solution technique for nonlinear finite element equations. We propose that an algorithm originally developed for numerical optimization can be adapted to the large sparse systems of nonlinear mechanics. It is a variation of Newton's method, constructed in such a way as to avoid the frequent recalculation of the Jacobian matrix of the nonlinear system (the tangent stiffness matrix). At the same time it aims to be more reliable, and to permit a larger loading increment, than the simplest modifications of Newton's method.

We write the nonlinear system as $F(u) = 0$. In a typical case the unknowns $u_1, \ldots, u_N$ represent the displacements at N nodal points within the structure, and the N equations impose conditions of equilibrium. The algorithm is not restricted to finite element applications, but we borrow the terminology of that field: the matrix of first derivatives $\partial F_i / \partial u_j$ is the stiffness matrix. If such a matrix K_i is computed at an approximation u_i, then the next Newton step is defined by

$$K_i(u_{i+1} - u_i) = -F(u_i) \tag{1}$$

In practice, the computation of each new K_i may be extremely expensive; the results achieved by computing and factoring the original matrix K_0 need to be reused. If K_i is replaced throughout the iterations by a fixed K_0, we have the modified Newton method. Its convergence is only linear, but if the K_i change slowly then it may be satisfactory. In the presence of plasticity, however, we cannot expect the changes in the matrix to remain small; the slope of the stress-strain curve is not at all steady. With unloading, which produces a sudden increase in slope, the modified Newton method can easily become divergent. Even for geometric nonlinearities the loading increment is severely restricted.

There are a number of algorithms which aim to yield an improvement in both speed and reliability. (We think of the latter as the more essential.) Among these is one whose substitution into an existing program we found to be straightforward. It is known as a quasi-Newton (or variable metric, or update) method, and it alters the matrix K in a simple way at each step. It has been successful in applications to optimization, where a cost function $\phi(u)$ is to be minimized--

and where it is normally the triangular factors of K (the L and U of Gaussian elimination) which are updated from one step to the next. In our experiments we adopted a different updating technique, intended for sparse systems, but the fundamental question is still the choice of update formula. We impose four conditions:

1. The new K_i should satisfy the quasi-Newton equation

$$K_i(u_i - u_{i-1}) = F(u_i) - F(u_{i-1}) \qquad (2)$$

This means that the information gained in the step just completed, from u_{i-1} to u_i, is correctly reflected in K_i. For one equation in one unknown we would have the standard secant method; with N unknowns, we can only know the secant <u>along the step</u>, but as these directions accumulate the matrices K_i should approach the exact $\partial F/\partial u$.

2. If K_{i-1} is symmetric then the new K_i should also be symmetric.

3. If K_{i-1} is positive definite then the new K_i should also be positive definite. This is to be expected in a pure displacement method. If constraints are introduced that lead to the inclusion of Lagrange multipliers among the unknowns, then the K_i become indefinite; we have not experimented with the modifications this would require.

4. The change from K_{i-1} to K_i must be easy to calculate. More precisely, it is the vector $d_i = K_i^{-1}F(u_i)$ which has to be computed as efficiently as possible, since d_i gives the "search direction" along which we move from u_i to u_{i+1}.

Among the update formulas which satisfy these conditions, one is by general agreement the most effective. It was derived independently by four authors (Broyden, Fletcher, Goldfarb, and Shanno) and is known as the <u>BFGS update</u>. We refer to the excellent survey by Dennis and Moré [1] for its derivation and for an analysis of convergence; they describe the history of several alternative updates, but we have restricted ourselves to BFGS. It is constructed from the vectors $\delta_i = u_i - u_{i-1}$ and $\gamma_i = F(u_i) - F(u_{i-1})$ which contain the information gained at the previous step. We list three different but equivalent expressions for the update:

a) Increment of rank two:

$$K_i = K_{i-1} + \frac{\gamma\gamma^T}{\gamma^T\delta} - \frac{(K_{i-1}\delta)(K_{i-1}\delta)^T}{\delta^T K_{i-1}\delta} \tag{3}$$

b) Product plus increment

$$K_i^{-1} = (I - \rho_i\delta_i\gamma_i^T)\, K_{i-1}^{-1}(I - \rho_i\gamma_i\delta_i^T) + \rho_i\delta_i\delta_i^T \tag{4}$$

Here ρ_i is the scalar $1/\delta_i^T\gamma_i$.

c) Product form

$$K_i^{-1} = (I + w_i v_i^T) K_{i-1}^{-1}(I + v_i w_i^T) \tag{5}$$

with vectors

$$v_i = \left[\frac{\delta_i^T\gamma_i}{\delta_i^T K_{i-1}\delta_i}\right]^{\frac{1}{2}} K_{i-1}\delta_i - \gamma_i; \quad w_i = \frac{1}{\delta_i^T\gamma_i}\quad \delta_i$$

It is this form, written for convenience as a correction to the inverse, that we used in our initial experiments [2].

The implementation of the three forms is quite different. The usual procedure in optimization is to carry out the updating on the Cholesky factors of K , converting the lower triangular matrix L_{i-1} into the updated L_i . This uses some variant of the incremental form, and it can be done more quickly than a fresh factorization of the matrix K_i . After L_i is computed, the differences δ_i and γ_i have done their part and are forgotten. This is reasonable when many updates are expected, and when the size and sparsity of the matrices do not dominate the situation; but our case is exactly the opposite. We cannot surrender sparsity, nor can we afford the $O(N^2)$ cost of matrix manipulations, and we hope not to make too many updates before achieving sufficient accuracy in u . Therefore we intend simply to remember all the updates, unless convergence is long delayed. Nocedal [3] has discovered what to do when our tolerance is exhausted: Forget the earliest update, shift the indices of all δ_i and γ_i (or v_i and w_i) from i to $i-1$, and give the new update its normal place.

The second form, product plus increment, deserves more attention than it has had. Like the others it preserves symmetry and positive definiteness; conditions 2 and 3 are automatically met, and the quasi-Newton equation (condition 1) is easy to verify. The key point is that a simple recurrence formula yields the new search direction $d_i = K_i^{-1} F(u_i)$. It starts with $q_i = F(u_i)$, multiplies by each $I - \rho_i \gamma_i \delta_i^T$, carries out back-substitution with K_0 , and then proceeds forward to the other multiplying factors $I - \rho_i \delta_i \gamma_i^T$ and the increments $\rho_i \delta_i \delta_i^T$:

Initialize $q_i = F(u_i)$.

For $j = i, \ldots, 1$:

Compute and store $\alpha_j = \rho_j \delta_j^T q_j$

Compute $q_{j-1} = q_j - \gamma_j \alpha_j$

Solve $K_0 r_0 = q_0$

For $j=1, \ldots, i$:

Compute $\beta_j = \rho_j \gamma_j^T r_{j-1}$

Compute $r_j = r_{j-1} + \delta_j(\alpha_j - \beta_j)$

End with new search direction $d_i = r_i$.

We intend to compare this recurrence with the pure product form which is already implemented. The operation counts in the following section are essentially unchanged.

2. Implementation of the method

The iterations begin with a matrix K_0, and the first search direction d_0 is the solution of $K_0 d_0 = F(u_0)$. The usual LDL^T factorization of K_0 is produced during Gaussian elimination, and the factors L and D are stored; D is diagonal, and L has the same band structure as K_0 (with some fill-in of the band).

At each succeeding step the updating factors are added, so that for example

$$d_2 = (I + w_2 v_2^T)(I + w_1 v_1^T) K_0^{-1} (I + v_1 w_1^T)(I + v_2 w_2^T) F(u_2)$$

In this case the program computes $b_2 = (I + v_2 w_2^T) F(u_2)$, then

$b_1 = (I + v_1 w_1^T) b_2$, then solves $K_0 c = b_1$ by back-substitution with the stored factors L and D , then $d_1 = (I + w_1 v_1^T) c$, and finally $d_2 = (I + w_2 v_2^T) d_1$. This is the third search direction.

It is essential to estimate the increase in computational work over the central back-substitution step, which requires 2mN multiplications if K_0 has half-bandwidth m . Each extra factor needs only 2N multiplications, half of them to compute the inner products like $w_2^T F(u_2)$, and the other half to multiply the vector v_2 by this inner product. The number of additions is also approximately 2mN for the central step and $2\ell N$ for ℓ extra factors. (There are also 3N multiplications to compute the pair v_i, w_i in the first place.) The program will impose an upper limit (say 15) on the number of factors to be stored. In most problems we expect to be within the required tolerance in a few iterations, so that the additional work is completely dominated by the back-substitution which is present already in modified Newton. Condition 4, on the computational effort, is satisfied.

There is one further ingredient of the algorithm. Given d_i , Newton's method would take a unit step in that direction to reach the new approximation $u_{i+1} = u_i - d_i$. Close to the solution this is entirely proper. In the initial iterations, however, when we are comparatively far from the solution, it is more prudent (and ultimately more efficient) to admit steps of different length: $u_{i+1} = u_i - s_i d_i$. The multiplying factor s_i is computed by a line search, aiming to annihilate the component of F in the search direction:

$$G_i(s) = d_i^T F(u_i - s d_i) = 0 \quad \text{at} \quad s = s_i \tag{6}$$

In an optimization problem this corresponds to finding the minimum of $\phi(u)$ along a ray. We cannot expect the ray to contain the exact global minimum of ϕ, or in our case the exact solution of $F(u) = 0$, but we do want to avoid an overshoot of the valley in which this solution lies. The line search is done only to modest accuracy; it is sufficient to find s_i such that

$$|d_i^T F(u_i - s_i d_i)| \leq \tfrac{1}{2} |d_i^T F(u_i)| \ , \quad \text{or} \quad |G_i(s_i)| \leq \tfrac{1}{2} |G_i(0)|$$

If $s_i = 1$ is satisfactory then the line search never starts; this was our experience after two or three iterations (see [2], which uses a modification of regula falsi to bracket the zero of G). We emphasize that the line search is a problem in one variable, and that it would be equally valuable for modified Newton.

The final step is to look again at the formula for v_i. The difference $\delta_i = u_i - u_{i-1}$ can be rewritten as $-s_{i-1}d_{i-1}$. Since $K_{i-1}d_{i-1} = F(u_{i-1})$, the expression $K_{i-1}\delta_i$ which appears twice in v_i does not require a matrix-vector multiplication. In fact we can rewrite v_i as

$$v_i = F(u_{i-1}) \left[1 + s_{i-1} \left[\frac{d_{i-1}^T \gamma_i}{\delta_i^T F(u_{i-1})} \right]^{\frac{1}{2}} \right] - F(u_i) \tag{7}$$

(The positive coefficient of s_{i-1} comes from choosing the negative of the square root.) In programming we can go further, replacing δ_i throughout by $-s_{i-1}d_{i-1}$ and then recognizing that the only two inner products needed within the square root are $d_{i-1}^T F(u_{i-1})$ and $d_{i-1}^T F(u_i)$. But these are exactly the $G_{i-1}(0)$ and $G_{i-1}(s_{i-1})$ found at the beginning and end of the line search just completed. A similar remark applies to $w_i = d_{i-1}/d_{i-1}^T \gamma_i$, and also to the quantity $w_i^T F(u_i)$ which is the first product needed at the far right of the formula for $d_i = K_i^{-1} F(u_i)$.

3. Programming and applications

We have tested the quasi-Newton method within the framework of the ADINA system developed by Bathe. It is important to emphasize that substituting the BFGS algorithm for the existing equilibrium iteration was completely straightforward. The other parts of the system were not affected. The program terminated when the vectors $u_i - u_{i-1}$ and $F(u_i)$ reached prescribed tolerances; it would also stop if a prescribed maximum number of updates were reached. We chose the tolerance as suitable multiples of a reference displacement and force, in order to measure relative error.

The tests themselves will be described at greater length elsewhere, but we do want to note two experimental observations:

1. *The algorithm gives convergence with a much larger loading increment than modified Newton.* For a uniformly loaded cantilever beam in plane stress, the force was increased until the end deflection

was comparable to the length. The geometric nonlinearity dominates the calculation. An efficient choice was to apply this load in five steps, solving at each step for the corresponding increment in displacement. (We followed Bathe [4] in using five 8-node isoparametric elements.) By comparison, modified Newton has rapidly increasing difficulty as the structure stiffens and can be used only for very small steps.

This was seen also in the dynamic case, with a large load imposed instantaneously at $t = 0$. Using the Wilson θ-method, with $\theta = 1.4$ and Δt chosen as 1/42 of the fundamental period, BFGS required a maximum of 5 updates and virtually no line searches; modified Newton diverged.

2. <u>The method remains reliable in the presence of plastic unloading</u>. This was tested both in the quasistatic case, for a crack problem, and also in the dynamic case for an elastic-plastic wave traveling along a beam. The oscillations of this wave led to rapid changes throughout the structure between loading and unloading. Again the quasi-Newton algorithm was successful; it seems able to cope with a wide variation of material properties.

The organization of the program is easy to describe. The user supplies two independent subroutines, one to carry out back-substitution using the computed factors of K_0, and the other to evaluate the out-of-balance forces F. (In the plastic case, the projection of elastic stresses back onto the current yield surface needs to be done efficiently; Krieg and Krieg [5] have supported a direct "radial return.") The main routine computes v_i and w_i, finds the new search direction d_i, and then either decides that $s = 1$ gives a satisfactory step or else carries out a line search. The updating vectors v_i and w_i, or δ_i and γ_i, can be stored along with their predecessors on a peripheral device; the program outlined in [2] requires space for only three additional vectors during the computations.

The algorithm is simple and reliable, but there is much more testing to do before it can be freely used in computational mechanics. It is now widely available in the ADINA system, and there it has already made a significant difference; at the recent ADINA users conference the importance of an improved iteration technique was absolutely clear, and in his report (as well as at the 1979 SMIRT conference in Berlin [6]) Bathe cited the results of additional tests of the method. The first author is joining also in numerical experiments with Hibbitt and Nickell, to study the application to nonlinear dynamics. We hope that implicit methods will become more feasible, if a very simple constant matrix can be given the role of K_0 at each step -- but it is too early to say. And there are alternative techniques, both in the acceleration of modified Newton and in quite different directions; Irons has tried a variant of conjugate gradients [7], and we know indirectly of a similar program developed by Sylvester's group at McGill; it is sure to be a very attractive option. But we believe that the quasi-Newton method is genuinely effective, and we hope very much that others will decide to test it within their own programs.

REFERENCES

[1] Dennis, J., and Moré, J., "Quasi-Newton methods, motivation and theory," SIAM Review, Vol. 19, 1977, pp. 46-89.

[2] Matthies, H., and Strang, G., "The solution of nonlinear finite element equations," International J. of Num. Meth. in Engineering, to appear.

[3] Nocedal, J., Updating quasi-Newton matrices with limited storage, Math. Comp., to appear.

[4] Bathe, K.-J., "An assessment of current finite element analysis of nonlinear problems," Numerical Solution of Partial Differential Equations III, B. Hubbard, ed., Academic Press, New York, 1976.

[5] Krieg, R., and Krieg, D., "Accuracies of numerical solution methods for the elastic-perfectly plastic model," J. of Pressure Vessel Technology, Vol. 99, No. 4, 1977, pp. 510-515.

[6] Bathe, K.-J., and Cimento, A., Some practical procedures for the solution of nonlinear finite element equations, preprint, SMIRT Conference, Berlin, 1979.

[7] Irons, B., and Elsawaf, A., "The conjugate Newton algorithm for solving finite element equations," Formulations and Computational Algorithms in Finite Element Analysis, K.-J. Bathe, J.T. Oden, and W. Wunderlich, eds., M.I.T. Press, Cambridge, 1977.

SESSION VIII

INVERSE PROBLEMS IN SEISMOLOGY

COMPUTING METHODS IN APPLIED SCIENCES AND ENGINEERING
R. Glowinski, J.L. Lions (editors)
North-Holland Publishing Company

AN OPTIMAL CONTROL SOLUTION OF THE INVERSE PROBLEM OF REFLECTION SEISMICS

by

A. BAMBERGER
C.M.A.
Ecole Polytechnique
91128 Palaiseau
FRANCE

G. CHAVENT - P. LAILLY
I.N.R.I.A.
B.P. 105
78150 Le Chesnay
FRANCE

ABSTRACT

We deal with the following inverse problem. We make an explosion at the surface of the earth (supposed to be a plane layered medium). The waves (supposed to be plane) as they go down are reflected and transmitted on each boundary between two layers. So we can record the waves coming back to the surface (seismogram). The inverse problem is to find out, from the seismogram, some characteristics of the medium as functions of the depth.

The inverse problem is set as an optimal control problem. We can expect, from some theoretical results [1], the stability of the optimum with respect to the noise corrupting the recording.

Some numerical results, obtained by this method, are given.

NOTATIONS

We shall use the following notations :

(1.1) z denotes the depth $(z \geq 0)$

(1.2) t denotes the time $(t \geq 0)$

(1.3) $y(z, t)$ denotes the displacement at depth z and time t

(1.4) $\rho(z)$ denotes the density at depth z

(1.5) $\mu(z)$ denotes the elasticity coefficient at depth z

(1.6) $\sigma(z) = (\rho(z)\ \mu(z))^{1/2}$ denotes the acoustical impedance at depth z

(1.7) $v(z) = (\mu(z)/\rho(z))^{1/2}$ denotes the sound celerity at depth z

(1.8) $x(z) = \int_0^z v^{-1}(z')dz'$ denotes the travel time ; it is the time required for a perturbation to propagate from the depth 0 to the depth z

(1.9) $\sigma(z)$ denotes the acoustical impedance at the travel time x.

1 - PROBLEM MODELING

1.1 - State equation

We suppose that the displacement is the solution of the 1-D wave equation :

(1.10) $$\rho(z) \frac{\partial^2 y}{\partial t^2} - \frac{\partial}{\partial z} (\mu(z) \frac{\partial y}{\partial z}) = 0$$

with the initial conditions :

(1.11) $$y(z, 0) = \frac{\partial y}{\partial t} (z, 0) = 0 \text{ (the medium is at rest at time } t = 0)$$

and the boundary condition :

(1.12) $$- \mu(0) \frac{\partial y}{\partial z} (0, t) = g(t)$$

$g(t)$ is the pressure pulse which is applied at the surface of the medium at time t.

This model is the simplest one since it is a 1-D model and the damping terms are neglected.

Changing the variable z (depth) into the variable x (travel time) the system (1.10), (1.11), (1.12) becomes :

(1.13) $$\sigma(x) \frac{\partial^2 y}{\partial t^2} - \frac{\partial}{\partial x} (\sigma(x) \frac{\partial y}{\partial x}) = 0$$

(1.14) $$y(x, 0) = \frac{\partial y}{\partial t} (x, 0) = 0$$

(1.15) $$- \sigma(0) \frac{\partial y}{\partial x} (0, t) = g(t)$$

We can see that the displacement at the surface y(0, t) depends only on the function σ ; so the inverse problem has been set in the following way :

1.2 - The inverse problem

We suppose that the pulse g(t) is given and we denote bt $Y(\sigma)$ the response of an acoustical impedance distribution to the pulse g :

(1.16) $$Y(\sigma) = y(0, t)$$

where y is the solution of (1.13), (1.14), (1.15) with the parameter σ.

The inverse problem consists in finding a distributed parameter $\hat{\sigma}$ from an observation of $Y(\hat{\sigma})$.

More precisely, we dispose of an observation $Y_d(t)$ of the displacement at the surface possibly corrupted by an additive noise b(t) for $t \in [0, T]$; T is the duration of the observation :

(1.17) $$Y_d(t) = Y(\hat{\sigma}, t) + b(t) \qquad t \in [0, T]$$

The inverse problem has been set as an optimization problem :

(1.18) $$\begin{cases} \text{To find an impedance distribution } \tilde{\sigma} \in \mathcal{T}_{ad} \text{ that the minimizes} \\ J(\sigma) = \int_0^T |Y(\sigma, t) - Y_d(t)|^2 dt \end{cases}$$

2 - MATHEMATICAL AND NUMERICAL CASE

This paragraph gives a summary of detailed papers [1], [2]. We refer to these papers for a proof of the results given here.

2.1 - Difficulties of the inverse problem

One of the difficulties in solving identification of distributed parameters problems is that these problems are generally ill posed (unstability of the inverse problem). We shall illustrate this point :

(2.1) Let $\Sigma = \{\sigma(x) \text{ s.t. } 0 < \sigma^- \le \sigma(x) \le \sigma^+ \quad \forall x \in [0, \frac{T}{2}]\}$

We define over the "distance" :

$$(2.2)\quad \begin{cases} d(\sigma', \sigma'') = \sup\limits_{x_1, x_2 \in [0, \frac{T}{2}]} \left| \int_0^{x_1} \sigma'(x)dx - \int_0^{x_2} \sigma''(x)dx \right| \\ \text{with } x_1, x_2 \text{ such that :} \\ \int_0^{x_1} \frac{1}{\sigma'(x)}\, dx = \int_0^{x_2} \frac{1}{\sigma''(x)}\, dx \end{cases}$$

Theorem :

The mapping $\sigma \to Y(\sigma)$ is continuous from Σ into $C^0(]0, T[)$ for the "distance" d defined by (1.20) if the pulse g verifies :

(2.3) $g(0) = g'(0) = 0 \qquad g \in W^{3,1}(]0, T[)$

Moreover :

$$\|Y(\sigma') - Y(\sigma'')\|_{C^0(]0, T[)} \le C(\Sigma)\ C(T) \|g\|_{W^{3,1}(]0, T[)}\ d(\sigma', \sigma'')$$

We can notice that the distance d defined by (2.2) is very weak with respect to an usual distance (L^2 for instance). This is illustrated in the following example :

Let σ' and σ'' the two following impedance distributions (figure 1) :

- σ'' is piecewise constant on intervals of length Δx :

  $\sigma'' = \sigma_1 = 1.$ on even intervals

  $\sigma'' = \sigma_2 = .25$ on odd intervals

  and is equal to 0. for $x > 1$ (*)

- σ' is constant = .5 up to the depth x = 1.25 and is zero after (*)

(*) so that there is a total reflection

It can be easily calculated that :

$$d(\sigma', \sigma'') = |\sigma_1 - \sigma_2| \Delta x$$

and then σ' and σ" are near with respect to the "distance" (2.2) when σ" is highly oscillating.

The computation of the seismograms Y(σ') and Y(σ") shows that they are the same, and this confirms the result of the theorem as it illustrates the unstability of the inverse problem with respect to a numerically significative distance.

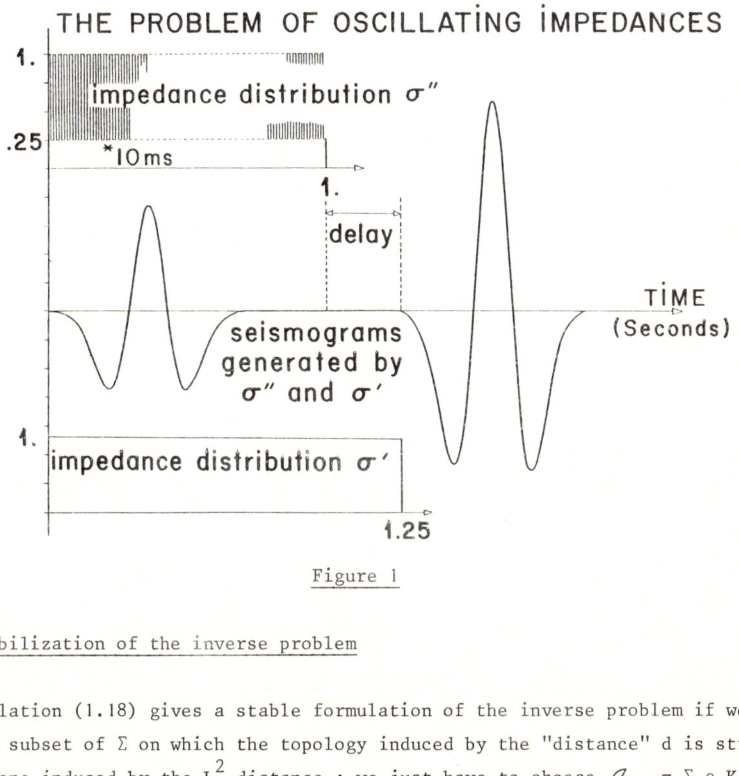

Figure 1

2.2 - Stabilization of the inverse problem

The formulation (1.18) gives a stable formulation of the inverse problem if we choose $\mathcal{a}_{ad}$ as a subset of Σ on which the topology induced by the "distance" d is stronger than the one induced by the L^2 distance : we just have to choose $\mathcal{a}_{ad} = \Sigma \cap K$ with K compact in L^2.

In order to be able to take account of discontinuous impedance distributions we have chosen :

(2.4)
$$\begin{cases} K = \{\sigma / \mathrm{var}(\sigma) \le M\} \\ \text{where var } \sigma = \text{total variation of the function } \sigma \text{ over } [0, \frac{T}{2}] \\ = \sup_{(x_i)} \sum_{i=0}^{N-1} |\sigma(x_{i+1}) - \sigma(x_i)| \\ 0 \le x_0 \le \quad \le x_i \le x_N \le \frac{T}{2} \end{cases}$$

Then we can only identify impedance distributions $\hat{\sigma}$ such that $\mathrm{var}(\hat{\sigma}) \le M$.

Remark :

We have eliminated highly oscillating impedance distributions and the situation of the previous example cannot happen.

2.3 - Approximated inverse problem

To solve numerically (1.18) an approximation of $\mathcal{a}_{ad}$ by a finite dimensional set $\mathcal{a}^h_{ad}$ is required.

Moreover the data $g(t)$ and $Y_d(t)$ are sampled data (sample step Δt)

(2.5)
$$\begin{cases} g_h(t) = \sum_{j=1}^{N_t} g_h^j X_j(t) & \text{with } X_j(t) = \begin{cases} 1 \text{ on } (j-1)\Delta t, j\Delta t \\ 0 \text{ elsewhere} \end{cases} \\ Y_{dh}(t) = \sum_{j=1}^{N_t} Y_{dh}^j X_j(t) & \text{and } N_t \Delta t = T \end{cases}$$

We have chosen $\mathcal{a}^h_{ad}$:

(2.6)
$$\mathcal{a}^h_{ad} = \{\sigma_h \text{ s.t. } \sigma_h(x) = \sigma_h^j = \text{cst. for }](j-1)h, jh[\text{ , } j=1,\dots,N \text{ and}$$
$$\mathrm{var}\sigma_h = \sum_{j=1}^{N-1} |\sigma_h^{j+1} - \sigma_h^j| \le M\}$$

We have chosen

$$k\Delta t = 2h \text{ with } k \in \mathbb{N}$$

so that the response $Y_h(\sigma_h)$ of an impedance distribution $\sigma_h \in \mathcal{a}^h_{ad}$ to the pulse g_h can be expressed by :

$$Y_h(\sigma_h) = \sum_{j=1}^{N_t} Y_h^j X_j(t)$$

Then the optimization problem (1.18) is approximated by :

(2.7) $$\begin{cases} \text{To find } \tilde{\sigma}_h \in \mathcal{A}^h_{ad} \text{ that minimizes} \\ J(\sigma_h) = \sum_{j=1}^{N_t} |Y^j_h(\sigma_h) - Y^j_{dh}|^2 = \|Y^j_h - Y^j_{dh}\|^2_{N_t} \end{cases}$$

Such an approximation is justified by the following lemma :

Lemma 1 :

With hypothesis (2.3) there exists two constants C_1 and C_2 such that :

$$\forall h \text{ and } \forall \sigma \in \mathcal{A}_{ad} \ \exists \ \sigma_h \in \mathcal{A}^h_{ad} \text{ such that}$$

$$\|\sigma - \sigma_h\|_{L^2} \leq C_1 h$$

$$\|Y(\sigma) - Y(\sigma_h)\|_{L^2} \leq C_2 h$$

The following theorem is a direct consequence of the previous and next lemmas.

Theorem :

With the hypothesis (2.3) and for a given h, if the noise b(t) is sufficiently small the problem (2.7) has unique solution such that :

$$\|\tilde{\sigma} - \tilde{\sigma}_h\|_{L^2(]0,\frac{T}{2}[)} \leq C(h + \|b(t)\|_{L^2(]0,T[)})$$

Lemma 2 :

We denote by $\mathcal{Y}$ the subset of R^{N_t} generated by $Y_h(\sigma_h)$ when σ_h moves in $\mathcal{A}^h_{ad}$.

There exists a neighbourhood $\mathcal{V}$ of the set $\mathcal{Y}$ in R^{N_t} and a constant $C_3(\mathcal{V})$ such that for every observed seismogram $Y_{dh} \in \mathcal{V}$ and every $\hat{\sigma} \in \mathcal{A}^h_{ad}$, the problem (2.7) has a unique solution σ_h depending on Y_{dh} in a (lipschitz) continuous way :

$$\|\hat{\sigma} - \tilde{\sigma}_h\|_N \leq C_3(\mathcal{V}) \| Y_{dh} - Y_h(\hat{\sigma}) \|_{N_t}$$

2.4 - Computation of the solution of the approximated inverse problem

The problem (2.7) has been solved by a gradient method. The following difficulties have appeared :

1 - The mapping $\sigma \to Y(\sigma)$ is not explicite. The gradient has been computed using adjoint equations which requires to solve twice the wave equation for a step in the monodimensional optimization of the gradient algorithm.

2 - The size of the problem is very large :

- many gradient iterations (~ 600) are required because the cost function $J(\sigma_h)$ is very flat and N (dimension of the vectorial space containing $\mathcal{A}^h_{ad}$) is large (~ 300).

- the wave equation must be solved very precisely (to obtain a descent direction for the flat function $J(\sigma_h)$) and in a large time x space domain (600 × 300).

The following tools have been used to overcome these difficulties :

- the method of characteristics which provides a very quick and exact way to solve the wave equation (1.13), (1.14), (1.15) when $\sigma \in \mathcal{A}^h_{ad}$.

- an efficient conjugate gradient algorithm to solve the optimization problem.

3 - The constraint $\mathrm{var}\,\sigma_h \leq M$ which appears in the definition of $\mathcal{A}^h_{ad}$ is not easy to handle. It has been introduced by adding a penalization term in the cost function $J(\sigma_h)$. Then the cost function becomes non differentiable (because $\sigma \to \mathrm{var}\,\sigma$ is not differentiable) and a special optimization algorithm must be used (LEMARECHAL [10]).

These tools have considerably reduced the computing time of our first attempts to solve this problem and now, the inversion of one usual seismogram (T = 2.4 sec., N = 300) requires about 3 minutes for a CDC 7600 computer.

3 - NUMERICAL RESULTS

In a first serie of runs using simulated datas, we test the ability of the method to handle noise corrupted seismograms ; then in a second serie of runs we apply our method to actual field recordings. The main characteristics of these runs are :

	Runs I	Runs II	Runs III
Nature of the data	Simulated seismogram	Actual field recordings	Actual field recordings
Number of runs	4	80	1
Seismogram length	2.4 sec.	6 sec. (including a 2.8 sec. water layer)	5 sec.
Layer thickness	4 msec.	4 msec.	5 msec.
Number of unknown layers	300	400	500
Difficulties	Noise level from 0 to 50 % but no model error	Noise corrupted seismograms and model error	Noise corrupted seismogram and model error greater than for runs II

Figure 2

3.1 - Inversion of simulated data

We give us an actual impedance distribution $\hat{\sigma}$ (in full line on the figures) recorded in the Sahara. Using $\hat{\sigma}$ and with the given pulse g we can compute the response $Y(\hat{\sigma})$; the seismogram Y_d simulating the field recording is then obtained by adding a noise having the same spectrum as the signal. Starting, as an initialization from a constant impedance distribution, the algorithm gives, after about 600 conjugate gradient iterations, the impedance distribution $\tilde{\sigma}$ (in dotted line on the figures).

As can be seen on figure 3, the efficiency of the algorithm is very good in the case of noiseless recordings. Only the mean value of $\hat{\sigma}$ has not been recovered when a zero mean pulse g is used, which is understandable.

We next show on figure 5 the results of two runs (with the zero mean pulse) starting from a seismogram Y_d having an exceptionnally strong noise level of 50 % (cf. fig. 4).

The top distribution $\tilde{\sigma}$ of figure 6 has been obtained without any bound VM on the total variation $\mathrm{Var}\sigma$ of the seeked impedance, and the result is not good : for instance the relative amplitudes of the peaks are not well recovered.

The bottom distribution $\tilde{\sigma}$ of figure 5 has been obtained using a bound VM = 28000 on the seeked impedance distribution (the total variation of the "exact" distribution $\hat{\sigma}$ is $\mathrm{Var}\,\hat{\sigma}$ = 25000), and the result is much improved. However, some peaks are added, especially as the depth grows, and more generally the greater the depth the worse the results. In particular the results are not very significant in the last quarter.

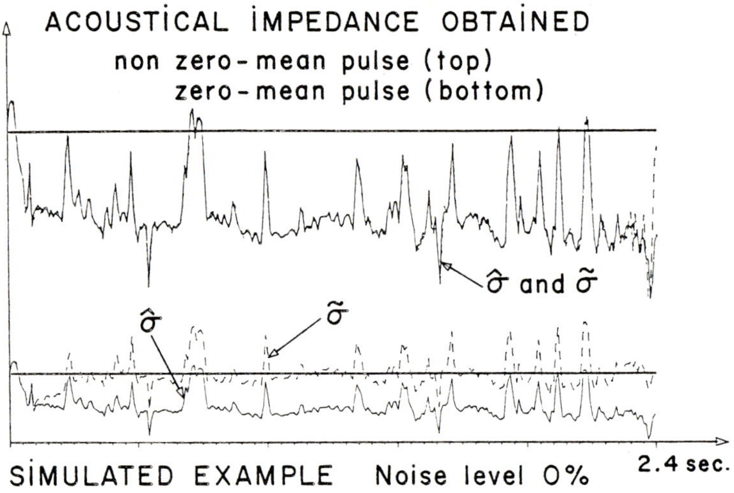

Figure 3

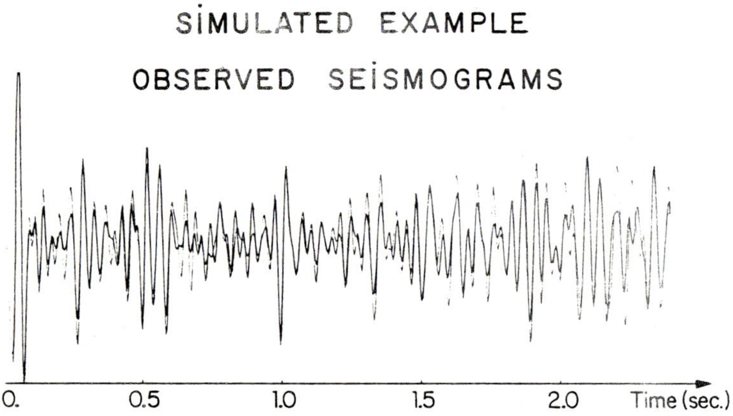

Figure 4

Simulated seismograms Y_d with a noise level of 0 % (thin line) and 50 % (thick line)

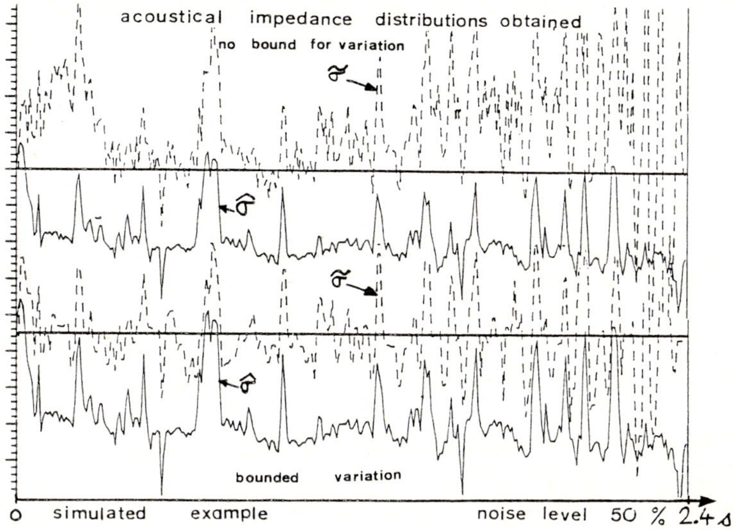

Figure 5

Acoustical impedance distribution obtained from a 50 % noise corrupted seismogram with (bottom) and without (top) bound on the total variation of the impedance.

As a conclusion for the runs on simulated datas, the test on the stability of the method with respect to the noise has been successfully passed, and the method gives good results, at least in the first half, up to 50 % noise.

3.2 - Inversion of field recordings

We have considered two examples (runs II and III) of increasing difficulties and interest for exploration.

3.2.1 - Runs II

The (simplest) wave propagation model (1.13), (1.14), (1.15) which has been used for this study is only valid when the following restrictive hypothesis are satisfied :

- horizontally stratified medium,
- plane waves propagating at normal incidence,
- no damping of the wave when it propagates in a homogeneous medium.

So we have chosen an example which verifies as much as possible these hypotheses the recordings have been obtained from offshore seismics with a very deep sea (about 2500 m.). At these depths the emitted spherical wave (point source) has nearly degenerated into a plane wave. In order to avoid useless computation in a medium which is homogeneous and known, we have eliminated in our model the greatest part of the water layer (1.4 sec. travel time). This is easily done by an adequate modification of the surface boundary condition (1.15).

We have inverted successively the 80 seismograms of the seismic section of fig. 6 (the sea layer is not represented). Each seismogram is the recording obtained when the source and the receiver are located at the current point (y, 0) of the surface where y is the lateral coordinate. So that the inversion of the seismic section gives a display of the ground in the (y, x) plane. This display (fig. 7) shows very accurate reflectors which gives to the geologist an information much easier to interpretate than the original data (fig. 6).

We can notice that, for each recorded seismogram Y_d of the seismic section, the computed seismogram $Y(\tilde{\sigma})$ is the same as Y_d (fig. 8).

3.2.2 - Run III

The previous example lacks some comparison of the computed impedance distribution with a recorded impedance (sonic log). The reason of this lack is that sonic logs are not available when the sea is so deep. So we have considered an other example where the impedance has been measured in a well (fig. 9). The seismic section (fig. 10) of this example shows that :

- the medium is nearly horizontally layered
- the water layer is very small (about 0.2 sec. travel time). Hence the recorded seismograms are far from the response to a plane wave excitation.

Two ways are often used to reconstruct plane waves seismograms from the actual recordings :

- the stack with normal moveout corrections which gives, for each value of the lateral coordinate y, an estimation of the response to a plane wave excitation for little depths.
- the stack without normal moveout corrections (slant stack) which gives an estimation of the response to a plane wave excitation for large depths.

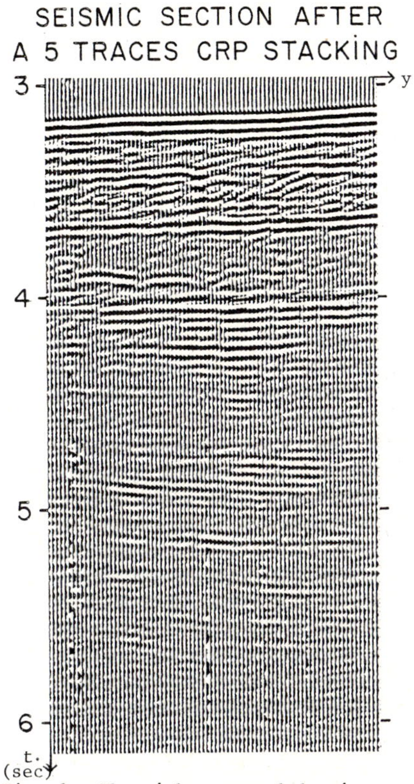

Fig. 6 : The eighty recorded seismograms Y_d

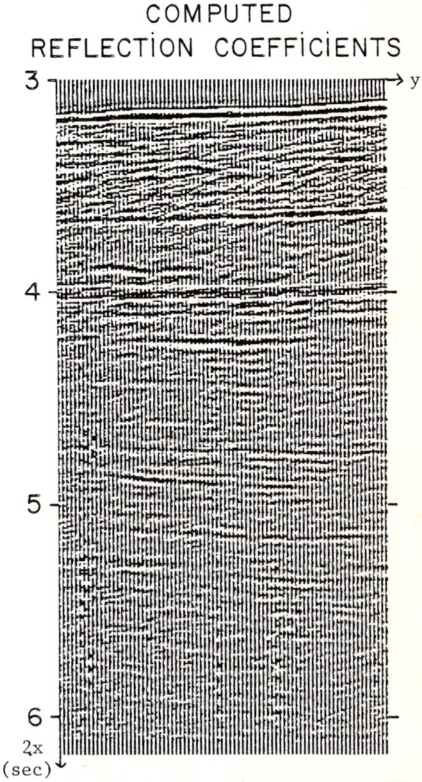

Fig. 7 : The eighty reflection coefficient distributions r computed from the data of figure 6. (The maximum admissible variation M = 14000).

Figure 8

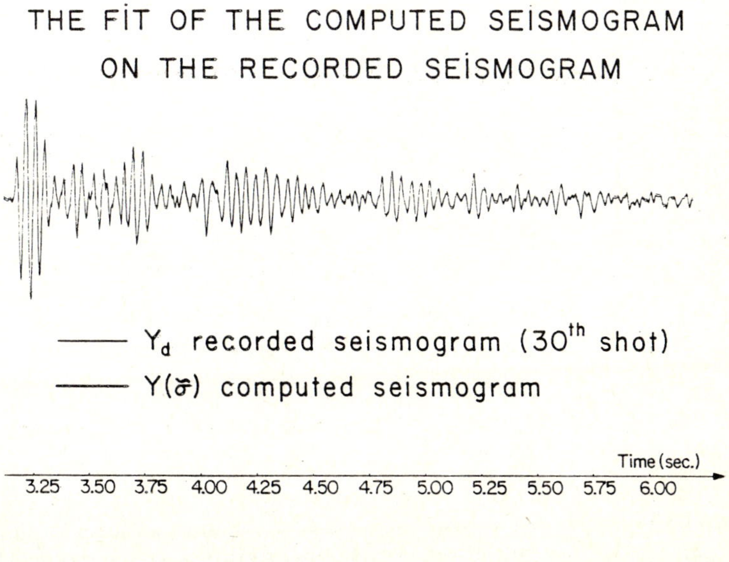

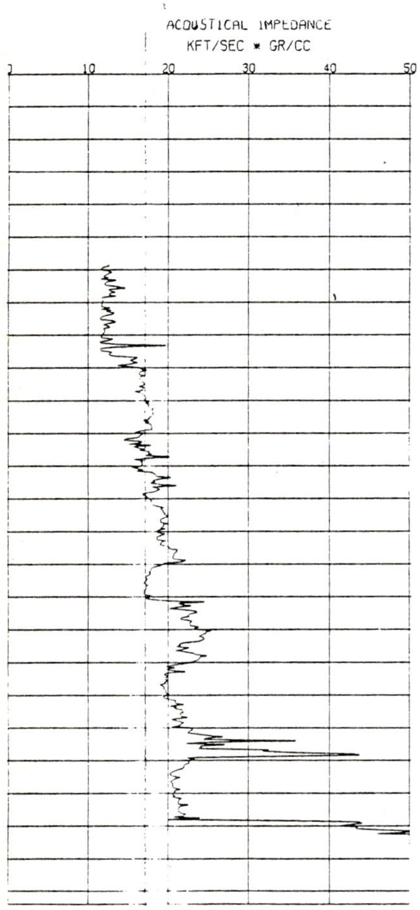

Figure 9 : acoustical impedance distribution recorded in the well.

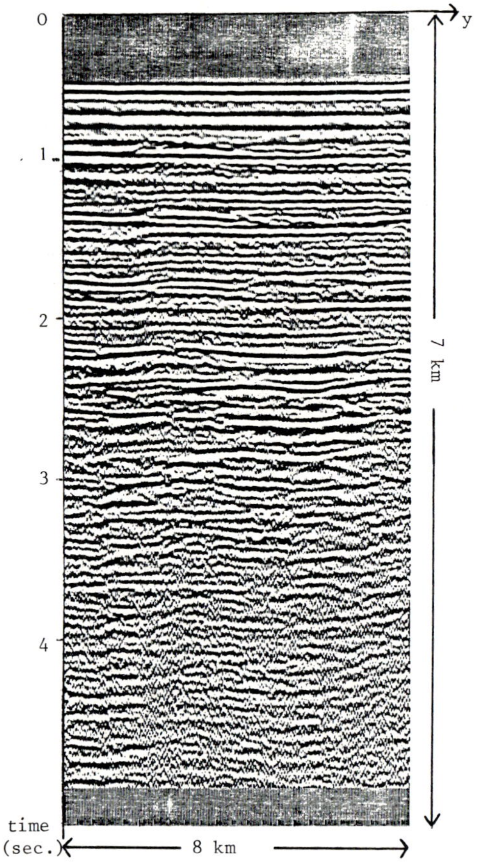

Figure 10 : seismic section for run III

Displaying the seismograms obtained by a stacking procedure for each value of the lateral coordinate y gives a seismic section : fig. 10 shows the seismic section obtained by the stack with normal moveout correction.

Let $Y_{d_1}(t)$ and $Y_{d_2}(t)$ denote respectively the seismograms extracted from the two stacked sections and corresponding to the location of the well. We define a new cost function (instead of the one appearing in (1.18)) :

$$J(\sigma) = \int_0^T \alpha(t)(Y(\sigma, t) - Y_{d_1}(t))^2 dt + \int_0^T (1 - \alpha(t))(Y(\sigma, t) - Y_{d_2}(t))^2 dt$$

where $\alpha(t)$ is the function :

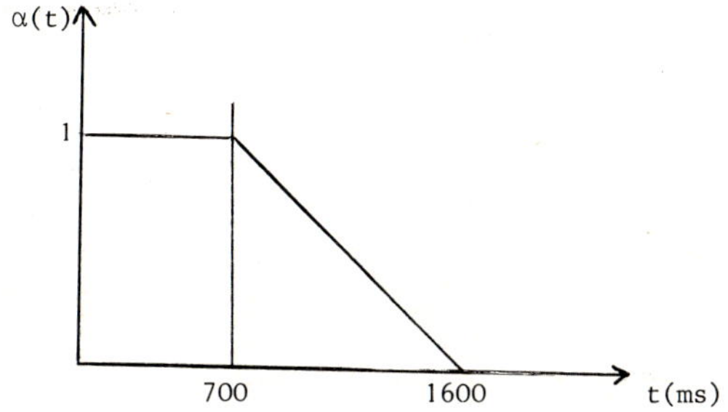

The comparison of the computed and recorded impedance distributions has been made by filtering these two functions by the pulse g(t) since, as we have seen in the previous examples, we can only obtain the components of the acoustical impedance which are in the bandwidth of the pulse g(t).

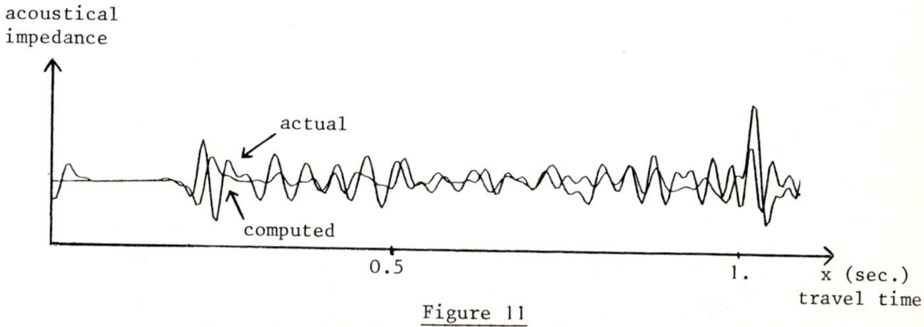

Figure 11

Comparison of the computed impedance distribution to the actual impedance distribution (recorded in the well).

We can see (fig. 11) that there is clearly some correlation between the computed and recorded impedance. As a matter of fact a good correlation is very rare for this kind of problem and the difficulties of such a comparison has been examined in a considerable amount of studies (see for instance [9], [13]). Briefly the 3 following problems still remain to solve :

- is the 1D wave equation an accurate enough model to simulate a plane wave propagating vertically in a horizontally layered medium ?
- have we a sufficient knowledge about the pulse g(t) ?
- can we reconstruct the response to a plane wave from the original experiment ?

CONCLUSION

The optimal control formulation and an adequate choice of the set of admissible parameters has lead to a method which gives a stable result when noise corrupts the data.

This has been checked successfully on simulated data. For inversion of field recordings, the interest of the method relies upon a positive answer to the 3 geophysical and technical problems previously mentionned.

Though a solution to these problems can be expected, our work at present time is studying 2D models as well for the interest of the direct problem as for a future solution to the 2D inverse problem.

REFERENCES

[1] A. BAMBERGER, G. CHAVENT, P. LAILLY - "Etude mathématique et numérique d'un problème inverse pour l'équation des ondes à une dimension" - Rapport LABORIA n° 226 - IRIA, BP 105, 78150 LE CHESNAY - Rapport Centre de Mathématiques Appliquées n° 14 - Ecole Polytechnique, 91128 PALAISEAU Cedex, France.

[2] A. BAMBERGER, G. CHAVENT, P. LAILLY - "About the stability of the inverse problem in the 1-D wave equation - Application to the interpretation of seismic profiles" - Journal of Applied Mathematics and Optimization - n° 5 - p. 1-47 - 1979.

[3] G. CHAVENT - "Deux résultats sur le problème inverse dans les équations aux dérivées partielles du deuxième ordre en t et sur l'unicité de la solution du problème inverse de la diffusion" - C.R. Acad. SC, Paris - 270 - 25-28 janvier 1970.

[4] D.M. DETCHMENDY - "Attenuation and inverse problems in wave propagation") (Swieeco 1968) - Esso production Research Company - Houston - Texas.

[5] M.L. GERVER - "The inverse problem for the vibrating string equation" - Izv. Acad. Sci. URSS - Physic Solid Earth - 1970-71.

[6] B. GJEVICK, A. NILSEN, J. HOYEN - "An attempt at the inversion of the reflection data " - Geophysical Propecting, 24 - 1976 - p. 492-505.

[7] B. GOPINATH, M.M. SONDHI - "Inversion of the telegraph equation and the synthesis of non uniform lines" - Proc. IEEE, 59 - 1971, 3.

[8] G. KUNETZ - "Quelques exemples d'analyse d'enregistrements sismiques" - Geophysical Prospecting, II - (1963), 4.

[9] M. LAVERGNE, C. WILLIM - "Inversion of seismograms and pseudo velocity logs" - Geophysical Prospecting - V. 25 - p. 231-250.

[10] C. LEMARECHAL - "Non differentiable optimization subgradient and ε-subgradient methods" - Lecture Notes in Econ. and Math. Systems - Vol. 117 - "Optimization and Operational Research" - Springer - Berlin - 1976.

[11] J.L. LIONS - "Contrôle optimal de systèmes gouvernés par des équations aux dérivées partielles" - Dunod - 1968.

[12] F. MURAT - "Contres exemples pour divers problèmes où le contrôle intervient dans les coefficients" - Annali Mat. ed applicata 4 - Vol. 112 - p. 49-68 - 1977.

[13] M. SCHOENBERGER, P.K. LEVIN - "The effect of subsurface sampling on 1-D synthetic seismograms" - Paper presented at the 48th Annual International SEG Meeting - November 1, 1978 - San Francisco - U.S.A.

[14] S. SPAGNOLO - "Convergence in energy for elliptic operators" - in Numerical solution of partial differential equations - III, Synspade 1975 - B. Hubbard, ed., Academic Press - New York - 1976.

COMPUTING METHODS IN APPLIED SCIENCES AND ENGINEERING
R. Glowinski, J.L. Lions (editors)
North-Holland Publishing Company

INVERSE IMAGING METHODS IN EXPLORATION SEISMOLOGY

Bjorn Engquist
Mathematics Department
University of California
Los Angeles, California 90024

1. Introduction. The purpose of seismic data processing is to determine geophysical properties of the earth from measurements of seismic waves at the surface. This process consists of many steps. In today's reflection seismology these steps include filtering, deconvolution, geometrical correction, velocity estimation, stacking and migration. See [1], [2] for general presentations of these methods. We will concentrate on the mathematical and numerical aspects of the last step - migration. In the migration process the location of reflecting and diffracting interfaces in the earth is determined.

The original data consists of measured quantities such as pressure or particle acceleration at the surface of the earth. The seismic wave field is initiated from a source and scattered by objects in the earth. We can assume that we know it at the surface. The basic step in wave equation migration is calculating the wave field in the interior of the earth from both this surface data and a rough estimate of the wave velocities. The calculated wave field is then used to map the interior scattering interfaces.

The pioneering work and most of the development of wave equation migration was done by Professor Jon Claerbout at Stanford University [3], [4], [5], [6]. Wave equation migration is now a standard technique in the numerical processing of data in reflection seismology exploration. The method is very cost effective which is of importance since the amount of data that needs processing is enormous.

For the calculation of the interior wave field we need to solve an acoustic or elastic wave equation with a space coordinate (down into the earth) as evolution direction. This is not a well posed problem as stated and approximations must be used.

In Section 3 we analyze these approximative wave equations and their numerical solutions. Our work in [7], [8], [9] is described in which a sequence of well-posed approximations is studied.

In Section 2 we shall briefly sketch the processing which is done before the migration step. We shall also discuss the imaging principles which are the foundations for migration.

2. Seismic data processing and inverse imaging principles. In order to have a background for our analysis we shall first consider a seismic experiment.

We let the source be the detonation of explosives which are located close to the surface of the earth. A pulse which mainly consists of P-waves travels down into the earth and passes through different layers of rocks and possibly through sea water. Parts of the downgoing wave are reflected back to the surface. Geophones measure, for a few seconds, the movements of the earth at a number of surface locations around the source.

Measured data from many such experiments will be processed in order to simulate the result from an idealized set of experiments. Below are some of the common steps in that process.

Special properties of the geophones must be compensated for. Filtering and stacking (superposition) of data reduce the noise level. Estimation of the wave form and then deconvolution compress the wave front to a sharp pulse. Geometrical compensation is made for the attenuation of the waves with time.

From now on we shall simplify the discussion to two space dimensions, one coordinate (x) along the surface and the other depth (z). This is no essential restriction from an analytic point of view and in fact most processing is today still done in two space dimension.

If processed data from different sources corresponding to a geophone at x are added together and plotted as a function of x and time (t) the plot will resemble a radar picture. The reflections from deeper interfaces arrive at later times. (In practice x is often chosen as the midpoint between the source and geophone, CDP-stack.)

This picture gives, however, a distortion of the location of the reflecting layers since the wave velocity is variable. Furthermore reflections from objects which are not exactly below the geophone will also be recorded. A point scatterer will look like a hyperbola in the x-t plot (see Figure 1) even in constant velocity medium.

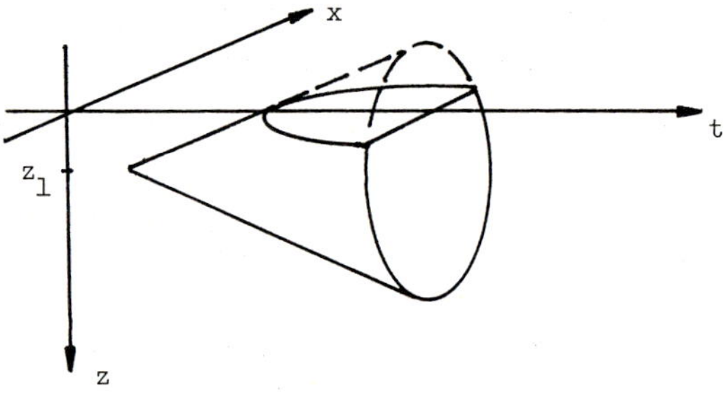

Figure 1: Scattered conical wave front set and its intersection with $z = 0$.

We shall use a wave equation to determine the field of upgoing waves between the reflector and the surface. Our principle involves displaying the wave field at the time of reflection. In [1] (p. 230) the imaging principle is described as "reflectors exist in the earth where the onset of the downgoing wave is time coincident with the upgoing wave." The imaging principle can also be thought of in terms of holography for a monocromatic source. The up and downgoing waves will be in phase at the reflector.

After a superposition of data from different experiments we get an idealized experiment with a downgoing plane wave pulse at the surface at $t = 0$. If we assume that P-waves dominate and the P-wave velocity is known we will also know where the downgoing wave front is located at different times $t > 0$.

The upgoing wave field can be determined by a wave equation using the data at the surface $z = 0$.

In Figure 1 the migration process consists of tracing the hyperbola at $z = 0$ back to $z = z_1$ where it is displayed as a point. This focusing effect can also be used to determine the P-wave velocity.

In industry practice often other superpositions are made but there is always the need for solving wave equations with data at the space time surface $z = 0$.

3. <u>One way wave equations and their numerical approximation</u>: We want to solve a scalar wave equation

$$\left(\frac{\partial^2}{\partial t^2} - v^2\left(\frac{\partial^2}{\partial x^2} + \frac{\partial^2}{\partial z^2}\right)\right)u(x,z,t) = 0, \qquad z > 0 \tag{1}$$

for which we begin by assuming constant velocity. We want to solve it as an initial value problem with

$$u(x,0,t) = u_0(x,t) \tag{2}$$

as initial data.

This problem is not well posed. The solution may contain infinitely fast growing modes. We are, however, only interested in upcoming traveling waves.

Fourier transform (1) with respect to x and t denote the duals ω and τ respectively

$$\left(-\tau^2 - v^2\left(-\omega^2 + \frac{d^2}{dz^2}\right)\right)\hat{u}(\omega,z,\tau) = 0 \tag{3}$$

The wave operator can now be factored for $\tau^2 > v^2\omega^2$

$$\left(\frac{d}{dz} + i\sqrt{(\tau/v)^2 - \omega^2}\right)\left(\frac{d}{dz} - i\sqrt{(\tau/v)^2 - \omega^2}\right)\hat{u} = 0.$$

The equation for a superposition of traveling upcoming waves of the form

$$u(x,z,t) = \iint e^{i(\omega x+\sqrt{(\tau/v)^2-\omega^2}\, z+\tau t)}\, \hat{u}$$

is then (for $\tau^2 > v^2\omega^2$)

$$\frac{\partial u}{\partial z} - \iint e^{i(\omega x+\tau t)}\, i\sqrt{(\tau/v)^2 - \omega^2}\, \hat{u}\, d\tau\, d\omega = 0. \tag{5}$$

Equation (5) corresponds to a differential equation if the square root is replaced by a rational approximation. Assuming $\tau^2 \gg v^2\omega^2$ which means waves traveling essentially in the z-direction we have

$$\sqrt{(\tau/v^2) - \omega^2} = \frac{\tau}{v}\sqrt{1-(v\omega/\tau)^2} \approx \frac{\tau}{v} - \frac{v\omega^2}{2\tau} .$$

The differential equation corresponding to the approximate version of (5) is

$$\left(\frac{\partial^2}{\partial z\partial t} - \frac{1}{v}\frac{\partial^2}{\partial t^2} + \frac{v}{2}\frac{\partial^2}{\partial x^2}\right) u = 0.$$

We can get rid of the $\partial^2/\partial t^2$-term via a coordinate transformation

$$\begin{aligned} x' &= x \\ z' &= z \\ t' &= t + z/v \\ \frac{\partial}{\partial z} &\to \frac{\partial}{\partial z'} + \frac{1}{v}\frac{\partial}{\partial t'} . \end{aligned} \tag{7}$$

Dropping the primes the transformed equation becomes

$$\left(\frac{\partial^2}{\partial z\partial t} + \frac{v}{2}\frac{\partial^2}{\partial x^2}\right) u = 0. \tag{8}$$

This is often called the parabolic approximation of the wave equation. The initial value problem approximating (1), (2) will have the form

$$w_{zt} + \frac{v}{2} w_{xx} = 0 \tag{9}$$

$$w(x,0,t) = u_0(x,t) . \tag{10}$$

Variable coefficients can be handled using the algebra of pseudo-differential operators [10], [11]. It is, however, often enough just to replace the constant v in (9) by $v(x,z)$.

If the earth contains strongly dipping layers the assumption $\tau^2 \gg v^2\omega^2$ is is not appropriate. A better approximation of the square root is needed. The Padé formula

$$\sqrt{1+\alpha} = 1 + \frac{\alpha}{2+\alpha/2} + O(|\alpha|^3) \tag{11}$$

gives the one way wave equation

$$w_{ztt} - \frac{v^2}{4} w_{zxx} + \frac{v}{2} w_{xxt} = 0. \tag{12}$$

In [9] and [11] we analyze different approximations of this type.

The pure initial value problems for (9) and (12) are well posed in L_2. This is also true for equations produced via higher order Padé-formulas.

Surprisingly a higher order Taylor expansion (13) of the square root gives rise to an ill posed differential equation (14)

$$\sqrt{1+\alpha} = 1 + \frac{\alpha}{2} - \frac{\alpha^2}{8} + O(|\alpha|^3) \tag{13}$$

$$w_{zttt} + \frac{v^3}{8} w_{xxx} + \frac{v}{2} w_{ttxx} = 0. \tag{14}$$

Corresponding formulas for the elastic wave equations are given in [9]. One way elastic wave equations will often be only conditionally well posed depending on the ratio of the Lamé parameters.

We use finite difference approximations of the one way wave equations for their numerical solutions. The function $w(x,z,t)$ is approximated by w^n_{jk} on the regular mesh (x_j, z^n, t_k). The following second order approximations of (9) are stable [7].

$$D^z_+ D^t_+ w^n_{jk} + \frac{v}{8} D^x_+ D^x_-(w^n_{jk} + w^{n+1}_{jk} + w^n_{j,k+1} + w^{n+1}_{j,k+1}) = 0 \tag{15}$$

$$D^z_+ D^t_+ w^n_{jk} + \frac{v}{4} D^x_+ d^x_-(w^n_{j,k+1} + w^{n+1}_{jk}) = 0. \tag{16}$$

The difference operators D_+ and D_- denote the forward and backward divided differencing respectively. The superscript denotes the differencing direction.

The implicit formula (15) is unconditionally stable and the explicit (16) is stable for $v\Delta t\, \Delta z/\Delta x^2 \leq 1$. In [7] we show that fully explicit difference approximations of (9) cannot have a higher order of approximation than two in the time direction.

We have not restricted the domain of computation in the x- and t-directions in our presentation. This must of course be done in practical calculations. In [8] we give side boundary conditions for wave equation migration which do not distort the solution close to the boundaries. An example of such a boundary condition for an x-boundary is

$$w_z + a w_x + \frac{b}{v} w_t = 0. \tag{17}$$

In the computations the derivatives are replaced by centered differences, a and b are constants.

References

1. Claerbout, J. F.: Fundamentals of Geophysical Data Processing, McGraw-Hill, 1976.

2. Waters, K. H.: Reflection Seismology, John Wiley and Sons, 1978.

3. Claerbout, J. F.: Coarse Grid Calculations of Waves in Inhomogeneous Media with Application to Delineation of Complicated Seismic Structure, Geophysics 35, No. 3 (1970), 407-418.

4. Claerbout, J. F.: Toward a Unified Theory of Reflector Mapping, Geophysics 36, No. 3 (1971), 467-481.

5. Claerbout, J. F. and A. G. Johnson: Extrapolation of Time Dependent Waveforms along Their Path of Propagation, Geophysics. J. R. Astron. Soc., 26, No.1-4 (1971), 285-295.

6. Claerbout, J. F. and S. M. Doherty: Downward Continuation of Moveout Corrected Seismograms, Geophysics 37, No. 5 (1972), 741-768.

7. Engquist, B.: High Order Difference Approximations to a Simplified Wave Equation, Uppsala Univ. Comp. Sciences Rep. (1975).

8. Clayton, R. and B. Engquist: Absorbing Side Boundary Conditions for Wave Equation Migration, to appear in Geophysics.

9. Clayton, R. and B. Engquist: Paraxial Elastic Wave Equations, to appear.

10. Nierenberg, L.: Lectures on Linear Partial Differential Equations, C.B.M.S. Regional Conference Ser. in Math., No. 17, A.M.S., Providence, Rhode Island, 1973.

11. Engquist, B. and A. Majda: Absorbing Boundary Conditions for the Numerical Simulation of Waves, Math. Comp., 31, No. 139 (1977) 629-651.

This work was supported by NSF Grant No. MCS-79-02735 and by the sponsors of the Stanford Exploration Project.

COMPUTING METHODS IN APPLIED SCIENCES AND ENGINEERING
R. Glowinski, J.L. Lions (editors)
North-Holland Publishing Company

TWO-POINT RAY TRACING IN HETEROGENEOUS MEDIA AND THE INVERSION OF TRAVEL TIME DATA

V. Pereyra
Escuela de Computacion
Facultad de Ciencias
Universidad Central de Venezuela
Apartado 59002, Caracas 104

1. Introduction

A general algorithm and its corresponding computer implementation has been devised to solve both direct and inverse problems involving ray tracing of elastic waves in heterogeneous, isotropic, piecewise smooth, two or three dimensional media.

The practical applications require the tracing of rays between a source (earthquake, explosion) and a receiver (seismometer, geophone). The velocity distributions we admit can be arbitrarily varying both in depth and laterally. Thus the resulting nonlinear two-point boundary value problem must be solved numerically. For this purpose we use a general adaptive finite difference code which can compute two-point rays to any desired accuracy. This code has provisions which facilitate the solution of the inverse problem.

Given arrival time data, we consider three kinds of inverse problems: calculation of velocity distributions, relocation of hypocenters, and determination of reflectors.

Some numerical examples using artificial data are presented.

2. Two-Point three-dimensional ray-tracing in heterogeneous, piecewise smooth media.

We are interested in studying the propagation of elastic waves in an heterogeneous, isotropic, three-dimensional medium. We assume that this medium is described by giving the velocity of propagation of either pressure (P) or shear (S) waves.

The geometry of the medium must also be given, specifying any interfaces across which the velocity is discontinuous. We allow also for switching from P to S waves, or viceversa, when passing across interfaces.

We assume finally that the phenomena under study is such that ray theory approximation is applicable, and seek to compute rays between a source and a receiver.

Rays can be described as stationary paths of certain functionals. The corresponding Euler-Lagránge equations for this calculus of variations problem are

$$\frac{d}{ds}\left[u(\underset{\sim}{\eta})\frac{d}{ds}\underset{\sim}{\eta}\right] - \nabla u = 0 , \tag{2.1}$$

where $\underset{\sim}{\eta} = (x,y,z)$ represents the ray and $u(\underset{\sim}{\eta}) \equiv 1/v(\underset{\sim}{\eta})$, with $v(\underset{\sim}{\eta})$ the velocity of propagation in the medium. We discuss first the case in which $v(\underset{\sim}{\eta})$ is smooth, say in C^4.

We add the constraint

$$\dot{x}^2 + \dot{y}^2 + \dot{z}^2 = 1 \tag{2.2}$$

which ensures that the independent variable s is arc-length along the ray.

The two-point ray tracing problem consists of solving these equations subject to the boundary conditions

$$\underset{\sim}{\eta}(0) = P_S , \quad \underset{\sim}{\eta}(S) = P_R ,$$

where P_S, P_R are the given Cartesian coordinates of the source and receiver respectively, and S is the total length of the arc of ray joining P_S and P_R. Obviously S is also unknown.

In the past, this problem has been solved mainly by shooting techniques (see [3,6,7]), taking advantage of available initial value problem software. Some researchers [8,16] have also looked into global methods based on finite differences.

However, it has not been until more recent times that software implementing versatile finite difference methods for nonlinear two-point boundary value problems have become available (see [12,14]). Is this software that has allowed us [15,17] to solve the two-point ray tracing problem routinarely and in a variety of situations. It will also allow us to attack the inverse problem in a more precise form than it has been done in the past, as we will show in Sections 3 and 4.

Our finite difference solver PASVA3 requires that the differential system be of first order. A few manipulations permit to transform (2.1) to the first order form

$$\dot{\omega}_i = \omega_{i+1}$$

(2.3) $$\dot{\omega}_{i+1} = v(-G(\underset{\sim}{\omega})\omega_{2i} + u_{\omega_{2i-1}})$$

$$i=1,2,3 \ ,$$

where $\underset{\sim}{\omega} = (x,\dot{x},y,\dot{y},z,\dot{z})$, and

$$G(\underset{\sim}{\omega}) = u_x\omega_2 + u_y\omega_4 + u_z\omega_6 \ .$$

Observe that these equations are very nonlinear as soon as $v(\omega_1,\omega_3,\omega_5)$ is non-constant.

Many other forms are possible, with various advantages and disadvantages. A very lucid and detailed account of this fascinating subject can be found in [3].

Unfortunately, system (2.3) is still not in a form appropriate for PASVA3, which requires a quite definite interval of integration to be given. We see that (2.3) is a free boundary problem, since the right end point S (in the independent variable s) is unknown. This is easily fixed by introducing the change of independent variable $s \to t \equiv s/S$, a new dependent variable $\omega_8 \equiv S$, and the artificial differential equation $S' = 0$. Now $'$ will denote differentiation with respect to t. The changes in (2.3) only involve multiplication of the right hand side of the equations by S, and the new interval of integration is simply $[0,1]$.

Finally, since the quantity we are really interested in is the travel time

(2.4) $$T = S\int_0^1 u\, dt \ ,$$

we introduce still one more variable $\omega_7(t)$, for the partial travel time, and the corresponding differential form of (2.4)

(2.5) $$\omega_7' = Su \ ,$$

together with the two extra boundary conditions $\omega_7(0) = 0$ and $\omega_2^2(0) + \omega_4^2(0) + \omega_6^2(0) = 1$. This latest condition is all what is necessary in order to enforce the constraint (2.2), as is shown in [15].

In the case of ray tracing in two space dimensions, further simplifications allow us to write the corresponding boundary value problem as

$$\begin{aligned} \omega_1' &= \omega_4 \cos \omega_3 \\ \omega_2' &= \omega_4 \sin \omega_3 \\ \omega_3' &= \omega_4\, v(\omega_1,\omega_2) \left[u_z \cos \omega_3 - u_x \sin \omega_3\right] \\ \omega_4' &= 0 \\ \omega_5' &= \omega_4\, u \end{aligned} \tag{2.6}$$

with the boundary conditions

$$\omega_1(0) = x_S, \quad \omega_2(0) = z_S, \quad \omega_5(0) = 0,$$
$$\omega_1(1) = x_R, \quad \omega_2(1) = z_R \ .$$

In order to consider the piecewise continuous case, let $\phi(x,y,z) = 0$ be a given interface. We will use Snell's law to derive the condition that the ray must satisfy when meeting this interface [3,11].

In general we will have two possibilities: either the ray is transmitted from one side to the other of the interface, or it is reflected by it.

Rays coming to the interface at a critical angle can be difracted and travel along the interface with the velocity of the fastest medium. We shall see that our formulation takes this fact into account automatically.

Let $\underset{\sim}{\eta}^*$ be the incident point, let $N(\eta^*)$ be the unit normal vector to the interface $\phi(\eta^*) = 0$, pointing from $\underset{\sim}{\eta}^*$ towards the first medium S_I (that contains the incident ray). Let $I(\underset{\sim}{\eta}^*)$ be the unit tangent vector to the ray, pointing from $\underset{\sim}{\eta}^*$ towards S_I. Finally, let α_I be the angle between N and I, let α_R be the angle between the reflected ray and N, and let α_T be the angle between the transmitted ray and N.

Finally, let n be the index of refraction at $\underset{\sim}{\eta}^*$: $n = v_T(\underset{\sim}{\eta}^*)/v_I(\underset{\sim}{\eta}^*)$, where the subindeces I,T indicate velocities in the first and second medium respectively.

Since the reflected or transmitted ray must lie in the plane determined by I and N, then it follows from the law of reflection and Snell's law that:

$$R = -I + 2 \cos \alpha_R N , \tag{2.8}$$

and

$$T = -n I + (n \cos \alpha_I + \cos \alpha_T) N . \tag{2.9}$$

These are actually three conditions each for the direction of the reflected and transmitted ray. But because of our equations, we already know that the tangent vector to the ray $(\dot{x},\dot{y},\dot{z})$ will be normalized, so it is enough to give two of these conditions in order to determine the proper exiting direction for the ray.

After some manipulation, these two pairs of conditions become:

$$\begin{aligned} \cos \alpha_I &= \cos \alpha_R \\ \langle I,R\rangle &= \cos 2 \alpha_I , \end{aligned} \tag{2.10}$$

and

$$\begin{aligned} -\{1-n^2\sin^2 \alpha_I\}^{\frac{1}{2}} &= \cos \alpha_T \\ \langle I,T\rangle &= -n \sin^2 \alpha_I + \cos \alpha_R \cos \alpha_I . \end{aligned} \tag{2.11}$$

Translating into our original variables we find that

$$I = -(\dot{x}_I,\dot{y}_I,\dot{z}_I) ,$$

$$N = (\phi_x,\phi_y,\phi_z)/\sqrt{\phi_x^2 + \phi_y^2 + \phi_z^2} ,$$

$$R = T = (\dot{x}_{II},\dot{y}_{II},\dot{z}_{II}) ,$$

$$n = v_{II}/v_I ,$$

where I, II indicate the two mediums and all functions are evaluated at $\tilde{\eta}^*$.

Since we want to use the solver PASVA3 without changes, we have to perform some transformations in order to accomodate this discontinuous problem. We will not spell out the details here, since that has already been done in [15], but we will briefly indicate the idea.

We consider the two smooth segments of the ray between the source and the point $\underset{\sim}{\eta}^*$ at the interface, and between $\underset{\sim}{\eta}^*$ and the receiver. For each segment we consider the ray equations and thus we end up with double the number of differential equations. We have in an obvious manner ten boundary conditions. The remaining six are provided by three continuity conditions that the ray must satisfy at $\underset{\sim}{\eta}^*$, plus the two discontinuity conditions on the ray direction that we discussed above, and finally the condition that $\underset{\sim}{\eta}^*$ be actually on the interface $\phi(\underset{\sim}{\eta}^*) = 0$.

Obviously this process can be repeated for any number of interface contacts, although it may become increasingly inefficient. The PASVA3 linear equation solver can be modified so that only one set of equations is solved for the whole ray, while the discontinuity conditions are worked in explicitly. The only thing one has to make sure is that the jumps occur always at mesh points [10].

3. The finite difference solver PASVA3

In this section we will briefly discuss some aspects of a genera program for solving nonlinear two point boundary value problems of the form

$$\text{(3.1)} \qquad \omega' = f(t,\omega) , \qquad g(\omega(a), \omega(b)) = 0 ,$$

where ω, f, g are smooth vector valued functions, and problem (3.1) is assumed to have an isolated solution $\omega^*(t)$.

PASVA3 [12,14,15] is a FORTRAN implementation of an adaptive finite difference method to solve problem (3.1). It has both the capability of adapting the mesh in a non uniform way and of changing the order of the approximating method by deferred corrections [13]. It produces asymptotic global error estimates and it has been extensively tested in a variety of applications.

Given a mesh $\pi = \{t_i: \ a=t_1 < \ldots < t_n = b\}$, the trapezoidal rule is used to discretize (3.1):

$$\text{(3.2)} \qquad \omega_{i+1} - \omega_i = \frac{h_i}{2}(f_i + f_{i+1}) , \quad i=1,\ldots,n-1, \qquad g(\omega_1,\omega_n) = 0 ,$$

where $h_i = t_{i+1} - t_i$, ω_i is sought to approximate $\omega^*(t_i)$, and $f_i = f(t_i, \omega_i)$.

This system of nonlinear equations is solved by a variant of Newton's method. Thus, the Jacobian matrix

$$f_\omega = \left(\frac{\partial f^i}{\partial \omega_j}\right)_{i,j=1,\ldots,m} ,$$

where f^i is the ith component of the vector function $f(t,\omega)$, is required. From it, a linearized form of (3.2) is constructed and solved by a stable LU factorization process that preserves sparseness by careful alternate pivoting. This linearized form amounts to a discretization of the variational equation associated with (3.1)

$$\begin{aligned} \delta' &= f_\omega(t,\omega)\delta + r(t) \ , \\ g_{\omega(a)}\delta(a) &+ g_{\omega(b)}\delta(b) = \beta \ . \end{aligned} \tag{3.3}$$

It is important to observe that once the Jacobian f_ω has been evaluated on the mesh π, and the matrix associated with the linearized discrete equations has been constructed and LU decomposed, the actual solution of any system with such a matrix amounts to the solution of one upper and one lower block bidiagonal system, which even have some further sparseness in their off-diagonal blocks.

This fact is taken full advantage of in PASVA3, since both the error estimation and the deferred corrections can be performed without recomputing the Jacobians, and therefore they cost very little extra computation. We will see in the following section that this feature is also very useful for computing partial derivatives of the travel time T with respect to parameters.

4. Inversion of travel time data.

Both in seismology and in the reflection method of seismic prospecting one models the interior of the Earth by using travel time data. In seismology the source will usually be a natural event, i.e. an earthquake. In such a case, it will be also of interest sometimes to calculate the position (hypocenter) and origin time of the event. In seismic prospecting the events are man-made explosions, and therefore source and origin time are known.

The modelling can involve both the determination of the velocity structure of a portion of the Earth interior, and that of any special material features that may produce strong discontinuities in the velocity field. These abrupt changes of material will be called "reflectors" in what follows.

We will consider three inverse problems:

(a) the modelling of the velocity of elastic waves, (b) location of hypocenters, and (c) detection of reflectors.

In a given application they may appear individually, by pairs, or all three together, and our program allows for any of these combinations to occur.

Each travel time observation (arrival time in some cases) should be accompanied by the coordinates of a source-receiver pair, and by a description of what kind of a ray may have produced the observation. For simplicity of the exposition, in what follows we would think of the model having none or at most one interface. If there are no interfaces the only possible rays are direct ones, and the only choice is between P and S waves. For long distances we may have reflections on the free surface, which is itself an interface. Thus we would have to indicate the number of reflections and if there is any switching between P and S waves or viceversa.

In the case of an interior interface, like a fault, a synclinal, or a salt dome, there are more possibilities that must be indicated. For more complicated structures or multiple reflected rays this task may become very difficult, and then the ray tracing code may be used to try to match one of the many possible rays with the observation. We will assume in what follows that the matching between rays and observations has been achieved, and that our task is only to compute accurate travel times with the direct ray tracing code. Having these computed travel times we will alter the parameters of the model in an iterative fashion, so that the whole set of observed times is best fitted in the least squares sense.

Therefore, let $\{T_i^0\}$, $i=1,\ldots,\ell$ be a set of observed travel times and let $\{T_i^c\}$ be the corresponding set of times computed with our ray tracing procedure. We recall (see (2.5)) that the travel time is actually one of the dependent variables in the ray tracing equations Although it could have been integrated independently once the ray was obtained, it is much more convenient to include it with the other equations.

Both the velocity $v(\underset{\sim}{\eta})$ and the interfaces $\phi(\underset{\sim}{\eta}) = 0$ will be modelled by a finite number of parameters.

This will be achieved by choosing an appropriate finite dimensional set of approximating functions. The choice of this representation is in itself a very important and delicate problem. Some examples can be found in the literature [1,2] , and we will not dwell here further on this topic, except to point out that splines [18] are probably fairly good candidates in many cases, unless some specific analytic form is indicated.

So, let $\underset{\sim}{\mu}$ be the vector of unknown parameters. Our problem can be readily stated as

$$\text{(4.1)} \qquad \min_{\underset{\sim}{\mu}} \sum_{i=1}^{\ell} (T_i^c(\underset{\sim}{\mu}) - T_i^0)^2$$

subject to

$$\text{(4.2)} \qquad \begin{aligned} &\dot{\omega}^i = f(t,\omega^i,\underset{\sim}{\mu}) \\ &g(\omega^i(0),\omega^i(1);\ \underset{\sim}{\mu}) = 0 , \end{aligned}$$

where (4.2) represents the ray tracing equations (2.3), (2.5), and where we have displayed explicitly the dependence upon the parameter vector $\underset{\sim}{\mu}$. This will come about either in the equations, because some of the parameters are used to represent an unknown velocity, or in the boundary conditions (2.10-2.11), when an unknown reflector has to be determined.

Problem (4.1)-(4.2) can be solved by more or less standard non-linear least squares techniques [4,5,9] . The most effective ones will require the partial derivatives of the computed times with respect to the parameters. These derivatives are easily obtained by observing that if we solve system (3.3), with the linearization being taken around the computed ray $\omega^i(t)$, and with

$$\text{(4.3)} \qquad r(t) = \frac{\partial f}{\partial \mu_j} \ , \quad \beta = \frac{\partial g}{\partial \mu_j} \ ,$$

then

$$\text{(4.4)} \qquad \frac{\partial T_i^c}{\partial \mu_j} = \delta_7(1) \ .$$

As we mentioned earlier, the discrete solution of (3.3) only involves the solution of two very simple triangular systems. Methods using shooting techniques for solving the ray equations require the independent setting and solution of these linearized equations, and thus usually avoid this calculation and resort to some gross approximations to the partial derivatives (4.4) [1,2] .

We point out in passing that this same observation is valid if one is interested in the computation of the geometrical spreading of the rays, which also requires the solution of the linearized system (3.3) with appropriate right hand sides. Thus PASVA3 is very economical in performing all these tasks as compared to shooting techniques.

We finally observe that hypocenters and origin times can be located by the same procedure, the unknown parameters appearing now in the boundary conditions and in the definition of the travel time itself

The nonlinear least squares problem (4.1) can be attacked by any of a number of available methods. However, it is necessary to recognize that the evaluation of the functional (4.1) is an expensive task, especially if the number of observation is large. Therefore, it is appropriate to choose a method that at each iteration takes full advantage of the evaluation of the functional and its gradient. After reviewing several candidates we have settled for a combination of methods due to Gill and Murray [5] and Jupp and Vozoff [4] .

The main task is to perform a singular value decomposition of the Jacobian matrix

$$J = \left(\frac{\partial T_i^c(\underset{\sim}{\mu})}{\partial \mu_j} \right)_{\substack{i=1,\ldots,\ell \\ j=1,\ldots,m}} .$$

This is the matrix of the linear least squares problems that result at each step of Gauss-Newton or Marquardt type methods. Although singular value decompositions are considered expensive, they provide a maximum of information which is specially useful in nearby rank deficient cases, a not uncommon occurrence in these inverse problems. Besides, compared to the ray tracing part, the cost is insignificant.

Our algorithm uses mainly the Gauss-Newton iteration with step control. If this fails to reduce the residual, then a Marquardt type correction is attained by damping some singular values and cutting off those below a pre-specified threshold.

5. Numerical examples.

We present now some numerical examples computed with the procedures just described. They are artificial examples, with the data generated by the ray tracing code and specific values of the modelling parameters. We show then details of the recovery of those parameters via the inversion code.

All the computations were performed in single precision on a Burroughs 6700 computer ($\approx$ 11 decimal digits).

i) Recovery of velocity in two-dimensions.

We consider a two-dimensional geometry with the velocity given by a linear function of the coordinates

$$v(x,z) = \mu_1 + \mu_2 x + \mu_3 z.$$

Thus the medium is both inhomogeneous vertically and laterally. The data is provided by five events at the known locations (measured in kms.):

Event	1	2	3	4	5
x	-30	-10	5	10	25
z	100	110	130	150	120

There are nine stations on the surface (z=0), at which the first arrival times of P waves (say) are recorded. Their positions are $x_j = -50 + 10j$, $j=1,\ldots,9$. Travel time data is generated by using the velocity $v^*(x,z) = 5 + .025(x+z)$. We show details of two runs, one starting with $\underset{\sim}{\mu}$ fairly close to the true solution (5, .025, .025), and a second one with a much poorer initial guess. In the column marked max. res. we show the maximum value of the difference $|T_i^c - T_i^o|$, while $(\Sigma\, res_i^2)^{1/2}$ gives the value of the L_2 residual (both measured in seconds):

It.	μ_1	μ_2	μ_3	max.res.	$(\Sigma\ res_i^2)^{\frac{1}{2}}$
0	5.100	.02600	.026	.508	1.68
1	4.998	.02497	.02496	.013	.043
2	5.000	.02500	.02500	$3.01\text{x}10^{-5}$	$6.87\text{x}10^{-5}$
0	4.0	.01	.01	8.92	28.43
1	4.857	.0189	.0178	2.17	6.38
2	5.009	.0243	.0239	.201	.533
3	5.000	.02499	.02499	.001	.003

ii) Simultaneous recovery of velocity and relocation of hypocenters.

With the same model of (i) we show now how we can recover the velocity parameters and the positions of the sources. To save computer time we use only 25 of the 45 available observations. The 10 additiona parameters for the source positions are:

$$\mu_{3+2j-1} = x_j \quad , \quad \mu_{3+2j} = z_j \quad ,$$

and the artificial data is the same as for example (i).

It\μ	μ_1	μ_2	μ_3	μ_4	μ_5	μ_6	μ_7
0	5.200	.0200	.03000	-32.00	105.	-12.0	105.0
1	4.918	.0253	.02539	-29.58	98.8	- 9.69	109.6
2	4.997	.0250	.02502	-30.02	100.	-10.0	110.0
3	4.998	.0250	.02503	-29.97	100.	- 9.97	110.0
Exact	5.000	.0250	.02500	-30.00	100.	-10.0	110.0

It	μ_8	μ_9	μ_{10}	μ_{11}	μ_{12}	μ_{13}	Max. Res.	Max.Rel Error
0	5.20	125.0	13.0	156.	22.0	125.0	6.77	.300
1	5.03	129.5	8.88	149.	24.4	118.5	.508	.117
2	4.96	130.0	9.90	150.	24.9	120.0	.026	.010
3	5.04	130.0	10.0	150	25.0	120.0	.001	.007
Exact	5.00	130.0	10.0	150	25.0	120.0		

In the last row we list the maximum relative error in the parameters for each iteration. The total CPU time for this run was 2 minutes.

iii) Simultaneous recovery of velocity and interfaces.

We consider the same velocity structure as before, but now we have stations and shooting points located in the surface (z=0) with $x_j = -5+j$, $j=1,\ldots,9$. We also have a reflector made up of three line segments:

$$z - \frac{4}{9}x - \frac{40}{9} = 0 \qquad -10 < x < -1 ,$$

$$z - 4 = 0 \qquad -1 < x < 1 ,$$

$$z + \frac{4}{9}x - \frac{40}{9} = 0 \qquad 1 < x < 10 .$$

Thus we have to determine two dipping reflectors and a horizontal one on a medium with a laterally inhomogeneous structure.

For economy, we take only 23 observations obtained by shooting from all the positions but receiving simple reflections only at the odd numbered ones. Again the first three parameters correspond to the velocity, while the last six serve to define the reflector segments: $z + \mu_{3+2j-1}x + \mu_{3+2j} = 0 \quad j=1,2,3.$

Param./Iter	0	1	2	3	Exact
μ_1	4.0	4.220	4.870	4.998	5.000
μ_2	0.0	0.0132	0.0208	0.02499	0.025
μ_3	0.0	0.2523	0.0678	0.02602	0.025
μ_4	-0.4	-0.4174	-0.446	-0.4453	-0.444
μ_5	-4.0	-4.072	-4.430	-4.451	-4.444
μ_6	0.1	0.072	-0.00076	-0.000066	0.0
μ_7	-3.8	-3.840	-3.989	-4.001	-4.000
μ_8	0.4	0.4203	0.446	0.4453	0.444
μ_9	-4.0	-4.060	-4.410	-4.451	-4.444
Max.Res.	1.4	0.284	0.0477	0.0013	
Rel. Error	0.2	0.227	0.043	0.001	
CPU time	317"				

Acknowledgement

This work was partly supported by Contract No. 14-08-0001-16777 with the U.S. Geological Survey (Earthquake Hazards Reduction Program).

REFERENCES

1. Aki, K. and W.H.K. Lee, "Determination of three-dimensional velocity anomalies under a seismic array using first P arrival times from local earthquakes. 1: A homogeneous initial model", J. Geophys. Res. 81 pp. 4381-4399 (1976).

2. Bois, P., M. La Porte, M. Lavergne, et G. Thomas, "Essai de determination automatique des vitesses sismiques par measures entre puits", Geophysical Prosp. 19 pp. 42-83 (1971).

3. Cerveny, V., I.A. Molotkov, and I. Psencik , Ray Method in Seismology. Univ. Karlova, Praha (1977).

4. Gay, D., "Modifying singular values: existence of solutions to systems of nonlinear equations having a possibly singular Jacobian matrix", Math. Comp. 31 pp. 962-973 (1977).

5. Gill, P.E., and W. Murray, "Algorithms for the solution of the nonlinear least squares problem", SIAM J. Numer. Anal. 15 pp. 977-992 (1978).

6. Jacob, K.H., "Three-dimensional seismic ray tracing in a laterally heterogeneous spherical earth", J. Geophys. Res. 75 pp. 6685-6689 (1970).

7. Julian, B.R., "Ray tracing in arbitrarily heterogeneous media", Tech. Note 1970-45, Lincoln Lab. (1970).

8. Julian, B.R. and D. Gubbins, "Three dimensional seismic ray tracing", J. Geophys. 43 pp. 95-114 (1977).

9. Jupp, D.L.B., and K. Vozoff, "Stable iterative methods for the inversion of Geophysical data", Geophys. J. R. Astro. Soc. 42 pp. 957-976 (1975).

10. Keller, H.B., "Accurate difference methods for nonlinear two-point boundary value problems", SIAM J. Numer. Anal. 11 pp. 305-320 (1974).

11. Keller, J.B. and H. B. Keller, "Determination of reflected and transmitted fields by geometrical optics", J. Opt. Soc. Amer. 40 pp. 48-52 (1950).

12. Lentini, M. and V. Pereyra, "An adaptive finite difference solver for nonlinear two-point boundary problems with mild boundary layers", SIAM J. Numer. Anal. 14 pp. 91-111 (1977).

13. Pereyra, V., "Iterated deferred corrections for nonlinear operator equations", Numer. Math. 10 pp. 316-323 (1967).

14. Pereyra, V., "PASVA3: An adaptive finite difference FORTRAN program for first order nonlinear, ordinary boundary problems", Proc. Working Conf. Codes for BVP in ODE's. Lecture Notes in Comp.Sci., Springer-Verlag. To appear.

15. Pereyra, V., W.H.K. Lee, and H. B. Keller, "Solving two-point seismic ray-tracing problems in a heterogeneous medium. Part 1. A general numerical method based on adaptive finite-differences". Submitted to Bull. Amer. Seismological Soc.

16. Wesson, R.L., "Travel time inversion for laterally inhomogeneous crustal velocity models", Bull. Seism. Soc. Am. 61 pp. 729-746 (1971).

17. Yang, J.P. and W.H.K. Lee, "Preliminary investigations on computational methods for solving two-point seismic ray-tracing problems in a heterogeneous and isotropic medium", USGS Open File Rep. 76-707 (1976).

18. de Boor, C., A practical guide to splines. Applied Math. Sc. 27, Springer-Verlag, New York (1979).

SESSION IX

BIOMATHEMATICS

COMPUTING METHODS IN APPLIED SCIENCES AND ENGINEERING
R. Glowinski, J.L. Lions (editors)
North-Holland Publishing Company

NUMERICAL SIMULATION OF NON-NEWTONIAN BLOOD FLOW MODELS BY A FINITE ELEMENT METHOD

by

M. Bercovier, M. Engelman
School of Applied Sciences, Hebrew University, Jerusalem
and
J. Borman
Hadassah University Hospital, Jerusalem

INTRODUCTION

Each year hundreds of heart valve replacements are performed around the world and there are many different types of valve prostheses available on the commercial market. Since hemodynamically favourable performance of an artifical valve is to a large degree characterised by non-turbulent or laminar flow patterns and small pressure gradients across the valve throughout the cardiac cycle, new valve prostheses are exhaustively tested in physical flow simulators in order to determine their characteristics, before their use in the human heart.

The motivation for and objective of this research was to develop a method of computer simulation of the blood flow past an artifical heart valve. The benefits to be gained from such a simulation are multifold. It offers advantages of adaptability, simplicity and reliability over a physical simulator which must produce pulsatile flow under realistic atrial, ventricular and aortic pressure wave conditions, with simulated blood. Pressure and velocity profiles and atrial and ventricular motion derived from in-vivo data taken in man can be introduced into the computer program; whereas in the physical simulator input signals furnished by a mechanical apparatus must be utilised. Of even greater importance, complete quantitative information on both the velocity and pressure fields is available from a computer simulation.

To date a number of computer simulations of blood flow in the heart have appeared in the literature. GREENFIELD and AU [10] treated the flow past artificial valves using a semi-infinite model and PESKIN has made an in-depth study and simulation of the flow past the natural leaflet heart valve [15] and past a Bjork - Shiley valve [16]. However, these simulations and the majority of the others that have been published use a Newtonian constitutive relation for blood whereas it is known that blood is a highly non-Newtonian fluid; its viscosity characteristics varying from point to point depending mainly on the cell concentration at that point.

In view of the complex boundaries which arise in the study of heart flows, even in the 2 D case, it was natural to use the Finite Element Method (FEM) for our numerical simulations.

The FEM is an established numerical technique which enjoys widespread use in solid and structural mechanics; however, existing FEM techniques for handling kinematic constraints, in particular the incompressibility condition of fluid flow, lead to significant computational difficulties; this being one of the reasons the FEM has not found wide acceptance in commercial fluid mechanics applications.

It is the purpose of this research to develop a method of dealing with the continuity equation in FEM applications to Newtonian and non-Newtonian viscous incompressible fluid flows. The basis of the method is a penalty function approach to the continuity equation, which allows a simple, effective FEM implementation of the incompressibility constraint, as well as allowing the elimination of the pressure unknown from the system of equations to be solved.

The penalty function approach, which is now widely used in structural mechanics was first applied to the Navier-Stokes equations by TEMAM [17] and later by ZIENKIEWICZ and GODBOLE [20] using an eight-node serendipity element. In this study we have employed a nine node isoparametric element the mathematical and numerical properties of which were thoroughly investigated beforehand [2], [3] [7].

An outline of this work is as follows. §1, 2 and 3 review the equations of motion for incompressible viscous fluid flow and their weak formulation. The penalty function formulation is introduced in §4. The formulation of the approximate (non Newtonian) problem with numerical integration is dealt with in §5. In §6 we give an example of a numerical simulation past two artificial valves.

1. Equations of Fluid Motion

Consider a viscous fluid in motion within a bounded region $\Omega \subset R^2$, with boundary Γ. Let $u_i(x,t)$ be the velocity of the fluid at a point $x \in \Omega$ at a time t, ρ the density of the fluid, $p(x,t)$ the hydrostatic pressure at a point x at time t, and f_i an external body force acting on the fluid then, if temperature effects are neglected, the motion of the fluid can be fully described by the following equations

equation of motion

$$(1.1) \qquad \left(\frac{\partial u_i}{\partial t} + u_j u_{i,j}\right) - \sigma_{ij,j} = \rho f_i \quad \text{in } \Omega$$

equation of continuity

$$(1.2) \qquad \frac{\partial \rho}{\partial t} - (\rho u_i)_{,i} \quad \text{in } \Omega$$

boundary conditions

$$(1.3) \qquad u_i(x,t) = g_i(x,t) \qquad x \in \Gamma \quad \text{and all} \quad t > 0.$$

$$(1.4) \qquad u_i(x,0) = h_i(x) \qquad x \in \Omega$$

where the stress tensor σ_{ij} is given by

$$(1.5) \qquad \sigma_{ij} = (-p + \lambda\theta)\,\delta_{ij} + \tau_{ij} \quad ; \quad \begin{cases} \theta = u_{i,i} \\ \lambda = \text{Lame constant} \end{cases}$$

In order that equations (1.1) - (1.5) represent a well-defined problem it is necessary to provide the constitutive relation for the fluid under consideration. This is usually of the form:

$$(1.6) \qquad \tau_{ij} = 2\,\mu(x, D_{\alpha\beta})D_{ij}$$

where $D_{ij} = \frac{1}{2}\,(u_{i,j} + u_{j,i})$ is the strain rate tensor, and μ is the viscosity of the fluid.

In the case that $\mu = \mu_0$, a constant, the fluid is known as a Newtonian fluid, otherwise it is called non-Newtonian.

Using (1.5), (1.1) becomes

$$(1.7) \qquad \rho(\frac{\partial u_i}{\partial t} + u_j\,u_{i,j}) + p_{,i} - \lambda\theta_{,i} - [\mu(u_{i,j} + u_{j,i})]_{,j} = \rho f_i$$

We will restrict our interest here to incompressible flows, that is, flows for which the density ρ is a constant. The continuity equation (1.2) thus reduces to

$$(1.8) \qquad \theta = u_{i,i} = 0$$

and (1.7) becomes

$$(1.9) \qquad \rho(\frac{\partial u_i}{\partial t} + u_i\,u_{i,j}) + p_{,i} - [\mu(u_{i,j} + u_{j,i})]_{,j} = \rho f_i$$

In order to reduce equations (1.8) and (1.9) to dimensionless equations let

$$u_i' = \frac{u_i}{U} \quad , \quad x_i' = \frac{x_i}{L}$$

where U,L are some characteristic velocity and length of the flow respectively. Carrying out the necessary substitutions the following equations are arrived at

$$(1.10) \qquad \frac{\partial u_i'}{\partial t'} + u_j'\,u_{i,j}' + p_{,i}' - [\frac{\mu}{\rho LU}\,(u_{i,j}' + u_{j,i}')]_{,j} = f_i'$$

$$(1.11) \qquad u_{i,i}' = 0$$

(differentiation being with respect to the tagged variables)

where

$$u_i = Uu'_i$$

$$x_i = L\, x'_i$$

$$t = L/U\, t'$$

$$p = \rho U^2 p'$$

$$f_i = U^2/L\, f'_i\ .$$

From this point on we will work with the equations (1.10), (1.11) and for convenience will drop the tags.

In the case where $\mu = \mu_0$, a constant, then equation (1.10), using (1.11) reduces to

$$\frac{\partial u_i}{\partial t} + u_j\, u_{i,j} = -\, p_{,i} + f_i + \frac{1}{Re}\, u_{i,jj} \tag{1.12}$$

where $Re = \frac{\rho LU}{\mu_0}$ is the Reynolds number of the flow.

2. The non Newtonian constitutive relation

In the present study we used a constitutive relation for blood developed by S. Moskowitz, it is presented in detail in MOSKOWITZ and ENGELMAN [14]. The formulation, in brief, is as follows. To each position x in the fluid we assign two absolute viscosities

$$\mu(x,\, D_{ij}) = \begin{cases} \lambda_0(x) & ,\ i = j \\ \lambda_1(x,\, D^*_{ij}) & ,\ i \neq j \end{cases} \tag{2.1}$$

where λ_0 and λ_1 are calculated by the following relations

$$|\tau_{ij}|^{\frac{1}{2}} = k_0(x) + k_1(x)\ \ |D_{ij}|^{\frac{1}{2}} \qquad i \neq j \tag{2.2}$$

$$k_1(x) = \left[\frac{2\mu_0}{(1 - c(x))^{\beta-1}}\right]^{\frac{1}{2}} \tag{2.3}$$

$$k_0(x) = \frac{\alpha}{\beta-1}\ \left[\frac{k_1(x)}{(2\mu_0)1/2} - 1\right] \tag{2.4}$$

where $c(x)$ is a given enthrocyte concentration function, α, β are parameters expressing the rod orientation and relation between axial ratio and shear rate and μ_0 is the viscosity of the suspending medium, assumed to be Newtonian.

Then

(2.5) $$\lambda_0(x) = k_1^2(x)/2$$

(2.6) $$\lambda_1(x) = \lambda_0(x) + k_0(x)\, k_1(x) \,/\, \sqrt{D_{12}^*} + k_0^2(x) \,/\, 2D_{12}^*$$

provided $|D_{ij}^*| > k$, otherwise $\lambda_1 = k$ a constant. D_{ij}^* denotes maximum shearing rates determined from principal stress directions i.e. direction i and j are not those of an arbitrary coordinate system.

The numerical values used above were $\alpha = 0.0498$, $\beta = 2.945$, $k = 0.4$ sec; these values being derived from the results of WALAWENDER et al [18]. We used the enthrocyte concentration function for a given flow radius R given in [8].

3. Weak formulation (Newtonian Fluids)

We first introduce the weak formulation for a Newtonian fluid. In this case the mathematical foundations of the finite element method cum penalty we are going to use are well defined. It allows us to give a (formal) extension of our method to the non-Newtonian case.

Notations We consider real valued functions defined on Ω. Let us denote by

(3.1) $$(u,v) = \int_\Omega u(x)v(x)dx$$

the scalar product in $L^2(\Omega)$ and by

(3.2) $$||v||_{0,\Omega} = (v,v)^{\frac{1}{2}}$$

the corresponding norm. Consider also the quotient space $L^2(\Omega)/R$ provided with the quotient norm

(3.3) $$||v||_{L^2(\Omega)/R} = \inf_{c\in R} ||v+c||_{0,\Omega}$$

where, for simplicity, we shall also denote by v any function in the class $v \in L^2(\Omega)/R$.

Given any integer $m \geq 0$, let

(3.4) $$H^m(\Omega) = \{v|\ v \in L^2(\Omega), \quad \partial^\alpha v \in L^2(\Omega), \quad |\alpha| \leq m\}.$$

be the usual Sobolev space provided with the norm

(3.5) $$||v||_{m,\Omega} = \Big(\sum_{|\alpha|\leq m} ||\partial^\alpha v||^2_{0,\Omega}\Big)^{\frac{1}{2}}$$

and the semi norm

(3.6) $$|v|_{m,\Omega} = \Big(\sum_{|\alpha|=m} ||\partial^\alpha v||^2_{0,\Omega}\Big)^{\frac{1}{2}}$$

In (3.4) - (3.6), α is a multi-index, $= (\alpha_1,\ldots,\alpha_N)$; $\alpha_i > 0$,

$$|\alpha| = \alpha_1 + \ldots + \alpha_N \quad \text{and } \partial^\alpha = \Big(\frac{\partial}{\partial x_1}\Big)^{\alpha_1} \ldots \Big(\frac{\partial}{\partial x_n}\Big)^{\alpha_N}$$

Let

(3.7) $$H_0^1(\Omega) = \{v \mid v \in H^1(\Omega), \quad v|\Gamma = 0\}$$

Let $(L^2(\Omega))^N$ (resp. $H^m(\Omega))^N$) be the space of vector functions $v = (v_1,\ldots,v_N)$ with components v_i in $L^2(\Omega)$ (resp. $H^m(\Omega)$). The scalar product in $(L^2(\Omega))^N$ is given by

(3.8) $$(u,v) = \int_\Omega u(x)\cdot v(x)dx = \sum_{i=1}^N \int_\Omega u_i(x) \; v_i(x) \; dx$$

We use the following norm and seminorm on the space $(H^m(\Omega))^N$

(3.9) $$||v||_{m,\Omega} = \Big(\sum_{i=1}^N ||v_i||^2_{m,\Omega}\Big)^{\frac{1}{2}}$$

(3.10) $$|v|_{m,\Omega} = \Big(\sum_{i=1}^N |v_i|^2_{m,\Omega}\Big)^{\frac{1}{2}}$$

Introduce now, the spaces

(3.11)
$$X = (H_0^1(\Omega))^N = \{v \mid v \in (H^1(\Omega))^N, \quad v|_\Gamma = 0\}$$
$$V = \{v \mid v \in X, \quad \operatorname{div} v = 0\}$$

Note that

(3.12) $$||v|| = |v|_{1,\Omega}$$

is a norm over X and V which is equivalent to the norm $||v||_{1,\Omega}$. We shall often use the notation $||\cdot||$ in place of $||\cdot||_{1,}$. We extend the scalar product in $(L^2(\Omega))^N$ to represent the duality between V and its dual space V'.

We provide V' with the dual norm

(3.13) $$||f||^* = \sup_{v \varepsilon V} \frac{(f,v)}{||v||}, \quad f \varepsilon V'$$

Let us define:

(3.14) $$a(u,v) = \sum_{i=1}^{N} \int_\Omega \frac{\partial u}{\partial x_i} \frac{\partial v}{\partial x_i} dx, \quad u,v \varepsilon (H^1(\Omega))^N$$

(3.15) $$\hat{b}(u,v,w) = \sum_{i=1}^{N} \int_\Omega u_i \frac{\partial v}{\partial x_i} \cdot w \, dx$$

(3.16) $$b(u,v,w) = \frac{1}{2} (\hat{b}(u,v,w) - \hat{b}(u,w,v))$$

By Sobolev's imbedding theorem, we have $(H_0^1(\Omega))^N \subset (L^4(\Omega))^N$, $(N \leq 3)$ so that the trilinear form $b(u,v,w)$ is defined and continuous on $X \times X \times X$. Moreover, we have

(3.17) $$b(u,v,w) = \hat{b}(u,v,w) \quad \text{for all} \quad u \varepsilon V \quad \text{and all} \quad v,w \varepsilon X$$

(3.18) $$b(u,v,v) = 0 \quad \text{for all} \quad u,v \varepsilon X$$

Mixed variational formulation

From a practical point of view, since the constraint $\text{div } u = 0$ leads to computational difficulties, we do not use a direct variational approach but rather what is known as a mixed variational formulation.

The mixed variational formulation of equations (1.12) and (1.11) with homogeneous boundary conditions, which we shall refer to as problem (P), is as follows:

Problem P

Given a function $f \varepsilon V'$, find functions $(u,p) \varepsilon X \times L^2(\Omega)/R$ such that

(3.19) $$\frac{d}{dt}(u,v) + b(u,u,v) - (p, \text{div } v) + \frac{1}{Re} a(u,v) = (f,v)$$

(3.20) $$(\text{div } u,q) = 0$$

for all $(v,q) \varepsilon X \times L^2(\Omega)/R$

We shall refer to the stationary problem associated with (P) as problem (PS). Then we have the following result (cf. LADYZHENSKAYA [12], LIONS [13]).

Theorem

Define

$$(3.21) \qquad \beta = \sup_{u,v,w \in V} \frac{|b(u,v,w)|}{||u||\,||v||\,||w||}$$

and assume that the function f *satisfies*

$$(3.22) \qquad \beta\, Re^2\, ||f||^* < 1 - \delta$$

for some constant $0 < \delta < 1$. *Then, there exists a unique pair of functions* $(u,p) \in X \times L^2(\Omega)/R$ *solution of problem (PS).*

From (2.18) and (2.20) letting $v = u$ we have the following estimate

$$(3.23) \qquad ||u|| < \frac{1}{Re\ \beta}$$

Note that the existence result remains valid without any restriction on f. (3.22) is required to ensure uniqueness of (u, p).

4. Penalty Function Approach

We consider the following perturbation of the problem described by (1.2), (1.3) and (1.4).

$$(4.1) \qquad \frac{\partial u_i}{\partial t} + u_j^\varepsilon u_{i,j}^\varepsilon + \frac{1}{2}(\nabla\cdot u^\varepsilon)u_i^\varepsilon = -\frac{1}{\varepsilon}\nabla(\mathrm{div}\ u^\varepsilon) + f_i + \frac{1}{Re} u_{i,jj}^\varepsilon \quad \text{in } \Omega$$

$$(4.2) \qquad u_i^\varepsilon = g_i(x,t)\ ,\ x \in \Gamma \quad ,\ t \geq 0$$

$$(4.3) \qquad u_i^\varepsilon(x,0) = h_i(x),\quad x \in \Omega$$

Introduction of the term $\frac{1}{2}$ (div u^ε) u_i^ε which defines (4.1) - (4.3) as a 'well-posed' problem is due to TEMAN [17], who showed that if this term is not present then the system is not a Cauchy-Kowaleska system, in which case uniqueness and convergence for the system is an open question.

Note that here the continuity equation $\operatorname{div} u = 0$ is replaced by the penalisation term $\frac{1}{\varepsilon} \nabla(\operatorname{div} u_\varepsilon)$ in (4.1); this in effect means that the restriction $\operatorname{div} u = 0$ has been weakened to $\operatorname{div} u = \varepsilon p$. This is not what is often called a slightly compressible flow, but it does belong to the same family of approximations introduced by CHORIN [5].

The weak formulation of equation (.1) with homogeneous boundary conditions, which we shall refer to as problem (PH^ε), is as follows

Problem (PH^ε)

Given a function $f \in V'$, find functions $u^\varepsilon \in X$ and p^ε $L^2(\Omega)$ such that

$$\frac{d}{dt}(u^\varepsilon,v) + b(u^\varepsilon,u^\varepsilon,v) - (p^\varepsilon,\operatorname{div} v) + \frac{1}{Re}\, a(u^\varepsilon,v) = (f,v) \tag{4.4}$$

$$+ \varepsilon(p^\varepsilon,q) + (\operatorname{div} u^\varepsilon,q) = 0$$

for all $v \in X$ and $q \in L^2(\Omega)$

Remark 4.1:

By using the second equation of (4.4) we can eliminate p^ε and restate Problem PH^ε as:

Given a function $f \in V'$, find a function $u^\varepsilon \in X$ such that

$$\frac{d}{dt}(u^\varepsilon,v) + b(u^\varepsilon,u^\varepsilon,v) + \frac{1}{\varepsilon}(\operatorname{div} u^\varepsilon, \operatorname{div} v) + \frac{1}{Re}\, a(u^\varepsilon,v) = (f,v) \tag{4.5}$$

for all $v \in X$.

The pressure p^ε can then be "recovered" by $p^\varepsilon = -\frac{1}{\varepsilon} \operatorname{div} u^\varepsilon$

The advantage of the penalty approach is that it eliminates the need for special techniques to handle the incompressibility and it allows the elimination of the pressure unknown, a great saving when the problem is to be solved numerically.

The above weak formulation also allows application of appropriate theorems for the existence and uniqueness of solutions to saddle point problems and it can be shown (cf. BERCOVIER [1]) that

$$||u - u^\varepsilon|| + ||p - p^\varepsilon|| \leq c\varepsilon$$

Thus we can solve problem (PH^ε) in place of problem (PH) without losing any significant accuracy provided ε is small enough. It can also be shown that this this same relationship holds between the numerical solutions of problem (PH) and (PH^ε), provided that one chooses the correct finite element method.

5. Non Newtonian Equations and FEM

Taking into account the fact that $\mu = \mu(u)$ we define (cf 1.10)

(5.1) $$a_u(u, v) = \sum_{i=1}^{N} \frac{1}{\rho LU} \int_\Omega \mu(u) \frac{\partial u}{\partial x_i} \frac{\partial v}{\partial x_i} \, dx$$

The critical question as to what space belongs $\mu(u)$ is not delt with in this report so that (5.1) and all what follows is formal only.

Instead of (4.5) we have to find a u^ε such that

(5.2) $$\frac{d}{dt}(u^\varepsilon,v) + b(u^\varepsilon,u^\varepsilon,v) + \frac{1}{\varepsilon}(\text{div } u^\varepsilon, \text{div } v) + a_{u^\varepsilon}(u^\varepsilon,v) = (f,v)$$

Again we do not define the test and trial spaces in the continuous case. We use (5.2) as the Galerkin formulation from which the finite element method is going to be constructed. Let us drop the index ε in u^ε (5.2) for the sake of clarity. To solve this non linear equation we use at each time step the Picard iterations:

given u_{n-1}, find u_n such that:

(5.3) $$\frac{d}{dt}(u_n,v) + b(u_{n-1},u_n, v) + \frac{1}{\varepsilon}(\text{div } u_n, \text{div } v) + a_{u_{n-1}}(u_n,v) = (f,v)$$

Remark 5.1 The non Newtonian law we are using is according to (2.1) ....,(2.6) of Bingham type, although more complex, so that the correct formulation of our problem would lead to a variational inequation . Nevertherless we have seen in the Bingham fluid case that iterations of type (5.3) converge (cf [3]). For a theoretical study of this kind of fluids see Cioranescu [6].

F.E.M.

Given a "triangulation" of Ω into 9 node isoparametric elements we take u and v in (5.3) from the corresponding FEM space of 9 node isoparametric basis functions. In order to have a consistant approximation of the constaint $\text{div } u = 0$ the penalty term on each element k

(5.4) $$\frac{1}{\varepsilon} \int_k \text{div } u \text{ div } v \, dx$$

has to be computed by a 2 × 2 by gaussian quadrature rule, dualy this corresponds to a bilinear pressure term in (3.19), (3.2). Details and justification of this so called "reduced integration technique" are given in [2], and a complete error analysis in [7].

On the other hand, because of its highly non linear nature it was found necessary to compute

(5.5) $$\sum_i \int_K \mu(u_{n-1}) \frac{\partial u}{\partial x_i} \frac{\partial v}{\partial x_i} \, dK$$

by a 4×4 gaussian quadrature rule and not by the 3×3 one that suffices to ensure uniform ellipticity in the Newtonian case. The fact that (2.6) is highly non linear and includes a free boundary $|D^*_{ij}| > k$ resulted in very poor or no convergence of (5.3) when (5.5) was computed by a 3×3 rule. An explanation of this can be found, in the case of Bingham fluids in [4].
Time integration was performed by a non linear Crank -Nicholson scheme equivalent to a modified Euler method (cf Fortin [9].) Finally, confirming the results of Jamet-Raviart [11] it was found that $b(u_{n-1},u_n,v)$ could be computed indifferently by a 2 x 2 or a 3 x 3 gaussian quadrature rule.

6. Numerical Simulations

To study the effectiveness of our approach a large FEM simulation comparing the hemodynamics of a Bjork-Shiley tilting disc valve with that of the Starr-Edwards caged ball valve was performed. The FEM triangulation of the domain for fully opened Bjork-Shiley mitral valve and the domain for a fully opened Starr-Edwards aortic valve are given in Fig. 1.
Fig. 2 gives the velocity vector plot and pressure contour plot for the first case. This figure was compared to pictures obtained with a physical simulator, WRIGHT [19]. The same major flow characteristics were observed.
Examination of the results for the Starr-Edwards aortic valve (Fig. 3) reveals high velocity and steep pressure gradient surrounding the ball, this result in shearing stresses, which can be a source of thromboembolic complications. Further results can be found in [8].

The present study included the development of a computer program by the second author; because of its out of core solver, for non-symmetric matrices the size of the problem is limited by available time only. Introducing a non-Newtonian model is easy, Bingham fluids, "monotone" fluids were studied using the same program. As such the present work can be used for 2D blood flows simulations as well as for testing new non-Newtonian models.

Acknowledgements: The authors are indebted to Professor S. Moskowitz for his participation in the modelisation steps, and mainly for providing us the constitutive law for blood suspension flows. Y. Hasbani provided considerable help with the computer programming. All figures are taken from the second author's Ph. D. Thesis.

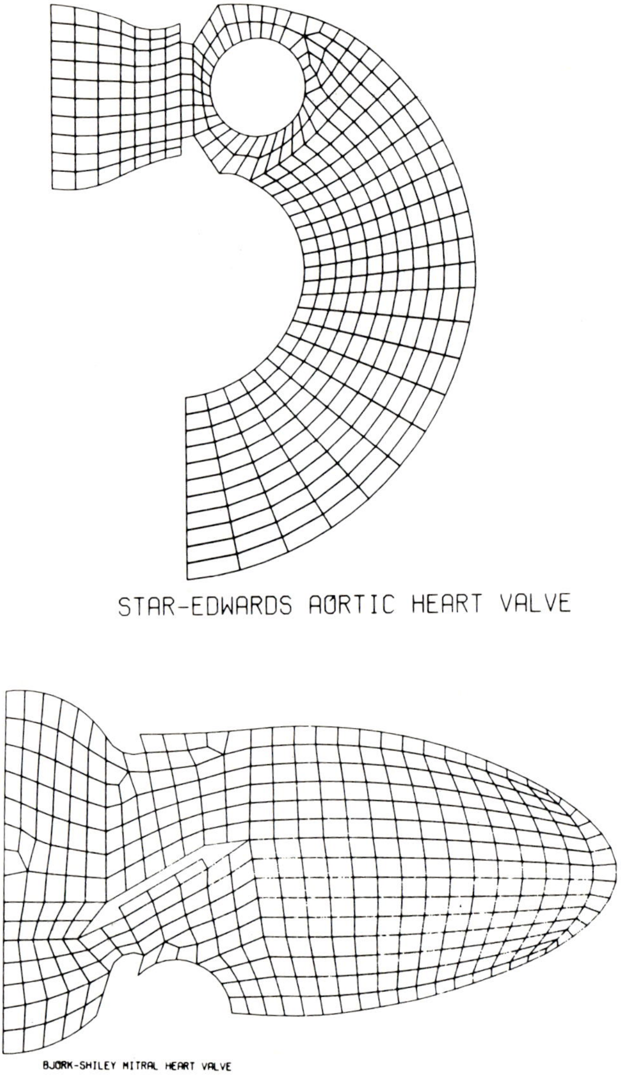
STAR-EDWARDS AØRTIC HEART VALVE
BJÖRK-SHILEY MITRAL HEART VALVE

Figure 1.

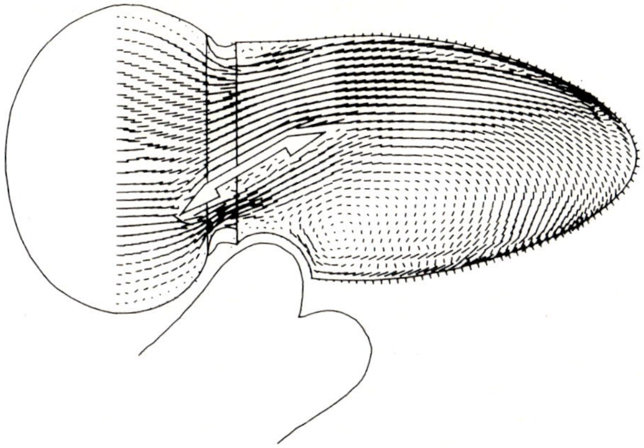

a) Velocity vector plot

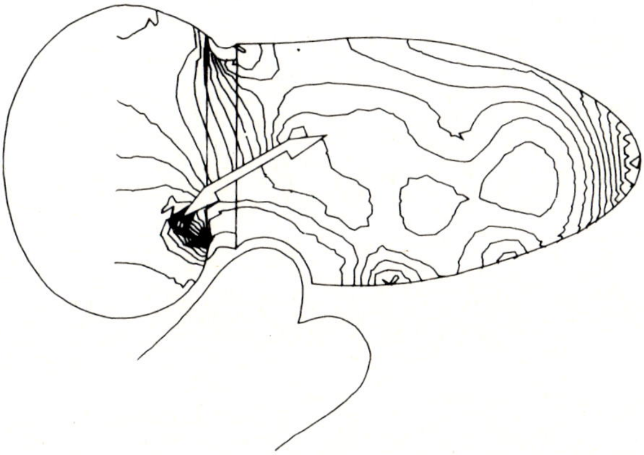

b) Pressure contour plot

Hemodynamics of Bjork-Shiley mitral valve

Figure 2.

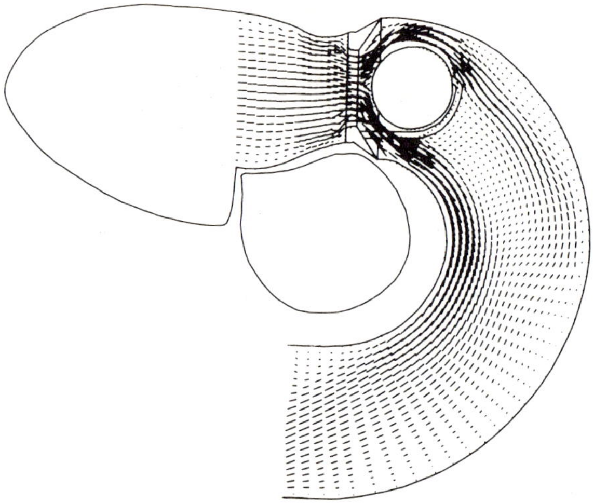

a) Velocity vector plot

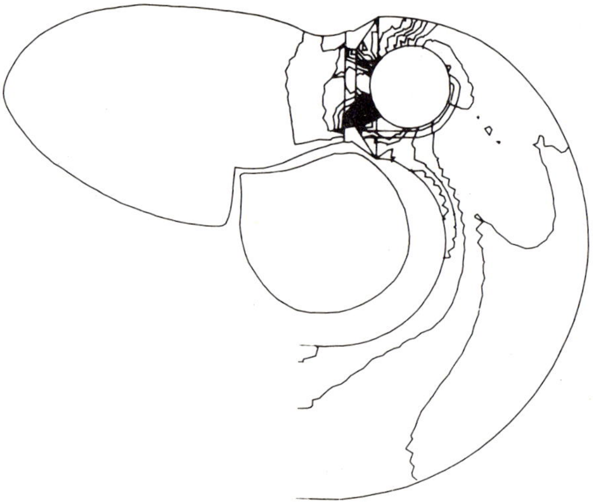

b) Pressure contour plot

Hemodynamics of Starr-Edwards aortic valve

Figure 3.

REFERENCES

[1] BERCOVIER, M. These de Doctoratd'Etat, Rouen, 1976.

[2] BERCOVIER, M. and ENGELMAN, M. A finite element for the numerical solution of viscous incompressible flows. J. Comp. Physics, 29, 181, 1979.

[3] BERCOVIER, M. and ENGELMAN, M. A finite element for incompressible non-Newtonian flows, to appear in J. Comp. Physics.

[4] BERCOVIER, M. ENGELMAN, M. and FORTIN, M. (To appear).

[5] CHORIN, A.J. A numerical method for solving incompressible viscous problems, J. Computational Physics, 2, 1967.

[6] CIORANESCU, D., Sur une Classe de Fluide Non Newtoniens, Applied Mathematics and Optimisation, 3, 263, 1977.

[7] ENGELMAN, M.S. BERCOVIER, M. Estimation de l'erreur pour la resolution du probleme de Stokes par un element finis quadrilateral conforme sur les vitesses. Comptes Rendues de l'Academie des Sciences. Serie A -555, 12 Mars 1979.

[8] ENGELMAN, M.S. MOSKOWITZ, S. BORMAN, J. Computer simulation, a diagnostic method in the comparative studies of prosthetic heart valves. To appear in J. Thorac. Cardiovascular Surgery.

[9] FORTIN, M. These, Paris, 1970.

[10] GREENFIELD, H. and AU. A. Computer visualisation of flow patterns for prosthetic heart valves. Scan J. Thorac Cardiovasc. Surg. 10, 197, 1976.

[11] JAMET, P. and RAVIART, P.A. Numerical solution of stationary Navier-Stokes equations by FEM's. Computing Methods in Appl. Sci. and Eng. ed. Glowinski, Lions. Springer-Verlag, Berlin, 1974.

[12] LADYZENSKAYA, O.A. The Mathematical Theory of Viscous Incompressible Flow, Gordon and Breach, 1964.

[13] LIONS, J.L. Quelques methodes de resolution de problemes aux limites et en controle optimal, Springer-Verlag, Berlin, 1973.

[14] MOSKOWITZ, S. and ENGELMAN, M.S. Navier-Casson-Stokes equations. In: First Mid-Atlantic conference on Bio-Fluid Mechanica Proceedings, Ed. D. J. Schneck, 1978.

[15] PESKIN, C.S. Flow patterns around heart valves: a numerical method. J. Comp Physics, 10, 252, 1972.

[16] PESKIN, C.S. Numerical Analysis of Blood Flow in the Heart, J. Comp. Physics, 25, 120, 1977.

[17] TEMAN, R. Navier Stokes equations, North Holland, Amsterdam, 1976.

[18] WALLAWENDER, W., CHEN, T. and CALA, D. An approximate Casson fluid model for tube flow of blood. Biorheology 12, 111, 1975.

[19] WRIGHT, J. The relative hydrodynamics performance of some mitral valve prostheses. In: Late results in Valvulator Replacements and Coronary Surgery. Ed's G. Stalpaert, R. Suy, F. Vermuelen, European Press, Ghent Belgium, 1976.

[20] ZIENKIEWICZ, O.C. and GODBOLE, P.N. Viscous incompressible flows with special reference to non-Newtonian (plastic) fluids, in Finite Element Method in Flow Problems, Wiley, New York, 1975.

COMPUTING METHODS IN APPLIED SCIENCES AND ENGINEERING
R. Glowinski, J.L. Lions (editors)
North-Holland Publishing Company

FLUID DYNAMICS OF THE HEART AND THE EAR

Charles S. Peskin
Courant Institute of Mathematical Sciences
251 Mercer St.
New York, N.Y. 10012

1. Introduction

The aim of this paper is to describe two intriguing problems in the field of physiological fluid dynamics: flow patterns around heart valves and wave propagation in the inner ear. These problems both involve an elastic boundary that is immersed in a viscous, incompressible fluid. These boundaries are the heart valve leaflets and the basilar membrane of the inner ear.

Despite this common physical basis, the fluid dynamics are very different in the two cases. In the heart we have to deal with vortex dynamics in flow at substantial Reynolds numbers, so the problem is highly nonlinear. In the ear, the amplitude of the motion is exceedingly small, so the convection terms of the Navier-Stokes equations can be neglected. Nevertheless, the problem is interesting because of the large variation in stiffness of the basilar membrane from one end to the other.

The research described in this paper has been carried out by the author and his colleagues over a period of several years, and this research program is still very active. Only an overview of this work will be given here, with emphasis on the formulation of the problem and with samples of numerical results. Details of the numerical methods are published elsewhere (see references).

2. Flow patterns around heart valves

a. Physiology

Each ventricle of the heart has an inflow and an outflow valve (see Fig. 2.1). We shall describe the function of these valves on the left side of the heart, but their function on the right side is similar. When the left ventricle is relaxed, its inflow valve (the mitral valve) is open and its outflow (aortic) valve is closed. In this situation the left ventricle forms a common chamber with the left atrium and it receives blood from the left atrium at low pressure (5 mm. Hg.). When the left ventricle contracts, the mitral valve closes and the aortic valve opens. The left ventricle then forms a common chamber with the aorta, and it pumps blood into this artery at high pressure (120 mm. Hg.).

The valves themselves are thin elastic membranes and their motions are primarily determined by the fluid that surrounds them. It is important to realize, however, that the valve leaflets also have a profound influence on the fluid in which they are immersed. This two-way interaction makes the heart-valve problem interesting from a mathematical point of view.

b. Equations of an immersed valve leaflet

In this section we state the equations of motion of a mechanical system that consists of an elastic boundary of zero mass immersed in a viscous incompressible fluid. The region of space occupied by the entire system will be called Ω ; the immersed boundary $B(t)$ moves about in the interior of Ω. We shall write equations of motion that apply on Ω as a whole; this avoids the difficulties of working with the fluid domain $\Omega - B(t)$.

The fluid has density ρ , viscosity μ , velocity $\underline{u}(\underline{x},t)$, and pressure $p(\underline{x},t)$. On account of viscosity, the velocity is continuous across the immersed boundary, and the restriction of $\underline{u}(\underline{x},t)$ to $B(t)$ gives the velocity of the material points of the boundary at time t.

The immersed boundary will be described in parametric form $\underline{X}(s,t)$. When Ω is three-dimensional, s will stand for the pair of parameters (s_1,s_2) and ds will stand for $ds_1 ds_2$. The domain of s will be denoted by B_o. Thus $B(t) = \{\underline{X}(s,t) | s \in B_o\}$. The choice of parameters is arbitrary except that fixed s must mark a material point.

Thus we use a Lagrangian description for the immersed boundary

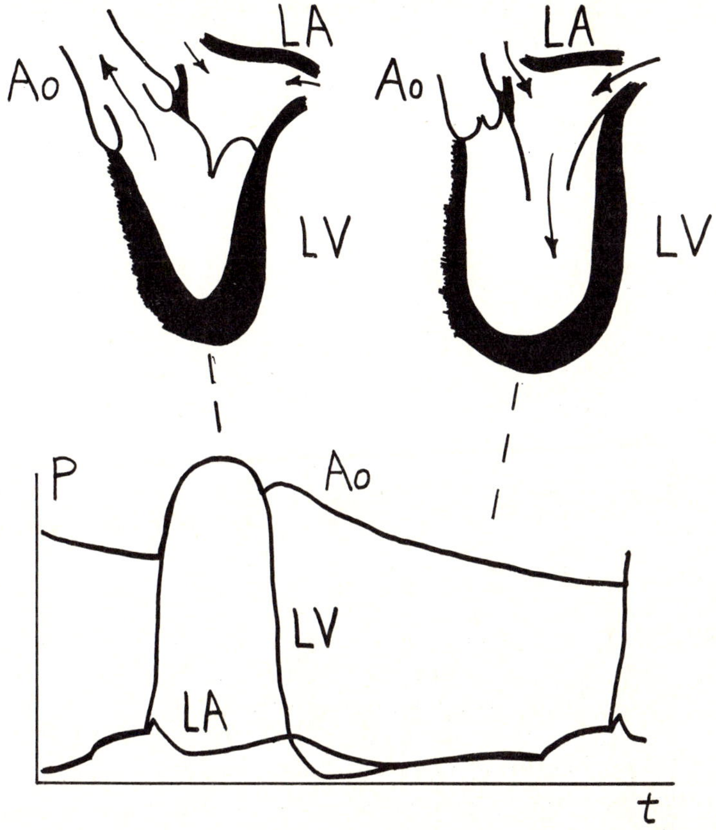

Fig. 2.1 Heart valves and pressures.
LV = Left Ventricle; LA = Left Atrium; Ao = Aorta;
P = Pressure; and t = time.

and an Eulerian description for the fluid. With this notation, the equation of motion may be stated as follows:

$$\rho\left(\frac{\partial \underline{u}}{\partial t} + \underline{u}\cdot\nabla\underline{u}\right) = -\nabla p + \mu\Delta\underline{u} + \underline{F} \tag{2.1}$$

$$\nabla\cdot\underline{u} = 0 \tag{2.2}$$

$$\underline{F}(\underline{x},t) = \int_{B_o} \underline{f}(s,t)\,\delta\big(\underline{x}-\underline{X}(s,t)\big)\,ds \tag{2.3}$$

$$\frac{\partial \underline{X}}{\partial t}(s,t) = \underline{u}\big(\underline{X}(s,t),t\big)$$

$$= \int_{\Omega} \underline{u}(\underline{x},t)\,\delta\big(\underline{x}-\underline{X}(s,t)\big)\,d\underline{x} \tag{2.4}$$

$$\underline{f}(\ ,t) = S\big(\underline{X}(\ ,t)\big) \tag{2.5}$$

This system of equations is not at all standard, and it requires considerable explanation. Equations (2.1-2.2) are the Navier-Stokes equations for a viscous incompressible fluid with an applied force-density $\underline{F}$. Here we use $\underline{F}$ to represent the force of the immersed boundary on the fluid. It follows that $\underline{F}(\ ,t)$ is a distribution with support on the immersed boundary $B(t)$. (Recall that $\underline{F}$ has units of force/volume, but the boundary exerts a finite force in zero volume.)

The explicit form of $\underline{F}$ is given in Eq. (2.3) in which $\delta(\underline{x})$ stands for $\delta(x)\delta(y)\delta(z)$. The integration in Eq. (2.3) does not completely remove the singularity since the dimension of the boundary is one less than the dimension of Ω. In our numerical method, therefore, $\underline{F}$ will be $O(h^{-1})$ and the volume of its support will be $O(h)$.

The exact meaning of $\underline{f}(s,t)$ can be found by integrating Eq. (2.3) over an arbitrary region $\Omega_1 \subset \Omega$. The result is

$$\int_{\Omega_1} \underline{F}(\underline{x},t)\,d\underline{x} = \int_{B_1} \underline{f}(s,t)\,ds \tag{2.6}$$

where $B_1 = \{s \mid \underline{X}(s,t) \in \Omega_1\}$.

Thus $\underline{f}$ is the density of the boundary force with respect to the boundary element ds , and Eq. (6) asserts that the force of the boundary on the region Ω_1 can be attributed to the part of the boundary that lies in Ω_1. In other words, the boundary force acts locally in the fluid.

Despite the local nature of the boundary force, its effects are felt instantaneously throughout the incompressible fluid. To see how this comes about, take the divergence of both sides of Eq. (2.1). The result is

$$\Delta p = -\rho \underline{u} \cdot \nabla \underline{u} + \nabla \cdot \underline{F} \tag{2.7}$$

Thus $\nabla \cdot \underline{F}$ acts as a source of the pressure field.

Eq. (2.4) simply asserts that the boundary moves at the local fluid velocity; here, we regard this as an equation of motion of the boundary and not as a constraint on the fluid velocity. The second equality in Eq. (2.4) is just the definition of the δ-function, which we write out to emphasize a certain symmetry with Eq. (2.3). Together, Eqs. (2.3) and (2.4) express the interaction between the boundary and the fluid, and the δ-function appears as a kernel in both cases. As remarked above, this expresses the local nature of the interaction.

Eq. (2.5) makes the important assertion that the boundary force is determined by the boundary configuration. This is a consequence of our assumptions that the boundary is elastic and massless. To illustrate this point, consider a two-dimensional example in which the boundary is an elastic curve (Fig. 2.2). Let s measure arc length in the unstressed configuration, let $\underline{\tau}$ be the unit tangent to the curve, and let T be the boundary tension. Then

$$\underline{\tau} = \frac{\partial \underline{X}/\partial s}{|\partial \underline{X}/\partial s|} \tag{2.8}$$

$$T = \sigma\left(|\partial \underline{X}/\partial s| - 1\right) \tag{2.9}$$

Consider an arc (a,b) of the boundary. Since the boundary has zero mass, the total force on this arc must be zero:

$$-\int_a^b \underline{f}\, ds + T\underline{\tau}\Big|_a^b = 0 \tag{2.10}$$

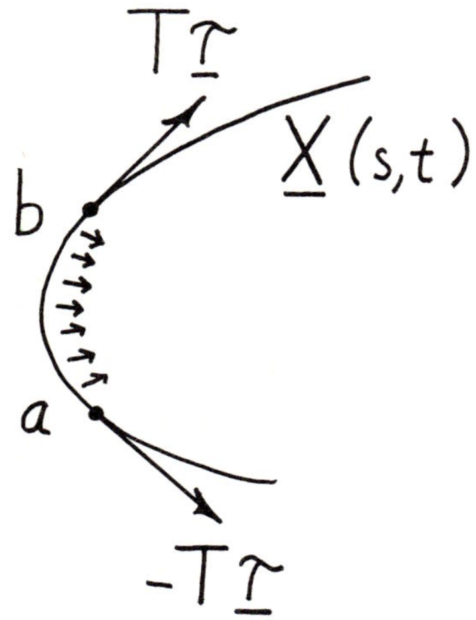

Fig. 2.2 Boundary force. T = Tension; $\underline{\tau}$ = unit tangent; $\underline{X}(s,t)$ = boundary configuration at time t. The forces at the ends of arc (a,b) are transmitted to the fluid along that arc.

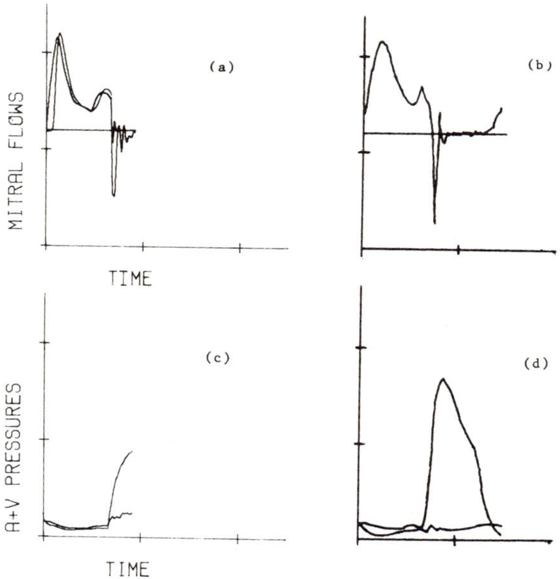

Fig. 2.3 Comparison of theory and experiment. Computed results are shown in (a,c) and experimental results from the laboratory of E. L. Yellin in (b,d). The two flow traces in (a) correspond to flow at the mitral ring and flow at the tips of the leaflets. The experimental flow (b) is measured at the ring only.

In Eq. (2.10), the first term is the force of the fluid on the arc (a,b) and the second term is the force of the rest of the boundary on this arc. Since a and b are arbitrary, this shows that

$$\underline{f} = \frac{d}{ds}(T\underline{\tau}) \tag{2.11}$$

Substituting (2.8) and (2.9) in (2.11), one can obtain an explicit example of the kind of relationship that is summarized by Eq. (2.5).

The system of equations (2.1-2.5) is remarkable in that the fluid stress tensor evaluated at the boundary never appears explicitly. As we have just seen, this is a direct consequence of the massless character of the boundary. If the boundary were massive, then we would need an equation for its acceleration. This equation would involve the sum of the elastic and fluid forces on an element of the boundary, and there would be no way to avoid explicit reference to the fluid stress tensor.

c. Computer test chamber for prosthetic mitral valves

We have discretized equations (2.1)-(2.5) to obtain a numerical method for the heart valve problem (Peskin, 1972, 1975, 1977; Peskin and McQueen, 1980). Since the details of the method have already been published, only its use will be considered here.

We use the method to solve the equations of motion of blood in a two-dimensional model of the left heart. The model has a left atrium and a left ventricle with contractile walls that have the physiological properties of heart muscle. The model heart is floating in fluid, so its walls as well as its valves are modeled as immersed boundaries. There is a source in the atrium that corresponds to pulmonary venous return, and there is a sink around the edges of the domain that accepts the volume displaced as the heart fills.

At the junction of the atrium and ventricle, we mount a model mitral valve, natural or artificial. With the natural valve in place, we adjust the physiological properties of heart walls until the results are in reasonable agreement with animal experiments as judged by records of presure and flow as functions of time. The comparison is shown in Fig. 2.3. In this figure all scales are the same except that the flow curves have been scaled arbitrarily on account of the difficulty of comparing three-dimensional flow (volume/time) with two-dimensional flow (area/time). The computed streamlines of the natural mitral valve are shown in Fig. (2.4).

Once we have established physiological conditions, we are in a position to test artificial valves in the computer and conduct parametric studies on the design of artificial valves. In Figs.(2.5)-(2.7) we show the computed flow patterns of a ball valve, and two pivoting disc valves.

The pivoting discs differ only in the position of the pivot point, and we do not impose any constraint on the maximum angle of opening. We find, however, that this angle depends on the position of the pivot point so that the valve in Fig. (2.7) opens much less than the valve in Fig. (2.6). This illustrates how the method can be used for parametric studies.

The principal limitations of this work are that the model is two-dimensional and that our numerical experiments are conducted at low Reynolds number. The Reynolds number appropriate for dog hearts (where heart valve experiments are often done) is about 500. Our numerical results were obtained at a Reynolds number of 20. As mentioned above, the computed and experimental results appear to agree despite this discrepancy. Nevertheless, it would be desirable to remove this limitation.

d. Vortex methods

A high Reynolds number method for the incompressible Navier-Stokes equation is the vortex method of A. J. Chorin (1973), and we conclude this discussion with a brief description of some papers that apply this method to the heart valve problem. In the thesis of Mendez (1977), a new method for the creation of vorticity at an immersed, elastic boundary is introduced. This method can be derived by taking the curl of Eq. (2.3) and noticing that each element of the boundary force acts as a source of a vortex dipole. In Peskin and Wolfe (1978), the vortex method is combined with conformal mapping and used to study the formation of the aortic sinus vortex. The use of conformal mapping avoids the difficulty of solving Laplace's equation numerically to compute the potential part of the flow. It also provides high resolution near corners where the slip velocity may be infinite. In McCracken and Peskin (1980), a vortex-grid method for the problem of blood flow in the heart is introduced. In this method, vorticity is stored either in the form of moving vortex blobs or on a fixed computational mesh. Tangential forces create vortex dipoles which are retained only for a fixed number of time steps; then their vorticity is transferred to the mesh.

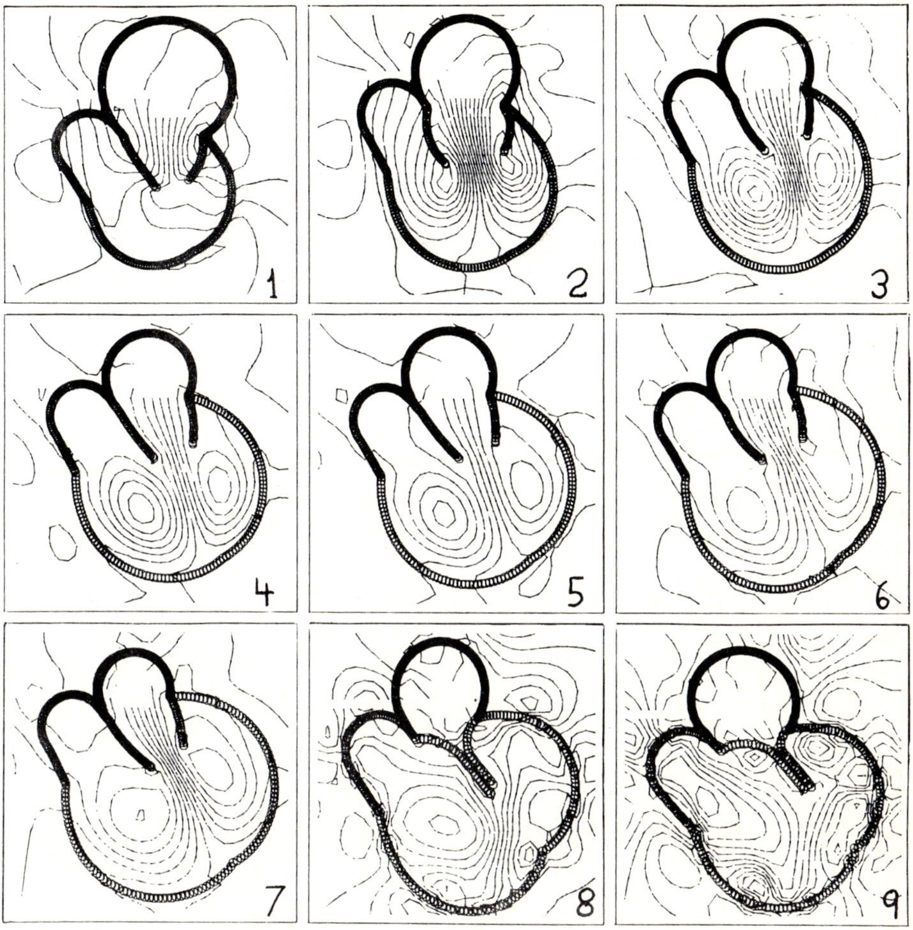

Fig. 2.4 Streamlines of the natural mitral valve at equally spaced times. (Note that streamlines cross moving boundaries.) In (1), the ventricle is relaxing and the valve is opening. Vortex formation occurs in (2) and establishes the characteristic flow pattern of ventricular filling. Contraction of the atrium strengthens the jet in (6-7). Valve closure has just begun in (7); it is completed by contraction of the ventricle in (8).

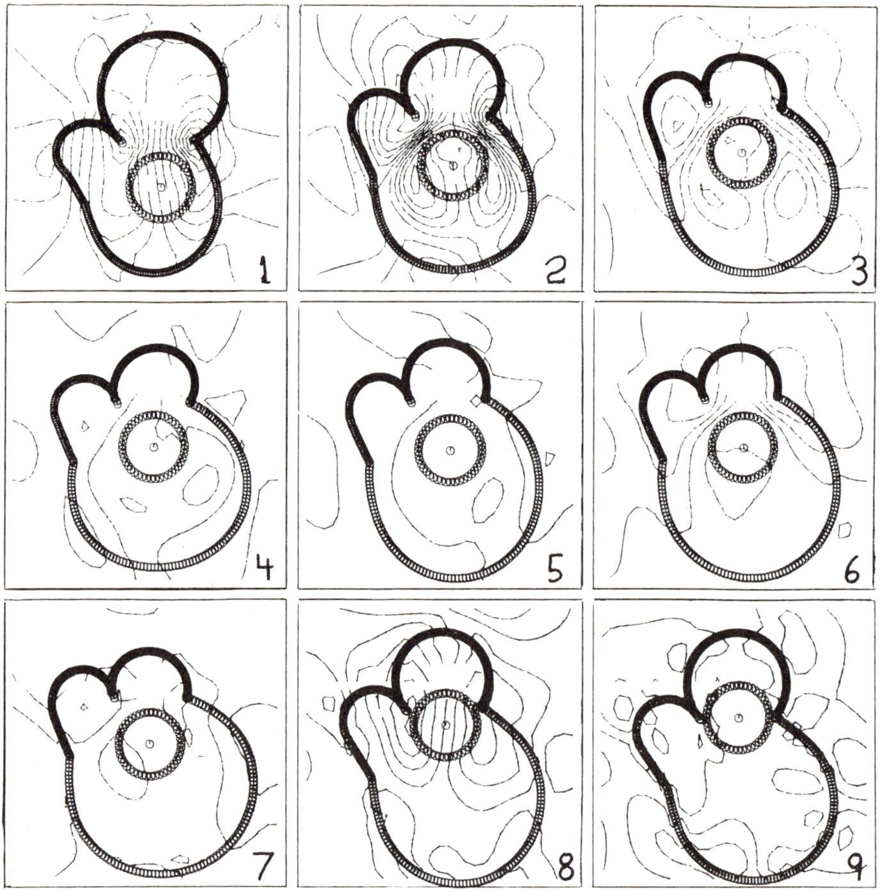

Fig. 2.5 Streamlines of a caged ball valve. Note the distinction between the flow pattern when the ball is opening (1) and after it has reached the open position (2). Similarly compare the closing flow pattern (8) with the closed flow pattern (9). As long as the ball is moving it is not an obstacle to the flow.

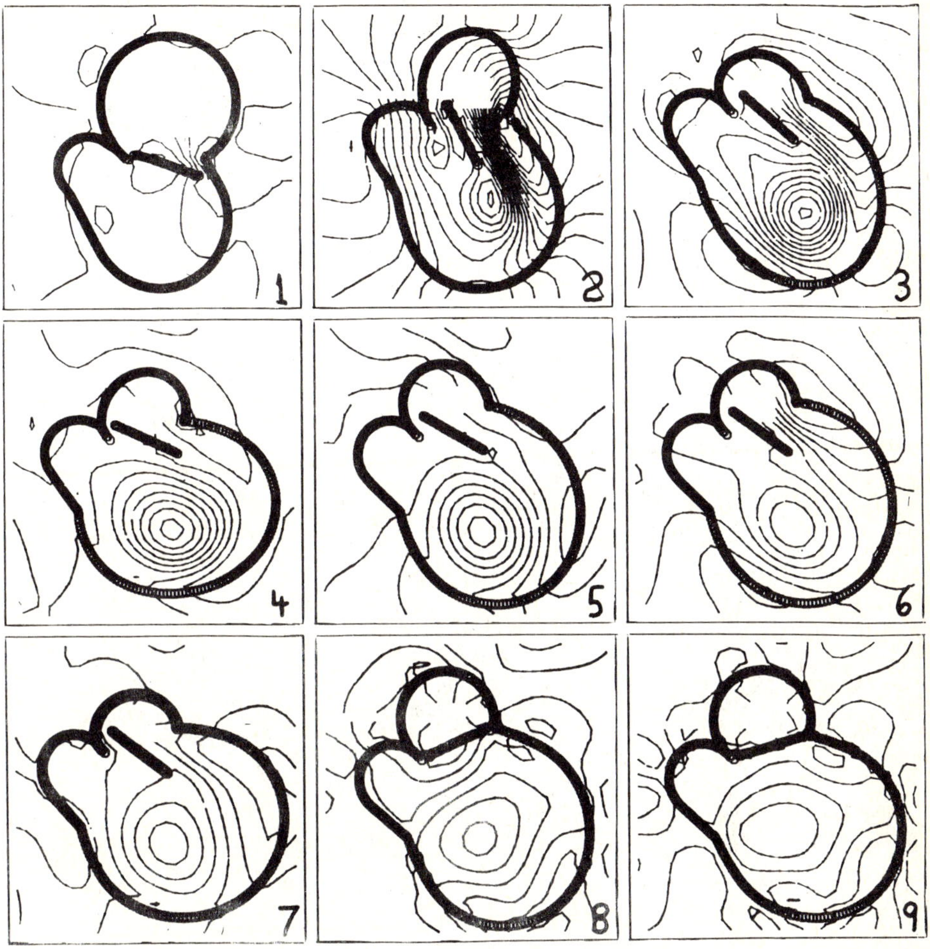

Fig. 2.6 Streamlines of a pivoting disc valve. In This computation, we impose no mechanical constraint on the angle of opening, which is set by the fluid dynamics. Our purpose is to study the effect of the position of the pivot point on the angle of opening (compare Fig. 2.7).

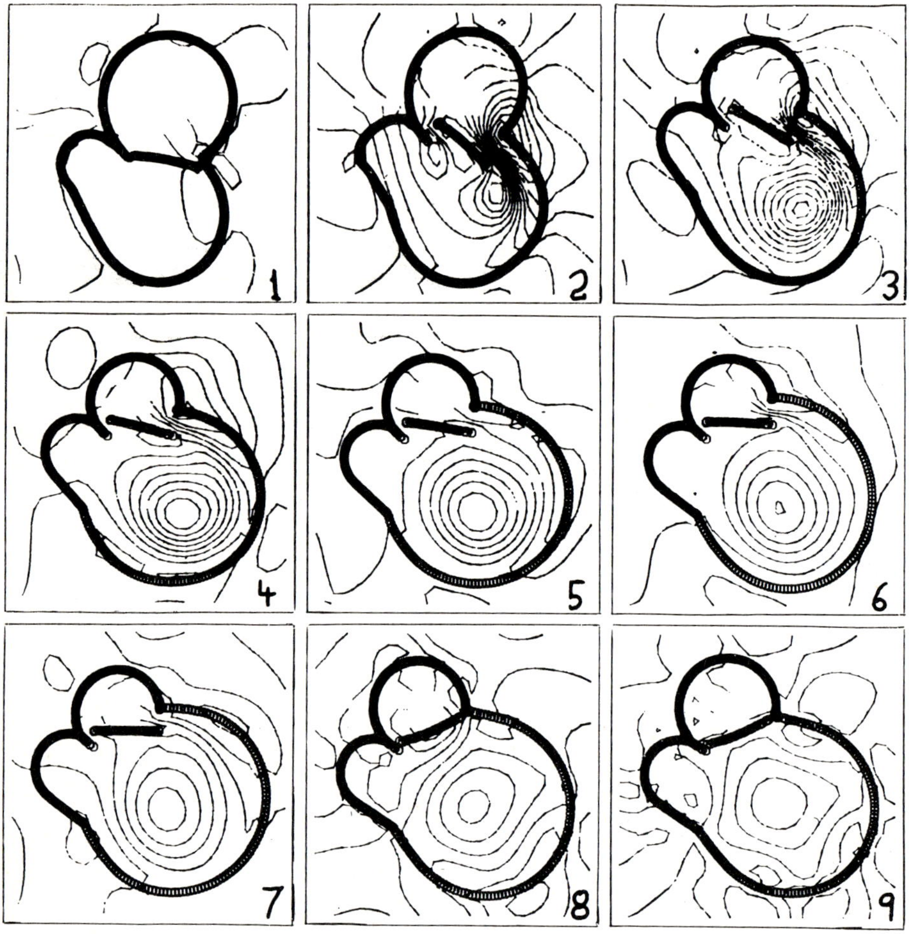

Fig. 2.7 Effect of moving the pivot point. The valve in this figure is identical to the valve in Fig. 2.6 except that the pivot point has been moved closer to the center of the valve. The angle of opening is substantially less than before.

3. Fluid dynamics of the inner ear

a. Physiology

The inner ear (cochlea) is a cavity in the temporal bone of the skull. Unlike the outer and middle ear, which are filled with air, the cochlea is filled with an essentially incompressible fluid. An elastic structure, the basilar membrane, runs along the length of the cochlea and divides it into two main parts. The wave motions that occur in the cochlea propagate along the basilar membrane. Although they occur in response to sound, these disturbances are not sound waves in the ordinary sense. Instead, they are vibrations in which the mass of an incompressible fluid is coupled to the elasticity of an immersed boundary. The kinetic energy of these waves is entirely a property of the fluid and the potential energy is entirely a property of the elastic boundary.

The physiology of the cochlea as we understand it today was first elucidated by George von Bekesy (1960). Using static tests with a constant pressure difference, von Bekesy discovered the important fact that the stiffness of the basilar membrane decreases exponentially with distance into the cochlea. Von Bekesy also studied the pattern of vibration of the basilar membrane in response to a pure tone (sine wave). His method was to observe the motions of the basilar membrane directly using a microscope with stroboscopic illumination. Von Bekesy found that the response of the ear to a steady pure tone takes the form of a wave. The points of constant phase propagate into the ear at a velocity that decreases with distance. The amplitude of the wave is a steady function of position that rises gradually to a unique maximum and then decays rapidly on the far side of this maximum. The point of maximum amplitude varies as the negative logarithm of the frequency of the stimulating sound. In fact, when the frequency of the sound is changed the whole pattern of vibration translates to a new position without much change in form. This correspondence between frequency and position (the cochlea map) is important because the fibers of the auditory nerve are distributed along the length of the basilar membrane. Thus each nerve fiber responds best to a particular frequency, and a complex sound is transmitted to the brain with its different frequency components carried along different nerve fibers. This is an important aspect of the mechanism of separating signals from noise in hearing.

It should be mentioned that the problem of constructing a theory to account for these observations has attracted the attention of many investigators (Lesser and Berkley, 1972; Siebert, 1974; Steele, 1974; Inselberg and Chadwick, 1976; Cole and Chadwick, 1977, Allen, 1977; Steele and Taber, 1979a, 1979b).

Nevertheless, the work that will be described here (see also Peskin, 1976, and Isaacson, 1979) is different in the following respects. First, we show how a simple conformal mapping can be used to reduce the inviscid cochlea problem to a standard water-wave problem. Second, we show how the corresponding viscous problem can be reduced to an integral equation on the basilar membrane, and we give two numerical methods for solving this integral equation.

b. Two-dimensional model

The model that we shall use is shown in Fig. 3.1. The boundaries at $y = \pm a$ are rigid, and the model is unbounded in the positive and negative x-direction. The moving boundary corresponding to the basilar membrane is described by the unknown function $y = h(x,t)$. We assume that the displacements of all fluid particles are small, so we neglect the non-linear terms in the Navier-Stokes equation, and we apply the boundary conditions appropriate for $y = h(x,t)$ to the undisturbed position of the basilar membrane which is $y = 0$.

We assume that the fluid is incompressible with density ρ and viscosity μ. The fluid velocity and pressure will be denoted (u,v) and p.

The equations of the model are as follows

$$\rho \frac{\partial u}{\partial t} + \frac{\partial p}{\partial x} = \mu \Delta u \qquad (3.1$$

$$\rho \frac{\partial v}{\partial t} + \frac{\partial p}{\partial y} = \mu \Delta v + f(x,t)\delta(y) \qquad (3.2)$$

$$\frac{\partial u}{\partial x} + \frac{\partial v}{\partial y} = 0 \qquad (3.3)$$

$$f(x,t) = -s_o e^{-\lambda x} h(x,t) \qquad (3.4)$$

$$\frac{\partial h}{\partial t}(x,t) = v(x,0,t) \qquad (3.5)$$

$$u(x,0,t) = 0 \qquad (3.6)$$

$$u(x,\pm a,t) = v(x,\pm a,t) = 0 \qquad (3.7)$$

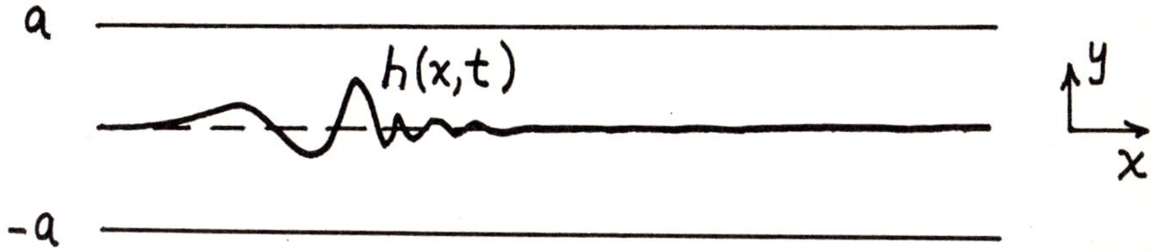

Fig. 3.1 The model cochlea

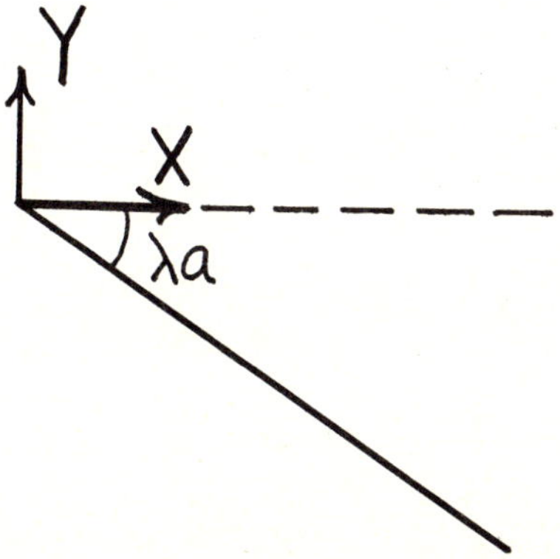

Fig. 3.2 The transformed problem. The conformal mapping given by Eqs. (3.14)-(3.16) takes the lower half-cochlea into the wedge shown here. The inviscid cochlea problem is equivalent to gravity waves on a sloping beach with the source at the beach and the waves going out.

In these equations $f(x,t)$ corresponds to the force per unit length exerted by the basilar membrane on the fluid. The resulting pressure difference across the membrane can be found by integrating (3.2) across $y=0$. The result is

$$[p] = \mu[\frac{\partial v}{\partial y}] + f \qquad (3.8)$$

when [ ] denotes the jump in a quantity across $y=0$. From (3.3) and the boundary condition $u=0$ at $y=0$, we see that $\partial v/\partial y = 0$ on both sides of the membrane. Therefore

$$[p] = f = -s_o e^{-\lambda x} h \qquad (3.9)$$

where we have used (3.4).

Equation (3.4) contains the physics of the model basilar membrane. According to this equation, the basilar membrane has zero mass, and it exhibits linear elastic behavior at each x with a spring stiffness of the form $s_o e^{-\lambda x}$. The exponential dependence of stiffness on position comes directly from the static measurements of von Bekesy (1960). The reader should notice that the different points of the basilar membrane are not coupled to each other by Eq. (3.4). That is, there are no terms involving the space derivatives of h. This comes about because the real basilar membrane is narrow and supported along its edges, see Peskin (1976).

Finally, we have to discuss how the model cochlea is driven. The real cochlea is excited at the stiff end by the piston-like action of two elastic membranes, one of which is connected to the eardrum by a chain of tiny bones. These two membranes are on opposite sides of the basilar membrane, and one moves out while the other moves in so that the total volume of the cochlea is conserved. In the model we assume that there is a sinusoidal source (with the appropriate antisymmetry) at $x = -\infty$. In a careful treatment of the problem, this would be built into the boundary conditions, but we shall be deliberately vague on this point since the details of how the cochlea is driven have very little effect on the form of the wave that results. Some reasons for this will appear below.

c. The inviscid case ($\mu=0$)

In this section we study the behavior of our model cochlea when $\mu=0$. In this case it is reasonable to look for potential flow solutions in which $(u,v) = \text{grad}\ \phi$. Because of the antisymmetry in the way that the cochlea is driven, it is also reasonable to look for solutions that satisfy

$$\phi(x,y,t) + \phi(x,-y,t) = 0 \tag{3.10}$$

Under these conditions, our equations reduce to the following system in the strip $-a \le y \le 0$:

$$\frac{\partial^2\phi}{\partial t^2} + \frac{s_o e^{-\lambda x}}{2\rho}\frac{\partial\phi}{\partial y} = 0 , \qquad y = 0 \tag{3.11}$$

$$\Delta\phi = 0 , \qquad -a < y < 0 \tag{3.12}$$

$$\frac{\partial\phi}{\partial y} = 0 \qquad y = -a \tag{3.13}$$

The boundary condition at the basilar membrane (3.11) was obtained from (3.9) by differentiating with respect to time and using the relations $\rho\partial\phi/\partial t + p = 0$, $\partial h/\partial t = (\partial\phi/\partial y)_{y=0}$, and $[\phi] = -2\phi_{y=0^-}$.

Equations (3.11)-(3.13) are very nearly the same as the equations for gravity waves of small amplitude in a channel of finite depth. The only difference between the two cases is that here the role of the gravitational constant g is played by the coefficient $s_o e^{-\lambda x}/2\rho$, which depends on x. We can get rid of this space-dependence, however, by applying the conformal mapping

$$X = e^{\lambda x} \cos \lambda y \tag{3.14}$$

$$Y = e^{\lambda x} \sin \lambda y \tag{3.15}$$

which can also be written

$$Z = e^{\lambda z} \tag{3.16}$$

where $z = x + iy$ and $Z = X + iY$. This mapping takes the strip $-a < y < 0$ into the wedge $-\lambda a < \arg(Z) < 0$. Along $y = 0$, we have

$$\frac{\partial\phi}{\partial y} = \lambda e^{\lambda x}\frac{\partial\Phi}{\partial Y} \tag{3.17}$$

so that our problem becomes

$$\frac{\partial^2\Phi}{\partial t^2} + \frac{s_o\lambda}{2\rho}\frac{\partial\Phi}{\partial Y} = 0 \qquad Y = 0, \quad X > 0 \tag{3.18}$$

$$\Delta\Phi = 0 \qquad -\lambda a < \arg(Z) < 0 \tag{3.19}$$

$$\frac{\partial\Phi}{\partial N} = 0 \qquad -\lambda a = \arg(Z) \tag{3.20}$$

where $\partial\Phi/\partial N$ stands for the normal derivative. Note that the factor $e^{-\lambda x}$ has disappeared and that we have transformed our problem into the problem of gravity waves on a sloping beach (see Fig. 3.2). In our case, the source is at the shoreline and the waves go out, because $x = -\infty$ maps into $X=Y=0$.

We do not need a detailed solution to this problem. Instead we can look at the behavior for large X. If the source has time-dependence $e^{i\omega t}$, then this will generate a wave that propagates away from the shoreline. For large X this wave looks like

$$\Phi(X,Y,t) = e^{i(\omega t-kX)}e^{kY} = e^{i(\omega t-kZ)} \tag{3.21}$$

where

$$k = \frac{2\rho\omega^2}{s_o\lambda} \tag{3.22}$$

Note that (3.21) satisfies (3.18) and (3.19) exactly, but not (3.20). Nevertheless (3.20) is satisfied approximately for large X because of the exponential decay as $Y \to -\infty$. Note also that our approximate solution is independent of the depth of the original cochlea model. It is also independent of the details of how the cochlea is driven.

Transforming back to the coordinates of the cochlea model, we find

$$\phi(x,y,t) = e^{i(\omega t-ke^{\lambda z})} \tag{3.23}$$

To find the vertical velocity we evaluate

$$\frac{\partial\phi}{\partial y}(x,y,t) = Ae^{i\theta} \tag{3.24}$$

where

$$A = k\lambda e^{(\lambda x+ke^{\lambda x}\sin\lambda y)} \tag{3.25}$$

$$\theta = \omega t - ke^{\lambda x}\cos\lambda y + \lambda y \tag{3.26}$$

For $y = 0$ we have simply

$$A = k\lambda e^{\lambda x} \tag{3.27}$$

$$\theta = \omega t - ke^{\lambda x} \tag{3.28}$$

According to these expressions, both the amplitude and the spatial frequency $(-\partial\theta/\partial x)$ of the basilar membrane velocity blow up exponentially as $x \to \infty$. This pathological behavior is a consequence of leaving viscosity out of the model. Nevertheless, we find something interesting if we examine the solution at a fixed depth $-\lambda y = \delta > 0$. Then the amplitude A has the form

$$A = k\lambda e^{(\lambda x - ke^{\lambda x}\sin\delta)} \tag{3.29}$$

which has a unique maximum at x_p given by

$$\lambda x_p = \log\frac{1}{k\sin\delta} \tag{3.30}$$

The expressions for A and θ can be written more simply in terms of $x - x_p$ as follows

$$A = \frac{\lambda}{\sin\delta}e^{(r-e^r)} \tag{3.31}$$

$$\theta = \omega t - \delta - \frac{e^r}{\tan\delta} \tag{3.32}$$

where

$$r = \lambda(x - x_p) \tag{3.33}$$

This result is sketched in Fig. 3.3. At any finite depth, then, the solution is very well-behaved. The amplitude A given by (3.31) rises exponentially for $r \ll 0$, achieves it maximum at $r=0$ $(x=x_p)$, and decays very rapidly $(\sim\exp(-\exp(r)))$ for $r \gg 0$. The spatial frequency still blows up exponentially but this is not a serious problem since the amplitude is decaying so rapidly. These results are consistent with the pathological behavior when $y=0$ because $x_p \to \infty$ as $y \to 0$.

Perhaps we can use this inviscid solution to get an idea how the model basilar membrane will behave when the fluid viscosity is small but not zero. By "small" fluid viscosity we mean that the boundary layer thickness $(\mu/\rho\omega)^{\frac{1}{2}}$ is small compared to λ^{-1}. In these circumstances we expect the inviscid solution to be valid outside the boundary layer and we can get a rough idea of what the basilar membrane is doing by evaluating the inviscid solution at the "edge" of the boundary layer. This motivates setting $y = -(\mu/\rho\omega)^{\frac{1}{2}}$ so that $\delta = \lambda(\mu/\rho\omega)^{\frac{1}{2}}$. By assumption, δ is small, so we set $\sin\delta = \delta$ in (3.30).

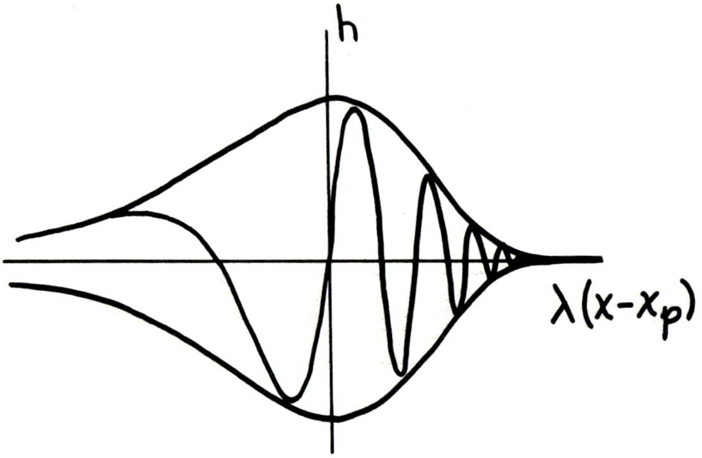

Fig. 3.3 Form of the cochlea wave as calculated by evaluating the inviscid solution at the edge of the viscous boundary layer. The direction of propagation is from left to right. Note the asymmetry of the envelope, which decays much more rapidly to the right than to the left. The spatial frequency increases exponentially from left to right.

Recalling the formula for k (3.22) we find the following expression for the cochlea map

$$\lambda x_p = -\frac{3}{2} \log \frac{\omega}{\omega_o} \qquad (3.34)$$

where $\omega_o = (s_o^2/4\rho\mu)^{1/3}$.

We are now in a good position to discuss the physics of the cochlea wave under conditions of small viscosity. The waves in the cochlea are concentrated near the basilar membrane with a depth of penetration into the fluid that is roughly equal to the local wavelength. The wavelength decreases as the wave propagates into the cochlea, however, so the depth of penetration decreases and the amplitude rises as the energy of the wave is compressed against the basilar membrane. When the wavelength becomes less than the boundary layer thickness, however, essentially all of the energy is in the viscous boundary layer and the amplitude of the wave decays rapidly. Thus the point of maximum amplitude occurs near the place where the wavelength and boundary layer thickness are equal. This condition determines the form of the cochlea map.

d. The viscous case: reduction to an integral equation

In this section we return to the viscous problem (3.1)-(3.7) and we consider solutions that are harmonic in time. These solutions satisfy a certain integral equation for the basilar membrane displacement. The derivation of this integral equation proceeds as follows:

First, express all quantities in terms of their complex amplitudes, e.g., let

$$h(x,t) = \text{Re}\left(H(x)e^{i\omega t}\right) \qquad (3.35)$$

Next, take the Fourier transforms in the x-direction, e.g., let

$$\hat{H}(\xi) = \frac{1}{\sqrt{2\pi}} \int_{-\infty}^{\infty} h(x)e^{-i\xi x}dx \qquad (3.36)$$

After these manipulations, the fluid equations (3.1-3.3 with boundary conditions 3.5-3.7) become a system of ordinary differential equations in y with ξ as a parameter. These equations can be solved explicitly for a relationship between $\hat{H}$ and $\hat{F}$ of the form

$$\hat{H}(\xi) = -\hat{K}(\xi)\hat{F}(\xi) \qquad (3.37)$$

where

$$\hat{K}(\xi) = \frac{1}{2}\frac{\xi}{\rho\omega^2\alpha}\frac{2\alpha\xi(1-\cosh\xi a\cosh\alpha a) + (\alpha^2+\xi^2)\sinh\xi a\sinh\alpha a}{\alpha\cosh\xi a\sinh\alpha a - \xi\sinh\xi a\cosh\alpha a} \tag{3.38}$$

$$\alpha = \sqrt{\xi^2 + \frac{i\omega\rho}{\mu}} \tag{3.38a}$$

When $a = \infty$, this expression simplifies as follows:

$$\hat{K}(\xi) = \frac{1}{2}\frac{|\xi|}{\rho\omega^2}\left(1 - \frac{1}{\sqrt{1+\frac{i\omega\rho}{\mu\xi^2}}}\right) \tag{3.38b}$$

which behaves like $|\xi|$ for small $|\xi|$ and like $|\xi|^{-1}$ for large $|\xi|$.

From (3.4) we also have

$$F(x) = -s_o e^{-\lambda x} H(x) \tag{3.39}$$

To combine these equations, we introduce the following operator notation

$$\mathcal{F} = \text{Fourier transform} \tag{3.40}$$

$$\mathcal{E} = \text{Multiplication by } e^{-\lambda x} \tag{3.41}$$

$$\hat{\mathcal{K}} = \text{Multiplication by } \hat{K}$$

Note that $\mathcal{F}$ is unitary, so $\mathcal{F}^* = \mathcal{F}^{-1}$. Taking the Fourier transform of (3.39) and substituting in (3.37), we find

$$\hat{H} = s_o\hat{\mathcal{K}}\mathcal{F}\mathcal{E}\mathcal{F}^*\hat{H} \tag{3.43}$$

which can also be written

$$H = s_o\mathcal{F}^*\hat{\mathcal{K}}\mathcal{F}\mathcal{E}H \tag{3.44}$$

we can think of (3.44) as an eigenproblem for the operator $\mathcal{F}^*\hat{\mathcal{K}}\mathcal{F}\mathcal{E}$, which is not self-adjoint. Suppose we have found an eigenfunction H corresponding to some eigenvalue s_o^{-1}. It is easy to check that the translates of H are also eigenfunctions corresponding to different values of s_o. In this way we can show that the spectrum of $F^*\hat{K}FE$ contains the positive real axis if it contains any point on that axis. We can therefore prescribe s_o arbitrarily and try to determine H. For numerical purposes it is useful to state this problem in least-squares form: Find H subject to

$$\int_{-\infty}^{\infty} |H|^2 dx = 1 \tag{3.45}$$

that minimizes

$$\int_{-\infty}^{\infty} |H - s_o F^*\hat{K}FEH|^2 \, dx \tag{3.46}$$

e. Two numerical methods for the viscous problem

The methods that will be outlined here were developed by Peskin (1976) and Isaacson (1979). Both are based on the least-squares formulation (3.45)-(3.46). Upon discretization, this leads to the problem of finding an eigenvector corresponding to the smallest eigenvalue of A^*A, where A is the matrix that arises from the discretization of $(I - s_o F^*\hat{K}FE)$. Such an eigenvector can be found by the inverse power method

$$A^*Aw^{n+1} = w^n / \|w^n\| \tag{3.47}$$

The two methods of this section use different discretizations, however, and they solve (3.47) in different ways.

In Peskin's method, the problem is discretized by introducing a mesh of N equally spaced points on the x-axis and a similar mesh on the ξ-axis. The mesh-widths are chosen as

$$\lambda \Delta x = \frac{\Delta \xi}{\lambda} = \left(\frac{2\pi}{N}\right)^{\frac{1}{2}} \tag{3.48}$$

so that $\Delta x \, \Delta \xi = 2\pi/N$, and

$$e^{-ix_j \xi_k} = e^{-i(j\Delta x)(k\Delta \xi)} = e^{-i\frac{2\pi}{N}jk} \tag{3.49}$$

Then the operator F is replaced by the discrete Fourier transform of order N, which has matrix elements, given by (3.49). The multiplication operators $\hat{K}$ and E are replaced by the appropriate diagonal matrices.

The matrix A*A that results from this process of discretization is not sparse, but this matrix is not needed explicitly, since (3.47) is solved using the subroutine package SYMMLQ (Paige and Saunders, 1973), which is a variant of the conjugate gradient method. This method needs the matrix of the system to be solved only in the form of a subroutine which can multiply this matrix by an arbitrary vector. In our case, multiplication by A*A can be broken down into a sequence of steps which are Fast Fourier Transforms or multiplication by diagonal matrices. Thus we can multiply by A*A in $O(N \log N)$ operations and solve our linear systems in $O(N^2 \log N)$ operations, since N multiplications by A*A are required. (In practice scaling is required to achieve the solution in N steps; see Peskin, 1976, for details.)

Some computational results obtained using this method are shown in Fig. 3.4. These results are for the simplest case which is $a = \infty$.

In Isaacson's method, discretization of (3.45)-(3.46) is achieved by means of the Rayleigh-Ritz procedure with the Hermite functions as an orthonormal basis. That is, the minimization is performed over the subspace of L_2 that is spanned by the first N Hermite functions. Note that the Hermite functions are eigenfunctions of the Fourier transformation.

It turns out that a good solution can be computed for moderate values of N, so Isaacson computes the matrix elements of A*A explicitly and uses the Cholesky factorization in the solution of (3.47). In computing the matrix elements, Isaacson uses Gauss-Hermite quadrature, and he takes advantage of the following identity

$$(FEH)(\xi) = \hat{H}(\xi - \lambda i) \qquad (3.50)$$

The expression $\hat{H}(\xi-\lambda i)$ always makes sense when $\hat{H}$ is a linear combination of Hermite functions, since the Hermite functions are entire functions.

In practice Isaacson's method is much faster than Peskin's because a comparable solution can be achieved with a much smaller value of N. Therefore his method was used to compute the form of the cochlea wave at several frequencies for a realistic finite depth and to construct a numerical cochlea map for the viscous problem. This map is plotted in Fig. (3.5).

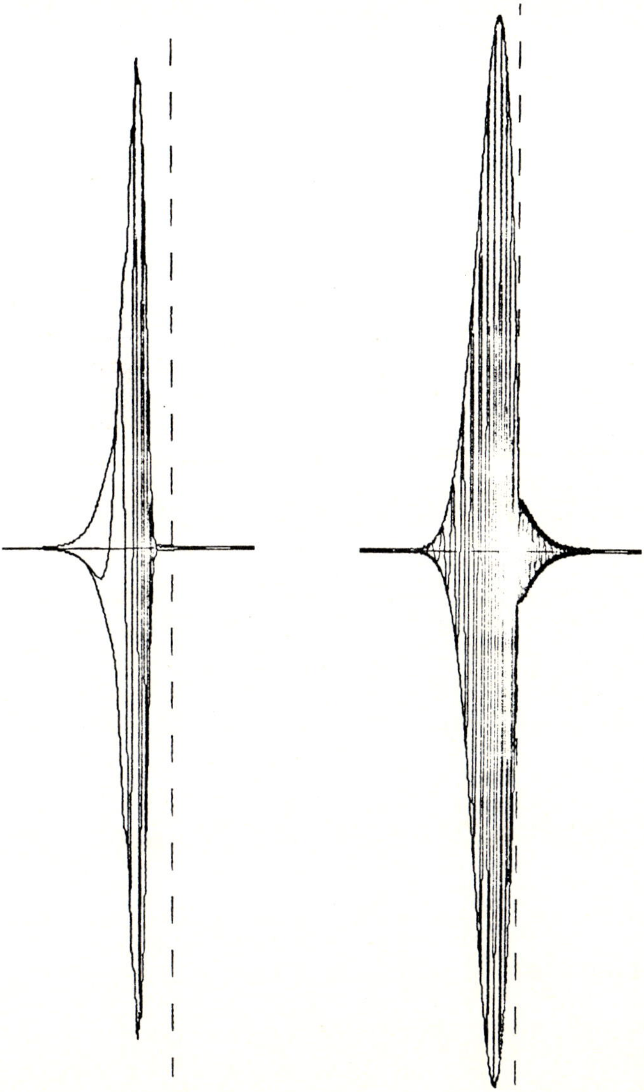

Fig. 3.4 The cochlea wave (left) and its Fourier transform (right). In the cochlea wave, note the gradual rise and rapid fall of the envelope as well as the increasing spatial frequency of the wave. In the Fourier transform note the predominance of negative frequencies which correspond to waves moving from left to right. This occurs despite the absence of any explicit reference to a source at $x = -\infty$ in the computation.

Acknowledgments

The author is indebted to the following individuals for their contributions to the work reported in this paper: Edward Yellin, Alexandre Chorin, Olof Widlund, David McQueen, Marjorie McCracken, Antoinette Wolfe, and Eli Isaacson.

The work on heart valves was supported by the National Institutes of Health (U.S.A.) under research grant HL-17859. Computation was also supported in part by the Department of Energy (U.S.A.) under contract EY-76-C-02-3077 at the Courant Mathematics and Computing Laboratory of New York University. Travel was supported by the NSF (U.S.A.).

References

Allen, J.B. (1977). "Two-dimensional cochlear fluid model: New results," J. Acoust. Soc. Am. 61, 110-119.

Bekesy, G. (1960) Experiments in Hearing (E.G. Weaver, trans.), McGraw Hill, New York.

Chorin, A.J. (1973). "Numerical study of slightly viscous flow," J. Fluid Mech. 57, 785-796.

Cole, J.D., and Chadwick, R.S. (1977). "An approach to mechanics of the Cochlea," ZAMP 28, 785-804.

Inselberg, A., and Chadwick, R.S. (1976). "Mathematical Model of the Cochlea," SIAM J. Appl. Math. 30, 149-179.

Isaacson, Eli (1979). A Numerical Method for a Finite-Depth, Two-Dimensional Model of the Inner Ear. Thesis, NYU (Mathematics), 1979.

Lesser, M.B., and Berkley, D.A. (1972). "Fluid mechanics of the cochlea. Part I," J. Fluid Mech. 51, 497-512.

McCracken, M.F., and Peskin, C.S. (1980). "A Vortex Method for Blood Flow Through Heart Valves," J. Comput Phys. (to appear).

Mendez, R. (1977). Numerical Study of Incompressible Flow in a Region Bounded by Elastic Walls. Thesis, U.C. Berkeley (Mathematics).

Paige, C.C., and Saunders, M.A. (1973). "Solutions of Sparse Indefinite Systems of Equations and Least Squares Problems" (report: STAN-CS-73-399) Stanford University (Computer Science).

Peskin, C.S. (1972). "Flow Patterns Around Heart Valves: A Numerical Method," J. Comput. Phys. 10, 252-271.

-- (1975). Mathematical Aspects of Heart Physiology Courant Institute Lecture Notes.

-- (1976). Partial Differential Equations in Biology Courant Institute Lecture Notes, Ch. 5.

-- (1977). "Numerical Analysis of Blood Flow in the Heart," J. Comput. Phys. 25, 220-252.

Peskin, C.S., and McQueen, D.M. (1980). "Modeling Prosthetic Heart Valves for Numerical Analysis of Blood Flow in the Heart," J. Comput. Phys. (to appear).

Peskin, C.S. and Wolfe, A.W. (1978). "The Aortic Sinus Vortex," Federation Proceedings 37, 2784-2792.

Siebert, W.M. (1974) "Ranke revisited -- a simple short-wave cochlear model," J. Acoust. Soc. Am. 56, 594-600.

Steele, C.R. (1974). "Behavior of the basilar membrane with pure tone excitation," J. Acoust Soc. Am. 55, 148-162.

Steele, C.R., and Taber, L. (1979a). "Comparison of WKB and Finite Difference Calculations for a Two-Dimensional Cochlear Model," J. Acoust. Soc. Am 65, 1001-1006.

-- (1979b). "Comparison of WKB Calculations and Experimental Results for Three-Dimensional Cochlear Models," J. Acoust. Soc. Am. 65, 1007-1018.

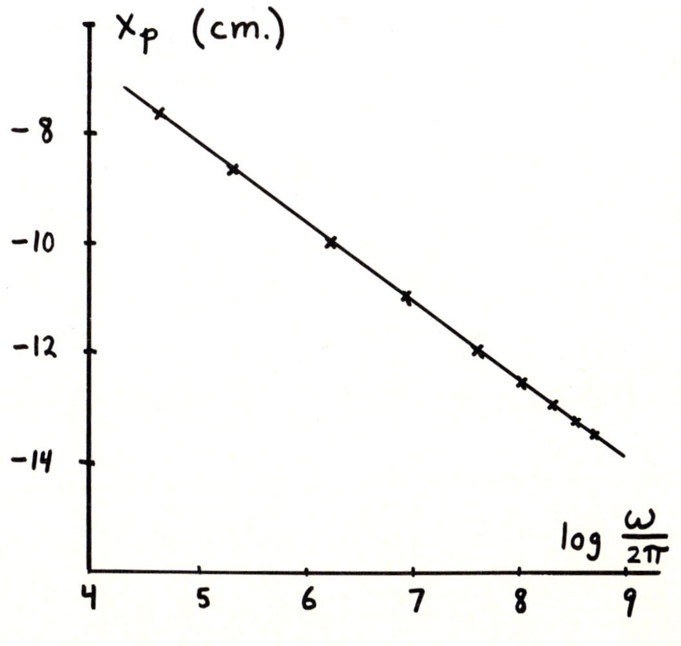

Fig. 3.5 The cochlea map (from the thesis of Eli Isaacson). The computations apply to the viscous, finite-depth case with the following parameters: $a = 0.2$ cm, $\mu = 0.02$ (cm^2/sec)·(gm/cm^3), $\rho = 1.0$ gm/cm^3, $\lambda = 1.4$ cm^{-1}.

COMPUTING METHODS IN APPLIED SCIENCES AND ENGINEERING
R. Glowinski, J.L. Lions (editors)
North-Holland Publishing Company

REGULARIZATION METHODS APPLIED TO AN INVERSE PROBLEM IN ELECTROCARDIOLOGY

P. Colli Franzone

Laboratorio di Analisi Numerica del C.N.R., Pavia

INTRODUCTION

During this last decade the technique of body surface mapping of the cardiac electric potential received great impulse in electrocardiology [1]. Compared to the usual technique of collecting 12 simultaneous electrocardiograms, the information content of a body surface map, which consists in collecting as many as 240 simultaneous electrocardiograms from evenly distributed locations on the torso, as proved to be a much more powerful tool to clinically discriminate between normal and abnormal cardiac states and among pathologies themselves. Moreover, surface maps furnish suitable information to attempt the solution of the "inverse" potential problem of electrocardiology i.e. the determination of "epicardial" from surface potential distributions in a volume conductor of known geometry [2,3]. The belief that the knowledge of the potential distributions over the "epicardium" would enhance clinical diagnosis was supported by experiments on the animals both in vitro [4] and in vivo [5].

As discussed in [6], the formulation of the inverse problem of electrocardiology leads to a Cauchy problem for a second order elliptic operator which, as it is well known, is an ill-posed problem in the usual functional spaces. To overcome the instability, the problem is here stated as a problem in control theory and it is approximated, via regularization techniques, by a family of stable problems which are numerically approximated using the finite element method. The performance of four different stabilization methods is tested on experimental data from isolated dog hearts [4]. The analysis of the accuracy attainable by the proposed stabilizations is preliminary to the attack of the inverse problem in the human case which is the real goal to be pursued.

Work performed in the frame of the Special Program on Biomedical Engineering of C.N.R.

1. THE MATHEMATICAL MODEL

As it is generally accepted [7], the human body is viewed as an isotropic, resistive, linear conducting medium excluding a region which contains all the primary sources (the heart) and including regions of different conductivities (the lungs and the bones, for instance) where no sources are present. The volume conductor is imbedded into an insulating medium (the air) and the electric field generated by the heart is considered quasi-static [7].

Under these assumptions, at any time, the electric cardiac potential satisfies an elliptic equation and its normal derivative vanishes at the boundary with the insulating medium.

Let $\Omega_1 \supset \Omega_0$ denote two bounded open nested sets of R^3, homeomorphic to a sphere, representing the human body and a fixed domain including the heart resp., and $\Gamma_1=\partial\Omega_1$, $\Gamma_0=\partial\Omega_0$ denote their boundaries. Γ_1 represents the body surface and Γ_0 represents a surface surrounding the heart and lying in proximity of the epicardial surface; in the following Γ_0 will be referred as "epicardial surface". Then define $\Omega=\Omega_1/\Omega_0$ with boundary $\partial\Omega=\Gamma=\Gamma_0 \cup \Gamma_1$, $\Gamma_0 \cap \Gamma_1=\phi$ and suppose Γ is a C^∞ manifold, fig. 1.1.

At any time t let $V(x)$ indicate the cardiac electric potential and $K(x) \geqslant K_0 > 0$, $K(x) \in C^\infty(\bar{\Omega})$ the electrical conductivity of the conducting medium Ω, $x \in \Omega$.

Then, if the potential values $u(x)$ are known on Γ_0, $V(x)$ in Ω satisfies the following problem:

(1.1) $\quad \operatorname{div} K(x) \operatorname{grad} V(x)=0 \text{ in } \Omega$

(1.2) $\quad V(x)=u(x) \qquad \text{on } \Gamma_0$

(1.3) $\quad \dfrac{V(x)}{\partial n}=0 \qquad \text{on } \Gamma_1$

where n indicates the outward normal to Γ_1.

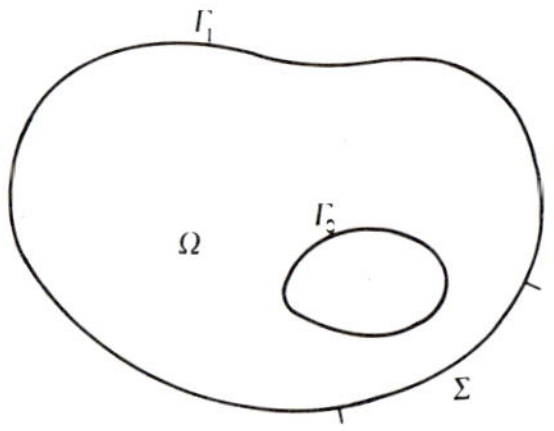

fig. 1.1

The solution of the "direct problem" which consists in evaluating $V(x)$ on Γ_1 is obtained by solving the mixed boundary value problem (1.1)-(1.3).

If no information is available about $V(x)$ on Γ_0 but it is possible to measure $V(x)$ on $\Sigma \subset \Gamma_1$ where Σ is an open subset on Γ_1 (in the real case $V(x)$ is measured in a finite set of locations on the torso), then $V(x)$ in Ω satisfies the following conditions:

(1.4) $\quad \operatorname{div} K(x) \operatorname{grad} V(x)=0 \qquad \text{in } \Omega$

(1.5) $\frac{\partial V(x)}{\partial_n} = 0$ on Γ_1

(1.6) $V(x) = z(x)$ on $\Sigma \subset \Gamma_1$

where $z(x)$ stands for measured potential values on Σ. The "inverse problem" of electrocardiology consists in estimating $V(x)$ on Γ_0 by solving the Cauchy problem for a second order elliptic operator (1.4)-(1.6).

If z were measured without errors, the Cauchy problem would uniquely define $V(x)$ but in a highly unstable fashion. In fact, it is well known that such a problem is an ill-posed problem i.e. the solution does not depend continuously on the data $z(x)$ in the usual functional spaces . This aspect is essential because, in the applications, $z(x)$ is always affected by a noise level. Hence, in order to solve the inverse problem, one must reduce it to the solution of a stable problem. The assumption of a distributed observation $z(x)$ on a surface Σ is not too restrictive since, by the technique of surface mapping |1| , the potential distribution on the torso can be recorded simultaneously in as many as 240 locations.

2. FORMULATION AS A CONTROL PROBLEM AND REGULARIZATIONS

The observation $z(x)$ is related to the surface Σ but condition (1.5) holds on the entire surface Γ_1; a way to take into account all the available information consists in restating the Cauchy problem as a control problem [8].

We shall use the following notations: $H^s(\Omega)$ with $s>0$ and $H^s(\Gamma_i)$, $i=0,1$, $H^s(\Sigma)$; $H^s(\Gamma)$ with s real, will denote the usual Sobolev spaces (see e.g. [9]). The corresponding norms will be denoted by $\|\cdot\|_{s,\Omega}$ and $|\cdot|_{s,\Gamma_i}$, $|\cdot|_{s,\Gamma}$, $i=0,1$, $|\cdot|_{s,\Sigma}$.

Define the state $y(x;v)=y(v)$ of the physical system by the following mixed boundary value problem:

$$\text{(2.1)} \qquad \begin{array}{ll} \operatorname{div} K(x) \operatorname{grad} y(v) = 0 & \text{in } \Omega \\ y(v) = v \text{ on } \Gamma_0, \ \dfrac{\partial y(v)}{\partial n} = 0 & \text{on } \Gamma_1 \end{array}$$

and introduce the following operator:

$$Av = y(v)_{|\Sigma}$$

For $z \in L^2(\Sigma)$ and $v \in L^2(\Gamma_0)$, we introduce the cost function

$$J(v) = \int_\Sigma |y(v) - z|^2 d\sigma = |Av - z|^2_{0,\Sigma}$$

which measures the distance between the measured surface data $z(x)$ and the predicted surface potential distribution and we consider the minimization problem:

(2.2) $\quad \underline{\text{Find}}\ u \in L^2(\Gamma_0):\ J(u) = \inf_{v \in L^2(\Gamma_0)} J(v)$

Assume that problem (1.4)-(1.6) admits a solution in Ω, then (2.2) admits a unique solution $u = V(x)|_{\Gamma}$ and $J(u) = 0$.

In this form the problem is still unstable because the operator A admits an unbounded inverse operator in the H^s space for all s. Hence in order to solve the problem (2.2) one must approximate it with stable problems via a stabilization procedure.

Among the methods reported in literature for the solution of ill-posed problems, especially for Fredhlom integral equations of the first kind, one may distinguish the two wide classes of regularization methods [see e.g. 10] and of statistical methods [see e.g. 11] which overcome the ill-conditioning by imposing smoothing constraints or utilizing a priori known deterministic or statistical information. As shown in [see e.g. 12], under reasonable hypotheses for the applications, regularization and statistical methods are related. A third class of methods is based on filtered singular value decomposition [see e.g. 13]. A different approach for the Cauchy problem is presented in [14].

Among these possible methods suited to attack the stabilization of the problem (2.2) we shall here develop some regularization techniques. We restrict the set of the admissible controls to a subspace U of $L^2(\Gamma_0)$ requiring the control $u = V(x)|_{\Gamma_0}$ to have regularity characteristics justified by the physical problem.

We consider the following cases:

(2.3) $\quad U = H^1(\Gamma_0) = \{v \in L^2(\Gamma_0):\ Bv \in L^2(\Gamma_0)\},\ B = \underset{\sim}{\text{grad}}$

(2.4) $\quad U = H^2(\Gamma_0) = \{v \in L^2(\Gamma_0):\ Bv \in L^2(\Gamma_0)\},\ B = \Delta|_{\Gamma_0}$, (see e.g. [9])

where $\underset{\sim}{\text{grad}}$ and $\Delta|_{\Gamma_0}$ are the gradient and the Laplace-Beltrami operator on Γ_0, respectively.

Then, the "inverse problem" (2.2) is approximated by a family of stable problems by means of the following optimal control problems for distributed systems [8] dependent on a smoothing parameter ε: for $\varepsilon > 0$,

(2.5) $\quad \text{Find}\ u_\varepsilon \in U:\ J_\varepsilon(u_\varepsilon) = \inf_{v \in U} J_\varepsilon(v)$

where

$$J_\varepsilon(v) = J(v) + \varepsilon \int_{\Gamma_0} |Bv|^2\, d\sigma$$

It is possible to show: [6]:

Theorem 2.1 If (2.3) or (2.4) holds, for any $\varepsilon>0$ the problem (2.5) admits a unique solution $u_\varepsilon \in U$ and the solution depends continuously on the data, i.e.: if $u^i_\varepsilon \in U$ is the solution of (2.5) for $z^i \in L^2(\Sigma)$ $i=1,2$, then:

(2.6) $$\|u^1-u^2\|_U \leq c(\varepsilon)\,|z^1-z^2|_{0,\Sigma}\ ,\ c(\varepsilon)=\eta(\varepsilon)^{-\frac{1}{2}},\eta(\varepsilon)\geq \min(c_1,\varepsilon c_2),c_1,c_2>0$$

Moreover if $z\in A(U)$ for $\varepsilon\to 0$, $u_\varepsilon \to u$ strongly in U where u is the unique solution of Au=z, and for $\hat{u}_\varepsilon$, solution of (2.5) for the data $\hat{z}$, holds:

(2.7) $$\|\hat{u}_\varepsilon-u\|_U \leq 2\ \sqrt{\frac{\varepsilon}{\eta(\varepsilon)}}\ |Bu|_{0,\Gamma_0} + \frac{|z-\hat{z}|_{0,\Sigma}}{\sqrt{\eta(\varepsilon)}}$$

Using the methods of the optimal control of distributed systems [8], it is easy to derive the following optimality system.

Theorem 2.2: $u\in U$ is a solution of (2.5) iff $(y_\varepsilon,p_\varepsilon,u_\varepsilon)\in H^1(\Omega)\times H^1(\Omega)\times U$ is the unique solution of the following variational system:

$$y_\varepsilon \in H^1(\Omega),\ y_{\varepsilon|\Gamma_0}=u_\varepsilon\ ,$$

(2.8) $$\int_\Omega K(x)\,\mathrm{grad}\,y_\varepsilon\cdot \mathrm{grad}\phi\,dx=0,\quad \forall\ \phi\in H^1(\Omega),\quad \phi_{|\Gamma_0}=0$$

$$p_\varepsilon\ \ H^1(\Omega),\ p_{\varepsilon|\Gamma_0}=0,$$

(2.9) $$\int_\Omega K(x)\,\mathrm{grad}\,p_\varepsilon\cdot \mathrm{grad}\phi\,dx=\int_\Sigma (y_\varepsilon-z)\phi\ d\sigma,\quad \forall \phi\in H^1(\Omega),\phi_{|\Gamma_0}=0$$

$$u_\varepsilon\ \ U,$$

(2.10) $$\varepsilon\int_\Gamma Bu_\varepsilon Bv\ d\sigma + \int_{\Gamma_0} K(x)\ \frac{\partial p_\varepsilon}{\partial n}\ v\ d\sigma=0,\qquad \forall\ v\in U$$

Further, $u_\varepsilon \in C^\infty(\Gamma_0)$

3. NUMERICAL APPROXIMATION

While in the approximation of case (2.3), only internal approximations of the space $H^1(\Gamma_0)$ are required, in the case (2.4), from a computational point of view, the approximation of the space $H^2(\Gamma_0)$ by a family of finite dimensional subspaces U_k, where k is the measure of the mesh size on Γ_0, is quite complex since it requires the use of conforming finite elements such that $U_k \subset C^1$. A way to avoid this difficulty consists in proceeding as in the mixed finite element method [15]. Set

$$\lambda_\varepsilon=\Delta_{|\Gamma_0} u_\varepsilon,\qquad G=\underset{\sim}{\mathrm{grad}}\ ,\qquad V=H^1(\Gamma_0)$$

where u_ε is the solution of (2.5) we have that $(u_\varepsilon,\lambda_\varepsilon)$ is the solution of the following equivalent variational system.

Find $(u_\varepsilon,\lambda_\varepsilon)\in H^2(\Gamma_0)\times L^2(\Gamma_0)$ such that:

$$\int_{\Sigma} Au_{\varepsilon} Av \, d\sigma + \varepsilon \int_{\Gamma_0} \lambda_{\varepsilon} \Delta v \, d\sigma = \int_{\Sigma} z \, Av \, d\sigma, \qquad \forall \, v \in H^2(\Gamma_0)$$

$$\int_{\Gamma} \lambda_{\varepsilon} \mu d - \int_{\Gamma_0} \Delta u_{\varepsilon} \mu d\sigma = 0, \qquad \forall \, \mu \in L^2(\Gamma_0)$$

From the regularity of u_{ε}, by means of the Green formula, we obtain that $(u_{\varepsilon}, \lambda_{\varepsilon})$ is a solution of the following problem:

Find $(u_{\varepsilon}, \lambda_{\varepsilon}) \in V \times V$ such that

$$\int_{\Sigma} Au_{\varepsilon} Av \, d\sigma - \varepsilon \int_{\Gamma_0} G\lambda_{\varepsilon} Gv \, d\sigma = \int_{\Sigma} z \, Av \, d\sigma, \qquad \forall \, v \in V \tag{3.1}$$

$$\int_{\Gamma_0} \lambda_{\varepsilon} \mu d\sigma + \int_{\Gamma_0} G\mu . Gu_{\varepsilon} d\sigma = 0, \qquad \forall \, v \in V \tag{3.2}$$

Moreover the solution of the variational problem (3.1)(3.2) is uniquely determined hence the variational system (3.1)(3.2) is an equivalent formulation of the problem (2.5) but its numerical approximation requires only the approximation of the space $V = H^1(\Gamma_0)$.

In order to find approximate solution of (2.5) it is necessary to introduce the approximation of the state equation. We shall now formulate a finite element method to construct an approximated operator A_h of A.

Let X_h be a sequence of finite dimensional subspaces of $H^1(\Omega)$ depending on the positive parameter h and we shall assume that the spaces X_h have the following approximation properties:

$$X_h \subset C^0(\bar{\Omega}) \cap H^1(\Omega) \quad \text{and} \quad X_{|\Gamma} \subset H^1(\Gamma)$$

where $X_{h|\Gamma}$ denote the functions defined on Γ which are restriction to of functions in X_h.

$$\inf_{\phi \in X_h} \|y - \phi\|_{s,\Omega} \leq C h^{3/2-s} \|y\|_{3/2,\Omega}, \ \forall \, y \in H^{3/2}(\Omega), \ \forall \, s \in [0,1]$$
$$\inf_{v \in X_{h|\Gamma}} |u - v|_{s,\Gamma} \leq C h^{1-s} |u|_{1,\Gamma}, \ \forall \, u \in H^1(\Gamma), \quad \forall \, s \in [0,1] \tag{3.3}$$

and the inverse assumption:

$$|v|_{1,\Gamma} < C h^{-1/2} \, |v|_{1,\Omega} \ , \ \forall v \in X_h \tag{3.4}$$

where C indicates a constant independent of y,u,h.

We introduce for $v \in V$ the approximated state equation:

$$\begin{cases} y_h(v) \in X_h, \ y_h(v) = \pi_h v \ \text{on} \ \Gamma_0 \\ \int_{\Omega} K(x) \operatorname{grad} y_h \cdot \operatorname{grad}\phi \, dx = 0, \quad \forall \, \phi \in X_h, \quad \phi_h = 0 \ \text{on} \ \Gamma_0 \end{cases} \tag{3.5}$$

where $\pi_h v$ denote the projection of $v \in H^1(\Gamma_0)$ onto $X_{h|\Gamma_0}$ and define

$A_h v = y_h(v)_{|\Gamma_1}$. We can now prove [6] the following error estimation:

Lemma 3.1: There exists a constant C independent of h such that, for $s \in [0,1]$:

(3.6) $$|(A-A_h)v|_{s,\Gamma_1} \leq Ch^{1-s}|v|_{1,\Gamma_0} \quad \forall\, v \in V$$

We shall now formulate a finite element method based on the variational formulation of (2.5) in the case we take as the regularization operator B=G.

In the following, since we are interested in the approximation and error estimation of the regularized problem (2.5) for fixed ε, we shall not indicate in the notation the dependence on ε.

Thus, let $\{V_k\}$ be a sequence of finite dimensional subspaces of $V=H^1(\Gamma_0)$ depending on the positive parameter k and we assume that the spaces V_k have the following approximation property:

(3.7) $$\inf_{v\ V_k} |u-v|_{1,\Gamma_0} \leq c\ k|u|_{2,\Gamma_0}, \qquad \forall\, u \in H^2(\Gamma_0)$$

For $z^h \in Y$, we shall consider the following finite element method:

(3.8) $$\text{Find } u_k^h \in V_k \text{ such that } a_h(u_k^h,v)=(f_h,v)_h, \ \forall\ v \in V_k$$

where for $u,v \in V_k$ and z^h

$$a_h(u,v)=(A_h u,A_h v)_Y+\varepsilon(Gu,Gv)_H$$

$$(f_h,v)_h=(z^h,A_h v)_Y$$

and we have set $Y=L^2(\Sigma)$, $H=L^2(\Gamma_0)$, $(\ ,\ )_Y$, $(\ ,\ )_H$ are the respective scalar products.

It is possible to prove [6] that for $\varepsilon>0$ there exists a unique solution of (3.8) and the following estimation holds:

Theorem 3.1 For $z^h \in Y$ there exists C independent of h and k such that:

$$|Au-A_h u_k^h|_{0,\Sigma}+|u-u_k^h|_{1,\Gamma_0} \leq C\ ((h+k)|u|_{2,\Gamma_0}+|z-z^h|_{0,\Sigma})$$

We shall now formulate a finite element method for problem (2.5) in the case of the regularization operator $B=\Delta_{|\Gamma_0}$ based on the variational formulation (3.1)(3.2).

$$\text{Find } (u_k^h,\lambda_k^h) \in V_k \times V_k \quad \text{such that:}$$

(3.9) $$(A_h u_k^h,A_h v)_Y-\varepsilon(G\lambda_k^h,Gv)_H=(z^h,A_h v)_Y, \ \forall\ v \in V_k$$

(3.10) $$(\lambda_k^h,\mu)_H + (G\mu,Gu_k^h)_H=0 \qquad , \ \forall\ \mu \in V_k$$

Then the following existence result and error estimation hold [6]:

Theorem 3.2 For $\varepsilon>0$, $z^h \in Y$, there exists a unique solution $(u_k^h, \lambda_k^h) \in V_k \times V_k$ satisfying (3.9)(3.10) and there exists a constant C indipendent of h and k such that:

$$|Au-A_h u_k^h|_{0,\Sigma}+|u-u_k^h|_{1,\Gamma_0}+|\lambda-\lambda_k^h|_{1,\Gamma_0}\leq C\left[(h+k)(|u_\varepsilon|_{2,\Gamma_0}+|\lambda_\varepsilon|_{2,\Gamma_0})+|z-z^h|_{0,\Sigma}\right]$$

Remark 3.1 For finite element spaces X_h and V_k satisfying (3.3)(3.4) (3.9) see [6].

4. NUMERICAL SOLUTION

The theoretical investigation of § 3 does not take into account the introduction of an approximate surface Γ^h of Γ which is needed to compute the numerical solution of problem (3.8) or (3.9)(3.10).

We shall give only an outline of the construction of this approximation, by means of isoparametric finite elements [15], which were used in the numerical computations. By means of a three dimensional mesh the domain Ω is approximated by a finite set of convex hexahdrhal elements, $\Omega_h=\bigcup_i C_i$, where each C_i is obtained by means of a polynomial mapping F_i linear in each variable defined on the unite reference cube $\hat{C}$ i.e. $C_i=F_i(\hat{C})$ and h is the maximum diameter of these elements.

Moreover we assume that the vertices of the elements belonging to the boundary $\Gamma^h=\Gamma_0^h \cup \Gamma_1^h$ of Ω_h lie also on the surface Γ. Then we consider the following spaces:

$$\tilde{X}_h = \{q:\ q\circ F_i^{-1} \in C^0(\bar{\Omega}_h) \text{ and linear on each variable on } \hat{C}\}$$

$$\tilde{U}_h = X_{|\Gamma_0^h} = \{\text{restriction of the functions of } X_h \text{ on } \Gamma_0^h\}$$

We note that the boundary Γ^h is composed by quadrilateral surface elements and the functions of $\tilde{X}_h(\tilde{U}_h)$ are univocally determined by the values they take on the node of the mesh $\Omega_h(\Gamma_0^h)$.

We can now formulate the following complete discrete form of (3.9)(3.10):

(4.1) find $(u_\varepsilon^h, \lambda_\varepsilon^h) \in \tilde{K}_h$:

$$J_\varepsilon^h(u_\varepsilon^h,\lambda_\varepsilon^h) = \inf_{(v^h,\mu^h)\in \tilde{K}_h} J_\varepsilon^h(v_\varepsilon^h,\mu_\varepsilon^h)$$

where:

$$\tilde{K}_h=\{(v^h,\mu^h)\in\tilde{U}_h\times\tilde{U}_h:\ \int_{\Gamma_0^h} v^h\nu^h d\sigma+\int_{\Gamma_0^h}\text{grad } v^h\cdot\text{grad } \nu^h d\sigma=0,\ \forall\, \nu^h\in\tilde{U}_h\}$$

$$J_\varepsilon^h(v^h,\nu^h)=\int_{\Sigma_h}|y^h(x;v^h)-z^h(x)|^2 d\sigma+\varepsilon\int_{\Gamma_0^h}|\nu^h|^2\, d\sigma$$

and $y^h(x;v^h)\in\tilde{Y}_h$, $y^h(x;v^h)=v^h$ on Γ_0^h

is solution of the following discrete state equation:

$$\int_{\Omega_h} k(x)\,\text{grad}\ y^h\cdot\text{grad}\ \phi^h\ dx=0$$
$$\forall\ \phi^h\in\tilde{Y}_h \quad \phi^h=0 \quad \text{on } \Gamma_0^h$$

Since for a fixed geometry (Ω,Γ,Σ) problem (4.1) must be solved repeatedly corresponding to the surface potential distribution z at different time instants of the heart beat and for each z several iterations are required for determing the "quasi optimal" smoothing parameter ε, it follows that the usual gradient techniques based on the use of the adjoint state variable to solve control problems is too time consuming and some more convenient alternative should be considered taking advantage of the linearity of the problem. This will be outlined below. Let $\{w_j^0(x),\ j=1,\dots,n\}$ and $\{w_j^1(x),\ j=1,\dots,r\}$ denote the basis of $\tilde{U}_h$ on Γ_0^h and of $\tilde{Y}_{h|\Sigma^h}$ on Σ_h respectively. Let $y_j^1(x)=y^h(x;w_j^0)_{|\Sigma h}$ i.e. $y_j^1(x)$ is the restriction to Σ_h of the solution $y^h(x;w_j^0)$ of problem (4.2) with Dirichlet data $w_j^0(x)$ on Γ_0^h. Then we have:

$y_j^1(x)=\sum_k \tau_{kj}w_k^1(x)$.

Given $u^h(x)\in U_h$ on Γ_0^h we have $u^h(x)=\sum_j u_j w_j^0(x)$ hence if $y^h(x;u^h)$ is the corresponding solution of (4.2) and $y\ (x)=y^h(x;u^h)_{|\Sigma_h}$ then:

$$y^1(x)=\sum_{jk}\tau_{kj}u_jw_k^1(x)$$

In the following if a,A,... are column vectors and matrices we denote by $a^*,A^*,\dots$ the transpose; in particular a^* is a row vector. Setting $T=(\tau_{kj})$ and considering the vectors $u=(u_1\dots u_n)^*$, $w^1(x)=(w_1^1(x)\dots w_r(x))^*$ we have $y^1(x)=w^{1*}Tu$. In particular if y^1 is the vector of the nodal values of $y^1(x)$ on Σ_h we have

$$y^1=T\ u,\ T \text{ is of order } (r,n)$$

i.e. T is the transfer matrix between the nodes of Γ_0^h and Σ_h.

If $\beta=(\beta_1\dots\beta_r)^*$ is the vector of the measured values of z in the nodes of Σ_h we shall define z(x) as the interpolated function:

$$z(x)=\sum_k\beta_kw_k^1(x) = \beta^*w^1=\beta\cdot w^1$$

Then we have:

$$\int_\Sigma |y^1(x)-z(x)|^2d\sigma=\int_{\Sigma_h}|w^1\cdot(Tu-\beta)|^2d\sigma=(Tu-\beta)^*\left(\int_{\Sigma_h}w^1w^{1*}d\sigma\right)(Tu-\beta)$$

Setting: $$\int_{\Sigma_h}w^1w^{1*}d\sigma = C^*C$$

with C upper triangular matrix (Choleski decomposition)

then: $\int_{\Sigma_h} |y^1(x)-z(x)|^2 d\sigma = ||Hu-b||^2$

where $H=CT$ and $b=C\beta$, H is of order (n,n)

If $(u^h,\mu^h)\in\tilde{K}_h$ then:

$$\int_{\Gamma_0^h} (\text{grad } u^h \cdot \text{grad } w_k^0 + \mu^h w_k^0)\, d\sigma = 0 \qquad ,\forall w_k^0 \in \tilde{U}_h$$

Introducing the vectors:

$$w^0(x) = (w_1^0(x) \ldots w_n^0(x))^*, \quad \mu = (\mu_1 \ldots \mu_n)^*, \quad u = (u_1 \ldots u_n)^*$$

we have $u^h(x) = \sum_j u_j w_j^0(x) = u \cdot w^0(x)$

$\mu^h(x) = \sum_j \mu_j w_j^0(x) = \mu \cdot w^0(x)$

and setting $\int_{\Gamma_0^h} w^0 w^{0*} d\sigma = P = D^* D$ (Choleski decomposition)

$$S = -\int_{\Gamma_0^h} \text{grad } w^0 \text{ grad } w^{0*}\, d\sigma$$

we obtain: $-Su+P\mu=0$ $\qquad \mu=P^{-1}S\,u$

Hence $\int_{\Gamma_0^h} |\mu^h|^2 d\sigma = u^* S P^{-1} \left(\int_{\Gamma_0^h} w^0 w^{0*} d\sigma\right) P^{-1} S u = u^* S P^{-1} S u = ||Ru||^2$

with $R=D^{*-1}S$.

Then problem (4.1) is equivalent to the n-dimensional least squares problem:

(4.3) $$\min_{u\in R^n} (||Hu-b||^2 + \varepsilon\, ||Ru||^2)$$

If in (4.1) the regularization term is $\varepsilon\int_{\Gamma_0^h} ||\text{grad } u^h||^2 d\sigma$ (G-reg) then we have (4.3) with $R=(-S)^{1/2}$.

Moreover we shall consider also the discrete problem (4.3) regularized by means of the euclidean norm (N-reg) i.e. $R=I$.

If instead of $\int_{\Sigma_h} |y^1(x)-z(x)|^2 d\sigma$ we consider $||y-z||^2$ where y and z are the vectors of the nodal values of $y(x)$ and $z(x)$ on Σ_h then (4.3) is approximated by:

(4.4) $$\min_{u\in R^n} (||Tu-z||^2 + \varepsilon\, ||Ru||^2)$$

we shall call (4.4) the discrete residual formulation.

For a mesh with a fairly high number of nodes uniformly distributed on Σ formulae (4.3) and (4.4) should have the same accuracy and this is borne out by the numerical tests. Therefore in the actual computations it is more advantageous to use (4.4) because it reduces computer time.

Any $u^h(x) \in \tilde{U}_h$ on Γ_0^h is a linear combination of $\{w_j^0(x)\}$ therefore $y^h(x;u^h(x))$ is a linear combination, with the same coefficients of $\{y^h(x;w_j^0(x))\}$.

The solution of problem (4.1) is achieved in two phases:

1) Numerical solution of n problems (4.2) with Dirichlet data $w_j^0(x)$ on Γ_0^h, computation of the transfer matrix T(H) and of the regularization matrix R.

2) Given z, computation of vector b and solution of problem (4.3).

Phase 1) is the most demanding in terms of computer time and memory occupation but for a given geometry is performed only once. In order to reduce the computer time and memory occupation a modified version of the frontal method suited to solve the n problem (4.2) which differ only for the Dirichlet data on Γ_0^h, was implemented.

As already mentioned the minimization problem of phase 2) must be solved for a sequence of z corresponding to different instants of the heart beat and for each z several values of ε must be considered. The first procedure to be used to solve this problem was the conjugate gradient method. However it was too time consuming and other more efficient procedures based on the generalized singular value decomposition have been applied. In particular the G-SVD of H and R has been developed following [16]. Two orthogonal matrices U,V of order n and one invertible matrix X of order n are determined such that: $U^*HX=D_H$ and $V^*RX=D_R$ with D_H,D_R non-negative diagonal matrices with diagonal elements arranged in decreasing order.

The condition number of D_H i.e. the ratio between the first and the last element of D_H is very high, thus revealing the ill-conditioned nature of H.

The solution u_ε of (4.3) is given by:

$$u_\varepsilon = Xq_\varepsilon \quad \text{where} \quad q_\varepsilon = \left(\frac{h_1 d_1}{h_1^2 + r_1^2}, \ldots, \frac{h_n d_n}{h_n^2 + \varepsilon r_n^2}\right)$$

with

$$d=U^*b, \quad D_H=\mathrm{diag}(h_1,\ldots,h_n), \quad D_R=\mathrm{diag}(r_1,\ldots,r_n)$$

The same results apply to (4.4) (discrete residual) with H=T and b=z. Since T and z are of order (r,n) and r respectively, U is of order (r,r) and D_H of order (r,n).

The computational advantage of this algorithm is that, given the discretization of the domain Ω and its boundary Γ, the factorisation of H and R is perfomed only once.

Another different way to stabilize the discrete inverse problem is that of truncated singular value decomposition (SVD) [13], i.e. to

solve in a least squares sense the linear system Tu-z=0. (u,z) vectors of nodal values).
This method, often referred to as the method of principal components, consists in replacing T by $T_{(k)}$ where $T_{(k)}=U\,D_{(k)}V$ and $D_{(k)}$ is the diagonal matrix of decreasing singular values of T truncated at the k-th element with the successive singular values set to zero. The choice of the truncation index k is somewhat equivalent to the choice of the regularization parameter ε. The solution $u_{(k)}$ is given by:

$$u_{(k)} = Vp_{(k)}$$

where

$$p^i_{(k)} = \begin{cases} g^i/h_i & , i=1, \quad ,k \\ 0 & , i=k+1, \ ,n \end{cases}$$

and $g=U^*z$ $\qquad D_{(k)} = \text{diag}\,(h_1,\ldots,h_k).$

5. APPLICATION

The stabilization methods described in the proceeding sections, i.e. Laplace (L), gradient (G), euclidean norm (N) and truncated singular value decomposition (SVD) were applied to a large set of data collected during experiments performed on isolated dog hearts [4] in order to extensively investigate the accuracy attainable in inverse calculations by each regularization operator.

The experimental data

The salient features which made these experiments most useful in testing an inverse procedure consisted in that they furnished almost simultaneous measurements within an "a priori" estimated accuracy of 7μV, of the potential distributions surrounding the epicardium and on the lateral surface of a tank. The tank contaning the isolated dog heart, a cylinder 15 cm high and 23 cm across, was filled with a highly conducting medium. The potential values were recorded from 600 locations in the bath of which 156 lying on the lateral surface Σ of the tank and the remaining in the bath so that most of them were quite uniformly distributed on ideal surfaces Γ_0, carrying 122 electrodes each, at an average distance from the epicardium of .5,1, and 7 cm, respectively.

The mesh

In the numerical calculations the domain Ω was that of the experiment, i.e. a cylinder 15 cm high and 23 cm across including a cavity 9 cm high and from 4 to 8 cm across, centered in the cylinder. The domain was discretized by a three dimensional mesh obtained subdividing the cylinder by 12 horizontal sections, each of which was divided into 12 sectors and each sector radially divided into 8 parts, thus yielding a mesh including 1200 hexahedral elements and 1446 nodes. Moreover, the nodes of the mesh lying on the inner and outer surfaces were, respecti vely, 122 and 156, i.e. there was correspondence between nodes in the mesh and sites of measurement, with no need of making interpolations for comparison with experimental data. Due to the homogeneity of the bath in the experiment, in the computations the electrical conductivity was assigned a constant unit value in the volume conductor: $k(x)=1$ in Ω.

Results

For sake of brevity, in describing the results, let:

z_{ex} and u_{ex} indicate the set of r=156 and n=122 potential values measu red, at time t, on the cylindical lateral surface Σ and on the cavity surface Γ_0, respectively;

z_{ct} the numerical solution on Σ obtained solving the direct problem of Eqs. (1.1)(1.2)(1.3) using $u=u_{ex}$ as input on Γ_0;

$u_{ct,ct}(u_{ct,ex})$ the inversely computed "epicardial" solutions obtained solving the inverse problem (1.4)(1.6)(1.7) using $z=z_{ct}(z=z_{ex})$ as input on Σ. If u and $\tilde{u}$ indicate experimental and calculated n-dimensional vector maps the comparison was performed by means of the two usual index: mean relative error (MRE) and correlation coefficient:

$$MRE = \frac{\|u-\tilde{u}\|}{\|u\|} \qquad CC = \frac{(u-u,\tilde{u}-\tilde{u})}{\|u-u\|\,\|\tilde{u}-\tilde{u}\|}$$

where $u,\tilde{u}$ are constant vectors of the arithmetic mean values. The stabi lization methods were applied to data for the entire beat in the geome try at .5 and 1 cm from the epicardium. Note that measured epicardial distributions u_{ex} in the two geometries were highly correlated. In fact, CC between the two distributions averaged .94 during the entire beat and never reached below .83.

Inverse calculations from simulated surface data

The numerical experiments hereafter reported on simulated surface data were performed to assess the accuracies of the different stabilization methods.

First, the surface potential distributions z_{ct} simulated from inner potential distributions u_{ex} collected at .5 cm from the epicardium in the normal beat, were used as input to the inverse procedures.

Note that, in the inverse calculations, the values of z_{ct} were rounded off to the closest integer, i.e. no fractions of μV were consi dered. Thus, the input data were affected by a white random noise of approximate standard deviation $\sigma=0.25$ μV.

Inverse distributions were computed for the entire normal beat by the stabilization methods previously described and Fig. 5.1 reports, for each time instant, the differences between MRE obtained by the L,N and SVD regularizations and MRE by the G method.

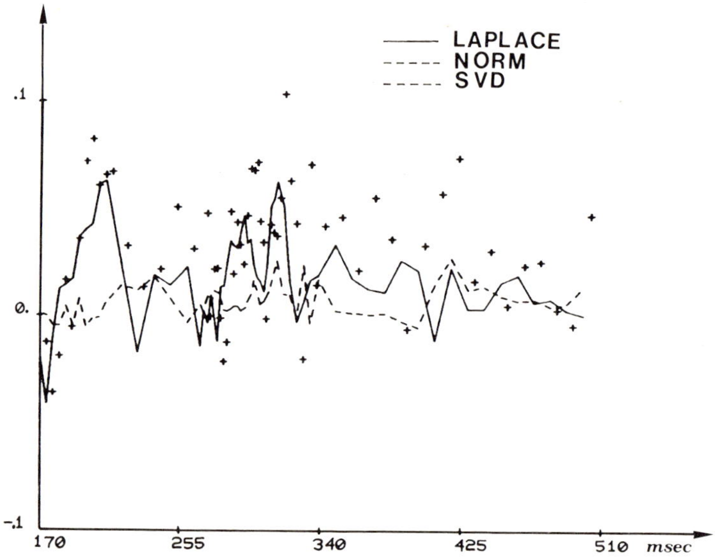

Fig. 5.1

The accuracies yielded by the four methods were all in the same range, with the SVD stabilization performing less well then the others and the regularization G giving the best results. The average values of MRE and CC over the entire beat, were, for each method, the following:

	N	L	SVD	G
MRE	0.45	0.44	0.46	0.43
CC	0.87	0.87	0.86	0.88

These results, as the following, were obtained minimizing the regularized cost function (2.5) considering the observation as discrete and no improvement was achieved in considering the case of a distributed observation, obtained from the discrete by suitable interpolations.

The regularization method by the G operator was then also applied to surface data simulated from measured inner potentials collected at 1 cm from the epicardium. By shortening the distance between inner and outer surfaces of only .5 cm, the average epicardial MRE over the entire nor mal beat lowered from .43 to .31 and, correspondingly, the average CC increased from .88 to .93.

All these results were obtained by computing at each time instant, the value of the regularization parameter $\varepsilon=r.10^{-10}$ (r=156 number of nodes on the surface) which minimized the MRE between computed $u_{ct,ct}$ and measured u_{ex}; we shall call this value of ε the "optimal" regularization parameter. It was observed that α remained quite constant over the heart beat waves, but at different values for each wave and over the entire beat, α averaged 2.88 and 3.45, in the two geometries, respectively. The surface residuals, computed as the root mean squares (RMS) differences between the input z_{ct} and the surface data simulated from the reconstructed $u_{ct,ct}$, i.e. $\frac{1}{\sqrt{r}} \|Tu_{ct}-z_{ct}\|$, showed that at all times oscillations of small amplitude about the average value which was 0.26 μV in both geometries in agreement with the standard deviation of the round off error which affected the data z_{ct}.

When simulated surface data with eight decimal digits (single precizion on a Honeywell 6040 System) were used in the inverse procedure, the accuracy of reconstructed $u_{ct,ct}$ greatly improved yielding, for instance, at 20 msec from QRS onset, a MRE=.09 versus the MRE=.26 obtained with z_{ct} rounded off at the closest integer. No better could be obtained due to the ill-posedness of the discrete inverse problem. Consider that the condition number of the discrete transfer matrix T relating surface to epicardial nodes was $2.2\cdot10^{10}$.

Inverse calculations from experimental surface data

All inverse numerical experiments performed on simulated surface data z_{ct} were also carried out on measured surface data z_{ex}. In this case, it was a priori known that surface data were affected by an "a priori" instrumental noise of approximate standard deviation σ=7μV and that a time-dependent error of unknown statistical characteristics had been superimposed on the measured data by the linear time-alignement [5]. Moreover, the matrice of transfer coefficients, evaluated on the basis of geometry measurements, were affected by the errors introduced by the numerical approximations to the originally continuous model. Note that, when dealing with experimental surface data, the inverse procedures estimated epicardial maps $u_{ct,ex}$ on the inner geometry deduced from the photographic prints taken after the experiment and the test of accuracy and pattern match was performed versus the measured epicardial u_{ex}, which had been collected on an epicardial geometry which differed slighly, but in an unknown fashion, from that assumed in building the transfer matrix. Therefore, the comparison of measured with inversely computed epicardial potentials could be only of qualitative type, yet the epicardial MRE was still used as index of accuracy. As observed for the simulated surface data, the accuracies yielded by the four methods were all in the same range, with the SVD stabilization performing less well then the others and the regularizations G and L almost equivalently, with the G method yielding slightly better accuracies than all others.

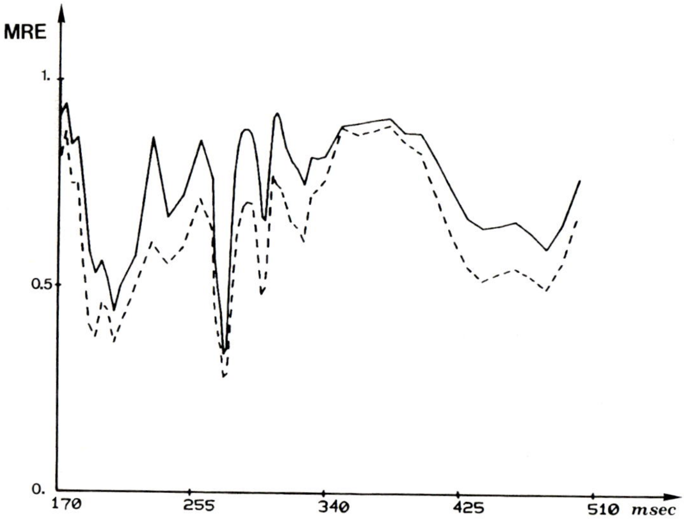

Fig. 5.2 Mean square root relative error between measured epicardial potentials and those inversely computed from experimental surface potentials. Solid line related to the geometry at .5 and dotted line to the geometry at 1 cm from the epicardium.

The time behavior of MRE obtained by the G regularization is reported in Fig. 5.2. By shortening the distance between inner and outer surfaces of only .5 cm, the average epicardial MRE over the entire normal beat decreased from .73 to .61 (see Fig 5.2) and, correspondingly, the average CC increased from .64 to .75. These results were obtained by computing, at each time instant, the value of the optimal regularization parameter $\varepsilon = r.10^{\alpha-10}$ (r=156; number of electrodes locations on the surface). The values of optimal α over the entire beat and for the two geometries showed high variability over the heart beat.
The surface residuals computed as the RMS differences between the input z_{ex} and the surface data simulated from the reconstructed $u_{ct,ex}$, i.e. $\frac{1}{\sqrt{r}} \| Tu_{ct,ex} - z_{ex} \|$, are reported in Fig. 5.3.
Residuals of surface fittings ranged between 2.5 and 12 μV, with average 5.2 μV, and were almost identical in the geometry at .5 and at 1 cm from the epicardium. The time-course of the residuals were suggestive of a time-dependent error on the surface data, which became more evident

over the waves, chiefly at times of rapid changes in epicardial potentials.

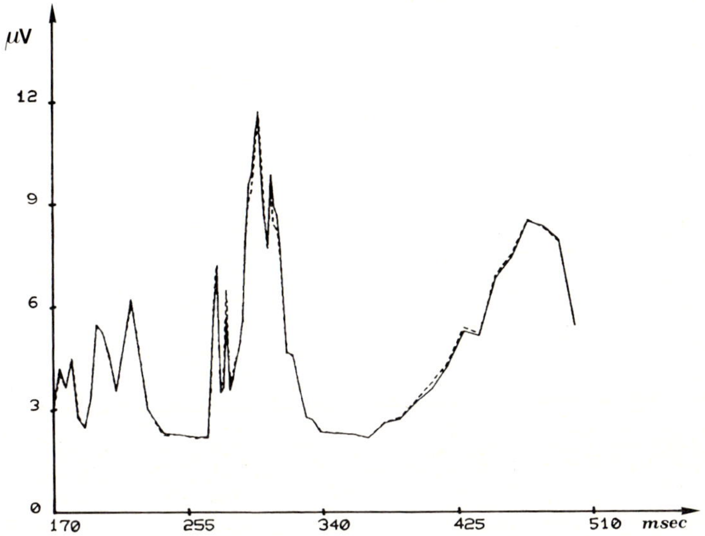

Fig. 5.3

Inverse calculations with the G regularization were also performed assigning α the costant average value attained by optimal α over the entire beat. When using average α, the main differences in accuracy and pattern match were in the time periods of very low surface data; it was, then, a posteriori concluded, that a mean value of the regularization parameter could be used for all time instants of the heart beat, thus notably reducing the cost of the inverse procedure and this conclusion was in agreement with the results reported in [3]. We also analized two methods proposed in literature for estimating the optimal regularization parameter ε, the discrepancy criterion(DC) [17] and the generalized cross validation (GCV) estimate [19] . This two methods were applied in inverse calculations from simulated data z_{ct} and it was observed that the GCV method, which did not utilize the a priori knowledge of σ, was successful in recognizing the best estimate $u_{ct,ct}$ even when a white noise level of standard deviation σ ranging between 1 and 100 μV, had artificially been added on z_{ct}. Both methods for choosing ε do not take into account the unavoidable errors coming from the numerical approximation of the mathematical model. The DC method had been theoretically extended to approximate model [18] but it is unpratical because its application requires the a priori knowledge of σ and the approximation error of the operator. In inverse calculations from experimental data z_{ex} the GCV method failed because the random

characteristic of the overall noise, due to measured surface data errors and model approximation errors, was not white. At present it is not completely clear which is the influence of each of these two error sources on the behaviour of the surface residual. The a posteriori analysis (fig.5.3) showed: a time dependent error, residual was the same in the two geometries and the average residual was of the magnitude of an "a priori" instrument noise level of 7 μV which affected the measured surface data. The understanding of the behavior of surface residual should be of help in finding new methods to estimate that ε which yields best average accuracy on the basis of surface information and model approximations. An example of $u_{ct,ex}$ reconstructed at 1 cm from the epicardium is reported in the bottom panel of the figure 5.4.
It may in general be remarked that inverse epicardial distributions correctly reproduced those features, like the number and approximate location of maxima and minima and coutour of equipotential lines which are relevant to the understanding of the activation sequence. The match between measured and inversely computed epicardial maps was accurate at times when epicardial potentials showed moderate gradients and less accurate when epicardial gradients were steep or surface data were at the noise level. In all cases, the reconstructed map was much more informative of cardiac events than it could be inferred from surface distributions.

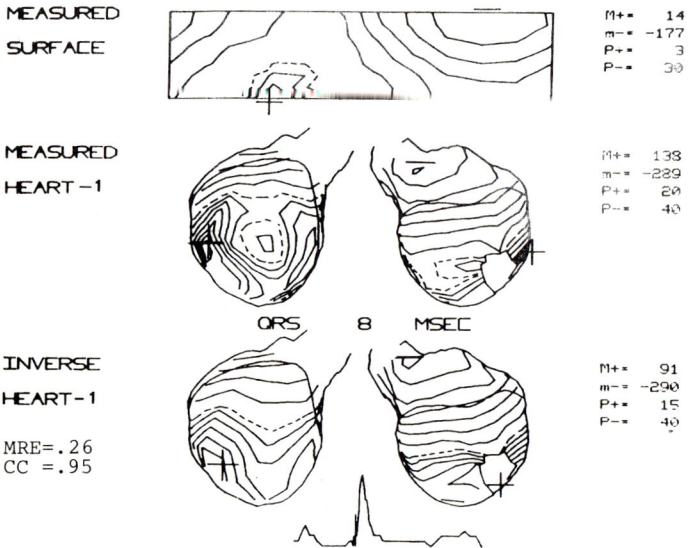

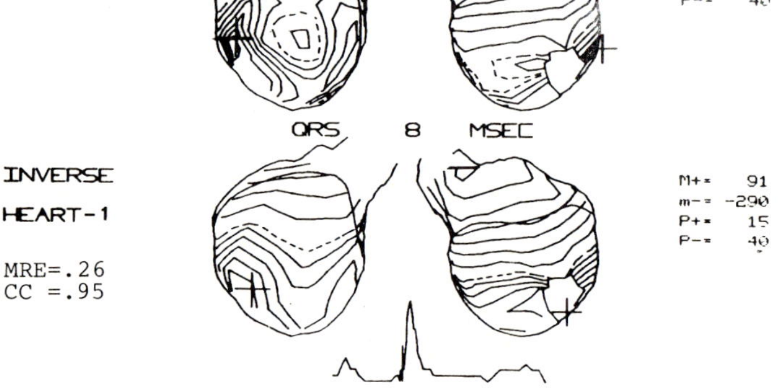

Fig. 5.4

BIBLIOGRAPHY

1 Taccardi B.,De Ambroggi L.,Viganotti C.: Body surface mapping of heart potentials, in The Theoretical Basis of Electrocardiology, edited by CV Nelson and DB Geselowitz, Oxford, Clarendon Press, 1976, pp. 436-466.

2 Colli Franzone P.,Taccardi B.,Viganotti C.: An approach to inverse calculation of epicardial potentials from body surface maps, in Adv. Cardiol. 21, Karger, Basel, 1977, 167-170.

3 Barr R.C.,Spach M.S.: Inverse Calculation of QRS-T Epicardial Potentials from Body Surface Potential Distributions for Normal and Ectopic Beats in the Intact Dog, Circ.Res., 45,5,1978, 661-675.

4 Taccardi B.,Viganotti C.,Macchi E.,De Ambroggi L.: Relationships between the current field surrounding an isolated dog heart and the potential distribution on the surface of the body, In Adv.Cardiol., 16, Karger, Basel, 1976, pp. 72-76.

5 Spach M.S., Barr R.C.,Lanning C.F.,Tucek P.C.: Origin of body surface QRS and T wave potentials from epicardial potential distributions in the intact chimpanzee. Circulation 55: 268-278, 1977.

6 Colli Franzone P.,Guerri L.,Taccardi B.,Viganotti C.: The direct and inverse potential problems in electrocardiology Report N.222 of L.A.N. C.N.R. Pavia 1979.

7 Plonsey R.,Fleming D.: Bioelectric Phenomena, Mac Graw-Hill,New York, 1969.

8 Lions J.L.: Optimal control of systems governed by partial differential equations, Grundlehren 170, Springer-Verlag, Berlin, 1971.

9 Lions J.L.,Magenes E.: Problèmes aux limites non homogènes et applications, Tome I, Dunod, Paris, 1968.

10 Tykhonov A.N.,Arsenine V.: Méthod de résolution de problèmes mal posée, Ed. Mir, Moscow, 1976.

11 Franklin N.J.: Well-posed stochastic extensions of ill-posed linear problems, J. Math. Anal. Appl., 31 (1970), pp. 682-716.

12 Stevens T.: A deterministic view of a statistical method for stable extension of unstable linear problems, J. Math. Anal. Appl., 64, (1978), pp. 251-262.

13 Lawson, C.L., Hanson R.J.: Solving least squares problems, New York, Prentice Hall Inc., 1974.

14 Lattes R.,Lions J.L.: The method of quasi-reversibility, American Elsevier, New York, 1969.

15 Ciarlet P.G.: The Finite Element Method for elliptic problems, Studies in Math. and Appl. Vol. 4, 1978 North-Holland.

16 Van Lôan C.: Generalizing the singular value decomposition, SIAM J. Numer. Anal. 13-1, 1976, 76-83.

17 Gordonova V.I.,Morozov V.A.: Numerical parameter selection algorithms in the regularization method, Zh. Vycisl. Mat. i Mat. Fiz.,13(1973), pp. 539-545.

18 Goncharskii A.V.,Leonov A.S.,Yagola A.G.: A generalized discrepancy principle, Zh. Vycisl. Mat. i Mat. Fiz., 13(1973) pp. 294-302.

19 Golub G.H.,Heath. M.,Wahba G.: Generalized cross-validation as a method for choosing a good ridge parameter, Report STAN-CS-77-622, Comp. Sc. Dept., Stanford University (1977).

SESSION X

ASYMPTOTIC AND PERTURBATION METHODS HOMOGENIZATION

COMPUTING METHODS IN APPLIED SCIENCES AND ENGINEERING
R. Glowinski, J.L. Lions (editors)
North-Holland Publishing Company

QUELQUES PROBLEMES DE Γ-CONVERGENCE.

Ennio De Giorgi
Scuola Normale Superiore, Pisa

Jusqu'à présent l'étude de la Γ-convergence et de la G-convergence a été développée surtout dans la théorie de l'homogénéisation. Cependant il y a beaucoup d'autres diréctions, à mon avis intéressantes, dans lesquelles cette étude est encore peu développée et je me propose de vous en indiquer quelques-unes pendant cette conférence.

Je pense que, comparés avec les développements possibles de la théorie, les résultats déjà connus ne représentent que le sommet d'un iceberg qui mériterait une exploration plus ample.

Mon exposé donnera seulement une indication de problèmes, pas une revue de résultats, ni une bibliographie. En particulier je tracerai les grandes lignes des problèmes sur les Γ-limites des obstacles, sur la convergence des minima locaux et des trajectoires de pente maximale, sur le rebond.

1. Γ-limites des obstacles.

La conjecture suivante joue un rôle central dans ce type de questions. Considérons les fonctionnelles

$$(1) \qquad F(u,\Omega) = \begin{cases} \int_\Omega f(x,u(x),Du(x))\,dx & \text{si } u \in W^{1,1}_{loc}(\Omega) \\ +\infty & \text{si } u \in L^1_{loc}(\Omega)\setminus W^{1,1}_{loc}(\Omega) \end{cases}$$

$$T_h(u,\Omega) = \begin{cases} 0 & \text{si } u \le \phi_h \text{ p.p. dans } \Omega \\ +\infty & \text{autrement} \end{cases}$$

où Ω est un ouvert de $\mathbb{R}^n$ et $\{\phi_h\}$ une suite de fonctions mesurables sur Ω. Supposons que

$$\Gamma(L^1_{loc}(\Omega)^-) \lim_{\substack{h\to+\infty \\ v\to u}} [F(v,\Omega) + T_h(v,\Omega)] = \bar{F}(u,\Omega) + T(u,\Omega) \;,$$

où $\bar{F}$ est la fonctionnelle semicontinue associée à F.

Nous faisons alors la conjecture que

$$(2) \qquad T(u,\Omega) = \int_{\bar{\Omega}} g(x,\bar{u}(x))\, d\mu(x)$$

où $g(x,t)$ est une fonction sur $\bar{\Omega}\times\bar{\mathbb{R}}$, μ une mesure de Borel et

$$\bar{u}(x) = \lim_{\rho\to 0^+} \rho^{-n} \int_{B(x,\rho)\cap\bar{\Omega}} u(y)\, dy \; ; \quad B(x,\rho) = \{y\in\mathbb{R}^n : |y-x|<\rho\} .$$

Il est très probable que cette conjecture est vraie dans le cas $f(x,u,p) = |p|^{\alpha}$ $(\alpha > 1)$ ainsi que la conjecture analogue pour le cas de l'obstacle bilatéral donné par

$$T_h(u,\Omega) = \begin{cases} 0 & \text{si } \psi_h \le u \le \phi_h \quad \text{p.p. dans } \Omega \\ +\infty & \text{autrement .} \end{cases}$$

Faisons observer qu'il est possible de répresenter par les fonctionnelles (2) les conditions au bord de Dirichlet en prenant

$$g(x,t) = \begin{cases} 0 & \text{si } x\in\partial\Omega \text{ et } t = \phi(x) \\ +\infty & \text{autrement} \end{cases}$$

$$\mu(B) = \mathcal{H}_{n-1}(\partial\Omega\cap B) \qquad (B \text{ borelien})$$

où $\mathcal{H}_{n-1}$ est la mesure de Hausdorff de dimension (n-1).

Dans quelques cas (peut-être $f(x,u,p) = |p|$) la conjecture devrait être modifiée: la fonctionnelle T, au lieu de la forme (2), pourrait avoir la forme

$$T(u,\Omega) = \int_{\bar{\Omega}} c(x)\, d\mu(x)$$

où $c(x) = \gamma(x,\bar{\bar{u}}_x)$, $\gamma : \bar{\Omega}\times L^1_{loc}(\mathbb{R}^n) \to \mathbb{R}$ et, pour tout $x\in\bar{\Omega}$, $\bar{\bar{u}}_x$ est la limite dans $L^1_{loc}(\mathbb{R}^n)$ quand $\rho\to 0^+$ des fonctions $\{W_{x,\rho}\}_{\rho>0}$ définies par

$$W_{x,\rho}(y) = \begin{cases} u(x+\rho y) & \text{si} \quad x+\rho y \in \Omega \\ 0 & \text{si} \quad x+\rho y \notin \Omega \end{cases} .$$

Autrement dit $\bar{\bar{u}}_x$ est la limite asymptotique (par rapport à la dilatation de centre x) de la fonction u et, naturellement, si $\bar{\bar{u}}_x$ est une fonction constante, elle coïncide avec $\bar{u}(x)$ multipliée par un nombre qui ne dépend pas de u .

Nous rappelons que des cas particuliers de limites d'obstacles ont été traités par H.Attouch, L.Boccardo et F.Murat, L.Carbone et F. Colombini. Ces derniers auteurs ont montré au moyen d'un exemple que, même dans le cas $f(x,u,p) = |p|^2$, T n'est pas toujours un "obstacle", c'est-à-dire que T peut prendre valeurs differentes de 0 et $+\infty$.

Quelques observations sur le problème de la limite des obstacles pour la fonctionnelle de l'aire ont été exposées pendant le seminaire "Superfici minime e questioni collegate" (Université de Trento (Italie), Mai 1979) en même temps que la construction d'un exemple de non-unicité pour le problème de minimum dans $BV(\Omega)$ pour la fonctionnelle

$$\int_B \sqrt{1+|Du|^2}\, dx + \int_{\partial B} |u-\phi|\, d\mathcal{H}_{n-1}$$

quand B est une boule ouverte du plan R^2 . Il est bien connu que ce dernier problème a une seule solution pourvu que la donnée au bord ϕ ait au moins un point de continuité dans ∂B et pour cette raison notre exemple concerne une fonction $\phi \in L^1(\partial B, \mathcal{H}_{n-1})$ qui n'est continue en aucun point de ∂B . Il me semble qu'un tel exemple de non-unicité ne se trouve pas dans la littérature, mais naturellement je serai reconnaissant à celui qui me signalera le contraire.

2. Convergence des minima locaux.

Nous dirons qu'une fonction $u \in L^1_{loc}(U)$ est un minimum local dans un ouvert U de $\mathbb{R}^n$ (pas nécessairement borné) d'une fonctionnelle F du type (1) si

$$F(u,\Omega) \le F(u+\phi,\Omega)$$

pour tout ouvert $\Omega \subset\subset U$ et pour toute fonction $\phi \in L^1_{loc}(U)$ dont le support est contenu dans Ω.

Considérons le problème suivant. Supposons que $\{F_h\}$ est une suite de fonctionnelles du type (1) et que

$$(3) \qquad F_\infty(u,\Omega) = \Gamma(L^1_{loc}(\Omega)^-) \lim_{\substack{h\to+\infty \\ v\to u}} F_h(v,\Omega)$$

pour tout ouvert Ω contenu dans un ouvert fixé U de $\mathbb{R}^n$. Supposons encore que $\{u_h\}$ est une suite qui converge dans $L^1_{loc}(U)$ vers une fonction u_∞, tel que toute u_h est un minimum local de F_h dans U. La fonction u_∞ est elle ou non un minimum local de F_∞ dans U ?

Une autre question: supposons que $\{F_h\}$ et F_∞ verifient (3) et que u_∞ est un minimum local de F_∞ dans U ; alors est-ce qu'il existe une suite $\{u_h\}$ qui converge dans $L^1_{loc}(U)$ vers u_∞, telle que chaque u_h est un minimum local de F_h dans U ?

Rappelons que des reponses positives à ces questions pour les minima absolus (au lieu de locaux) sont à la base de l'introduction de la Γ-convergence.

Un premier résultat positif pour la première de ces questions a été obtenu par L.Modica dans le cas des fonctionnelles

$$F_h(u,\Omega) = \begin{cases} \int_\Omega [\frac{|Du|^2}{h} + h(1-u^2)^2]\,dx & \text{si } u \in W^{1,1}_{loc}(\Omega) \\ +\infty & \text{si } u \in L^1_{loc}(\Omega)\setminus W^{1,1}_{loc}(\Omega) \end{cases}$$

$$F_\infty(u,\Omega) = \begin{cases} 8/3 \ \mathrm{Per}_\Omega\{x \in \Omega : u(x)=1\} & \text{si } |u| = 1 \text{ p.p. dans } \Omega \\ +\infty & \text{autrement}. \end{cases}$$

($\mathrm{Per}_\Omega A$ est le périmètre de A dans Ω) et appliqué pour prouver un intéressant comportement asymtotique des minima locaux bornés de F_1 dans $\mathbb{R}^n$ qui est lié à la théorie des surfaces minimales. Plus récemment, un article de G.Dal Maso et L.Modica a donné une réponse positive à la première question quand $\{F_h\}$ est une suite de fonctionnelles du type (1) satisfaisant certaines majorations: la classe de ces fonctionnelles comprend presque toutes les fonctionnelles dont on a étudié jusqu'à aujourd'hui la Γ-convergence, comme par exemple les formes quadratiques, y compris dans le cas de l'homogénéisation, et les fonctionelles du type de l'aire.

3. Convergence des trajectories de pente maximale.

Commençons par donner la définition de trajectoire de pente maximale introduite par A.Marino, M.Tosques et moi-même.

Soit $f : M \to R$ une fonction définie dans un espace métrique (M,d). Posons

$$|\nabla_- f|(x) = \max \{0, \max_{y \to x} \lim \frac{f(x)-f(y)}{d(x,y)} \}$$

$$|\nabla_+ f|(x) = \max \{0, \max_{y \to x} \lim \frac{f(y)-f(x)}{d(x,y)} \} .$$

Nous appelons $|\nabla_- f|$ la pente descendante, $|\nabla_+ f|$ la pente ascendante; observons que, si $M = \mathbb{R}^n$ et si f est fonction dérivable, alors $|\nabla_- f| = |\nabla_+ f| = |Df|$.

Une fonction $\phi :]a,b[\to M$ définie sur un intervalle de $\mathbb{R}$ sera dite une trajectoire de pente descendente maximale pour f si la fonction $u(t) = f(\phi(t))$ verifie les deux égalités suivantes pour tout $t \in]a,b[$

$$\lim_{\varepsilon \to 0^+} \frac{u(t+\varepsilon)-u(t)}{\varepsilon} = [\, |\nabla_- f|(\phi(t))]^2 \tag{4}$$

$$\lim_{\varepsilon \to 0^+} \frac{d(\phi(t+\varepsilon),\phi(t))}{\varepsilon} = |\nabla_- f|(\phi(t)) . \tag{5}$$

Si $M = \mathbb{R}^n$ et si f est une fonction dérivable, alors (4) et (5) sont équivalentes à l'équation

$$\frac{d\phi}{dt} = (\operatorname{grad} f)\,(\phi(t))$$

Le cas échéant on peut considérer des formes plus faibles de (4) et

(5), en supposant par exemple que (4) et (5) sont vérifiées seulement presque partout sur $]a,b[$.

Il est maintenant possible de poser le problème suivant: si $\{f_h\}$ est une suite de fonctions définies sur M qui $\Gamma(M^-)$-converge vers une fonction f_∞ et si $\{\phi_h\}$ est une suite formée par des trajectoires de pente maximale descendante pour f_h qui converge uniformement sur les compacts de $]a,b[$ vers une fonction ϕ_∞ , alors la fonction ϕ_∞ est elle ou non une trajectoire de pente maximale descendante pour f_∞ ?

Le problème parait avoir une réponse positive assez facile quand M est un espace vectoriel et quand f est convexe, moins facile avec des hypothèses plus generales.

4. Le rebond.

Quelques cas multidimensionnels intéressants de problèmes de rebond ont été considérés par L.Amerio, C.Citrini, G.Prouse et M.Schatzman qui ont mis en evidence la grande délicatesse du problème de l'unicité.

Récemment M.Carriero et E.Pascali ont etudié le cas plus simple d'un point matériel qui se meut sur une droite en étant soumis à une force $f(t)$ dépendant du temps et à une loi de rebond parfaitement élastique.

Un problème de ce genre peut être obtenu, par exemple, comme la limite, quand $h \to +\infty$, de l'équation différentielle suivante

$$\frac{d^2u}{dt^2} = f(t) - hg'(u(t)) \tag{6}$$

où g est une fonction de la classe C^2 qui vérifie les hypothèses

$$\begin{cases} g(u) = 0 & \text{si} \quad u \leq 0 \\ g(u) > 0 & \text{si} \quad u > 0 \\ \lim\limits_{u\to 0^+} \dfrac{g'(u)}{g(u)} = +\infty & . \end{cases} \tag{7}$$

En faisant seulement l'hypothèse $f \in L^1_{loc}(\mathbb{R})$, on démontre que, si t_0, a, b sont fixés dans $\mathbb{R}$ et $a \leq 0$, il est possible d'extraire de la suite des solutions du problème de Cauchy $u(t_0) = a$, $u'(t_0) = b$

pour l'équation (6) une sous-suite qui converge dans $L^1_{loc}(]a,b[)$, tandis qu'il semble beaucoup plus difficile de démontrer la convergence de la suite sauf dans des cas particuliers, comme $f \leq 0$, ou f fonction en escalier.

Il est intéressant observer que la pénalisation introduite par g n'est pas nécessairement convexe: par example g peut coïncider sur la demi-droite positive avec un polynôme positif dont les dérivées première et seconde sont nulles à l'origine.

Cet exemple fait penser que, même dans beaucoup d'autres cas, les pénalisations convexes peuvent être remplacées par des pénalisations rapidement croissantes, c'est-à-dire satisfaisant une condition analogue à la troisième condition de (7). D'ailleurs les résultats de L.Amerio, C.Citrini, G.Prouse, M.Schatzman montrent déjà que dans l'étude des équations

$$\frac{d^2u}{dt^2} = (\mathrm{grad}\, f)(u(t))$$

la convexité de f ne fournit pas une aide comparable à celle qu'on obtient dans l'étude de l'equation correspondante parabolique.

COMPUTING METHODS IN APPLIED SCIENCES AND ENGINEERING
R. Glowinski, J.L. Lions (editors)
North-Holland Publishing Company

HOMOGENIZATION AND PERTURBATION PROBLEMS

N.S. Bakhvalov
Moscow State University

The study of many problems of science and engineering demands to research the characteristic of media with periodic structures [1-4]. The examples of such media are crystal, polymer, fabric, antenna, radiator, brick wall. Last time the interest to this problems increased and a lot of mathematicians work in this field [5-13].

Let us suppose, that the region of regularity is transformed in cube $0 \leqslant x_1, \ldots, x_s \leqslant \varepsilon$. In some cases the problems are interesting, where the original region is parallelepiped $0 \leqslant x_j \leqslant \varepsilon_j = \varepsilon^{\lambda_j}$; in this case it is also suitable to transform this region in cube with the side ε.

In some cases it is nesessary to study the structures with characteristics slow change from one such cube to another, or structures with different characteristics in different parts of space (when some bodies with different periodical structures are in contact).

§1 Abstract formulation and methods of averaging

Let U^ε, F^ε be linear spaces and the imprints $u(X)$, $f(X)$ are definided for their elements u, f for almost everything $X = (x_1, \ldots, x_s)$. For convenience we identify the point X and vector X. Let S_m be the displacement operator:

$$S_m\, g(X) = g(X + m\varepsilon),\ m = (m_1, \ldots, m_s).$$

The operator p reflecting U^ε in F^ε corespond to periodical structure if

$$p(S_m\, u) = S_m\,(pu) \tag{1.1}$$

for any vector m with all whole m_i. Let P_o^ε be the set of operators for which (1.1) is true.

The condition (1.1) in elasticity theory means: the structure is periodical if the shift of the displacement field to m leads to the same shift of the forces. Let P^{ε} be the set of operators $p(X,u)$, which transforme $R \times U$ in F^{ε} and have the characteristic: if Y is fixed, then operator $p(Y,u)$ belong to P_0^{ε}. Such operator p correspond to structure with slowly changing characteristics. We consider the equation

$$p(X,u) = 0, \; p \in P^{\varepsilon}. \tag{1.2}$$

By analogy with P_0^{ε} and P^{ε} we determine the operators sets $\mathbf{Q}_0^{\varepsilon}$ and $\mathbf{Q}^{\varepsilon}$ with periodical characteristics, which transform U^{ε} in U^{ε} and $R \times U$ in U^{ε} accordingly. We need in some smoothness $p(X,u)$ with respect to X, else we recieve usually only zero approach. The general recomendation about averaging of equation (1.2) is to use the standart methodics of asymptotic expansion. We seek an asymptotic series of u in the form

$$u \sim \sum_{\ell=0}^{\infty} \varepsilon^{a_\ell} N_l(X,Z), \tag{1.3}$$

where $a_l < a_{l+1}$, $Z = (\zeta_1, \ldots, \zeta_s)$, $\zeta_j = x_j/\varepsilon$ and $N_l(X,Z)$ are periodical on ζ_j; the values a_l will be determined below. If some element depends on Z we mean usually that it is periodical on all of ζ_j and period is equal 1. We expand $p(X,u)$ in an asymptotic series and choose a_l and N_l to satisfy the asymptotical equality $p(X,u) \sim \sum \varepsilon^{b_\ell} H_l(X,Z) \sim 0$. The condition that $p(X,u)$ expand in asymptotic series leads in nonlinear case to some equation which containing first of a_l and N_l. There are some examples when using the form (1.3) does not lead to the aim: we must find the solution in more general form

$$u \sim \varepsilon^{a} \int_0^{\infty} \varepsilon^{\lambda} N(\lambda, X, Z)\, dF(\lambda), \tag{1.4}$$

where F have a local limited variation. By analysing the equations $H_l(X,Z) = 0$ we find more definite form of asymptotical series

$$u \sim \sum \varepsilon^{a_\ell} N_l(X,v), \; N_l \in \mathbf{Q}^{\varepsilon}, \tag{1.5}$$

or more general form

$$u \sim \varepsilon^{a} \int_0^{\infty} \varepsilon^{\lambda} N(\lambda, X, v)\, dF(\lambda).$$

Here v is unknown smooth function of x and ε. We consider further the form (1.5). We expand $p(X,u)$ in an asymptotic series

$$p(X,u) \sim \sum \varepsilon^{b_\ell} H_1(X,v). \qquad (1.6)$$

We will choose a_1 and N_1 from conditions that such asymptotical series exist and operators H_1 do not depend on Z:

$$H_1(Y, S_q\, v) = S_q\, H_1(Y,v)$$

for any vector q. Then the equation

$$\bar{p}(X,v) \sim \sum \varepsilon^{b_\ell} H_1(X,v) \sim 0 \qquad (1.7)$$

does not contain fast variables ξ_j. We can find $v(X, \varepsilon)$ as the regular asymptotic series in powers of ε. The first method is more universal; in that time the second methods is more convenient for physical interpretation. The form of asymptotic expansion (1.5) may be concretized [6-8]. We denote $v^{(j_1,\dots,j_\ell)} = \frac{\partial}{\partial x_{j_1}} \cdots \frac{\partial}{\partial x_{j_\ell}} v$, $v^{(q;j_1,\dots,j_\ell)} = \frac{\partial^q}{\partial t^q} v^{(j_1,\dots,j_\ell)}$. In constructing the averaged equation, the order of the differentiations with respect to different variables x_i cannot be interchanged. We recieve more information about problems, if we impose this restriction [8,12].

For many linear equations $pu = f(X)$, where $p \in P_0^\varepsilon$, the form of asymptotical expansion is suitable:

$$u \sim \sum \varepsilon^{p+1} N^p_{j_1,\dots,j_1}(Z)\, v^{(j_1,\dots,j_\ell)}\ . \qquad (1.8)$$

Often we recieve at first some asymptotical expansion

$$p(X,u) \sim \sum \varepsilon^{c_\ell} p_1(X,u),$$

where $p_1 \in P_0$. Here P_0 is set of operators $p \in P_0^\varepsilon$ for which the following condition is execute: if $u = u(Z) \in U^\varepsilon$ does not explicity depend on ε, then $pu = h(Z)$ does not explicity depend on ε. For example we represent the operator p: $pu = \frac{d}{dx}(a_2(\xi)\frac{du}{dx}) + a_0 u$ in the form $p = \varepsilon^{-2} p_{-2} + p_0$, where

$p_{-2}u = \varepsilon^2 \frac{d}{dx}(a_2 \frac{du}{dx})$, $p_0 u = a_0 u$, $p_{-2}, p_0 \in P_0$.

§2. The formal averaging of elliptic and hyperbolic equations

We consider the system of equations

$$p_0 u = -Lu = f(X), \quad Lu = \frac{\partial}{\partial x_i}(A_{ij}(Z) \frac{\partial u}{\partial x_j}), \qquad (2.1)$$

$$p_1 u = R(Z)\frac{\partial^2 u}{\partial t^2} - Lu = f(t,X), \qquad (2.2)$$

where R, A_{ij} are matrices, u, f are vectors, the summation is understood to be from 1 to s.

In many cases the coefficients do not depend on one of the fast variables, for example in (2.2) on $\tau = t/\varepsilon$. We can introduce the fictitions dependence from this variable, or [12] consider total combination $u(t,X)$ for all t as one element $u(X)$. We seek a solution of (2.1), (2.2) in a form

$$u \sim \sum_{l \le t,q} \varepsilon^{q+l} N^q_{j1,\dots,j_l} v^{(q;j_1,\dots,j_l)};\ N^0 = E.$$

Let us consider the system (2.2). The system (2.1) may be considered as a special case of systems (2.2) where u and v do not depend on t. We substitute u in (2.2) and recieve

$$p_1 u \sim \sum \varepsilon^{q+l-2} H^q_{j_1\dots j_l}(Z)\, v^{(q;j_1,\dots,j_l)}.$$

Here $H^q_{j_1,\dots,j_l} = -L_o(N^q_{j_1,\dots,j_l}) + T^q_{j_1,\dots,j_l}$,

$$L_o(N) = \frac{\partial}{\partial \xi_i}\left(A_{ij}(Z)\frac{\partial N}{\partial \xi_j}\right),\quad T^q_{j_1\dots j_l} = -\left(A_{j_1 j}\frac{\partial N^q_{j_2\dots j_l}}{\partial \xi_j} + \right.$$

$$+ \frac{\partial}{\partial \xi_j}\left(A_{jj_1} N^q_{j_2\dots j_l}\right) + A_{j_1 j_2} N^q_{j_3\dots j_l}\Big) + R\, N^{q-2}_{j_1,\dots,j_l},$$

$N^q_{j_1\dots j_l} \equiv 0$ in the case (2.1) if $q > 0$. We order the collection $(q;j_1,\dots,j_l)$ in some way keeping the parameters $q+l$ in lexicographic order. We introduce the notation

$$\left\{A(X,Z)\right\} = \varepsilon^{-s}\int_{x_1-\frac{\varepsilon}{2}-0}^{x_1+\frac{\varepsilon}{2}-0} \cdots \int_{x_s-\frac{\varepsilon}{2}-0}^{x_s+\frac{\varepsilon}{2}-0} A\left(Y, \frac{Y}{\varepsilon}\right) dy_1\dots dy_s;$$

and $\left[A(Z)\right] = \int_{0-0}^{1-0} \cdots \int_{0-0}^{1-0} A(Z)\, d\xi_1 \dots d\xi_s$. Let us suppose, that

$$0 < \alpha \le \int_{R_s} \left(A_{ij}(Z)\frac{\partial u}{\partial x_j}, \frac{\partial u}{\partial x_i}\right) dX \Big/ \int_{R_s}\left(\frac{\partial u}{\partial x_j}, \frac{\partial u}{\partial x_j}\right) dX \le \beta < \infty \qquad (2.3)$$

for all finite $u(X) \ne 0$. Each time we find $N^q_{j_1\dots j_l}$ from the condition that $H^q_{j\dots j}$ do not depend from fast variables ξ_j:

$$H^q_{j_1\dots j_l}(Z) \equiv h^q_{j_1\dots j_l}.$$

In the case (2.3) the equation $L_o(N(Z)) = G(Z)$ is solvable for $[G] = 0$, and its general solution is expressible in the form $N = \overline{N}(Z) + C$, where C is an arbitrary constant matrix. So we take each time

$$h^q_{j_1\cdots j_l} = \left[T^q_{j_1\cdots j_l} \right].$$

The final averaged equations have a form

$$\overline{p}_o v \sim \sum_{1<\ell} \varepsilon^{l-2}\, h^o_{j_1\cdots j_l}\, v^{(j_1,\ldots,j_\ell)} \sim f, \qquad (2.4)$$

$$\overline{p}_1 v \sim \sum_{1<q+\ell} \varepsilon^{q+l-2}\, h^q_{j_1\cdots j_l}\, v^{(q;j_1,\ldots,j_\ell)} \sim f. \qquad (2.5)$$

The coefficients in these equations are not uniquely defined. The choice of matrixes C influence on the values of some of the following $h^q_{j_1\cdots j_l}$, and is equivalent to some substitution

$$v \sim \sum \varepsilon^{q+l}\, g^q_{j_1\cdots j_l}\, w^{(q;j_1,\ldots,j_\ell)}, \quad g^o = E,$$

$g^q_{j_1,\ldots,j_\ell}$ are constant matrixes. We will specify some canonical form of equation (2.4) and (2.5). We choose each time the constant term in $N^q{}_{j_1,\ldots,j_\ell}$, $q+l > 0$ from condition $[N] = 0$.

Theorem 1. Then

$$\overline{p}_o \sim \sum_{1<\ell} \varepsilon^{l-2}\, p_l, \quad \overline{p}_1 \sim \sum_{1<2q+\ell} \varepsilon^{2q+l-2} \frac{\partial^q}{\partial t^q} p_{q,l}$$

where p_l, $p_{q,l}$ are symmetrical homogeneous differential operators of order l with respect to the variables x_j. The averaged equation $\overline{p}_1 v = f$ contain differentiation with respect to variable t of order higher then the second and therefore is not natural.

Theorem 2. There is some regular procedure to recieve the operator $\mathbf{Q}$ with constant coefficients

$$\mathbf{Q}w = w + \sum_{0<q,\,0\leq\ell} \varepsilon^{2q+l}\, g^{2q}_{j_1\cdots j_l}\, w^{(2q;j_1,\ldots,j_\ell)}$$

with following characteristic:

$$\mathbf{Q}^*\overline{p}_1\mathbf{Q}w = \overline{\overline{p}}_1 w = \overline{p}_o w + \frac{\partial^2}{\partial t^2} \sum_{0\leq\ell} \varepsilon^l\, \overline{p}_{2,l},$$

where $\overline{p}_o$ is the same operator as in Theorem 1 and $\overline{p}_{2,l}$ are symmetrical homogeneous differentias operators of order l with respect to the variables x_j. So we have the averaged equation $\overline{\overline{p}}_1 w = \mathbf{Q}^* f$.

Let w be finite. Then we recieve from there the integral of energy

$$\frac{dE}{dt} = \int_{R_s} (Q^* f, \frac{\partial w}{\partial t})\, dx.$$

The main term of E is positive by conditions $R > 0$ and (2.3). Let $\bar{L}$ be the main part of $\bar{p}_0$:

$$\bar{L}v = \frac{\partial}{\partial x_i}(\bar{A}_{ij} \frac{\partial}{\partial x_j}).$$

Lemma. If $A^*_{ij} = A_{ji}$, $\forall$ i,j, then $\bar{A}^*_{ij} \equiv \bar{A}_{ji}$. If the inequality (2.3) is true for the operator L, then it is true for the operator L.

Let a^{kl}_{ij} be elements of matrix A_{ij} and dimension of vector u be s.

Lemma. It $a^{kq}_{ij} = a^{kj}_{iq}$, $\forall$ i,j,k,q, then $a^{kq}_{ij} \equiv a^{kj}_{iq}$.

Consider now the first approach $u \sim u^1 = v + \varepsilon N_j \frac{\partial v}{\partial x_j}$. We have [6,13] for basic boundary problems $\| u^1 - u \|_{W_2^1} \to 0$ if $\varepsilon \to 0$. Consequently

$$\frac{\partial u}{\partial x_k} \sim \frac{\partial v}{\partial x_k} + \frac{\partial N_j}{\partial \xi_k} \frac{\partial v}{\partial x_j}. \tag{2.6}$$

Thats way, deformations and stresses oscillate rapidly. In the theory of elasticity $\sigma_i = A_{ij} \frac{\partial u}{\partial x_j}$ are stresses and $e = (\sigma_i, \frac{\partial u}{\partial x_i})$ - "energy".

Lemma. Next relations are true [12]:

$$\frac{\partial u}{\partial t} \sim \frac{\partial v}{\partial t}, \left\{\frac{\partial u}{\partial x_j}\right\} \sim \frac{\partial v}{\partial x_j}, \left\{\sigma_i\right\} \sim \bar{\sigma_i}, \left\{e\right\} \sim \bar{e}.$$

The relation (2.6) is true in norm L_2 and we can expect, that it is true inside the domain in uniform metric. The asymptotic near plane boundary is recieved in [14]. We have from [14] that (2.6) is not true near boundary.

§3. The changing of equations type.

We give some examples, where the averaged equation have another type than original equation.

1. We consider the system of integro-differential equation with distributed parameters [10]:

$$R(Z) \frac{\partial^2 u}{\partial t^2} = \varepsilon^{-s} \int_{-\infty}^{\infty} \cdots \int_{-\infty}^{\infty} K(Z, \Lambda)\, u(Y)\, dY,$$

where $Y = (y_1, \ldots, y_s)$, $\Lambda = (\lambda_1, \ldots, \lambda_s)$, $\lambda_j = y_j / \varepsilon$, and K(Z,) have the following properties:

$$K(Z+m, \Lambda+m) = K(Z, \Lambda)$$

if all vectors m coordinates m_i are whole numbers, $\| K(Z, \Lambda)\| \; \| Z-\Lambda\|^q \to 0$ if $\| Z-\Lambda\| \to \infty \;\; \forall q$. This problem is special case of problem (1.2). By using decomposition (1.8) we recieve the averaged equation which have the form (2.4).

2. We consider the equations system

$$R(Z)\frac{\partial^2 u}{\partial t^2} - L_o u - L_1 \frac{\partial u}{\partial t} = 0, \quad L_k u = \frac{\partial}{\partial x_i}(A^k_{ij}(Z)\frac{\partial u}{\partial x_j}).$$

By averaging we recieve [13,15] the system

$$\bar{R}\frac{\partial^2 v}{\partial t^2} - \bar{L}_o v - \bar{L}_1 \frac{\partial v}{\partial t} - \int_{-\infty}^{t} \Omega_{ij}(t-\tau)\frac{\partial^2 v(\tau, x)}{\partial x_i \partial x_j} d\tau = 0$$

$$\bar{L}_k v = \frac{\partial}{\partial x_i}(\bar{A}^k_{ij}\frac{\partial v}{\partial x_j}).$$

3. Let $X = n\varepsilon$ and $u(x)$ be the solution of next problem:

$$k\frac{d^2u}{dx^2} = f(x)$$

in each interval $(j\varepsilon, (j+1)\varepsilon)$, $0 \leqslant j < n$, where $k = \text{const}$, $u(o+o) = u_o$, $u(X-0) = u_1$.

$$k\left.\frac{du}{dx}\right|_{x=j\varepsilon+0} = k\left.\frac{du}{dx}\right|_{x=j\varepsilon-0} = g(u(j\varepsilon+0), u(j\varepsilon-0)), \; 0 < j < n.$$

We consider for definity the case $g(a,b) = h(b) - h(a)$, which corresponds to radiation heat transfer. Then our methodic leads to averaged equation

$$\varepsilon\frac{d}{dx}(h'(v)\frac{dv}{dx}) - \frac{\varepsilon^2}{k}\frac{d}{dx}((h'(v))^2\frac{dv}{dx}) + \ldots \sim f(x). \tag{3.3}$$

There is next peculiarity in this equation: there is not term of zero order.

We consider case

$$g(a,b) = (h(b) - h(a)) + d(a,b)(b-a)^{q(a,b)}\,\text{sign}(b-a),$$

$d(a,b) > 0$, $q(a,b) \geqslant c > 1$. The second term correspond to heat exchange by convection.

If $q(a,b) \equiv c$ we can find the asymptotic solution in the form (1.5). If $q(a,b) \not\equiv$ const we find the asymptotic solution in the form (1.4):

$$u \sim v - \varepsilon \xi \frac{dv}{dx} + \int_C^{\infty} \varepsilon^{\lambda} N(\lambda, \xi, v)\, dF(\lambda).$$

The final averaged equation is

$$\varepsilon \frac{d}{dx} \left(h'(v) \frac{dv}{dx}\right) + \int_C^{\infty} \varepsilon^{\lambda} H(\lambda, v)\, dF_1(\lambda) \sim f(x).$$

§4. About velocity of sound in mixtures.

We consider one example [14], where the asymptotical theory substantiate and precise the result which is recieved from physical reason. There is the fact of small sound velocity in mixtures. The sound velocity in air is near 300 m/sec, in water in near 1500 m/sec but in some air-water mixtures it is less then 100 m/sec.

By some conditions the process of the sound waves spreading in periodical mixture of gas and fluid is described by system of equations of the elasticity theory where the second Lame coefficient is equal to zero:

$$\rho(Z) \frac{\partial^2 u}{\partial t^2} - \text{grad}\,(\lambda(Z)\,\text{div}\, u) = 0 \qquad (4.1)$$

where $\rho(Z)$ is density, $\lambda(Z)$ the first Lame coefficient.

The physics explanation of small sound velocity is: the value λ^{-1} define the compressibility; the compressibility and the density are additive function. So the equation for middle values is

$$\rho_0 \frac{\partial^2 v}{\partial t^2} = \text{grad}\,(\lambda_0 \text{div}\, v), \qquad (4.2)$$

where $\rho_0 = [\rho]$ and $(\lambda_0)^{-1} = [\lambda^{-1}]$. Thats way the middle sound velosity is

$$c_0 = \sqrt{\lambda_0 / \rho_0}\,. \qquad (4.3)$$

Be definite concentration air and water we recieve from there small sound velocity. We consider this problem in detail. Our methodic lead [12] to the same averaged system (4.2). The system (4.1) is not huperbolic and then we can not estimate the difference between solutions of systems (4.1) and (4.2).

We will see further that u is not close to v. We apply to (4.1) the operator $\mathbf{Q}$: $\mathbf{Q}\omega = \mathrm{div}\,(\rho^{-1}\omega)$. Then we recieve the equation

$$\lambda^{-1}\frac{\partial^2 p}{\partial t^2} - Lp = 0, \quad Lp = \mathrm{div}\,(\rho^{-1}\,\mathrm{grad}\, p), \tag{4.4}$$

where $p = \lambda(Z)\,\mathrm{div}\, u$. The averaging methodics leads to the equation

$$(\lambda^{\circ})^{-1}\frac{\partial^2 q}{\partial t^2} - \bar{L}q = 0, \quad \bar{L}q = \frac{\partial}{\partial x_i}\left(\bar{a}_{ij}\frac{\partial q}{\partial x_j}\right). \tag{4.5}$$

The operator L is elliptical and so there is nearness between the solutions of systems (4.4) and (4.5). The sound velocity for (4.5) may be different in various direction. The sound velocity in x_1-direction is $c_1 = \sqrt{\lambda^{\circ}\bar{a}_{11}}$. Let $W(m,i,k)$ be the class of vectors $n(Z)$ with size m which belong W_2^1 and satisfy the following conditions: n is periodic in each of the arguments ξ_j exept ξ_i,

$$n|_{\xi_i=1} - n|_{\xi_i=0} = e_k,$$

where e_k is ort of axis x_k. There is the relation [4,15]

$$\bar{a}_{11} = \inf_{n\in W(1,1,1)}\left[\rho^{-1}\frac{\partial n}{\partial \xi_j}\frac{\partial n}{\partial \xi_j}\right]$$

We have

$$\bar{a}_{11} \geqslant \bar{a}_{11}^{-} = \inf_{n\in W(1,1,1)}\left[\rho^{-1}\left(\frac{\partial n}{\partial \xi_1}\right)^2\right] = \left[\left[\left[\rho\right]_1^{-1}\right]_2\right]_3,$$

where $[h]_k$ means the value of integral $\int_{0-0}^{1-0} h\, d\xi_k$. In the case $\rho(Z) \equiv \rho(\xi_1)$ we have $\bar{a}_{11} = \rho_0^{-1}$ and $c_1 = c_0$, else $\bar{a}_{11} > \bar{a}_{11}^{-} > \rho_0^{-1}$ and so $c_1 > c_0$. Let us consider the following structure: the spherical bubbles of air are disposed periodically with centrums in the points (j_1, j_2, j_3) and the other part of the space is filled by water, air and water occupy the equal volumes. In this case $\bar{a}_{ij} = a\,\delta_{i-j}$ and the sound velocities in all direction are the same. We have calculated the value of $\bar{a}_{11}$ numerically and have recieved the value of sound velocity $c_1 \approx 1.2\, c_0$. If the phases exchange their places then $c_1 \approx 11.5\, c_0$.

However the formulae (4.3) has some sense. Let us consider the system of equations of the elasticity theory in local isotropic medium

$$\rho(Z)\frac{\partial^2 u}{\partial t^2} = \frac{\partial}{\partial x_i}\left(A_{ij}(Z)\frac{\partial u}{\partial x_j}\right) \text{ in the case } s = 3. \tag{4.6}$$

We have the relation

$$\bar{a}_{jj}^{kk} = \inf_{W(3,j_2,k)} [e(n)],$$

where $e(n) = \frac{\mu(Z)}{2}\left(\frac{\partial n_i}{\partial \xi_q} + \frac{\partial n_q}{\partial \xi_i}\right)^2 + \lambda(Z)(\operatorname{div} n)^2$, $n = (n_1, n_2, n_3)$.

Let $\eta(Z)$ be the periodical solution of the equation $\operatorname{div}\eta = \lambda^{\circ}\lambda^{-1} - 1$ with minimal norm W_2^1 on period, and let $n_j^o = (\eta_1 + \delta_{1-j}\xi_1, \eta_2 + \delta_{2-j}\xi_2, \eta_3 + \delta_{3-j}\xi_3)$. We have $[e(n_j^o)] = \lambda^{\circ} + O(\bar{\mu})$, $\bar{\mu} = \sup \mu(Z)$. On the basis of definition $\bar{a}_{jj}^{kk}$ we have

$$\bar{a}_{jj}^{jj} \geqslant \inf_{W(3,j,j)} \left[\lambda (\operatorname{div} n)^2\right] = \lambda^{\circ}.$$

So $\bar{a}_{jj}^{jj} = \lambda^{\circ} + O(\bar{\mu})$. If $j \neq k$ then $\bar{a}_{jj}^{kk} \leqslant [e(\xi_j e_k)] = O(\bar{\mu})$. We suppose that matrixes A_{ij} are invariant conserning the transformations $x_i \Longleftrightarrow -x_i$. The coefficient by $\frac{\partial^2 v}{\partial t^2}$ in averaged equation is ρ^c. So we recieve that the sound velocities in any direction have the following form. One of them $c^1 = c^o + O(\bar{\mu})$ and two others are $c^2, c^3 = O(\bar{\mu})$. Now we give resume. The formulae (4.3) is not true for the system (4.1). However it has some sense, as it gives the limit value of sound velocity for a system (4.6) if $\bar{\mu} \to 0$.

§5. The equation with random coefficients.

The differential equation with random coefficients arises often in applications. Let us consider an expanded model of equations with random coefficients. Let the space R_s be divided into the parts Π_n : $n_j\varepsilon \leqslant x_j < (n_j+1)\varepsilon$, $n = (n_1,\dots,n_s)$, and let ξ_i be bounded ranbom variables with $]\xi_i[= 1$, where $][$ is the notation for the mean. Let $a_i(X)$ be some bounded function. As the value of coefficient a_i on Π_n we take $a_i(X)\xi_i^n$, where ξ_i^n with different n are independent realizations of the random variable ξ_i.

Let us consider the case $s=1$ and $a_i(X) = a_i = \text{const}$. We introduce some notation. If $v = (v^1,\dots,v^m)$, v^i are defined on $I = [0,X]$ and belong to $L_p(I)$, then we write $v \in L_p(I)$, $\|v\|_p = \|\,|v|\,\|_{L_p}$, $|v|^2 = v^i v^i$. If $v, v_x \in L_p(I)$, $v(0) = v(X) = 0$, then we write $v \in \overset{\circ}{W}{}_p^1(I)$, $\|v\|_p^{(1)} = \|v_x\|_p$. We set

$$\|v\|_p^{(-1)} = \sup \frac{|(v,\eta)|}{\|\eta\|_{p'}^{(1)}}, \quad (v,\eta) = \int_I v^i \eta^i \, dx,$$

sup ranges over all $\eta \in \overset{\circ}{W}{}^1_{p'}$, $p^{-1} + (p')^{-1} = 1$. The notation is analogous for matrices.

Lemma. The inequality

$$\left] \sup_{0 \leq x \leq X} \left\| \int_0^x (a_i(\tau) -]a_i(\tau)[)\, d\tau \right\| \right[\leq \text{const} \sqrt{\varepsilon}$$

is valid.

From the lemma follows

$$\left] \| a_i(x) -]a_i(x)[\|_p^{(-1)} \right[\leq \text{const} \sqrt{\varepsilon}\,, \quad 1 \leq p \leq \infty .$$

Thus, estimates of how near the solutions of equations with random coefficients are reduce to similar estimates for the solutions in terms of negative norms of the differences of coefficients. If the coefficients of one of the equations are smooth, then to estimate proximity we can apply a method analogous to the one used in the case of periodic coefficients [16]. We seek an approximation to u_1 in the form

$$u_1 \sim w = u_2 + \sum_{j=1}^{2} \sum_{k,l=0}^{j} N_{kl}{}^{j}(x)\, u_{2_{t^k x^l}} ;$$

to the derivative of u_2 we assign a formal order Δ^0 and to the derivative $(N^j_{kl})_{x^r}$ an order Δ^{j-r}, where Δ is small. We substitute w into the equation corresponding to u_1 and at the expense of choosing N^j_{kl} we achieve proximity with the equation corresponding to u_2.

As an example we consider the boundary-value problem

$$Lu = (Ku_x)_x + Bu_x + (Du)_x - Qu = f, \quad x \in I; \quad u(0)=u(X)=0.$$

Let

$$(Kv_x + Dv, v_x) - (Bv_x - Qv, v) \geq \nu \| v_x \|_2^2 \qquad \forall v \in W_2^1(I),$$

where K, B, D, Q are matrices; f is a vector with random coefficients of the type being examined. By the method described above we can show that the solution of this system is close to the solution of a system of equations with constant coefficients

$$]K^{-1}[^{-1} v_{xx} + (]BK^{-1}[\,]K^{-1}[^{-1} +]K^{-1}[^{-1}]K^{-1}D[)\, v_x$$
$$+ (]BK^{-1}[\,]K^{-1}[^{-1}]K^{-1}D[-]BK^{-1}D[-]Q[)\, v = f, \quad x \in I, \; v(0)=v(X)=0.$$

Analogous estimates for the difference between the solutions are recieved for the boundary-value problem for the systems

$$\sum_{j=0}^{m} \frac{d^j}{dx^j}\left(A^i_j \frac{d^j u}{dx^j}\right) = f_i + \frac{d\omega_i}{dx}, \quad u_i \in \overset{\circ}{W}{}^m_2 (I), \quad m > 1.$$

We consider further the following problems:

I. $L_i u_i = (K_i u_{i_x})_x - Q_i u_i = g_i = f_i \qquad x \in I = (0,X)$; $u_i(0) = u_i(X) = 0$.

II. $C_i u_{i_t} - L_i u_i = g_i(x,t), (x,t) \in \Pi = I \times (0,T)$; $u_i|_{x=0,X} = 0$, $0 < t < T$; $u_i|_{t=0} = \varphi_i$, $0 < x < X$.

III. $C_i u_{i_{tt}} - L_i u_i = g_i(x,t), (x,t) \in \Pi$; $u_i|_{x=0,X} = 0$, $0 < t < T$; $u_i|_{t=0} = \varphi_i$, $u_{i_t}|_{t=0} = \psi_i$, $0 < x < X$.

Here C_i, K_i, Q_i are $m \times m$ matrices depending on x; u_i, f_i, ω_i, φ_i, ψ_i are vectors with values in R^N, $i=1,2$.

Everywhere below $C_i + C_i^* \geqslant \nu E$, $K_i + K_i^* \geqslant \nu E$, $\|C_i\|_\infty$, $\|K_i\|_\infty \leqslant \mu$, where E is the identity matrix of order N, and $\nu, \mu > 0$. We give some estimates of that kinds from [16]. We consider problem I. Let $(K_i v_x, v_x) + (Q_i v, v) \geqslant \nu \|v_x\|_2^2 \quad \forall\ v \in W_2^1 (I)$.

Theorem 3. If $\|Q_1\|_2, \|Q_2\|_1, \|f_1\|_1, \|f_2\|_2 \leqslant \mu$, then

$$\|\mathfrak{F} u\|_2 \leqslant c\left(\|\mathfrak{F} Y\|_2 + |\mathfrak{F} Y(X)| + \|\mathfrak{F} Q\|_2^{(-1)} + \|\mathfrak{F} f\|_1^{(-1)}\right).$$

Here $\mathfrak{F} v = v_2 - v_1$, $Y_i(x) = \int_0^x K_i^{-1}(y)\, dy$.

Theorem 4. If $G_i(x,\tau)$ is the Green's function for problem I, then

$$\| |\mathfrak{F} G| \|_C \leqslant c\left(\|\mathfrak{F} Y\|_\infty + \|\mathfrak{F} Q\|_1^{(-1)}\right).$$

Theorem 5. If $K_i = K_i^*$, $Q_i = Q_i^*$, $\|Q_i\|_2 \leqslant \mu$, then

$$|\mathfrak{F} \lambda_n^{-1}| \leqslant c\left(\|\mathfrak{F} Y\|_2 + |\mathfrak{F} Y(X)| + \|\mathfrak{F} Q\|_1^{(-1)}\right)$$

uniformly in n, for the eigenvalues $\lambda_{n,i}$ of operator L_i (with boundary condition $u(0) = u(X) = 0$). Let us estimate the proximity of the quantities $s_i = K_i u_{i_x}$, which are strains in the case of problems in elasticity theory.

Theorem 6. If $\|Q_i\|_1, \|f_i\|_1 \leqslant \mu$, then

$$\|\mathfrak{F} s\|_\infty \leqslant c\left(\|\mathfrak{F} Y\|_\infty + \|\mathfrak{F} Q\|_\infty^{(-1)} + \|\mathfrak{F} f\|_\infty^{(-1)}\right)$$

The next theorem is valid for problem II.

Theorem 7. If $K_i = K_i^*$, $\|Q_i\|_2$, $\|f_1\|_{2,1}$, $\|f_2\|_2$, $\|\varphi_1\|_2$, $\|\varphi_2\|_2^1 \leq \mu$, then

$$\|\mathfrak{S}u\|_2 + \|\mathfrak{S}Cu\|_{2,\infty}^{(-1)} \leq C\,(\|\mathfrak{S}Y\|_2 + |\mathfrak{S}Y(X)| + \|\mathfrak{S}Q\|_2^{(-1)} +$$

$$+ \|\mathfrak{S}C\|_\infty^{(-1)} + \|\mathfrak{S}f\|_{2,1}^{(-1)} + \|\mathfrak{S}(C\varphi)\|_2^{(-1)}.$$

Here $\|v\|_{p,q} = \|\,\|v(\cdot,t)\|_p\|_q$, $\|v^{(-1)}\|_{p,q} = \|\,\|v(\cdot,t)\|_p^{(-1)}\|_q$.

The next theorem is valid for problem III.

Theorem 8. If $C_i = C_i^*$, $K_i = K_i^*$, $\|f_1\|_{2,1}$, $\|\varphi_1\|_2$, $\|\psi_1\|_2$, $\|f_2\|_{2,1}$, $\left\|\frac{\partial f_2}{\partial t}\right\|_{2,1}$, $\|\varphi_2\|_2^1$, $\|L_2\varphi_2\|_2$, $\|\psi_2\|_2^1 \leq \mu$ then

$$\left\|\int_0^t \mathfrak{S}u\,d\tau\right\|_{2,\infty} + \|\mathfrak{S}Cu\|_{2,\infty}^{(-1)} \leq C\,(\|\mathfrak{S}Y\|_2 + |\mathfrak{S}Y(X)| +$$

$$+ \|\mathfrak{S}Q\|_2^{(-1)} + \|\mathfrak{S}C\|_2^{(-1)} + \|\mathfrak{S}f\|_{2,1}^{(-1)} + \|\mathfrak{S}(C\varphi)\|_2^{(-1)} + \|\mathfrak{S}(C\psi)\|_1^{(-1)})$$

These theorems are recieved by the energy method. In the case of Theorem 7 and 8 we integrate first the equation for $\mathfrak{S}u$ with respect to t once or several times and then multiply by $\Lambda_1 C_1 \mathfrak{S}u$.

In multidimensional case more particular result is known. Let us consider the diffusion equation

$$\frac{\partial}{\partial x_1}\left(p\frac{\partial u}{\partial x_1}\right) + \frac{\partial}{\partial x_2}\left(p\frac{\partial u}{\partial x_2}\right) = 0$$

with the boundary condition $u|_\Gamma = \varphi$. Let ξ be the random variables such that $\ln\xi$ and $\ln\xi^{-1}$ have the same distributions. As the value of coefficient p on Π_n we take ξ^n, where ξ^n are independent realization of the random variable ξ. Then the solution of the boundary problem (5.4) converges [17] in probability to the solution of the boundary problem

$$\Delta v = 0,\; v|_\Gamma = \varphi, \text{ when } \varepsilon \to 0.$$

We remark, that this result have been known in physics for a long time[18].

References

1. Борн М., Хуан Кунь. Динамическая теория решеток. ИЛ, 1958.

2. Бриллюэн Л., Пароди М. Распространение волн в периодических структурах. М., ИЛ, 1959.

3. Григолюк Э.И., Фильштинский Л.А. Перфорированные пластинки и оболочки. М., Наука, 1970.

4. Бреховских Л.М. Волны в слоистых средах. М., Наука, 1973.

5. Sanchez-Palencia E. Compartements local et macroscopic d'un type de milieux physiques hétéreogènes. Intern. J. Engrn. Sci., 1974, 12, 331-351.

6. De Giorgi, Spagnolo S. Sulla convergenza degli integrali dell energia. Boll. Un. Mat. Ital., 1973, 8, № 3, 391-411.

7. Бахвалов Н.С. Осредненные характеристики тел с периодической структурой. Докл. АН СССР, 1974, 218, № 5, 1046-1048.

8. Бахвалов Н.С. Осреднение дифференциальных уравнений с частными производными с быстро осциллирующими коэффициентами. Докл. АН СССР, 1975, 221, № 3, 516-519.

9. Бахвалов Н.С. Осреднение нелинейных уравнений с частными производными с быстро осциллирующими коэффициентами. Докл. АН СССР, 1975, 225, № 2, 249-252.

10. Кунин И.А. Теория упругих сред с микроструктурой. М., Наука, 1975.

11. Bensoussan A., Lions J.L. et Papanicolaou. Sur quelques phénomènes asymptotiques d'evolution. Comptes rendus Acad Sci., 1975, 281, série A, 317-322.

12. Бахвалов Н.С. Осреднение уравнений с часными производными с быстро осциллирующими коэффициентами. Сб. Проблемы вычислительной математики и математической физики. М., 1977, 34-51.

13. Bensoussan A., Lions J.L. et Papanicolaou. Asymptotic methods for media with periodic structures. North Holland, 1978.

14. Панасенко Г.П. Асимптотики высших порядков решений уравнений с быстро осциллирующими коэффициентами. Докл. АН СССР, 1978, 240, № 6, 1293-1296.

15. Бахвалов Н.С. К вопросу о скорости звука в смесях. Докл. АН СССР, 1979, 245, № 6, 1345-1348.

16. Бахвалов Н.С., Злотник А.А. Коэффициентная устойчивость и осреднение уравнений со случайными коэффициентами. Докл. АН СССР, 1978, 242, № 4, 745-748.

17. Козлов С.М. Осреднение случайных структур. Докл. АН СССР, 1978, 241, № 5, 1016-1019.

18. Дыхне А.М. Проводимость двумерной двухфазной системы. ЖЭТФ, 1970, 59, № 7, 110-115.

COMPUTING METHODS IN APPLIED SCIENCES AND ENGINEERING
R. Glowinski, J.L. Lions (editors)
North-Holland Publishing Company

STOCHASTICALLY PERTURBED BIFURCATION

George C. Papanicolaou *
Courant Institute of Mathematical Sciences
New York University
New York, New York 10012/USA

1. Introduction

Let $\mathcal{O} \subset R^d$ be a bounded open set and consider the nonlinear eigenvalue problem

$$\Delta u + \lambda F(u,x) = 0, \quad x \in \mathcal{O}, \quad \lambda \in R^1, \tag{1.1}$$
$$u = 0, \quad x \in \partial\mathcal{O}$$

Suppose that $F(0,x) \equiv 0$ so that $u \equiv 0$ is a solution of (1.1) for all λ. Consider the variational equations about $u = 0$.

$$(\Delta + \lambda V(x))z = 0, \quad x \in \mathcal{O} \tag{1.2}$$
$$z = 0, \quad x \in \partial\mathcal{O}, \quad V(x) \equiv F_u(0,x) .$$

Since $V(x)$ is bounded there is a number $\bar{\lambda}$ (assumed $\neq 0$) such that for all $\lambda < \bar{\lambda}$, (1.2) has only the trivial solution $z \equiv 0$. At $\lambda = \bar{\lambda}$ we assume that there is a one-dimensional nullspace spanned by the function $\phi(x)$, normalized so that

$$(\phi,\phi) = \int_{\mathcal{O}} \phi^2(x)\, dx = 1 . \tag{1.3}$$

For $\lambda < \bar{\lambda}$, (1.1) will have a unique solution in a neighborhood of $u \equiv 0$, namely the identically zero function itself. At $\lambda = \bar{\lambda}$ uniqueness is lost and other solution branches will appear (bifurcation). Near $\lambda = \bar{\lambda}$ the nature of these new solutions branches can be determined by perturbation calculations (cf. for example [1] and [2]). We recapitulate this quickly in the next section.

In many physical problems (1.1) is not quite the proper equation. It may be more appropriate to have F replaced by $F^{\varepsilon}(u,x)=F(u,x)+\varepsilon F_1(u,x)$ where ε is a small number called the imperfection parameter. Now $u \equiv 0$ is not a solution of (1.1) (since $F_1(u,x) \not\equiv 0$ in general) and the question arises as to how the bifurcation picture near $\lambda = \bar{\lambda}$ is affected by the ε perturbation. This is the subject of the so called perturbed bifurcation [3]-[8] which is also analyzed by perturbation calculations.

One frequently encounters the question: what happens in the bifurcation problem (1.1) when F is perturbed by a random function of x? The answer is simple enough when the random perturbation of

* Research supported by the U. S. Air Force Office of Scientific Research under Grant No. AFOSR-78-3668.

F is of the form εF_1 , as above, with F_1 a random function of x with finite moments. Namely, the deterministic perturbed bifurcation analysis carries right over and the statistics play no role at all.

A more interesting stochastic perturbation is the replacement of F in (1.1) by $F(u,x) + F_1(u,\frac{x}{\varepsilon})$ where $F_1(u,y)$ is a stationary random process (depending parametrically on u, like $uF_1(y)$ for example) with mean zero and finite moments. Now the perturbation is small only in a weak or average sense while the perturbed version of (1.1) is a nonlinear stochastic partial differential equation so it is not immediately obvious what happens to the bifurcation picture.

In this paper we shall show that near $\lambda = \bar{\lambda}$ the stochastic bifurcation is described asymptotically (after suitable scaling) by a random algebraic equation of low order (usually quadratic or cubic) with gaussian coefficients. The gaussian nature of the coefficients is independent of the statistical properties of F_1 except for some general ones that are usually necessary for the validity of the central limit theorem. Although one can derive our limit theorem rather quickly, as we do here, the detailed mathematical treatment is rather lengthy. Limit theorems similar to ours but for ordinary differential equations are given in [9].

This work was done jointly with W. Day (cf. [10] for some more examples). It is my pleasure to thank E. L. Reiss for getting me interested in stochastic bifurcation and S.R.S. Varadhan for a number of discussions on the present problem.

2. Deterministic Bifurcation. Notation.

We review briefly the deterministic analysis of (1.1). We write

$$\lambda = \bar{\lambda} + \nu, \quad u = \mu\phi + \psi , \tag{2.1}$$

$$(\psi,\phi) = 0 \tag{2.2}$$

where μ and ν are parameters and $(u,\phi) = \mu$ because of (2.2). Inserting into (1.1) and recalling that $(\Delta + \bar{\lambda}V)\phi = 0$, we obtain

$$\begin{gathered}(\Delta+\bar{\lambda}V)\psi+\bar{\lambda}[F(\mu\phi+\psi,x)-V(x)(\mu\phi+\psi)]+\nu F(\mu\phi+\psi,x) = 0 , \\ \psi = 0, \quad x \in \partial\mathcal{O}.\end{gathered} \tag{2.3}$$

To simplify notation we shall frequently omit the x argument and write for example $F(\mu\phi+\psi)$ instead of $F(\mu\phi(x)+\psi(x),x)$. With this in mind we define

$$G(\mu,\nu,\psi) = (\bar{\lambda}(F(\mu\phi+\psi)-V(\mu\phi+\psi))+\nu F(\mu\phi+\psi),\phi) \tag{2.4}$$

and note that (2.3) is equivalent to

$$(2.5)\qquad \begin{aligned} &(\Delta+\bar{\lambda}V)\psi+\bar{\lambda}(F(\mu\phi+\psi)-V(\mu\phi+\psi))+\nu F(\mu\phi+\psi)-G(\mu,\nu,\psi)\phi = 0 \\ &\psi = 0 \quad \text{on} \quad \partial\mathcal{O} \end{aligned}$$

$$(2.6)\qquad G(\mu,\nu,\psi) = 0 \ .$$

In (2.5) we can apply $(\Delta+\bar{\lambda}V)^{-1}$ since the component along ϕ has been removed (Fredholm alternative). For ν and μ small the equation (2.5) (in integral form) can be solved for $\psi = \psi(x;\mu,\nu)$ and the solution ψ satisfies (2.2) and depends smoothonly on μ and ν. Inserting this into (2.6) we obtain

$$(2.7)\qquad U(\mu,\nu) = G(\mu,\nu,\psi(\cdot,\mu,\nu)) = 0$$

which is the bifurcation equation, that is, the functional equation that relates the amplitude $\mu = (u,\phi)$ of the solution branch to the bifurcation parameter $\nu = \lambda - \bar{\lambda}$. Of course (2.7) is to be solved in a neighborhood of (0,0) and clearly $U(0,0) = 0$.

The object now is to classify all solutions $\mu = \mu(\nu)$ of (2.7) for ν small and to find what these solution branches look like.

A way to do this that generalizes well in the stochastic case is to introduce another parameter, say ε, and replace ν and μ by $\varepsilon\nu$ and $\varepsilon\mu$ respectively. Then $\varepsilon^{-2}U(\varepsilon\mu,\varepsilon\nu)$ will tend to a limit $U_0(\mu,\nu)$ as $\varepsilon \to 0$ and the function $U_0(\mu,\nu)$ will be the canonical form of the local bifurcation picture. A simple computation shows ψ is proportional to ε^2 and hence

$$G(\varepsilon\mu,\varepsilon\nu,\psi(\cdot,\varepsilon\mu,\varepsilon\nu)) = \varepsilon^2(\bar{\lambda}\ \tfrac{1}{2}\ F_{uu}(0)\mu^2\phi^2+\nu\mu F_u(0)\phi,\phi)+ O(\varepsilon^3)$$

so that

$$(2.8)\qquad U_0(\mu,\nu) = \frac{\bar{\lambda}}{2}\left(F_{uu}(0)\phi^2,\phi\right)\mu^2 + \left(F_u(0)\phi,\phi\right)\nu\mu \ .$$

Here

$$(2.9)\qquad \left(F_{uu}(0)\phi^2,\phi\right) = \int F_{uu}(0,x)\ \phi^3(x)\ dx \ ,$$

and similarly for $\left(F_u(0)\phi,\phi\right)$.

If $\left(F_{uu}(0)\phi^2,\phi\right) \neq 0$ then the canonical bifurcation equation is quadratic and the two solution branches are $\mu = 0$ and

$$(2.10)\qquad \mu = -\frac{2\left(F_u(0)\phi,\phi\right)}{\bar{\lambda}\left(F_{uu}(0)\phi^2,\phi\right)}\ \nu$$

If $\left(F_{uu}(0)\phi^2,\phi\right) = 0$ then another scaling is necessary to obtain a nontrivial bifurcation picture namely $\nu \to \varepsilon^2\nu$, $\mu \to \varepsilon\mu$ so that, after a simple computation

$$\varepsilon^{-3}U(\varepsilon\mu,\varepsilon^2\nu) \to U_0(\mu,\nu) \tag{2.11}$$

where

$$U_0(\mu,\nu) = \frac{\bar{\lambda}}{6}\left(F_{uuu}(0)\phi^3,\phi\right)\mu^3 + [\bar{\lambda}(F_{uu}(0)\psi_0\phi,\phi) + \nu(F_u(0)\phi,\phi)]\mu \tag{2.12}$$

and $\psi_0 = -(\Delta+\bar{\lambda}V)^{-1}(\frac{\bar{\lambda}}{2} F_{uu}(0)\phi^2)$. Thus in this case the canonical bifurcation equation is a cubic. More general situations can be dealt with in the same way.

Regarding detailed proofs of the above, under reasonable assumptions on F like smoothness as a function of u, they can be obtained easily as in [2].

3. Stochastically Perturbed Bifurcation

Let $(\Omega,\mathcal{F},\mathcal{P})$ be a probability space and let $S(x) = S(x,\omega)$ be a stationary random field on R^d with mean zero and finite moments. Stationarity means that the joint distribution of $S(x_1), S(x_2)\ldots S(x_n)$ and of $S(x_1+h)$, $S(x_2+h)$, ..., $S(x_n+h)$, for any $x_1,x_2,\ldots,x_n$ and h, is the same.

We shall consider the perturbed bifurcation problem

$$\begin{aligned} &\Delta u(x,\omega) + \lambda F(u(x,\omega)) + S(\tfrac{x}{\varepsilon},\omega) = 0, \quad x \in \mathcal{O}, \\ &u(x,\omega) = 0, \quad x \in \partial\mathcal{O}, \end{aligned} \tag{3.1}$$

in which F(u) does not depend on x for simplicity. One may think of S as a random source term due to spatial noise and, as usual, to articulate its rapidly fluctuating nature we introduce the small parameter ε. For any finite region D

$$\int_D S(\tfrac{x}{\varepsilon},\omega)\,dx \to 0$$

as $\varepsilon \to 0$ with probability one by the ergodic theorem. The question is what happens when λ is near $\bar{\lambda}$, when the deterministic problem undergoes bifurcation, as $\varepsilon \to 0$ in (3.1)?

We shall assume that

$$F(0) = 0, \quad F_u(0) \neq 0, \quad F_{uu}(0) \neq 0 \tag{3.2}$$

so that the quadratic bifurcation picture is valid when the S term in (3.1) is absent. We proceed formally to obtain the limiting bifurcation equation. Write

$$u = \mu\phi + \psi, \quad (\psi,\phi) = 0 \tag{3.3}$$

as before. Using the definition (2.4) we obtain

(3.4) $(\Delta+\bar{\lambda}V)\psi + \bar{\lambda}[F(\mu\phi+\psi)-V(\mu\phi+\psi)] + \nu F(\mu\phi+\psi) - G(\mu,\nu,\psi)\phi + S^{\varepsilon}-(S^{\varepsilon},\phi)\phi=0,$

(3.5) $$G(\mu,\nu,\psi) + (S^{\varepsilon},\phi) = 0$$

with $S^{\varepsilon} = S(\frac{x}{\varepsilon},\omega)$.

The random variable (S^{ε},ϕ) behaves like $\varepsilon^{d/2}$ times a gaussian random variable under suitable conditions (which are detailed in the next section). In view of (3.2) the right scaling for μ and ν is therefore

(3.6) $$\mu \to \varepsilon^{d/4}\mu , \qquad \nu \to \varepsilon^{d/4}\nu$$

Formally now one expects that ψ^{ε} is of order $\varepsilon^{d/2}$ (this is *not* correct but it does not affect the results when $3 \leq d < 8$) and hence

(3.7) $$\varepsilon^{-d/2}G\left(\varepsilon^{d/4}\mu,\varepsilon^{d/4}\nu,\psi^{\varepsilon}\right) + \varepsilon^{-d/2}(S^{\varepsilon},\phi) \to U_0(\mu,\nu) + \gamma$$

where U_0 is given by (2.8) and γ is a gaussian random variable with mean zero and variance σ^2 where

(3.8) $$\sigma^2 = \int_{R^d} R(x)\,dx, \qquad R(x) = E\{S(x+y)\,S(y)\}$$

The limiting variance is obtained by noting that

$$E\{\varepsilon^{-d}(S^{\varepsilon},\phi)^2\} = \varepsilon^{-d}\int_0\int_0 \phi(x)\phi(y)R(\tfrac{x-y}{\varepsilon})dx\,dy \xrightarrow{\varepsilon\downarrow 0} \sigma^2\int_0 \phi^2(x)dx = \sigma^2$$

The very rough analysis above leads actually to the following correct statement: The random, scaled bifurcation function

$$\varepsilon^{-d/2}G\left(\varepsilon^{d/4}\mu,\varepsilon^{d/4}\nu,\psi^{\varepsilon}\right) - \varepsilon^{-d/2}(S^{\varepsilon},\phi)$$

converges weakly as $\varepsilon \to 0$ to the gaussian random process $U_0(\mu,\nu)+\gamma$. This means that statistical properties of solution branches (zeros of bifurcation function) behave asymptotically like the corresponding objects computed from $U_0(\mu,\nu)+\gamma$.

Let

(3.9) $$\alpha = \frac{\bar{\lambda}}{2}F_{uu}(0)(\phi^2,\phi) , \qquad \beta = F_u(0) .$$

The limiting bifurcation function is then $\alpha\mu^2 + \beta\mu\nu + \gamma$ and its zeros are

(3.10) $$\mu_{\pm}(\nu) = -\frac{\beta\nu}{2\alpha} \pm \frac{1}{2\alpha}\left((\beta\nu)^2 - 4\alpha\gamma\right)^{1/2}$$

To fix ideas suppose that α and β are positive. We see then that the effect of γ is to eliminate bifurcation when γ is negative and cause bifurcation to occur at a value of ν less than zero when γ is positive. Since γ is random, a quantity of interest is the expected value of the value of ν at which a solution branch changes multiplicity. Denote this random value of ν by $\bar{\nu}$. Then

$$E(\bar{\nu}) = -\frac{2\sqrt{\alpha}}{\beta}\int_0^\infty \sqrt{\gamma}\,\frac{e^{-\gamma^2/2\sigma^2}}{\sqrt{2\pi\ \sigma^2}}\,d\gamma \tag{3.11}$$

This is probably the most interesting quantity to compute for the following reason. When α and β are positive, simple analysis shows that the deterministic solution branch $\mu = 0$, $\nu < 0$ is (infinitesimally) stable. Therefore $\nu = 0$ is the value of the bifurcation parameter at which stability is lost. In the perturbed bifurcation case the value of ν at which stability is lost is $-2\sqrt{\alpha\gamma}/\beta$ $(\gamma > 0)$. If $\rho(\nu)$, $\nu < 0$ is the probability that at this value of ν stability has not been lost, we have from (3.10)

$$\rho(\nu) = \frac{1}{\sqrt{2\pi\sigma^2}}\int_0^{(\beta\nu)^2/4\alpha} e^{-\gamma^2/2\sigma^3}\,d\gamma \tag{3.12}$$

From (3.10) many other computations can be carried out. In addition, the cubic bifurcation case can be studied in a similar manner ($F_{uu}(0)=0$, $F_{uuu}(0) \neq 0$) and the scaling is now $\mu \to \varepsilon^{d/6}\mu$, $\nu \to \varepsilon^{d/3}\nu$.

4. Stochastically Perturbed Bifurcation. Analysis.

In this section, we shall give a method for making precise the heuristics of Section 3.

Clearly one must find a way to solve (3.4) and estimate ψ for μ, ν and ε small. One runs into difficulties however because S^ε is not small, only its averages are small so that norm estimates are too crude to work. But this is a familiar situation in stochastic problems and in homogenization and can be overcome by a device we introduce after stating some hypotheses.

First we simplify the problem to avoid cumbersome notation. We shall consider

$$\Delta u + \lambda u + au^2 + S^\varepsilon = 0, \quad x \in \mathcal{O}, \tag{4.1}$$

$$u = 0 \quad \text{for} \quad x \in \partial\mathcal{O}.$$

Here $\mathcal{O}$ is a bounded open subset of R^d with smooth boundary, a is a positive number and $S^\varepsilon = S(\frac{x}{\varepsilon},\omega)$ where $S(y,\omega)$ is a zero mean, stationary stochastic process with finite moments of all orders.* We denote the correlation function of S by

$$R(z) = E\{S(x+z)S(x)\} \tag{4.2}$$

---

* It will be clear from the following that many hypotheses can be weakened considerably.

and its Fourier transform by $\hat{R}(k)$

$$R(z) = \int_{R^d} e^{ik\cdot z}\, \hat{R}(k)\, dk \ . \tag{4.3}$$

We assume that $\hat{R}(k)$ is a smooth function (it is always positive and integrable since $E\{(S(x))^2\} < \infty$) that decays rapidly at infinity. We will need in addition higher order correlation functions, for example,

$$R_4(z_1,z_2,z_3,z_4) = E\{S(z_1)S(z_2)S(z_3)S(z_4)\} \tag{4.4}$$

which is translation invariant again

$$R_4(z_1,z_2,z_3,z_4) = R_4(0,z_2-z_1,z_3-z_1,z_4-z_1) \ . \tag{4.5}$$

We also define ρ_4 by

$$\begin{aligned} R_4(z_1,z_2,z_3,z_4) = {} & R(z_1-z_2)R(z_3-z_4) + R(z_1-z_3)R(z_2-z_4) \\ & + R(z_1-z_4)R(z_2-z_3) + \rho_4(z_1,z_2,z_3,z_4) \end{aligned} \tag{4.6}$$

which is called a cluster function because it has the property that it is uniformly small for any configuration of the four points with mutual distances bigger than a fixed number. The definitions for higher order correlation and cluster functions go similarly. The (uniform) decay rate for the cluster functions, and of $R(z)$ in particular, will be assumed to be exponential.

The results that we shall obtain *depend on the dimension* d . We shall work in detail the case $d = 3$ and comment briefly about the other cases. It will be clear how to carry over our estimates.

To begin the analysis we define a process

$$\chi(y,\omega) = \frac{1}{4\pi} \int_{R^3} \left(\frac{1}{|y-z|} - \frac{1}{|z|} \right) S(z,\omega)\, dz \ . \tag{4.7}$$

The integral in (4.7) converges for almost all ω and is continuous in y, but the process χ is not stationary while $\chi(0,\omega) = 0$. To see the convergence we note that

$$\begin{aligned} E\{\chi^2(y)\} &= \frac{1}{(4\pi)^2} \int\int \left|\frac{1}{|y-z_1|} - \frac{1}{|z_1|}\right| \left|\frac{1}{|y-z_2|} - \frac{1}{|z_2|}\right| R(z_1-z_2)\,dz_1 dz_2 \\ &= \int_{R^3} \frac{|e^{ik\cdot y}-1|^2}{|k|^4} R(k)\, dk < \infty. \end{aligned} \tag{4.8}$$

Actually χ satisfies the Poisson equation

$$\Delta\chi + S = 0 \ , \tag{4.9}$$

almost everywhere in ω. In addition we need the following two estimates for

(4.10) $$\chi^\varepsilon(x,\omega) = \varepsilon^2\chi(\tfrac{x}{\varepsilon},\omega) \; .$$

The first estimates are

(4.11) $$\sup_{x\in\mathcal{O}} E\{(\chi^\varepsilon(x))^2\} \le C_1\varepsilon^{5/2}, \qquad \sup_{x\in\mathcal{O}} E\{(\nabla\chi^\varepsilon(x))^2\} \le C_2\varepsilon^2$$

and the second

(4.12) $$\sup_{x\in\mathcal{O}} E\{(\chi^\varepsilon(x))^4\} \le C_3\varepsilon^5$$

where C_1, C_2 and C_3 are constants independent of ε.

Let us prove the first in (4.11). As in (4.8) we have that

$$E\{(\chi^\varepsilon(x))^2\} = \int_{R^3} \frac{|e^{ik\cdot x/\varepsilon}-1|^2}{|k/\varepsilon|^4}\hat{R}(k)\,dk$$

$$= \int_{|k|<\varepsilon^{3/2}} + \int_{|k|\ge\varepsilon^{3/2}}$$

For $|k| < \varepsilon^{3/2}$ we estimate the numerator by $|k|^2|x|^2/\varepsilon^2$ so the integral is less than a constant times $\varepsilon^{5/2}$. In the second integral the numerator is estimated by 2 and, since R decays rapidly at infinity the resulting estimate is $O(\varepsilon^{5/2})$ proving (4.11). To prove (4.12) we use (4.6), (4.6), the decay of ρ_4 and (4.11). Using the notation

(4.13) $$\Gamma(x,z) = \frac{1}{4\pi}\left|\frac{1}{|x-z|} - \frac{1}{|z|}\right|$$

we have

(4.14) $$E\{(\chi^\varepsilon(x))^4\} = \frac{\varepsilon^8}{(4\pi)^4}\iiiint \Gamma(\tfrac{x}{\varepsilon},z_1)\Gamma(\tfrac{x}{\varepsilon},z_2)\Gamma(\tfrac{x}{\varepsilon},z_3)\Gamma(\tfrac{x}{\varepsilon},z_4)$$

$$\cdot R_4(z_1,z_2,z_3,z_4)\;dz_1dz_2dz_3dz_4$$

$$= 3\left(\frac{\varepsilon^4}{(4\pi)^2}\iint \Gamma(\tfrac{x}{\varepsilon},z_1)\Gamma(\tfrac{x}{\varepsilon},z_2)R(z_1-z_2)\;dz_1\;dz_2\right)^2$$

$$+ \frac{\varepsilon^8}{(4\pi)^4}\iiiint \Gamma(\tfrac{x}{\varepsilon},z_1)\Gamma(\tfrac{x}{\varepsilon},z_2)\Gamma(\tfrac{x}{\varepsilon},z_3)\Gamma(\tfrac{x}{\varepsilon},z_4)\rho_4(0,z_2-z_1,z_3-z_1,z_4-z_1)$$

$$\cdot\; dz_1dz_2dz_3dz_4$$

The first term on the right is $O(\varepsilon^5)$. The second term is actually smaller (of order ε^9) as can be seen by using the fact that $\rho_4(0,z_2,z_3,z_4)$ decays to zero rapidly (exponentially, say) as $|z_1|+|z_2|+|z_3| \to \infty$. This proves (4.12).

Let $\theta(s)$ be a C^∞ cutoff function defined by

$$\theta(s) = \begin{cases} 1\ , & s \geq 2 \\ 0\ , & s \leq 1 \end{cases} \tag{4.15}$$

Recall that the eigenfunction ϕ

$$(\Delta+\bar{\lambda})\phi = 0, \quad x \in \mathcal{O}, \quad \phi = 0 \text{ on } \partial\mathcal{O} \tag{4.16}$$

is a smooth function that is positive in the interior of $\mathcal{O}$. We now define

$$\tilde{\chi}^{\varepsilon}(x,\omega) = \chi^{\varepsilon}(x,\omega)\theta^{\varepsilon}(x)\ , \qquad \theta^{\varepsilon}(x) = \theta\left(\frac{\phi(x)}{\sqrt{\varepsilon}}\right), \quad x \in \mathcal{O}\ , \tag{4.16}$$

so that $\tilde{\chi}^{\varepsilon} \equiv 0$ for x near $\partial\mathcal{O}$.

The decisive step in our analysis is the ansatz

$$u = \mu\phi + \tilde{\chi}^{\varepsilon} + \psi\ , \qquad (\psi,\phi) = 0 \tag{4.17}$$

for the problem (4.1). Inserting (4.17) in (4.1), setting $\lambda = \bar{\lambda} + \nu$ and collecting terms we find that ψ must satisfy

$$\begin{aligned}(\Delta+\bar{\lambda})\psi + \nu(\mu\phi+\psi) + a(\mu\phi+\psi)^2 + 2a(\mu\phi+\psi)\tilde{\chi}^{\varepsilon} \\ + a(\tilde{\chi}^{\varepsilon})^2 + S^{\varepsilon}(1-\theta^{\varepsilon}) + 2\nabla\theta^{\varepsilon}\cdot\nabla\chi^{\varepsilon} + \Delta\theta^{\varepsilon}\chi^{\varepsilon} + (\bar{\lambda}+\nu)\tilde{\chi}^{\varepsilon} = 0\ .\end{aligned} \tag{4.18}$$

Let

$$\zeta^2 = a(\tilde{\chi}^{\varepsilon})^2 + S^{\varepsilon}(1-\theta^{\varepsilon}) + 2\nabla\theta^{\varepsilon}\cdot\nabla\chi^{\varepsilon} + \Delta\theta^{\varepsilon}\chi^{\varepsilon} + \nu\tilde{\chi}^{\varepsilon} \tag{4.19}$$

and

$$\begin{aligned}G(\mu,\nu,\varepsilon,\psi) = \nu\mu + a\mu^2(\phi^2,\phi) + 2a\mu(\phi\psi,\phi) + a(\psi^2,\phi) \\ + 2a\mu(\phi\tilde{\chi}^{\varepsilon},\phi) + 2a(\psi\tilde{\chi}^{\varepsilon},\phi) + (\zeta^{\varepsilon},\phi) + \bar{\lambda}(\tilde{\chi}^{\varepsilon},\phi)\end{aligned} \tag{4.20}$$

Now the problem at hand is this: find a random function $\psi(x,\omega)$ for μ, ν and ε sufficiently small such that

$$\begin{aligned}(\Delta+\bar{\lambda})\psi + \nu(\mu\phi+\psi) + a(\mu\phi+\psi)^2 + 2a(\mu\phi+\psi)\tilde{\chi}^{\varepsilon} \\ + \zeta^{\varepsilon} + \bar{\lambda}\tilde{\chi}^{\varepsilon} - G(\mu,\nu,\varepsilon,\psi)\phi = 0\ , \qquad (\phi,\psi) = 0\ ,\end{aligned} \tag{4.21}$$

$$G(\mu,\nu,\varepsilon,\psi) = 0\ . \tag{4.22}$$

Lemma 1. For any $\eta > 0$ and for all μ, ν and ε sufficiently small there is a set $\Omega_{\varepsilon} \subset \Omega$ such that (4.21) has a unique solution $\psi(x,\omega;\mu,\nu,\varepsilon)$ in $W^{2,2}(\mathcal{O}) \cap W_0^{1,2}(\mathcal{O})$ for each $\omega \in \Omega_{\varepsilon}$ such that *

$$\sup_{\omega\in\Omega_{\varepsilon}} \|\psi(\cdot,\omega)\|_{W^{2,2}} \leq \eta\ . \tag{4.23}$$

Moreover

$$\lim_{\varepsilon\downarrow 0} P(\Omega_{\varepsilon}) = 1\ . \tag{4.24}$$

---

* $W^{\ell,\rho}(\mathcal{O})$ is the Sobolev space of functions with distribution derivatives of order ℓ in L^{ρ}.

Remark. Since we are in 3 dimensions (4.23) implies that

$$\sup_{\omega\in\Omega_\varepsilon}\ \sup_{\in\mathcal{O}} |\psi(x,\omega)| \le C\eta \tag{4.25}$$

where C is some constant. This is used in Lemma 2.

Proof: First we recall that the operator $\Delta+\bar\lambda$ is an isomorphism from L^2_ϕ = orthogonal complement of ϕ in $L^2(\mathcal{O})$ to $W^{2,2}(\mathcal{O}) \cap W_0^{1,2}(\mathcal{O}) \cap L^2_\phi(\mathcal{O})$. The inverse of $\Delta+\bar\lambda$ on the range is denoted by $K = -(\Delta+\bar\lambda)^{-1}$. So we may write (4.21) in fixed-point form

$$\psi = K[\nu(\mu\phi+\psi) + a(\mu\phi+\psi)^2 + 2a(\mu\phi+\psi)\tilde\chi^\varepsilon + \zeta^\varepsilon + \bar\lambda\tilde\chi^\varepsilon] \equiv T(\psi) \tag{4.26}$$

where we omit the $G\phi$ term that makes the terms in the bracket be in L^2_ϕ since that is understood as part of the definition of K.

We shall show that the operator $T(\psi)$ defined by the right side of (4.26) is a contraction in $W^{2,2}(\mathcal{O})$ for ν,μ,ε small and for ω in a suitably defined set Ω_ε.

Suppose ψ is a fixed function of x such that $\|\psi\|_{W^{2,2}} \le \eta$ (we are only interested in small η fixed). Then by the isomorphism there is a constant C independent of ν,μ,ε,ψ and ω such that *

$$\|T(\psi)\|_{W^{2,2}} \le C[\nu\|\psi\|_{L^2} + a\mu^2\|\phi^2\|_{L^2} + 2a\mu\|\phi\psi\|_{L^2} + a\|\psi^2\|_{L^2} + 2a\|(\mu\phi+\psi)\tilde\chi^\varepsilon\|_{L^2} + \|\zeta^\varepsilon\|_{L^2} + \bar\lambda\|\tilde\chi^\varepsilon\|_{L^2}]$$

But $\|\psi\|_{L^\infty(\mathcal{O})} \le C\eta$ so that

$$\|T(\psi)\|_{W^{2,2}} \le C[\nu\eta + a\mu^2 C_1 + 2a\mu C_2\eta + a\eta^2 + C_3\|(\tilde\chi^\varepsilon)^2\|_{L^2} + \|\zeta^\varepsilon\|_{L^2}] \tag{4.27}$$

From (4.19) we see that ζ^ε contains $(\tilde\chi^\varepsilon)^2$ so that it suffices now to show that $\|\zeta^\varepsilon\|_{L^2}$ can be made small with ε. We shall construct a set Ω_ε such that

$$\|\zeta^\varepsilon\|_{L^2} \le C_4\eta^2 \quad \text{for} \quad \omega\in\Omega_\varepsilon\ . \tag{4.28}$$

The set Ω_ε depends on η but for each η fixed and positive $P(\Omega_\varepsilon) \to 1$ as $\varepsilon \to 0$.

Returning to (4.27) we see that for η small and ν,μ small, $\|T(\psi)\|_{W^{2,2}} \le \eta$. From this our Lemma 1 follows easily. It therefore remains to construct Ω_ε and to show (4.28).

From (4.19) it follows that for some constant C

$$\|\zeta^\varepsilon\|^2_{L^2} \le C\Big[\int_{\mathcal{O}} (\chi^\varepsilon(x))^4 dx + \int_{\mathcal{O}_\varepsilon} (S^\varepsilon(x))^2 dx + \frac{1}{\varepsilon}\int_{\mathcal{O}_\varepsilon} (\nabla\chi^\varepsilon)^2 dx + \frac{1}{\varepsilon^2}\int_{\mathcal{O}_\varepsilon} (\chi^\varepsilon(x))^2 dx\Big], \tag{4.29}$$

---

* We use C to denote such constants, even different ones. But we sometimes use subscripts on C for clarity.

where

$$\mathcal{O}_\varepsilon = \left\{ x \in \mathcal{O} \mid \text{distance}(x, \partial\mathcal{O}) \le C_5\sqrt{\varepsilon} \right\} . \tag{4.30}$$

We now take

$$\Omega_\varepsilon = \left\{ \omega \in \Omega \;\Big|\; \int_{\mathcal{O}} (\chi^\varepsilon)^4\, dx + \int_{\mathcal{O}_\varepsilon} (S^\varepsilon)^2 dx + \frac{1}{\sqrt{\varepsilon}} \int_{\mathcal{O}} (\nabla\chi^\varepsilon)^2 dx + \frac{1}{\varepsilon}\int_{\mathcal{O}_\varepsilon} (\chi^\varepsilon)^2\, dx \le \eta^4 \right\} \tag{4.31}$$

so that (4.28) is immediately valid. It remains to show that $P(\Omega_\varepsilon^c) \to 0$ as $\varepsilon \to 0$ which is done by Chebyshev's inequality term by term as follows.

$$P\left\{ \int_{\mathcal{O}} (\chi^\varepsilon)^4\, dx > \eta^4 \right\} \le \frac{1}{\eta^4}\int_{\mathcal{O}} E(\chi^\varepsilon)^4\, dx \le \frac{C_6}{\eta^4}\,\varepsilon^5 \quad \text{(by (4.12))}; \tag{4.32}$$

$$P\left\{ \int_{\mathcal{O}_\varepsilon} (S^\varepsilon)^2\, dx > \eta^4 \right\} \le \frac{1}{\eta^4}\int_{\mathcal{O}_\varepsilon} E(S^\varepsilon)^2\, dx \le \frac{C_7}{\eta^4}\sqrt{\varepsilon} \tag{4.33}$$

$$P\left\{ \frac{1}{\varepsilon}\int_{\mathcal{O}_\varepsilon} (\nabla\chi^\varepsilon)^2\, dx > \eta^4 \right\} \le \frac{1}{\eta^4}\int_{\mathcal{O}_\varepsilon} \frac{1}{\varepsilon} E(\nabla\chi^\varepsilon)^2 dx \le \frac{C_8}{\eta^4}\frac{1}{\varepsilon} \cdot \sqrt{\varepsilon} \cdot \varepsilon^2 \tag{4.34}$$

$$P\left\{ \frac{1}{\varepsilon^2}\int_{\mathcal{O}_\varepsilon} (\chi^\varepsilon)^2 dx > \eta^4 \right\} \le \frac{1}{\eta^4}\int_{\mathcal{O}_\varepsilon} \frac{1}{\varepsilon^2} E(\chi^\varepsilon)^2 dx \le \frac{C_9}{\eta^4}\frac{1}{\varepsilon^2}\sqrt{\varepsilon} \cdot \varepsilon^{5/2}. \tag{4.35}$$

In (4.33)-(4.35) we have used (4.10)-(4.12) as well as the fact that $\text{vol}(\mathcal{O}_\varepsilon) \le C_{10}\cdot\sqrt{\varepsilon}$. The proof of the lemma is complete.

Now it can be seen that one can actually get rates of decay to zero of ψ in the $W^{2,2}$ norm on Ω_ε as $\varepsilon \to 0$. But this is not enough for our limit theorem. Before going to our next estimate we recall some facts about $-(\Delta+\bar{\lambda})^{-1}$. This operator can be represented by a kernel $K(x,y)$ that has the form

$$K(x,y) = \frac{\cos\sqrt{\bar{\lambda}}\,|x-y|}{4\pi\,|x-y|} + \kappa(x,y) , \qquad x,y \in \mathcal{O} , \tag{4.36}$$

with $\kappa(x,y)$ a bounded function. We also have the estimates

$$|K(x,y)| \le \frac{C}{|x-y|} , \qquad |\nabla K(x,y)| \le \frac{C}{|x-y|^2} \tag{4.37}$$

where C is some constant. The function K is called the generalized Green's function and, as is well known, it is not positive.

Let $\psi(x,\omega;\mu,\nu,\varepsilon)$ be the solution of Lemma 1, with $\omega \in \Omega_\varepsilon$ and put

(4.38) $$\psi^\varepsilon(x,\omega) = \psi(x,\omega,\varepsilon^{3/4}\mu,\varepsilon^{3/4}\nu,\varepsilon)\ , \qquad \omega \in \Omega_\varepsilon$$

Lemma 2. We have that uniformly over compact (μ,ν) sets

(4.39) $$\lim_{\varepsilon \downarrow 0} \varepsilon^{-3/2}\, E\left\{\|\psi^\varepsilon\|^2_{L^2(\mathcal{O})}, \Omega_\varepsilon\right\} = 0$$

where $E\{f,\Omega\}$ denotes the integral of f over Ω with respect to the probability measure P.

Proof: From (4.21) we have, omitting the ω, that

(4.40) $$\psi^\varepsilon(x) = \int_{\mathcal{O}} K(x,y)[\varepsilon^{3/4}\nu\ \psi(y) + a\left(\varepsilon^{3/4}\mu\ \phi(y) + \psi^\varepsilon(y)\right)^2 + 2a\left(\varepsilon^{3/4}\mu\ \phi(y) + \psi^\varepsilon(y)\right)\tilde{\chi}^\varepsilon(y) + \zeta^\varepsilon(y) + \bar{\lambda}\tilde{\chi}^\varepsilon(y)]\, dy$$

where again we omit the term $G\phi$ since it is understood that K annihilates it. We estimate $\|\psi^\varepsilon\|^2_{L^2(\mathcal{O})}$ by a constant times the sum of squares of the L^2 norms on the right. The first term on the right is estimated by

$$\int_{\mathcal{O}} dx \int_{\mathcal{O}}\int_{\mathcal{O}} K(x,y)K(x,z)\,\varepsilon^{3/2}\nu^2\ \psi^\varepsilon(y)\ \psi^\varepsilon(z)\ dy\, dz$$
$$< \varepsilon^{3/2}\nu \int_{\mathcal{O}} dx \int_{\mathcal{O}}\int_{\mathcal{O}} \frac{c^2}{|x-y|\,|x-z|}\ |\psi^\varepsilon(y)|\ |\psi^\varepsilon(z)|$$
$$\leq \varepsilon^{3/2}\nu\ C_1\|\ \psi^\varepsilon\|^2_{L^2(\mathcal{O})}\ , \qquad \omega \in \Omega_\varepsilon$$

where C_1 is another constant. The other terms go the same way and only the following requires comment:

$$\int_{\mathcal{O}} dx \int_{\mathcal{O}}\int_{\mathcal{O}} K(x,y)\ K(x,z)\ \left(\psi^\varepsilon(y)\right)^2\ \left(\psi^\varepsilon(z)\right)^2\ dy\ dz \leq C_2\eta^2\|\psi^\varepsilon\|^2_{L^2(\mathcal{O})}$$
$$\omega \in \Omega_\varepsilon$$

To get this we use Lemma 1. In fact this is the reason we must prove Lemma 1 first.

Collecting estimates we have that

$$\|\psi^\varepsilon\|^2_{L^2(\mathcal{O})} \leq C_3[\varepsilon^{3/2}\nu^2\|\psi^\varepsilon\|^2_{L^2(\mathcal{O})} + a^2\mu^4\varepsilon^3 + a^2\varepsilon^{3/2}\mu^2\|\psi^\varepsilon\|^2_{L^2(\mathcal{O})} + \eta^2\|\psi^\varepsilon\|^2_{L^2(\mathcal{O})} + \|(\tilde{\chi}^\varepsilon)^2\|^2_{L^2(\mathcal{O})} + \|K\zeta^\varepsilon\|^2_{L^2(\mathcal{O})}]$$

Therefore since $\eta > 0$ is small and ν, μ are in a compact set we have that (omitting the set Ω_ε in E{ }),

$$(4.41)\quad \lim_{\varepsilon\downarrow 0} \varepsilon^{-3/2} E\left\{\|\psi^\varepsilon\|^2_{L^2(\mathcal{O})}\right\} \leq C_4 \lim_{\varepsilon\downarrow 0} \varepsilon^{-3/2}\Big[E\big(\|(\tilde{\chi}^\varepsilon)^2\|^2_{L^2(\mathcal{O})}\big) + E\big(\|K\zeta^\varepsilon\|^2_{L^2(\mathcal{O})}\big)\Big]$$

Now

$$\varepsilon^{-3/2} E\left\{\|(\tilde{\chi}^\varepsilon)^2\|^2_{L^2(\mathcal{O})}\right\} = \varepsilon^{-3/2} \int_{\mathcal{O}} E\left\{\big(\tilde{\chi}^\varepsilon(x)\big)^4\right\} dx \leq C\, \varepsilon^{-3/2}\varepsilon^5 \to 0$$

by (4.12). Therefore we must look at the second term on the right in (4.41). This requires estimating several terms.

From (4.19) we have

$$(4.42)\quad E\left\{\|K\zeta^\varepsilon\|^2_{L^2(\mathcal{O})}\right\} = E\left\{ \int_{\mathcal{O}} dx \left| \int_{\mathcal{O}} K(x,y)\,[a(\tilde{\chi}^\varepsilon(y))^2 + s^\varepsilon(y)(1-\theta^\varepsilon(y)) + 2\nabla\theta^\varepsilon\cdot\nabla\chi^\varepsilon + \Delta\theta^\varepsilon\chi^\varepsilon + \varepsilon^{3/4}\nu\,\tilde{\chi}^\varepsilon]\, dy\right|^2\right\}$$

We shall omit the tedious verification that this goes to zero faster than $\varepsilon^{3/2}$ as $\varepsilon \to 0$. Only one term requires comment. It is

$$(4.43)\quad \int_{\mathcal{O}} dx \int_{\mathcal{O}}\int_{\mathcal{O}} K(x,y)K(x,z)\ \Delta\theta^\varepsilon(y)\Delta\theta^\varepsilon(z)\ E\{\chi^\varepsilon(y)\chi^\varepsilon(z)\}\, dy\, dz\ .$$

To handle this we integrate by parts once on y and once on z. Then we use the gradient estimates (4.37) and (4.11) and the result follows. The proof of Lemma 2 is complete.

We move closer to our main objective with the following lemma.

Lemma 3. Let $G^\varepsilon(\mu,\nu,\omega) = \varepsilon^{-3/2}G(\varepsilon^{3/4}\mu, \varepsilon^{3/4}\nu, \psi^\varepsilon)$ with G defined by (4.20) and ψ^ε by (4.38) while $\omega \in \Omega_\varepsilon$. Then for any $\delta > 0$ and any compact set τ of (μ,ν) we have that

$$(4.44)\quad \lim_{\varepsilon\downarrow 0} P\left\{ \sup_{(\mu,\nu)\in\tau} |G^\varepsilon(\mu,\nu,\omega) - (\nu\mu + a\mu^2(\phi^2,\phi) + \varepsilon^{-3/2}(s^\varepsilon,\phi))| > \delta, \Omega_\varepsilon\right\} = 0\ .$$

Proof of Lemma 3: By inspection after Chebyshev's inequality is applied to the probability in (4.44) we see that Lemma 2 is just the right thing to knock out most terms. What remains to be shown is

$$(4.45)\quad \lim_{\varepsilon\downarrow 0} \varepsilon^{-3/2} E\left\{|(\zeta^\varepsilon + \bar{\lambda}\tilde{\chi}^\varepsilon - \delta^\varepsilon, \phi)|\right\} = 0\ .$$

But from (4.19) we see that

$$\zeta^\varepsilon + \bar\lambda\tilde\chi^\varepsilon - S^\varepsilon = a(\tilde\chi^\varepsilon)^2 + \Delta\tilde\chi^\varepsilon + \varepsilon^{3/4}\nu\,\tilde\chi^\varepsilon + \bar\lambda\tilde\chi^\varepsilon$$

$$= a(\tilde\chi^\varepsilon)^2 + (\Delta+\bar\lambda)\tilde\chi^\varepsilon + \varepsilon^{3/4}\tilde\chi^\varepsilon \quad .$$

Thus

$$(\zeta^\varepsilon + \bar\lambda\tilde\chi^\varepsilon - S^\varepsilon,\ \phi) = a\big((\tilde\chi^\varepsilon)^2,\phi\big) + \varepsilon^{3/4}\nu(\tilde\chi^\varepsilon,\phi)$$

since $(\Delta+\bar\lambda)\phi = 0$ and integration by parts is permissible because $\tilde\chi^\varepsilon$ vanishes near $\partial\mathcal{O}$. From this and (4.11) the result (4.45) and hence Lemma 3 follow.

We are now in a position to state the main result of this paper.

Theorem. Assume that $S(y,\omega)$ satisfies the properties that are stated at the beginning of this section and, in addition that it is α-mixing [11] with, say, an exponential rate. Then the scaled bifurcation function $G^\varepsilon(\mu,\nu,\omega) = \varepsilon^{-3/2}G(\varepsilon^{3/4}\mu,\varepsilon^{3/4}\nu,\psi^\varepsilon)$ converges weakly (as a process on $C(R^2;R^1)$) to the Gaussian process $a\mu^2(\phi^2,\phi) + \nu\mu + \gamma$ where γ is a Gaussian random variable with mean zero and variance given by (3.8).

Remark. It follows that statistical properties of solution branches of $G^\varepsilon(\mu,\nu,\omega)$, say $\mu = \mu^\varepsilon(\nu,\omega)$, converge to statistical properties of the corresponding branches of the limit quadratic bifurcation equation.

Proof of Theorem: We have shown in Lemma 3 that G and $a\mu^2(\phi^2,\phi) + \nu\mu + \varepsilon^{-3/2}(S^\varepsilon,\phi)$ have the same weak limit, if any, as processes. But this means that it suffices to find the weak limit of the random variable $\varepsilon^{-3/2}(S^\varepsilon,\phi)$ and this is just the central limit theorem [11].

The proof of the theorem is complete.

References

[1] J. B. Keller and S. Antman, "Bifurcation Theory and Nonlinear Eigenvalue Problems," W. A. Benjamin, New York, 1969.

[2] D. Sattinger, Topics in Stability and Bifurcation Theory, Lecture Notes in Math. 309 (1973), Springer-Verlag.

[3] E. Reiss, Imperfect Bifurcation, in Applications of Bifurcation Theory, Academic Press, New York, 1977, pp. 37-65.

[4] B. Matkowski and E. Reiss, Singular Perturbation of Bifurcation, SIAM J. Appl. Math. 33 (1977) pp. 230-255.

[5] M. Golubitski and D. Schaffer, A Theory for Imperfect Bifurcation via Singularity Theory, Comm. Pure Appl. Math. 32 (1979), pp. 21-98.

[6] J. Hale, Generic Bifurcation with Applications, in Nonlinear Analysis and Mechanics: Heriot-Watt Symposium, Vol. I, R. J. Knaps, Editor, Pitman, London, 1977.

[7] J. Keener and H. B. Keller, Perturbed Bifurcation Theory, Arch. Rat. Mech. Anal. 50 (1973), pp. 159-175.

[8] D. S. Cohen, Multiple Solutions of Nonlinear Partial Differential Equations, Springer Lecture Notes in Math. 322 (1973) pp. 15-77.

[9] B. S. White and J. V. Franklin, A Limit Theorem for Stochastic Two-Point Boundary-Value Problems of Ordinary Differential Equations, Comm. Pure Appl. Math. 32 (1979) pp. 253-276.

[10] W. Day and G. Papanicolaou, Stochastically Perturbed Bifurcation, SIAM J. Appl. Mathematics, to appear.

[11] I. A. Ibragimov and Yu. V. Linnik, Independent and Stationary Sequences of Random Variables, Groningen, Walters-Noordoff, 1971.

COMPUTING METHODS IN APPLIED SCIENCES AND ENGINEERING
R. Glowinski, J.L. Lions (editors)
North-Holland Publishing Company

SOME TYPES OF PERTURBATION FOR EVOLUTION EQUATIONS

Sergio Spagnolo (Pisa)

L'objet de cet exposé est la stabilité de certaines classes d'équations hyperboliques par rapport à la convergence des solutions (pour ce qui concerne les équations paraboliques, on renvoie à [1],[3],[10],[8] et [9]). Les équations ici considérées seront toutes du type suivant

(1) $$u_{tt} - \sum_{i,j}^{1,n} (a_{ij}(x,t)\, u_{x_i})_{x_j} = 0 \qquad (0 \leqslant t \leqslant T).$$

Les *coefficients* a_{ij} sont des fonctions mesurables sur $R^n \times [0,T]$ qui vérifient les conditions usuelles d'ellipticité (où $\lambda > 0$) :

(2) $$\begin{cases} a_{ij} = a_{ji} \\ \lambda|\xi|^2 \leqslant \Sigma\, a_{ij}\, \xi_i \xi_j \leqslant \Lambda|\xi|^2 , \quad \forall\, \xi \in \mathbb{R}^n. \end{cases}$$

Outre à cette condition, les a_{ij} seront soumises, à mesure que la necessité se présentera, à des hypothèses supplémentaires de régularité en t ou en x.

Le problème aux limites qu'on fait correspondre habituellement à l'équation (1) est le *problème de Cauchy*: on assigne les valeurs de u et de u_t à l'instant initial $t = 0$, et l'on cherche une solution u pour $t \in [0,T]$. Quant aux conditions aux limites par rapport à la variable x, on en peut considérer de deux types différents : ayant fixé un ouvert Ω de $\mathbb{R}^n_x$, chercher une solution qui soit nulle sur $\partial\Omega$ pour tout t (*problème de Cauchy-Dirichlet*); ou bien chercher des solution définies sur $\mathbb{R}^n_x$ tout entier, sans imposer aucune restriction pour $|x| \to \infty$.

On se bornera ici à ce dernier type de problème. En tout cas beaucoup des résultats (positifs ou négatifs) exposés ici se transposent au problème de Cauchy-Dirichlet.

On considère donc le problème

$$(3) \qquad \begin{cases} u_{tt} - \Sigma (a_{ij}(x,t) u_{x_i})_{x_j} = 0 & (0 \leqslant t \leqslant T) \\ u(x,0) = \varphi(x) , \quad u_t(x,0) = \psi(x) \end{cases}$$

où φ et ψ sont deux fonctions définies sur $\mathbb{R}^n$.

Dans les questions de stabilité on a une suite de problèmes :

$$(3)_k \qquad \begin{cases} u^k_{tt} - \Sigma (a^k_{ij}(x,t) u_{x_i})_{x_j} = 0 & (0 \leqslant t \leqslant T) \\ u^k(x,0) = \varphi(x) , \quad u^k_t(x,0) = \psi(x) \end{cases}$$

où $k = 1,2,3,\ldots$ et où les données initiales φ et ψ ne dépendent pas de k. On supposera toujours que les coefficients a^k_{ij} vérifient (2) de façon uniforme par rapport à k.

Tout résultat de stabilité s'appuie sur quelque résultat d'existence et unicité des solutions et sur des estimations convenable de la solution par rapport aux données initiales et aux coefficients. Or, à différence de ce qui se passe pour les équations paraboliques (pour lesquelles on ne considère que le problème de Cauchy Dirichlet), le problème (3) est en général depourvu de solutions, à moins que les coefficients a_{ij} ne soient suffisamment réguliers en t ou en x. Pour des contre-exemples à ce propos, on renvoie à [2].

Dans cet exposé seront utilisés essentiellement les deux théorèmes d'existence suivants (l'un ou l'autre des deux, selon que les $a^k_{ij}(x,t)$ soient équiréguliers en t ou bien en x).

<u>Th.1</u> (*existence lorsque les coefficients sont lipschitziens en* t *;* voir [6])

Considérons le problème (3) avec des $a_{ij}(x,t)$ qui vérifient, outre à (2), la condition

$$(4) \qquad |a_{ij}(x,t_2) - a_{ij}(x,t_1)| \leqslant L|t_2 - t_1| \quad , \quad \forall\, x,t_1,t_2 .$$

Alors le problème admet, pour tout $\varphi \in H^1_{loc}$ et tout $\psi \in L^2_{loc}$, une et une seule solution $u \in L^{\infty}([0,T],H^1_{loc})$.

Considérons la suite des problèmes $(3)_k$, avec des a^k_{ij} qui vérifient (2) et (4) uniformément par rapport à k. Alors les solutions u^k restent bornées dans $L^{\infty}([0,T],H^1_{loc})$.

Th.2 (*existence lorsque les coefficients sont analytiques en* x *; voir* [5])

Considérons le problème (3) avec des $a_{ij}(x,t)$ qui vérifient, outre a (2), les conditions

$$|D^h_x a_{ij}(x,t)| \leqslant \Lambda_r A_r^{|h|} h! \qquad , \forall h \in \mathbb{N}^n , \tag{5}$$

pour tout $x \leqslant |r|$ et tout $r \geqslant 0$.

Alors le problème admets, pour tout φ et ψ analytique sur R^n, une et une seule solution $u \in C^1([0,T],A(\mathbb{R}^n))$, où $A(\mathbb{R}^n)$ est l'espace des fonctions analytiques sur R^n.

Considérons la suite des problèmes $(3)_k$, avec des a^k_{ij} qui vérifient (2) et (5) uniformément par rapport à k. Alors les solutions u^k restent bornées dans $C^1([0,T],A(\mathbb{R}^n))$ pourvu que les a^k_{ij} restent dans un compacte de $L^1_{loc}(\mathbb{R}^n \times [0,T])$.

Remarque 1 Il est important de signaler que, soit dans le th.1 que dans le th.2, le *domaine de dépendence* de la solution par les données initiales φ et ψ sur quelque ouvert Ω de $\mathbb{R}^n$, contient toujours le cône

$$\Gamma_{\Omega} = \{(x,t) : \mathrm{dist}(x, \complement\Omega) > t\sqrt{\Lambda}\} .$$

Ça signifie que si φ et ψ sont nulles sur Ω alors la solution u est nulle sur Γ_{Ω}, ou (dans le cas du th.2) que si $\{\varphi^k\}$ et $\{\psi^k\}$ sont deux suites convergeant à zéro dans $A(\Omega)$ alors $\{u^k\}$ converge à zéro uniformément sur les compactes de Γ_{Ω}.

En correspondence aux théorèmes d'existence 1 et 2, on obtients les deux résultats suivants de stabilité.

Th.3 (*convergence dans le cas de coefficients équilipschitziens en* t ; voir [4])

Considérons les problèmes $(3)_k$ et (3), avec $\varphi \in H^1_{loc}$ et $\psi \in L^2_{loc}$, et avec des coefficients a^k_{ij} qui vérifient (2) et (4) uniformément par rapport à k.

Supposons en outre qu'on ait, au sens de la *G-convergence elliptique* (voir [7]),

$$\{-\Sigma D_{x_i}(a^k_{ij}(\circ,t)\,D_{x_j})\} \overset{G}{\rightarrow} -\Sigma D_{x_i}(a_{ij}(\circ,t)\,D_{x_j})$$

sur $\mathbb{R}^n_x$, pour tout $t \in [0,T]$ (et $k \to \infty$).

Alors les fonctions a_{ij} sont lipschitziennes en t, et on a

$$\{u^k\} \rightarrow u \qquad \text{dans } L^\infty([0,T], L^2_{loc}) .$$

Th.4 (*convergence dans le cas de coefficients équianalytiques en* x ; voir [5])

Considérons les problèmes $(3)_k$ et (3), avec φ et ψ analytiques sur $\mathbb{R}^n$, et avec des coefficients a^k_{ij} qui vérifient (2) et (5) uniformément par rapport à k.

Supposons en outre que, lorsque $k \to \infty$, on ait

$$\{a^k_{ij}\} \rightarrow a_{ij} \qquad \text{dans } L^1_{loc}(\mathbb{R}^n \times [0,T]) .$$

Alors les fonctions a_{ij} sont analytiques en x pour tout t, et

$$\{u^k\} \rightarrow u \qquad \text{dans } C^1([0,T], A(\mathbb{R}^n)).$$

On peut montrer par des exemples (voir [2]) que les résultats des théorèmes 2 et 4 tombent en défaut si les données initiales ne sont pas des fonctions analytiques, même s'ils sont des fonctions indéfiniment différentiables. Plus précisément, il existent des coefficients $a_{ij} \equiv a_{ij}(t)$ hölderiens (mais non lipschitziens, puisque autrement on retomberais dans le th.1) et des données initiales φ et ψ de classe C^∞ (mais non analytiques) tels que le problème (3) n'a pas de solutions dans l'espace des distributions.

Il existent en outre des suites $a_{ij}^{k}(t)$ de coefficients tel que $\{a_{ij}^{k}\} \to \delta_{ij}$ (symbole de Kronecker) en L^{∞} pour $k \to \infty$ et des données initiales φ et ψ de classe C^{∞}, de façon que la suite $\{u^{k}\}$ des solutions de $(3)_k$ n'est pas bornée dans l'espace des distributions.

Des contre-exemples encore plus forts se présenterons très probablément si l'on utilise des coefficients qui dépendent aussi de x.

On relève donc ici une différence très nette avec les équations elliptiques ou paraboliques, dans lesquelles une petite perturbation dans L^{1}_{loc} des coefficients ne produit qu'une petite perturbation des solutions correspondentes.

Un cas particulier de problème de stabilité est celui de l'*homogénéisation*, où l'on assigne des fonctions périodiques (et suffisamment régulières) $\alpha_{ij}(y,\tau)$ et on définit

$$a_{ij}^{k}(x,t) = \alpha_{ij}(k^{p}x,\ k^{q}t) \qquad ,\ p \text{ et } q \geqslant 0 .$$

On s'attend alors qu'il existent des coefficients-*limite* a_{ij} (constants) tels que la suite $\{u^{k}\}$ des solutions de $(3)_k$ converge dans quelque espace vers la solution de (3). C'est ça, en effet, qui arrive pour les équations paraboliques (voir [1],[8] et [9]), où l'on peut calculer les coefficients-limite a_{ij} par trois procédés différents, selon que soit $p < 2q$, $p = 2q$ ou $p > 2q$.

Malhereusement, toutefois, ni le th.3 ni le th.4 peuvent être utilisés dans le cas de l'homogénéisation (sauf dans le cas simple où $q = 0$ qui rentre dans le th.3) des équations hyperboliques.

De plus, les exemples ci-dessous montrent que la situation de l'homogénéisation en t est presque sans espérance.

<u>Exemple 1</u> (*homogénéisation en t* ;voir [4])

Considérons les problèmes (dans une seule variable d'espace)

$$(6)_k \quad \begin{cases} u_{tt}^{k} - (\alpha(kt)\, u_{x}^{k})_{x} = 0 & (t \geqslant 0) \\ u^{k}(x,0) = \varphi(x) \quad , \quad u_{t}^{k}(x,0) = \psi(x) \end{cases}$$

avec

$$\alpha(\tau) = 1 - 4\epsilon \sin 2\tau - \epsilon^2 (1 - \cos 2\tau)^2 \qquad (0 < \epsilon \leqslant \frac{1}{10})$$

et

$$\varphi(x) = 0 \quad , \quad \psi(x) = \sum_{h=0}^{\infty} e^{-\delta h} \sin(hx) \qquad (\delta > 0)$$

Alors $\psi(x)$ est une fonction analytique sur $\mathbb{R}^n$, mais les solutions u^k de $(6)_k$ ne sont pas bornées (au sens des distributions) au déhors de la bande fermée $\{(x,t) : 0 \leqslant t \leqslant \delta/\epsilon\}$.

<u>Remarque 2</u> Si, avec les mêmes $\alpha(\tau)$ et $\varphi(x)$, on prend $\psi(x) = \Sigma \exp(-\delta \sqrt{h}) \sin(hx)$, où $\delta > 0$, alors ψ est de classe C^∞ et les solutions u^k ne sont pas bornées (au sens des distributions) dès que $t > 0$.

<u>Remarque 3</u> On peut construire d'autres contre-exemples en considérant des équations du type suivant (*homogénéisation en* x *et* t):

$$(7)_k \qquad u^k_{tt} - (\alpha(kx, kt)\, u^k_x)_x = 0$$

avec $\alpha(y, \tau) = \alpha_1(y)\, \alpha_2(\tau)$.

Malgré ces exemples, on peut prover quelques résultats (positifs) de stabilité dans des cas très particuliers qui sont assez proches au cas de l'homogénéisation. Un de ces résultats est le suivant:

<u>Th. 5</u> Considérons, dans le cas $n = 1$, les problèmes $(3)_k$ avec

$$a^k(x, t) = \alpha(e^{\delta k} x, kt) \qquad (\delta > 0),$$

où $\alpha(y,\tau)$ est une fonction périodique de classe C^1, comprise entre les deux constantes postives λ et Λ , et supposons que la fonction réciproque $1/\alpha(y,\tau)$ ait une moyenne par rapport à y indépendente de τ , i.e. que

$$(8) \qquad ⨍ \frac{1}{\alpha(y,\tau)}\, dy = \frac{1}{a} \quad , \text{ avec } a \text{ constante.}$$

On a alors, pour $\varphi = 0$, $\psi \in H^1_{loc}$ et $T \leqslant c(\lambda)\,\delta$:

$$\{u^k\} \rightarrow u \qquad \text{dans} \quad L^\infty([0,T], L^2_{loc}) \ ,$$

où $u(x,t)$ est la solution du problème (3) avec coefficient a donné par la (8) .

Remarque 4 L'hypothèse que $\alpha(y,t)$ soit périodique en y et τ peut être substituée par la condition plus faible que $\alpha(y,t)$ soit périodique en y pour tout τ et que α et α_τ soient bornées.

Remarque 5 Un cas simple où l'hypothèse (8) est remplie est celui où

$$\alpha(y,\tau) = \beta(y+\tau)$$

avec β périodique.

Remarque 6 Le résultat du th.5 est encore valable si $a^k(x,t) = \alpha(\mu_k x, \nu_k t)$, avec $\alpha(y,\tau)$ vérifiant les hypothèses du théorème, pourvu que (pour $k \rightarrow \infty$) on ait $\{\nu_k\} \rightarrow +\infty$ et $\{\mu_k e^{-\delta\nu_k}\} \rightarrow +\infty$ avec $\delta > 0$. Des résultats analogues (voir [1],[3],[8] et [9]) sont valables pour les équations paraboliques du type $u_t - (\alpha(\mu_k x, \nu_k t)u_x)_x = 0$, pourvu que $\{\mu_k \nu_k^{-1/2}\} \rightarrow +\infty$.

Remarque 7 Le th.5 peut probablement s'étendre au cas général n-dimensionnel, à condition de remplacer l'hypothèse (8) par la condition que l'opérateur qu'on obtient à partir de $-\Sigma D_{y_i}(\alpha_{ij}(y,\tau)D_{y_j})$ par homogénéisation elliptique en y, soit indépendent du paramètre τ , i.e.

$$\{-\Sigma D_{x_i}(\alpha(kx,\tau) D_{x_j})\} \overset{G}{\rightarrow} -\Sigma a_{ij} D_{x_i} D_{x_j} \quad , \forall \tau \, ,$$

pour $k \rightarrow \infty$, avec a_{ij} constantes.

Remarque 8 Il est possible que, en utilisant le th.2, on puisse arriver à prover des résultats de convérgence des solutions des problèmes $(3)_k$ dans le cas où

$$a^k_{ij}(x,t) = a_{ij}(x,t) + \epsilon_k \, \alpha_{ij}(\mu_k x, \nu_k t)$$

avec $a_{ij}(x,t)$ analytique en x , $\alpha(y,\tau)$ analytique an y et périodique, et avec $\{\epsilon_k\} \to 0$, $\{\mu_k\} \to +\infty$, $\{\nu_k\} \to +\infty$, à condition que la suite $\{\epsilon_k \circ \mu_k\}$ converge à zéro assez rapidement.

BIBLIOGRAPHIE

1 A.BENSOUSSAN, J.L.LIONS et G.PAPANICOLAOU *Asymptotic Analysis for periodic Structures.* North Holland, 1978.

2 F.COLOMBINI, E.DE GIORGI et S.SPAGNOLO *Sur les équations hyperboliques avec des coefficients qui ne dépendent que du temps.* Ann. Scu. Norm. Pisa, 6 (1979), p.511.

3 F.COLOMBINI et S.SPAGNOLO *Sur la convergence de solutions d'équations paraboliques.* J.Math. pures et appl. , 56 (1977), p.263.

4 F.COLOMBINI et S.SPAGNOLO *On the convergence of solutions of hyperbolic equations.* Comm. in Part. Diff. Eq., 3 (1978), p.77.

5 F.COLOMBINI et S.SPAGNOLO *Second order hyperbolic equations with coefficients real analytic in space variables and discontinuous in time.* A paraître.

6 R.COURANT et D.HILBERT *Methods of Mathematical Physics.* Interscience, New York 1962.

7 E.DE GIORGI et S.SPAGNOLO *Sulla convergenza degli integrali dell'energia per operatori ellittici del secondo ordine.* Boll. Un. Mat. It. , 8(1973).

8 A.PROFETI et B.TERRENI *Uniformità per una convergenza di operatori parabolici nel caso dell'omogeneizzazione.* A paraître sur Boll. Un. Mat. It., 1979.

9 A.PROFETI et B.TERRENI *Su alcuni metodi per lo studio della convergenza di equazioni paraboliche.* A paraître sur Boll. Un. Mat. It. ,1980.

10 S.SPAGNOLO *Convergence of parabolic equations.* Boll. Un. Mat. It. , 14-B (1977), p.547.

COMPUTING METHODS IN APPLIED SCIENCES AND ENGINEERING
R. Glowinski, J.L. Lions (editors)
North-Holland Publishing Company

PERTURBATION OF SPECTRAL PROPERTIES FOR A CLASS OF STIFF PROBLEMS

E. SANCHEZ - PALENCIA *, M. LOBO - HIDALGO **

* (Laboratoire de Mécanique Théorique, associé au CNRS. Université Paris VI, 4, place Jussieu, T. 66 ; 75230 Paris)

** (Departamento de Matematicas, Universidad de Santiago de Compostela, España).

1.- INTRODUCTION - Stiff problems (in french "problèmes raides", c.f. Lions [3]) are boundary value problems in domains Ω such that the coefficients of the partial differential equation are multiplied by a parameter ε on a part Ω^1 of Ω. One then studies the asymptotic behaviour as ε (real and positive) tends to zero. As a rule, the solution in the region Ω^1 tends to infinity, and the asymptotic behaviour may be given by a chain of boundary value problems in the regions Ω^1 and Ω^0 ($\Omega^0 = \Omega \setminus \overline{\Omega}^1$).

We consider here asymptotic properties of the spectral properties of such problem. In physical applications, the interesting spectral eigenvalue problems are of the kind

$$\mathcal{A}^\varepsilon u_\varepsilon = \lambda \mathcal{B}^\varepsilon u_\varepsilon \tag{1.1}$$

i.e., eigenvalue problems of the operator $\mathcal{A}^\varepsilon$ with respect to another operator $\mathcal{B}^\varepsilon$ which also depends on ε, and which is, in some sense, also a "stiff operator".

Generally speaking, the spectral properties of the stiff problem are associated with the spectral properties of certain boundary value problems in the domains Ω^0 and Ω^1.

We have two methods to study such problems. We introduce them in a model problem in sections 2 and 3. Nevertheless, each method applies, with minor modifications, to a variety of problems.

The first method (sect. 2), is based on analyticity properties of the resolvent, as in the "analytic perturbation theory" of Kato [1] . This method uses the classical asymptotic expansion of Lions [3] in a sharper form, which needs regularity properties of the solutions. Moreover, this first method is only fitted for pointwise spectra, but it applies to non-selfadjoint operators.

The second method (sect. 3) is based on the study of an auxiliar "hyperbolic" stiff problem and the Fourier transform ($t \Rightarrow \lambda$) of its solution. This method was used in Lobo and Sanchez [2] for the study of other perturbation problems. It uses essentially the sefladjointness of the problems, but it gives information about the spectral family and works for continuous as well as pointwise spectra.

Sect. 4 is devoted to complements about problems in unbounded domains, in particular the problem of the vibration of an elastic body in air. It appears that, as ε (the density of the air) tends to zero, the continuous spectral family tends in some sense to the piecewise constant spectral family corresponding to the vibration of the elastic body in vacuum. Earlier results about the scattering frequencies are recalled.

Remark about notation.

Standard notations are used for Sobolev spaces.

$\rho(A)$ and $\sigma(A)$ denote the resolvent set and the spectrum of the operator A .

If u is a function defined on Ω, u^0 and u^1 are its restrictions to the subdomains Ω^0 and Ω^1. These restrictions are sometimes denoted $u\big|_{\Omega^0}$, $u\big|_{\Omega^1}$.

$$L^\infty (R , V) = L^\infty(-\infty , +\infty ; V)$$

$$(u , v)_{\Omega^k} = \int_{\Omega^k} u\bar{v} \, dx$$

2.- METHOD OF THE EXPANSION IN POWERS OF ε.

We consider the following model problem. Let Ω^0, Ω^1 be two connected bounded domains of R^N with smooth boundaries Γ and S, located as shown in fig. 1. We also consider the "total domain"

$$\Omega = \Omega^0 \cup S \cup \Omega^1$$

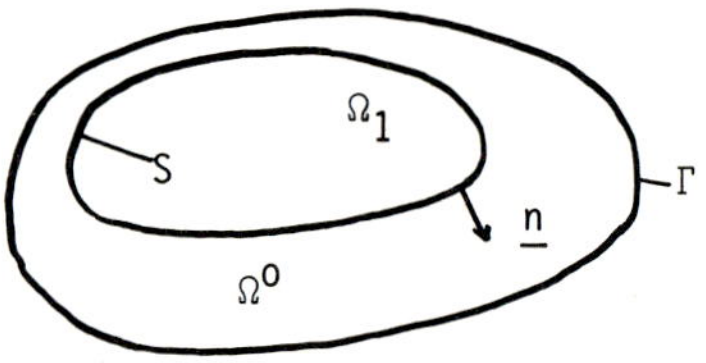

fig. 1

We define the real coefficients

$$a^k_{ij}(x) = a^k_{ji}(x) \quad , \quad k = 0,1 \quad ; \quad i,j = 1 \ldots N$$

which are smooth functions defined on Ω^k resp. They satisfy

$$(2.1) \qquad a^k_{ij}\, \xi_i\, \xi_j \geqslant \delta\, |\xi|^2 \quad ; \quad \delta > 0 \quad , \quad \xi \in R^N$$

We then define the hermitian forms :

$$(2.2) \qquad a^k(u\ ,\ v) = \int_{\Omega^k} a^k_{ij} \frac{\partial u}{\partial x_i} \frac{\partial \bar{v}}{\partial x_j}\, dx \qquad k = 0,1$$

$$(2.3) \qquad b^k(u\ ,\ v) = \int_{\Omega^k} u\, \bar{v}\, dx$$

Moreover, we consider the forms (depending on the real positive parameter ε):

$$(2.4) \qquad a^\varepsilon(u\ ,\ v) = a^0(u\ ,\ v) + \varepsilon\, a^1(u\ ,\ v)$$

$$(2.5) \qquad b^\varepsilon(u\ ,\ v) = b^0(u\ ,\ v) + \varepsilon\, b^1(u\ ,\ v\ )$$

Let $f \in L^2(\Omega)$ be a given function, and ζ a "spectral parameter" which is, for the time being, any complex number. We then define :

Problem Π_ε - Find $u_\varepsilon \in H^1_0(\Omega)$ (which also depends on ζ and f) such that

$$(2.6) \qquad a^\varepsilon(u_\varepsilon\ ,\ v) - \zeta\, b^\varepsilon(u_\varepsilon\ ,\ v) = (f\ ,\ v)_\Omega \qquad \forall\, v \in H^1_0(\Omega)$$

Remark 2.1 - It is clear that, for fixed $\varepsilon > 0$, the forms a^ε and b^ε are hermitian, continuous and coercive on $H^1_0(\Omega)$ and $L^2(\Omega)$ resp. Consequently, b^ε is a scalar product on L^2 with associated norm equivalent to the standard one. The problem Π_ε is then a standard spectral problem for a selfadjoint operator with compact resolvent. The spectrum is formed by the real positive eigenvalues ζ^ε_m with finite multiplicities, which tend to infinity as $m \to \infty$. For ζ out of the spectrum, the problem Π_ε has a unique solution. ∎

In order to give an interpretation of problem Π_ε as boundary value problem, we introduce the following notations :

$$(2.7) \qquad A^k\, u = -\frac{\partial}{\partial x_i}\left(a^k_{ij} \frac{\partial u}{\partial x_j}\right) \quad ; \quad \frac{\partial u}{\partial \nu^k} = a^k_{ij} \frac{\partial u}{\partial x_j}\, n_i$$

where n is the outer unit normal to S. We then have the following Green formulae for functions which are zero on Γ(otherwise there is a supplementary term on Γ) :

$$(2.8) \qquad a^0(u\ ,\ v) = \int_{\Omega^0} (A^0\, u)\, \bar{v}\, dx - \int_S \frac{\partial u}{\partial \nu^0}\, \bar{v}\, dS$$

$$a^1(u\ ,\ v) = \int_{\Omega^1} (A^1 u)\ \overline{v}\ dx + \int_S \frac{\partial u}{\partial \nu^1}\ \overline{v}\ dS \tag{2.9}$$

Then, Π_ε is equivalent to the transmission problem :

$$A^0 u_\varepsilon^0 - \zeta u_\varepsilon^0 = f^0 \text{ in } \Omega^0 \quad ; \quad \varepsilon A^1 u_\varepsilon^1 - \zeta \varepsilon u_\varepsilon^1 = f^1 \text{ in } \Omega^1 \tag{2.10}$$

$$u_\varepsilon \Big|_\Gamma = 0 \quad ; \quad u_\varepsilon^0 \Big|_S = u_\varepsilon^1 \Big|_S \quad ; \quad \frac{\partial u_\varepsilon^0}{\partial \nu^0} = \varepsilon \frac{\partial u_\varepsilon^1}{\partial \nu^1} \tag{2.11}$$

Remark 2.2 - We recall that regularity theory holds for this problem. If f is given in $L^2(\Omega)$ and ζ is not in the spectrum, $u_\varepsilon \in H_0^1(\Omega)$; $u_\varepsilon^k \in H^2(\Omega^k)$ and consequently, formulae (2.8), (2.9) make sense. On the other hand, the necessary and sufficient condition for a function w to belong to $H^1(\Omega)$ is that

$$w^k \in H^1(\Omega^k) \text{ and } w^0 \Big|_S = w^1 \Big|_S \ . \ \blacksquare$$

Finally, we introduce the following elliptic boundary value operators (in $L^2(\Omega^k)$) :

$$\left\{ \begin{array}{l} \mathcal{A}^0 \Leftrightarrow A^0 \text{ in } \Omega^0 \text{, with } u \Big|_\Gamma = 0 \quad ; \quad \dfrac{\partial u}{\partial \nu^0}\Big|_S = 0 \\ \mathcal{A}^1 \Leftrightarrow A^1 \text{ in } \Omega^1 \text{, with } u \Big|_S = 0 \end{array} \right. \tag{2.12}$$

and the sets

$$\left\{ \begin{array}{l} X = \{v \ ; \ v \in H^1(\Omega^0) \ , \ v \Big|_\Gamma = 0 \} \\ Y = \{v \ ; \ v \in H_0^1(\Omega) \ , \ v^0 = 0 \} \cong \{ v \ ; \ v^1 \in H_0^1(\Omega^1) \ , \ v^0 = 0 \} \end{array} \right. \tag{2.13}$$

We introduce the standard expansion :

$$u_\varepsilon = \varepsilon^{-1} u_{-1} + u_0 + \varepsilon u_1 + \varepsilon^2 u_2 + \dots \quad ; \quad u_i \in H_0^1(\Omega) \tag{2.14}$$

and we have the following conditions :

$$\varepsilon^{-1}) \qquad a^0(u_{-1}\ ,\ v) - \zeta\, b^0(u_{-1}\ ,\ v) = 0 \qquad \forall v \in H_0^1(\Omega) \tag{2.15}$$

$$\varepsilon^{0}) \qquad a^0(u_0\ ,\ v) - \zeta\, b^0(u_0\ ,\ v) + a^1(u_{-1}\ ,\ v) - \zeta\, b^1(u_{-1}\ ,\ v) = (f\ ,\ v)_\Omega \qquad \forall v \in H_0^1(\Omega) \tag{2.16}$$

$$\varepsilon^{j}, j > 0) \qquad a^0(u_j\ ,\ v) - \zeta\, b^0(u_j\ ,\ v) + a^1(u_{j-1}\ ,\ v) - \zeta\, b^1(u_{j-1}\ ,\ v) = 0 \qquad \forall v \in H_0^1(\Omega) \tag{2.17}$$

(2.18) $\left\{ \begin{array}{l} \text{Let } \zeta \text{ be a point in the resolvent sets of } \mathcal{A}^0, \mathcal{A}^1, \text{ i.e.} \\ \zeta \in \rho(\mathcal{A}^0) \cap \rho(\mathcal{A}^1) \end{array} \right.$

Then, the terms of (2.14) may be obtained from (2.15) - (2.17).

Study of u_{-1} - If we take $v \in X$ in (2.15) (which is equivalent to $v \in H_0^1(\Omega)$) we obtain $u_{-1}^0 \in X \;;\; a^0(u_{-1}^0 , v) - \zeta\, b^0(u_{-1}^0 , v) = 0 \quad \forall v \in X$
and, by virtue of (2.18) :

(2.19) $$u_{-1}^0 = 0$$

Next, from (2.16) with $v \in Y$,

$$u_{-1}^1 \in Y \quad ; \quad a^1(u_{-1}^1 , v) - \zeta\, b^1(u_{-1}^1 , v) = (f^1 , v)_{\Omega^1} \qquad \forall v \in Y$$

which is equivalent to

(2.20) $$\mathcal{A}^1 u_{-1}^1 - \zeta u_{-1}^1 = f^1$$

and (2.18) shows that u_{-1}^1 exists and is unique. Moreover, by standard regularity theory,

(2.21) $$\| u_{-1}^1 \|_{H^2(\Omega^1)} \leq C_1 \|f^1\|_{L^2(\Omega^1)} \qquad \blacksquare$$

Study of u_0 - If we multiply (2.20) by $\bar{v}$ (not necessarily zero on S) and we use (2.9), we have :

(2.22) $$a^1(u_{-1}^1 , v) - \zeta\, b^1(u_{-1}^1 , v) = (f^1 , v)_{\Omega^1} + \int_S \frac{\partial u_{-1}^1}{\partial \nu^1} \bar{v}\, dS$$

and (2.17) gives for u_0^0 :

(2.23) $$u_0^0 \in X \quad ; \quad a^0(u_0^0 , v) - \zeta\, b^0(u_0^0 , v) = (f^0 , v)_{\Omega^0} - \int_S \frac{\partial u_{-1}^1}{\partial \nu^1} \bar{v}\, dS \quad \forall v \in X$$

and this with (2.18) gives u_0^0 in a unique manner ; in fact, this is the variational formulation of

(2.24) $$A^0 u_0^0 - \zeta u_0^0 = f^0 \quad ; \quad u_0^0\big|_\Gamma = 0 \quad ; \quad \left.\frac{\partial u_0^0}{\partial \nu^0}\right|_S = \left.\frac{\partial u_{-1}^1}{\partial \nu^1}\right|_S$$

and from standard regularity :

(2.25) $$\| u_0^0 \|_{H^2(\Omega^0)} \leq C_2 \|f^0\|_{L^2(\Omega^0)} + C_3' \left\|\frac{\partial u_{-1}^1}{\partial \nu^1}\right\|_{H^{1/2}(S)} \leq$$
$$\leq C_2 \|f^0\|_{L^2(\Omega^0)} + C_3 \|u_{-1}^1\|_{H^2(\Omega^1)}$$

Next, to obtain u_o^1, we know that u_o^1 is equal to u_o^o on S ; on the other hand, (2.17) with j = 1 and $v \in Y$ gives

$$a^1(u_o^1\ ,\ v) - \zeta\, b^1(u_o^1\ ,\ v) = 0 \qquad \forall v \in Y \tag{2.26}$$

and u_o^1 is the unique solution of

$$\mathcal{A}^1\, u_o^1 - \zeta\, u_o^1 = 0 \quad ; \quad u_o^1\Big|_S = u_o^o\Big|_S \Longrightarrow$$

$$\| u_o^1 \|_{H^2(\Omega^1)} \leq C_4 \, \| u_o^o \|_{H^2(\Omega^o)} \qquad \blacksquare \tag{2.27}$$

<u>Study of u_m , $m > 0$</u> - It is analogous to the precedent one, but with f = 0. We obtain

$$\begin{cases} \| u_m^o \|_{H^2(\Omega^o)} \leq C_3 \, \| u_{m-1}^1 \|_{H^2(\Omega^1)} \\ \| u_m^1 \|_{H^2(\Omega^1)} \leq C_4 \, \| u_m^o \|_{H^2(\Omega^o)} \end{cases} \tag{2.28}$$

where the constants C_3, C_4 <u>are the same</u> as in (2.25) and (2.27). ■

Now, we may state the following result :

<u>Theorem 2.1</u> - Under the hypothesis (2.1), let K be a compact set contained in the intersection of $\rho(\mathcal{A}^0)$ and $\rho(\mathcal{A}^1)$ (which are the operators defined by (2.12)). Then, for sufficiently small ε ($\varepsilon < \mu(K)$, say), the problem Π_ε (defined by (2.6)) has, for any $\zeta \in K$ and $f \in L^2(\Omega)$ a unique solution u_ε given by the series (2.14), which is uniformly convergent (in the norm of $H_0^1(\Omega)$) for $\zeta \in K$, $\varepsilon < \mu(K)$.

<u>Proof</u> - From the preceeding calculation of the terms u_i, it is clear that the series (2.14) may be formally written for $\zeta \in K$. Moreover, the constants C_1, C_2, C_3, C_4 in (2.21), (2.25), (2.27), (2.28) may be taken independent of ζ for $\zeta \in K$. Next, from (2.21), (2.25), (2.27), (2.28) it follows that

$$\| u_m \|_{H_0^1(\Omega)} \leq C_5 (C_3\, C_4)^m \, \| f \|_{L^2(\Omega)} \tag{2.29}$$

and the conclusion follows. ■

In particular, if ζ is a fixed point contained in $\rho(\mathcal{A}^0) \cap \rho(\mathcal{A}^1)$, for sufficiently small ε, ζ is in the resolvent set of the problem Π_ε (see remark 2.1).

Now, we study some properties which are, in some sense, the "converse" of the preceeding one : a point of the spectra of either $\mathcal{A}^0$ or $\mathcal{A}^1$ is near a point of the spectrum of Π_ε for small ε.

Lemma 2.1 - Let γ be a closed simple curve in the plane of the variable ζ. For fixed $\varepsilon > 0$, let $u_\varepsilon(\zeta)$ be the solution of (2.6) (for ζ in the resolvent set of problem Π_ε). Then, if

(2.30) $$\int_\gamma u_\varepsilon(\zeta)\, d\zeta \neq 0$$

there exists at least a point of the spectrum of Π_ε in the region of the plane enclosed by γ.

Proof - It is immediate (see remark 2.1) because $u_\varepsilon(\zeta)$ is a holomorphic function of ζ in the resolvent set of Π^ε. Then, (2.30) implies that γ encloses a singularity of this function. ∎

Theorem 2.2 - Let ζ^* be a point of the spectrum of either $\mathcal{A}^0$ or $\mathcal{A}^1$. Let γ be a simple closed curve enclosing ζ^* and contained in $\rho(\mathcal{A}^0) \cap \rho(\mathcal{A}^1)$. Then, for sufficiently small ε, there is at least an eigenvalue of the problem Π_ε in the region enclosed by γ .

Proof - First, let us assume $\zeta^* \in \sigma(\mathcal{A}^0)$. We take in (2.6) f such that $f^1 = 0$, $f^0 \neq 0$ equal to an eigenvector of $\mathcal{A}^0$ associated with the eigenvalue ζ^*. From standard theory of the resolvent (in $L^2(\Omega^0)$) we have :

(2.31) $$\frac{-1}{2\pi i} \int_\gamma (\mathcal{A}^0 - \zeta)^{-1} f^0\, d\zeta = f^0$$

On the other hand, for $\zeta \in \gamma$, we construct $u_\varepsilon(\zeta)$; the term $u_{-1}(\zeta)$ is zero, and (2.14) becomes :

(2.32) $$u_\varepsilon(\zeta) = u_0(\zeta) + \varepsilon\, u_1(\zeta) + \dots \qquad ; \qquad \zeta \in \gamma$$

which is uniformly convergent for $\zeta \in \gamma$. Then :

(2.33) $$\left(\int_\gamma u_\varepsilon(\zeta)\, d\zeta \ , f\right)_\Omega$$

may be calculated term by term from (2.32). Moreover, by taking the limit value as $\varepsilon \searrow 0$, we see that (2.33) is different from zero for sufficiently small ε if

(2.34) $$\left(\int_\gamma u_0(\zeta)\, d\zeta \ , f\right)_\Omega \neq 0$$

but from (see (2.23) with $u_{-1} = 0$)

$$f^1 = 0 \qquad ; \qquad u_0^0 = (\mathcal{A}^0 - \zeta)^{-1} f^0$$

we see that (2.34) is a consequence of (2.31). Then, the conclusion follows from Lemma 2.1.

If $\zeta^* \in \sigma(\mathcal{A}^1)$, the proof is analogous, by taking $f^0 = 0$, f^1 an eigenvector of $\mathcal{A}^1$. We then have

$$u_{-1}^1 = (\mathcal{A}^1 - \zeta)^{-1} f^1$$

On the other hand, instead of (2.32), we have

$$\varepsilon\, u_\varepsilon(\zeta) = u_{-1}(\zeta) + \varepsilon\, u_0 + \ldots$$

and the proof is the same. ■

Remark 2.3 - As we said in the introduction, the preceeding method applies to non selfadjoint problems. As an example, we may take

$$a^k(u\ ,\ v) = \int_{\Omega^k} \left(a_{ij}^k \frac{\partial u}{\partial x_i} \frac{\partial \bar{v}}{\partial x_j} + c_i^k \frac{\partial u}{\partial x_i} \bar{v}\right) dx$$

$$A^k\, u = - \frac{\partial}{\partial x_j}\left(a_{ij}^k \frac{\partial u}{\partial x_i}\right) + c_i^k \frac{\partial u}{\partial x_i}$$

where c_i^k are smooth functions defined on Ω^k. The problem Π_ε is non longer selfadjoint, but it is elliptic and with compact resolvent. Theorems 2.1 and 2.2 again hold, but the eigenvalues are complex in general. ■

3.- METHOD OF THE FOURIER TRANSFORM.

We first make some remarks about the solution of a class of "hyperbolic problems" and their Fourier transforms.

Let V and H be two Hilbert spaces, V contained in H, with dense and continuous embedding. H is identified to its dual. Let V' be the dual of V :

(3.1) $$V \subset H \subset V'$$

Let $a(u\ ,\ v)$ be a continuous and symmetric form on V, such that

(3.2) $$a(v,\ v) \geq \delta \| v \|_V^2 \qquad (\delta > 0) \qquad \forall\, v \in V$$

We denote by A the selfadjoint operator of H associated with the form a in the standard framework :

(3.3) $$a(u\ ,\ v) = (Au\ ,\ v)_H \qquad \forall\, u \quad D(A),\quad v \in V$$

and of course

(3.4) $$D(A^{1/2}) = V \ .$$

We consider the

Initial value hyperbolic problem - For given $\alpha \in V$, $\beta \in H$, find $u(t)$ such that (' denotes derivative with respect to time) :

(3.5) $(u\ ,\ u') \in L^{\infty}(R\ ;\ V) \times L^{\infty}(R\ ;\ H)$

(3.6) $u'' + A\,u = 0$

(3.7) $u(0) = \alpha$

(3.8) $u'(0) = \beta$

According to standard semigroup theory, this problem has a unique solution. In fact, u, u' form a Stone's (unitary) group in $V \times H$ equipped with the norm

$$\|(u\ ,\ u')\|^2_{V \times H} = a\,(u\ ,\ u) + \|u'\|^2_H \tag{3.9}$$

The solution of (3.5) - (3.8) may be written :

$$\begin{cases} u(t) = \cos t\,A^{1/2}\,\alpha + A^{-1/2}\sin t\,A^{1/2}\,\beta \\ u'(t) = -\,A^{1/2}\sin t\,A^{1/2}\,\alpha + \cos t\,A^{1/2}\beta \end{cases} \tag{3.10}$$

On the other hand, let Φ be the set of the functions of class C^1 which are zero, as well as their first derivatives for $t = T$. Let $\tilde{V}$ be a dense subset of V. As in Lions - Magenes [4] we have

Proposition 3.1 - The solution (3.10) of the initial value hyperbolic problem is the (unique) function u which satisfies (3.5), (3.7) and :

$$\begin{cases} \int_0^T [a(u\ ,\ \psi) - (u'\ ,\ \psi')_H]\ dt = (\beta\ ,\ \psi(0))_H \\ \forall\ \psi = \phi \otimes v\quad ;\quad \phi \in \Phi\ ,\quad v \in \tilde{V} \end{cases} \tag{3.11}$$

where T is arbitrarily fixed.

Moreover, u'(t) is a temperated distribution. By taking the Fourier transform of (3.10) and using

$$\cos t\,A^{1/2} = \frac{1}{2}\int_0^{\infty}(e^{i\lambda t} + e^{-i\lambda t})\ dE(A^{1/2}\ ,\ \lambda)$$

where $E(A^{1/2}\ ,\ \lambda)$ is the spectral family of the operator $A^{1/2}$, we obtain (see Lobo and Sanchez [2] for details) :

Proposition 3.2 - If u is the solution of the initial value hyperbolic problem (3.5) - (3.8) with $\alpha = 0$, $\beta \neq 0$, we have :

$$\mathcal{F}(u'(t)\ ,\ v)_H = \sqrt{\frac{\pi}{2}}\,\frac{d}{d\lambda}\,([\,E(A^{1/2}\ ,\ \lambda) - E(A^{1/2}\ ,\ -\lambda)]\ \beta\ ,\ v)_H \tag{3.12}$$

for any $\beta\ ,\ v \in H$. Here $\mathcal{F}$ and $d/d\lambda$ denote the Fourier transform $(t \Rightarrow \lambda)$ and the distributional derivative.

Remark 3.1 - From (3.2) we see that $E(A^{1/2}, \lambda)$ (resp. $E(A^{1/2}, -\lambda)$) has its support strictly contained in the positive (resp. negative) half axis.■

After these remarks, we consider again the forms a^{ε}, b^{ε} of the preceeding section (2.2), (2.3). By taking into account remark 2.1, we denote by $\mathcal{A}^{\varepsilon}$ the self-adjoint operator of $L^2(\Omega)$ associated with the form $a^{\varepsilon}(u, v)$ when $L^2(\Omega)$ is equipped with the scalar product $b^{\varepsilon}(u, v)$ and the corresponding norm. We then consider the problem :

Problem P^{ε} - Find $u_{\varepsilon}(t)$ such that, for given $\beta \in L^2(\Omega)$

(3.13) $$(u_{\varepsilon}, u_{\varepsilon}') \in L^{\infty}(R ; H_0^1(\Omega)) \times L^{\infty}(R ; L^2(\Omega))$$

(3.14) $$u_{\varepsilon}'' + \mathcal{A}^{\varepsilon} u_{\varepsilon} = 0$$

(3.15) $$u_{\varepsilon}(0) = 0 \quad ; \quad u_{\varepsilon}'(0) = \beta$$

Moreover, we define w^0, w^1 as the solutions of the following problems (see (2.13), (2.14)) :

Problem P^0 - Find w^0 such that

(3.16) $$(w^0, w^{0\prime}) \in L^{\infty}(R ; X) \times L^{\infty}(R ; L^2(\Omega^0))$$

(3.17) $$w^{0\prime\prime} + \mathcal{A}^0 w^0 = 0$$

(3.18) $$w^0(0) = 0 \quad ; \quad w^{0\prime}(0) = \beta^0$$

Problem P^1 - Find w^1 such that

(3.19) $$(w^1, w^{1\prime}) \quad L^{\infty}(R ; H_0^1(\Omega^1)) \times L^{\infty}(R ; L^2(\Omega^1))$$

(3.20) $$w^{1\prime\prime} + \mathcal{A}^1 w^1 = 0$$

(3.21) $$w^1(0) = 0 \quad ; \quad w^{1\prime}(0) = \beta^1$$

We then have :

Lemma 3.1 - Let u_{ε}, w^0, w^1 be the solutions of the problems P^{ε}, P^0, P^1. Then :

i) If $\beta^1 = 0$ (i.e., if the support of β is in Ω^0) we have :

(3.22) $$u_{\varepsilon}^{0\prime} \rightarrow w^{0\prime} \quad \text{in } L^{\infty}(R ; L^2(\Omega^0)) \text{ weakly } *$$

(it is clear that u_{ε}^0 is the restriction of u_{ε} to Ω^0).

ii) If $\beta^0 = 0$ (i.e., if the support of β is in Ω^1) we have :

(3.23) $$u_\varepsilon^{1\prime} \to w^{1\prime} \quad \text{in } L^\infty(R\ ;\ L^2(\Omega^1)) \quad \text{weakly} * \quad .$$

(it is clear that u_ε^1 is the restriction of u_ε to Ω^1).

Remark 3.1 - The preceeding Lemma immediately follows from general results about hyperbolic stiff problems (Lions [3] chap. IV, th. 8.3). Nevertheless, we give hereafter another proof which also applies to other problems where regularity properties are not known. ■

Proof of lemma 3.1 - Part i) It is clear that (3.14) may be written

(3.24) $$b^\varepsilon(u_\varepsilon'' , v) + a^\varepsilon(u_\varepsilon , v) = 0 \qquad \forall\ v \in H_0^1(\Omega)$$

and we have the a priori estimate

(3.25) $$b^\varepsilon(u_\varepsilon' , u_\varepsilon') + a^\varepsilon(u_\varepsilon , u_\varepsilon) = b^\varepsilon(\beta , \beta) \leq C$$

and we can extract a subsequence such that

(3.26) $$\begin{cases} u_\varepsilon^0 \to u^* & \text{in } L^\infty(R\ ;\ X) \text{ weakly } * \\ u_\varepsilon^{0\prime} \to u^{*\prime} & \text{in } L^\infty(R\ ;\ L^2(\Omega^0)) \text{ weakly } * \end{cases}$$

For the problem P^ε, (3.11) becomes

(3.27) $$\int_0^T [a^\varepsilon(u_\varepsilon , \psi) - b^\varepsilon(u_\varepsilon' , \psi)]\ dt = b^\varepsilon(\beta , \psi(0))$$

and (3.25), (3.26) suffice to pass to the limit $\varepsilon \searrow 0$ in (3.27). We then see that u^* is the solution of the problem P^0. The part i) is proved (note that the hypothesis $\beta^1 = 0$ has not been used).

Part ii) The estimate (3.25) becomes in this case

(3.28) $$b^\varepsilon(u_\varepsilon' , u_\varepsilon') + a^\varepsilon(u_\varepsilon , u_\varepsilon) = \varepsilon\,(\beta^1 , \beta^1)_{\Omega^1} \leq \varepsilon\, C$$

and consequently, after extraction of a subsequence :

(3.29) $$u_\varepsilon^0 \to 0 \quad \text{in } L^\infty(R\ ;\ H^1(\Omega^0)) \quad \text{strongly}$$

(3.30) $$u_\varepsilon^{0\prime} \to 0 \quad \text{in } L^\infty(R\ ;\ L^2(\Omega^0)) \quad \text{strongly}$$

(3.31) $$u_\varepsilon^1 \to \tilde{u} \quad \text{in } L^\infty(R\ ;\ H^1(\Omega^1)) \quad \text{weakly} *$$

(3.32) $$u_\varepsilon^{1\prime} \to \tilde{u}' \quad \text{in } L^\infty(R\ ;\ L^2(\Omega^1)) \quad \text{weakly} *$$

Then, (3.29) and (3.31) show that $\tilde{u}$ takes its values in $H_0^1(\Omega^1)$; moreover, we write (3.27) with $\psi = \phi \otimes v$, $v \in H_0^1(\Omega^1)$ and we pass to the limit by virtue of (3.31), (3.32). We see that $\hat{\mathcal{U}}$ is the solution of P^1. Lemma 3.1 is proved. ■

We then have the following result about the convergence of the derivatives of the spectral families :

<u>Proposition 3.3</u> - In the conditions of lemma 3.1, we have, as $\varepsilon \to 0$:

i) If $\beta^1 = 0$ (and any $\beta^0 \in L^2(\Omega^0)$),

(3.33) $$([\frac{d}{d\lambda} E(\mathcal{A}^{\varepsilon 1/2}, \lambda)\beta]\Big|_{\Omega^0}, v)_{\Omega^0} \to (\frac{d}{d\lambda} E(\mathcal{A}^{0 1/2}, \lambda)\, \beta^0, v)_{\Omega^0}$$

in $\mathcal{D}'(0, \infty)$ for any $v \in L^2(\Omega^0)$.

ii) If $\beta^0 = 0$ (and any $\beta^1 \in L^2(\Omega^1)$),

(3.34) $$([\frac{d}{d\lambda} E(\mathcal{A}^{1/2}, \lambda)\beta]\Big|_{\Omega^1}, v)_{\Omega^1} \to (\frac{d}{d\lambda} E(\mathcal{A}^{0 1/2}, \lambda)\, \beta^1, v)_{\Omega^1}$$

in $\mathcal{D}'(0, \infty)$, for any $v \in L^2(\Omega^1)$.

<u>Proof</u> - It suffices to apply proposition 3.2 to the results of lemma 3.1 (which holds in the topology of temperated distributions, for which the Fourier transform is continuous. Moreover, by remark 3.1, the parts corresponding to $E(\mathcal{A}^{1/2}, -\lambda)$ are disregarded if we consider $\lambda \in (0, \infty)$. ■

<u>Remark 3.2</u> - Theorem 2.2 of the preceeding section is a consequence of proposition 3.3. It suffices to remark that if $\mathcal{A}$ is anticompact, $\frac{d}{d\lambda} E(\mathcal{A}^{1/2}, \lambda)$ is a sum of Dirac distributions located at the points $\lambda_j^{1/2}$, where λ_j are the eigenvalues of $\mathcal{A}$ (see Lobo - Sanchez [2] for details and other analogous problems). In addition, prop. 3.3 contains some (little !) information about the asymptotic behaviour of the eigenvectors, which is not given in theorem 2.2. ■

Finally, it is possible to obtain convergence properties for the spectral families (instead of their derivatives). We have

<u>Theorem 3.1</u> - In the conditions of lemma 3.1, we have, as $\varepsilon \to 0$

i) If $\beta^1 = 0$ (and any $\beta^0 \in L^2(\Omega^0)$) :

(3.35) $$([E(\mathcal{A}^{\varepsilon}, \lambda)\beta]\Big|_{\Omega^0}, v)_{\Omega^0} \to (E(\mathcal{A}^0, \lambda)\beta^0, v)_{\Omega^0}$$

in $L^\infty(0, \infty)$ weakly $*$, for any $v \in L^2(\Omega^0)$.

ii) If $\beta^0 = 0$ (and any $\beta^1 \in L^2(\Omega^1)$) :

(3.36) $([E(\mathcal{A}^{\varepsilon}, \lambda)\beta]\big|_{\Omega^1}, v)_{\Omega^1} \to (E(\mathcal{A}^1, \lambda)\beta^1, v)_{\Omega^1}$

in $L^{\infty}(0, \infty)$ weakly $*$, for any $v \in L^2(\Omega^1)$.

Proof - We first prove (3.35), (3.36) with $\mathcal{A}^{1/2}$ instead of $\mathcal{A}$. It is clear that the left hand sides of (3.35), (3.36) are bounded in $L^{\infty}(0, \infty)$ and then, precompact for the weak star topology. Moreover, we consider the difference with the right hand side, and any limit must be a constant function (by virtue of prop. 3.3); moreover, this constant is zero because it is so in a neighbourhood of $\lambda = 0$ (see remark 3.1). Finaly, we pass from the spectral families of $\mathcal{A}^{1/2}$ to those of $\mathcal{A}$ by changing the variable $\lambda = \mu^2$. ∎

Remark 3.3 - It is noticeable that theorem 3.1 implies some sort of convergence of $E(\mathcal{A}^{\varepsilon})$ to $E(\mathcal{A}^0)$, and $E(\mathcal{A}^1)$ when applied to appropriate functions. But it is clear that $E(\mathcal{A}^{\varepsilon})$ is a family of projectors of $L^2(\Omega)$ equipped with scalar product $b^{\varepsilon}(u, v)$. On the other hand, $E(\mathcal{A}^0)$, $E(\mathcal{A}^1)$ are families of projectors of $L^2(\Omega^0)$ and $L^2(\Omega^1)$ resp. ∎

4.- COMPLEMENTS AND CONCLUDING REMARKS.

The method of the Fourier transform applies to problems in unbounded domains. For instance, we may consider a problem analogous to that of sect. 2 and 3 but with $\Omega = R^3$, Ω^1 bounded and Ω^0 of course unbounded. If the coefficients a^k_{ij} are constant in Ω^0, it is known that the spectra of $\mathcal{A}^{\varepsilon}$ and $\mathcal{A}^0$ are continuous (in fact, absolutely continuous and they fill the positive real semi-axis). $\mathcal{A}^1$ has of course a purely point spectrum.

Theorem 3.3 holds for such problem. The proof is essentially the same as the preceeding one. In fact, the coerciveness hypothesis (3.2) does not hold, but $a(u, v) + (u, v)_H$ satisfies such inequality. Consequently, we may study the operator $A + I$ instead of A. The spectral family of A is then obtained by a unit translation in the variable λ. The part i) of theorem 3.1 shows in this case that the continuous spectral family $E(\mathcal{A}^{\varepsilon}, \lambda)$, "restricted to Ω^0 " (such is to say, operating on a function with support in Ω^0 and taking the restriction of the result to Ω^0) "converges" to the continuous spectral family $E(\mathcal{A}^0, \lambda)$. On the other hand, part ii) of theorem 3.1 shows that the continuous family $E(\mathcal{A}^{\varepsilon}, \lambda)$ "restricted" to Ω^1 "converges" to the piecewise constant family $E(\mathcal{A}^1, \lambda)$ associated with the operator ^{1}.

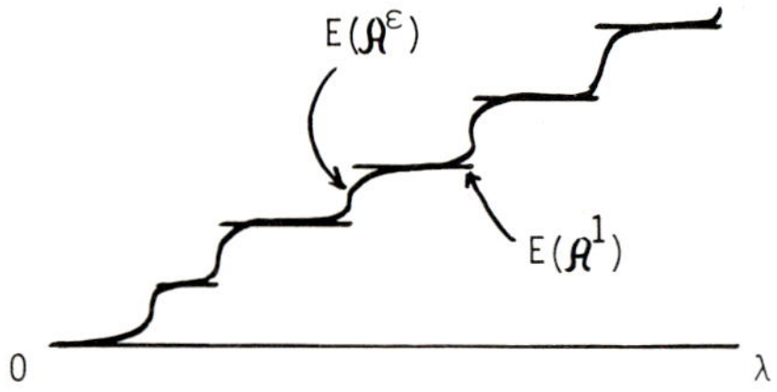

fig. 2

Analogous results are obtained if the roles of Ω^0 and Ω^1 are exchanged (i.e. if the coefficients tend to zero in the outer region). This problem is analogous (c.f. Sanchez - Palencia [5] for details) to the problem of the vibration of an elastic body Ω^0 in air Ω^1. The small parameter ε is the ratio of the densities of the air and the body.

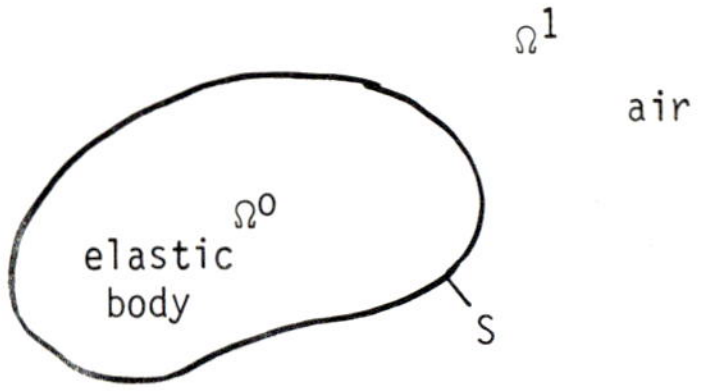

fig. 3

The continuous spectral family of $\mathcal{A}^\varepsilon$ when restricted to the "air" Ω^1 converges to the continuous spectral family of $\mathcal{A}^1$ (which is in this case the problem of the vibration of the air with the body Ω^0 at rest). On the other hand, $E(\mathcal{A}^\varepsilon, \lambda)$ when restricted to Ω^0 tends to the piecewise constant family of $\mathcal{A}^1$ (which is in this case, the problem of the vibration of the elastic body in vacuum).

The preceeding results are to the compared with those obtained in Sanchez - Palencia [5] where the asymptotic behaviour as $\varepsilon \searrow 0$ of the scattering frequencies of this problem were studied. If ζ^* is either an eigenvalue of $\mathcal{A}^0$ or a scattering frequency of $\mathcal{A}^1$, there exists a scattering frequency ζ^ε of $\mathcal{A}^\varepsilon$ which converges to ζ^* as $\varepsilon \searrow 0$.

REFERENCES

[1] Kato, T "Perturbation theory for linear operators", Springer, Berlin (1966)

[2] Lobo - Hidalgo et Sanchez - Palencia "Sur certaines propriétés spectrales des perturbations du domaine dans les problèmes aux limites" Comm. Part. Diff. Eq. (to be published).

[3] Lions, J.L. "Perturbations singulières dans les problèmes aux limites et en contrôle optimal" Springer, Berlin, Lect. Not. Math. 323 (1973)

[4] Lions, J.L. et Magenes, E. "Problèmes aux limites non homogènes et applications, vol. I " Dunod, Paris (1968)

[5] Sanchez - Palencia, E. "Perturbations spectrales liées à la vibration d'un corps élastique dans l'air" in "Singular perturbations and boundary layer theory", Springer, Berlin, Lect. Not. Math. 594 (1977) (p. 437 - 455).

COMPUTING METHODS IN APPLIED SCIENCES AND ENGINEERING
R. Glowinski, J.L. Lions (editors)
North-Holland Publishing Company

SINGULAR PERTURBATIONS AND FREE BOUNDARY VALUE PROBLEMS

C.M. BRAUNER (*) and B. NICOLAENKO (**)

0. INTRODUCTION

We present here various aspects of the interplay between Free Boundary Value Problems (F.B.V.P.) and singular perturbations for nonlinear elliptic partial differential equations (P.D.E.). In a wide range of situations, it is possible to show that F.B.V.P. arise as singular limits of such singularly perturbed P.D.E., as opposed to classical boundary layer situations. There appears some kind of internal layer, centered on the unknown limiting free boundary ; with a whole gamut of asymptotic expansions, ranging from only a limited loss of regularity (e.g. from $C^k(\bar{\Omega})$ for some $k \geq 2$ to the $W^{2,p}(\Omega)$ optimal estimates for F.B.V.P.) to extreme situations of unilateral stiffness on one side of the free boundary.

In previous work within the context of enzyme kinetics and chemical heterogeneous catalysts [1-5] we have first evidenced that some nonlinear eigenvalue problems can indeed be extended into F.B.V.P. ; specifically we considered in some bounded domain Ω in $\mathbb{R}^n$, with a smooth boundary Γ :

$$(0.1) \quad \begin{cases} \Delta u = \lambda u^m \left(\dfrac{\varepsilon+1}{\varepsilon+u} \right)^{m+k} & \text{in } \Omega , \\ u = 1 \quad \text{on } \Gamma, \\ u \geq 0 ; \lambda \geq 0 ; m \geq 1 ; -1 < k < +1 ; \end{cases}$$

(*) Laboratoire de Mathématiques - Informatique - Systèmes, Ecole Centrale de Lyon, 69130 Ecully, France.

(**) University of California, Mathematical Analysis Group T7, Los Alamos Scientific Laboratory, 87545 Los Alamos, New Mexico, USA.

λ is the eigenvalue parameter, and ε the singular perturbation parameter. For λ large enough, we have established that a solution $u(\lambda, \varepsilon)$ of (0.1) converges as $\varepsilon \to 0$, to $u_o(\lambda)$ in $H^1(\Omega)$ strong, such that :

$$(0.2)\quad \begin{cases} - u_o^{m+k} \Delta u_o + \lambda u_o^m = 0, \\ u_o \geq 0 \text{ in } \Omega, \\ u_o = 1 \text{ on } \Gamma, \\ \text{meas } \Omega_o = \{x \in \Omega, u_o(x) = 0\} > 0 \text{ strictly.} \end{cases}$$

However a Variational Inequality (V.I.) formulation of (0.2) and smoothness of the free boundary can be established only in the case $-1 < k \leq 0$. For $0 < k < 1$, even C^0 - smoothness of the free boundary is lacking!

In the present article, we survey similar phenomena associated with classical V.I. and smooth free boundaries, and their implications for numerical algorithms for V.I. Stiffness problems for related numerical schemes will be discussed.

We first present a simple, yet enlightening example of a F.B.V.P. which is a singular limit of a nonlinear eigenvalue problem.

In Part 2, we exhibit a new class of phenomena : a singular perturbation problem for a nonlinear P.D.E., where the nonlinearity is "incompatible" with a "forcing function" term. This problem is highlighted with dichotomic asymptotic expansions, unilaterally stiff on one side of the (unknown) free boundary.

Finally, in Part 3, we establish a new approximation scheme for a wide class of V.I., based on the simple example discussed below. Related numerical algorithms are discussed.

1. A FREE BOUNDARY VALUE PROBLEM WHICH IS A SINGULAR LIMIT OF A NONLINEAR EIGENVALUE PROBLEM

Let Ω be a bounded open set in $\mathbb{R}^n$, with a smooth boundary Γ. Consider the nonlinear eigenvalue problem :

$$(1.1)\quad \begin{cases} \Delta u_\varepsilon = \lambda \dfrac{u_\varepsilon}{\varepsilon + u_\varepsilon} \text{ in } \Omega \\ u_\varepsilon = 1 \text{ on } \Gamma, \\ \lambda \geq 0, \ \varepsilon > 0. \end{cases}$$

We look for positive solutions $u_\varepsilon \geq 0$ of (1.1).

Lemma 1.1 : For $\varepsilon > 0$ fixed, (1.1) has an unique positive solution $u_\varepsilon(\lambda)$, $\forall \lambda \geq 0$, such that $u_\varepsilon(\lambda) \in C^j(\bar{\Omega})$, $\forall j \geq 0$ and $0 \leq u_\varepsilon(\lambda) \leq 1$.

Proof : remark that 0 is a subsolution and 1 a supersolution for (1.1). Monotonicity arguments insure uniqueness.

■

As ε is formally set equal to zero, the formal limit problem associated to (1.1) is :

(1.2) $\Delta u = \lambda$ in Ω, $u = 1$ on Γ.

Corollary 1.1 : whenever $u(\lambda) \geq 0$, then $u(\lambda) \leq u_\varepsilon(\lambda) \leq 1$.

Proof : indeed $u(\lambda)$ is a subsolution of (1.1) if $u(\lambda) \geq 0$. ■

But we immediately see that for λ large enough, $u(\lambda)$ is locally < 0 (i.e. < 0 on a set of measure > 0). More precisely, let ξ be the solution of

(1.3) $\Delta\xi = 1$ in Ω, $\xi = 0$ on Γ.

Clearly $\xi < 0$, $\forall x \in \Omega$ and $u(\lambda) = \lambda\,\xi + 1$.

Let $M = |\xi|_{L^\infty(\Omega)}$. The minimum of $u(\lambda)$ is $1 - \lambda M$; thus $u(\lambda)$ is locally < 0 as soon as $\lambda > \lambda_* = 1/_M$.

As a consequence, $u(\lambda)$ cannot be a limit of $u_\varepsilon(\lambda)$ for $\lambda > \lambda_*$, $\varepsilon \to 0$, since $u_\varepsilon(\lambda) \geq 0$.

Theorem 1.1 : There exists an unique $u_o(\lambda) \in W^{2,p}(\Omega)$, $\forall p \geq 1$, such that $u_\varepsilon(\lambda) \searrow u_o(\lambda)$ in $W^{2,p}(\Omega)$ weak, hence in $C^{1,\alpha}(\bar{\Omega})$ strong, $\forall \alpha$, $0 < \alpha < 1$. We have $u_o = 1$ on Γ, $u_o(\lambda) \geq 0$ in Ω.

Proof : Sub - and supersolution arguments easily yield $u_{\varepsilon_1}(\lambda) < u_{\varepsilon_2}(\lambda)$ if $\varepsilon_1 < \varepsilon_2$. As

$$0 \leq \frac{u_\varepsilon}{u_\varepsilon + \varepsilon} \leq 1 ,$$

we immediately deduce the estimate $0 \leq \Delta u_\varepsilon(\lambda) \leq \lambda$, which yields that the sequence u_ε is bounded in $W^{2,p}(\Omega)$, $\forall p \geq 1$.

■

Corollary 1.2 : Let λ verify $0 < \lambda < \lambda_*$. Then $u_o(\lambda) = u(\lambda)$ and $u_\varepsilon \searrow u$ in $C^\infty(\bar{\Omega})$.

Proof : If $\lambda < \lambda_*$, $u(\lambda) \geqslant 1 - \lambda M > 0$; hence $u_\varepsilon \geqslant - \lambda M + 1$ and $\dfrac{u_\varepsilon}{\varepsilon + u_\varepsilon} \to 1$.

We conclude with Lebesgue's Theorem.

■

Remark 1.1 : the above proof still applies when $\lambda = \lambda_*$, for the set of points where u reaches its minimum has null measure ; hence $\dfrac{u_\varepsilon}{\varepsilon + u_\varepsilon} \to 1$ a.e.

Corollary 1.3 : Let $\lambda > \lambda_*$. Then $u(\lambda)$ is null on a set of positive measure.

Proof : Suppose that $u_o > 0$ a.e. in Ω. Then $\dfrac{u_\varepsilon}{\varepsilon + u_\varepsilon} \to 1$ a.e., and $u_o(\lambda) \equiv u(\lambda)$. But $u < 0$ locally on a set of measure > 0 , and we reach a contradiction.

■

Thus the limit $u_o(\lambda)$ satisfies a F.B.V.P., where the free boundary $\partial\Omega_o$ is the frontier of $\Omega_o = \left\{ x \in \Omega \ / \ u_o(\lambda) = 0 \right\}$. For $\lambda > \lambda_*$, $\Omega_o \subset \Omega$ (strictly). More precisely, we shall demonstrate that u_o is solution to a V.I.

Theorem 1.2 : Let $K = \left\{ v \in H^1(\Omega) \ , \ v \geqslant 0, \ v/\partial\Omega = 1 \right\}$. Then $u_o(\lambda) \in K$, and is the solution of the V.I. :

(1.4) $$(\nabla u_o, \nabla(v-u_o)) + (\lambda, v-u_o) \geqslant 0 \qquad \forall v \in K.$$

Proof : Let $v \in K$. Multiply (1.1) by $(v-u_\varepsilon)$:

$$(-\Delta u_\varepsilon, v-u_\varepsilon) + \lambda \left(\frac{u_\varepsilon}{\varepsilon + u_\varepsilon}, v-u_\varepsilon \right) = 0 ;$$

hence

$$(-\Delta u_\varepsilon, v-u_\varepsilon) + (\lambda, v-u_\varepsilon) + \lambda \left(\frac{\varepsilon}{\varepsilon + u_\varepsilon}, u_\varepsilon \right) = \lambda \left(\frac{\varepsilon}{\varepsilon + u}, v \right) \geqslant 0$$

and

$$\left(\frac{\varepsilon}{\varepsilon + u_\varepsilon}, u_\varepsilon \right) = \varepsilon \int_\Omega \frac{u_\varepsilon}{\varepsilon + u_\varepsilon} \, dx \to 0.$$

In the limit, we have indeed :

$$(- \Delta u_o, v-u_o) + (\lambda, v-u_o) \geqslant 0 \qquad \forall v \in K.$$

This is, of course, equivalent to :

(1.4)' $$u_o \geqslant 0, \ - \Delta u_o + \lambda \geqslant 0, \ u_o(- \Delta u_o + \lambda) = 0 \text{ in } \Omega, \ u_o = 1 \text{ on } \Gamma.$$

■

Theorem 1.3 : we have the following error estimate :

(1.5) $$||u_o - u_\varepsilon||_{H^1(\Omega)} \leq \text{Cst} \sqrt{\varepsilon}$$

Proof : Multiply (1.1) by $u_o - u_\varepsilon$, and take $v = u_\varepsilon$ into (1.4) :

$$(\nabla u_\varepsilon, \nabla(u_o - u_\varepsilon)) = (-\lambda, u_o - u_\varepsilon) + \lambda\left(\frac{\varepsilon}{\varepsilon + u_\varepsilon}, u_o - u_\varepsilon\right)$$

$$(\nabla u_o, \nabla(u_\varepsilon - u_o)) \geq (-\lambda, u_\varepsilon - u_o)$$

hence

$$-|\nabla(u_\varepsilon - u_o)|^2 \geq \lambda\left(\frac{\varepsilon}{\varepsilon + u_\varepsilon}, u_o - u_\varepsilon\right)$$

and

$$|\nabla(u_\varepsilon - u_o)|^2 \leq -\lambda\left(\frac{\varepsilon}{\varepsilon + u_\varepsilon}, u_o\right) + \lambda\left(\frac{\varepsilon}{\varepsilon + u_\varepsilon}, u_\varepsilon\right) \leq \lambda\, \varepsilon \text{ meas } \Omega.$$

■

Remark 1.2 : the above V.I. may be recast into a more familiar form by setting :

(1.6) $$w_o = 1 - u_o,\ w_\varepsilon = 1 - u_\varepsilon ;$$

the convex K becomes

(1.7) $$\tilde{K} = \{w \in H_o^1(\Omega),\ w \leq 1\},$$

and the V.I. is

(1.8) $$(-\Delta w_o, w - w_o) \geq (\lambda, w - w_o),\quad \forall w \in \tilde{K}$$

with the approximation scheme.

(1.9) $$-\Delta w_\varepsilon = \lambda \frac{(1 - w_\varepsilon)}{\varepsilon + (1 - w_\varepsilon)},\ w_\varepsilon \in H_o^1(\Omega)$$

Remark 1.3 : the V.I. (1.4) defines in fact the continuation of positive solutions of (1.2) beyond λ_*. In this respect, λ_* is an end - point of regular, C^∞, positive solutions of (1.2), since beyond λ_* we have at best $W^{2,p}(\Omega)$ regularity. This concept of end - point of a regular arc of solutions was emphasized in [4] [5].

Remark 1.4 : (1.6) - (1.9) suggest a new approximation scheme for a wide class of V.I., which will be systematically developed in Part 3.

2. SINGULAR PERTURBATIONS WHICH ARISE FROM INCOMPATIBLE NONLINEARITIES AND FORCING FUNCTIONS

2.0 INTRODUCTION

From now on, Ω is a bounded open set in $\mathbb{R}^n$, with a C^∞ boundary Γ.

Consider the perturbation problem

$$(0.1) \qquad \begin{cases} -\varepsilon \, \Delta u_\varepsilon + \dfrac{u_\varepsilon}{1+u_\varepsilon} = f \ , \ f \geq 0 \\ u_{\varepsilon/\Gamma} = 0 \end{cases}$$

whose nonlinearity becomes, as $\varepsilon \to 0$, "incompatible" with the R.H.S. "forcing function" f whenever the latter is not < 1. Indeed, if $u_\varepsilon \to u_o$ for some topology, and if we can take the limit in (0.1) when $\varepsilon \to 0$, we have

$$(0.2) \qquad \frac{u_o}{1+u_o} = f$$

which is impossible if $f \not< 1$.

We are then led to consider the function $w_\varepsilon = \varepsilon \, u_\varepsilon$ which verifies

$$(0.3) \qquad -\Delta w_\varepsilon + \frac{w_\varepsilon}{\varepsilon + w_\varepsilon} = f \ , \ w_{\varepsilon/\Gamma} = 0 \ ;$$

This problem is similar to the one considered in Part 1, with the exception of the non-homogeneity f . We expect w_ε to converge to the solution of a F.B.V.P., characterized by a V.I.

<u>This remark seems quite general</u> : we may also consider the singular perturbation problem

$$(0.4) \qquad -\varepsilon \, \Delta u_\varepsilon - u_\varepsilon^- = f, \ u_\varepsilon/_\Gamma = 0$$

The same problem of "incompatibility" between the nonlinearity and the R.H.S. term arises if f is not ≤ 0. The function $w_\varepsilon = \varepsilon \, u_\varepsilon$ is now solution of

$$(0.5) \qquad -\Delta w_\varepsilon - \frac{w_\varepsilon^-}{\varepsilon} = f \ , \ w_\varepsilon/_\Gamma = 0$$

which is the <u>penalized</u> equation classically associated to the V.I.

$$(0.6) \qquad (-\Delta w_o, v - w_o) \geq (f, v - w_o)$$

$\forall v \in K = \{ v \in H_o^1(\Omega) , v \geq 0 \}$.

Rewriting Eq. (0.3) as

$$(0.7) \qquad - \Delta w_\varepsilon - \frac{\varepsilon}{\varepsilon + w_\varepsilon} = f - 1$$

we see that the term $- \frac{\varepsilon}{\varepsilon + w_\varepsilon}$ is a penalization analogous to $- \frac{w_\varepsilon^-}{\varepsilon}$. This fact will be fully exploited in the 3rd part of this article.

We may construct a whole array of problems similar to (0.1) or (0.4), for instance

$$(0.8) \qquad - \varepsilon \Delta u_\varepsilon + e^{u_\varepsilon} = f \, , \; u_\varepsilon /_\Gamma = 0$$

in the case where f is not > 0 ; but we shall not initiate a systematical study of such problems.

2.1 THE CASE $0 \leqslant f < 1$

Consider the problem

$$(1.1) \qquad - \varepsilon \Delta u_\varepsilon + \frac{u_\varepsilon}{1 + u_\varepsilon} = f \, , \; u_\varepsilon /_\Gamma = 0$$

under the hypothesis

$$(1.2) \qquad f \in L^\infty(\Omega), \; f \geqslant 0, \; ||f||_\infty < 1.$$

Because of the monotonicity of the nonlinear operator (cf. LIONS [10]), problem (1.1) admits, for $\varepsilon > 0$ fixed, an unique solution $\geqslant 0$ in $H^1_o(\Omega) \cap W^{2,p}(\Omega)$, $1 \leqslant p < + \infty$.

With the hypothesis $||f||_\infty < 1$, there is no incompatibility between the nonlinearity and f, and the situation is relatively simple (the monotonicity of the operator plays a crucial role).

From the weak maximum principle, we have

$$(1.3) \qquad 0 \leqslant u_\varepsilon \leqslant M = \frac{||f||_\infty}{1 - ||f||_\infty} < + \infty.$$

Multiply (1.1) by u_ε and integrate over Ω :

$$(1.4) \qquad \varepsilon \int_\Omega (\nabla u_\varepsilon)^2 \, dx + \int_\Omega \frac{u_\varepsilon^2}{1 + u_\varepsilon} \, dx = \int_\Omega f \, u_\varepsilon \, dx$$

hence

$$(1.5) \qquad \int_\Omega \frac{u_\varepsilon^2}{1 + M} \, dx \leqslant M \int_\Omega f \, dx$$

and the sequence u_ε remains bounded in $L^2(\Omega)$. We extract a subsequence u_ε such that

(1.6) $$u_\varepsilon \to u_o \quad \text{in} \quad L^2(\Omega) \text{ weak.}$$

We now use a monotonicity method (cf LIONS [11] p. 393) ; denote by (,) the scalar product in $L^2(\Omega)$ and the duality between $H^1_o(\Omega)$ and $H^{-1}(\Omega)$:

(1.7) $$(-\varepsilon\Delta(u_\varepsilon - v), u_\varepsilon - v) + \left(\frac{u_\varepsilon}{1+u_\varepsilon} - \frac{v}{1+|v|}, u_\varepsilon - v\right) \geqslant 0 \quad \forall v \in H^1_o(\Omega)$$

and

$$\left(-\varepsilon\,\Delta u_\varepsilon + \frac{u_\varepsilon}{1+u_\varepsilon}, u_\varepsilon - v\right) = (f, u_\varepsilon - v) \to (f, u_o - v)$$ hence, after taking the limit in (1.7),

(1.8) $$\left(f - \frac{v}{1+|v|}, u_o - v\right) \geqslant 0 \qquad \forall v \in L^2(\Omega).$$

In (1.8), set $v = u_o - \lambda w$, $\lambda > 0$, $w \in L^2(\Omega)$,

(1.9) $$\left(f - \frac{u_o - \lambda w}{1 + |u_o - \lambda w|}, w\right) \geqslant 0$$

Then let $\lambda \to 0$: we obtain, almost everywhere in Ω,

(1.10) $$f = \frac{u_o}{1+u_o}, \quad \text{i.e.} \quad u_o = \frac{f}{1-f}.$$ ■

To obtain strong convergence, remark that

(1.11) $$\begin{cases} \dfrac{u_\varepsilon}{1+u_\varepsilon} \to f \quad \text{in } L^\infty(\Omega) \text{ weak}* \\ \dfrac{1}{1+u_\varepsilon} \to 1 - f \quad \text{in } L^\infty(\Omega) \text{ weak}* \end{cases}$$

and we rewrite (1.4) as

(1.12) $$\begin{cases} \varepsilon \displaystyle\int_\Omega (\nabla u_\varepsilon)^2\,dx + \int_\Omega \frac{(u_\varepsilon - u_o)^2}{1+u_\varepsilon}\,dx \\ = -2\displaystyle\int_\Omega \frac{u_\varepsilon u_o}{1+u_\varepsilon}\,dx + \int_\Omega \frac{u_o^2}{1+u_\varepsilon}\,dx + \int_\Omega f\,u_\varepsilon\,dx \end{cases}$$

hence

(1.13) $$\begin{cases} \limsup \displaystyle\int_\Omega \frac{(u_\varepsilon - u_o)^2}{1+M} \leqslant -2\int_\Omega f\,u_o\,dx + \int_\Omega u_o^2(1-f)\,dx \\ + \displaystyle\int_\Omega f\,u_o\,dx = -\int_\Omega \frac{u_o^2}{1+u_o}\,dx + \int_\Omega \frac{u_o^2}{1+u_o}\,dx = 0\,. \end{cases}$$ ■

As $f_{/\Gamma} \neq 0$ in general, $u_{o/\Gamma} \neq 0$, and there appears the usual boundary layer phenomenon in a neighborhood of the boundary Γ.

2.2 THE CASE $f > 1$

2.2.1 Preliminary results

We shall need the following theorem :

Theorem 2.1 : Let $\rho(x) = d(x, \Gamma)$. Then the mapping $\frac{1}{\rho} : v \to \frac{v}{\rho}$ is bounded linear from $H^1_o(\Omega)$ in $L^2(\Omega)$, and there exists a constant $c > 0$ such that

$$(2.1) \qquad \left\| \frac{v}{\rho} \right\|_{L^2(\Omega)} \leqslant c \, \|v\|_{H^1_o(\Omega)} \qquad \forall\, v \in H^1_o(\Omega)$$

Proof : Inequality (2.1) is established the following way (cf LIONS [10] p. 104) : using a partition of unity and local maps, everything follows from the inequality

$$\int_o^\infty \left|\frac{1}{x}\, \phi\,(x)\right|^2 dx \leqslant 2 \int_o^\infty |\phi'(x)|^2\, dx, \quad \phi \in \mathcal{D}(]0, \infty[).$$

(We write $\frac{1}{x}\,\phi(x) = \frac{1}{x}\int_o^x \phi'(y)\, dy$ and we apply the Hardy - Littlewood inequality). ■

Remark 2.1 : A more general result is given in LIONS - MAGENES [12] , I, p. 76 : let s integer > 0. Then $u \in H^s_o(\Omega)$ iff $u \in \mathcal{D}'(\Omega)$ and $\rho^{-s+|\alpha|} D^\alpha u \in L^2(\Omega)$, $\forall\, \alpha$ with $|\alpha| \leqslant s$. This result with $\alpha = 0$, $s = 2$, will be crucial in the F.B.V.P. case, where f is both > 1 and < 1 in Ω.

2.2.2 The singular problem

We still consider the problem

$$(2.2) \qquad -\varepsilon\,\Delta u_\varepsilon + \frac{u_\varepsilon}{1 + u_\varepsilon} = f, \quad u_{\varepsilon/\Gamma} = 0$$

with, now, the following hypothesis

$$(2.3) \qquad f \in C^o(\bar{\Omega}),\ f - 1 > 0 \text{ on } \bar{\Omega}.$$

Remark 2.2 : $\varepsilon\,\Delta u_\varepsilon = \frac{u_\varepsilon}{1 + u_\varepsilon} - f < 0$

Lemma 2.1 : When $\varepsilon \to 0$, the sequence u_ε is increasing.

Proof : Let $\varepsilon_1 < \varepsilon_2$, and set $u_i = u_{\varepsilon_i}$; $u_1 - u_2$ verifies

$$(2.4) \qquad -\varepsilon_1\,\Delta(u_1 - u_2) + \frac{u_1 - u_2}{(1 + u_1)(1 + u_2)} = \Delta u_2\,(\varepsilon_1 - \varepsilon_2)$$

and the second term of (2.4) is > 0, hence $u_1 > u_2$. ∎

2.2.3 Renormalization

Let us define the function w_ε as

$$(2.5) \qquad w_\varepsilon = \varepsilon\, u_\varepsilon$$

and w_ε is solution of the problem

$$(2.6) \qquad - \Delta w_\varepsilon + \frac{w_\varepsilon}{\varepsilon + w_\varepsilon} = f \;,\; w_{\varepsilon/\Gamma} = 0$$

or

$$(2.6)' \qquad - \Delta w_\varepsilon = f - 1 + \frac{\varepsilon}{\varepsilon + w_\varepsilon} \;,\; w_{\varepsilon/\Gamma} = 0$$

Lemma 2.2 : When $\varepsilon \to 0$, the sequence w_ε is decreasing.

Proof : Let $\varepsilon_1 < \varepsilon_2$, and set $w_i = w_{\varepsilon_i}$; $w_1 - w_2$ verifies

$$(2.7) \qquad - \Delta(w_1 - w_2) + \varepsilon_2 \frac{w_1 - w_2}{(\varepsilon_1 + w_1)(\varepsilon_2 + w_2)} = \frac{w_2\,(\varepsilon_1 - \varepsilon_2)}{(\varepsilon_1 + w_1)(\varepsilon_2 + w_2)}$$

and the second term of (2.7) is < 0, hence $w_1 < w_2$. ∎

We now introduce the function u_{-1} solution of

$$(2.8) \qquad - \Delta u_{-1} = f - 1 \;,\; u_{-1/\Gamma} = 0$$

Lemma 2.3 : With the hypothesis (2.3), $u_{-1} \in C^{1,\alpha}(\bar\Omega) \cap C^2(\Omega)$, $0 < \alpha < 1$, $u_{-1} > 0$ in Ω and $\frac{\partial u_{-1}}{\partial \nu} < 0$ on Γ where ν is the normal to Γ oriented outwards.

Proof : $u_{-1} > 0$ from the maximum principle and $\frac{\partial u_{-1}}{\partial \nu} < 0$ on Γ from a lemma of HOPF (see [6] , II, p. 321). ∎

Corollary 2.1 : There exists a constant $K > 0$ such that $u_{-1} \geqslant K\rho$, where $\rho(x) = d(x, \Gamma)$.

Proof : $\frac{\partial u_{-1}}{\partial \nu} \in C^\alpha(\Gamma)$ and reaches its negative maximum on Γ (which is a compact manifold). ∎

Corollary 2.2 : $\frac{1}{u_{-1}} \in H^{-1}(\Omega)$.

Proof : Let $v \in H^1_o(\Omega)$,

$$\left| \left(\frac{1}{u_{-1}}, v \right) \right| = \left| \left(\frac{\rho}{u_{-1}}, \frac{v}{\rho} \right) \right| \leqslant \frac{\sqrt{\text{mes } \Omega}}{K} \cdot \left\| \frac{v}{\rho} \right\|_{L^2(\Omega)} \leqslant \text{Cst} \, \|v\|_{H^1_o(\Omega)}$$

from Theorem 2.1.

■

Theorem 2.2 : We have $\|w_\varepsilon - u_{-1}\|_{H^1_o(\Omega)} \leqslant \text{Cst}.\ \varepsilon$

Proof : $- \Delta(w_\varepsilon - u_{-1}) = \frac{\varepsilon}{\varepsilon + w_\varepsilon} \geqslant 0$, hence

$w_\varepsilon \geqslant u_{-1}$ and $- \Delta(w_\varepsilon - u_{-1}) \leqslant \frac{\varepsilon}{u_{-1}}$, consequently

$$\|\nabla (w_\varepsilon - u_{-1})\|^2_{L^2(\Omega)} \leqslant \left(w_\varepsilon - u_{-1}, \frac{1}{u_{-1}}\right) \leqslant \varepsilon \, \|w_\varepsilon - u_{-1}\|_{H^1_o(\Omega)} \cdot \left\| \frac{1}{u_{-1}} \right\|_{H^{-1}(\Omega)}$$

and we conclude with POINCARE'S Inequality.

■

Corollary 2.3 : $w_\varepsilon \searrow u_{-1}$ in $W^{2,p}(\Omega)$ weak and $C^{1,\alpha}(\bar{\Omega})$, $1 \leqslant p < \infty$, $0 < \alpha < 1$.

Proof : Simply notice that Δw_ε remains bounded in $L^\infty(\Omega)$ when $\varepsilon \to 0$.

■

From this corollary and lemma 2.3, we deduce the

Theorem 2.3 : As $\varepsilon \to 0$, $u_\varepsilon(x) \to +\infty \quad \forall x \in \Omega$.

Proof : Indeed, $\forall x \in \Omega$, $\varepsilon u_\varepsilon(x) = w_\varepsilon(x) \to u_{-1}(x) > 0$.

■

2.2.4 Asymptotic expansion

Theorem 2.3 shows the singular behaviour of u_ε when $\varepsilon \to 0$. As in LIONS [11], we look for an asymptotic expansion of u_ε as follows

$$u_\varepsilon = \frac{u_{-1}}{\varepsilon} + u_o + \varepsilon u_1 + \ldots \tag{2.9}$$

Carrying over (2.9) into (2.2), we find that u_{-1} verifies (2.8), u_o being given by

$$- \Delta u_o = \frac{1}{u_{-1}}, \quad u_{o/\Gamma} = 0 \tag{2.10}$$

and u_1 by

$$(2.11) \qquad -\Delta u_1 = -\frac{1+u_o}{(u_{-1})^2}, \quad u_{1/\Gamma} = 0, \text{ etc } \ldots$$

Lemma 2.4 : Problem (2.10) admits an unique solution in $H^1_o(\Omega) \cap L^\infty(\Omega)$. ∎

Proof : u_o exists and is unique in $H^1_o(\Omega)$ since $\frac{1}{u_{-1}} \in H^{-1}(\Omega)$. That it belongs to $L^\infty(\Omega)$ from the regularity of u_o inside Ω and the condition $u_{o/\Gamma} = 0$.

Remark 2.3 : Using Remark 2.1, one can see that the 2^d member of (2.11) belongs to $H^{-2}(\Omega)$. The existence of u_1 is an open question (see § 2.3.3).

Through the knowledge of u_o, we can complete theorem 2.2 by a pointwise estimate.

Theorem 2.4 : $||w_\varepsilon - u_{-1}||_{L^\infty(\Omega)} \leqslant \varepsilon ||u_o||_{L^\infty(\Omega)}$

Proof : Set

$$(2.12) \qquad \tilde{w}_\varepsilon = u_{-1} + \varepsilon u_o$$

and compute $-\Delta(\tilde{w}_\varepsilon - w_\varepsilon) = \frac{\varepsilon}{u_{-1}} - \frac{\varepsilon}{\varepsilon + w_\varepsilon} \geqslant 0$, hence $\tilde{w}_\varepsilon \geqslant w_\varepsilon$ and

$$(2.13) \qquad u_{-1} \leqslant w_\varepsilon \leqslant \tilde{w}_\varepsilon = u_{-1} + \varepsilon u_o$$

consequently

$$(2.14) \qquad 0 \leqslant w_\varepsilon - u_{-1} \leqslant \varepsilon u_o.$$ ∎

From the above results, we deduce the

Theorem 2.4 : As $\varepsilon \to 0$, $u_\varepsilon - \frac{u_{-1}}{\varepsilon} \to u_o$ in $H^1_o(\Omega)$ weak and in $L^\infty(\Omega)$ weak $*$.

Proof : The sequence $u_\varepsilon - \frac{u_{-1}}{\varepsilon}$ remains bounded in $H^1_o(\Omega)$ (th. 2.2) and in $L^\infty(\Omega)$ (th. 2.4). We can extract a subsequence converging to χ in $H^1_o(\Omega)$ weak and in $L^\infty(\Omega)$ weak $*$. $\forall \phi \in \mathcal{D}(\Omega)$, we have

$$(2.15) \qquad \int_\Omega -\Delta\left(u_\varepsilon - \frac{u_{-1}}{\varepsilon}\right)\phi\, dx = \int_\Omega \frac{\phi}{\varepsilon + w}\, dx$$

and, on the support of ϕ, $\frac{1}{\varepsilon + w_\varepsilon} \nearrow \frac{1}{u_{-1}}$, hence

$$(2.16) \qquad \int_\Omega -\Delta\chi\, \phi\, dx = \int_\Omega \frac{\phi}{u_{-1}}\, dx \quad \forall \phi \in \mathcal{D}(\Omega)$$

and $- \Delta\chi = \dfrac{1}{u_{-1}}$ in $\mathcal{D}'(\Omega)$. As $\chi \in H_o^1(\Omega)$, $\chi = u_o$. ∎

2.3 THE GENERAL CASE : THE FREE BOUNDARY

We consider the same problem as in § 2.1 and 2.2

$$(3.1) \qquad - \varepsilon\, \Delta u_\varepsilon + \frac{u_\varepsilon}{1 + u_\varepsilon} = f \;, \quad u_{\varepsilon|\Gamma} = 0$$

but now with the hypothesis

$$(3.2) \qquad f \in C^\infty(\bar{\Omega}),\; f \geqslant 0 \;.$$

We suppose that f can take values inferior and superior to 1, but with the restriction

$$(3.3) \qquad \left\{ \begin{array}{l} \text{The manifold } \Sigma_1 = \{x,\ f(x) = 1\} \text{ is a simply connected } C^\infty\text{- mani-} \\ \text{fold of dimension } n - 1 \text{ which bisects } \Omega \text{ into 2 simply connected open} \\ \text{sets } F_- = \{x \in \Omega,\ f(x) < 1\} \text{ and } F_+ = \{x \in \Omega,\ f(x) > 1\} \end{array} \right.$$

Indeed, the case where f takes the value 1 on a set of measure > 0 is much more complex and shall not be dealt here (cf. §2.4 for the case $f \equiv 1$). Other hypothesis than (3.3) will be considered in a forthcoming article.

2.3.1 The free boundary value problem

Let u_{-1} be the solution to the variational inequality

$$(3.4) \qquad \left| \begin{array}{l} (-\Delta u_{-1},\, v - u_{-1}) \geqslant (f - 1,\, v - u_{-1}) \\ \forall\, v \in K = \{v \in H_o^1(\Omega),\ v \geqslant 0\} \end{array} \right.$$

It is well known (cf. e.g. LIONS [10]) that the solution of (3.4) belongs to $W^{2,p}(\Omega)$, $1 \leqslant p < \infty$, hence to $C^{1,\alpha}(\bar{\Omega})$, $0 < \alpha < 1$.

We set

$$(3.5) \qquad \Omega_o = \{x \in \Omega,\ u_{-1}(x) = 0\}$$

(if $x \in \Omega_o$, $\nabla u_{-1}(x) = 0$)

$$(3.6) \qquad \Omega_+ = \{x \in \Omega,\ u_{-1}(x) > 0\}$$

Ω_o and Ω_+ are separated by a <u>free boundary</u> S which is not Σ_1. However S is determinated by Σ_1 and F_+.

Lemma 3.1 : For a.e. $x \in \Omega_o$, $f(x) < 1$.

Proof : Since u_{-1} is solution of the V.I. (3.4), we have

(3.8) $$- \Delta u_{-1}(x) \geqslant f(x) - 1 \quad \text{a.e. } x \in \Omega$$

For a.e. $x \in \Omega_o$, $\Delta u_{-1}(x) = 0$ and $f(x) - 1 \leqslant 0$. Then $f(x) < 1$ a.e. from hyp. (3.3) ■

More precisely, we have the

Theorem 3.1 : $\Omega_o \subset F_- = \{ x \in \Omega,\ f(x) < 1 \}$.

Proof : i) Let us prove that $\Omega_o \cap F_+ = \emptyset$. Suppose $\exists\, x_o \in \Omega_o \cap F_+$; there exists a ball B, of center x_o, $\subset F_+$. In B, $- \Delta u_{-1} = f - 1$, and by the maximum principle, u_{-1} reaches its minimum on ∂B : that is inconsistent with $u_{-1}(x_o) = 0$.

ii) Suppose $\exists\, x_o \in \Omega_o \cap \Sigma_1$. There exists a ball B strictly contained in F_+ and tangent to Σ_1 in x_o. In B, $- \Delta u_{-1} = f - 1$. By the maximum principle u_{-1} reache its minimum on B at $x_o \in \partial B$ and $\frac{\partial u_{-1}}{\partial \nu}(x_o) < 0$. But $\frac{\partial u_{-1}}{\partial \nu}(x_o) = \nabla u_{-1}(x_o) . \nu$ and we reach a contradition because $\nabla u_{-1}(x_o) = 0$ as $x_o \in \Omega_o$. ■

Remark 3.1 . $\bar{\Omega}_o \cap \Sigma_1$ can be empty or reduced to $\Sigma_1 \cap \Gamma$.

To simplify, we shall assume from now on that

(3.9) S is a manifold of dimension n - 1

Then, from the regularity estimates for (3.4), S is at worst a $C^{1,\alpha}$ manifold, $\forall\, \alpha$, $0 < \alpha < 1$.

Transmission conditions across S are then defined

(3.10) $$u_{-1} /_S = \frac{\partial u_{-1}}{\partial \nu} /_S = 0$$

Corollary 3.1 : The manifold S is in $\bar{F}_-$, and its intersection with Σ_1 is empty, except may be at exceptional points $\in \Sigma_1 \cap \Gamma$. Moreover S cannot be a closed manifold strictly inside F_-.

Proof : Suppose S is a closed manifold $\subset F_-$. Then $\int_{\Omega_+} -\Delta u_{-1}\,dx = -\int_S \frac{\partial u_{-1}}{\partial \nu}\,dx = 0$, but $-\Delta u_{-1} = f - 1 < 0$ in Ω_+. ■

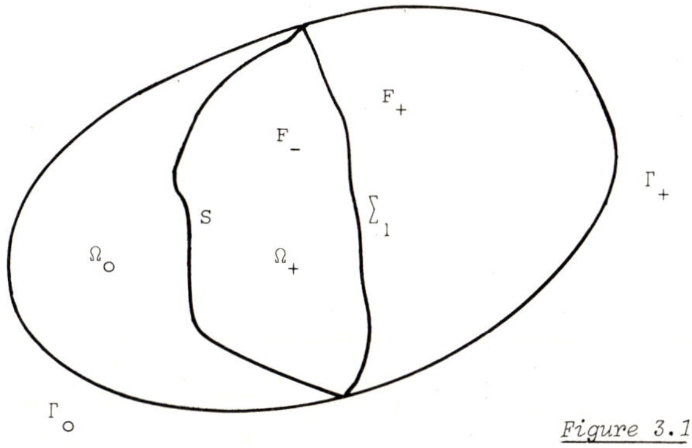

Figure 3.1

2.3.2 Convergence results

As in § 2.2, we renormalize u_ε as follows :

(3.11) $$w_\varepsilon = \varepsilon\, u_\varepsilon$$

(3.12) $$-\Delta w_\varepsilon + \frac{w_\varepsilon}{\varepsilon + w_\varepsilon} = f\ ,\ w_{\varepsilon/\Gamma} = 0$$

Theorem 3.2 : $||w_\varepsilon - u_{-1}||_{H^1_o(\Omega)} \leqslant \text{Cst.}\sqrt{\varepsilon}$

Proof : From (3.12), we deduce

(3.13) $$(-\Delta w_\varepsilon, v) = (f - 1, v) + \left(\frac{\varepsilon}{\varepsilon + w_\varepsilon}, v\right) \quad \forall v \in H^1_o(\Omega)$$

we take $v = w_\varepsilon$ in (3.4), $v = u_{-1} - w_\varepsilon$ in (3.13).

Then $-\int_\Omega (\nabla(u_{-1} - w_\varepsilon))^2\,dx \geqslant \varepsilon \int_\Omega \frac{u_{-1} - w_\varepsilon}{\varepsilon + w_\varepsilon}\,dx$, hence

(3.14) $$\int_\Omega (\nabla(u_{-1} - w_\varepsilon))^2\,dx \leqslant -\varepsilon \int_\Omega \frac{u_{-1}}{\varepsilon + w_\varepsilon}\,dx + \varepsilon \int_\Omega \frac{w_\varepsilon}{\varepsilon + w}\,dx$$

$$\leqslant \varepsilon \text{ meas } \Omega.$$

■

Remark 3.2 : In theorem 2.2 (case f > 1), we obtained a better estimate in "ε".

Corollary 3.2 : $w_\varepsilon \searrow u_{-1}$ in $W^{2,p}(\Omega)$ weak, $1 \leqslant p < \infty$, and $C^{1,\alpha}(\bar{\Omega})$, $0 < \alpha < 1$ (as in § 2.2).

Corollary 3.3 : $\dfrac{\partial w_\varepsilon}{\partial \nu}\Big|_S \to 0$ in $L^2(S)$ strong.

Proof : If $\overset{\circ}{w}_\varepsilon = w_\varepsilon\,|_{\Omega_o}$, we have $\overset{\circ}{w}_\varepsilon \to 0$ in $W^{2,p}(\Omega_o)$ weak and in particular in $H^{3/2}(\Omega_o)$ strong. The result follows then from a trace theorem (cf. e.g. LIONS - MAGENES [12], I).

Lemma 3.2 : $\Delta w_\varepsilon \to 0$ in $L^2(\Omega_o)$ strong

Proof : Multiply (3.12) by Δw_ε and integrate over Ω_o

(3.15) $$||\Delta w_\varepsilon||^2_{L^2(\Omega_o)} - \int_{\Omega_o} \Delta w_\varepsilon \cdot \frac{w_\varepsilon}{\varepsilon + w_\varepsilon}\,dx = -\int_{\Omega_o} f\,\Delta w_\varepsilon\,dx$$

hence, from Green's formula,

(3.16) $$||\Delta w_\varepsilon||^2_{L^2(\Omega_o)} + \varepsilon \int_{\Omega_o} \frac{(\nabla w_\varepsilon)^2}{(\varepsilon + w_\varepsilon)}\,dx = -\int_{\Omega_o} f\,\Delta w_\varepsilon\,dx + \int_S \frac{\partial w_\varepsilon}{\partial \nu}\,\frac{w_\varepsilon}{\varepsilon + w_\varepsilon}\,dS$$

and the R.H.S. of (3.16) goes to 0, because of corollaries 3.2 and 3.3. ∎

Theorem 3.3 : As $\varepsilon \to 0$,

(i) $u_\varepsilon(x) \to \dfrac{f(x)}{1 - f(x)}$ a.e. in Ω_o

(ii) $u_\varepsilon(x) \to +\infty$ $\forall$ x in Ω_+

Proof : (i) Since $\dfrac{u_\varepsilon}{1 + u_\varepsilon} = f + \Delta w_\varepsilon$, $\dfrac{u_\varepsilon}{1 + u_\varepsilon} \to f$ in $L^2(\Omega_o)$ strong and a.e., hence the result with th. 3.1.

(ii) $\varepsilon\,u_\varepsilon(x) = w_\varepsilon(x) \to u_{-1}(x) > 0$ $\forall\, x \in \Omega_+$. ∎

Now we are going to prove that theorem 3.2 is not optimal, and also that $u_\varepsilon \to \dfrac{f}{1 - f}$ in some L^1 space.

Theorem 3.4 : As $\varepsilon \to 0$,

(i) $\left(\sqrt{\varepsilon}\, u_\varepsilon - \frac{u_{-1}}{\sqrt{\varepsilon}}\right) \to 0$ in $H_0^1(\Omega)$ strong

(ii) $\int_{\Omega_0} (1 - f)\, u_\varepsilon \, dx \to \int_{\Omega_0} f \, dx$

Proof : We set $\psi_\varepsilon = w_\varepsilon - u_{-1}$. It is easy to check that ψ_ε verifies in Ω

$$- \Delta\psi_\varepsilon + \chi_{\Omega_0} \cdot (1 - f) = \frac{\varepsilon}{\varepsilon + w_\varepsilon} \tag{3.17}$$

where χ_{Ω_0} is the charateristic function of Ω_0.

Multiply (3.17) by ψ_ε and integrate over Ω :

$$\begin{aligned} \int_\Omega (\nabla\psi_\varepsilon)^2 \, dx + \int_{\Omega_0} (1 - f)\psi_\varepsilon \, dx &= \int_\Omega \frac{\varepsilon\, \psi_\varepsilon}{\varepsilon + w_\varepsilon} \, dx \\ &= \varepsilon \int_{\Omega_0} \frac{u_\varepsilon}{1 + u_\varepsilon} \, dx + \varepsilon \int_{\Omega_+} \frac{w_\varepsilon - u_{-1}}{\varepsilon + w_\varepsilon} \, dx \end{aligned} \tag{3.18}$$

Divide (3.18) by ε :

$$\begin{aligned} \int_\Omega \left(\frac{\nabla\psi_\varepsilon}{\sqrt{\varepsilon}} \right)^2 dx + \int_{\Omega_0} (1 - f)\, u_\varepsilon \, dx \\ = \int_{\Omega_0} \frac{u_\varepsilon}{1 + u_\varepsilon} \, dx + \int_{\Omega_+} \frac{w_\varepsilon - u_{-1}}{\varepsilon + w_\varepsilon} \, dx \end{aligned} \tag{3.19}$$

It follows from th. 3.3, and Lebesgue's theorem that

$$\int_{\Omega_0} \frac{u_\varepsilon}{1 + u_\varepsilon} \, dx \to \int_{\Omega_0} f \, dx \tag{3.20}$$

Also by Lebesgue's theorem (using $w_\varepsilon \geq u_{-1}$),

$$\int_{\Omega_+} \frac{w_\varepsilon - u_{-1}}{\varepsilon + w_\varepsilon} \, dx \to 0 \tag{3.21}$$

hence

$$0 \leq \overline{\lim} \int_{\Omega_0} (1 - f)\, u_\varepsilon \, dx \leq \int_{\Omega_0} f \, dx \tag{3.22}$$

But we know after Fatou's theorem that

$$\int_{\Omega_0} (1 - f) \lim u_\varepsilon \, dx \leq \underline{\lim} \int_{\Omega_0} (1 - f)\, u_\varepsilon \, dx \tag{3.23}$$

As $u_\varepsilon \to \frac{f}{1 - f}$ a.e. in Ω_0,

$$(3.24)\qquad \int_{\Omega_o} f\,dx \leqslant \underline{\lim} \int_{\Omega_o} (1-f)\,u_\varepsilon\,dx \leqslant \overline{\lim} \int_{\Omega_o} (1-f)\,u_\varepsilon\,dx \leqslant \int_{\Omega_o} f\,dx$$

and (ii) is proved.

Finally

$$(3.25)\qquad \int_\Omega \left(\frac{\nabla\psi_\varepsilon}{\varepsilon}\right)^2 dx \to 0$$

which demonstrates (i).

■

Corollary 3.3 : $(1-f)\,u_\varepsilon \to f$ in $L^1(\Omega_o)$ strong

Proof : We use the following result : If $g_j \geqslant 0$, $g_j \to f$ a.e., $\int_{\Omega_o} g_j\,dx \to \int_{\Omega_o} f\,dx$, then $g_j \to f$ in $L^1(\Omega_o)$ strong (see in [5] , Prop. 5.1, a simple proof due to H. Brezis).

■

Corollary 3.4 : $u_\varepsilon \to \dfrac{f}{1-f}$ in $L^1(\tilde{\Omega}_o)$ strong, where $\tilde{\Omega}_o = \Omega_o \setminus \theta$, θ being a fixed neighborhood, as small as we wish, of $\Gamma \cap \sum_1 \cap S$.

Proof : On $\tilde{\Omega}_o$, $\min(1-f) \geqslant \beta > 0$.

■

2.3.3 Asymptotic expansion

As in § 2.2, we look for an asymptotic expansion of u_ε as follows

$$(3.26)\qquad u_\varepsilon = \frac{u_{-1}}{\varepsilon} + u_o + \varepsilon\,u_1 + \ldots$$

where u_{-1} is given by (3.4), that is

$$(3.27)\qquad \begin{cases} u_{-1} = 0 \text{ in } \Omega_o, \quad -\Delta u_{-1} = f-1 \text{ in } \Omega_+, \quad u_{-1}|_{\Gamma_+} = 0, \\ u_{-1}|_S = \dfrac{\partial u_{-1}}{\partial \nu}\Big|_S = 0 \end{cases}$$

Remark 3.1 : as a matter of fact, the asymptotic expansion $u_\varepsilon \sim u_{-1}/\varepsilon + u_o$ has been established in the domain $\tilde{\Omega}_o$ with $u_{-1} \equiv 0$, $u_o = f/1-f$, and convergence in $L^1(\tilde{\Omega}_o)$.

Formally, u_o is given by

$$(3.28)\quad \begin{cases} u_o = \dfrac{f}{1-f} & \text{in } \Omega_o \\ -\Delta u_o = \dfrac{1}{u_{-1}} & \text{in } \Omega_+ \\ u_o|_{\Gamma_+} = 0 & \\ u_o|_S = \dfrac{f}{1-f}\Big|_S & \end{cases}$$

But we see that $\frac{1}{u_{-1}} \notin H^{-1}(\Omega_+)$ since $u_{-1}|_S = \frac{\partial u_{-1}}{\partial \nu}\Big|_S = 0$. Indeed, by local maps, we reduce the problem to the half - plane $x_n > 0$ where $\frac{1}{u_{-1}}$ has a behaviour as $\frac{1}{x_n^2}$ in a neighborhood of $x_n = 0$. The existence of u_o is then an open question. (In some particular cases, we can establish that $\frac{1}{u_{-1}} \in H^{-2}(\Omega_+)$ using Remark 2.1).

2.4 THE CASE $f \equiv 1$

In the case $f \equiv 1$, the singular perturbation problem becomes

$$(4.1)\qquad -\varepsilon\, \Delta u_\varepsilon + \frac{u_\varepsilon}{1+u_\varepsilon} = 1 \quad , \quad u_\varepsilon|_\Gamma = 0$$

Or, equivalently,

$$(4.1)'\qquad -\varepsilon\, \Delta u_\varepsilon = \frac{1}{1+u_\varepsilon} \, , \quad u_\varepsilon|_\Gamma = 0$$

We remark that the solution u_{-1} of (3.4) is identically null ; yet it is obvious that $u_\varepsilon \to +\infty$. The renormalization (3.10) is not adequate anymore. We set

$$(4.2)\qquad \hat{w}_\varepsilon = \sqrt{\varepsilon}\, u_\varepsilon$$

and $\hat{w}_\varepsilon$ is solution of

$$(4.3)\qquad -\sqrt{\varepsilon}\, \Delta \hat{w}_\varepsilon = \frac{\sqrt{\varepsilon}}{\sqrt{\varepsilon} + \hat{w}_\varepsilon}$$

hence

$$(4.4)\qquad -\Delta\, \hat{w}_\varepsilon = \frac{1}{\sqrt{\varepsilon} + \hat{w}_\varepsilon} \, , \quad \hat{w}_\varepsilon|_\Gamma = 0$$

Theorem 4.1 : $\hat{w}_\varepsilon \to \hat{u}_{-1}$ in $H_o^1(\Omega)$ weak, where $\hat{u}_{-1}$ is the unique solution in $C^2(\Omega) \cap C^o(\bar{\Omega})$ of

$$(4.5) \qquad -\Delta\hat{u}_{-1} = \frac{1}{\hat{u}_{-1}} \quad , \quad \hat{u}_{-1}|_{\Gamma} = 0$$

Problem (4.5) is a special case of a more general problem studied by CRANDALL - RABINOWITZ - TARTAR [7] . The estimate in $H^1_o(\Omega)$ is simply obtained by multiplying (4.4) by $\hat{w}_\varepsilon$.

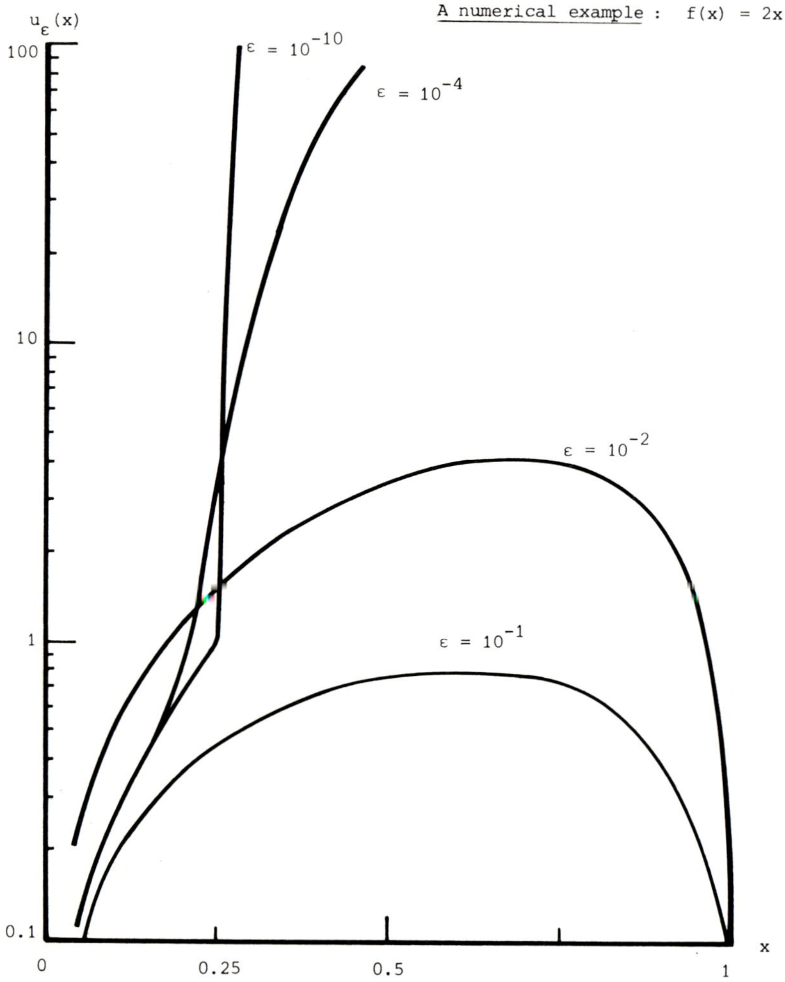

$$\Omega =]0,1[, \ \Sigma_1 = \left\{\frac{1}{2}\right\}, \ S = \left\{\frac{1}{4}\right\}$$

3. A NEW APPROXIMATION SCHEME FOR VARIATIONAL INEQUALITIES

3.1 HIGHLIGHTS OF THE NEW SCHEME

We present here a new approximation to a wide classe of Variational Inequalities (V.I.), which offers many advantages over classical penalization schemes. It is directly inspired from the free boundary value problem (F.B.V.P.) studied in part 1 as a singular perturbation limit of a class of nonlinear elliptic eigenvalue problems. The new scheme is based on the properties of the bounded, monotone, homographic application :

(1.1) $$x \longrightarrow \frac{x}{\varepsilon + x}, \; x \in R^1_+ , \; \varepsilon \in R^1_+ .$$

To clarify the presentation, we consider the simplified problem with Dirichlet conditions :

(1.2) $$A u - f \geqslant 0 , \; u - \psi \geqslant 0 , \; (A u - f)(u - \psi) = 0 \text{ in } \Omega$$

(1.3) $$u = 0 \text{ on } \Gamma , \; u \in H^1_o(\Omega)$$

(1.4) $$A u = - \sum \frac{\partial}{\partial x_i} \left(a_{ij} \frac{\partial u}{\partial x_j} \right) + \sum a_j \frac{\partial u}{\partial x_j} + a_o u$$

where

(1.5) $$a_o, \; a_j \in L^\infty(\Omega), \; a_{ij} \in C^1(\bar{\Omega}).$$

Ω is a bounded domain in $\mathbb{R}^n$, with a smooth boundary Γ. The following bilinear form is associated to A, for u, v $\in H^1_o(\Omega)$:

(1.6) $$a(u, v) = \sum \int_\Omega a_{ij} \frac{\partial u}{\partial x_j} \frac{\partial v}{\partial x_i} dx + \sum \int_\Omega a_j \frac{\partial u}{\partial x_j} v \, dx + \int_\Omega a_o u v \, dx$$

We assume the following coercivity hypothesis :

(1.7) $$a(v, v) \geqslant \alpha ||v||^2, \; \forall v \in H^1_o(\Omega), \; \alpha > 0.$$

In this part 3, $||u||$ will denote the norm of u in $H^1_o(\Omega)$. (1.5) insures that the solution u of the V.I. (1.2) (1.3) belongs to $H^2(\Omega)$, if $f \in L^2(\Omega)$.

We will first suppose that :

(1.8) $$f \in L^2(\Omega), \; \psi \in H^2(\Omega)$$
$$(f - A\psi)^- \in L^\infty(\Omega)$$

Remark 1.1 : These conditions can be relaxed to $\psi \in H^1(\Omega)$, $f - A\psi$ is a measure, and $(f - A\psi)^- \in L^2(\Omega)$.

We choose $g \in L^\infty(\Omega)$ such that

(1.9) $$g \geqslant 0,\ f + g - A\psi \geqslant 0 \text{ in } \Omega$$

(From the hypothesis (1.8), such a choice is possible).

For each $\varepsilon > 0$, let u_ε be the solution of :

(1.10) $$A u_\varepsilon + g \frac{u_\varepsilon - \psi}{u_\varepsilon - \psi + \varepsilon} = f + g, \text{ in } \Omega,$$

(1.11) $$u_\varepsilon \in H^1_o(\Omega).$$

Verify the

Lemma 1.1 : Problem (1.10) (1.11) has an unique solution such that $u_\varepsilon \in H^1_o(\Omega) \cap H^2(\Omega)$ and :

(1.12) $$\psi \leqslant u_\varepsilon$$

Remark 1.2 : We will show that u_ε is an <u>approximation</u> of u, solution of the V.I. (1.2), (1.3); (1.12) shows that this approximation satisfies the constraints.

1) Consider a priori the equation :

(1.13) $$A w + g \frac{(w - \psi)^+}{w - \psi + \varepsilon} = f + g,\ w \in H^1_o(\Omega) ;$$

it admits an <u>unique</u> solution, from the theory of <u>monotone operators</u>.

2) Verify that

(1.14) $$w \geqslant \psi$$

(This implies that (1.13) coincides with (1.10) (1.11), taking $u_\varepsilon = w$).

Indeed, multiplying (1.13) by $v \in H^1_o(\Omega)$, we have :

(1.15) $$a(w - \psi, v) + \left(g \frac{(w - \psi)^+}{w - \psi + \varepsilon}, v \right) = (f + g - A\psi, v).$$

Taking $v = (w - \psi)^-$, we obtain

(1.16) $$a\big((w - \psi)^-, (w - \psi)^-\big) + \big(f + g - A\psi, (w - \psi)^-\big) = 0 ;$$

from (1.9), the last term in (1.16) is ≥ 0 ; hence (1.16) implies $(w - \psi)^- = 0$, hence (1.14).

∎

We proceed to demonstrate :

Theorem 1.1 : Under conditions (1.8) (1.9), let u_ε (resp. u) be the solution of (1.10) (1.11) (resp. of the V.I. (1.2) (1.3)). Then, as $\varepsilon \to 0$, we have : $u_\varepsilon \to u$ in $H_o^1(\Omega)$ and more precisely :

$$(1.17) \qquad ||u_\varepsilon - u|| < C\,\varepsilon^{1/2}.$$

Remark 1.3 : As we underlined it, the approximation of u by u_ε satisfies the constraints of the V.I.

Proof : We may rewrite (1.10) (1.11) under the shape :

$$A u_\varepsilon - \frac{\varepsilon g}{\varepsilon + u_\varepsilon - \psi} = f, \quad u_\varepsilon \in H_o^1(\Omega), \text{ hence}$$

$$(1.18) \qquad a(u_\varepsilon, v) = (f, v) + \left(\frac{\varepsilon g}{\varepsilon + u_\varepsilon - \psi}, v\right) \qquad \forall v \in H_o^1(\Omega).$$

Moreover

$$(1.19) \qquad a(u, v - u) \geq (f, v - u), \quad \forall v \in H_o^1(\Omega), \; v \geq \psi.$$

Take $v = u_\varepsilon$ in (1.19) (this is possible from (1.13)), and $v = -(u_\varepsilon - u)$ in (1.18) ; we obtain

$$- a(u_\varepsilon - u, u_\varepsilon - u) \geq \left(\frac{\varepsilon g}{\varepsilon + u_\varepsilon - \psi}, u - u_\varepsilon\right)$$

$$\geq -\left(\frac{\varepsilon g}{\varepsilon + u_\varepsilon - \psi}, u_\varepsilon - \psi\right), \text{ hence}$$

$$(1.20) \qquad \alpha\, ||u_\varepsilon - u||^2 \leq \varepsilon \left(\frac{u_\varepsilon - \psi}{\varepsilon + u_\varepsilon - \psi}, g\right),$$

α defined in (1.7). As $0 \leq \dfrac{u_\varepsilon - \psi}{\varepsilon + u_\varepsilon - \psi} \leq 1$, this implies :

$$(1.21) \qquad \alpha\, ||u_\varepsilon - u||^2 \leq \varepsilon \int_\Omega g\, dx,$$

hence (1.17).

Remark 1.4 : The optimal choice of g with (1.9) is :

(1.22) $g = (f - A\,\psi)^-$, hence

(1.23) $$\alpha\, \|u_\varepsilon - u\|^2 \leqslant \varepsilon \int_\Omega (f - A\,\psi)^-\, dx$$

Finally, we deduce from our new approximation, the following result due to H. LEWY and G. STAMPACCHIA [9], J.L. JOLY [8] :

Theorem 1.2 : Within the hypothesis of Theorem 1.1, we have :

(1.24) $$f \leqslant A\,u_\varepsilon \leqslant \operatorname{Sup}\,(f,\, A\,\psi)$$

(1.25) $$f \leqslant A\,u \leqslant \operatorname{Sup}\,(f,\, A\,\psi)$$

Proof : We remark that $0 \leqslant \dfrac{\varepsilon\, g}{u_\varepsilon - \psi + \varepsilon} \leqslant g$, so $f \leqslant A\,u_\varepsilon \leqslant f + g$; choosing g as in (1.22), we obtain (1.24). Hence (1.25), since $u_\varepsilon \to u$ in $H^1_o(\Omega)$.

Remark 1.5 : (1.25) remains valid whenever $\psi \in H^1(\Omega)$, via an approximation and limit scheme.

Remark 1.6 : Naturally, we recover from Theorem (1.2) the classical regularity result : if $f \in L^p(\Omega)$, $\psi \in W^{2,p}(\Omega)$, then

(1.26) $$u \in W^{2,p}(\Omega) \cap H^1_o(\Omega).$$

Remark 1.7 : The method is easily extended to higher order coercive elliptic operators, and to constraints of the type $u \leqslant \psi$. Moreover, if we relax conditions on ψ as in remark 1.1, we can choose $g \in L^2(\Omega)$. Finally, the parabolic case can be easily treated.

Remark 1.8 : Dropping the Dirichlet conditions, the method is easily extended to the V.I :

(1.27) $$A\,u - f \geqslant 0\ ,\ u - \psi \geqslant 0\ ,\ (A\,u - \psi)\ (u - \psi) = 0 \text{ in } \Omega\ ;$$

(1.28) $$\frac{\partial u}{\partial \nu_A} \geqslant 0\ ,\ u - \psi \geqslant 0,\ \frac{\partial u}{\partial \nu_A}\ (u - \psi) = 0 \text{ on } \Gamma.$$

We need only to suppose that $\psi \in H^1(\Omega)$, $A\,\psi \in L^2(\Omega)$, $\dfrac{\partial \psi}{\partial \nu_A}$ is a measure on Γ, and $\left(\dfrac{\partial \psi}{\partial \nu_A}\right)^+ \in L^2(\Gamma)$.

Remark 1.9 : The method can be adapted to the game theory case, with constraints $\psi_1 \leqslant u \leqslant \psi_2$. However, in full generality, for the adapted scheme, u_ε will satisfy only one constraint ($u_\varepsilon \leqslant \psi_2$ or $u_\varepsilon \geqslant \psi_1$) for every ε.

3.2 A NUMERICAL ALGORITHM

To simplify, we take $\psi \equiv 0$, and g is a given constant satisfying (1.9).

We define u^o by

(2.1) $$- \Delta u^o = f + g \, , \; u^o_{|\Gamma} = 0$$

Suppose we have construct $u^1, u^2, \ldots, u^n$ such that

(2.2) $$u^j \in W^{2,p}(\Omega) \, , \; 1 \leqslant p < \infty, \; u^j > 0 \text{ in } \Omega, \; u^{j+1} \leqslant u^j, \; j = 0, \ldots, n-1$$

Then define u^{n+1} by

(2.3) $$- \Delta u^{n+1} + g \frac{u^{n+1}}{u^n} = f + g, \; u^{n+1}_{|\Gamma} = 0 \quad (^1)$$

Lemma 2.1 : u^{n+1} is uniquely defined in $W^{2,p}(\Omega)$ by (2.3), $u^{n+1} > 0$ in Ω and $u^{n+1} \leqslant u^n$.

Proof : Introduce u_ε^{n+1} solution of

(2.4) $$- \Delta u_\varepsilon^{n+1} + g \frac{u_\varepsilon^{n+1}}{\varepsilon + u^n} = f + g, \; u_\varepsilon^{n+1}{}_{|\Gamma} = 0$$

Using $u^n \in C^{1,\alpha}(\bar{\Omega})$, $u^n > 0$ in Ω, it is easy to see that $u_\varepsilon^{n+1} \searrow u^{n+1}$ in $H_o^1(\Omega)$ strong as $\varepsilon \to 0$. On the other hand, $u^{n+1} - u^n$ is solution of

(2.5) $$- \Delta (u^{n+1} - u^n) + \frac{g}{u^n} (u^{n+1} - u^n) = \frac{g}{u^{n-1}} (u^n - u^{n-1}) \leqslant 0$$

hence $u^{n+1} \leqslant u^n$, and Δu^{n+1} belongs to $L^\infty(\Omega)$. ■

Lemma 2.2 : There exists $u^* \geqslant 0$ such that $u^n \searrow u^*$ in $W^{2,p}(\Omega)$ weak as $n \to +\infty$.

Proof : Remark that Δu^{n+1} is bounded in $L^\infty(\Omega)$ independently of n, because $\frac{u^{n+1}}{u^n} \leqslant 1$ in Ω. ■

Lemma 2.3 : u^* is solution of the V.I.

(2.6) $$(- \Delta u^*, v - u^*) \geqslant (f, v - u^*)$$
$$\forall v \in K = \{ v \in H_o^1(\Omega), \; v \geqslant 0 \}.$$

(^{1}) The idea of taking $\varepsilon = 0$ in (2.3) is due to A. MARROCO, who also performed the numerical treatment (see [13]).

Proof : $(- \Delta u^{n+1}, v - u^{n+1}) = (f, v - u^{n+1}) + g (1 - \frac{u^{n+1}}{u^n}, v)$

$$- g (1 - \frac{u^{n+1}}{u^n}, u^{n+1}) \geqslant (f, v - u^{n+1}) - g (1 - \frac{u^{n+1}}{u^n}, u^{n+1})$$

and $(1 - \frac{u^{n+1}}{u^n}, u^{n+1}) = \int_\Omega (u^n - u^{n+1}) \frac{u^{n+1}}{u^n} dx \to 0$ as $n \to +\infty$. ■

ACKNOWLEDGMENTS : The authors are glad to thank Professor J.L. LIONS for many fruitful discussions. They also thank A. MORROCCO for the communication of his numerical results (see [13]).

REFERENCES

[1] C.M. BRAUNER and B. NICOLAENKO, Perturbation singulière, solutions multiples et hystérésis dans un problème de biochimie, C.R.Acad. Sc. Paris, Série A, 283, p. 775-778 (1976).

[2] C.M. BRAUNER and B. NICOLAENKO, Singular perturbation, multiple solutions and hysteresis in a nonlinear problem, Lect. Notes in Math., 594, Springer - Verlag, p. 50-76 (1977).

[3] C.M. BRAUNER and B. NICOLAENKO, Sur une classe de problèmes elliptiques non linéaires, C.R. Acad. Sc. Paris, Série A, 286, p. 1007-1010 (1978).

[4] C.M. BRAUNER and B. NICOLAENKO, Sur des problèmes aux valeurs propres non linéaires qui se prolongent en problèmes à frontière libre, C.R. Acad. Sc. Paris, Série A, 287, p. 1105-1108 (1978), and 288, p. 125-127 (1979).

[5] C.M. BRAUNER and B. NICOLAENKO, On nonlinear eigenvalue problems which extend into free boundaries problems, to appear in Lect. Notes in Maths (volume edited by C. BARDOS), Preprint Lyon 1979.

[6] R. COURANT and D. HILBERT, Methods of Mathematical Physics, Interscience.

[7] M.G. CRANDALL, P.H. RABINOWITZ and L. TARTAR, On a Dirichlet problem with a singular nonlinearity, Comm. in Partial Differential Equations, 2, p. 193-222 (1977).

[8] J.L. JOLY, Thèse, Grenoble (1970).

[9] H. LEWY and G. STAMPACCHIA, On the smoothness of superharmonics which solve a minimum problem, J. Anal. Math. 23, p. 227-236 (1970).

[10] J.L. LIONS, Quelques méthodes de résolution des problèmes aux limites non linéaires, Dunod (1969).

[11] J.L. LIONS, Perturbations singulières dans les problèmes aux limites et en contrôle optimal, Lect. Notes in Math., 323, Springer-Verlag (1973).

[12] J.L. LIONS and E. MAGENES, Problèmes aux limites non homogènes et applications, I, II, Dunod (1968).

[13] A. MARROCCO, to appear.